Probability and Random Processes with Applications to Signal Processing

Third Edition

Henry Stark
Illinois Institute of Technology

John W. Woods
Rensselaer Polytechnic Institute

PRENTICE HALL, Upper Saddle River, New Jersey 07458

Library of Congress Cataloging-in-Publication Data

Stark, Henry, 1938-
    Probability and random processes with applications to signal processing/Henry Stark
and John W. Woods.–3rd ed.
        p. cm.
    Rev. ed.: Probability, random processes, and estimation theory for engineers. 2nd ed. 1994.
    Includes bibliographical references and index.
    ISBN 0-13-020071-9
    1. TK5102.9 S717 2001 2. Signal processing–Mathematics. 3. Probabilities. 4.
    Stochastic processes. I. Woods, John W. (John William), 1943-II. Stark, Henry, 1938-
    Probability, random processes, and estimation theory for engineers. III. Title.

    519.2–dc21   2001021253

Vice President and Editorial Director, ECS: *Marcia J. Horton*
Vice President and Director of Production and Manufacturing, ESM: *David W. Riccardi*
Publisher: *Tom Robbins*
Acquisitions Editor: *Eric Frank*
Associate Editor: *Alice Dworkin*
Executive Managing Editor: *Vince O'Brien*
Managing Editor: *David A. George*
Production Editor: *Scott Disanno*
Director of Central Services: *Paul Belfanti*
Creative Director: *Carole Anson*
Art Director: *Jayne Conte*
Art Editor: *Adam Velthaus*
Manufacturing Manager: *Trudy Pisciotti*
Manufacturing Buyer: *Lisa McDowell*
Senior Marketing Manager: *Holly Stark*
Editorial Assistant: *Jody McDonnell*

 © 2002 Prentice-Hall
Prentice-Hall, Inc.
Upper Saddle River, New Jersey 07458

The author and publisher of this book have used their best efforts in preparing this book. These
efforts include the development, research, and testing of the theories and programs to determine their
effectiveness. The author and publisher make no warranty of any kind, expressed or implied, with regard to
these programs or the documentation contained in this book. The author and publisher shall not be liable
in any event for incidental or consequential damages in connection with, or arising out of, the furnishing,
performance, or use of these programs.

Printed in the United States of America
10 9 8 7 6 5 4 3 2 1

ISBN 0-13-020071-9

Prentice-Hall International (UK) Limited, *London*
Prentice-Hall of Australia Pty. Limited, *Sydney*
Prentice-Hall Canada Inc., *Toronto*
Prentice-Hall Hispanoamericana, S.A., *Mexico*
Prentice-Hall of India Private Limited, *New Delhi*
Prentice-Hall of Japan, Inc., *Tokyo*
Prentice-Hall Asia Pte. Ltd., *Singapore*
Editora Prentice-Hall do Brasil, Ltda., *Rio de Janeiro*

To my father P.D. Stark (in memoriam)
From darkness to light (1941–1945)

To Harriet.

J. W. Woods

Contents

Preface

The first edition of this book (1986) grew out of a set of notes used by the authors to teach two one-semester courses on probability and random processes at Rensselaer Polytechnic Institute (RPI). At that time the probability course at RPI was required of all students in the Computer and Systems Engineering Program and was a highly recommended elective for students in closely related areas. While many undergraduate students took the course in the junior year, many seniors and first-year graduate students took the course for credit as well. Then, as now, most of the students were engineering students. To serve these students well, we felt that we should be rigorous in introducing fundamental principles while furnishing many opportunities for students to develop their skills at solving problems.

There are many books in this area and they range widely in their coverage and depth. At one extreme are the very rigorous and authoritative books that view probability from the point of view of measure theory and relate probability to rather exotic theorems such as the Radon-Nikodym theorem (see for example *Probability and Measure* by Patrick Billingsley, Wiley, 1978). At the other extreme are books that usually combine probability and statistics and largely omit underlying theory and the more advanced types of applications of probability. In the middle are the large number of books that combine probability and random processes, largely avoiding a measure theoretic approach, preferring to emphasize the axioms upon which the theory is based. It would be fair to say that our book falls into this latter category. Nevertheless this begs the question: why write or revise another book in this area if there are already several good texts out there that use the same approach and provide roughly the same coverage? Of course back in 1986 there were few books that emphasized the engineering applications of probability and random processes and that integrated the latter into one volume. Now there are several such books.

Both authors have been associated (both as students and faculty) with colleges and universities that have demanding programs in engineering and applied science. Thus their experience and exposure have been to superior students that would not be content with a text that furnished a shallow discussion of probability. At the same time, however, the authors wanted to write a book on probability and random processes for engineering and applied science students. A measure-theoretic book, or one that avoided the engineering applications of probability and the processing of random signals, was regarded not suitable for such students. At the same time the authors felt that the book should have enough depth so that students taking 2nd year graduate courses in advanced topics such as estimation and detection, pattern recognition, voice and image processing, networking and queuing, and so forth would not be handicapped by insufficient knowledge of the fundamentals and applications of random phenomena. In a nutshell we tried to write a book that combined rigor with accessibility and had a strong self-teaching orientation. To that end we included a large number of worked-out examples, MATLAB codes, and special appendices that include a review of the kind of basic math needed for solving problems in probability as well as an introduction to measure theory and its relation to probability. The MATLAB codes, as well as other useful material such as multiple choice exams that cover each of the book's sections, can be found at the book's web site http : //www.prenhall.com/stark.

The normal use of this book would be as follows: for a first course in probability at, say the junior or senior year, a reasonable goal is to cover Chapters 1 through 4. Nevertheless we have found that this may be too much for students not well prepared in mathematics. In that case we suggest a load reduction in which *combinatorics* in Chapter 1 (parts of Section 1.8), *failure rates* in Chapter 2 (Section 2.7), *more advanced density functions* and *the Poisson transform* in Chapter 3 are lightly or not covered the first time around. The proof of the *Central Limit Theorem, joint characteristic functions*, and Section 4.8, which deals with statistics, all in Chapter 4, can, likewise, also be omitted on a first reading.

Chapters 5 to 9 provide the material for a first course in random processes. Normally such a course is taken in the first year of graduate studies and is required for all further study in signal processing, communications, computer and communication networking, controls, and estimation and detection theory. Here what to cover is given greater latitude. If pressed for time, we suggest that the pattern recognition applications and simultaneous diagonalization of two covariance matrices in Chapter 5 be given lower preference than the other material in that chapter. Chapters 6 and 7 are essential for any course in random processes and the coverage of the topics therein should be given high priority. Chapter 9 on signal processing should, likewise be given high priority, because it illustrates the applications of the theory to current state-of-art problems. However, within Chapter 9, the instructor can choose among a number of applications and need not cover them all if time pressure becomes an issue. Chapter 8 dealing with advanced topics is critically important to the more advanced students, especially those seeking further studies toward the Ph.D. Nevertheless it too can be lightly covered or omitted in a first course if time is the critical factor.

Readers familiar with the 2nd edition of this book will find significant changes in the 3rd edition. The changes were the result of numerous suggestions made by lecturers and students alike. To begin with, we modified the title to *Probability and Random Processes with Applications to Signal Processing*, to better reflect the contents. We removed the two chapters on estimation theory and moved some of this material to other chapters where it naturally fitted in with the material already there. Some of the material on parameter estimation e.g., the Gauss-Markov Theorem has been removed, owing to the need for finding space for new material. In terms of organization, the major changes have been in the random processes part of the book. Many readers preferred seeing discrete-time random phenomena in one chapter and continuous-time phenomena in another chapter. In the earlier editions of the book there was a division along these lines but also a secondary division along the lines of stationary versus non-stationary processes. For some this made the book awkward to teach from. Now all of the material on discrete-time phenomena appears in one chapter (Chapter 6); likewise for continuous-time phenomena (Chapter 7). Another major change is a new Chapter 9 that discusses applications to signal processing. Included are such topics as: the orthogonality principle, Wiener and Kalman filters, The Expectation-Maximization algorithm, Hidden Markov Models, and simulated annealing. Chapter 8 (Advanced Topics) covers much of the same ground as the old Chapter 9 e.g., stochastic continuity, mean-square convergence, Ergodicity etc. and material from the old Chapter 10 on representation of random processes.

There have been significant changes in the first half of the book also. For example, in Chapter 1 there is an added section on the misuses of probability in ordinary life. Here we were helped by the discussions in Steve Pinker's excellent book *How the Mind Works* (Norton Publishers, New York, 1997). Chapter 2 (*Random Variables*) now includes discussions on more advanced distributions such as the Gamma, Chi-square and the Student-t. All of the chapters have many more worked-out examples as well as more homework problems. Whenever convenient we tried to use MATLAB to obtain graphical results. Also, being a book primarily for engineers, many of the worked-out example and homework problems relate to real-life systems or situations.

We have added several new appendices to provide the necessary background mathematics for certain results in the text and to enrich the reader's understanding of probability. An appendix on Measure Theory falls in the latter category. Among the former are appendices on the delta and gamma functions, probability-related basic math, including the principle of proof-by-induction, Jacobians for n-dimensional transformations, and material on Fourier and Laplace inversion.

For this edition, the authors would like to thank Geoffrey Williamson and Yongyi Yang for numerous insightful discussions and help with some of the MATLAB programs. Also we thank Nikos Galatsanos, Miles Wernick, Geoffrey Chan, Joseph LoCicero, and Don Ucci for helpful suggestions. We also would like to thank the administrations of Illinois Institute of Technology and Rensselaer Polytechnic Institute for their patience and support while this third edition was being prepared. Of course, in the end, it is the reaction of the students that is the strongest driving force for improvements. To all our students and readers we owe a large debt of gratitude.

Henry Stark
John W. Woods

1

Introduction to Probability

1.1 INTRODUCTION: WHY STUDY PROBABILITY?

One of the most frequent questions posed by beginning students of probability is: "Is anything truly random and if so how does one differentiate between the truly random and that which, because of a lack of information, is treated as random but really isn't?" First, regarding the question of truly random phenomena: "Do such things exist?" A theologian might state the case as follows: "We cannot claim to know the Creator's mind, and we cannot predict His actions because He operates on a scale too large to be perceived by man. Hence there are many things we shall never be able to predict no matter how refined our measurements."

At the other extreme from the cosmic scale is what happens at the atomic level. Our friends the physicists speak of such things as the *probability* of an atomic system being in a certain state. The uncertainty principle says that, try as we might, there is a limit to the accuracy with which the position and momentum can be simultaneously ascribed to a particle. Both quantities are fuzzy and indeterminate.

Many, including some of our most famous physicists, believe in an essential randomness of nature. Eugen Merzbacher in his well-known textbook on quantum mechanics [1-1] writes:

> The probability doctrine of quantum mechanics asserts that the indetermination, of which we have just given an example, is a property inherent in nature and not merely a profession of our temporary ignorance from which we expect to be relieved by a future better and more complete theory. The conventional interpretation thus denies the possibility of an ideal theory which would encompass the present quantum mechanics

1

but would be free of its supposed defects, the most notorious "imperfection" of quantum mechanics being the abandonment of strict classical determinism.

But the issue of determinism versus inherent indeterminism need never even be considered when discussing the validity of the probabilistic approach. The fact remains that there is, quite literally, a nearly uncountable number of situations where we cannot make any categorical deterministic assertion regarding a phenomenon because we cannot measure all the contributing elements. Take, for example, predicting the value of the current $i(t)$ produced by a thermally excited resistor R. Conceivably, we might accurately predict $i(t)$ at some instant t in the future if we could keep track, say, of the 10^{23} or so excited electrons moving in each other's magnetic fields and setting up local field pulses that eventually all contribute to producing $i(t)$. Such a calculation is quite inconceivable, however, and therefore we use a probabilistic model rather than Maxwell's equations to deal with resistor noise. Similar arguments can be made for predicting weather, the outcome of a coin toss, the time to failure of a computer, and many other situations.

Thus to conclude: Regardless of which position one takes, that is, determinism versus indeterminism, we are forced to use probabilistic models in the real world because we do not know, cannot calculate, or cannot measure all the forces contributing to an effect. The forces may be too complicated, too numerous, or too faint.

Probability is a mathematical model to help us study physical systems in an *average sense*. Thus we cannot use probability in any meaningful sense to answer questions such as: "What is the probability that a comet will strike the earth tomorrow?" or "What is the probability that there is life on other planets?"[†]

R. A. Fisher and R. Von Mises, in the first third of the twentieth century, were largely responsible for developing the groundwork of modern probability theory. The modern axiomatic treatment upon which this book is based is largely the result of the work by Andrei N. Kolmogorov [1-2].

1.2 THE DIFFERENT KINDS OF PROBABILITY

There are essentially four kinds of probability. We briefly discuss them here.

A. Probability as Intuition

This kind of probability deals with judgments based on intuition. Thus "She will probably marry him," and "He probably drove too fast," are in this category. Intuitive probability can lead to contradictory behavior. Joe is still likely to buy an imported Itsibitsi, world famous for its reliability, even though his neighbor Frank has a 19-year-old Buick that has never broken down and Joe's other neighbor, Bill, has his Itsibitsi in the repair shop. Here Joe may be behaving "rationally," going by the statistics and ignoring, so-to-speak, his

---

[†]Nevertheless, certain evangelists and others have dealt with this question rather fearlessly. However, whatever probability system these people use, it is not the system that we shall discuss in this book.

personal observation. On the other hand, Joe will be wary about letting his nine-year-old daughter Jane swim in the local pond, if Frank reports that Bill thought that he might have seen an alligator in it. This despite the fact that no one has ever reported seeing an alligator in this pond, and countless people have enjoyed swimming in it without ever having been bitten by an alligator. To give this example some credibility, assume that the pond is in Florida. Here Joe is ignoring the statistics and reacting to, what is essentially, a rumor. Why? Possibly because the *cost* to Joe "just-in-case" there is an alligator in the pond would be too high [1-3].

People buying lottery tickets intuitively believe that certain number combinations like month/day/year of their grandson's birthday are more likely to win that, say, 06–06–06. How many people will bet even odds that a coin that, heretofore has behaved "fairly," that is in an unbiased fashion, will come up heads on the next toss, *if in the last seven tosses it has come up heads?* Many of us share the belief that the coin has some sort of memory and that, after seven heads, that coin must "make things right" by coming up with more tails.

A mathematical theory dealing with intuitive probability was developed by B. O. Koopman [1-4]. However, we shall not discuss this subject in this book.

B. Probability as the Ratio of Favorable to Total Outcomes (Classical Theory)

In this approach, which is not experimental, the probability of an event is computed *a priori*[†] by counting the number of ways N_E that E can occur and forming the ratio N_E/N where N is the number of all possible outcomes, that is, the number of all alternatives to E plus N_E. An important notion here is that all outcomes are equally likely. Since equally likely is really a way of saying equally probable, the reasoning is somewhat circular. Suppose we throw a pair of unbiased dice and ask what is the probability of getting a seven? We partition the outcome into 36 equally likely outcomes as shown in Table 1.2-1 where each entry is the sum of the numbers on the two dice.

Table 1.2-1 Outcomes of Throwing Two Dice

		1st die				
2nd die	1	2	3	4	5	6
1	2	3	4	5	6	7
2	3	4	5	6	7	8
3	4	5	6	7	8	9
4	5	6	7	8	9	10
5	6	7	8	9	10	11
6	7	8	9	10	11	12

[†] *A priori* means relating to reasoning from self-evident propositions or presupposed by experience. *A posteriori* means relating to reasoning from observed facts.

The total number of outcomes is 36 if we keep the dice distinct. The number of ways of getting a seven is $N_7 = 6$. Hence

$$P[\text{getting a seven}] = \frac{6}{36} = \frac{1}{6}.$$

Example 1.2-1

Throw a fair coin twice (note that since no physical experimentation is involved, there is no problem in postulating an ideal "fair coin"). The possible outcomes are HH, HT, TH, TT. The probability of getting at least one tail T is computed as follows: With E denoting the event of getting at least one tail, the event E is the set of outcomes

$$E = \{HT, TH, TT\}.$$

Thus, E occurs whenever the outcome is HT or TH or TT. The number of elements in E is $N_E = 3$; the number of all outcomes, N, is four. Hence

$$P[\text{at least one } T] = \frac{N_E}{N} = \frac{3}{4}.$$

The classical theory suffers from at least two significant problems: (1) It cannot deal with outcomes that are not equally likely; and (2) it cannot handle uncountably infinite outcomes without ambiguity (see the example by Athanasios Papoulis [1-5]). Nevertheless, in those problems where it is impractical to actually determine the outcome probabilities by experimentation and where, because of symmetry considerations, one can indeed argue equally likely outcomes the classical theory is useful.

Historically, the classical approach was the predecessor of Richard Von Mises' [1-6] relative frequency approach developed in the 1930s.

C. Probability as a Measure of Frequency of Occurrence

The relative-frequency approach to defining the probability of an event E is to perform an experiment n times. The number of times that E appears is denoted by n_E. Then it is tempting to define the probability of E occurring by

$$P[E] = \lim_{n \to \infty} \frac{n_E}{n}. \tag{1.2-1}$$

Quite clearly since $n_E \leq n$ we must have $0 \leq P[E] \leq 1$. One difficulty with this approach is that we can never perform the experiment an infinite number of times so we can only estimate $P[E]$ from a finite number of trials. Secondly, we *postulate* that n_E/n approaches a limit as n goes to infinity. But consider flipping a fair coin 1000 times. The likelihood of getting exactly 500 heads is very small; in fact, if we flipped the coin 10,000 times, the likelihood of getting exactly 5000 heads is even smaller. As $n \to \infty$, the event of observing exactly $n/2$ heads becomes vanishingly small. Yet our intuition demands that $P[\text{head}] = \frac{1}{2}$

for a fair coin. Suppose we choose a $\delta > 0$; then we shall find experimentally that if the coin is truly fair, the number of times that

$$\left| \frac{n_E}{n} - \frac{1}{2} \right| > \delta \qquad (1.2\text{-}2)$$

as n becomes large, becomes very small. Thus although it is very unlikely that at any stage of this experiment, especially when n is large, n_E/n is exactly $\frac{1}{2}$, this ratio will nevertheless hover around $\frac{1}{2}$, and the number of times it will make significant excursion away from the vicinity of $\frac{1}{2}$ according to Equation 1.2-2 becomes very small indeed.

Despite these problems with the frequency definition of probability, the relative-frequency concept is essential in applying probability theory to the physical world.

D. Probability Based on an Axiomatic Theory

This is the approach followed in most modern textbooks on the subject. To develop it we must introduce certain ideas, especially those of a random experiment, a sample description space, and an event. Briefly stated, a random experiment is simply an experiment in which the outcomes are nondeterministic, that is, probabilistic. Hence the word *random* in *random experiment*. The *sample description space* is the set of all outcomes of the experiment. An *event* is a subset of the sample description space that satisfies certain constraints. In general, however, almost any subset of the sample description space is an event.

These notions are refined in Sections 1.4 and 1.5.

1.3 MISUSES, MISCALCULATIONS, AND PARADOXES IN PROBABILITY

The misuse of probability and statistics in everyday life is quite common. Many of the misuses are illustrated by the following examples. Consider a defendant in a murder trial who pleads not guilty to murdering his wife. The defendant has on numerous occasions beaten his wife. His lawyer argues that, yes, the defendant has beaten his wife but that among men who do so, the probability that one of them will actually murder his wife is only 0.001, that is, only one in a thousand. Let us assume that this statement is true. It is meant to sway the jury by implying that the fact of beating one's wife is no indicator of murdering one's wife. Unfortunately, unless the members of the jury have taken a good course in probability, they might not be aware that a far more significant question is the following: *Given that a battered wife is murdered, what is the probability that the husband is the murderer?* Statistics show that this probability is, in fact, *greater than one-half.*

In the 1996 presidential race, Senator Bob Dole's age became an issue. His opponents claimed that a 72-year-old white male has a 27 per cent risk of dying in the next five years. Thus it was argued, were Bob Dole elected, the probability that he would fail to survive his term was greater than one-in-four. The trouble with this argument is that the probability of survival, as computed, is not conditioned on additional pertinent facts. As it happens, if a 72-year-old male is still in the work force and, additionally, happens to be rich, then taking these additional facts into consideration, the average 73-year-old (the age at which

Dole would have assumed the presidency) *has only a one-in-eight chance of dying in the next four years* [1-3].

Misuse of probability appears frequently in predicting life elsewhere in the universe. In his book *Probability* 1 (Harcourt Brace & Company, 1998), Amir Aczel assures us that we can be certain that alien life forms are out there just waiting to be discovered. However, in a cogent review of Aczel's book, John Durant of London's Imperial College writes:

> Statistics are extremely powerful and important, and Aczel is a very clear and capable exponent of them. But statistics cannot substitute for empirical knowledge about the way the universe behaves. We now have no plausible way of arriving at robust estimates about the way the universe behaves. We now have no plausible way of arriving at robust estimates for the probability of life arriving spontaneously when the conditions are right. So, until we either discover extraterrestial life or understand far more about how at least one form of life—terrestrial life—first appeared, we can do little more than guess at the likelihood that life exists elsewhere in the universe. And as long as we're guessing, we should not dress up our interesting speculations as mathematical certainties.

The computation of probabilities based on relative frequency can lead to paradoxes. An excellent example is found in [1-3]. We repeat the example here:

> In a sample of American women between the ages of 35 and 50, 4 out of 100 develop breast cancer within a year. Does Mrs. Smith, a 49-year-old American woman, therefore have a 4% chance of getting breast cancer in the next year? There is no answer. Suppose that in a sample of women between the ages of 45 and 90—a class to which Mrs. Smith also belongs—11 out of 100 develop breast cancer in a year. Are Mrs. Smith's chances 4%, or are they 11%? Suppose that her mother had breast cancer, and 22 out of 100 women between 45 and 90 whose mothers had the disease will develop it. Are her chances 4%, 11%, or 22%? She also smokes lives in California, had two children before the age of 25 and one after 40 is of Greek descent ... What group should we compare her with to figure out the "true" odds? You might think, the more specific the class, the better—but the more specific the class, the smaller its size and the less reliable the frequency. If there were only two people in the world very much like Mrs. Smith, and one developed breast cancer, would anyone say that Mrs. Smith's, chances are 50%? In the limit, the only class that is truly comparable with Mrs. Smith in all her details is the class containing Mrs. Smith herself. But in a class of one "relative frequency" makes no sense.

The previous example should not leave the impression that the study of probability, based on relative frequency, is useless. For one, there are a huge number of engineering and scientific situations that are not as nearly complex as the case of Mrs. Smith's likelihood of getting cancer. Also, it is true that if we refine the class and thereby reduce the class size, our estimate of probability based on relative frequency becomes less stable. But exactly how much less stable is deep within the realm of study of probability and its offspring *statistics* (e.g., see the Law of Large Numbers in Section 4.4). Also, there are many situations where the required conditioning, that is, class refinement, is such that the class size is sufficiently large for excellent estimates of probability. And finally returning to Mrs. Smith, if the class size starts to get too small, then stop adding conditions and learn to live with a probability estimate associated with a larger, less refined class. This estimate may be sufficient for all kinds of actions, that is, planning screening tests, and the like.

1.4 SETS, FIELDS, AND EVENTS

A set is a collection of objects, either concrete or abstract. An example of a concrete set is the set of all New York residents whose height equals or exceeds 6 feet. A subset of a set is a collection that is contained within the larger set. Thus the set of all New York City residents whose height is between 6 and $6\frac{1}{2}$ feet is a subset of the previous set. In the study of probability we are particularly interested in the set of all outcomes of a random experiment and subsets of this set. We sometimes denote the experiment by the symbol $\mathscr{H}$ and the set of all outcomes by Ω. The set Ω is called the *sample space* or *sample description space* of the random experiment $\mathscr{H}$. Subsets of Ω are called *events*. Since any set is a subset of itself, Ω is itself an event. In particular Ω is called the *certain event*. Thus Ω is used to denote two objects: the set of all elementary outcomes of a random event and the certain event. We shall see that using the same symbol for both objects is entirely consistent. A little later we shall be somewhat more precise as to our definition of events.

Examples of Sample Spaces

Example 1.4-1
The experiment consists of flipping a coin once. Then $\Omega = \{H, T\}$ where H is a head and T is a tail.

Example 1.4-2
The experiment consists of flipping a coin twice. Then $\Omega = \{HH, HT, TH, TT\}$. A typical subset of Ω is $\{HH, HT, TH\}$; it is the event of getting at least one head in two flips.

Example 1.4-3
The experiment consists of choosing a person at random and counting the hairs on his or her head. Then
$$\Omega = \{0, 1, 2, \ldots, 10^7\},$$
that is, the set of all nonnegative integers up to 10^7, it being assumed that no human head has more than 10^7 hairs.

Example 1.4-4
The experiment consists of determining the age to the nearest year of each member of a married couple chosen at random. Then with x denoting the age of the man and y denoting the age of the woman, Ω is described by

$$\Omega = \{\text{2-tuples } (x, y)\colon x \text{ any integer in } 10 - 200;\ y \text{ any integer in } 10 - 200\}.$$

Note that in Example 1.4-4 we have assumed that no human lives beyond 200 years and that no married person is ever less than ten years old. Similarly, in Example 1.4-1, we assumed that the coin never lands on edge. If the latter is a possible outcome, it must be included in Ω in order for it to denote the set of *all* outcomes as well as the certain event.

Example 1.4-5

The experiment consists of observing the angle of deflection of a nuclear particle in an elastic collision. Then

$$\Omega = \{\theta: -\pi \leq \pi \leq \pi\}.$$

An example of a subset of Ω is

$$E = \left\{ \frac{\pi}{4} \leq \theta \leq \frac{\pi}{4} \right\} \subset \Omega.$$

Example 1.4-6

The experiment consists of measuring the instantaneous power, P, across a thermally agitated resistor. Then

$$\Omega = \{P: P \geq 0\}.$$

Since power cannot be negative, we leave out negative values of P in Ω. A subset of Ω is $E = \{P > 10^{-3} \text{ watts}\}$.

Note that in Examples 1.4-5 and 1.4-6, the number of elements in Ω is uncountably infinite. Therefore there are an uncountably infinite number of subsets. When, as in Example 1.4-4, the number of outcomes is finite, the number of distinct subsets is also finite, and each represents an event. Thus if $\Omega = \{\zeta_1, \ldots, \zeta_N\}$, the number of subsets of Ω is 2^N. We can see this by noting that each descriptor ζ_i, $i = 1, \ldots, N$ either is or is not present in any given subset of Ω. This gives rise to 2^N distinct events, one of which being the one in which none of the ζ_i appear. This event, involving none of the $\zeta \in \Omega$, is called the *impossible* event and is denoted by ϕ. It is the event that cannot happen and is only included for completeness. The set ϕ is also called the empty set.

Set algebra.   The *union* (sum) of two sets E and F, written $E \cup F$ or $E + F$, is the set of all elements that are in at least one of the sets E or F. Thus with $E = \{1, 2, 3, 4\}$ and $F = \{1, 3, 4, 5, 6\}$[†]

$$E \cup F = \{1, 2, 3, 4, 5, 6\}.$$

If E is a subset of F, we indicate this by writing $E \subset F$. Clearly for $E \subset F$ it follows that $E \cup F = F$. We indicate that ζ is an element of Ω or "belongs" to Ω by writing $\zeta \in \Omega$. Thus we can write

$$E \cup F = \{\zeta: \zeta \in E \text{ or } \zeta \in F \text{ or } \zeta \text{ lies in both}\}. \tag{1.4-1}$$

The *intersection* or set product of two sets E and F, written $E \cap F$ or EF is the set of elements common to both E and F. Thus in the preceding example

$$EF = \{1, 3, 4\}.$$

[†]The order of the elements in a set is not important.

Formally, $EF \triangleq \{\zeta : \zeta \in E \, and \, \zeta \in F\}$. The *complement* of a set E, written E^c, is the set of *all elements not in E*. From this it follows that if Ω is the sample description space or, more generally, the universal set, then

$$E \cup E^c = \Omega. \tag{1.4-2}$$

Also $EE^c = \phi$. The *difference* of two sets or, more appropriately, the *reduction* of E by F, written $E - F$ is the set of elements in E that are not in F. It should be clear that

$$E - F = EF^c$$

$$F - E = FE^c$$

and, in general $E - F \neq F - E$. The *exclusive-or* of two sets, written $E \oplus F$, is the set of all elements in E *or* F but not both. It is readily shown that[†]

$$E \oplus F = (E - F) \cup (F - E). \tag{1.4-3}$$

The operation of unions, intersections, and so forth, can be symbolically represented by Venn diagrams, which are useful as aids in reasoning and in establishing probability relations. The various set operations $E \cup F$, EF, E^c, $E - F$, $F - E$, $E \oplus F$ are shown in Figure 1.4-1 in hatch lines.

Two sets E, F are said to be disjoint if $EF = \phi$; that is, they have no elements in common. Given any set E, an *n-partition* of E consists of a sequence of sets E_i, $i = 1, \ldots, n$ such that $E_i \subset E$, $\bigcup_{i=1}^{N} E_i = E$ and $E_i E_j = \phi$ all $i \neq j$. Thus given two sets E, F a 2-partition of F is

$$F = FE \cup FE^c. \tag{1.4-4}$$

It is easy to demonstrate, using Venn diagrams, the following results:

$$(E \cup F)^c = E^c F^c \tag{1.4-5}$$

$$(EF)^c = E^c \cup F^c \tag{1.4-6}$$

and, by induction,[‡] given sets $E_1, \ldots, E_n$:

$$\left[\bigcup_{i=1}^{n} E_i \right]^c = \bigcap_{i=1}^{n} E_i^c \tag{1.4-7}$$

$$\left[\bigcap_{i=1}^{n} E_i \right]^c = \bigcup_{i=1}^{n} E_i^c. \tag{1.4-8}$$

[†]Equation 1.4-3 shows why $\cup$ is preferable, at least initially, to $+$ to indicate union. The beginning student might—in error—write $(E - F) + (F - E) = E - F + F - E = 0$, which is meaningless.

[‡]See Appendix A for the meaning of induction.

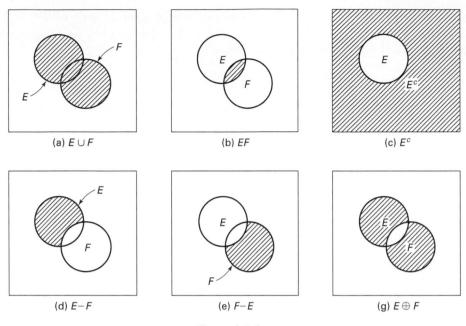

Figure 1.4-1

The relations are known as De Morgan's laws after the English mathematician Augustus De Morgan (1806–1871).

Two sets E and F are said to be equal if every element in E is in F and vice versa. Equivalently,

$$E = F \quad \text{if} \quad E \subset F, \qquad F \subset E. \tag{1.4-9}$$

Sigma fields.   Consider a universal set Ω and a collection of subsets of Ω. Let $E, F, \ldots$ denote subsets in this collection. This collection of subsets forms a *field* $\mathcal{M}$[†] if

1. $\phi \in \mathcal{M}$, $\Omega \in \mathcal{M}$.
2. If $E \in \mathcal{M}$ and $F \in \mathcal{M}$, then $E \cup F \in \mathcal{M}$, and $EF \in \mathcal{M}$.[‡]
3. If $E \in \mathcal{M}$, then $E^c \in \mathcal{M}$.

A *sigma* (σ) *field*[§] $\mathcal{F}$ is a field that is closed under any countable set of unions, intersections, and combinations. Thus if $E_1, \ldots, E_n, \ldots$ belong to $\mathcal{F}$ so do

$$\bigcup_{i=1}^{\infty} E_i \quad \text{and} \quad \bigcap_{i=1}^{\infty} E_i,$$

[†] Also sometimes called an algebra.
[‡] From this it follows that if $E_1, \ldots, E_n$ belongs to $\mathcal{M}$ so do $\bigcup_{i=1}^{n} E_i \in \mathcal{M}$ and $\bigcap_{i=1}^{N} E_i \in \mathcal{M}$.
[§] Also sometimes called a σ-algebra.

where these are defined as

$$\bigcup_{i=1}^{\infty} E_i \triangleq \{\text{the set of all elements in } at\ least\ one\ E_i\}$$

and

$$\bigcap_{i=1}^{\infty} E_i \triangleq \{\text{the set of all elements in } every\ E_i\}$$

Events.   Consider an experiment $\mathscr{H}$ with sample description space Ω. If Ω has a countable number of elements, then every subset of Ω may be assigned a probability in a way consistent with the axioms given in the next section. Then the class of all subsets make up a σ-field and each subset is an *event*. Thus we speak of the σ-field $\mathscr{F}$ of events. However, when Ω is not countable, that is, when $\Omega = R =$ the real line, then not every subset of Ω can be assigned a probability that will be consistent with the axiomatic theory. Only those subsets to which a probability can be assigned consistent with the axioms will be called events. The collection of those subsets is smaller than the collection of all possible subsets that one can define on Ω. This smaller collection forms a σ-field. On the real line R the σ-field is sometimes called the Borel field of events and, as a practical matter, includes all subsets of engineering and scientific interest.[†]

At this stage of our development, we have two of the three objects required for the axiomatic theory of probability, namely, a sample description space Ω and a σ-field $\mathscr{F}$ of events defined on Ω. We still need a probability measure P. The three objects $(\Omega, \mathscr{F}, P)$ form a triplet called a *probability space $\mathscr{P}$.*

1.5 AXIOMATIC DEFINITION OF PROBABILITY

Probability is a set function $P[\cdot]$ that assigns to every event $E \in \mathscr{F}$ a number $P[E]$ called the probability of E such that

(1) $P[E] \geq 0.$ (1.5-1)

(2) $P[\Omega] = 1.$ (1.5-2)

(3) $P[E \cup F] = P[E] + P[F]$ if $EF = \phi.$ (1.5-3)

These axioms are sufficient[‡] to establish the following basic results, all but one of which we leave as exercises for the reader:

(4) $P[\phi] = 0.$ (1.5-4)

(5) $P[EF^c] = P[E] - P[EF]$ where $E \in \mathscr{F}, F \in \mathscr{F}.$ (1.5-5)

[†]For two-dimensional Euclidean sample spaces, the Borel field of events would be subsets of $R \times R$; for three-dimensional sample spaces it would be subsets of $R \times R \times R$.

[‡]A fourth axiom: $P\left[\bigcup_{i=1}^{\infty} E_i\right] = \sum_{i=1}^{\infty} P[E_i]$ if $E_i E_j = \phi$ all $i \neq j$ must be included to enable one to deal rigorously with limits and countable unions. This axiom is of no concern to us here but will be in later chapters.

(6) $P[E] = 1 - P[E^c]$.   (1.5-6)

(7) $P[E \cup F] = P[E] + P[F] - P[EF]$.   (1.5-7)

From Axiom 3 we can establish by mathematical induction that

$$P\left[\bigcup_{i=1}^{n} E_i\right] = \sum_{i=1}^{n} P[E_i] \text{ if } E_i E_j = \phi \quad \text{all} \quad i \neq j \tag{1.5-8}$$

From this result and Equation 1.5-7 we establish the general result that $P\left[\bigcup_{i=1}^{n} E_i\right] \leq \sum_{i=1}^{n} P[E_i]$. This result is sometimes known as the *union bound*.

Example 1.5-1

We wish to prove result (7). First we decompose the event $E \cup F$ into three disjoint events as follows:

$$E \cup F = EF^c \cup E^c F \cup EF.$$

By Axiom 3

$$P[E \cup F] = P[EF^c \cup E^c F] + P[EF]$$
$$= P[EF^c] + P[E^c F] + P[EF], \text{ by axiom (3) again}$$
$$= P[E] - P[EF] + P[F] - P[EF] + P[EF]$$
$$= P[E] + P[F] - P[EF]. \tag{1.5-9}$$

In obtaining line three from line two, we used result (5).

Example 1.5-2

The experiment consists of throwing a coin once. Hence

$$\Omega = \{H, T\}.$$

The σ-field of events consists of the following sets: $\{H\}$, $\{T\}$, Ω, ϕ. With the coin assumed fair, we have[†]

$$P[\{H\}] = P[\{T\}] = \tfrac{1}{2}, \qquad P[\Omega] = 1, \qquad P[\phi] = 0.$$

Example 1.5-3

The experiment consists of throwing a die once. The outcome is the number of dots n_i appearing on the upface of the die. The set Ω is given by $\Omega = \{1, 2, 3, 4, 5, 6\}$. The σ-field of events consists of 2^6 elements. Some are

$$\phi, \Omega, \{1\}, \{1, 2\}, \{1, 2, 3\}, \{1, 4, 6\}, \{1, 2, 4, 5\},$$

[†]To be clear we must distinguish between *outcomes* ζ and *elementary events* $\{\zeta\}$. The outcomes ζ are elements of Ω. The elementary events $\{\zeta\}$ are subsets of Ω. Probability is a set function; it assigns a number to every event. Thus should we write $P[\{\zeta\}]$ rather than $P[\zeta]$ and more generally $P[\{\zeta_1, \zeta_2, \ldots, \zeta_n\}]$ rather than $P[\zeta_1, \zeta_2, \ldots, \zeta_n]$. However, we shall frequently dispense with this notational complication and hope that the meaning of the equation will be clear from the context.

and so forth. We assign

$$P[\{n_i\}] = \tfrac{1}{6} \qquad i = 1, \dots, 6.$$

All probabilities can now be computed from the basic axioms and the assumed probabilities for the elementary events. Thus with $A = \{1\}$ and $B = \{2, 3\}$ we obtain $P[A] = \tfrac{1}{6}$. Also $P[A \cup B] = P[A] + P[B]$, since $AB = \phi$. Furthermore, $P[B] = P[\{2\}] + P[\{3\}] = \tfrac{2}{6}$ so that

$$P[A \cup B] = \tfrac{1}{6} + \tfrac{2}{6} = \tfrac{1}{2}.$$

Example 1.5-4

The experiment consists of picking at random a numbered ball from 12 balls numbered 1 to 12 from an urn:

$$\Omega = \{1, \dots, 12\}.$$

Let

$$A^\dagger = \{1, \dots, 6\} \qquad B = \{3, \dots, 9\}$$

$$A \cup B = \{1, \dots, 9\} \qquad AB = \{3, 4, 5, 6\}, \qquad AB^c = \{1, 2\}$$

$$B^c = \{1, 2, 10, 11, 12\}, \qquad A^c = \{7, \dots, 12\}, \qquad A^c B^c = \{10, 11, 12\}$$

$$(AB)^c = \{1, 2, 7, 8, 9, 10, 11, 12\}.$$

Hence

$$P[A] = P[\{1\}] + P[\{2\}] + \dots + P[\{6\}]$$

$$P[B] = P[\{3\}] + \dots + P[\{9\}]$$

$$P[AB] = P[\{3\}] + \dots + P[\{6\}].$$

If $P[\{1\}] = \dots = P[\{12\}] = \tfrac{1}{12}$, then $P[A] = \tfrac{1}{2}$, $P[B] = \tfrac{7}{12}$, $P[AB] = \tfrac{4}{12}$, and so forth.

We point out that a theory of probability could be developed from a different set of axioms [1-7]. However, whatever axiom is used and whatever theory is developed, for it to be useful in solving problems in the physical world, it must model our empirical concept of probability as a relative frequency and the consequences that follow from it.

The extension of Equation 1.5-7 to the case of three events is straightforward but somewhat tedious. We consider the three events, E_1, E_2, E_3, and wish to compute that the probability, $P[\bigcup_{i=1}^{3} E_i]$ that at least one of these events occurs. From the Venn diagram in Figure 1.5-1, we see that there are seven disjoint regions in $\bigcup_{i=1}^{3} E_i$ which we label as σ_i, $i = 1, \dots, 7$. Then $P[\bigcup_{i=1}^{3} E_i] = P[\bigcup_{i=1}^{7} \sigma_i] = \sum_{i=1}^{7} P[\sigma_i]$, from Axiom 3.

---

†Thus A occurs whenever any of the numbers 1 through 6 is an outcome. AB occurs whenever any of the 3 through 6 occurs, and so forth, for the other events.

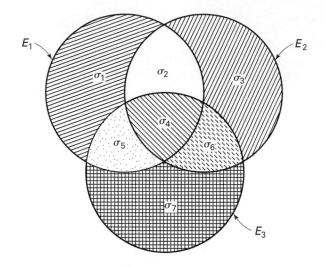

Figure 1.5-1   Partitioning $\bigcup_{i=1}^{3} E_i$ into seven disjoint regions $\sigma_1, \ldots, \sigma_7$.

In terms of the original events, the seven disjoint regions are identified as

$$\sigma_1 = E_1 E_2^c E_3^c = E_1 (E_2 \bigcup E_3)^c;$$

$$\sigma_2 = E_1 E_2 E_3^c;$$

$$\sigma_3 = E_1^c E_2 E_3^c = E_2 (E_1 \cup E_3)^c;$$

$$\sigma_4 = E_1 E_2 E_3;$$

$$\sigma_5 = E_1 E_2^c E_3;$$

$$\sigma_6 = E_1^c E_2 E_3;$$

$$\sigma_7 = E_1^c E_2^c E_3 = E_3 (E_1 \cup E_2)^c.$$

The computations of the probabilities $P[\sigma_i]$, $i = 1, \ldots, 7$, follow from Equations 1.5-5 and 1.5-7. Thus we compute

$$P[\sigma_1] = P[E_1] - P[E_1 E_2 \cup E_1 E_3]$$
$$= P[E_1] - \{P[E_1 E_2] + P[E_1 E_3] - P[E_1 E_2 E_3]\}.$$

In obtaining the first line, we used Equation 1.5-5. In obtaining the second line, we used Equation 1.5-7. The computations of $P[\sigma_i]$, $i = 3, 7$ are quite similar to the computation $P[\sigma_1]$ and involve the same sequence of steps. Thus

$$P[\sigma_3] = P[E_2] - \{P[E_1 E_2] + P[E_2 E_3] - P[E_1 E_2 E_3]\};$$
$$P[\sigma_7] = P[E_3] - \{P[E_1 E_3] + P[E_2 E_3] - P[E_1 E_2 E_3]\};$$

The computations of $P[\sigma_2]$, $P[\sigma_2]$, and $P[\sigma_6]$ are also quite similar and involve applying Equation 1.5-5. Thus

$$P[\sigma_2] = P[E_1 E_2] - P[E_1 E_2 E_3];$$

$$P[\sigma_5] = P[E_1 E_3] - P[E_1 E_2 E_3];$$

$$P[\sigma_6] = P[E_2 E_3] - P[E_1 E_2 E_3];$$

and, finally

$$P[\sigma_4] = P[E_1 E_2 E_3].$$

Now, recalling that $P[\bigcup_{i=1} E_i] = \sum_{i=1}^{7} P[\sigma_i]$, we merely add all the $P[\sigma_i]$ to obtain the desired result. This gives

$$P\left[\bigcup_{i=1}^{3} E_i\right] = \sum_{i=1}^{3} P[E_i] - \{P[E_1 E_2] + P[E_1 E_3] + P[E_2 E_3]\} + P[E_1 E_2 E_3]. \qquad (1.5\text{-}10)$$

If we adopt the notation $P_i \triangleq P[E_i]$, $P_{ij} \triangleq P[E_i E_j]$, and $P_{ijk} \triangleq P[E_i E_j E_k]$, where $1 \leq i < j < k \leq 3$, we can rewrite Equation 1.5-10 as

$$P\left[\bigcup_{i=1}^{3} E_i\right] = \sum_{i=1}^{3} P_i - \sum_{1 \leq i < j \leq 3} P_{ij} + \sum_{1 \leq i < j < k \leq 3} P_{ijk}.$$

The last sum contains only one term, namely P_{123}. Denote now each sum by the symbol S_l, where the l denotes the number of subscripts associated with the terms in that sum. Then

$$P\left[\bigcup_{i=1}^{3} E_i\right] = S_1 - S_2 + S_3, \quad \text{where } S_1 \triangleq \sum_{i=1}^{3} P_i, S_2 \triangleq \sum_{1 \leq i < j \leq 3} P_{ij}, \text{ and } S_3$$

$$\triangleq \sum_{1 \leq i < j < k \leq 3} P_{ijk}.$$

Why this introduction of new notation? Using the symbols $S_l, l = 1, \ldots$, we can extend Equation 1.5-10 to the general case.

Theorem 1.5-1   The probability P that at least one among the events E_1, E_2, E_n occurs in a given experiment is given by

$$P = S_1 - S_2 + \ldots \pm S_n,$$

where $S_1 \triangleq \sum_{i=1}^{n} P_i$, $S_2 \triangleq \sum_{1 \leq i < j \leq n} P_{ij}, \ldots, S_n \triangleq \sum_{1 \leq i < j < k < \ldots < l \leq n} P_{ijk\ldots l}$. The last sum has n subscripts and contains only one term. ∎

The proof of this theorem is given in [1-8, p. 89]. It can also be proved by induction that is, assume that $P = S_1 - S_2 + \ldots \pm S_n$ is true. Then show that for the case $n + 1$, $P = S_1 - S_2 + \ldots \mp S_{n+1}$. We leave this exercise for the braver reader.

1.6 JOINT, CONDITIONAL, AND TOTAL PROBABILITIES; INDEPENDENCE

Assume that we perform the following experiment: We are in a certain U.S. city and wish to collect weather data about it. In particular we are interested in three events, call them A, B, C, where

A is the event that on any particular day, the temperature equals or exceeds 10°C;

B is the event that on any particular day, the amount of precipitation equals or exceeds 5 millimeters;

C is the event that on any particular day A and B both occur, that is, $C \triangleq AB$.

Since C is an event, $P[C]$ is a probability that satisfies the axioms. But $P[C] = P[AB]$; we call $P[AB]$ the *joint probability of the events A and B*. This notion can obviously be extended to more than two events, that is, $P[EFG]$ is the joint probability of events E, F, and G.[†] Now let n_i denote the number of days on which event i occurred. Over a thousand-day period ($n = 1000$), the following observations are made: $n_A = 811$, $n_B = 306$, $n_{AB} = 283$. By the relative frequency interpretation of probability

$$P[A] \simeq \frac{n_A}{n} = \frac{811}{1000} = 0.811$$

$$P[B] \simeq \frac{n_B}{n} = 0.306$$

$$P[AB] \simeq \frac{n_{AB}}{n} = 0.283$$

Consider now the ratio n_{AB}/n_A. This is the relative frequency with which event AB occurs when event A occurs. Put into words, it is the fraction of time that the amount of precipitation equals or exceeds 5 millimeters on those days *when the temperature equals or exceed* 10°C. Thus we are dealing with the frequency of an event, given that or *conditioned upon the fact that another event has occurred.* Note that

$$\frac{n_{AB}}{n_A} = \frac{n_{AB}/n}{n_A/n} \simeq \frac{P[AB]}{P[A]}. \tag{1.6-1}$$

This empirical concept suggests that we introduce in our theory a conditional probability measure $P[B|A]$ defined by

$$P[B|A] \triangleq \frac{P[AB]}{P[A]}, \qquad P[A] > 0 \tag{1.6-2}$$

and described in words as the probability that B occurs given that A has occurred, and similarly,

$$P[A|B] \triangleq \frac{P[AB]}{P[B]}, \qquad P[B] > 0. \tag{1.6-3}$$

[†] E, F, G are any three events defined on the same probability space.

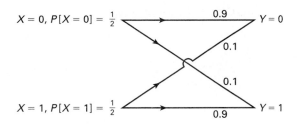

Figure 1.6-1   A binary communication system.

Definitions 1.6-2 and 1.6-3 can be used to compute the joint probability of AB. We illustrate this with an example.

Example 1.6-1

In a binary communication system (Figure 1.6-1), that is, one in which a two-symbol alphabet is used to communicate, the two symbols are "zero" and "one." In this system let Y stand for the received symbol and X for the transmitted symbol. The sample description space Ω for this experiment is $\Omega = \{(X,Y): X = 0$ or $1, Y = 0$ or $1\} = \{(0,0),(0,1),(1,0),(1,1)\}$. The event $\{X\}$ is given by $\{X\} = \{(X,0),(X,1)\}$. The probability function is $P[\{(X,Y) = (i,j)\}] = P[\{X = i\}]P[\{Y = j|X = i\}]$ $i,\ j = 0,1$. We shall dispense with the curly-bracket notation when writing joint probabilities. Thus $P[\{(X,Y) = (i,j)\}]$ will be written $P[X = i, Y = j]$. The reader should understand that we are speaking of the joint event $\{X = i\} \cap \{Y = j\}$ and not of the event $\{X = i\} \cup \{Y = j\}$.

Because of noise a transmitted zero sometimes gets decoded as a received one and vice versa. From measurements it is known that

$$P[Y = 1|X = 1] = 0.9 \qquad P[Y = 1|X = 0] = 0.1$$
$$P[Y = 0|X = 1] = 0.1 \qquad P[Y = 0|X = 0] = 0.9$$

and by design[†] $P[X = 0] = P[X = 1] = 0.5$. The various joint probabilities are then

$$P[X = 0, Y = 0] = P[Y = 0|X = 0]P[X = 0] = 0.45$$
$$P[X = 0, Y = 1] = P[Y = 1|X = 0]P[X = 0] = 0.05$$
$$P[X = 1, Y = 0] = P[Y = 0|X = 1]P[X = 1] = 0.05$$
$$P[X = 1, Y = 1] = P[Y = 1|X = 1]P[X = 1] = 0.45.$$

The introduction of conditional probabilities raises the intriguing question of whether conditional probabilities satisfy Axioms 1 to 3. In other words, given any two events E, F such that $EF = \phi$ and a third event A, all belonging to the σ-field of events $\mathscr{F}$ in the probability space $(\Omega, \mathscr{F}, P)$ does

$$P[E|A] \geq 0?$$

[†]It is good practice to design a code in which the zeros and ones appear at the same rate.

$$P[\Omega|A] = 1?$$

$$P[E \cup F|A] = P[E|A] + P[F|A] \quad \text{for } EF = \phi?$$

The answer is yes. We leave the details as an exercise to the reader.

The next problem illustrates the use of joint and conditional probabilities.

Example 1.6-2[†]

Assume that a beauty contest is being judged by the following rules: (1) There are N contestants not seen by the judges before the contest, and (2) the contestants are individually presented to the judges in a random sequence. Only one contestant appears before the judges at any one time. (3) The judges must decide on the spot whether the contestant appearing before them is the most beautiful. If they decide in the affirmative, the contest is over but the risk is that a still more beautiful contestant is in the group as yet not displayed. In that case the judges would have made the wrong decision. On the other hand, if they pass over the candidate, the contestant is disqualified from further consideration even if it turns out that all subsequent contestants are less beautiful. What is the probability of correctly choosing the most beautiful contestant? What is a good strategy to follow?

Solution   To make the problem somewhat more quantitative assume that all the virtues of each contestant are summarized into a single "beauty" number. Thus the most beautiful contestant is associated with the highest number, the least beautiful has the lowest number. The numbers, unseen by the judges, are placed in a bag and the numbers are drawn individually from the bag. Imagine that the draws are ordered along a line as shown in Figure 1.6-2. Thus the first draw is number 1, the second is 2, and so forth. At each draw, a number appears; is it the largest of all N numbers?

Assume that the following "wait-and-see" strategy is adopted: We pass over the first α draws (that is, we reject the first α contestants) but record the highest number (that is, the most beautiful contestant) observed within this group of α. Call this number ξ.

Now we continue drawing numbers (that is, call for more contestants to appear) but *don't reject them out of hand*. The first draw (contestant) after the α passed-over draws that yields a number exceeding ξ is taken to be the winner.

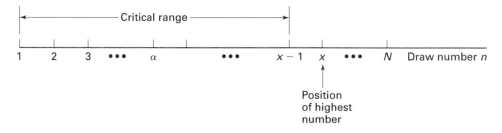

Figure 1.6-2   The numbers along the axis represent the chronology of the draw, *not* the number actually drawn from the bag.

[†]Thanks are due to Geof Willianson and Jerry Tiemann for valuable discussions regarding this problem.

Now define $C_m(\alpha)$ as the event that the *largest number* that is drawn from the first m draws occurs in the group of first α draws. Then $C_m = \Omega$ (the certain event) for $m \leq \alpha$ and $C_m \subset \Omega$ for $m > \alpha$. Let X denote the draw that will contain the largest number among the N numbers in the bag. Then two events must occur jointly for the correct decision to be realized. (1) (obvious) $\{X > \alpha\}$; and (2) (subtle) $C_m(\alpha)$ for all values of m leading up to X. Thus for a correct decision to happen, the subevent $\{X = \alpha + j + 1\}$ must occur jointly with the event $C_{\alpha+j}(\alpha)$. The event $\{X > \alpha\}$ can be resolved into disjoint subevents as

$$\{X > \alpha\} = \{X = \alpha + 1\} \cup \{X = \alpha + 2\} \cup \ldots \cup \{X = N\}.$$

Thus

$$\{X > \alpha\} \cap C_m(\alpha) = \{X = \alpha + 1, C_\alpha(\alpha)\} \cup$$

$$\{X = \alpha + 2, C_{\alpha+1}(\alpha)\} \cup \ldots \cup \{X = N, C_{N-1}(\alpha)\}$$

and the probability of a correct decision is

$$P[D] = \sum_{j=\alpha+1}^{N} P[X = j, C_{j-1}(\alpha)]$$

$$= \sum_{j=\alpha+1}^{N} P[C_{j-1}(\alpha)|X = j,]P[X = j]$$

$$= \frac{1}{N} \sum_{j=\alpha+1}^{N} \frac{\alpha}{j-1}.$$

By Euler's summation formula[†] for large N

$$P[D] = \frac{1}{N} = \sum_{j=\alpha+1}^{N} \frac{\alpha}{j-1} \simeq \frac{\alpha}{N} \int_{\alpha}^{N} \frac{dx}{x} = -\frac{\alpha}{N} \ln \frac{\alpha}{N}.$$

The best choice of α, say, α_0 is found by differentiation. Thus by setting

$$\frac{dP[D]}{d\alpha} = 0$$

we find that

$$\alpha_0 = \frac{N}{e}$$

and the maximum probability, P_0, of a correct decision is

$$P_0 \simeq 0.37 \quad (N \text{ large}).$$

Thus we let approximately the first third of the contestants pass by before beginning to judge the contestants in earnest.

[†]See, for example, G. F. Carrier et al., *Functions of a Complex Variable* (New York: McGraw-Hill, 1966), p. 246.

Unconditional probability. In many problems in engineering and science we would like to compute the unconditional probability, $P[B]$, of an event B in terms of the sum of weighted conditional probabilities. Such a computation is easily realized through the following theorem.

Theorem 1.6-1 Let $A_1, A_2, \ldots, A_n$ be n mutually exclusive events such that $\bigcup_{i=1}^n A_i = \Omega$ (the A_i's are *exhaustive*). Let B be any event defined over the probability space of the A_i's. Then, with $P[A_i] \neq 0$ all i,

$$P[B] = P[B|A_1]P[A_1] + \ldots + P[B|A_n]P[A_n]. \qquad (1.6\text{-}4)$$

Sometimes $P[B]$ is called the "average" or "total" probability of B because the expression on the right is reminiscent of the operation of averaging and $P[B]$ is seem to be the sum of parts.

Proof We have $A_i A_j = \phi \; i \neq j$ and $\bigcup_{i=1}^n A_i = \Omega$. Also $B\Omega = B = B\bigcup_{i=1}^n A_i = \bigcup_{i=1}^n BA_i$. But by definition of the intersection operation, $BA_i \subset A_i$; hence $(BA_i)(BA_j) = \phi$ for all $i \neq j$. Thus from Axiom 3 (generalized to n events):

$$P[B] = P\left[\bigcup_{i=1}^n BA_i\right] = P[BA_1] + P[BA_2] + \ldots + P[BA_n]$$

$$= P[B|A_1]P[A_1] + \ldots + P[B|A_n]P[A_n]. \qquad (1.6\text{-}5)$$

The last line follows from Equation 1.6-2. ■

Example 1.6-3
For the binary communication system shown in Figure 1.6-1 compute $P[Y = 0]$ and $P[Y = 1]$.

Solution We use Equation 1.6-5 as follows:

$$P[Y = 0] = P[Y = 0|X = 0]P[X = 0] + P[Y = 0|X = 1]P[X = 1]$$

$$= (0.9)(0.5) + (0.1)(0.5)$$

$$= 0.5.$$

We can compute $P[Y = 1]$ in a similar fashion or by noting that $\{Y = 0\} \cup \{Y = 1\} = \Omega$ and $\{Y = 0\} \cap \{Y = 1\} = \phi$. Hence $P[Y = 0] + P[Y = 1] = 1$ or

$$P[Y = 1] = 1 - P[Y = 0] = 0.5.$$

Independence. This is an important concept and full justice to all its implications cannot be done until we discuss random variables in Chapter 2.

Definitions (A) Two events $A \in \mathscr{F}$, $B \in \mathscr{F}$ with $P[A] > 0$, $P[B] > 0$ are said to be independent if and only if (iff)

$$P[AB] = P[A]P[B]. \qquad (1.6\text{-}6)$$

Since, in general, $P[AB] = P[B|A]P[A] = P[A|B]P[B]$ it follows that for independent events

$$P[A|B] = P[A] \qquad\qquad\qquad (1.6\text{-}7a)$$

$$P[B|A] = P[B]. \qquad\qquad\qquad (1.6\text{-}7b)$$

Thus the definition satisfies our intuition: If A and B are independent, the outcome B should have no effect on the probability of A and vice versa.

(B) Three events A, B, C defined on $\mathscr{P}$ and having nonzero probabilities are said to be independent if

$$P[ABC] = P[A]P[B]P[C] \qquad\qquad\qquad (1.6\text{-}8a)$$

$$P[AB] = P[A]P[B] \qquad\qquad\qquad (1.6\text{-}8b)$$

$$P[AC] = P[A]P[C] \qquad\qquad\qquad (1.6\text{-}8c)$$

$$P[BC] = P[B]P[C]. \qquad\qquad\qquad (1.6\text{-}8d)$$

This is an extension of (A) above and suggests the pattern for the definition of n independent events $A_1, \ldots, A_n$. Note that it is *not enough* to have just $P[ABC] = P[A]P[B]P[C]$. Pairwise independence must also be shown.

(C) Let A_i, $i = 1, \ldots, n$ be n events defined on $\mathscr{P}$. The $\{A_i\}$ are said to be independent if

$$P[A_i A_j] = P[A_i]P[A_j]$$

$$P[A_i][A_j][A_k] = P[A_i]P[A_j]P[A_k]$$

$$\vdots$$

$$P[A_1 \ldots A_n] = P[A_1]P[A_2] \ldots P[A_n]$$

for all combination of indices such that $1 \leq i < j < k < \ldots \leq n$. ∎

Example 1.6-4 _____

(Testing for Independence.) An urn contains 10 numbered black balls (some even, some odd) and 20 numbered white balls (some even, some odd). Some of the balls of each color are lighter in weight than the others. The composition of the urn is shown in the diagram associated with this example.

Let A denote the event of picking a black ball, B denote the event of picking one of the lighter weight balls, and C denote the event of picking a even-numbered-ball. Are A, B, C independent events?

Solution   We first test whether $P[ABC] = P[A]P[B]P[C]$. Since $P[A] = 1/3$, $P[B] = 1/2$, $P[C] = 2/5$ and $P[ABC] = 2/30$, it follows that $P[ABC] = P[A]P[B]P[C]$. However to show independence we must also show that $P[AB] = P[A]P[B]$, $P[AC] = P[A]P[C]$, $P[BC] = P[B]P[C]$. Note that $P[AC] = 2/30$ while $P[A]P[C] = 1/3 \times 12/30 = 2/15 \neq 2/30$. Hence A, B, C are not independent events.

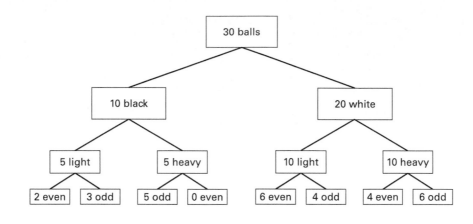

1.7 BAYES' THEOREM AND APPLICATIONS

The previous results enable us now to write a fairly simple formula known as Bayes' theorem.[†] Despite its simplicity, this formula is widely used in biometrics, epidemiology, and communication theory.

Bayes' Theorem   Let A_i, $i = 1, \ldots, n$ be a set of disjoint and exhaustive events defined on a probability space $\mathscr{P}$. Then $\bigcup_{i=1}^{n} A_i = \Omega$, $A_i A_j = \phi$ for $i \neq j$. With B any event defined on $\mathscr{P}$ with $P[B] > 0$ and $P[A_i] \neq 0$ all i

$$P[A_j|B] = \frac{P[B|A_j]P[A_j]}{\sum\limits_{i=1}^{n} P[B|A_i]P[A_i]}. \qquad \blacksquare \tag{1.7-1}$$

Proof   The denominator is simply $P[B]$ by Equation 1.6-5 and the numerator is simply $P[A_j B]$. Thus Bayes' Theorem is merely an application of the definition of conditional probability.   ∎

Remark   In practice the terms in Equation 1.7-1 are given various names: $P[A_j|B]$ is known as the *a posteriori* probability of A_j given B; $P[B|A_j]$ is called the *a priori* probability of B given A_j; and $P[A_i]$ is the *causal* or *a priori* probability of A_j. In general *a priori* probabilities are estimated from past measurements or presupposed by experience while *a posteriori* probabilities are measured or computed from observations.   ∎

Example 1.7-1 _____

In a communication system a zero or one is transmitted with $P[X = 0] = P_0$, $P[X = 1] = 1 - P_0 \overset{\Delta}{=} P_1$, respectively. Due to noise in the channel, a zero can be received as a one with probability β and a one can be received as a zero also with probability β. A one is observed. What is the probability that a one was transmitted?

_____
[†]Named after Thomas Bayes, English mathematician/philosopher (1702–1761).

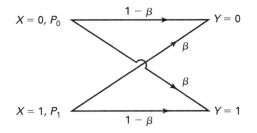

Figure 1.7-1   Representation of a binary communication channel subject to noise.

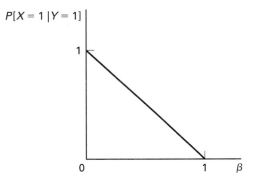

Figure 1.7-2   *A posteriori* probability versus β.

Solution   The structure of the channel is shown in Figure 1.7-1. We write

$$P[X = 1|Y = 1] = \frac{P[X = 1, Y = 1]}{P[Y = 1]}$$

$$= \frac{P[Y = 1|X = 1]P[X = 1]}{P[Y = 1|X = 1]P[X = 1] + P[Y = 1|X = 0]P[X = 0]}$$

$$= \frac{P_1(1 - \beta)}{P_1(1 - \beta) + P_0\beta}.$$

If $P_0 = P_1 = \frac{1}{2}$, the *a posteriori* probability $P[X = 1|Y = 1]$ depends on β as shown in Figure 1.7-2.

The channel is said to be noiseless if $\beta = 1$ or $\beta = 0$.

Example 1.7-2 _____

(Parzen [1-9, p. 119].) Suppose there exists a (fictitious) test for cancer with the following properties. Let

A = event that the test states that tested person has cancer.

B = event that person has cancer.

A^c = event that test states person is free from cancer.

B^c = event that person is free from cancer.

It is known that $P[A|B] = P[A^c|B^c] = 0.95$ and $P[B] = 0.005$. Is the test a good test?

Solution   To answer this question we should like to know the likelihood that a person actually has cancer if the test so states, that is, $P[B|A]$. Hence

$$P[B|A] = \frac{P[B]P[A|B]}{P[A|B]P[B] + P[A|B^c]P[B^c]} = \frac{(0.005)(0.95)}{(0.95)(0.005) + (0.05)(0.995)}$$
$$= 0.087.$$

Hence in only 8.7% of the cases where the tests are positive will the person actually have cancer. This test has a very high false-alarm rate and in this sense cannot be regarded as a good test. The fact that, initially, the test seems like a good test is not surprising given that $P[A|B] = P[A^c|B^c] = 0.95$. However, when the scarcity of cancer in the general public is considered, the test is found to be vacuous.

1.8 COMBINATORICS[†]

Before proceeding with our study of basic probability we introduce a number of counting formulas of importance in probability. Some of the results presented here will have immediate application in Section 1.9; others will be useful later.

A *population of size n* will be taken to mean a collection of n elements without regard to order. Two populations are considered different if one contains at least one element not contained in the other. A *subpopulation* of size r from a population of size n is a subset of r elements taken from the original population. Likewise, two subpopulations are considered different if one has at least one element different from the other.

Next, consider a population of n elements $a_1, a_2, \ldots, a_n$. Any ordered arrangement $a_{k_1}, a_{k_2}, \ldots, a_{k_r}$ of r symbols is called an *ordered sample* of size r.

Consider now the generic urn containing n distinguishable numbered balls. Balls are removed one by one. How many different ordered samples of size r can be formed? There are two cases:

(*i*) *Sampling with replacement.*   Here after each ball is removed, its number is recorded and it is returned to the urn. Thus for the first sample there are n choices, for the second there are again n choices, and so on. Thus we are led to the following result: For a population of n elements, there are n^r different ordered samples of size r that can be formed with replacement.

(*ii*) *Sampling without replacement.*   After each ball is removed, it is not available anymore for subsequent samples. Thus n balls are available for the first sample, $n - 1$

[†]This material closely follows that of William Feller [1-8].

for the second and so forth. Thus we are now led to the result: For a population of n elements there are

$$(n)_r \overset{\Delta}{=} n(n-1)(n-2)\ldots(n-r+1)$$

$$= \frac{n!}{(n-r)!} \tag{1.8-1}$$

different ordered samples of size r that can be formed without replacement.[†]

The Number of Subpopulations of Size r in a Population of Size n. A basic problem that often occurs in probability is the following: How many groups, that is, subpopulations of size r can be formed from a population of size n? For example, consider 6 balls numbered 1 to 6. How many groups of size 2 can be formed? The following table shows that there are 15 groups of size 2 that can be formed:

12	23	34	45	56
13	24	35	46	
14	25	36		
15	26			
16				

Note that this is different from the number of ordered samples that can be formed without replacement. These are $(6 \cdot 5 = 30)$:

12	21	31	41	51	61
13	23	32	42	52	62
14	24	34	43	53	63
15	25	35	45	54	64
16	26	36	46	56	65

Also it is different from the number of samples that can be formed with replacement $(6^2 = 36)$:

11	21	31	41	51	61
12	22	32	42	52	62
13	23	33	43	53	63
14	24	34	44	54	64
15	25	35	45	55	65
16	26	36	46	56	66

A general formula for the number of subpopulations, C_r^n of size r in a population of size n can be computed as follows: Consider an urn with n distinguishable balls. We already know that the number of ordered samples of size r that can be formed is $(n)_r$. Now consider a specific subpopulation of size r. For this subpopulation there are $r!$ arrangements and therefore $r!$ different ordered samples. Thus for C_r^n subpopulations there must be $C_r^n \cdot r!$ different ordered samples of size r. Hence

$$C_r^n \cdot r! = (n)_r$$

[†]Different samples will often contain the same subpopulation but with a different *ordering*. For this reason we sometimes speak of $(n)_r$ ordered samples that can be formed without replacement.

or

$$C_r^n = \frac{(n)_r}{r!} = \frac{n!}{(n-r)!r!} \triangleq \binom{n}{r}. \tag{1.8-2}$$

Equation 1.8-2 is an important result, and we shall apply it in the next section. The symbol

$$C_r^n \triangleq \binom{n}{r}$$

is called a *binomial coefficient.* Clearly

$$\binom{n}{r} = \frac{n!}{r!(n-r)!} = \frac{n!}{(n-r)!r!} \triangleq \binom{n}{n-r} = C_{n-r}^n. \tag{1.8-3}$$

We already know from Section 1.4 that the total number of subsets of a set of size n is 2^n. The number of subsets of size r is $\binom{n}{r}$. Hence we obtain that

$$\sum_{r=0}^{n} \binom{n}{r} = 2^n.$$

A result which can be viewed as an extension of C_r^n is given by the following.

Theorem 1.8-1   Let $r_1, \ldots, r_k$ be a set of nonnegative integers such that $r_1 + r_2 + \ldots + r_k = n$. Then the number of ways in which a population of n elements can be partitioned into k subpopulations of which the first contains r_1 elements, the second r_2 elements, and so forth, is

$$\frac{n!}{r_1! r_2! \ldots r_k!} \tag{1.8-4}$$

This coefficient is called the *multinomial coefficient.* Note that the order of the subpopulation is essential in the sense that ($r_1 = 7$, $r_2 = 10$) and ($r_1 = 10$, $r_2 = 7$) represent different partitions. However, the order within each group does not receive attention. For example, suppose we have five distinguishable balls (1,2,3,4,5) and we ask how many subpopulations can be made with three balls in the first group and two in the second. Here $n = 5$, $r_1 = 3$, $r_2 = 2$, and $r_1 + r_2 = 5$. The answer is $5!/3!2! = 10$ and the partitions are

Group 1:	1,2,3	2,3,4	3,4,5	4,5,1	5,1,2	2,4,5	2,3,5	1,3,5	1,3,4	1,2,4
Group 2:	4,5	5,1	1,2	2,3	3,4	1,3	1,4	2,4	2,5	3,5

Note that the order is important in that had we set $r_1 = 2$ $r_2 = 3$ we would have gotten a *different* partition, for example,

Group 1:	4,5	5,1	1,2	2,3	3,4	1,3	1,4	2,4	2,5	3,5
Group 2:	1,2,3	2,3,4	3,4,5	4,5,1	5,1,2	2,4,5	2,3,5	1,3,5	1,3,4	1,2,4

The partition (4,5), (1,2,3) is, however, identical with (5,4), (2,1,3).

While Equation 1.8-4 seems somewhat esoteric, its proof is actually quite straight-forward. Note that we can rewrite Equation 1.8-4 as

$$\frac{n!}{r_1} \cdot \frac{1}{r_2!} \cdot \cdots = \frac{1}{r_k!}$$

$$= \frac{n!}{r_1!(n-r_1)!} \cdot \frac{(n-r_1)!}{r_2!(n-r_1-r_2)!} \frac{(n-r_1-r_2)!}{r_3!(n-r_1-r_2-r_3)!} \cdots \frac{\left(n - \sum\limits_{j=1}^{k-1} r_j\right)!}{r_k! \left(n - \sum\limits_{j=1}^{k} r_j\right)!}$$

Recalling that $0! = 1$, we see that the last term is unity. Then the multinomial formula is written as

$$\frac{n!}{r_1!r_2!\ldots r_k!} = \binom{n}{r_1}\binom{n-r_1}{r_2}\binom{n-r_1-r_2}{r_3}\cdots\binom{n-r_1-r_2-\ldots-r_{k-2}}{r_{k-1}} \quad (1.8\text{-}5)$$

To affect a realization of r_1 elements in the first subpopulation, r_2 in the second, and so on, we would select r_1 elements from the given n, r_2 from the remaining $n - r_1$, r_3 from the remaining $n - r_1 - r_2$, etc. But there are $\binom{n}{r_1}$ ways of choosing r_1 elements out of n, $\binom{n-r_1}{r_2}$ ways of choosing r_2 elements out of the remaining $n - r_1$, and so forth. Thus, the total number of ways of choosing r_1 from n, r_2 from $n - r_1$, and so on, is simply the product of the factors on the right-hand side of Equation 1.8-5 and the proof is complete. ■

Example 1.8-1

[1-8, p. 36] Suppose we throw 12 dice; since each die throw has six outcomes there are 6^{12} outcomes. Consider now the event E that each face appears twice. There are, of course, many ways in which this can happen. Two outcomes in which this happens are shown below:

Dice I.D. Number	1	2	3	4	5	6	7	8	9	10	11	12
Outcome 1	3	1	3	6	1	2	5	4	4	6	2	5
Outcome 2	6	1	3	2	6	3	4	5	5	1	4	2

The total number of ways that this event can occur is the number of ways 12 dice ($n = 12$) can be arranged into six groups ($k = 6$) of two each ($r_1 = r_2 = \ldots = r_6 = 2$). Assuming that all outcomes are equally likely we compute

$$P[E] = \frac{N_E}{E} = \frac{\text{no. of ways } E \text{ can occur}}{\text{total no. of outcomes}}$$

$$= \frac{12!}{(2!)^6 6^{12}} = 0.003438.$$

The binomial and multinomial coefficients appear in the binomial and multinomial probability laws discussed in the next sections. The multinomial coefficient is also important in a class of problems called *occupancy problems* that occur in theoretical physics.

Occupancy Problems

Occupancy problems are generically modeled as the random placement of r balls into n cells. For the first ball there are n choices, for the second ball there are n choices, and so on, so that there are n^r possible distributions of r balls in n cells and each has a probability of n^{-r}. If the balls are distinguishable, then each of the distributions is distinguishable; if the balls are not distinguishable, then there are fewer than n^r distinguishable distributions. For example, with three distinguishable balls ($r = 3$) labeled "1," "2," "3" and two cells ($n = 2$) we get eight (2^3) distinguishable distributions:

Cell no. 1	1	2	3	1,2	1,3	2,3	1,2,3	—
Cell no. 2	2,3	2,3	1,2	3	2	1	—	1,2,3

When the balls are not distinguishable (each ball is represented by a $*$), we obtain four distinct distributions:

Cell no. 1	***	**	*	—
Cell no. 2	—	*	**	***

How many distinguishable distributions can be formed from r balls and n cells? An elegant way to compute this is furnished by William Feller [1-8, p. 38] using a clever artifice. This artifice consists of representing the n cells by the spaces between $n + 1$ bars and the balls by stars. Thus,

$$| \quad | \quad | \quad |$$

represents three empty cells, while

$$|**| \quad | \quad |*|**| \quad |*****|$$

represents two balls in the first cell, zero balls in the second and third cells, one in the fourth, two in the fifth, and so on. Indeed, with $r_i \geq 0$ representing the number of balls in the ith cell and r being the total number of balls, it follows that

$$r_1 + r_2 \ldots r_n = r.$$

The n-tuple $(r_1, r_2, \ldots, r_n)$ is called the *occupancy* and the r_i are the *occupancy numbers*; two distribution of balls in cells are said to be indistinguishable if their corresponding occupancies are identical. The occupancy of

$$|**| \quad | \quad |*|**| \quad |*****|$$

is (2,0,0,1,2,0,5). Note that n cells require $n + 1$ bars but since the first and last symbols must be bars, only $n - 1$ bars and r stars can appear in any order. Thus, we are asking for the number of subpopulations of size r in a population of size $n - 1 + r$. The result is, by Equation 1.8-2

$$\binom{n+r-1}{r} = \binom{n+r-1}{n-1}. \tag{1.8-6}$$

Example 1.8-2 _____

Show that the number of distinguishable distributions in which no cell remains empty is $\binom{r-1}{n-1}$. Here we require that no bars be adjacent. Therefore, n of the r stars must occupy spaces between the bars but the remaining $r - n$ stars can go anywhere. Thus $n - 1$ bars and $r - n$ stars can appear in any order. The number of distinct distributions is then equal to the number of ways of choosing $r - n$ places in $(n - 1)$ bars $+(r - n)$ stars or $r - n$ out of $n - 1 + r - n = r - 1$. This is, by Equation 1.8-2

$$\binom{r-1}{r-n} = \binom{r-1}{n-1}.$$

_____

In statistical mechanics, a six-dimensional space called *phase space* is defined which consists of three position and three momentum coordinates. Because of the uncertainty principle which states that the uncertainty in position times the uncertainty in momentum cannot be less than Planck's constant, h, phase space is quantized into tiny cells of volumes $v = h^3$. In a system that contains atomic or molecular size particles, the distribution of these particles among the cells constitute the *state* of the system. In Maxwell–Boltzman statistics, all distributions of r particles among in cells are equally likely. It can be shown (see, for example, *Concepts of Modern Physics* by A. Beiser, McGraw-Hill, 1973) that this leads to the famous Boltzmann law

$$n(\varepsilon) = \frac{2\pi N}{(\pi k)^{3/2}} \sqrt{\varepsilon}\, e^{-\varepsilon/kT}, \tag{1.8-7}$$

where $n(\varepsilon)d\varepsilon$ is the number of particles with energy between ε and $\varepsilon + d\varepsilon$, N is the total number of particles, T is absolute temperature, and k is the Boltzmann constant. The Maxwell–Boltzmann law holds for identical particles that, in some sense, can be distinguished. It is argued that the molecules of a gas are particles of this kind. It is not difficult to show that Equation 1.8-7 integrates to N.

In contrast to the Maxwell–Boltzmann statistics where all n^r arrangements are equally likely, Bose-Einstein statistics considers only distinguishable arrangements of indistinguishable identical particles. For n cells and r particles, the number of such arrangements is Equation 1.8-6

$$\binom{n+r-1}{r}$$

and each arrangement is assigned a probability

$$\binom{n+r-1}{r}^{-1}.$$

It is argued that Bose-Einstein statistics are valid for photons, nuclei, and particles of zero or integral spin that do not obey the exclusion principle. The exclusion principle, discovered by Wolfgang Pauli in 1925, states that for a certain class of particles (e.g., electrons) no two particles can exist in the same quantum states (e.g., no two or more balls in the same cell).

To deal with particles that obey the exclusion principle, a third assignment of probabilities is construed. This assignment, called Fermi-Dirac statistics, assumes

1. the exclusion principle (no two or more balls in the same cell); and
2. all distinguishable arrangements satisfying (1) are equally probable.

Note that for Fermi-Dirac statistics, $r \leq n$. The number of distinguishable arrangements under the hypothesis of the exclusion principle is the number of subpopulations of size $r \leq n$ in a population of n elements of $\binom{n}{r}$. Since each in equally likely, the probability of any one state is $\binom{n}{r}^{-1}$.

The above discussions should convince the reader of the tremendous importance of probability in the basic sciences as well as its limitations: No amount of pure reasoning based on probability axioms could have determined which particles obey which probability laws.

Extensions and Applications

Theorem 1.5-1 can be used to solve problems of engineering interest. First we note that the *number* of individual probability terms in the sum S_i is $\binom{n}{i}$. Why? There are a total of n indices and in S_i, all terms have i indices. For example with $n = 5$ and $i = 2$, S_2 will consist of the sum of the terms P_{ij}, where the indices ij are 12; 13; 14; 15; 23; 24; 25; 34; 35; 45. Each set of indices in S_i never repeats, that is, they are all different. Thus, the number of indices and, therefore, the number of terms in S_i is the number of subpopulations of size i in a population of size n which is $\binom{n}{i}$ from Equation 1.8-2. Note that S_n will have only a single term.

Example 1.8-3

We are given r balls and n cells. The balls are indistinguishable and are to be randomly distributed among the n cells, assuming that each arrangement is equally likely. Compute the probability that all cells are occupied. Note that the balls may represent data packets and the cells buffers. Or, the balls may represent air-dropped food rations and the cells, people in a country in famine.

Solution   Let E_i denote the event that cell i is empty $(i = 1, \ldots, n)$. Then the r balls are placed among the remaining $n - 1$ cells. For each of the r balls there are $n - 1$ cells to choose from. Hence there are $A(r, n - 1) \triangleq (n - 1)^r$ ways of arranging the r balls among the $n - 1$ cells. Obviously, since the balls are indistinguishable, not all arrangements will be distinguishable. Indeed there are only $\binom{n + r - 1}{n - 1}$ distinguishable distributions and these are not, typically, equally likely. The total number of ways of distributing the r balls among the n cells is n^r. Hence

$$P[E_i] = (n-1)^r/n^r = \left(1 - \frac{1}{n}\right)^r \triangleq P_i.$$

Next assume that cells i and j are empty. Then $A(r, n-2) = (n-2)^r$ and

$$P[E_i E_j] \triangleq P_{ij} = \left(1 - \frac{2}{n}\right)^r.$$

In a similar fashion, it is easy to show that $P[E_i E_j E_k] = \left(1 - \frac{3}{n}\right)^r \triangleq P_{ijk}$, and so on. Note that the right-hand side expressions for P_i, P_{ij}, P_{ijk}, and so on, do not contain the subscripts i, ij, ijk, and so on. Thus each S_i contains $\binom{n}{i}$ identical terms and their sum amounts to

$$S_i = \binom{n}{i}\left(1 - \frac{i}{n}\right)^r.$$

Let E denote the event that at least one cell is empty. Then

$$P[E] = P\left[\bigcup_{i=1}^{n} E_i\right]$$

$= S_1 - S_2 + \ldots + S_n$ (by Theorem 1.5-1). Substituting for S_i from two lines above, we get

$$P[E] = \sum_{i=1}^{n} \binom{n}{i}(-1)^{i+1}\left(1 - \frac{i}{n}\right)^r. \tag{1.8-8}$$

The event that all cells are occupied is E^c. Hence $P[E^c] = 1 - P[E]$ which can be written as

$$P[E^c] = \sum_{i=0}^{n} \binom{n}{i}(-1)^i\left(1 - \frac{i}{n}\right)^r. \tag{1.8-9}$$

We leave the details as an exercise for the reader.

Example 1.8-4 _____

Use Equation 1.8-9 to compute $P_m(r, n)$, the probability that exactly m out of the n cells are empty after the r balls have been distributed.

Solution   Using the $P_m(r, n)$ notation, we write $P[E^c] = P_0(r, n)$. Now assume that exactly m cells are empty and $n - m$ cells are occupied. Next, let's fix the m cells that are empty, for example, cells numbers $\underbrace{2, 4, 5, 7, \ldots, l}_{m \text{ terms}}$. Let $B(r, n-m)$ be the number of ways of distributing r balls among the remaining $n - m$ cells such that no cell remains empty and let $A(r, n-m)$ denote the number of ways of distributing r balls among $n - m$ cells. Then $P_0(r, n-m) = B(r, n-m)/A(r, n-m)$ and, since $A(r, n-m) = (n-m)^r$, we get that $B(r, n-m) = (n-m)^r P_0(r, n-m)$. There are $\binom{n}{m}$ ways of placing m empty cells among n cells. Hence the total number of arrangements of r balls among n cells such that m remain empty is $\binom{n}{m} (n-m)^r P_0(r, n-m)$. Finally, the number of ways of distributing r balls among n cells is n^r. Thus

$$P_m(r, n) = \binom{n}{m} (n-m)^r P_0(r, n-m)/n^r$$

or, after simplifying

$$P_m(r, n) = \binom{n}{m} \sum_{i=0}^{n} \binom{n}{i} (-1)^i \left(1 - \frac{i+m}{n}\right)^r. \qquad (1.8\text{-}10)$$

1.9 BERNOULLI TRIALS—BINOMIAL AND MULTINOMIAL PROBABILITY LAWS

Consider the very simple experiment consisting of a single trial with a binary outcome: A success $\{\zeta_1 = s\}$ with probability p or a failure $\{\zeta_2 = f\}$ with probability $q = 1 - p$. Thus $P[\{s\}] = p$, $P[\{f\}] = q$ and the sample description space is $\Omega = \{s, f\}$. The σ-field of events $\mathscr{F}$ is ϕ, Ω, $\{s\}$, $\{f\}$.

Suppose we do the experiment twice. The new sample description space Ω_2, written $\Omega_2 = \Omega \times \Omega$ is the set of all ordered 2-tuples

$$\Omega_2 = \{ss, sf, fs, ff\}.$$

$\mathscr{F}$ contains $2^4 = 16$ events. Some are ϕ, Ω, $\{ss\}$, $\{ss, ff\}$, and so forth.

The product $\Omega \times \Omega$ is called the *Cartesian product*. If we do n independent trials, the sample space is

$$\Omega_n = \underbrace{\Omega \times \Omega \times \ldots \times \Omega}_{n \text{ times}}$$

and contains 2^n elementary outcomes, each of which is an ordered n-tuple. Thus

$$\Omega_n = \{a_1, \ldots, a_M\} \quad \text{where} \quad M = 2^n$$

and $a_i \triangleq z_{i_1} \ldots z_{i_n}$, an ordered n-tuple, where $z_{i_j} = s$ or f. Since each outcome z_{i_j} is independent of any other outcome, the joint probability $P[z_{i_1} \ldots z_{i_n}] = P[z_{i_1}]P[z_{i_2}] \ldots P[z_{i_n}]$. Thus the probability of a given ordered set of k successes and $n-k$ failures is simply $p^k q^{n-k}$. For example, suppose we throw a coin three times with $p = P[\{H\}]$ and $q = P[\{T\}]$. The probability of the event $\{HTH\}$ is $pqp = p^2 q$. The probability of the event $\{THH\}$ is also $p^2 q$. The different events leading to two heads and one tail are listed here:

$$E_1 = \{HHT\}$$
$$E_2 = \{HTH\}$$
$$E_3 = \{THH\}.$$

If F denotes the event of getting two heads and one tail without regard to order, then $F = E_1 \cup E_2 \cup E_3$. Since $E_i E_j = \phi$ for $i \neq j$, we obtain $P[F] = P[E_1] + P[E_2] + P[E_3] = 3p^2 q$.

Let us now generalize the previous result by considering an experiment consisting of n Bernoulli trials. The sample description space Ω_n contains $M = 2^n$ outcomes $a_1, a_2, \ldots, a_M$, where each a_i is a string of n symbols, and each symbol represents a success s or a failure f. Consider the event $A_k \triangleq \{k \text{ successes in } n \text{ trials}\}$ and let the primed outcomes, that is a_i', denote strings with k successes and $n - k$ failures. Then, with K denoting the number of ordered arrangements involving k successes and $n - k$ failures, we write

$$A_k = \bigcup_{i=1}^{K} \{a_i'\}.$$

To determine how large K is, we use an artifice similar to that used in proving Equation 1.8-6. Let bars represent failures and stars represent successes. Then, as an example,

$$| \; * \; * \; | \; * \; * \; | \; | \; *$$

represents five successes in nine tries in the order $fssfssffs$. How many such arrangements are there? The solution is given by Equation 1.8-6 with $r = k$ and $(n-1) + r$ replaced by $(n - k) + k = n$. (Note that there is no restriction that the first and last symbols must be bars.) Thus

$$K = \binom{n}{k}$$

and, since the $\{a_i\}$ are disjoint, that is, $\{a_i\} \cap \{a_j\} = \phi$ for $i \neq j$, we obtain

$$P[A_k] = P\left[\bigcup_{i=1}^{K} \{a_i'\}\right] = \sum_{i=1}^{K} P[\{a_i'\}].$$

Finally, since $P[\{a_i'\}] = p^k q^{n-k}$ regardless of the ordering of the s and f's, we obtain

$$P[A_k] = \binom{n}{k} p^k q^{n-k}$$

$$\triangleq b(k; n, p). \tag{1.9-1}$$

The symbol $b(k; n, p)$ denotes the *binomial probability law*, defined in Equation 1.9-1, which is the probability of getting k successes in n independent tries with individual Bernoulli trial success probability p.

The binomial coefficient

$$C_k^n = \binom{n}{k}$$

was introduced in the previous section and is the number of subpopulations of size k that can be formed from a population of size n. In the preceding example the population has size 3 (three tries) and the subpopulation has size 2 (two heads), and we are interested in getting two heads in three tries with order being irrelevant. Thus the correct result is

$C_2^3 = 3$. Note that had we asked for the probability of getting two heads on the first two tosses followed by a tail, that is, $P[E_1]$, we would *not have used the coefficient C_2^3 there is only one way that this event can happen.*

Example 1.9-1

Suppose $n = 4$; that is, there are four balls numbered 1 to 4 in the urn. The number of distinguishable, ordered samples of size 2 that can be drawn without replacement is 12, that is, $\{1, 2\}; \{1, 3\}; \{1, 4\}; \{2, 1\}; \{2, 3\}; \{2, 4\}; \{3, 1\}; \{3, 2\}; \{3, 4\}; \{4, 1\}; \{4, 2\}; \{4, 3\}$. The number of distinguishable unordered sets is 6, that is,

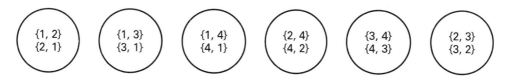

From Equation 1.8-2 we obtain this result directly; that is $(n = 4, k = 2)$

$$\binom{n}{k} = \frac{4!}{2!2!} = 6$$

Example 1.9-2

Ten independent binary pulses per second arrive at a receiver. The error (that is, a zero received as a one or vice versa) probability is 0.001. What is the probability of at least one error/second?

Solution

$$P[\text{at least one error/sec}] = 1 - P[\text{no errors/sec}]$$

$$= 1 - \binom{10}{0}(0.001)^0(0.999)^{10} = 1 - (0.999)^{10} \simeq 0.01.$$

Observation. Note that

$$\sum_{k=0}^{n} b(k; n, p) = 1. \quad \text{Why?}$$

Example 1.9-3

An odd number of people want to play a game that requires two teams made up of even numbers of players. To decide who shall be left out to act as umpire, each of the N persons tosses a fair coin with the following stipulation: If there is one person whose outcome (be it heads or tails) is different from the rest of the group, that person will be the umpire. Assume that there are 11 players. What is the probability that a player will be "odd-man out," that is, will be the umpire on the first play?

Solution Let $E \triangleq \{10H, 1T\}$, where $10H$ means $H, H, \ldots, H$ ten times, and $F \triangleq \{10T, 1H\}$. Then $EF = \phi$ and

$$P[E \cup F] = P[E] + P[F]$$

$$= \binom{11}{10} \left(\frac{1}{2}\right)^{10} \left(\frac{1}{2}\right) + \binom{11}{1} \left(\frac{1}{2}\right) \left(\frac{1}{2}\right)^{10}$$

$$\simeq 0.01074.$$

Example 1.9-4

In Example 1.9-3 derive a formula for the probability that the odd-man out will occur for the first time on the nth play. (*Hint*: Consider each play as an independent Bernoulli trial with success if an odd-man out occurs and failure otherwise.)

Solution   Let E be the event of odd-man out for first time on the nth play. Let F be the event of no odd-man out in $n-1$ plays and let G be the event of an odd-man out on the nth play. Then

$$E = FG.$$

Since it is completely reasonable to assume F and G are independent events, we can write

$$P[E] = P[F]P[G]$$

$$P[F] = \binom{n-1}{0} (0.0107)^0 (0.9893)^{n-1} = (0.9893)^{n-1}$$

$$P[G] = 0.0107.$$

Thus $P[E] = (0.0107)(0.9893)^{n-1}$.

Further discussion of the binomial law.   We write down some self-evident formulas for further use. The probability $B(k; n, p)$ of k or fewer *successes in n tries* is given by

$$B(k; n, p) = \sum_{i=0}^{k} b(i; n, p) = \sum_{i=0}^{k} \binom{n}{i} p^i q^{n-i}. \tag{1.9-2}$$

The symbol $B(k; n, p)$ is called the *binomial distribution function*. The probability of at least k successes in n tries is

$$\sum_{i=k}^{n} b(i; n, p) = 1 - B(k-1; n, p).$$

The probability of more than k successes but no more than j successes is

$$\sum_{k+1}^{j} b(i; n, p).$$

We illustrate this with an example.

Example 1.9-5 _____

Five missiles are fired against an aircraft carrier in the ocean. It takes at least two direct hits to sink the carrier. All five missiles are on the correct trajectory but must get through the "point defense" guns of the carrier. It is known that the point defense guns can destroy a missile with probability $p = 0.9$. What is the probability that the carrier will still be afloat after the encounter?

Solution   Let E be the event that the carrier is still afloat and let F be the probability of a missile getting through the point-defense guns. Then

$$P[F] = 0.1$$

and

$$P[E] = 1 - P[E^c]$$

$$= 1 - \sum_{i=2}^{5} \binom{5}{i} (0.1)^i (0.9)^{5-i} \simeq 0.92.$$

Multinomial Probability Law

The multinomial probability law is a generalization of the binomial law. The binomial law is based on Bernoulli trials in which only two outcomes are possible. The multinomial law is based on a generalized Bernoulli trial in which k outcomes are possible. Thus, consider an elementary experiment consisting of a single trial with k elementary outcomes $\zeta_1, \zeta_2, \ldots, \zeta_k$. Let the probability of outcome ζ_i be $p_i (i = 1, \ldots, k)$. Then

$$p_i \geq 0, \quad \text{and} \sum_{i=1}^{k} p_i = 1. \tag{1.9-3}$$

Assume that this generalized Bernoulli trial is repeated n times and consider the event consisting of a *prescribed, ordered* string of elementary outcomes in which ζ_1 appears r_1 times, ζ_2 appears r_2 times and so on until ζ_k appears r_k times. What is the probability of this event? The key here is that the order is prescribed *a priori*. For example, with $k = 3$ (three possible outcomes) and $n = 6$ (six tries), a prescribed string might be $\zeta_1 \zeta_3 \zeta_2 \zeta_2 \zeta_1 \zeta_2$ so that $r_1 = 2$, $r_2 = 3$, $r_3 = 1$. Observe that $\sum_{i=1}^{k} r_i = n$. Since the outcome of each trial is an independent event, the probability of observing a prescribed ordered string is $p_1^{r_1} p_2^{r_2} \ldots p_k^{r_k}$. Thus for the string $\xi_1 \zeta_3 \zeta_2 \zeta_2 \zeta_1 \zeta_2$ the probability is $p_1^2 p_2^3 p_3$.

A different (greater) probability results when order is not specified. Suppose we perform n repetitions of a generalized Bernoulli trial and consider the event in which ζ_1 appears r_1 times, ζ_2 appears r_2 times, and so forth, *without regard to order*. Before computing the probability of this event we furnish an example.

Example 1.9-6 _____

In calling the Sav-Yur-Life health care facility to report an emergency one of three things can happen:

1. the line is busy (event E_1);
2. you get the wrong party (event E_2); and
3. you get through to the triage nurse (event E_3).

Assume $P[E_i] = p_i$. What is the probability that in five disparate emergencies at different times, initial calls are met with four busy signals and one wrong number?

Solution   Let F denote the event of getting four busy signals and one wrong number. Then

$$F = F_1 \cup F_2 \cup F_3 \cup F_4 \cup F_5 \text{ where } F_1 = \{E_1 E_1 E_1 E_1 E_2\},$$

$$F_2 = \{E_1 E_1 E_1 E_2 E_1\}, \ F_3 = \{E_1 E_1 E_2 E_1 E_1\}, \ F_4 = \{E_1 E_2 E_1 E_1 E_1\}$$

and

$$F_5 = \{E_2 E_1 E_1 E_1 E_1\}.$$

Since $F_i F_j = \phi$, $P[F] = \sum_{i=1}^{5} P[F_i]$. But $P[F_i] = p_1^4 p_2^1 p_3^0$ independent of i. Hence

$$P[F] = 5 p_1^4 p_2^1 p_3^0.$$

With the assumed $p_1 = 0.3$, $p_2 = 0.1$, $p_3 = 0.6$, we get

$$P[F] = 5 \times 8.1 \times 10^{-3} \times 0.1 \times 1 = 0.004.$$

In problems of this type we must count all the strings of length n in which ζ_1 appears r_1 times, ζ_2 appears r_2 times, and so on. In the example just considered, there were five such strings. In the general case of n trials with r_1 outcomes of ζ_1, r_2 outcomes of ζ_2, and so on, there are

$$\frac{n!}{r_1! r_2! \ldots r_k!} \tag{1.9-4}$$

such strings. In Example 1.9-6, $n = 5$, $r_1 = 4$, $r_2 = 1$, $r_3 = 0$ so that

$$\frac{5!}{4! 1! 0!} = 5.$$

The number in Equation 1.9-4 is recognized as the multinomial coefficient. To demonstrate that it is the appropriate coefficient, consider the r_1 outcomes ζ_1. The number of ways of placing the r_1 outcomes ζ_1 among the n trials is identical with the number of subpopulations of size r_1 in a population of size n which is $\binom{n}{r_i}$. That leaves $n - r_1$ trials among which we wish to place r_2 outcomes ζ_2. The number of ways of doing that is $\binom{n - r_1}{r_2}$. Repeating this process we obtain for the total number of distinguishable arrangements

$$\binom{n}{r_1}\binom{n-r_1}{r_2}\cdots\binom{n-r_1-r_2\ldots-r_{k-1}}{r_k} = \frac{n!}{r_1!r_2!\ldots r_k!}$$

Example 1.9-7

Consider four repetitions of a generalized Bernoulli experiment in which the outcomes are $*$, $\bullet$, 0. What is the number of ways of getting two $*$, one $\bullet$, and one 0.

Solution   The number of ways of getting two $*$ in four trials is $\binom{4}{2} = 6$. If we let the spaces between bars represent a trial then we can denote the outcomes as

$$|*|*|\ |\ |;\ |*|\ |*|\ |;\ |\ |*|*|\ |;\ |\ |*|\ |*|;\ |\ |\ |*|*|;\ |*|\ |\ |*|.$$

The number of ways of placing $\bullet$ among the two remaining cells is $\binom{2}{1} = 1$. The number of ways of placing 0 among the remaining cell is $\binom{1}{1} = 1$. Hence the total number of arrangements is $6 \cdot 2 \cdot 1 = 12$. They are

*	0	•	*
*	•	0	*
*	*	0	•
*	*	•	0
*	0	*	•
*	•	*	0
0	*	*	•
•	*	*	0
0	*	•	*
•	*	0	*
0	•	*	*
•	0	*	*

We can now state the multinomial probability law. Consider a generalized Bernoulli trial with outcomes $\zeta_1, \zeta_2, \ldots, \zeta_k$ and let the probability of observing ζ_i be p_i, $i = 1, \ldots, k$ where $p_i \geq 0$ and $\sum_{i=1}^{k} p_i = 1$. The probability that in n trials ζ_1 occurs r_1 times, ζ_2 occurs r_2 times, and so on, is

$$P(\boldsymbol{r}; n, \boldsymbol{p}) = \frac{n!}{r_1!r_2!\ldots r_k!}p_1^{r_1}p_2^{r_2}\cdots p_k^{r_k}, \tag{1.9-5}$$

where $\boldsymbol{r}$ and $\boldsymbol{p}$ are k-tuples defined by

$$\boldsymbol{r} = (r_1, r_2, \ldots, r_k), \quad \boldsymbol{p} = (p_1, p_2, \ldots, p_k) \text{ and } \sum_{i=1}^{k} r_i = n.$$

Observation.   With $r = 2$, Equation 1.9-5 becomes the binomial law with $p_1 \triangleq p$, $p_2 \triangleq$ $1 - p$, $r_1 \triangleq k$, and $r_2 \triangleq n - k$. Functions such as Equations 1.9-1 and 1.9-5 are often called *probability mass functions.*

Example 1.9-8
(Emergency calls.) In the United States, 911 is the all-purpose number used to summon an ambulance, the police, or the fire department. In the rowdy city of Nirvana in upstate New York, it has been found that 60 percent of all calls request the police, 25 percent request an ambulance, and 15 percent request the fire department. We observe the next ten calls. What is the probability of the event that six will ask for the police, three for ambulances, and one for the fire department?

Solution   Using Equation 1.9-5 we get

$$P(6, 3, 1; 10, 0.6, 0.25, 0.15)$$

$$= \frac{10!}{6!3!1!}(0.6)^6(0.25)^3(0.15)^1 \simeq 0.092.$$

A numerical problem appears if n gets large. For example, suppose we observe 100 calls and consider the event of 60 calls for the police, 30 for ambulances, and 10 for the fire department; clearly computing numbers such as 100!, 60!, 30! requires some care. An important result that helps in evaluating such large factorials is Stirling's[†] formula:

$$n! \simeq (2\pi)^{1/2}n^{n+(1/2)}e^{-n},$$

where the approximation improves as n increases, for example,

n	$n!$	Stirling's formula	Percent error
1	1	0.922137	8
10	3,628,800	3,598,700	0.8

As stated earlier, the binomial law is a special case, perhaps the most important case, of the multinomial law. When the parameters of the binomial law attain extreme values, the binomial law can be used to generate another important probability law. This is explored next.

1.10 ASYMPTOTIC BEHAVIOR OF THE BINOMIAL LAW: THE POISSON LAW

Suppose that in the binomial function $b(k; n, p)$, $n \gg 1$, $p \ll 1$, but np remains constant, say $np = a$. Recall that $q = 1 - p$. Hence

$$\binom{n}{k} p^k (1 - p)^{n-k} \simeq \frac{1}{k!}a^k \left(1 - \frac{a}{n}\right)^{n-k},$$

[†] James Stirling, eighteenth-century mathematician.

where $n(n-1)\ldots(n-k+1) \simeq n^k$ if n is allowed to become large enough and k is held fixed. Hence in the limit as $n \to \infty$, $p \to 0$, and $k << n$, we obtain

$$b(k; n, p) \simeq \frac{1}{k!} a^k \left(1 - \frac{a}{n}\right)^{n-k} \to \frac{a^k}{k!} e^{-a}. \tag{1.10-1}$$

Thus in situations where the binomial law applies with $n >> 1$, $p << 1$ but $np = a$ is a finite constant, we can use the approximation

$$b(k; n, p) \simeq \frac{a^k}{k!} e^{-a}. \tag{1.10-2}$$

Example 1.10-1

A computer contains 10,000 components. Each component fails independently from the others and the yearly failure probability per component is 10^{-4}. What is the probability that the computer will be working one year after turn-on? Assume that the computer fails if one or more components fail.

Solution

$$p = 10^{-4} \qquad n = 10{,}000, \qquad k = 0, \qquad np = 1.$$

Hence

$$b(0; 10{,}000, 10^{-4}) = \frac{1^0}{0!} e^{-1} 1 \frac{1}{e} = 0.368.$$

Example 1.10-2

Suppose that n independent points are placed at random in an interval $(0, T)$. Let $0 < t_1 < t_2 < T$ and $t_2 - t_1 \triangleq \tau$. Let $\tau/T << 1$ and $n >> 1$. What is the probability of observing exactly k points in τ seconds? (Figure 1.10-1.)

Solution   Consider a single point placed at random in $(0, T)$. The probability of the point appearing in τ is τ/T. Let $p = \tau/T$. Every other point has the same probability of being in τ seconds. Hence, the probability of finding k points in τ seconds is the binomial law

$$P[k \text{ points in } \tau \text{ sec.}] = \binom{n}{k} p^k q^{n-k}. \tag{1.10-3}$$

With $n >> 1$, we use the approximation in Equation 1.10-1 to give

$$b(k; n, p) \simeq \left(\frac{n\tau}{T}\right)^k \frac{e^{-(n\tau/T)}}{k!}, \tag{1.10-4}$$

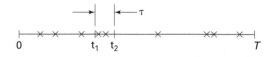

Figure 1.10-1   Points placed at random on a line. Each point is placed with equal likelihood anywhere along the line.

where n/T can be interpreted as the "average" number of points per unit interval. Equations 1.10-1 and 1.10-4 are examples of the *Poisson probability law*.

The Poisson law with parameter a $(a > 0)$ is defined by[†]

$$P[k \text{ points}] = e^{-a}\frac{a^k}{k!} \tag{1.10-5}$$

where $k = 0, 1, 2, \ldots$. With $a \triangleq \lambda\tau$, where λ is the average number of events per unit time and τ is the length of the interval $(t, t + \tau)$, the probability of k events in τ is

$$P(k; t, t + \tau) = e^{-\lambda\tau}\frac{(\lambda\tau)^k}{k!}. \tag{1.10-6}$$

In Equation 1.10-6 we assumed that λ was independent of t. If λ depends on t, the product $\lambda\tau$ gets replaced by the integral $\int_t^{t+\tau} \lambda(\xi)\, d\xi$, and the probability of k events in the interval $(t, t + \tau)$ is

$$P(k; t, t + \tau) = \exp\left[-\int_t^{t+\tau} \lambda(\xi)\, d\xi\right]\frac{1}{k!}\left[\int_t^{t+\tau} \lambda(\xi)d\xi\right]^k. \tag{1.10-7}$$

The Poisson law $P[k \text{ events in } \Delta x]$ or more generally $P[k \text{ events in } (x, x + \Delta x)]$ where x is time, volume, distance, and so forth, and Δx is the interval associated with x is widely used in engineering and sciences. Some typical situations in various fields where the Poisson law is applied are listed below.

Physics. In radioactive decay—$P[k$ alpha particles in τ seconds] with λ the average number of emitted alpha particle per second.

Operations Research. In planning the size of a switchboard—$P[k$ telephone calls in τ seconds] with λ the average number of calls per second.

Biology. In water pollution monitoring—$P[k$ coliform bacteria in 1000 cubic centimeters] with λ the average number of coliform bacteria per cubic centimeter.

Transportation. In planning the size of a highway toll facility—$P[k$ automobiles arriving in τ minutes] with λ the average number of automobiles per minute.

Optics. In designing an optical receiver—$P[k$ photons per second over a surface of area A] with λ the average number of photons-per-second-per-unit area.

Communications. In designing a fiber optical transmitter-receiver link—$P[k$ photoelectrons generated at the receiver in one second] with λ the average number of photoelectrons per second.

The parameter λ is often called, in this context, the Poisson *rate parameter*. Its dimensions are events per unit interval, the interval being time, distance, volume, and so forth. When the form of the Poisson law that we wish to use is as in Equation 1.10-6 or 1.10-7, we speak of the Poisson law with *rate parameter* λ or *rate function* $\lambda(t)$.

[†]The term "points" here is a generic term. Equally appropriate would be "arrivals," "occurrences" and the like.

Example 1.10-3

(Misuse of probability.) (a) "Prove" that there must be life in the universe, other than that on our own Earth, by using the following numbers[†]: average number of stars per galaxy, 300×10^9; number of galaxies, 100×10^9; probability that a star has a planetary system, 0.5; average number of planets per planetary system, 9; probability that a planet can sustain life, 1/9; probability of life emerging on a life-sustaining planet, 10^{-12}.

Solution   First we compute, n_{LS}, the number of planets that are life-sustaining:

$$n_{LS} = 300 \times 10^9 \times 100 \times 10^9 \times 0.5 \times 9 \times 1/9$$

$$= 1.5 \times 10^{22}.$$

Next we use the Poisson approximation to the binomial with $a = n_{LS}\, p = 1.5 \times 10^{22} \times 10^{-12}$, for computing the probability of *no life outside of Earth's* and obtain

$$b(0, 1.5 \times 10^{22}, 10^{-12}) = \frac{(1.5 \times 10^{10})^0}{0!} e^{-1.5 \times 10^{10}} \approx 0.$$

Hence we have just "shown" that the probability of life outside Earth has a probability of unity, that is, a sure bet. Note that the number for life emerging on other planets, 10^{-12}, is impressively low.

(b) Now show that life outside Earth is extremely unlikely by using the same set of numbers except that the probability of life emerging on a life-sustaining planet has been reduced to 10^{-30}.

Solution   Using the Poisson approximation to the binomial, with $a = 1.5 \times 10^{22} \times 10^{-30} = 1.5 \times 10^{-8}$, we obtain for the probability of no life outside Earth's:

$$b(0, 1.5 \times 10^{22}, 10^{-30}) = \frac{(1.5 \times 10^{-8})^0}{0!} e^{-1.5 \times 10^{-8}}$$

$$\approx 1 - (1.5 \times 10^{-8}) \approx 1$$

where we have used the approximation $e^{-x} \approx 1 - x$ for small x.

Thus by changing only one number, we have gone from "proving" that the universe contains extraterrestrial life to proving that, outside of ourselves, the universe is lifeless. The reason that this is a misuse of probability is that, at present, we have no idea as to the factors that lead to the emergence of life from nonliving material. While the calculation is technically correct, this examples illustrates the use of contrived numbers to either prove or disprove what is essentially a belief or faith.

Example 1.10-4

A switchboard receives on the average 16 calls per minute. If the switchboard can handle at most 24 calls per minute, what is the probability that in any one minute the switchboard will saturate?

Solution   Saturation occurs if the number of calls in a minute exceeds 24. The probability of this event is:

[†]All the numbers have been quoted at various times by proponents of the idea of extraterrestrial life.

$$P[\text{saturation}] = \sum_{k=25}^{\infty} [\lambda\tau]^k \frac{e^{-\lambda\tau}}{k!} \tag{1.10-8}$$

$$= \sum_{k=25}^{\infty} [16]^k \frac{e^{-16}}{k!} \simeq 0.017 \simeq 1/60. \tag{1.10-9}$$

Thus about once in every 60 minutes (on the "average") will a caller experience a delay.

Given the numerous applications of the Poisson law in the physical and biological sciences, one would think that its origin is of somewhat more noble birth than "merely" as a limiting form of the binomial law. Indeed this is the case, and the Poisson law can be derived once three assumptions are made. Obviously these three assumptions should reasonably mirror the characteristics of the underlying physical process; otherwise our results will be of only marginal interest. Fortunately, in many situations these assumptions seem to be quite valid.

In order to be concrete, we shall talk about events[†] taking place in *time* (as opposed to, say, length or volume). The Poisson law is based on the following assumptions:

1. The probability, $P(1; t, t+\Delta t)$, of a single event occurring in $(t, t+\Delta t)$ is proportional to Δt, that is,

$$P(1; t, t + \Delta t) \simeq \lambda(t)\Delta t \qquad \Delta t \to 0. \tag{1.10-10}$$

In Equation 1.10-10 $\lambda(t)$ is the Poisson rate parameter.
2. The probability of $k(k > 1)$ events in $(t, t + \Delta t)$ goes to zero:

$$P(k; t, t + \Delta t) \simeq 0 \qquad \Delta t \to 0, \qquad k = 2, 3, \ldots. \tag{1.10-11}$$

3. Events in nonoverlapping time intervals are statistically independent.

Starting with these three simple physical assumptions, it is a straightforward task to obtain the Poisson probability law. We leave this derivation to Chapter 7 but merely point out that the clever use of the assumptions leads to a set of elementary, first-order differential equations whose solution is the Poisson law. The general solution is furnished by Equation 1.10-7 but, fortunately, in a large number of physical situations the Poisson rate parameter $\lambda(t)$ can be approximated by a constant, say, λ. In that case Equation 1.10-6 can be applied. We conclude this section with a final example.

Example 1.10-5 _____
A manufacture of computer tape finds that the defect density along the length of tape is not uniform. After a careful compilation of data, it is found that for tape strips of length

---

[†]Examples of events that apply here: the emission of an α-particle, a car arriving at a toll gate, a telephone starting to ring, and the like.

D, the defect density $\lambda(x)$ along the tape length x varies as

$$\lambda(x) = \lambda_0 + \frac{1}{2}(\lambda_1 - \lambda_0)\left(1 + \cos\left[\frac{2\pi x}{D}\right]\right), \qquad \lambda_1 > \lambda_0$$

for $0 \leq x \leq D$ due to greater tape contamination at the edges $x = 0$ and $x = D$.

(a) What is the meaning of $\lambda(x)$ in this case?
(b) What is the average number of defects for a tape strip of length D?
(c) What is an expression for k defects on a tape strip of length D?
(d) What are the Poisson assumptions in the case?

Solution

(a) Bearing in mind that $\lambda(x)$ is a *defect density*, that is, the average number of defects per unit length at x, we conclude that $\lambda(x)\Delta x$ is the average number of defects in the tape from x to $x + \Delta x$.

(b) Given the definition of $\lambda(x)$, we conclude that the average number of defects along the whole tape is merely the integral of $\lambda(x)$, that is,

$$\int_0^D \lambda(x)dx = \int_0^D \left[\lambda_0 + \frac{1}{2}(\lambda_1 - \lambda_0)\left(1 + \cos\frac{2\pi x}{D}\right)\right]dx$$

$$= \frac{\lambda_0 + \lambda_1}{2}D$$

$$\triangleq \Omega.$$

(c) Assuming the Poisson law holds, we use Equation 1.10-7 with x and Δx (distances) replacing t and τ (times). Thus

$$P(k; x, x + \Delta x) = \exp\left[-\int_x^{x+\Delta x} \lambda(\zeta)d\zeta\right] \cdot \frac{1}{k!}\left[\int_x^{x+\Delta x} \lambda(\zeta)d\zeta\right]^k.$$

In particular, with $x = 0$ and $x + \Delta x = D$, we obtain

$$P(k; 0, D) = \Omega^k \frac{e^{-\Omega}}{k!}.$$

(d) The Poisson assumptions become

(i) $P[1; x, x + \Delta x] \simeq \lambda(x)\Delta x$, as $\Delta x \to 0$.
(ii) $P[k; x, x + \Delta x] = 0$ $\quad \Delta x \to 0$, for $k = 2, 3, \ldots$ that is, the probability of there being more than one defect in the interval $(x, x + \Delta x)$ as Δx becomes vanishingly small, is zero.
(iii) the occurrences of defects (events) in nonoverlapping sections of the tape are independent.

1.11 NORMAL APPROXIMATION TO THE BINOMIAL LAW

In this section we give, without proof, a numerical approximation to binomial probabilities and binomial sums. Let S_n denote the number of successes in n Bernoulli trials. Then S_n has a binomial distribution and

$$P[S_n = k] = \binom{n}{k} p^k q^{n-k} = b(k; n, p), \quad 0 \le k \le n. \tag{1.11-1}$$

For large values of n and k, Equation 1.11-1 may be difficult to evaluate numerically. Also, the probability of the event $\{\alpha \le S_n \le \beta\}$, where α and β are integers, may involve many terms, making a direct evaluation of $P[\alpha \le S_n \le \beta]$ impractical. Fortunately, when n is large, we can use approximate methods for evaluating both $b(k; n, p)$ and $P[\alpha \le S_n \le \beta]$. These approximate methods involve the *Normal distribution*.

The Normal distribution and its significance will be discussed in greater detail in Chapter 2 and subsequent chapters in this book. Here we use it only to help evaluate binomial probabilities. For the present, define the function $\phi(x)$, known as the *standard Normal density*, by

$$\phi(x) = \frac{1}{\sqrt{2\pi}} \exp\left[-\frac{1}{2}x^2\right] \tag{1.11-2}$$

and its integral, known as the *standard Normal distribution* by

$$\Phi(x) = \frac{1}{\sqrt{2\pi}} \int_{-\infty}^{x} \exp\left[-\frac{1}{2}y^2\right] dy. \tag{1.11-3}$$

Then, when n is large it can be shown [1-8] that

$$b(k; n, p) \approx \frac{1}{\sqrt{npq}} \phi\left(\frac{k - np}{\sqrt{npq}}\right). \tag{1.11-4}$$

The approximation becomes better when $npq \gg 1$. We reproduce the results from [1-8] in Table 1.11-1. Even in this case, $npq = 1.6$, the approximation is quite good.

The approximation for sums, when $n \gg 1$ and a and b are fixed integers, takes the form

$$P[\alpha \le S_n \le \beta] \approx \Phi\left[\frac{\beta - np + 0.5}{\sqrt{npq}}\right] - \Phi\left[\frac{\alpha - np - 0.5}{\sqrt{npq}}\right]. \tag{1.11-5}$$

Some results, for various values of n, p, α, β are furnished in Table 1.11-2 which uses the results in [1-8].

In using the Normal approximation one should refer to Table 2.4-1. In Table 2.4-1 a function called erf(x) is given rather than $\Phi(x)$. The erf(x) is defined by

$$\text{erf}(x) = \frac{1}{\sqrt{2\pi}} \int_{0}^{x} e^{-\frac{y^2}{2}} dy.$$

Table 1.11-1   Normal Approximation to the Binomial for Selected Numbers

k	$b(k; 10, 0.2)$	Normal approximation
0	0.1074	0.0904
1	0.2864	0.2307
2	0.3020	0.3154
3	0.2013	0.2307
4	0.0880	0.0904
5	0.0264	0.0189
6	0.0055	0.0021

Table 1.11-2   Event Probabilities Using the Normal Approximation (Adapted from [1-8])

n	p	α	β	$P[\alpha \leq S_n \leq \beta]$	Normal approximation
200	0.5	95	105	0.5632	0.5633
500	0.1	50	55	0.3176	0.3235
100	0.3	12	14	0.00015	0.00033
100	0.3	27	29	0.2379	0.2341
100	0.3	49	51	0.00005	0.00003

However, since it is easy to show that

$$\Phi(x) = \tfrac{1}{2} + \mathrm{erf}(x), \quad x > 0 \tag{1.11-6}$$

and

$$\Phi(x) = \tfrac{1}{2} - \mathrm{erf}(|x|), \quad x < 0, \tag{1.11-7}$$

we can easily compute Equation 1.11-5 in terms of the table values. Thus with $a' \triangleq \frac{\alpha - np - 0.5}{\sqrt{npq}}$ and $b' \triangleq \frac{\beta - np + 0.5}{\sqrt{npq}}$ and $b' > a'$, we can use the results in Table 1.11-3.

The Normal approximation is also useful in evaluating Poisson sums. For example, a sum such as in Equation 1.10-9 is tedious to evaluate if done directly. However if $\lambda\tau >> 1$, we can use the *Normal approximation to the Poisson law*, which is merely an extension of the Normal approximation to the binomial law. This extension is expected since we have seen that the Poisson law is itself an approximation to the binomial law under certain circumstances. From the results given above we are able to justify the following approximation.

Table 1.11-3   Use of the Error Function to Approximate Binomial Sums

If a'	If b'	Then to compute $P[\alpha \leq S_n \leq \beta]$ use
>0	>0	$\mathrm{erf}(b') - \mathrm{erf}(a')$
<0	>0	$\mathrm{erf}(b') + \mathrm{erf}(|a'|)$
<0	<0	$\mathrm{erf}(|a'|) - \mathrm{erf}(|b'|)$

$$\sum_{k=\alpha}^{\beta} e^{-\lambda\tau} \frac{[\lambda\tau]^k}{k!} = \frac{1}{\sqrt{2\pi}} \int_{l_1}^{l_2} \exp\left[-\frac{1}{2}y^2\right] dy, \qquad (1.11\text{-}8)$$

where

$$l_2 \triangleq \frac{\beta - \lambda\tau + 0.5}{\sqrt{\lambda\tau}}$$

and

$$l_1 \triangleq \frac{\alpha - \lambda\tau - 0.5}{\sqrt{\lambda\tau}}$$

Another useful approximation is

$$e^{-\lambda\tau} \frac{[\lambda\tau]^k}{k!} = \frac{1}{\sqrt{2\pi}} \int_{l_3}^{l_4} \exp\left[-\frac{1}{2}y^2\right] dy, \qquad (1.11\text{-}9)$$

where

$$l_4 \triangleq \frac{k - \lambda\tau + 0.5}{\sqrt{\lambda\tau}}$$

and

$$l_3 \triangleq \frac{k - \lambda\tau - 0.5}{\sqrt{\lambda\tau}}.$$

For example, with $\lambda\tau = 5$, and $k = 5$, the error in using the Normal approximation of Equation 1.11-9 is less than 1 percent.

1.12   SUMMARY

In this, the first chapter of the book, we have reviewed some different definitions of probability. We developed the axiomatic theory and showed that for a random experiment three important objects were required: the sample space Ω, the sigma field of events $\mathscr{F}$, and a probability measure P. The triplet $(\Omega, \mathscr{F}, \mathscr{P})$ is called the probability space $\mathscr{P}$.

We discussed some important formulas from combinatorics and briefly illustrated how important they were in theoretical physics. We then discussed the binomial probability law and its generalization, the multinomial law. We saw that the binomial law could, when certain limiting conditions were valid, be approximated by the Poisson law. The Poisson law, one of the central laws in probability theory, was shown to have application in numerous branches of science and engineering. We stated, but deferred verification until Chapter 7, that the Poisson law can be derived directly from simple and entirely reasonable physical assumptions.

Approximations for the binomial and Poisson laws, based on the Normal distribution, were furnished. Several occupancy problems of engineering interest were discussed. In Chapter 4 we shall revisit these problems.

PROBLEMS

1.1.   In order for a statement such as "Ralph is probably guilty of theft," to have meaning in the relative frequency approach to probability, what kind of data would one need?

1.2.   Problems in applied probability (a branch of mathematics called statistics) often involve testing P $\rightarrow$ Q (P implies Q) type statements, for example, if she smokes, she will probably get sick; if he is smart he will do well in school. You are given a set of four cards that have a letter on one side and a number on the other. You are asked to test the rule "If a card has a D on one side, it has a three on the other." Which of the following cards should you turn over to test the veracity of the rule:

| Card one | Card two | Card three | Card four |

1.3.   In a spinning-wheel game, the spinning wheel contains the numbers 1 to 9. The contestant wins if an *even* number shows. What is the probability of a win? What are your assumptions?

1.4.   A fair coin is flipped three times. The outcomes on each flip are heads (H) or tails (T). What is the probability of obtaining two heads and one tail?

1.5.   An urn contains three balls numbered 1, 2, 3. The experiment consists of drawing a ball at random, recording the number, and replacing the ball before the next ball is drawn. This is called sampling with replacement. What is the probability of drawing the same ball twice in two tries?

1.6.   An experiment consists of drawing two balls *without* replacement from an urn containing six balls numbered 1 to 6. Describe the sample description space Ω. What is Ω if the ball is replaced before the second is drawn?

1.7.   The experiment consists of measuring the heights of each partner of a randomly chosen married couple. (a) Describe Ω in convenient notation; (b) let E be the event that the man is shorter than the woman. Describe E in convenient notation.

1.8. An urn contains ten balls numbered 1 to 10. Let E be the event of drawing a ball numbered no greater than 5. Let F be the event of drawing a ball numbered greater than 3 but less than 9. Evaluate E^c, F^c, EF, $E \cup F$, EF^c, $E^c \cup F^c$, $EF^c \cup E^c F$, $EF \cup E^c F^c$, $(E \cup F)^c$, and $(EF)^c$. Express these events in words.

1.9. An experiment consists of drawing two balls at random, with replacement from an urn containing five balls numbered 1 to 5. Three students "Dim," "Dense," and "Smart" were asked to compute the probability p that the sum of numbers appearing on the two draws equals 5. Dim computed $p = \frac{2}{15}$, arguing that there are 15 distinguishable unordered pairs and only 2 are favorable, that is, $(1, 4)$ and $(2, 3)$. Dense computed $p = \frac{1}{9}$, arguing that there are 9 distinguishable sums (2 to 10), of which only 1 was favorable. Smart computed $p = \frac{4}{25}$, arguing that there were 25 distinguishable ordered outcomes of which 4 were favorable, that is, $(4, 1)$, $(3, 2)$, $(2, 3)$, and $(1, 4)$. Why is $p = \frac{4}{25}$ the correct answer? Explain what is wrong with the reasoning of Dense and Dim.

1.10. Use the axioms given in Equations 1.5-1 to 1.5-3 to show the following: ($E \in \mathscr{F}$, $F \in \mathscr{F}$) (a) $P[\phi] = 0$ (b) $P[EF^c] = P[E] - P[EF]$; (c) $P[E] = 1 - P[E^c]$.

1.11. Use the "exclusive-or" operator in Equation 1.4-3 to show that $P[E \oplus F] = P[EF^c] + P[E^c F]$.

1.12. Show that $P[E \oplus F]$ in Problem 1.11 can be written as $P[E \oplus F] = P[E] + P[F] - 2P[EF]$.

1.13. (a) Let the sample space $\Omega = \{\text{cat, dog, goat, pig}\}$. Then assume that only the following probability information is given:

$$P[\{\text{cat, dog}\}] = 0.9,$$

$$P[\{\text{goat, pig}\}] = 0.1,$$

$$P[\{\text{pig}\}] = 0.05,$$

$$P[\{\text{dog}\}] = 0.5.$$

For this given set of probabilities, find the appropriate field of events $\mathscr{F}$ so that the overall probability space $(\Omega, \mathscr{F}, P)$ is well defined. Specify the field $\mathscr{F}$ by listing all the events in the field, along with their corresponding probabilities.

(b) Repeat part (a), but without the information that $P[\{\text{pig}\}] = 0.05$.

1.14. An urn contains eight balls. The letters a and b are used to label the balls. Two balls are labeled a two are labeled b and the remaining balls are labeled with both letters, that is, a, b. Except for the labels, all the balls are identical. Now a ball is drawn at random from the urn. Let A and B represent the events of observing letters a and b, respectively. Find $P[A]$, $P[B]$, and $P[AB]$. Are A and B independent?

1.15. A fair die is tossed twice (a die is said to be fair if all outcomes $1, \ldots, 6$ are equally likely). Given that a 3 appears on the first toss, what is the probability of obtaining the sum 7 in two tosses?

1.16. A card is selected at random from a standard deck of 52 cards. Let A be the event of selecting an ace and let B be the event of selecting a red card. There are 4 aces and 26 red cards in the normal deck. Are A and B independent?

1.17. A random-number generator generates integers from 1 to 9 (inclusive). All outcomes are equally likely; each integer is generated independently of any previous integer. Let Σ denote the sum of two consecutively generated integers; that is, $\Sigma = N_1 + N_2$. Given that Σ is odd, what is the conditional probability that Σ is 7? Given that $\Sigma > 10$, what is the conditional probability that at least one of the integers is > 7? Given that $N_1 > 8$, what is the conditional probability that Σ will be odd?

1.18. The following problem was given to 60 students and doctors at the famous Hevardi Medical School (HMS): Assume there exists a test to detect a disease, say D, whose prevalence is 0.001, that is, the probability, $P[D]$, that a person picked at random is suffering from D, is 0.001. The test has a false positive rate of 0.005 and a correct detection rate of one. The correct detection rate is the probability that if you have D, the test will say that you have D. Given that you test positive for D, what is the probability that you actually have it? Many of the HMS experts answered 0.95 and the average answer was 0.56. Show that your knowledge of probability is greater than that of the HMS experts by getting the right answer of 0.17.

1.19. Henrietta is 29 years old and physically very fit. In college she majored in geology. During her student days, she frequently hiked in the national forests and biked in the national parks. She participated in anti-logging and anti-mining operations. Now, Henrietta works in an office building in downtown Nirvana. Which is greater: the probability that Henrietta's occupation is that of office manager; or the probability that Henrietta is an office manager who is active in nature-defense organizations like the Sierra Club?

1.20. In the ternary communication channel shown in Figure P1.20 a 3 is sent three times more frequently than a 1, and a 2 is sent two times more frequently than a 1. A 1 is observed; what is the conditional probability that a 1 was sent?

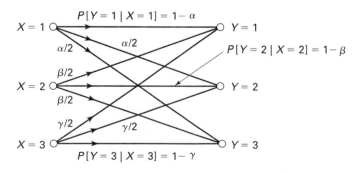

Figure P1.20

$$P(\overline{T}/C) = \frac{P(TC)}{P(C)} = \frac{P(C/\overline{T})P(\overline{T}) + P(C/\overline{?})}{P(C)} \quad P(\overline{?})$$

1.21. A large class in probability theory is taking a multiple-choice test. For a particular question on the test, the fraction of examinees who know the answer is p; $1 - p$ is the fraction that will guess. The probability of answering a question correctly is unity for an examinee who knows the answer and $1/m$ for a guessee; m is the number of multiple-choice alternatives. Compute the probability that an examinee knew the answer to a question given that he or she has correctly answered it.

1.22. In the beauty-contest problem, Example 1.5-2, what is the probability of picking the most beautiful contestant if we decide *a priori* to choose the ith ($1 \leq i \leq N$)?

1.23. Assume there are 3 machines, A, B, and C in a semiconductor manufacturing facility that make chips. They manufacture, respectively, 25, 35, and 40 percent of the total semiconductor chips there. Of their outputs, respectively, 5, 4, and 2 percent of the chips are defective. A chip is drawn randomly from the combined output of the three machines and is found defective. What is the probability that this defective chip was manufactured by machine A? by machine B? by machine C?

1.24. In Example 1.6-2, plot the probability of making a correct decision versus α/N, assuming that the "wait-and-see" strategy is adopted. In particular, what is $P[D]$ when $\alpha/N = 0.5$. What does this suggest about the sensitivity of $P[D]$ vis-a-vis α when α is not too far from α_0 and N is large?

1.25. Consider r indistinguishable balls (particles) and n cells (states) where $n > r$. The r balls are placed at random into the n cells (multiple occupancy is possible). What is the probability P that the r balls appear in r preselected cells (one to a cell)?

1.26. Assume that we have r indistinguishable balls and n cells. The cells can at most hold only one ball. As in Problem 1.25 $r < n$. What is the probability P that the r balls appear in r preselected cells?

1.27. Some digital communication networks use *packet switching* to create virtual circuits between two users, even though the users are sharing the same physical channel with others. In packet switching, the data stream is broken up into *packets* that travel different paths and are reassembled in the proper chronological order and at the correct address. Suppose the *order* information is missing. Compute the probability that a data stream broken up into N packets will reassemble itself correctly, even without the order information.

1.28. In the previous problem assume that $N = 3$. A lazy engineer decides to omit the order information in favor of repeatedly sending the data stream until the packets re-order correctly for the first time. Derive a formula that the correct re-ordering occurs for the first-time on the nth try. How many repetitions should be allowed before the cumulative probability of a correct re-ordering for the first-time is at least 0.95?

1.29. Prove that $\sum_{k=0}^{n} b(k; n, p) = 1$, where $b(k; n, p)$ is the binomial law.

1.30. War-game strategists make a living by solving problems of the following type. There are six incoming ballistic missiles (BM's) against which are fired twelve antimissile missiles (AMM's). The AMM's are fired so that two AMM's are directed against each BM. The single-shot-kill probability (SSKP) of an AMM is 0.8. The SSKP is simply the probability that an AMM destroys a BM. Assume that the AMM's don't interfere with each other and that an AMM can, at most, destroy only the BM against which it is fired. Compute the probability that (a) all BM's are destroyed,

1.41. In a particular computer communication network, the host computer broadcasts a packet of data (say L bytes long) to N receivers. The host computer then waits to receive an acknowledgement message from each of the N receives before proceeding to broadcast the next packet. If the host does not receive all the acknowledgments within a certain time period, it will rebroadcast (retransmit) the same packet. The host computer is then said to be in the "retransmission mode." It will continue retransmitting the packet until all N acknowledgments are received. Then it will proceed to broadcast the next packet.

Let $p \triangleq P[\text{successful transmission of a single packet to a single receiver along with successful acknowledgment}]$. Assume that these events are independent for different receivers or separate transmission attempts. Due to random impairments in the transmission media and the variable condition of the receivers (terminals or PCs), we have that $p < 1$.

(a) In a fixed protocol or method of operation, we require that all N of the acknowledgments be received in response to a given transmission attempt for that packet transmission to be declared successful. Let the event $S(m)$ be defined as follows: $S(m) \triangleq \{$a successful transmission of one packet to all N receivers in m or fewer attempts$\}$. Find the probability

$$P(m) \triangleq P[S(m)].$$

[*Hint*: Consider the complement of the event $S(m)$.]

(b) An improved system operates according to a dynamic protocol as follows. Here we relax the acknowledgment requirement on retransmission attempts, so as to only require acknowledgments from those receivers that have not yet been heard from on previous attempts to transmit the current packet. Let $S_D(m)$ be the same event as in part (a) but using the dynamic protocol. Find the probability

$$P_D(m) \triangleq P[S_D(m)].$$

[*Hint*: First consider the probability of the event $S_D(m)$ for an individual receiver, and then generalize to the N receivers.]

Note: If you try $p = 0.9$ and $N = 5$ you should find that $P(2) < P_D(2)$.

1.42. Toss two unbiased dice (each with six faces: 1 to 6), and write down the sum of the two face numbers. Repeat this procedure 100 times. What is the probability of getting 10 readings of value 7? What is the Poisson approximation for computing this probability?

[*Hint*: Consider the event $A = \{sum = 7\}$ on a single toss and let p in Equation 1.9-1 be $P(A)$.]

1.43. On behalf of your tenants you have to provide a laundry facility. Your choices are

1. lease two inexpensive "Clogger" machines at \$50.00/month each; or
2. lease a single "NeverFail" at \$100/month.

The Clogger is out of commission 40 percent of the time while the NeverFail is out of commission only 20 percent of the time.

    (a) From the tenant's point, which is the better alternative?

    (b) From your point of view as landlord which is the better alternative?

1.44. In the politically unstable country of Eastern Borduria, it is not uncommon to find a bomb on-board passenger aircraft. The probability that on any given flight, a bomb will be on-board is 10^{-2}. A nervous passenger always flies with an unarmed bomb in his suitcase, reasoning that the probability of there being *two bombs on-board* is 10^{-4}. By this maneuver, the nervous passenger believes that he has greatly reduced the airplane's chances of being blown up. Do you agree with his reasoning? If not why not?

1.45. In a ring network consisting of eight links as shown in Figure P1.45, there are two paths connecting any two terminals. Assume that links fail independently with probability q, $0 < q < 1$. Find the probability of successful transmission of a packet from terminal A to terminal B. (Note: Terminal A transmits the packet in *both* directions on the ring. Also, terminal B removes the packet from the ring upon reception. Successful transmission means that terminal B received the packet from either direction.)

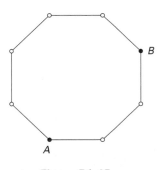

Figure P1.45

1.46. A union directive to the executives of the telephone company demands that telephone operators receive overtime payment if they handle more than 5760 calls in an eight-hour day. What is the probability that Curtis, a unionized telephone operator, will collect overtime on a particular day where the occurrence of calls during the eight-hour day, follows the Poisson law with rate parameter $\lambda = 720$ calls/hour?

1.47. An aging professor, desperate to finally get a good review for his course on probability, hands out chocolates to his students. The professor's short-term memory is so bad that he can't remember which students have already received a chocolate. Assume that, for all intents and purposes, the chocolates are distributed randomly. There are 10 students and 15 chocolates. What is the probability that each student received at least one chocolate?

1.48. Let us assume that two people have their birthdays on the same day if both the month and the day are same for each (not necessarily the year). How many people would you need to have in a room before the probability is $\frac{1}{2}$ or greater that at least two people have their birthdays on the same day?

1.49. Research Problem: Percolating Fractals

Consider a square lattice with N^2 cells, that is, N cells per side. Write a program that does the following: With probability p you put an electrically conducting element

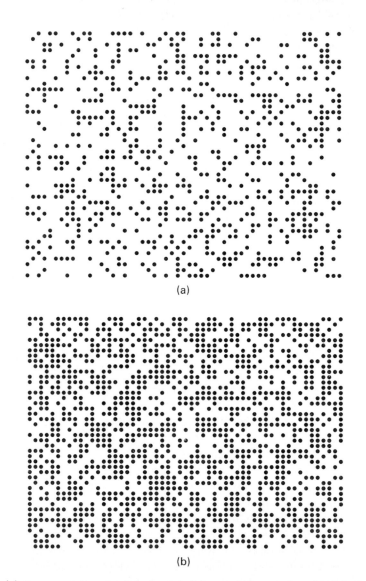

(a)

(b)

Figure P1.49 (a) Nonpercolating random fractal; (b) percolating random fractal. Can you find a percolating path? (From *Fractals, Chaos, Power Laws* by M. Schroeder, W.H. Freeman, New York, 1991. With permission.)

in a cell and with probability $q = 1 - p$, you leave the cell empty. Do this for every cell in the lattice. When you are done, does there exist a continuous path for current to flow from the bottom of the lattice to the top? If yes, the lattice is said to *percolate*. Percolation models are used in the study of epidemics and forest fires among others. The lattice is called a *random fractal* because of certain invariant properties that it possesses. Try $N = 10, 20, 50$; $p = 0.1, 0.3, 0.6$. You will need a random number generator. MATLAB has the function *rand* (uniform) which generator random numbers x_i in the interval $(0.0, 1.0)$. If the number $x_i \leq p$, make the cell electrically conducting; otherwise leave it alone. Repeat the procedure as often as time permits in order to estimate the probability of percolation for different p's.

A non-percolating lattice is shown in Figure P1.49(a); a percolating lattice is shown in (b). For more discussion of this problem see M. Schroeder, *Fractals, Chaos, Power Laws* (New York: W.H. Freeman, 1991).

*1.50. Research Problem: Counterintuitive strategies

You are a contestant on a TV game show. There are three identical closed doors leading to three rooms. Two of the rooms contain nothing, but the third contains a \$100,000 Rexus luxury automobile which is yours if you pick the right door. You are asked to pick a door by the master of ceremonies (MC) who knows which room contains the Rexus. After you pick a door, the MC opens a door (not the one you picked) to show a room *not* containing the Rexus. Show that even without any further knowledge, you will greatly increase your chances of winning the Rexus if you *switch* your choice from the door you originally picked to the one remaining closed door.

REFERENCES

1-1. E. Merzbacher, *Quantum Mechanics*. New York: John Wiley, 1961.

1-2. A. Kolmogorov, *Foundations of the Theory of Probability*. New York: Chelsea, 1950.

1-3. S. Pinker, *How the Mind Works*. New York: Norton, 1997.

1-4. B. O. Koopman, "The Axioms of Algebra and Intuitive Probability," *Annals of Mathematics* (2), Vol. 41, pp. 269–292.

1-5. A. Papoulis, *Probability, Random Variables, and Stochastic Processes*. New York: McGraw-Hill, 1965, p. 11.

1-6. R. Von Mises, *Wahrscheinlichkeit, Statistic and Wahrheit*. Vienna: Springer-Verlag, 1936.

1-7. W. B. Davenport, Jr., *Probability and Random Processes*. New York: McGraw-Hill, 1970.

1-8. W. Feller, *An Introduction to Probability Theory and Its Applications*, Vol. 1, (2nd edition). New York: John Wiley, 1950, Chapter 2.

1-9. E. Parzen, *Modern Probability Theory and Its Applications*. New York: John Wiley, 1960, p. 119.

2

Random Variables

2.1 INTRODUCTION

Many random phenomena have outcomes that are sets of real numbers: the voltage $v(t)$, at time t, across a noisy resistor, the arrival time of the next customer at a movie theatre, the number of photons in a light pulse, the brightness level at a particular point on the TV screen, the number of times a light bulb will switch on before failing, the lifetime of a given living person, the number of people on a New York to Chicago train, and so forth. In all these cases the sample description spaces are sets of numbers on the real line.

Even when a sample space Ω is not numerical, we might want to generate a new sample space from Ω that is numerical, that is, converting random speech, color, gray tone, and so forth, to numbers, or converting the physical fitness profile of a person chosen at random into a numerical "fitness" vector consisting of weight, height, blood pressure, heart rate, and so on, or describing the condition of a patient afflicted with, say, black lung disease by a vector whose components are the number and size of lung lesions and the number of lung zones affected.

In science and engineering, we are in almost all instances interested in numerical outcomes, whether the underlying experiment $\mathcal{H}$ is numerical-valued or not. To obtain numerical outcomes, we need a rule or *mapping* from the original sample description space Ω to the real line R. Such a mapping is what a random variable fundamentally is and we discuss it in some detail in the next several sections.

Let us, however, make a remark or two. The concept of a random variable will enable us to replace the original probability space with one in which events are sets of numbers. Thus

58

on the induced probability space of a random variable every event is a subset of R. But is every subset of R always an event? Are there subsets of R that could get us into trouble via violating the axioms of probability? The answer is yes, but fortunately these subsets are not of engineering or scientific importance. We say that they are *nonmeasurable*.[†] Sets of practical importance are of the form $\{x = a\}$, $\{x : a \leq x \leq b\}$, $\{x : a < x \leq b\}$, $\{x : a \leq x < b\}$, $\{x : a < x < b\}$ and their unions and intersections. These five sets are usually abbreviated, respectively, as $[a]$, $[a, b]$, $(a, b]$, $[a, b)$, and (a, b). Intervals that include the end points are said to be *closed*; those that leave out end points are said to be *open*. Intervals can be half-open, and so forth.

We can define more than one random variable on the same underlying sample space Ω. For example, suppose that Ω consist of a large, representational group of people in the United States. Let the experiment consist of choosing a person at random. Let X denote the person's lifetime and Y denote that person's daily consumption of cigarettes. We can now ask: Are X and Y related? That is, can we predict X from observing Y? Suppose we define a third random variable Z that denotes the person's weight. Is Z related to X or Y?

The main advantage of dealing with random variables is that we can define certain probability functions that make it both convenient and easy to compute the probabilities of various events. These functions must naturally be consistent with the axiomatic theory. For this reason we must be a little careful in defining events on the real line. Elaboration of the ideas introduced in this section is given next.

2.2 DEFINITION OF A RANDOM VARIABLE

Consider an experiment $\mathscr{H}$ with sample description space Ω. The elements or points of Ω, ζ are the *random* outcomes of $\mathscr{H}$. If to every ζ we assign a real number $X(\zeta)$, we establish a correspondence rule between ζ and R, the real line. Such a rule, subject to certain constraints, is called a *random variable*. Thus, a random variable $X(\cdot)$ or simply X is not really a variable but a function whose domain is Ω and whose range is some subset of the real line. Being a function, every ζ generates a specific $X(\zeta)$ although for a particular $X(\zeta)$ there may be more than one outcome ζ that produced it. Now consider an event $E_B \subset \Omega (E_B \in \mathscr{F})$.

Through the mapping X, such an event maps into points on the real line Figure 2.2-1. In particular, the event $\{\zeta : X(\zeta) \leq x\}$, often abbreviated $\{X \leq x\}$, will denote an event of unique importance, and we should like to assign a probability to it. The probability $P[X \leq x] \triangleq F_X(x)$ is called the *probability distribution function* (PDF) of X: It is shown in more advanced books [2-1], [2-2] that in order for $F_X(x)$ to be consistent with the axiomatic definition of probability, the function X must satisfy the following: For every Borel set of numbers B, the set $\{\zeta : X(\zeta) \in B\}$ must correspond to an event $E_B \in \mathscr{F}$, that is, it must be in the domain of the probability function $P(\cdot)$. Stated somewhat more mathematically, this requirement demands that X can be a random variable only if the *inverse image* under

---

[†]See Appendix D for a brief discussion on measure.

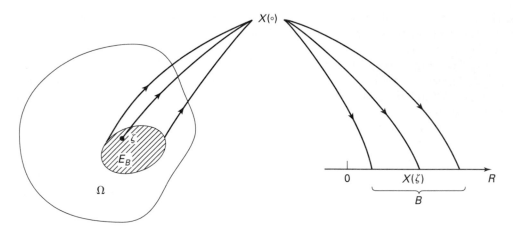

Figure 2.2-1   Symbolic representation of the action of the random variable X.

X of all Borel subets in R, making up the field $\mathscr{B}^\dagger$ are events. What is an inverse image? Consider an arbitrary Borel set of real numbers B; the set of points E_B in Ω for which $X(\zeta)$ assumes values in B is called the inverse image of the set B under the mapping X. Finally, all sets of engineering interest can be written as countable unions or intersections of events of the form $(-\infty, x]$. The event $\{\zeta : X(\zeta) \leq x\} \in \mathscr{F}$ gets mapped under X into $(-\infty, x] \in \mathscr{B}$. Thus if X *is a random variable, the set of points* $(-\infty, x]$ *is an event.*

In many if not most scientific and engineering applications, we are not interested in the actual form of X or the specification of the set Ω. For example, we might conceive of an underlying experiment that consists of heating a resistor and observing the position and velocities of the electrons in the resistor. The set Ω is then the totality of positions and velocities of all N electrons present in the resistor. Let X be the thermal noise current produced by the resistor; clearly $X \colon \Omega \to R$ although the form of X, that is, the exceedingly complicated equations of quantum electrodynamics that map from electron positions and velocity configurations to current, is not specified. What we are really interested in is the behavior of X. Thus although an underlying experiment with sample description space Ω may be implied, it is the real line R and its subsets that will hold our interest and figure in our computations. Under the mapping X we have, in effect, generated a new probability space $(R, \mathscr{B}, P_X)$ where R is the real line, $\mathscr{B}$ is the Borel σ-algebra of all subsets of R generated by countable unions and intersections of sets of the form $(-\infty, x)$, and P_X is a set function assigning a number $P_X[A] \geq 0$ to each set $A \in \mathscr{B}.^\ddagger$

In order to assign certain desirable continuity properties to the function $F_X(x)$ at $x = \pm\infty$, we require that the events $\{X = \infty\}$ and $\{X = -\infty\}$ have probability zero. With

---

†The σ-algebra of events defined on Ω are denoted by $\mathscr{F}$. The family of Borel subsets of points on R is denoted by $\mathscr{B}$. For definitions see Section 1.4 in Chapter 1.

‡The extraordinary advantage of dealing with random variables is that a single pointwise function, that is, the distribution function $F_X(x)$ can replace the set function $P_X(\cdot)$ that may be extremely cumbersome to specify, since it must be specified for every event (set) $A \in \mathscr{B}$. See Section 2.3.

the latter our specification of a random variable is complete, and we can summarize much of the above discussion in the following definition.

Definition 2.2-1 Let $\mathcal{H}$ be an experiment with sample description space Ω. Then the real random variable X is a function whose domain is Ω that satisfies the following: (i) For every Borel set of numbers B, the set $E_B \triangleq \{\zeta \in \Omega, X(\zeta) \in B\}$ is an event and (ii) $P[X = -\infty] = P[X = +\infty] = 0$.

Loosely speaking, when the range of X consists of a countable set of points, X is said to be a discrete random variable; and if the range of X is a continuum, X is said to be continuous. This is a somewhat inadequate definition of discrete and continuous random variables for the simple reason that we often like to take for the range of X the whole real line R. Points in R not actually reached by the transformation X with a nonzero probability are then associated with the impossible event.[†] ■

Example 2.2-1

A person, chosen at random in the street, is asked if he or she has a younger brother. If the answer is *no*, the data is encoded as *zero*; if the answer is *yes*, the data is encoded as *one*. The underlying experiment $\mathcal{H}$ has sample description space $\Omega = \{no, yes\}$, $\mathscr{F} = [\phi, \Omega, \{no\}, \{yes\}]$, probabilities $P[\phi] = 0$, $P[\Omega] = 1$, $P[no] = \frac{3}{4}$ (an assumption), $P[yes] = \frac{1}{4}$. The associated probabilities for X are $P[\phi] = 0$, $P[X \leq \infty] = P[\Omega] = 1$, $P[X = 0] = P[no] = \frac{3}{4}$, $P[X = 1] = P[yes] = \frac{1}{4}$. Take any x_1, x_2 and consider, for example, the probabilities that X lies in sets of the type $[x_1, x_2]$ $[x_1, x_2)$ or $(x_1, x_2]$. Thus

$$P[3 \leq X \leq 4] = P[\phi] = 0$$

$$P[0 \leq X < 1] = P[no] = \tfrac{3}{4}$$

$$P[0 \leq X \leq 2] = P[\Omega] = 1$$

$$P[0 < X \leq 1] = P[yes] = \tfrac{1}{4},$$

and so on. Thus every set $\{X = x\}$, $\{x_1 \leq X < x_2\}\{X \leq x_2\}$, and so forth, is related to an event defined on Ω. Hence X is a random variable.

Example 2.2-2

A bus arrives at random in $[0, T]$; let t denote the time of arrival. The sample description space Ω is $\Omega = \{t : t \in [0, T]\}$. A random variable X is defined by

$$X(t) = \begin{cases} 1, & t \in \left[\dfrac{T}{4}, \dfrac{T}{2}\right] \\ 0, & \text{otherwise.} \end{cases}$$

Assume that the arrival time is uniform over $[0, T]$. We can now ask and compute what is $P[X(t) = 1]$ or $P[X(t) = 0]$ or $P[X(t) \leq 5]$.

[†]An alternative definition is the following: X is discrete if $F_X(x)$ is a staircase-type function, and X is continuous if $F_X(x)$ is a continuous function. Some random variables cannot be classified as discrete or continuous; they are discussed in Section 2.5.

Example 2.2-3

An urn contains three colored balls. The balls are colored white (W), black (B), and red (R), respectively. The experiment consists of choosing a ball at random from the urn. The sample description space is $\Omega = \{W, B, R\}$. The random variable X is defined by

$$X(\zeta) = \begin{cases} \pi, & \zeta = W \text{ or } B \\ 0, & \zeta = R. \end{cases}$$

We can ask and compute the probability $P[X \leq x_1]$ where x_1 is any number. Thus $\{X \leq 0\} = \{R\}$, $\{2 \leq X < 4\} = \{W, B\}$. The computation of the associated probabilities are left as an exercise.

Example 2.2-4

A spinning wheel and pointer has 50 sectors numbered $n = 0, 1, \ldots, 49$. The experiment consists of spinning the wheel. Because the players are interested only in *even* or *odd* outcomes, they choose $\Omega = \{even, odd\}$ and the only events in the σ-field are $\{\phi, \Omega, even, odd\}$. Let $X = n$, that is, if n shows up, X assumes that value. Is X a random variable? Note that the inverse image of the set $[2, 3]$ is not an event. Hence X is not a valid random variable on this probability space because it is not a function on Ω.

2.3 PROBABILITY DISTRIBUTION FUNCTION

In Example 2.2-1 the induced event space under X includes $\{0, 1\}$, $\{0\}$, $\{1\}$, ϕ, for which the probabilities are $P[X = 0 \text{ or } 1] = 1$, $P[X = 0] = \frac{3}{4}$, $P[X = 1] = \frac{1}{4}$, and $P[\phi] = 0$. From these probabilities, we can infer any other probabilities such as, for example, $P[X \leq 0.5]$. In many cases it is awkward to write down $P[\cdot]$ for every event. For this reason we introduce a pointwise probability function called the *probability distribution function* (PDF). The PDF[†] is a function of x, which contains all the information necessary to compute $P[E]$ for any E in the Borel field of events. The PDF, $F_X(x)$, is defined by

$$F_X(x) = P[\{\zeta : X(\zeta) \leq x\}] = P_X[(-\infty, x)]. \tag{2.3-1}$$

Equation 2.3-1 is read as "the set of all outcomes ζ in the underlying sample description space such that the function $X(\zeta)$ assumes values less than or equal to x." Thus there is a subset of outcomes $\{\zeta : X(\zeta) \leq x\} \subset \Omega$ that under the mapping $X(\cdot)$ generates the set $[-\infty, x]$. The sets $\{\zeta : X(\zeta) \leq x\} \subset \Omega$ and $[-\infty, x]$ are equivalent events. We shall frequently leave out the dependence on the underlying sample space and write merely $P[X \leq x]$ or $P[a < X \leq b]$.

For the present we shall denote random variables by capital letters, that is, X, Y, Z and the values they can take by lowercase letters x, y, z. The subscript X on $F_X(x)$ associates it with the random variable for which it is the PDF. Thus writing $F_X(y)$ is perfectly consistent

---

[†]PDF should not be confused with pdf which will be the abbreviation for *probability density function*, a function to be introduced shortly.

notation and refers to $P[X \leq y]$. If $F_X(x)$ is discontinuous at a point, say, x_o then $F_X(x_o)$ will be taken to mean the value of the PDF immediately to the right of x_o (continuity from the right property) while $F_X(x_o^-)$ will denote the value immediately to the left.

Properties[†] of $F_X(x)$

(i) $F_X(\infty) = 1$, $F_X(-\infty) = 0$.

(ii) $x_1 \leq x_2 \rightarrow F_X(x_1) \leq F_X(x_2)$, that is, $F_X(x)$ is a nondecreasing function of x.

(iii) $F_X(x)$ is continuous from the right, that is,

$$F_X(x) = \lim_{\varepsilon \to 0} F_X(x + \varepsilon) \qquad \varepsilon > 0.$$

Proof of (ii)   Consider the event $\{x_1 < X \leq x_2\}$ with $x_2 > x_1$. The set $[x_1, x_2]$ is nonempty and $\in \mathscr{B}$. Hence

$$0 \leq P[x_1 < X \leq x_2] \leq 1.$$

But

$$\{X \leq x_2\} = \{X \leq x_1\} \cup \{x_1 < X \leq x_2\}$$

and

$$\{X \leq x_1\}\{x_1 < X \leq x_2\} = \phi.$$

Hence

$$F_X(x_2) = F_X(x_1) + P[x_1 < X \leq x_2]$$

or

$$\text{\Large ⚐} \quad \underline{P[x_1 < X \leq x_2] = F_X(x_2) - F_X(x_1) \geq 0 \text{ for } x_2 > x_1.} \qquad (2.3\text{-}2)$$

We leave it to the reader to establish the following results:

$$\text{\Large ⚐} \left[\begin{array}{l} P[a \leq X \leq b] = F_X(b) - F_X(a) + P[X = a]; \\[4pt] P[a < X < b] = F_X(b) - P[X = b] - F_X(a); \\[4pt] P[a \leq X < b] = F_X(b) - P[X = b] - F_X(a) + P[x = a]. \end{array} \right.$$

Example 2.3-1

The experiment consists of observing the voltage X of the parity bit in a word in computer memory. If the bit is *on*, then $X = 1$; if *off* then $X = 0$. Assume that the *off* state has probability q and the *on* state has probability $1 - q$. The sample space has only two points: $\Omega = \{\text{off, on}\}$.

[†]Properties (i) and (iii) require proof. This is furnished with the help of extended axioms in Chapter 7. Also see Wilbur F. Davenport [2-3, Chapter 4].

Computation of $F_X(x)$

   (i) $x < 0$: The event $\{X \le x\} = \phi$ and $F_X(x) = 0$.

  (ii) $0 \le x < 1$: The event $\{X \le x\}$ is equivalent to the event $\{\text{off}\}$ and excludes the event $\{\text{on}\}$.

$$X(\text{on}) = 1 > x$$
$$X(\text{off}) = 1 \le x.$$

       Hence $F_X(x) = q$.

 (iii) $x \ge 1$: The event $\{X \le x\} = $ is the certain event since

$$X(\text{on}) = 1 \le x$$
$$X(\text{off}) = 0 \le x.$$

      The solution is shown in Figure 2.3-1.

Example 2.3-2 _____

A bus arrives at random in $(0, T]$. Let the random variable X denote the time of arrival. Then clearly $F_X(t) = 0$ for $t \le 0$ and $F_X(T) = 1$ because the former is the probability of the impossible event while the latter is the probability of the certain event. Suppose it is known that the bus is equally likely or *uniformly* likely to come at any time within $(0, T]$. Then

$$F_X(t) = \begin{cases} 0 & t \le 0 \\ \dfrac{t}{T} & 0 < t \le T \\ 1 & t > T. \end{cases} \tag{2.3-3}$$

Actually Equation 2.3-3 defines "equally likely," not the other way around. The PDF is shown in Figure 2.3-2. In this case we say that X is *uniformly* distributed.

_____

If $F_X(x)$ is a continuous function of x, then

$$F_X(x) = F_X(x^-). \tag{2.3-4}$$

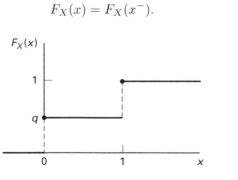

Figure 2.3-1 Probability distribution function associated with the parity bit observation experiment.

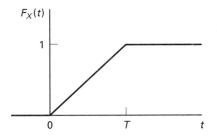

Figure 2.3-2 Probability distribution function of the uniform random variable X of Example 2.3-2.

However, if $F_X(x)$ is discontinuous at the point x then, from Equation 2.3-2

$$F_X(x) - F_X(x^-) = P[x^- < X \le x]$$
$$= \lim_{\varepsilon \to 0} P[x - \varepsilon < X \le x]$$
$$\triangleq P[X = x]. \tag{2.3-5}$$

Typically $P[X = x]$ is a discontinuous function of x; it is zero whenever $F_X(x)$ is continuous and nonzero only at discontinuities in $F_X(x)$.

Example 2.3-3 _____
Compute the PDF for a binomial random variable with parameters (n, p).

Solution Since X takes on only discrete values, that is, $X \in \{0, 1, 2, \ldots, n\}$, the event $\{X \le x\}$ is the same as $\{X \le [x]\}$, where $[x]$ is the largest integer equal to or smaller than x. Then $F_X(x)$ is given by

$$F_X(x) = \sum_{k=0}^{[x]} \binom{n}{k} p^k (1 - p)^{n-k}.$$

For $p = 0.6$, $n = 4$, the PDF has the appearance of a staircase function as shown in Figure 2.3-3

Example 2.3-4 _____
Using the results of Example 2.3-3, compute the following:

(a) $P[1.5 < X < 3]$;
(b) $P[0 \le X \le 3]$;
(c) $P[1.2 < X \le 1.8]$;
(d) $P[1.99 \le X < 3]$.

Solution

(a) $P[1.5 < X < 3] = F_X(3) - P[X = 3] - F_X(1.5)$
$\qquad = 0.8704 - 0.3456 - 0.1792 = 0.3456$;

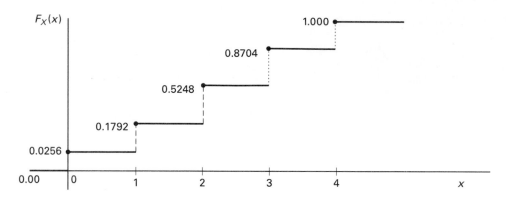

Figure 2.3-3 Probability distribution function for a binomial r.v. with $n = 4$, $p = 0.6$.

(b) $P[0 \leq X \leq 3] = F_X(3) - F_X(0) + P[X = 0]$

$\qquad = 0.8704 - 0.0256 + 0.0256 = 0.8704;$

(c) $P[1.2 < X \leq 1.8] = F_X(1.8) - F_X(1.2)$

$\qquad = 0.1792 - 0.1792 = 0;$

(d) $P[1.99 \leq X < 3] = F_X(3) - P[X = 3] - F_X(1.99) + P[X = 1.99]$

$\qquad = 0.8704 - 0.3456 - 0.1792 + 0 = 0.3456$

2.4 PROBABILITY DENSITY FUNCTION (pdf)

If $F_X(x)$ is continuous and differentiable, the pdf is computed from

$$f_X(x) = \frac{dF_X(x)}{dx}. \tag{2.4-1}$$

Properties.   If $f_X(x)$ exists, then

(i) $f_X(x) \geq 0.$ (2.4-2)

(ii) $\displaystyle\int_{-\infty}^{\infty} f_X(\xi)d\xi = F_X(\infty) - F_X(-\infty) = 1.$ (2.4-3)

(iii) $\displaystyle F_X(x) = \int_{-\infty}^{x} f_X(\xi)d\xi = P[X \leq x].$ (2.4-4)

(iv) $\displaystyle F_X(x_2) - F_X(x_1) = \int_{-\infty}^{x_2} f_X(\xi)d\xi - \int_{-\infty}^{x_1} f_X(\xi)d\xi$

$$= \int_{x_1}^{x_2} f_X(\xi)d\xi = P[x_1 < X \leq x_2]. \tag{2.4-5}$$

Interpretation of $f_X(x)$.

$$P[x < X \leq x + \Delta x] = F_X(x + \Delta x) - F_X(x).$$

If $F_X(x)$ is continuous in its first derivative then, for sufficiently small Δx,

$$F_X(x + \Delta x) - F_X(x) = \int_x^{x+\Delta x} f(\xi)d\xi \simeq f_X(x)\Delta x.$$

Hence for small Δx

$$P[x < X \leq x + \Delta x] \simeq f_X(x)\Delta x. \tag{2.4-6}$$

Observe that if $f_X(x)$ exists, then $F_X(x)$ is continuous and therefore, from Equation 2.3-5, $P[X = x] = 0$.

The univariate Normal (Gaussian[†]) pdf.   The pdf is given by

$$f_X(x) = \frac{1}{\sqrt{2\pi\sigma^2}} e^{-\frac{1}{2}\left[\frac{x-\mu}{\sigma}\right]^2}. \tag{2.4-7}$$

There are two independent parameters: μ, the mean, and σ, the standard deviation (σ^2 is called the variance). For the Gaussian random variable and, for that matter, for any other random variable with a well-defined pdf, we can compute the mean and variance (the sequence of the standard deviation) from the following formulas:

$$\mu = \int_{-\infty}^{\infty} x f_X(x)dx \tag{2.4-8}$$

and

$$\sigma^2 = \int_{-\infty}^{\infty} (x - \mu)^2 f_X(x)dx. \tag{2.4-9}$$

For random variables that take on discrete values such as the Bernoulli, Poisson, and others, we compute the mean and standard deviation from

$$\mu = \sum_{x_i : P[X=x_i]>0} x_i P[X = x_i] \tag{2.4-10}$$

and

$$\sigma^2 = \sum_{x_i : P[X=x_i]>0} (x_i - \mu)^2 P[X = x_i]. \tag{2.4-11}$$

While the last two equations are correct, in many places it is the convention to write $\mu = \sum_{i=-\infty}^{\infty} x_i P[X = x_i]$ and, likewise, for the summation to compute σ^2. Another possible convention is to write, merely, $\mu = \sum_i x_i P[X = x_i]$ and $\sigma^2 = \sum_i (x_i - \mu)^2 P[X = x_i]$. If in Equations 2.4-8 and 2.4-9 we substitute on the right-hand side (RHS) the Gaussian pdf

---

[†]After the German mathematician/physicist Carl F. Gauss (1777–1855).

with parameters a, b, that is, $f_X(x) = (2\pi b^2)^{-1/2}\exp\{-\frac{1}{2}(\frac{x-a}{b})^2\}$, we would indeed find that $\mu = a$, $\sigma = b$. These calculations are postponed until Chapter 4. In the meanwhile we furnish some simple examples.

Example 2.4-1

Let $f_X(x) = 1$, for $0 < x \le 1$ and zero elsewhere. This pdf is a special case of the *uniform law* discussed below. The mean is computed as

$$\mu = \int_{-\infty}^{\infty} x f_X(x)dx = \int_0^1 x\,dx = 0.5$$

and the variance is computed as

$$\sigma^2 = \int_{-\infty}^{\infty} (x-\mu)^2 f_X\,dx = \int_0^1 (x-0.5)^2 dx = 1/12.$$

Example 2.4-2

Suppose we are given that $P[X = 0] = P[X = 2] = 0.25$ and $P[X = 1] = 0.5$. For this discrete random variable (r.v.) we use Equations 2.4-10 and 2.4-11 to obtain

$$\mu = 0 \times 0.25 + 1 \times 0.5 + 2 \times 0.25 = 1$$

and

$$\sigma^2 = (0-1)^2 \times 0.25 + (1-1)^2 \times 0.5 + (2-1)^2 \times 0.25 = 0.5$$

In Chapter 4 we shall discuss what these quantities actually mean. When we want to say that a random variable X obeys the Normal probability law with mean μ and standard deviation σ, we shall use the symbols $X: N(\mu, \sigma^2)$. The Normal pdf is shown in Figure 2.4-1.

The Normal pdf is widely encountered in all branches of science and engineering as well as in social and demographic studies. For example, the IQ of children, the heights of men (or women), the noise voltage produced by a thermally agitated resistor, all are postulated to be approximately Normal over a large range of values.

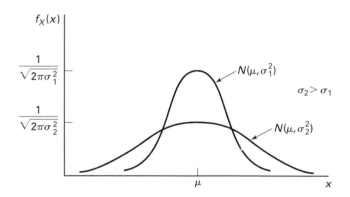

Figure 2.4-1   The Normal pdf.

Conversion of the Gaussian pdf to the standard Normal.   Suppose we are given $X: N(\mu, \sigma^2)$ and must evaluate $P[a < X \le b]$. We have

$$P[a < X \le b] = \frac{1}{\sqrt{2\pi\sigma^2}} \int_a^b e^{-\frac{1}{2}\left[\frac{x-u}{\sigma}\right]^2} dx.$$

$X = b, \quad \beta = \frac{b-\mu}{\sigma}$

$X = a, \quad \alpha' = \frac{a-\mu}{\sigma}$

With $\beta \triangleq (x - \mu)/\sigma$, $d\beta = (1/\sigma)dx$, $b' \triangleq (b - \mu)/\alpha$, $a' \triangleq (a - \mu)/\alpha$, we obtain

$$P[a < X \le b] = \frac{1}{\sqrt{2\pi}} \int_{a'}^{b'} e^{-\frac{1}{2}x^2} dx, \quad \checkmark$$

$\beta = \frac{x-\mu}{\sigma}$

$d\beta = \frac{dx}{\sigma}$

$$= \frac{1}{\sqrt{2\pi}} \int_0^{b'} e^{-\frac{1}{2}x^2} dx - \frac{1}{\sqrt{2\pi}} \int_0^{a'} e^{-\frac{1}{2}x^2} dx.$$

The function

$$\mathrm{erf}(x) \triangleq \frac{1}{\sqrt{2\pi}} \int_0^x e^{-\frac{1}{2}t^2} dt \qquad (2.4\text{-}12)$$

is sometimes called the error function $[\mathrm{erf}(x)]$ although other definitions of $\mathrm{erf}(x)$ exist.[†] The $\mathrm{erf}(x)$ is tabulated in Table 2.4-1 and is plotted in Figure 2.4-2.

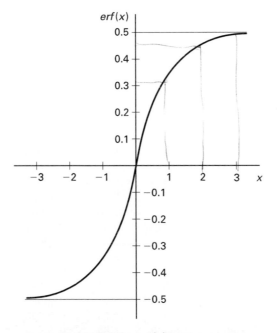

Figure 2.4-2 $\mathrm{erf}(x)$ versus x.

[†]For example, a widely used definition of $\mathrm{erf}(x)$ is $\mathrm{erf}(x) \triangleq (2/\sqrt{\pi}) \int_0^x e^{-t^2} dt$. If we call this $\mathrm{erf}(x) \triangleq \mathrm{erf}_2(x)$ and we call $\mathrm{erf}(x)$ of Equation 2.4-12 $\mathrm{erf}_1(x)$, then $\mathrm{erf}_1(x) = \frac{1}{2}\mathrm{erf}_2(x/\sqrt{2})$.

Table 2.4-1   Selected Values of erf(x)

$$\text{erf}(x) = \frac{1}{\sqrt{2\pi}} \int_0^x \exp\left(-\frac{1}{2}t^2\right) dt$$

x	erf(x)	x	erf(x)
0.05	0.01994	2.05	0.47981
0.10	0.03983	2.10	0.48213
0.15	0.05962	2.15	0.48421
0.20	0.07926	2.20	0.48609
0.25	0.09871	2.25	0.48777
0.30	0.11791	2.30	0.48927
0.35	0.13683	2.35	0.49060
0.40	0.15542	2.40	0.49179
0.45	0.17364	2.45	0.49285
0.50	0.19146	2.50	0.49378
0.55	0.20884	2.55	0.49460
0.60	0.22575	2.60	0.49533
0.65	0.24215	2.65	0.49596
0.70	0.25803	2.70	0.49652
0.75	0.27337	2.75	0.49701
0.80	0.28814	2.80	0.49743
0.85	0.30233	2.85	0.49780
0.90	0.31594	2.90	0.49812
0.95	0.32894	2.95	0.49840
1.00	0.34134	3.00	0.49864
1.05	0.35314	3.05	0.49884
1.10	0.36433	3.10	0.49902
1.15	0.37492	3.15	0.49917
1.20	0.38492	3.20	0.49930
1.25	0.39434	3.25	0.49941
1.30	0.40319	3.30	0.49951
1.35	0.41149	3.35	0.49958
1.40	0.41924	3.40	0.49965
1.45	0.42646	3.45	0.49971
1.50	0.43319	3.50	0.49976
1.55	0.43942	3.55	0.49980
1.60	0.44519	3.60	0.49983
1.65	0.45052	3.65	0.49986
1.70	0.45543	3.70	0.49988
1.75	0.45993	3.75	0.49990
1.80	0.46406	3.80	0.49992
1.85	0.46783	3.85	0.49993
1.90	0.47127	3.90	0.49994
1.95	0.47440	3.95	0.49995
2.00	0.47724	4.00	0.49996

Hence if $X: N(\mu, \sigma^2)$, then

$$P[a < X \le b] = \text{erf}\left(\frac{b - \mu}{\sigma}\right) - \text{erf}\left(\frac{a - \mu}{\sigma}\right). \qquad (2.4\text{-}13)$$

Example 2.4-3 ──

Suppose we choose a resistor with resistance R from a batch of resistors with parameters $\mu = 1000$ ohms with $\sigma = 200$ ohms. What is the probability that R will have a value between 900 and 1100 ohms?

Solution   Assuming that $R: N[1000, (200)^2]$ we compute from Equation 2.4-13

$$P[900 < R \le 1100] = \text{erf}(0.5) - \text{erf}(-0.5).$$

But $\text{erf}(-x) = -\text{erf}(x)$ (deduced from Equation 2.4-12). Hence

$$P[900 < R \le 1100] = 0.38.$$

Using Figure 2.4-3 as an aid in our reasoning, we readily deduce the following for $X: N(0, 1)$. Assume $x > 0$, then

$$P[X \le x] = \tfrac{1}{2} + \text{erf}(x) \qquad (2.4\text{-}14a)$$

$$P[X > -x] = \tfrac{1}{2} + \text{erf}(x) \qquad (2.4\text{-}14b)$$

$$P[X > x] = \tfrac{1}{2} - \text{erf}(x) \qquad (2.4\text{-}14c)$$

$$P[-x < X \le x] = 2\,\text{erf}(x) \qquad (2.4\text{-}14d)$$

$$P[|X| > x] = 1 - 2\,\text{erf}(x). \qquad (2.4\text{-}14e)$$

Note that since $\text{erf}(-x) = -\text{erf}(x)$, the first three formulas remain valid for $x < 0$ also.

Four Other Common Density Functions

1.   Rayleigh $(\sigma > 0)$:

$$f_X(x) = \frac{x}{\sigma^2} e^{-x^2/2\sigma^2} u(x). \qquad (2.4\text{-}15)$$

The function $u(x)$ is the unit step, that is, $u(x) = 1$, $x \ge 0$, $u(x) = 0$, $x < 0$. Thus $f_X(x) = 0$ for $x < 0$. Examples of where the Rayleigh pdf shows up are in rocket-landing errors, random fluctuations in the envelope of certain waveforms, and radial distribution of misses around the bull's-eye at a rifle range.

2.   Exponential $(\mu > 0)$:

$$f_X(x) = \frac{1}{\mu} e^{-x/\mu} u(x). \qquad (2.4\text{-}16)$$

The exponential law occurs, for example, in waiting-time problems, lifetime of machinery, and in describing the intensity variations of incoherent light.

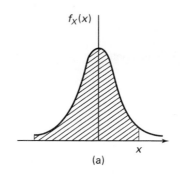

(a)

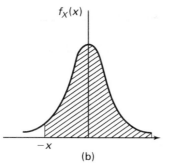

(b)

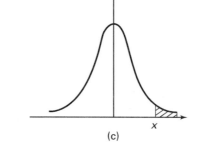

(c)

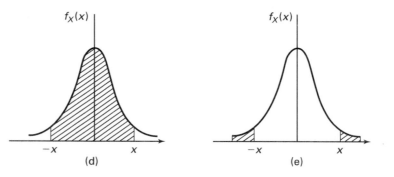

(d)

(e)

Figure 2.4-3   The areas of the shaded region under curves are (a) $P[X \leq x]$; (b) $P(X > -x)$; (c) $P(X > x)$; (d) $P(-x < X \leq x)$; and (e) $P(|X| > x)$.

3. Uniform $(b > a)$:

$$f_X(x) = \frac{1}{b - a} \qquad a < x < b$$

$$= 0 \qquad \text{otherwise.} \qquad (2.4\text{-}17)$$

The uniform pdf is used in communication theory, in queueing models, and in situations where we have no *a priori* knowledge favoring the distribution of outcomes except for the

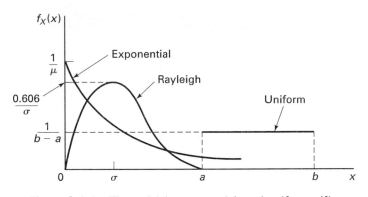

Figure 2.4-4   The rayleigh, exponential, and uniform pdf's.

end points; that is, we don't know when a business call will come but it must come, say, between 9 A.M. and 5 P.M.

The three pdf's are shown in Figure 2.4-4.

4.   Another pdf that has assumed importance since the introduction of computer analysis of speech and images is the Laplacian. The pdf is defined by

$$f_X(x) = \frac{c}{2}e^{-c|x|}, \quad c > 0. \tag{2.4-18}$$

The Laplacian is widely used to model speech sources and image gray levels. The Laplacian r.v. in these cases is called the *adjacent-sample difference* and is the difference in signal level from a sample point and its neighbor. Since the level of the sample point and its neighbor are often the same, the Laplacian peaks at zero. The Laplacian pdf is sometime written as

$$f_X(x) = \frac{1}{\sqrt{2}\sigma} \exp[-\sqrt{2}|x|/\sigma], \quad \sigma > 0, \tag{2.4-19}$$

where σ is the standard deviation of the Laplacian r.v. X. Precisely what this means will be explained in Chapter 4. The Laplacian pdf is shown in Figure 2.4-5.

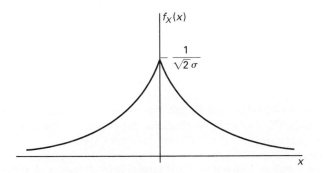

Figure 2.4-5   The Laplacian pdf used in computer analysis of speech and images.

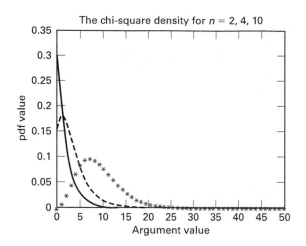

Figure 2.4-6   The Chi-square probability density function for $n = 2$ (solid), $n = 4$ (dashed), and $n = 10$ (stars). Note that for larger values of n, the shape approaches that of a Normal pdf with a positive mean-parameter μ.

More Advanced Density Functions

5.   Chi-square (n an integer)

$$f_X(x) = K_\chi x^{\left(\frac{n}{2}\right)-1} e^{-\frac{x}{2}} u(x), \tag{2.4-20}$$

where the normalizing constant K_χ is computed as $K_\chi = \frac{1}{2^{n/2}\Gamma(n/2)}$ and $\Gamma(\cdot)$ is the *Gamma function* discussed in Appendix B. The Chi-square pdf is shown in Figure 2.4-6.

6.   Gamma: ($b > 0, c > 0$)

$$f_X(x) = K_\gamma x^{b-1} e^{-cx} u(x), \tag{2.4-21}$$

where $K_\gamma = c^b / \Gamma(b)$.

7.   Student-t: (n an integer)

$$f_X(x) = K_{st} \left(1 + \frac{x^2}{n}\right)^{-\left(\frac{n+1}{2}\right)}, \tag{2.4-22}$$

where

$$K_{st} = \frac{\Gamma[(n+1)/2]}{\Gamma(n/2)\sqrt{\pi n}}$$

The Chi-square and Student-t densities are widely used in statistics.[†] We shall encounter these densities later in the book. The Gamma density is mother to other densities. For

---

[†]The Student-t distribution is so named because its discoverer W. S. Gossett (1876–1937) published his papers under the name "Student." Gossett, E. S. Pearson, R. A. Fisher, and J. Neyman are regarded as the founders of modern statistics.

example with $b = 1$, there results the exponential density; and with $b = n/2$ and $c = 1/2$, there results the Chi-square density.

There are other pdf's of importance in engineering and science, and we shall encounter some of them as we continue our study of probability. They all, however, share the properties that

$$f_X(x) \geq 0 \tag{2.4-23}$$

$$\int_{-\infty}^{\infty} f_X(x)dx = 1. \tag{2.4-24}$$

When $F_X(x)$ is not continuous, strictly speaking, its finite derivative does not exist and, therefore, the pdf doesn't exist. The question of what probability function is useful in describing X depends on the classification of X. We consider this next.

2.5 CONTINUOUS, DISCRETE, AND MIXED RANDOM VARIABLES

If $F_X(x)$ is continuous for every x and its derivative exists everywhere except at a countable set of points, then we say that X is a *continuous random variable* (r.v.). At points x where $F'_X(x)$ exists, the pdf is $f_X(x) = F'_X(x)$. At points where $F_X(x)$ is continuous, but $F'_X(x)$ is discontinuous, we can assign any positive number to $f_X(x)$; $f_X(x)$ will then be defined for every x, and we are free to use the following important formulas:

$$F_X(x) = \int_{-\infty}^{x} f_X(\xi)d\xi \tag{2.5-1}$$

$$P[x_1 < X \leq x_2] = \int_{x_1}^{x_2} f_X(\xi)d\xi, \tag{2.5-2}$$

and

$$P[B] = \int_{\xi:\xi \in B} f_X(\xi)d\xi, \tag{2.5-3}$$

where, in Equation 2.5-3, $B \in \mathscr{B}$, that is, B is an event. Equation 2.5-3 follows from the fact that for a continuous random variable, events can be written as a union of disjoint intervals in R. Thus, for example, let $B = \{\xi : \xi \in \cup_{i=1}^{n} I_i, I_i I_j = \phi \text{ for } i \neq j\}$, where $I_i = (a_i, b_i]$. Then clearly,

$$P[B] = \int_{a_1}^{b_1} f_X(\xi)d\xi + \int_{a_2}^{b_2} f_X(\xi)d\xi + \ldots + \int_{a_n}^{b_n} f_X(\xi)d\xi$$

$$= \int_{\xi:\xi \in B} f_X(\xi)d\xi. \tag{2.5-4}$$

A discrete *random variable* has a staircase type of distribution function (Figure 2.5-1).

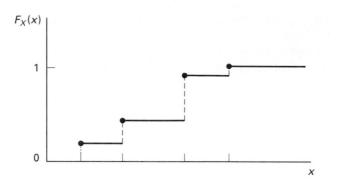

Figure 2.5-1   The probability distribution function for a discrete random variable.

A probability measure for discrete r.v. is the *probability mass function*[†] (PMF). The probability mass function $P_X(x)$ of a (discrete) random variable X is defined as

$$P_X(x) = P[X \leq x] - P[X < x]. \tag{2.5-5}$$

Thus $P_X(x) = 0$ everywhere where $F_X(x)$ is continuous and has finite values only where there is a discontinuity, that is, jump, in the PDF. If we denote $P[X < x]$ by $F_X(x^-)$, then at the jumps x_i, $i = 1, 2, \ldots$, the finite values of $P_X(x_i)$ can be computed from $P_X(x_i) = F_X(x_i) - F_X(x_i^-)$.

The probability mass function is used when there are at most a countable set of outcomes of the random experiment. Indeed $P_X(x_i)$ lends itself to the following frequency interpretation: Perform an experiment n times and let n_i be the number of tries that x_i appears as an outcome. Then, for n large,

$$P_X(x_i) \simeq \frac{n_i}{n}. \tag{2.5-6}$$

Because the PMF is so closely related to the frequency notion of probability, it is sometimes called the *frequency function*.

Since for a discrete r.v. $F_X(x)$ is not continuous $f_X(x)$, strictly speaking, does not exist. Nevertheless, with the introduction of Dirac delta functions,[‡] we shall be able to assign pdf's to discrete r.v.'s as well. The PDF for a discrete r.v. is given by

$$F_X(x) \triangleq P[X \leq x] = \sum_{\text{all } x_i \leq x} P_X(x_i). \tag{2.5-7}$$

and, more generally, for any event B when X is discrete:

$$P[B] = \sum_{\text{all } x_i \in B} P_X(x_i). \tag{2.5-8}$$

[†]Like mass, probability is nonnegative and conserved. Hence the term "mass" in probability *mass* function.

[‡]Also called impulses or impulse functions. Named after the English physicist Paul A. M. Dirac (1902–1984). Delta functions are discussed in Section B.2 of Appendix B.

Examples of Probability Mass Functions

1. Bernoulli $(p > 0, q > 0, p + q = 1)$:

$$P_X(0) = p \qquad P_X(1) = q$$
$$P_X(x) = 0, \qquad x \neq 0, 1. \tag{2.5-9}$$

The Bernoulli law applies in those situations where the outcome is one of two possible states, that is, whether a particular bit in a digital sequence is "one" or "zero." A r.v. that has the Bernoulli PMF is said to be a Bernoulli r.v. The Bernoulli PMF can be conveniently written as $P_X(x) = p^{1-x}(1-p)^x$ for $x = 0$, 1 and zero elsewhere.

2. Binomial $(n = 1, 2, \ldots; 0 < p < 1)$:

$$P_X(k) = \binom{n}{k} p^k q^{n-k}, \qquad k = 0, 1, 2, \ldots, n$$
$$= 0, \qquad \text{otherwise.} \tag{2.5-10}$$

The binomial law applies in games of chance, military defense strategies, failure analysis, and many other situations. A binomial r.v. has a PMF as in Equation 2.5-10. The Bernoulli law is often called a *point binomial*.

3. Poisson $(a > 0)$:

$$P_X(k) = e^{-a} \frac{a^k}{k!} \qquad k = 0, 1, 2, \ldots$$
$$= 0, \qquad \text{otherwise.} \tag{2.5-11}$$

The Poisson law is widely used in every branch of science and engineering (see Section 1.10). A r.v. whose PMF is given by Equation 2.5-11 is said to be a Poisson r.v.

Sometimes a r.v. is neither purely discrete nor purely continuous. We call such a r.v. a *mixed r.v.* The PDF of a mixed r.v. is shown in Figure 2.5-2. Thus $F_X(x)$ is discontinuous but not a staircase type-function.

The distinction between continuous and discrete r.v.'s is somewhat artificial. Continuous and discrete r.v.'s are often regarded as different objects even though the only real difference

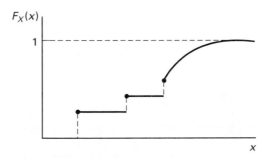

Figure 2.5-2 The PDF of a mixed r.v.

between them is that for the former the PDF is continuous while for the latter it is not. By introducing delta functions we can, to a large extent, treat them in the same fashion and compute probabilities for both continuous and discrete r.v.'s by using pdf's.

Returning now to Equation 2.5-7, which can be written as

$$F_X(x) = \sum_{i=-\infty}^{\infty} P_X(x_i)u(x - x_i), \qquad (2.5\text{-}12a)$$

and using the results from the section on delta functions in Appendix B enables us to write for a discrete r.v.

$$f_X(x) = \frac{dF_X(x)}{dx} = \sum_{i=-\infty}^{\infty} P_X(x_i)\delta(x - x_i), \qquad (2.5\text{-}12b)$$

where we recall that $P_X(x) \triangleq F_X(x_i) - F_X(x_i^-)$ and the unit step assures that the summation is over all i such that $x_i \le x$.

Example 2.5-1

Let X be a discrete r.v. with distribution function as shown in Figure 2.5-3(a). The pdf of X is

$$f_X(x) = \frac{dF_X}{dx} = 0.2\delta(x) + 0.6\delta(x - 1) + 0.2\delta(x - 3)$$

and is shown in Figure 2.5-3(b). To compute probabilities from the pdf for a discrete r.v. great care must be used in choosing the interval of integration. Thus

$$F_X(x) = \int_{-\infty}^{x^+} f_X(\xi)d\xi,$$

which includes the delta function at x if there is one there.

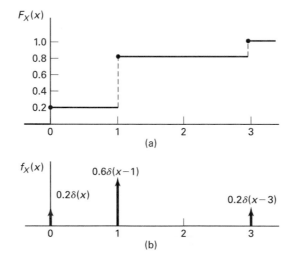

Figure 2.5-3   (a) PDF of a discrete r.v. X; (b) pdf of X using delta functions.

Similarly $P[x_1 < X \le x_2]$ involves the interval

and includes the impulse at x_2 (if there is one there) but excludes what happens at x_1. On the other hand $P[x_1 \le X < x_2]$ involves the interval

and therefore

$$P(x_1 \le X < x_2) = \int_{x_1^-}^{x_2^-} f_X(\xi)d\xi.$$

Applied to the foregoing example, these formulas give

$$P[X \le 1.5] = F_X(1.5) = 0.8$$
$$P[1 < X \le 3] = 0.2$$
$$P[1 \le X < 3] = 0.6$$

Example 2.5-2 _____

The pdf associated with the Poisson law with parameter a

$$f_X(x) = e^{-a} \sum_{k=0}^{\infty} \frac{a^k}{k!} \delta(x - k).$$

Example 2.5-3 _____

The pdf associated with the binomial law $b(k; n, p)$

$$f_X(x) = \sum_{k=0}^{n} \binom{n}{k} p^k q^{n-k} \delta(x - k).$$

Example 2.5-4 _____

The pdf of a mixed r.v. is shown in Figure 2.5-4. (1) What is the constant K? (2) Compute $P[X \le 5]$, $P[5 \le X < 10]$. (3) Draw the distribution function.

Solution   (1) Since

$$\int_{-\infty}^{\infty} f_X(\xi)d\xi = 1,$$

we obtain $10K + 0.25 + 0.25 = 1 \Rightarrow K = 0.05$.

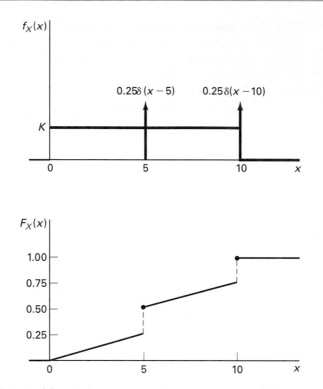

Figure 2.5-4   (a) pdf of a mixed r.v. for Example 2.5-4; (b) computed PDF.

(2) Since $P[X \leq 5] = P[X < 5] + P[X = 5]$, the impulse at $x = 5$ must be included. Hence

$$P[X \leq 5] = \int_0^{5^+} [0.05 + 0.25\delta(\xi - 5)]d\xi$$

$$= 0.5.$$

To compute $P(5 \leq X < 10)$, we leave out the impulse at $x = 10$ but include the impulse at $x = 5$. Thus

$$P[5 \leq X < 10] = \int_{5^-}^{10^-} [0.05 + 0.25\delta(\xi - 5)]d\xi$$

$$= 0.5.$$

2.6 CONDITIONAL AND JOINT DISTRIBUTIONS AND DENSITIES

Consider the event C consisting of all outcomes $\zeta \in \Omega$ such that $X(\zeta) \leq x$ and $\zeta \in B \subset \Omega$ where B is another event. The event C is then the set intersection of the two events

$\{\zeta : X(\zeta) \leq x\}$ and $\{\zeta : \zeta \in B\}$. We define the *conditional distribution function of X given the event B* as

$$F_X(x|B) \triangleq \frac{P[C]}{P[B]} = \frac{P[X \leq x, B]}{P[B]}, \tag{2.6-1}$$

where $P[X \leq x, B]$ is the probability of the joint event $\{X \leq x\} \cap B$ and $P[B] \neq 0$. If $x = \infty$, the event $\{X \leq \infty\}$ is the certain event Ω and since $\Omega \cap B = B$, $F_X(\infty|B) = 1$. Similarly, if $x = -\infty$, $\{X \leq -\infty\} = \phi$ and since $\Omega \cap \phi = \phi$, $F_X(-\infty|B) = 0$. Continuing in this fashion, it is not difficult to show that $F_X(x|B)$ has all the properties of an ordinary distribution, that is, $x_1 \leq x_2 \to F_X(x_1|B) \leq F_X(x_2|B)$.

For example, consider the event $\{X \leq x_2, B\}$ and write (assuming $x_2 \geq x_1$)

$$\{X \leq x_2, B\} = \{X \leq x_1, B\} \cup \{x_1 < X \leq x_2, B\}.$$

Since the two events on the right are disjoint, their probabilities add and we obtain

$$P[X \leq x_2, B] = P[X \leq x_1, B] + P[x_1 < X \leq x_2, B]$$

or

$$P[X \leq x_2|B]P[B] = P[X \leq x_1|B]P[B] + P[x_1 < X \leq x_2|B]P[B].$$

Thus when $P[B] \neq 0$, we obtain after rearranging terms

$$P[x_1 < X \leq x_2|B] = P[X \leq x_2|B] - P[X \leq x_1|B]$$

$$= F_X(x_2|B) - F_X(x_1|B). \tag{2.6-2}$$

Generally the event B will be expressed on the probability space $(R, \mathscr{B}, P_X)$ rather than the original space $(\Omega, \mathscr{F}, P)$. The conditional pdf is simply

$$f_X(x|B) \triangleq \frac{dF_X(x|B)}{dx}. \tag{2.6-3}$$

Following are some examples.

Example 2.6-1 _____

Let $B \triangleq \{X \leq 10\}$. We wish to compute $F_X(x|B)$.

(i) For $x \geq 10$, the event $\{X \leq 10\}$ is a subset of the event $\{X \leq x\}$. Hence $P[X \leq 10, X \leq x] = P[X \leq 10]$ and use of Equation 2.6-1 gives

$$F_X(x|B) = \frac{P[X \leq x, X \leq 10]}{P[X \leq 10]} = 1.$$

(ii) For $x \leq 10$, the event $\{X \leq x\}$ is a subset of the event $\{X \leq 10\}$. Hence $P[X \leq 10, X \leq x] = P[X \leq x]$ and

$$F_X(x|B) = \frac{P[X \leq x]}{P[X \leq 10]}.$$

The result is shown in Figure 2.6-1. We leave as an exercise to the reader to compute $F_X(x|B)$ when $B = \{b < X \leq a\}$.

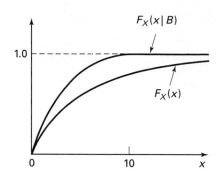

Figure 2.6-1   Conditional and unconditional PDF of X.

Example 2.6-2

Let X be a Poisson r.v. with parameter a. We wish to compute the conditional PMF and pdf of X given $B \triangleq \{X(\text{is}) \text{ even}\} = \{X = 0, 2, 4, \ldots\}$. First observe that $P[X \text{ even}]$ is given by

$$P[X = 0, 2, \ldots] = \sum_{k=0,2,\ldots}^{\infty} e^{-a} \frac{a^k}{k!}$$

while for X odd, we have

$$P[X = 1, 3, \ldots] = \sum_{k=1,3,\ldots}^{\infty} e^{-a} \frac{a^k}{k!}.$$

From these relations, we obtain

$$\sum_{k \text{ even}} e^{-a} \frac{a^k}{k!} - \sum_{k \text{ odd}} e^{-a} \frac{a^k}{k!} = \sum_{k=0}^{\infty} e^{-a} \frac{a^k}{k!}(-1)^k$$

$$= e^{-2a}$$

and

$$\sum_{k \text{ even}} e^{-a} \frac{a^k}{k!} + \sum_{k \text{ odd}} e^{-a} \frac{a^k}{k!} = 1.$$

Hence $P[X = 0, 2, \ldots] = \frac{1}{2}(1 + e^{-2a})$. Using the definition of a conditional PMF, we obtain

$$P_X(k|X \text{ is even}) = \frac{P[X = k, X \text{ is even}]}{P[X \text{ even}]}.$$

If k is even, then $\{X = k\}$ is a subset of $\{X \text{ even}\}$. If k is odd, $\{X = k\} \cap \{X \text{ even}\} = \phi$. Hence

$$P_X(k|X \text{ is even}) = \begin{cases} \dfrac{2e^{-a}a^k}{k!(1 + e^{-2a})}, & k \text{ even} \\ 0, & k \text{ odd}. \end{cases}$$

The conditional pdf is

$$f_X(x|X \text{ is even}) = \sum_{k=0,2,\ldots}^{\infty} \frac{2e^{-a}a^k}{k!(1+e^{-2a})}\delta(x-k)$$

and the conditional PDF is

$$F_X(x|X \text{ is even}) = \sum_{\text{all } k \leq x} P_X(k|X \text{ even}).$$

Let us next derive some important formulas involving conditional PDF's and pdf's.

The distribution function written as a weighted sum of conditional distribution functions.   Equation 1.6-4 in Chapter 1 gave the probability of the event B in terms of n mutually exclusive and exhaustive events $\{A_i\}i = 1, \ldots, n$ defined on the same probability space as B. With $B \triangleq \{X \leq x\}$, we immediately obtain from Equation 1.6-4:

$$F_X(x) = \sum_{i=1}^{n} F_X(x|A_i)P[A_i]. \tag{2.6-4}$$

Equation 2.6-4 describes $F_X(x)$ as a weighted sum of conditional distribution functions. One way to view Equation 2.6-4 is an "average" over all the conditional PDF's.[†] Since we haven't yet made concrete the notion of average (this will be done in Chapter 4), we ask only that the reader recall the nomenclature since it is in use in the technical literature.

Example 2.6-3 _____
In the automated manufacturing of computer memory chips, company Z produces one defective chip for every five good chips. The defective chips (DC) have a time of failure X that obeys the PDF

$$F_X(x|DC) = (1 - e^{-x/2})u(x) \quad (x \text{ in months})$$

while the time of failure for the good chips (GC) obeys the PDF

$$F_X(x|GC) = (1 - e^{-x/10})u(x) \quad (x \text{ in months}).$$

The chips are visually indistinguishable. A chip is purchased. What is the probability that the chip will fail before six months of use?

Solution   The unconditional PDF for the chip is, from Equation 2.6-4,

$$F_X(x) = F_X(x|DC)P[DC] + F_X(x|GC)P[GC],$$

[†]For this reason, when $F_X(x)$ is written as in Equation 2.6-4, it is sometimes called the *average* distribution function.

where $P[DC]$ and $P[GC]$ are the probabilities of selecting a defective and good chip, respectively. From the given data $P[DC] = 1/6$ and $P[GC] = 5/6$. Thus

$$F_X(6) = [1 - e^{-3}]\tfrac{1}{6} + [1 - e^{-0.6}]\tfrac{5}{6}$$

$$= 0.158 + 0.376 = 0.534.$$

Bayes' formula for probability density functions.   Consider the events B and $\{X = x\}$ defined on the same probability space. Then from the definition of conditional probability, it seems reasonable to write

$$P[B|X = x] = \frac{P[B, X = x]}{P[X = x]}. \tag{2.6-5}$$

The problem with Equation 2.6-5 is that if X is a continuous r.v., then $P[X = x] = 0$. Hence Equation 2.6-5 is undefined. Nevertheless, we can compute $P[B|X = x]$ by taking appropriate limits of probabilities involving the event $\{x < X \le x + \Delta x\}$. Thus consider the expression

$$P[B|x < X \le x + \Delta x] = \frac{P[x < X \le x + \Delta x|B]P[B]}{P[x < X \le x + \Delta x]}.$$

If we (i) divide numerator and denominator of the expression on the right by Δx, (ii) use the fact that $P[x < X \le x + \Delta x|B] = F(x + \Delta x|B) - F(x|B)$, and (iii) take the limit as $\Delta x \to 0$, we obtain

$$P[B|X = x] = \lim_{\Delta x \to 0} P[B|x < X \le x + \Delta x]$$

$$= \frac{f_X(x|B)P[B]}{f_X(x)}, \qquad f(x) \ne 0. \tag{2.6-6}$$

The quantity on the left is sometimes called the *a posteriori* probability (or *a posteriori* density) of B given $X = x$. Multiplying both sides of Equation 2.6-6 by $f_X(x)$ and integrating enables us to obtain the important result

$$P[B] = \int_{-\infty}^{\infty} P[B|X = x]f_X(x)dx. \tag{2.6-7}$$

In line with the terminology used in this section, $P[B]$ is sometimes called the *average probability* of B, the usage being suggested by the form of Equation 2.6-7.

Example 2.6-4

A signal, X, can come from one of three different sources designated as A, B, or C. The signal from A is $N(-1, 4)$; the signal from B is $N(0, 1)$; and the signal from C is $N(1, 4)$. In order for the signal to reach its destination at R, the switch in the line must be closed. Only one switch can be closed when the signal X is observed at R, but it is not known which switch it is. However it is known that switch a is closed twice as often as switch b, which is closed twice as often as switch c (Figure 2.6-2).

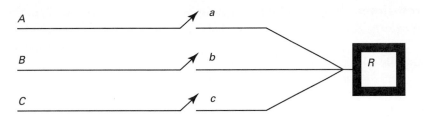

Figure 2.6-2 Based upon observing the signal, the receiver R must decide which switch was closed or, equivalently, which of the sources A, B, C was responsible for the signal. Only one switch can be closed at the time the receiver is on.

(a) Compute $P[X \leq -1]$;
(b) Given that we observe the event $\{X > -1\}$ from which source was this signal most likely?

Solution   (a) Let $P[A]$ denote the probability that A is responsible for the observation at R, that is, switch a is closed. Likewise for $P[B]$, $P[C]$. Then from the information about the switches we get $P[A] = 2P[B] = 4P[C]$ and $P[A] + P[B] + P[C] = 1$. Hence $P[A] = 4/7$, $P[B] = 2/7$, $P[C] = 1/7$. Next we compute $P[X \leq -1]$ from

$$P[X \leq -1] = P[X \leq -1|A]P[A] + P[X \leq -1|B]P[B] + P[X \leq -1|C]P[C]$$

where

$$P[X \leq -1|A] = 1/2$$

$$P[X \leq -1|B] = 1/2 - \text{erf}(1) = 0.159$$

$$P[X \leq -1|C] = 1/2 - \text{erf}(1) = 0.159.$$

Hence $P[X \leq -1] = 1/2 \times 4/7 + 0.159 \times 2/7 + 0.159 \times 1/7 \approx 0.354$.

(b) We wish to compute $\max\{P[A|X > -1], P[B|X > -1], P[C|X > -1]\}$. To enable this computation, we note that $P[X > -1|A] = 1 - P[X \leq -1|A]$, and so on, for B and C. Concentrating on source A, and using Bayes' rule, we get

$$P[A|X > -1] = \frac{\{1 - P[X \leq -1|A]\} \times P[A]}{1 - P[X \leq -1]}$$

which, using the values already computed, yields $P[A|X > -1] = 0.44$.

Repeating the calculation for the other sources, we obtain

$$P[B|X > -1] = 0.372,$$

$$P[C|X > -1] = 0.186.$$

Hence, since the maximum *a posteriori* probability favors A, source A was the most likely cause of the event $\{X > -1\}$.

Poisson transform. An important specific example of Equation 2.6-7 is the so-called *Poisson transform* in which B is the event that a random variable Y takes on integer values $k = 0, 1, \ldots$, that is, $B \triangleq \{Y = k\}$ and X is the Poisson parameter, treated here as a random variable with pdf $f_X(x)$. The ordinary Poisson law

$$P[Y = k] = e^{-a} \frac{a^k}{k!}, \qquad k \geq 0 \tag{2.6-8}$$

where a is the average number of events in a given interval (time, distance, volume, and so forth), treats the parameter a as a constant. But in many situations the underlying phenomenon that determines a is itself random and a must be viewed as a random outcome, that is, the outcome of a random experiment. Thus there are two elements of randomness: the random value of a and the random outcome $\{Y = k\}$. When a is random it seems appropriate to replace it by the notation of a random variable, say X. Thus for any given outcome $\{X = x\}$ the probability $P[Y = k | X = x]$ is Poisson; but the *unconditional* probability of the event $\{Y = k\}$ is not necessarily Poisson. Because both the number of events and the Poisson parameter are random, this situation is sometimes called *doubly stochastic*. From Equation 2.6-7 we obtain for the unconditional PMF of Y

$$P_Y(k) = \int_0^\infty \frac{x^k}{k!} e^{-x} f_X(x) dx, \qquad k \geq 0. \tag{2.6-9}$$

Equation 2.6-9 is known as the Poisson transform and can be used to obtain $f_X(x)$ if $P_Y(k)$ is obtained by experimentation. The mechanism by which $F_X(x)$ is obtained from $P_Y(k)$ is the *inverse Poisson transform*. The derivation of the latter is as follows. Let

$$F(\omega) \triangleq \frac{1}{2\pi} \int_0^\infty e^{j\omega x} e^{-x} f_X(x) dx \tag{2.6-10}$$

that is, the inverse Fourier transform of $e^{-x} f_X(x)$. Since

$$e^{j\omega x} = \sum_{k=0}^\infty [j\omega x]^k / k! \tag{2.6-11}$$

we obtain

$$F(\omega) = \frac{1}{2\pi} \sum_{k=0}^\infty (j\omega)^k \int_0^\infty \frac{x^k}{k!} e^{-x} f_X(x) dx$$

$$= \frac{1}{2\pi} \sum_{k=0}^\infty (j\omega)^k P_Y(k) \text{ from Equation 2.6-9.} \tag{2.6-12a}$$

Thus $F(\omega)$ *is known if* $P_Y(k)$ *is known.* Taking the forward Fourier transforms of $F(\omega)$ yield

$$e^{-x} f_X(x) = \int_{-\infty}^\infty F(\omega) e^{-j\omega x} d\omega$$

or

$$f_X(x) = e^x \int_{-\infty}^{\infty} F(\omega)e^{-j\omega x}d\omega. \tag{2.6-12b}$$

Equation 2.6-12b is the inverse relation we have been seeking. Thus to summarize: If we know $P_Y(k)$, we can compute $F(\omega)$. Knowing $F(\omega)$ enables us to obtain $f_X(x)$ by a Fourier transform. We illustrate the Poisson transform with an application from optical communication theory.

Example 2.6-5

In an optical communication system, light from the transmitter strikes a photodetector, which generates a photocurrent consisting of valence electrons having become conduction electrons (Figure 2.6-3).

It is known from physics that if the transmitter uses coherent laser light of constant intensity the Poisson parameter X has pdf

$$f_X(x) = \delta(x - x_0) \qquad x_o > 0 \tag{2.6-13}$$

where x_o, except for a constant, is the laser intensity. On the other hand, if the transmitter use thermal illumination, then the Poisson parameter X obeys the exponential law:

$$f_X(x) = \frac{1}{\mu}e^{-(1/\mu)x}u(x), \tag{2.6-14}$$

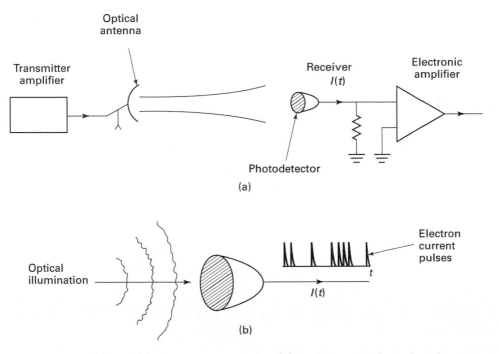

Figure 2.6-3   (a) Optical communication system; (b) output current from photodetector.

where $\mu > 0$ is a parameter called the mean value of X. Compute the PMF for the electron-count variable Y.

Solution   For coherent laser illumination we obtain from Equation 2.6-9

$$P_Y(k) = \int_0^\infty \frac{x^k}{k!} e^{-x} \delta(x - x_0) dx \qquad (2.6\text{-}15)$$

$$= \frac{x_o^k}{k!} e^{-x_0}, k \geq 0. \qquad (2.6\text{-}16)$$

Thus for coherent laser illumination, the photoelectrons *obey* the Poisson law. For thermal illumination, we obtain[†]

$$P_Y(k) = \int_0^\infty \frac{x^k}{k!} e^{-x}, \frac{1}{\mu} e^{-x/\mu} dx$$

$$= \frac{\mu^k}{[1 + \mu]^{k+1}}, k \geq 0. \qquad (2.6\text{-}17)$$

This PMF law is known as the geometric distribution and is sometimes called *Bose-Einstein statistics* [2-4]. It obeys the interesting recurrence relation

$$P_Y(k+1) = \frac{\mu}{1 + \mu} P_Y(k). \qquad (2.6\text{-}18)$$

Depending on which illumination applies, the statistics of the photocurrents are widely dissimilar.

Joint distributions and densities.   As stated in Section 2.1, it is possible to define more than one random variable on a probability space. For example, consider a probability space $(\Omega, \mathscr{F}, P)$ involving an underlying experiment consisting of the simultaneous throwing of two fair coins. Here the ordering is not important and the only elementary outcomes are $\zeta_1 = HH, \zeta_2 = HT, \zeta_3 = TT$, the sample description space is $\Omega = \{HH, HT, TT\}$, the σ-field of events is $\phi, \Omega, \{HT\}, \{TT\}, \{HH\}, \{TT \text{ or } HT\}, \{HH \text{ or } HT\}$, and $\{HH \text{ or } TT\}$. The probabilities are easily computed and are, respectively, 0, 1, 1/2, 1/4, 1/4, 3/4, 3/4, and 1/2. Now define two random variables

$$X_1(\zeta) = \begin{cases} 0, & \text{if at least one } H \\ 1, & \text{otherwise} \end{cases}$$

$$X_2(\zeta) = \begin{cases} -1, & \text{if one } H \text{ and one } T \\ +1, & \text{otherwise.} \end{cases}$$

Then $P[X_1 = 0] = 3/4$, $P[X_1 = 1] = 1/4$, $P[X_2 = -1] = 1/2$, $P[X_2 = 1] = 1/2$. Also we can easily compute the probability of joint events, for example, $P[X_1 = 0, X_2 = 1] = P[\{HH\}] = 1/4$.

[†]The evaluation of the integral, while not difficult, requires a little work and is left as an exercise.

In defining more than one random variable on a probability space, it is possible to define degenerate random variables. For example suppose the underlying experiment consists of observing the number ζ that is pointed to when a spinning wheel, numbered 0 to 100, comes to rest. Suppose we let $X_1(\zeta) = \zeta$ and $X_2(\zeta) = e^\zeta$. This situation is degenerate because observing one random variable completely specifies the other. In effect the uncertainty is associated with only one random variable, not both; we might as well forget about observing the other one. If we define more than one random variable on a probability space, degeneracy can be avoided if the underlying experiment is complex enough, or rich enough in outcomes. In the example we considered at the beginning, observing that $X_1 = 0$ doesn't specify the value of X_2 while observing $X_2 = 1$ doesn't specify the value of X_1.

The event $\{X \leq x,\ Y \leq y\} \triangleq \{X \leq x\} \cap \{Y \leq y\}$ consists of all outcomes $\zeta \in \Omega$ such that $X(\zeta) \leq x$ and $Y(\zeta) \leq y$. The point set induced by the event $\{X \leq x, Y \leq y\}$ is the shaded region in the $x'y'$ plane shown in Figure 2.6-4. In the diagram the numbers x, y are shown positive. In general they can have any value. The *joint distribution function* of X and Y is defined by

$$F_{XY}(x, y) = P[X \leq x, Y \leq y]. \tag{2.6-19}$$

By definition $F_{XY}(x, y)$ is a probability; thus it follows that $F_{XY}(x, y) \geq 0$ for all x, y. Since $\{X \leq \infty, Y \leq \infty\}$ is the certain event, $F_{XY}(\infty, \infty) = 1$. The point set associated with the certain event is the whole x', y' plane. The event $\{X \leq -\infty, y \leq -\infty\}$ is the impossible event and therefore $F_{XY}(-\infty, -\infty) = 0$. The reader should consider the events $\{X \leq x, Y \leq -\infty\}$ and $\{X \leq -\infty, Y \leq y\}$; are they impossible events also?

Since $\{X \leq \infty\}$ and $\{Y \leq \infty\}$ are certain events, and for any event B, $B \cap \Omega = B$, we obtain

$$\{X \leq x, Y \leq \infty\} = \{X \leq x\} \tag{2.6-20a}$$

$$\{X \leq \infty, Y \leq y\} = \{Y \leq y\} \tag{2.6-20b}$$

so that

$$F_{XY}(x, \infty) = F_X(x) \tag{2.6-21a}$$

$$F_{XY}(\infty, y) = F_Y(y). \tag{2.6-21b}$$

If $F_{XY}(x, y)$ is continuous and differentiable, the joint pdf can be obtained from

$$f_{XY}(x, y) = \frac{\partial^2}{\partial x\, \partial y}[F_{XY}(x, y)]. \tag{2.6-22}$$

It is left as a homework problem (Problem 2.29) to show that

$$f_{XY}(x, y)dx\, dy = P[x < X \leq x + dx, y < Y \leq y + dy]$$

and hence that $f_{XY}(x, y) \geq 0$ for all (x, y).

By twice integrating Equation 2.6-22 we obtain

$$F_{XY}(x, y) = \int_{-\infty}^{x} d\xi \int_{-\infty}^{y} d\eta f_{XY}(\xi, \eta). \tag{2.6-23}$$

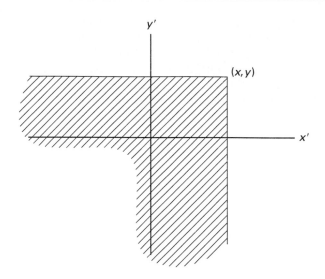

Figure 2.6-4 Point set associated with the event $\{X \leq x, Y \leq y\}$.

Equation 2.6-23 says that $F_{XY}(x,y)$ is the integral of the nonnegative function $f_{XY}(x,y)$ over the surface shown in Figure 2.6-4. It follows that integrating $f_{XY}(x,y)$ over a larger surface will generally yield a larger probability (never a smaller one!) than integrating over a smaller surface. From this we can deduce some obvious but important results. Thus if (x_1, y_1) and (x_2, y_2) denote two pairs of numbers and if $x_1 \leq x_2$, $y_1 \leq y_2$, then $F_{XY}(x_1, y_1) \leq F_{XY}(x_2, y_2)$. In general, $F_{XY}(x,y)$ increases as (x,y) moves up and to the right and decreases as (x,y) moves down and to the left. Also F_{XY} is continuous from above and from the right, that is, at a point of discontinuity, say x_0, y_0, with ε, $\delta > 0$:

$$F_{XY}(x_0, y_0) = \lim_{\substack{\varepsilon \to 0 \\ \delta \to 0}} F_{XY}(x_0 + \varepsilon, y_0 + \delta).$$

Thus at a point of discontinuity, F_{XY} assumes the value immediately to the right and above the point.

Properties of $F_{XY}(x,y)$

(i) $F_{XY}(\infty, \infty) = 1$; $F_{XY}(-\infty, y) = F_{XY}(x, -\infty) = 0$; Also $F_{XY}(x, \infty) = F_X(x)$; $F_{XY}(\infty, y) = F_Y(y)$.

(ii) if $x_1 \leq x_2$, $y_1 \leq y_2$, then $F_{XY}(x_1, y_1) \leq F_{XY}(x_2, y_2)$.

(iii) $F_{XY}(x, y) = \lim_{\substack{\varepsilon \to 0 \\ \delta \to 0}} F_{XY}(x + \varepsilon, y + \delta)$ $\varepsilon, \delta > 0$.

The next result of interest is the computation of the probability of the event $\{x_1 < X \leq x_2, y_1 < Y \leq y_2\}$. The point set induced by this event is shown in Figure 2.6-5.

The key to this computation is to observe that the set $\{X \leq x_2, Y \leq y_2\}$ lends itself to the following decomposition into disjoint sets:

$$\{X \leq x_2, Y \leq y_2\} = \{x_1 < X \leq x_2, y_1 < Y \leq y_2\}$$

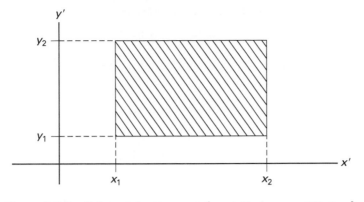

Figure 2.6-5   Point set for the event $\{x_1 < X \le x_2, y_1 < Y \le y_2\}$.

$$\cup \{x_1 < X \le x_2, Y \le y_1\{\cup\}X \le x_1, y_1 < Y \le y_2\}$$
$$\cup \{X \le x_1, Y \le y_1\}.$$

Now using the induced result from Axiom 3 (Equation 1.5-3), we obtain

$$F_{XY}(x_2, y_2) = P[x_1 < X \le x_2, y_1 < Y \le y_2]$$
$$+ P[x_1 < X \le x_2, Y \le y_1] + P[X \le x_1, y_1 < Y \le y_2] \qquad (2.6\text{-}24)$$
$$+ F_{XY}(x_1, y_1).$$

According to the elementary properties of the definite integral, the second and third terms on the right-hand side of Equation 2.6-24 can be written, respectively, as

$$\int_{x_1}^{x_2} \int_{-\infty}^{y_1} f_{XY}(\xi, \eta)d\xi\, d\eta = \int_{-\infty}^{x_2} \int_{-\infty}^{y_1} f_{XY}(\xi, \eta)d\xi\, d\eta$$
$$- \int_{-\infty}^{x_1} \int_{-\infty}^{y_1} f_{XY}(\xi, \eta)d\xi\, d\eta \qquad (2.6\text{-}25)$$

$$\int_{-\infty}^{x_1} \int_{y_1}^{y_2} f_{XY}(\xi, \eta)d\xi\, d\eta = \int_{-\infty}^{x_1} \int_{-\infty}^{y_2} f_{XY}(\xi, \eta)d\xi\, d\eta$$
$$- \int_{-\infty}^{x_1} \int_{-\infty}^{y_1} f_{XY}(\xi, \eta)d\xi\, d\eta. \qquad (2.6\text{-}26)$$

But the terms on the right-hand sides of these equations are all distributions; thus Equations 2.6-25 and 2.6-26 become

$$\int_{x_1}^{x_2} \int_{-\infty}^{y_1} f_{XY}(\xi, \eta)d\xi\, d\eta = F_{XY}(x_2, y_1) - F_{XY}(x_1, y_1), \qquad (2.6\text{-}27)$$

$$\int_{-\infty}^{x_1} \int_{y_1}^{y_2} f_{XY}(\xi, \eta)d\xi\, d\eta = F_{XY}(x_1, y_2) - F_{XY}(x_1, y_1). \qquad (2.6\text{-}28)$$

Now going back to Equation 2.6-24 and using Equations 2.6-27 and 2.6-28 yields

$$F_{XY}(x_2, y_2) = P[x_1 < X \le x_2, y_1 < Y \le y_2]$$
$$+ F_{XY}(x_2, y_1) - F_{XY}(x_1, y_1) + F_{XY}(x_1, y_2) - F_{XY}(x_1, y_1)$$
$$+ F_{XY}(x_1, y_1).$$

After simplifying and rearranging term so that the desired quantity appears on the left-hand side, we finally get

$$P[x_1 < X \le x_2, y_1 < Y \le y_2] = F_{XY}(x_2, y_2) - F_{XY}(x_2, y_1)$$
$$- F_{XY}(x_1, y_2) + F_{XY}(x_1, y_1). \qquad (2.6\text{-}29)$$

Equation 2.6-29 is generally true for any random variables X, Y independent or not. Some caution must be used in applying Equation 2.6-29. For example, Figure 2.6-6(a) and (b) show two regions A, B involving excursions on random variables X, Y such that $\{x_1 < X \le x_2\}$ and $\{y_1 < Y \le y_2\}$. However the use of Equation 2.6-29 would not be appropriate here since neither region is a rectangle with sides parallel to the axes. In the case of the event shown in Figure 2.6-6(a), a rotational coordinate transformation might save the day but this would involve some knowledge of transformation of random variables, a subject covered in the next chapter. The events whose point sets are shown in Figure 2.6-6 can still be computed by integration of the probability density function (pdf) provided that the integration is done over the appropriate region. We illustrate with the following example.

Example 2.6-6

We are given $f_{XY}(x, y) = e^{-(x+y)} u(x)u(y)$ and wish to compute $P[(X, Y) \in \mathscr{A}]$, where $\mathscr{A}$ is the shaded region shown in Figure 2.6-7. The region $\mathscr{A}$ is described by $\mathscr{A} = \{(x, y) : 0 \le x \le 1, |y| \le x\}$. We obtain

$$P[(X, Y) \in \mathscr{A}] = \int_{x=0}^{x=1} \int_{y=-x}^{x} e^{-(x+y)} u(x)u(y)dx\, dy$$

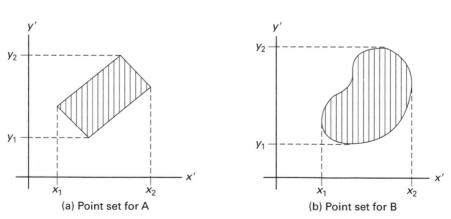

(a) Point set for A   (b) Point set for B

Figure 2.6-6   Points sets of events A and B whose probabilities are not given by Equation 2.6-29.

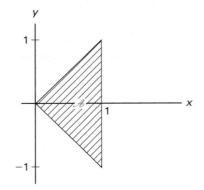

Figure 2.6-7   The region $\mathscr{A}$ for Example 2.6-5.

$$= \int_{x=0}^{x=1} dx e^{-x} u(x) \int_{y=-x}^{x} e^{-y} u(y) dy$$

$$= 0.27.$$

Example 2.6-7 _____

Let X, Y be two random variables with joint pdf $f_{XY}(x, y) = 1$ for $0 < x < 1$, $0 < y < 1$, and zero elsewhere. The support for the pdf is shown in gray; the support for the event $(-\infty, x) \times (-\infty, y)$ for $0 < x < 1$, $0 < y < 1$ is shown bounded by the heavy black line.

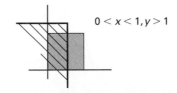

0 < x < 1, 0 < y < 1

For this case $F_{XY}(x, y) = \int_0^y \int_0^x 1 dx' \, dy' = xy$.

0 < x < 1, y > 1

When $0 < x < 1$, $y > 1$, we obtain $F_{XY}(x, y) = \int_0^x dx' \int_0^1 dy = x$. Proceeding in this way we obtain a complete characterization of the PDF as

$$F_{XY}(x, y) = 0, x < 0, \text{ any } y$$

$$= 0, y < 0, \text{ any } x$$

$$= x, 0 < x < 1, y > 1$$

$$= y, x > 1, 0 < y < 1$$

$$= xy, 0 < x < 1, 0 < y < 1.$$

$$= 1, x \geq 1, y \geq 1.$$

As Examples 2.6-6 and 2.6-7 illustrate for specific cases, the probability of any event of the form $\{(X, Y) \in \mathcal{A}\}$ can be computed by the formula

$$P[(X, Y) \in \mathcal{A}] = \iint_{\mathcal{A}} f_{XY}(x, y) dx\, dy \qquad (2.6\text{-}30)$$

provided $f_{XY}(x, y)$ exists. While Equation 2.6-30 seems entirely reasonable, its veracity requires demonstration. One way to do this is to decompose the arbitrarily shaped region into a (possibly very large) number of tiny disjoint rectangular regions $\mathcal{A}_1, \mathcal{A}_2, \ldots, \mathcal{A}_N$. Then the event $\{X, Y \in \mathcal{A}\}$ is decomposed as

$$\{(X, Y) \in \mathcal{A}\} = \bigcup_{i=1}^{N} \{(X, Y) \in \mathcal{A}_i\}$$

with the consequence that (by induced Axion 3):

$$P[(X, Y) \in \mathcal{A}] = \sum_{i=1}^{N} P[(X, Y) \in \mathcal{A}_i]. \qquad (2.6\text{-}31)$$

But the probabilities on the right-hand side can be expressed in terms of distributions and hence in terms of integrals of densities (Equation 2.6-29). Then, taking the limit as N becomes large and the $\mathcal{A}_i$ become infinitesimal, we would obtain Equation 2.6-30.

The functions $F_X(x)$ and $F_Y(y)$ are called *marginal* distributions if they are derived from a joint distribution as in Equation 2.6-21. Thus

$$F_X(x) = F_{XY}(x, \infty) = \int_{-\infty}^{x} d\xi \int_{-\infty}^{\infty} dy\, f(\xi, y) \qquad (2.6\text{-}32)$$

$$F_Y(y) = F_{XY}(\infty, y) = \int_{-\infty}^{y} d\eta \int_{-\infty}^{\infty} dx\, f(x, \eta). \qquad (2.6\text{-}33)$$

Since the marginal densities are given by

$$f_X(x) = \frac{dF_X(x)}{dx}$$

$$f_Y(y) = \frac{dF_Y(y)}{dy}$$

we obtain by differentiating Equations 2.6-32 and 2.6-33:

$$f_X(x) = \int_{-\infty}^{\infty} f_{XY}(x,y)dy \tag{2.6-34}$$

$$f_Y(y) = \int_{-\infty}^{\infty} f_{XY}(x,y)dx. \tag{2.6-35}$$

We point out in passing that the joint pdf $f_{XY}(x,y)$ shares important properties with its one-dimensional relative $f_X(x)$, namely,

(i) $f_{XY}(x,y) \geq 0$ all x, y;

(ii) $\int_{-\infty}^{\infty} \int_{-\infty}^{\infty} f_{XY}(x,y)dx\,dy = 1$ (the certain event);

(iii) While $f_{XY}(x,y)$ is not a probability, $f_{XY}(x,y)dx\,dy$ is. Indeed $f_{XY}(x,y)dx\,dy = P[x < X \leq x + dx, y < Y \leq y + dy]$. We leave the demonstration of these results as a homework problem.

For discrete random variables we obtain similar results. Given the joint probability mass function $P_{XY}(x_i, y_k)$ for all x_i, y_k, we compute the marginal probability mass functions from

$$P_X(x_i) = \sum_{\text{all } y_k} P_{XY}(x_i, y_k) \tag{2.6-36}$$

$$P_Y(y_k) = \sum_{\text{all } x_i} P_{XY}(x_i, y_k). \tag{2.6-37}$$

Independent random variables.   Two r.v.'s X and Y are said to be *independent* if the events $\{X \leq x\}$ and $\{Y \leq y\}$ are independent for every combination of x, y. In Section 1.5 two events A and B were said to be independent if $P[AB] = P[A]P[B]$. Taking $AB \triangleq \{X \leq x\} \cap \{Y \leq y\}$, where $A \triangleq \{X \leq x\}, B = \{Y \leq y\}$, and recalling that $F_X(x) \triangleq P[X \leq x]$, and so forth for $F_Y(y)$, it then follows immediately that we can write

$$F_{XY}(x,y) = F_X(x)F_Y(y) \tag{2.6-38}$$

for every x, y if and only if X and Y are independent. Also

$$f_{XY}(x,y) = \frac{\partial^2 F_{XY}(x,y)}{\partial x\,\partial y} \tag{2.6-39}$$

$$= \frac{\partial F_X(x)}{\partial x} \cdot \frac{\partial F_Y(y)}{\partial y}$$

$$= f_X(x)f_Y(y). \tag{2.6-40}$$

From the definition of conditional probability we obtain for independent X, Y:

$$F_X(x|Y \leq y) = \frac{F_{XY}(x,y)}{F_Y(y)}$$

$$= F_X(x), \tag{2.6-41}$$

and so forth, for $F_Y(y|X \leq x)$. From these results it follows (by differentiation) that for independent events the conditional pdf's are equal to the marginal pdf's, that is,

$$f_X(x|Y \leq y) = f_X(x) \tag{2.6-42}$$

$$f_Y(y|X \leq x) = f_Y(y). \tag{2.6-43}$$

It is easy to show from Equation 2.6-29 that the events $\{x_1 < X \leq x_2\}$ and $\{y_1 < Y \leq y_2\}$ are independent if X and Y are independent random variables, that is,

$$P[x_1 < X \leq x_2, y_1 < Y \leq y_2] = P[x_1 < X \leq x_2]P[y_1 < Y \leq y_2] \tag{2.6-44}$$

if $F_{XY}(x,y) = F_X(x)F_Y(y)$. Indeed, using Equation 2.6-29

$$P[x_1 < X \leq x_2, y_1 < Y \leq y_2]$$
$$= F_{XY}(x_2, y_2) - F_{XY}(x_2, y_1) - F_{XY}(x_1, y_2) + F_{XY}(x_1, y_1)$$
$$= F_X(x_2)F_Y(y_2) - F_X(x_2)F_Y(y_1) - F_X(x_1)F_Y(y_2) + F_X(x_1)F_Y(y_1)$$
$$= (F_X(x_2) - F_X(x_1))(F_Y(y_2) - F_Y(y_1))$$
$$= P[x_1 < X \leq x_2]P[y_1 < Y \leq y_2].$$

Example 2.6-8 _____

$$f_{XY}(x,y) = \frac{1}{2\pi\sigma^2}e^{-(1/2\sigma^2)(x^2+y^2)}$$

$$= \frac{1}{\sqrt{2\pi\sigma^2}}e^{-\frac{1}{2}(x^2/\sigma^2)}\frac{1}{\sqrt{2\pi\sigma^2}}e^{-\frac{1}{2}(y^2/\sigma^2)}. \tag{2.6-45}$$

Hence X and Y are independent r.v.'s.

Example 2.6-9 _____

The joint pdf of two random variables is given by $f_{XY}(x,y) = [2\pi]^{-1}\exp[-\frac{1}{2}(x^2+y^2)]$ for $-\infty < x$, $y < \infty$. Compute the probability that $\{X,Y\}$ are restricted to (a) the 2×2 square; and (b) the unit circle.

Solution   (a) Let $\Re_1$ denote the surface of the square. Then

$$P[\zeta:(X,Y) \in \Re_1] = \iint_{\Re_1} f_{XY}(x,y)dxdy$$

$$= \frac{1}{\sqrt{2\pi}} \int_{-1}^{1} \exp\left[-\frac{1}{2}x^2\right] dx \times \frac{1}{\sqrt{2\pi}} \int_{-1}^{1} \exp\left[-\frac{1}{2}y^2\right] dy$$

$$= 2\mathrm{erf}(1) \times 2\mathrm{erf}(1) = 0.466.$$

(b) Let $\Re_2$ denote the surface of the unit circle. Then

$$P[\zeta\colon (X,Y) \in \Re_2] = \iint_{\Re_2} f_{XY}(x,y)dxdy$$

$$= \iint_{\Re_2} [2\pi]^{-1} \exp\left[-\frac{1}{2}(x^2 + y^2)\right] dxdy.$$

With $r \triangleq \sqrt{x^2 + y^2}$ and $\tan\theta \triangleq y/x$, the infinitesimal area $dxdy \to rdrd\theta$, and we obtain

$$P[\zeta\colon (X,Y) \in \Re_2] = \frac{1}{2\pi} \iint_{\Re_2} \exp\left[-\frac{1}{2}r^2\right] r\,dr\,d\theta$$

$$= \frac{1}{2\pi} \int_0^{2\pi} \left\{ \int_{r=0}^{1} r\exp\left[-\frac{1}{2}r^2\right] dr \right\} d\theta = 0.393.$$

Joint densities involving nonindependent r.v.'s.   Lest the reader think that all joint PDF's or pdf's factor, we next consider a case involving nonindependent random variables.

Example 2.6-10

Let $f_{XY}(x,y) = A(x+y) \qquad 0 < x \le 1, \qquad 0 < y \le 1$

$$= 0, \qquad \text{otherwise.}$$

(i) What is A? We know that

$$\int_{-\infty}^{\infty} \int_{-\infty}^{\infty} f_{XY}(x,y)dx\,dy = 1.$$

Hence

$$A \int_0^1 dy \int_0^1 x\,dx + A \int_0^1 dx \int_0^1 y\,dy = 1 \Rightarrow A = 1.$$

(ii) What are the marginal pdf's?

$$f_X(x) = \int_{-\infty}^{\infty} f_{XY}(x,y)dy = \int_0^1 (x+y)dy = (xy + y^2/2)\Big|_0^1$$

$$= \begin{cases} x + \frac{1}{2} & 0 < x \le 1 \\ 0, & \text{otherwise.} \end{cases}$$

Similarly,

$$f_Y(y) = \int_{-\infty}^{\infty} f_{XY}(x,y)dx$$

$$= \begin{cases} y + \frac{1}{2} & 0 < y \le 1 \\ 0, & \text{otherwise.} \end{cases}$$

(iii) What is $F_{XY}(x,y)$? $F_{XY}(x,y) \triangleq P[X \le x, Y \le y]$, so we must integrate over the infinite rectangle with vertices (x,y), $(x,-\infty)$, $(-\infty,-\infty)$, and $(-\infty,y)$. However, only where this rectangle actually overlaps with the region over which $f_{XY}(x,y) \ne 0$, will there be a contribution to the integral

$$F_{XY}(x,y) = \int_{-\infty}^{x} dx' \int_{-\infty}^{y} dy' f_{XY}(x',y').$$

(a) $x \ge 1$, $y \ge 1$ [Figure 2.6-8(a)]

$$F_{XY}(x,y) = \int_0^1 \int_0^1 f_{XY}(x',y')dx'\,dy' = 1.$$

(b) $0 < x \le 1$, $y \ge 1$ [Figure 2.6-8(b)]

$$F_{XY}(x,y) = \int_{y'=0}^{1} dy' \left(\int_{x'=0}^{x} dx'(x' + y') \right)$$

$$= \int_{y'=0}^{1} dy' \left(\int_{x'=0}^{x} x'\,dx \right) + \int_{y'=0}^{1} dy'\,y' \left(\int_{x'=0}^{x} dx' \right)$$

$$= \frac{x}{2}(x + 1).$$

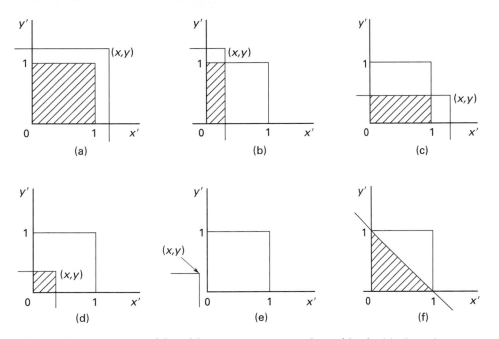

Figure 2.6-8   Shaded region in (a) to (e) is the intersection of supp(f_{XY}) with the point set associated with the event $\{-\infty < X \le x, -\infty < Y \le y\}$. In (f), the shaded region is the intersection of supp(f_{XY}) with $\{X + Y \le 1\}$.

(c) $0 < y \le 1,\ x \ge 1$ [Figure 2.6-8(c)]

$$F_{XY}(x, y) = \int_{y'=0}^{y} \int_{x'=0}^{1} (x' + y')dx'\, dy' = \frac{y}{2}(y + 1).$$

(d) $0 < x \le 1,\ 0 < y \le 1$ [Figure 2.6-8(d)]

$$F_{XY}(x, y) = \int_{y'=0}^{y} \int_{x'=0}^{x} (x' + y')dx'\, dy' = \frac{yx}{2}(x + y).$$

(e) $x < 0$, any y; or $y < 0$, any x [Figure 2.6-8(e)]

$$F_{XY}(x, y) = 0.$$

(f) Compute $P[X + Y \le 1]$. The point set is the half-space separated by the line $x + y = 1$ or $y = 1 - x$. However, only where this half-space intersects the region over which $f_{XY}(x, y) \ne 0$, will there be a contribution to the integral

$$P[X + Y \le 1] = \iint_{x'+y'\le 1} f_{XY}(x', y')dx'\, dy'.$$

[See Figure 2.6-8(f).] Hence

$$P[X + Y \leq 1] = \int_{x'=0}^{1} \int_{y'=0}^{1-x'} (x' + y')dy'\, dx'$$

$$= \int_{x'=0}^{1} x'(1 - x')dx' + \int_{x'=0}^{1} \frac{(1 - x')^2}{2} dx'$$

$$= \frac{1}{3}.$$

In the previous example we dealt with a pdf that was not factorable. Another example of a joint pdf that is not factorable is

$$f_{XY}(x, y) = \frac{1}{2\pi\sigma^2 \sqrt{1 - \rho^2}} \exp\left(\frac{-1}{2\sigma^2(1 - \rho^2)}(x^2 + y^2 - 2\rho xy)\right) \qquad (2.6\text{-}46)$$

when $\rho \neq 0$. Thus X and Y are not independent.

In the special case when $\rho = 0$ in Equation 2.6-46, $f_{XY}(x, y)$ factors as $f_X(x)f_Y(y)$ and X and Y become independent random variables. A picture of $f_{XY}(x, y)$ under these circumstances is shown in Figure 2.6-9 for $\sigma = 1$.

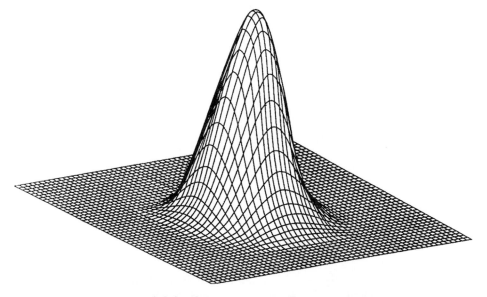

Figure 2.6-9   Graph of the joint Gaussian density

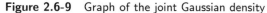

$$f_{XY}(x, y) = (2\pi)^{-1} \exp\left[-\frac{1}{2}(x^2 + y^2)\right].$$

As we shall see in Chapter 4, Equation 2.6-46 is a special case of the *jointly Gaussian probability density* of two r.v.'s. We defer a fuller discussion of this important pdf until we discuss the meaning of the parameter ρ. This we do in Chapter 4.

Example 2.6-11

Consider again the problem considered in Example 2.6-9, part (b) except let

$$f_{XY}(x,y) = [2\pi\sqrt{1-\rho^2}]^{-1} \exp\left[-\frac{1}{2(1-\rho^2)}(x^2 + y^2 - 2\rho xy)\right].$$

As before let $\Re_2$ denote the surface of the unit circle. Then

$$P[\zeta\colon (X,Y) \in \Re_2] = \iint_{\Re_2} f_{XY}(x,y)dxdy$$

$$= \iint_{\Re_2} [2\pi\sqrt{1-\rho^2}]^{-1} \exp\left[-\frac{1}{2(1-\rho^2)}(x^2 + y^2 - 2\rho xy)\right] dx\, dy.$$

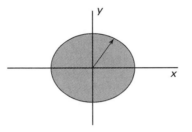

As before we use a polar coordinate transformation as $r = \sqrt{x^2 + y^2}$, $\tan\theta = y/x$ and obtain

$$P[\zeta\colon (X,Y) \in \Re_2] = \frac{1}{4\pi K} \int_0^{2\pi} \left(\frac{1 - \exp[-2K^2(1 - \rho\sin 2\theta)]}{1 - \rho\sin 2\theta}\right) d\theta,$$

where $K \triangleq \frac{1}{2\sqrt{1-\rho^2}}$. We have skipped several steps that the reader should have no trouble in reproducing. For $\rho = 0$, we get for this probability 0.393, that is, the same as in Example 2.6-9. However when $\rho \neq 0$, this probability can be computed numerically since the integral is generally not available in closed form. A MATLAB.m file that enables the computation of $P[\zeta\colon (X,Y) \in \Re_2]$ is furnished below. The result is shown in Figure 2.6-10.

MATLAB.m file for computing. $P[\zeta\colon (X,Y) \in \Re_2]$

```
function[Pr]=corrprob
p=[0:100]/100.;
q=p*2*pi;
Pr=zeros(1,100);
```

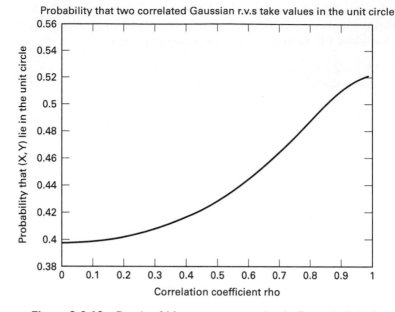

Figure 2.6-10   Result of MATLAB computation in Example 2.6-11.

```
K=.5./sqrt(1-p.^2);

for i=1:100
    f=(1-exp(-2*K(i)^2*(1-p(i)*sin(2*q)))))./(1-p(i)*sin(2*q));
    Pr(i)=sum(f)/(4*pi)/K(i)*(2*pi/100);
end

plot(p(1:100),Pr)
title('Probability that two correlated Gaussian r.v.s take values in the
    unit circle')
xlabel('correlation coefficient rho')
ylabel('Probability that X,Y) lie in the unit circle')
```

In Section 4.3 of Chapter 4 we demonstrate the fact that as $\rho \to 1$
$$f_{XY}(x,y) \to \frac{1}{2\pi} e^{-\frac{x^2}{2}} \delta(y-x). \text{ Hence}$$

$$P[\zeta : (X,Y) \in \Re_2] = \iint_{\Re_2} \frac{1}{\sqrt{2\pi}} e^{-0.5x^2} \delta(x-y)dxdy = \int_{-0.707}^{0.707} \frac{1}{\sqrt{2\pi}} e^{-0.5x^2} dx$$

$$= 0.52.$$

This is the result that we observe in Figure 2.6-10.

Conditional densities. We shall now derive a useful formula for conditional densities involving two r.v.'s. The formula is based on the definition of conditional probability given in Equation 1.6-2. From Equation 2.6-29 we obtain

$$P[x < X \le x + \Delta x, y < Y \le y + \Delta y]$$
$$= F_{XY}(x + \Delta x, y + \Delta y) - F_{XY}(x, y + \Delta y) - F_{XY}(x + \Delta x, y) + F_{XY}(x, y).$$

Now dividing both sides $\Delta x \, \Delta y$, taking limits, and subsequently recognizing that the right-hand side, by definition, is the second partial derivative of F_{XY} with respect to x and y enables us to write that

$$\lim_{\substack{\Delta x \to 0 \\ \Delta y \to 0}} \frac{P[x < X \le x + \Delta x, y < Y \le y + \Delta y]}{\Delta x \, \Delta y} = \frac{\partial^2 F_{XY}}{\partial x \, \partial y} \triangleq f_{XY}(x, y).$$

Hence for Δx, Δy small

$$P[x < X \le x + \Delta x, y < Y \le y + \Delta y] \simeq f_{XY}(x, y)\Delta x \, \Delta y, \qquad (2.6\text{-}47)$$

which is the two-dimensional equivalent of Equation 2.4-6. Now consider

$$P[y < Y \le y + \Delta y | x < X \le x + \Delta x] = \frac{P[x < X \le x + \Delta x, y < Y \le y + \Delta y]}{P[x < X \le x + \Delta x]}$$
$$\simeq \frac{f_{XY}(x, y)\Delta x \, \Delta y}{f_X(x)\Delta x}.$$

But the quantity on the left is merely

$$F_{Y|B}(y + \Delta y | x < X \le x + \Delta x) - F_{Y|B}(y | x < X \le x + \Delta x),$$

where $B \triangleq \{x < X \le x + \Delta x\}$. Hence

$$\lim_{\substack{\Delta x \to 0 \\ \Delta y \to 0}} \frac{F_{Y|B}(y + \Delta y | x < X \le x + \Delta x) - F_{Y|B}(y | x < X \le x + \Delta x)}{\Delta y}$$

$$= \frac{f_{XY}(x, y)}{f_X(x)}$$

$$= \frac{\partial F_{X|Y}(y | X = x)}{\partial y}$$

$$= f_{Y|X}(y | x) \qquad (2.6\text{-}48)$$

by Equation 2.6-3. The notation $f_{Y|X}(y|x)$ reminds us that it is the conditional pdf of Y given the event $\{X = x\}$. We thus obtain the important formula

$$f_{Y|X}(y|x) = \frac{f_{XY}(x, y)}{f_X(x)}, \qquad f_X(x) \ne 0. \qquad (2.6\text{-}49)$$

If we use Equation 2.6-49 in Equation 2.6-35 we obtain the useful formula

$$f_Y(y) = \int_{-\infty}^{\infty} f_{Y|X}(y|x)f_X(x)dx. \tag{2.6-50}$$

Also

$$f_{X|Y}(x|y) = \frac{f_{XY}(x,y)}{f_Y(y)}, \qquad f_Y(y) \neq 0. \tag{2.6-51}$$

The quantity $f_{X|Y}(x|y)$ is called the conditional pdf of X given the event $\{Y = y\}$. From Equations 2.6-49 and 2.6-50 it follows that

$$f_{X|Y}(x|y) = \frac{f_{Y|X}(y|x)f_X(x)}{f_Y(y)}, \tag{2.6-52}$$

We illustrate with an example.

Example 2.6-12 _____
(Laser coherence.) Suppose we observe the light field $U(t)$ being emitted from a laser. Laser light is said to be *temporally coherent*, which means that the light at any two times t_1 and t_2 is statistically dependent if $t_2 - t_1$ is not too large [2-5]. Let $X \triangleq U(t_1)$, $Y \triangleq U(t_2)$ and $t_2 > t_1$. Suppose X and Y are jointly Gaussian as in Equation 2.6-46 with $\sigma^2 = 1$. For $\rho \neq 0$, they are dependent. Still it is easy to show from Equations 2.6-34 and 2.6-35 that the marginal pdf's $f_X(x)$ and $f_Y(y)$ are given by

$$f_X(x) = \frac{1}{\sqrt{2\pi}}e^{-\frac{1}{2}x^2}$$

$$f_Y(y) = \frac{1}{\sqrt{2\pi}}e^{-\frac{1}{2}y^2}$$

and that both are centered about zero. Now suppose that we measure the light at t_1, that is, X and find that $X = x$. Is the pdf of Y, conditioned upon this new knowledge, still centered at zero, that is, is the average[†] value of Y still zero?

Solution   We wish to compute $f_{Y|X}(y|x)$.
Applying Equation 2.6-49,

$$f_{Y|X}(y|x) = \frac{f_{XY}(x,y)}{f_X(x)}$$

yields

$$f_{Y|X}(y|x) = \frac{1}{\sqrt{2\pi(1-\rho^2)}} \exp\left\{\left[-\frac{1}{2(1-\rho^2)}(x^2+y^2-2\rho xy)\right]+\frac{1}{2}x^2\right\}.$$

_____
[†]A concept to be fully developed in Chapter 4.

If we multiply and divide the isolated $\frac{1}{2}x^2$ term in the far right of the exponent by $1 - \rho^2$, we simplify the above as

$$f_{Y|X}(y|x) = \frac{1}{\sqrt{2\pi(1-\rho^2)}} \exp\left\{-\left[\frac{x^2 + y^2 - 2\rho x y - x^2(1-\rho^2)}{2(1-\rho^2)}\right]\right\}.$$

Further simplifications result when quadratic terms in the exponent are combined into a perfect square:

$$f_{Y|X}(y|x) = \frac{1}{\sqrt{2\pi(1-\rho^2)}} \exp\left\{-\frac{(y-\rho x)^2}{2(1-\rho^2)}\right\}.$$

Thus when $X = x$, the pdf of Y is centered at $y = \rho x$ and not zero as previously. If $\rho x > 0$, Y is more likely to take on positive values and if $\rho x < 0$, Y is more likely to take on negative values. This is in contrast to what happens when X is not observed: The most likely value of Y is then zero!

A major application of conditioned events and conditional probabilities occurs in the science of estimating failure rates. This is discussed in the next section.

2.7 FAILURE RATES

In modern industrialized society where planning for equipment replacement, issuance of life insurance, and so on, are important activities there is a need to keep careful records of the failure rates of objects, be they machines or humans. For example consider the cost of life insurance: Clearly it wouldn't make much economic sense to price a five-year term-life insurance policy for a 25-year-old woman at the same level as, say, for a 75-year-old man. The "failure" probability (i.e., death) for the older man is much higher than for the young woman. Hence, sound pricing policy will require the insurance company to insure the older man at a higher price. How much higher? This is determined from actuarial tables which are estimates of life expectancy conditioned on many factors. One important condition is "that you have survived until (a certain age)." In other words, the probability that you will survive to age 86, given that you have survived to age 85, is much higher than the probability that you will survive to age 86 if you are an infant.

Let X denote the time of failure or, equivalently, the failure time. Then by Bayes' theorem, the probability that failure will occur in the interval $[t, t+dt]$ *given that* the object has survived to t can be written as

$$P[t < X \leq t + dt | X > t] = \frac{P[t < X \leq t + dt, X > t]}{P[X > t]}. \tag{2.7-1}$$

But since the event $\{X > t\}$ is subsumed by the event $\{t < X \leq t + dt\}$, it follows that $P[t < X \leq t + dt, X > t] = P[t < X \leq t + dt]$. Hence

$$P[t < X \leq t + dt | X > t] = \frac{P[t < X \leq t + dt]}{P[X > t]}. \tag{2.7-2}$$

By recalling that $P[t < X \le t + dt] = F_X(t + dt) - F_X(t)$, we obtain

$$P[t < X \le t + dt | X > t] = \frac{F_X(t + dt) - F_X(t)}{1 - F_X(t)}. \tag{2.7-3}$$

A Taylor series expansion of $F_X(t + dt)$ about the point t yields (we assume that F_X is differentiable)

$$F_X(t + dt) = F_X(t) + f_X(t)\, dt.$$

When this result is used in Equation 2.7-3, we obtain at last

$$P[t < X \le t + dt | X > t] = \frac{f_X(t) dt}{1 - F_X(t)} \tag{2.7-4}$$

$$\stackrel{\Delta}{=} \alpha(t)\, dt,$$

where

$$\alpha(t) \stackrel{\Delta}{=} \frac{f_X(t)}{1 - F_X(t)}. \tag{2.7-5}$$

The object $\alpha(t)$, is called the *conditional failure rate* although it has other names such as the hazard rate, force of mortality, intensity rate, instantaneous failure rate, or simply failure rate. If the conditional failure rate at t is large, then an object surviving to time t will have a higher probability of failure in the next Δt seconds than another object with lower conditional failure rate. Many objects, including humans, have failure rates that vary with time. During the early life of the object, failure rates may be high due to inherent or congenital defects. After this early period the object enjoys a useful life characterized by a near-constant failure rate. Finally, as the object ages and parts wear out, the failure rate increases sharply leading quickly and inexorably to failure or death.

The pdf of the random variable X can be computed explicitly from Equation 2.7-3 when we observe that $F_X(t + dt) - F_X(t) = F_X'(t)dt = dF_X$. Thus we get

$$\frac{dF_X}{1 - F_X} = \alpha(t)\, dt \tag{2.7-6}$$

which can be solved by integration. First recall from calculus that

$$\int_{y_0}^{y_1} \frac{dy}{1 - y} = -\int_{1-y_0}^{1-y_1} \frac{dy}{y} = \int_{1-y_1}^{1-y_0} \frac{dy}{y} = \ln\frac{1 - y_0}{1 - y_1};$$

Second, use the facts that

(i) $F_X(0) = 0$ since we assume that the object is working at $t = 0$ (the time that the object is turned on);

(ii) $F_X(\infty) = 1$ since we assume that object must ultimately fail. Then

$$\int_{F_X(0)}^{F_X(t)} \frac{dF_X}{1 - F_X} = -\ln[1 - F_X(t)] = \int_0^t \alpha(t')dt'$$

from which we finally obtain

$$F_X(t) = 1 - \exp\left[-\int_0^t \alpha(t')dt'\right]. \tag{2.7-7}$$

Since $F_X(\infty) = 1$ we must have

$$\int_0^\infty \alpha(t')\,dt' = \infty. \tag{2.7-8}$$

Equation 2.7-7 is the PDF for the failure time X. By differentiating Equation 2.7-7 we obtain for the pdf

$$f_X(t) = \alpha(t)\exp\left[-\int_0^t \alpha(t')dt'\right]. \tag{2.7-9}$$

Different pdf's result from different models for the conditional failure rate $\alpha(t)$.

Example 2.7-1
Assume that X obeys the exponential probability law, that is, $F_X(t) = (1 - e^{-\lambda t})u(t)$. We find

$$\alpha(t) = \frac{f_X(t)}{1 - F_X(t)} = \frac{\lambda e^{-\lambda t}}{e^{-\lambda t}} = \lambda.$$

Thus the conditional failure rate is a constant. Conversely, if $\alpha(t)$ is a constant, the failure time obeys the exponential probability law.

An important point to observe is that the conditional failure rate is not a pdf (see Equation 2.7-8). The conditional density of X, given $\{X \geq t\}$ can be computed from the conditional distribution by differentiation. For example,

$$F_X(x|X > t) \triangleq P[X \leq x|X > t]$$
$$= \frac{P[X \leq x, X > t]}{P(X > t)}.$$

The event $\{X \leq x, X > t\}$ is clearly empty if $t > x$. If $t < x$, then $\{X \leq x, X > t\} = \{t < X \leq x\}$. Thus

$$F_X(x|X > t) = \begin{cases} 0, & t > x \\ \dfrac{F_X(x) - F_X(t)}{1 - F_X(t)}, & x \geq t. \end{cases} \tag{2.7-10}$$

Hence

$$f_X(x|X \geq t) = \begin{cases} 0, & t > x \\ \dfrac{f_X(x)}{1 - F_X(t)}, & x \geq t. \end{cases} \tag{2.7-11}$$

and the connection between $\alpha(t)$ and $f_X(x|X \geq t)$ is obtained by comparing Equation 2.7-11 with Equation 2.7-5, that is,

$$f_X(t|X \geq t) = \alpha(t). \tag{2.7-12}$$

Example 2.7-2 _____

Oscar, a college student, has a nine-year-old Itsibitsi, an import car famous for its reliability. The conditional failure rate, based on field data, is $\alpha(t) = 0.06tu(t)$ assuming a normal usage of 10,000 mile/year. To celebrate the end of the school year, Oscar begins a 30-day cross-country motor trip. What is the probability that Oscar's Itsibitsi will have a first breakdown during his trip?

Solution   First, we compute the pdf $f_X(t)$ as

$$f_X(t) = 0.06te^{-\int_0^t 0.06t' \, dt'} u(t)$$

$$= 0.06te^{-0.03t^2} u(t).$$

Next, we convert 30 days into 0.0824 years. Finally, we note that

$$P[9.0 < X \le 9.0824 | X > 9] = \frac{P[9.0 < X \le 9.0824]}{1 - F_X(9)},$$

where we have used Bayes' rule and the fact the event $\{9 < X \le 9.0824\} \cap \{X > 9\} = \{9 < X \le 9.0824\}$.

Since

$$P[9.0 < X \le 9.0824] = 0.06 \int_{9.0}^{9.0824} te^{-0.03t^2} dt$$

$$\simeq 0.0038$$

and

$$1 - F_X(9) = 0.088,$$

Oscar's car has a $3.8 \times 10^{-3}/8.8 \times 10^{-2}$ or 0.043 probability of suffering a breakdown in the next 30 days.

Incidentally, the probability that a newly purchased Itsibitsi will have at least one breakdown in 10 years is 0.95.

_____

2.8 SUMMARY

The material discussed in this chapter is central to the concept of the whole book. We began by defining a real random variable as a mapping from the sample description space Ω to the real line R. We then introduced a point function $F_X(x)$ called the probability distribution function (PDF), which enabled us to compute the probabilities of events of the type $\{\zeta : \zeta \in \Omega, X(\zeta) \le x\}$. The probability density function pdf and probability mass function (PMF) were derived from the PDF, and a number of useful and specific probability laws were discussed. We showed how, by using Dirac delta functions, we could develop a unified theory for both discrete and continuous random variables. We then discussed joint distributions, the Poisson transform, and its inverse and the application of these concepts to physical problems.

We discussed the important concept of conditional probability and illustrated its application in the area of conditional failure rates. The conditional failure, often high at the outset, constant during mid-life, and high at old age, is fundamental in determining the probability law of time-to-failure.

PROBLEMS

2.1. The event of k successes in n tries regardless of the order is the binomial law $b(k, n; p)$ (Equation 1.9-1). Let $n = 10$, $p = 0.3$. Define the r.v. (random variable) X by

$$X(k) = \begin{cases} 1, & \text{for } 0 \le k \le 2 \\ 2, & \text{for } 2 < k \le 5 \\ 3, & \text{for } 5 < k \le 8 \\ 4, & \text{for } 8 < k \le 10 \end{cases}.$$

Compute $P[X = j]\, j = 1, \dots, 4$. Plot $F_X(x) = P[X \le x]$ for all x.

2.2. In a restaurant known for its unusual service, the time X, in minutes, that a customer has to wait before he captures the attention of a waiter is specified by the following distribution function:

$$F_X(x) = \begin{cases} \left(\dfrac{x}{2}\right)^2 & \text{for } 0 \le x \le 1, \\[2mm] \dfrac{x}{4} & \text{for } 1 \le x \le 2, \\[2mm] \dfrac{1}{2} & \text{for } 2 \le x \le 10, \\[2mm] \dfrac{x}{20} & \text{for } 10 \le x \le 20, \\[2mm] 1 & \text{for } x \ge 20. \end{cases}$$

(a) Sketch $F_X(x)$. (b) Compute and sketch the pdf $f_X(x)$. Verify that the area under the pdf is indeed unity. (c) What is the probability that the customer will have to wait (1) at least 10 minutes, (2) less than 5 minutes, (3) between 5 and 10 minutes, (4) exactly 1 minute?

2.3. Compute the probabilities of the events $\{X < a\}$, $\{X \le a\}$, $\{a \le X < b\}$, $\{a \le X \le b\}$, $\{a < X \le b\}$, and $\{a < X < b\}$ in terms of $F_X(x)$ and $P[X = x]$ for $x = a, b$.

2.4. In the following pdf's, compute the constant B required for proper normalization:

(a) Cauchy ($\alpha < \infty, \beta > 0$):

$$f_X(x) = \frac{B}{1 + [(x - \alpha)/\beta]^2}, \qquad -\infty < x < \infty.$$

(b) Maxwell ($\alpha > 0$):

$$f_X(x) = \begin{cases} Bx^2 e^{-x^2/\alpha^2}, & x > 0 \\ 0, & \text{otherwise.} \end{cases}$$

(c) Beta ($b > -1, c > -1$):

$$f_X(x) = \begin{cases} Bx^b(1-x)^c, & 0 \le x \le 1 \\ 0, & \text{otherwise.} \end{cases}$$

(See formula 6.2-1 on page 258 of [2-6].)

(d) Chi-square ($\sigma > 0, n = 1, 2, \ldots$):

$$f_X(x) = \begin{cases} Bx^{(n/2)-1} \exp(-x/2\sigma^2), & x > 0 \\ 0, & \text{otherwise.} \end{cases}$$

2.5. A noisy resistor produces a voltage $v_n(t)$. At $t = t_1$, the noise level $X \triangleq v_n(t_1)$ is known to be a Gaussian r.v. with pdf

$$f_X(x) = \frac{1}{\sqrt{2\pi\sigma^2}} \exp\left[-\frac{1}{2}\left(\frac{x}{\sigma}\right)^2\right].$$

Compute and plot the probability that $|X| > k\sigma$ for $k = 1, 2, \ldots$.

2.6. Compute $F_X(k\sigma)$ for the Rayleigh pdf (Equation 2.4-15) for $k = 0, 1, 2, \ldots$.

2.7. Write the *probability density functions* (using delta functions) for the Bernoulli, binomial, and Poisson PMF's.

2.8. The pdf of a r.v. X is shown in Figure P2.8. The numbers in parentheses indicate area.

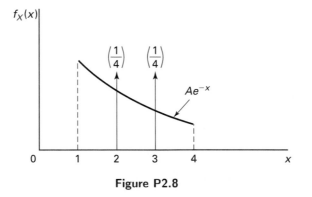

Figure P2.8

(a) Compute the value of A; (b) sketch the PDF; (c) compute $P[2 \le X < 3]$; (d) compute $P[2 < X \le 3]$; (e) compute $F_X(3)$.

2.9. The PDF of a random variable X is given by $F_X(x) = (1-e^{-x})u(x)$. The probability of the event $\{\zeta : X(\zeta) < 1 \text{ } or \text{ } X(\zeta) > 2\}$ is: (a) 0.914; (b) 0.767; (c) 0.632; (d) 0.135; (e) 0.041.

2.10. Consider a binomial r.v. with PMF $b(k; 4, \frac{1}{2})$. Compute $P[X = k | X \text{ even}]$ for $k = 0, \ldots, 4$.

2.11. The time-to-failure in months, X, of light bulbs produced at two manufacturing plants A and B obey, respectively, the following PDF's

$$F_X(x) = (1 - e^{-x/5})u(x) \text{ for plant } A$$

$$F_X(x) = (1 - e^{-x/2})u(x) \text{ for plant } B.$$

Plant B produces three times as many bulbs as plant A. The bulbs, indistinguishable to the eye, are intermingled and sold. What is the probability that a bulb purchased at random will burn at least (a) two months; (b) five months; (c) seven months?

2.12. Show that the conditioned distribution of X given the event $A = \{b < X \leq a\}$ is

$$F_X(x|A) = \begin{cases} 0, & x \leq b \\ \dfrac{F_X(x) - F_X(b)}{F_X(a) - F_X(b)}, & b < x \leq a \\ 1, & x \geq a. \end{cases}$$

2.13. It has been found that the number of people Y waiting in a queue in the bank on payday obeys the Poisson law as

$$P[Y = k | X = x] = e^{-x}\frac{x^k}{k!}, \quad k \geq 0, x > 0$$

given that the normalized serving time of the teller x (that is, the time it takes the teller to deal with a customer) is constant. However, the serving time is more accurately modeled as an r.v. X. For simplicity let X be a uniform r.v. with

$$f_X(x) = \tfrac{1}{5}[u(x) - u(x - 5)].$$

Then $P[Y = k | X = x]$ is still Poisson but $P[Y = k]$ is something else. Compute $P[Y = k]$ for $k = 0, 1, 2, \ldots$. (*Hint*: Use Equation 2.6-9.)

2.14. Consider the joint pdf of X and Y:

$$f_{XY}(x, y) = \frac{1}{3\pi}e^{-\frac{1}{2}[(x/3)^2 + (y/2)^2]}u(x)u(y).$$

Are X and Y independent r.v.? Compute the probability of $\{0 < X \leq 3, 0 < Y \leq 2\}$.

2.15. Consider the random variable X with pdf $f_X(x)$ given by

$$f_X(x) = \begin{cases} A(1 + x), & -1 < x \leq 0 \\ A(1 - x), & 0 < x < 1 \\ 0, & \text{elsewhere.} \end{cases}$$

(a) Find A and plot $f_X(x)$;
(b) Plot $F_X(x)$, the PDF;
(c) Find point b such that

$$P[X > b] = \tfrac{1}{2}P[X \le b].$$

2.16. Consider a communication channel corrupted by noise. Let X be the value of the transmitted signal and Y be the value of the received signal. Assume that the conditional density of Y given $\{X = x\}$ is Gaussian, that is,

$$f_{Y|X}(y|x) = \frac{1}{\sqrt{2\pi\sigma^2}} \exp\left(\frac{-(y-x)^2}{2\sigma^2}\right),$$

and X is uniformly distributed on $[-1, 1]$. What is the conditional probability density of X given Y (i.e., $f_{X|Y}(x|y)$?

2.17. The arrival time of a professor to his office is a continuous r.v. uniformly distributed over the hour between 8 A.M. and 9 A.M. Define the events:

$$A = \{\text{The prof. has not arrived by 8.30 A.M.}\},$$

$$B = \{\text{The prof. will arrive by 8:31 A.M.}\}.$$

Find

(a) $P[B|A]$.
(b) $P[A|B]$.

2.18. Let X be a random variable with *pdf*

$$f_X(x) = \begin{cases} 0, & x < 0 \\ ce^{-2x}, & x \ge 0 \end{cases} \qquad (c > 0).$$

(a) Find c;
(b) Let $a > 0$, $x > 0$, find $P[X \ge x + a]$;
(c) Let $a > 0$, $x > 0$, find $P[X \ge x + a | X \ge a]$.

2.19. To celebrate getting a passing grade in a course on probability, Wynette invites her Professor, Dr. Chance, to dinner at the famous French restaurant "C'est Tres Chere." The probability of getting a reservation if you call y days in advance is given by $1 - e^{-y}$, where $y \ge 0$. What is the minimum numbers of days that Wynette should call in advance in order to have a probability of at least 0.95 of getting a reservation?

2.20. A U.S. defense radar scans the skies for unidentified flying objects (UFOs). Let M be the event that a UFO is present and M^c the event that a UFO is absent. Let $f_{X/M}(x|M) = \frac{1}{\sqrt{2\pi}} \exp(-0.5[x - r]^2)$ be the conditional pdf of the radar return signal X when a UFO is actually there and let $f_{X/M}(x/M^c) = \frac{1}{\sqrt{2\pi}} \exp(-0.5[x]^2)$ be the conditional pdf of the radar return signal X when there is no UFO. To be specific, let $r = 1$ and let the *alert level* be $x_A = 0.5$. Let A denote the event of an alert, that is, $\{X > x_A\}$. Compute $P[A|M]$, $P[A^c|M]$, $P[A|M^c]$, $P[A^c|M^c]$.

2.21. In Problem 2.20 assume that $P[M] = 10^{-3}$. Compute

$$P[M|A], P[M|A^c], P[M^c|A], P[M^c|A^c]. \text{ Repeat for } P[M] = 10^{-6}.$$

Note: By assigning drastically different numbers to $P[M]$, this problem attempts to illustrate the difficulty of using probability in some types of problems. Because a UFO appearance is so rare (except in Roswell, New Mexico), it may be considered a *one-time event* for which accurate knowledge of the prior probability $P[M]$ is near impossible. Thus, in the surprise attack by the Japanese on Pearl Harbor in 1941, while the radar clearly indicated a massive cloud of incoming objects, the signals were ignored by the commanding officer (CO). Possibly the CO assumed that the prior probability of an attack was so small that a radar failure was more likely.

2.22. Research Problem: Receiver Operating Characteristics.

In Problem 2.20, $P[A|M^c]$ is known as α, the *probability of a false alarm*, while in Problem 2.21, $P[M|A]$ is known as β, the *probability of a correct detection*. Clearly $\alpha = \alpha(x_A)$, $\beta = \beta(x_A)$. Write a MATLAB program to plot β *versus* α for a fixed value of r. Choose $r = 0, 1, 2, 3$. The curves so obtained are known among radar people as the *receiver-operating characteristic* (ROC) for various values of r.

2.23. A sophisticated house security system uses an infrared beam to complete a circuit. If the circuit is broken, say by a robber crossing the beam, a bell goes off. The way the system works is as follows: The photodiode generates a beam of infrared photons at a Poisson rate of 9×10^6 photons per second. Every microsecond a counter counts the total number of photons collected at the detector. If the count drops below 2 photons in the counting interval (10^{-6} sec.), it is assumed that the circuit is broken and the bell rings. Assuming the Poisson PMF, compute the probability of a false alarm during a one-second interval.

2.24. A traffic light can be in one of three states: *green* (G), *red* (R), and *yellow* (Y). The light changes in a random fashion (e.g., the light at the corner of Hoosick and Doremus in Nirvana, New York). At any one time the light can be in only one state. The experiment consists of observing the state of the light.

    (a) Give the sample description space of this experiment and list five events.
    (b) Let a random variable $X(\cdot)$ be defined as follows: $X(\text{G}) = -1$; $X(\text{R}) = 0$; $X(\text{Y}) = \pi$. Assume that $P[\text{G}] = P[\text{Y}] = 0.5 \times P[\text{R}]$. Plot the PDF of X. What is $P[X \leq 3]$?

2.25. A *token-based, multi-user* communication system works as follows: say that nine user-stations are connected to a ring and an electronic signal, called a *token*, is passed around the ring in, say, a clockwise direction. The token stops at each station and allows the user (if there is one) up to five minutes of signaling a message. The token waits for a maximum of one minute at each station for a user to initiate a message. If no user appears at the station at the end of the minute, the token is passed on to the next station. The five-minute window *includes* the waiting time of the token at the station. Thus a user who begins signaling at the end of the token waiting period has only four minutes of signaling left.

(a) Assume that you are a user at a station. What are the minimum and maximum waiting times you might experience? The token is assumed to travel instantaneously from station to station.

(b) Let the probability that a station is occupied be p. If a station is occupied, the "occupation time" is a random variable that is uniformly distributed in $(0,5)$ minutes. Using MATLAB, write a program that simulates the waiting time at your station. Assume that the token has just left your station. Pick various values of p.

2.26. Let X and Y be jointly Gaussian r.v.'s with pdf

$$f_{XY}(x,y) = \frac{1}{2\pi} \exp\left[-\frac{1}{2}(x^2 + y^2)\right].$$

What is the smallest value of c such that $P[X^2 + Y^2 \le c^2] \ge 0.95$. Hint: Use polar coordinates.

2.27. A laser used to scan the bar code on supermarket items is assumed to have a constant conditional failure rate μ. What minimum value of μ will ensure that the probability of a first breakdown in 100 hours of operation is less than 0.05?

2.28. Compute the pdf of the failure time X if the conditional failure rate $\alpha(t)$ is as shown in Figure P2.28.

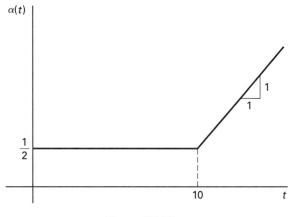

Figure P2.28

2.29. Use the basic properties of the joint PDF $F_{XY}(x,y)$ to show

(a) $f_{XY}(x,y)dx\,dy = P[x < X \le x + dx, y < Y \le y + dy]$;

(b) $\displaystyle\int_{-\infty}^{\infty}\int_{-\infty}^{\infty} f_{XY}(x,y)dx\,dy = 1$; and

(c) $f_{XY}(x,y) \ge 0$.

2.30. Show that Equation 2.6-46 factors as $f_X(x)f_Y(y)$ when $\rho = 0$. What are $f_X(x)$ and $f_Y(y)$? For $\sigma = 1$ and $\rho = 0$, what is $P[-\frac{1}{2} < X \le \frac{1}{2}, -\frac{1}{2} < Y \le \frac{1}{2}]$?

REFERENCES

2-1. M. Loeve, *Probability Theory*. New York: Van Nostrand, Reinhold, 1962.

2-2. W. Feller, *An Introduction to Probability Theory and Its Applications*, 2 vols. New York: John Wiley, 1950, 1966.

2-3. W. F. Davenport, *Probability and Random Processes: An Introduction for Applied Scientists and Engineers*. New York: McGraw-Hill, 1970.

2-4. B. Saleh, *Photoelectron Statistics*. New York: Springer-Verlag, 1978.

2-5. M. Born and E. Wolf, *Principles of Optics*. New York: Pergamon Press, 1965.

2-6. M. Abramowitz and I. A. Stegun, *Handbook of Mathematical Functions*. New York: Dover, 1965.

ADDITIONAL READING

Cooper, G. R. and C. D. McGillem, *Probabilistic Methods of Signal and System Analysis*. 3rd edition. New York: Holt, Rinehart and Winston, 1999.

Garcia, L.-G., *Probability and Random Processes for Electrical Engineering*, 2nd edition. Reading, MA: Addison-Wesley, 1994.

Helstrom, C. W., *Probability and Stochastic Processes for Engineers*, 2nd edition. New York: Macmillan, 1991.

Papoulis, A., *Probability, Random Variables, and Stochastic Processes*, 3rd edition. New York: McGraw-Hill, 1991.

Pebbles, P. Z. Jr., *Probability, Random Variables, and Random Signal Principles*, 4th edition. New York: McGraw-Hill, 2001.

Scheaffer, R. L., *Introduction to Probability and Its Applications*. Belmont, CA: Duxbury 1990.

Yates, R. D. and D. J. Goodman, *Probability and Stochastic Processes*. New York: Wiley, 1999.

Ziemer, R. E., *Elements of Engineering Probability & Statistics*. Upper Saddle River, NJ: Prentice Hall, 1997.

3

Functions of Random Variables

3.1 INTRODUCTION

A classic problem in engineering is the following: We are given the input to a system and we must calculate the output. If the input to a system is random, the output will generally be random as well. To put this somewhat more formally, if the input at some instant t or point x is a random variable, the output at some corresponding instant t' or point x' will be a random variable. Now the question arises, what if we know the PDF, PMF, or pdf of the input r.v. can we compute these functions for the output r.v.? In many cases we can, while in other cases the computation is too difficult and we settle for descriptors of the output r.v. which contains less information than the PDF. Such descriptors are called *averages* or *expectations* and are discussed in Chapter 4. In general for systems with memory, that is, systems in which the output at a particular instant of time depends on past values of the input (possibly an infinite number of such past values), it is much more difficult (if not impossible) to calculate the PDF of the output. To a large degree, the whole second half of this book (Chapters 6–9) is devoted to a study of such situations. In this chapter, we study much simpler situations. We illustrate with some examples.

Example 3.1-1
As is well known from electric-circuit theory, the current I flowing through a resistor R (Figure 3.1-1) dissipates an amount of power W given by

$$W \stackrel{\Delta}{=} W(I) = I^2 R. \tag{3.1-1}$$

116

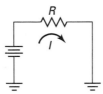

Figure 3.1-1   Ohmic power dissipation in a resistor.

Equation 3.1-1 is an explicit rule that generates for every value of I a number $W(I)$. This rule or correspondence is called a *function* and is denoted by $W(\cdot)$ or merely W or sometimes even $W(I)$—although the latter notation confuses the difference between the rule and the actual number. Clearly, if I were a random variable, the rule $W = I^2 R$ generates a new random variable W whose probability distribution function might be quite different from that of I.[†] Indeed, this alludes to the heart of the problem: Given a rule $g(\cdot)$, and a random variable X with pdf $f_X(x)$, what is the pdf $f_Y(y)$ of the random variable $Y = g(X)$?

The computation of $f_Y(y)$, $F_Y(y)$ or the PMF of Y, that is, $P_Y(y_i)$, can be very simple or quite complex. We illustrate such a computation with a second example, one that comes from communication theory.

Example 3.1-2
A two-level waveform is made analog because of the effect of additive Gaussian noise (Figure 3.1-2). A decoder samples the analog waveform $x(t)$ at t_0 and decodes according to the following rule:

Input to Decoder x	Output of Decoder y
If $x(t_0)$:	Then y is assigned:
$\geq \frac{1}{2}$	1
$< \frac{1}{2}$	0

What is the PMF or pdf of Y?

Solution   Clearly with Y (an r.v.) denoting the output of the decoder, we can write the following events:

$$\{Y = 0\} = \{X < 0.5\} \tag{3.1-2a}$$

$$\{Y = 1\} = \{X \geq 0.5\} \tag{3.1-2b}$$

[†]This is assuming that the composite function $I^2(\zeta)R$ satisfies the required properties of a r.v. (see Section 2.2)

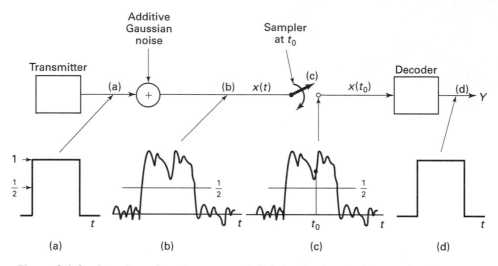

Figure 3.1-2   Decoding of a noise-corrupted digital pulse by sampling and hard clipping.

where $X \triangleq x(t_0)$. Hence if we assume $X : N(1, 1)$, we obtain the following:

$$P[Y = 0] = P[X < 0.5] = \frac{1}{\sqrt{2\pi}} \int_{-\infty}^{0.5} e^{-\frac{1}{2}(x-1)^2} dx$$

$$= 0.31. \tag{3.1-3}$$

In arriving at Equation 3.1-3 from the previous line, we use the normalization procedure explained in Section 2.4 and the fact that for $X : N(0, 1)$ and any $x < 0$ $F_X(x) = \frac{1}{2} - \text{erf}(|x|)$. The area under the normal curve $N(1, 1)$ associated with $P[Y = 0]$ is shown in Figure 3.1-3.

In a similar fashion we compute $P[Y = 1] = 0.69$. Hence the PMF of Y is

$$P_Y(0) = 0.31$$

$$P_Y(1) = 0.69 \tag{3.1-4}$$

$$P_Y(y) = 0, \quad \text{for} \quad y \neq 0, 1.$$

Using delta functions, we can write for the pdf of Y:

$$f_Y(y) = 0.31\,\delta(y) + 0.69\,\delta(y - 1). \tag{3.1-5}$$

Unfortunately not all function-of-a-random-variable (FRV) problems are this easy to evaluate. To gain a deeper insight into the FRV problem we take a closer look at the underlying concept of FRV. The gain in insight will be useful for us in a practical sense when we discuss random sequences and processes beginning in Chapter 6.

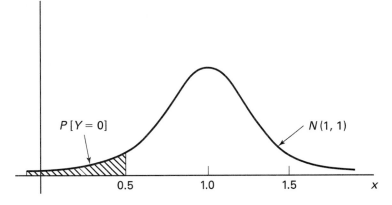

Figure 3.1-3   The area associated with $P[Y = 0]$ in Example 3.1-2.

Functions of a Random Variable (Several Views)

There are several different but essentially equivalent views of a FRV. We will now present two of them. The differences between them are mainly ones of *emphasis*.

Assume as always an underlying probability space $(\Omega, \mathscr{F}, P)$ and a random variable X defined on it. Recall that X is a rule that assigns to every $\zeta \in \Omega$ a number $X(\zeta)$. X transforms the σ-field of events $\mathscr{F}$ into the Borel σ-field $\mathscr{B}$ of sets of numbers on the real line. If R_X denotes the subset of the real line actually reached by X as ζ roams over Ω, then we can regard X as an ordinary function with domain Ω and range R_X. Now, additionally, consider a measurable real function $g(x)$ of the real variable x.

First view ($Y:\Omega \to R_Y$).   For every $\zeta \in \Omega$, we generate a number $g[X(\zeta)] \triangleq Y(\zeta)$. The rule Y, which generates the numbers $\{Y(\zeta)\}$ for random outcomes $\{\zeta \in \Omega\}$, is a r.v. with domain Ω and range $R_Y \subset R$. Finally for every Borel set of real numbers B_Y, the set $\{\zeta : Y(\zeta) \in B_Y\}$ is an event. In particular the event $\{\zeta : Y(\zeta) \leq y\}$ is equal to the event $\{\zeta : g[X(\zeta)] \leq y\}$.

In this view, the stress is on Y as a mapping from Ω to R_Y. The intermediate role of X is suppressed.

Second view (input/output systems view).   For every value of $X(\zeta)$ in the range R_X we generate a new number $Y = g(X)$ whose range is R_Y. The rule Y whose domain is R_X and range is R_Y is a function of the random variable X. In this view the stress is on viewing Y as a mapping from one set of real numbers to another. A model for this view is to regard X as the input to a system with transmittance $g(\cdot)$.[†] For such a system, an input x gets transformed to an output $y = g(x)$ and an input function X gets transformed to an output function $Y = g(X)$. (See Figure 3.1-4.)

[†]g can be any *measurable* function, that is, if R_Y is the range of Y, then the inverse image of every subset in R_Y generated by countable unions and intersections of sets of the form $\{Y \leq y\}$ is an event.

Figure 3.1-4   Input/output view of a function of a random variable.

The input–output viewpoint is the one we stress, partly because it is particularly useful in dealing with random processes where the input consists of waveforms or sequence of random variables. The central problem in computations involving FRV's is: Given $g(x)$ and $F_X(x)$, find the point set C_y such that the following events are equal:

$$\{\zeta : Y(\zeta) \le y\} = \{\zeta : g[X(\zeta)] \le y\}$$
$$= \{\zeta : X(\zeta) \in C_y\}. \tag{3.1-6a}$$

In general we shall submerge the role of ζ and write $\{Y \le y\} = \{X \in C_y\}$, and so forth. For C_y so determined it follows that

$$P[Y \le y] = P[X \in C_y] \tag{3.1-6b}$$

since the underlying event is the same. If C_y is empty, then the probability of $\{Y \le y\}$ is zero.

In dealing with the input-output model it is generally convenient to omit any references to an abstract underlying experiment $\mathcal{H}$ and deal, instead, directly with the r.v.'s X and Y. In this approach the underlying experiments are the observations on X, events are Borel subsets of the real line R, and the set function $P[\cdot]$ is replaced by the distribution function $F_X(\cdot)$. Then Y is a mapping (an r.v.) whose domain is the range, R_X, of X and whose range is a subset, R_Y, or R. The functional properties of X are ignored in favor of viewing X as a mechanism that gives rise to numerically valued random phenomena. In this view the domain of X is irrelevant.

For additional discussion on the various views of a function of a random variable, see Wilbur Davenport [3-1, p. 174].

3.2 SOLVING PROBLEMS OF THE TYPE $Y = g(X)$

We shall now demonstrate how to solve problems of the type $Y = g(X)$. Eventually we shall develop a formula that will enable us to solve problems of this type very rapidly. However, use of the formula at too early a stage of the development will tend to mask the underlying principles that we shall need later to deal with more difficult problems.

Example 3.2-1 _____

Let X be a uniform r.v. in $(0,1)$ and let $Y = 2X + 3$. Then we need to find the point set C_y in Equation 3.1-6b to compute $F_Y(y)$. Clearly

$$\{Y \le y\} = \{2X + 3 \le y\} = \left\{X \le \tfrac{1}{2}(y - 3)\right\}.$$

Hence C_y is the interval $(-\infty, \frac{1}{2}(y - 3))$ and

$$F_Y(y) = F_X\left(\frac{y - 3}{2}\right).$$

The pdf of Y is

$$f_Y(y) = \frac{dF_Y(y)}{dy} = \frac{d}{dy}\left[F_X\left(\frac{y - 3}{2}\right)\right]$$

$$= \frac{1}{2}f_X\left(\frac{y - 3}{2}\right).$$

The solution is shown in Figure 3.2-1.

Generalization.   Let $Y = aX + b$ with X a continuous r.v. with pdf $f_X(x)$. Then for $a > 0$ the outcomes $\{\zeta\} \subset \Omega$ that produce the event $\{aX + b \le y\}$ are identical with the outcomes $\{\zeta\} \subset \Omega$ that produce the event $\{X \le \frac{y-b}{a}\}$. Thus

$$\{Y \le y\} = \{aX + b \le y\} = \left\{X \le \frac{y - b}{a}\right\}.$$

From the definition of the PDF:

$$F_Y(y) = F_X\left(\frac{y - b}{a}\right), \tag{3.2-1}$$

and

$$f_Y(y) = \frac{1}{a}f_X\left(\frac{y - b}{a}\right). \tag{3.2-2}$$

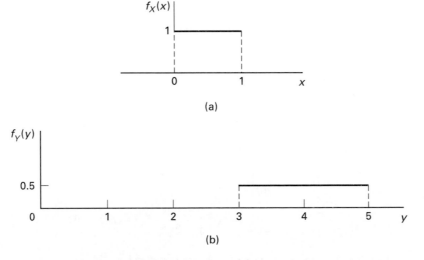

(a)

(b)

Figure 3.2-1   (a) Original pdf of X; (b) the pdf of $Y = 2X + 3$.

For $a < 0$, the following events are equal[†]

$$\{Y \le y\} = \{aX + b \le y\} = \left\{ X \ge \frac{y - b}{a} \right\}.$$

Since the events $\left\{ X < \frac{y-b}{a} \right\}$ and $\left\{ X \ge \frac{y-b}{a} \right\}$ are disjoint and their union is the certain event, we obtain from Axiom 3

$$P\left[X < \frac{y - b}{a} \right] + P\left[X \ge \frac{y - b}{a} \right] = 1.$$

Finally for a continuous r.v.

$$P\left[X < \frac{y - b}{a} \right] = P\left[X \le \frac{y - b}{a} \right]$$

and

$$P\left[X \ge \frac{y - b}{a} \right] = P\left[X > \frac{y - b}{a} \right].$$

Thus for $a < 0$

$$F_Y(y) = 1 - F_X\left(\frac{y - b}{a} \right), \tag{3.2-3}$$

and

$$f_Y(y) = \frac{1}{|a|} f_X\left(\frac{y - b}{a} \right), \qquad a \ne 0. \tag{3.2-4}$$

We leave the computation of $F_Y(y)$ when X is not necessarily continuous as an exercise for the reader.

Example 3.2-2

(Square-law detector.) Let X be an r.v. with continuous PDF $F_X(x)$ and let $Y = X^2$. Then

$$\{Y \le y\} = \{X^2 \le y\} = \{-\sqrt{y} \le X \le \sqrt{y}\}$$
$$= \{-\sqrt{y} < X \le \sqrt{y}\} \cup \{X = -\sqrt{y}\}. \tag{3.2-5}$$

The probability of the union of disjoint events is the sum of their probabilities. Using the result of Equation 2.4-4 we obtain

$$F_Y(y) = F_X(\sqrt{y}) - F_X(-\sqrt{y}) + P[X = -\sqrt{y}]. \tag{3.2-6}$$

If X is continuous, $P[X = -\sqrt{y}] = 0$. Then for $y > 0$,

$$f_Y(y) = \frac{d}{dy}[F_Y(y)]$$

$$= \frac{1}{2\sqrt{y}} f_X(\sqrt{y}) + \frac{1}{2\sqrt{y}} f_X(-\sqrt{y}). \tag{3.2-7}$$

[†]Recall that we mean the event $\{\zeta : Y(\zeta) \le y\} = \left\{ \zeta : X(\zeta) \ge \frac{y-b}{a} \right\}$.

For $y < 0$, $f_Y(y) = 0$. How do we know this? Recall from Equation 3.1-6a that if C_y is empty, then $P[Y \le y] = 0$ and hence $f_Y(y) = 0$. For $y < 0$, there are no values of the r.v. X on the real line that satisfy

$$\{-\sqrt{y} \le X \le \sqrt{y}\}.$$

Hence $f_Y(y) = 0$ for $y < 0$. If $X: N(0,1)$, then from Equation 3.2-7,

$$f_Y(y) = \frac{1}{\sqrt{2\pi y}} e^{-\frac{1}{2}y} u(y) \tag{3.2-8}$$

where $u(y)$ is the standard unit step, that is, $u(y) = 1, y \ge 0$, and zero elsewhere.

Example 3.2-3 ~~Problem #1~~ Find the PDF of $Y = g(X)$ if $X: N(0,1)$
(Half-wave rectifier.) A half-wave rectifier has the transfer characteristic $g(x) = xu(x)$ and (Figure 3.2-2). $g(\cdot)$ is g
   Thus

$$F_Y(y) = P[Xu(X) \le y] = \int_{\{x: xu(x) \le y\}} f_X(x)\,dx \tag{3.2-9}$$

(i) Let $y > 0$; then $\{x: xu(x) \le y\} = \{x: x > 0; x \le y\} \cup \{x: x < 0; x \le y\} = \{x: x \le y\}$. Thus $F_Y(y) = \int_{-\infty}^{y} f_X(x)\,dx = F_X(y)$.
(ii) Next let $y = 0$. Then $P[Y = 0] = P[X \le 0] = F_X(0)$.
(iii) Finally let $y < 0$. Then $\{x: xu(x) \le y\} = \phi$ (the empty set).

Thus

$$F_Y(y) = \int_{\phi} f_X(x)\,dx = 0.$$

If $X: N(0,1)$, then $F_Y(y)$ has the form in Figure 3.2-3.

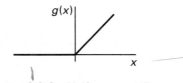

Figure 3.2-2   Half-wave rectifier.

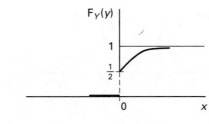

Figure 3.2-3   The PDF of Y when $X: N(0,1)$ for the half-wave rectifier.

The pdf is obtained by differentiation. Because of the discontinuity of $y = 0$, we obtain an impulse in the pdf at $y = 0$, that is,

$$f_Y(y) = \begin{cases} 0, & y < 0 \\ F_X(0)\delta(y), & y = 0 \\ F_X(y), & y > 0. \end{cases} \tag{3.2-10}$$

Example 3.2-4

Let X be a Bernoulli r.v. with $P[X = 0] = p$ and $P[X = 1] = q$. Then

$$f_X(x) = p\delta(x) + q\delta(x - 1)$$
$$F_X(x) = pu(x) + qu(x - 1),$$

where $u(x)$ is the unit step.

Let $Y \triangleq X - 1$. Then (Figure 3.2-4)

$$\begin{aligned} F_Y(y) &= P[X - 1 \le y] \\ &= P[X \le y + 1] \\ &= F_X(y + 1) \\ &= pu(y + 1) + qu(y). \end{aligned}$$

The pdf is

$$f_Y(y) = \frac{d}{dy}[F_Y(y)] = p\delta(y + 1) + q\delta(y). \tag{3.2-11}$$

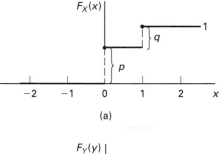

(a)

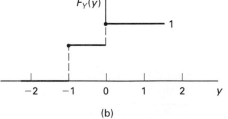

(b)

Figure 3.2-4   (a) PDF of X; (b) PDF of $Y = X - 1$.

Example 3.2-5

(Transformation of PDF's.) Let X have a continuous PDF $F_X(x)$, which is a strict monotone increasing function† of x. Let Y be an r.v. formed from X by the transformation

$$Y = F_X(X). \tag{3.2-12}$$

To compute $F_Y(y)$, we proceed as usual:

$$\{Y \le y\} = \{F_X(X) \le y\}$$
$$= \{X \le F_X^{-1}(y)\}.$$

Hence

$$F_Y(y) = P[F_X(X) \le y]$$
$$= P[X \le F_X^{-1}(y)]$$
$$= \int_{\{x: F_X(x) \le y\}} f_X(x)dx.$$

1. Let $y < 0$. Then since $0 \le F_X(x) \le 1$ for all $x \in [-\infty, \infty]$, the set $\{x: F_X(x) \le y\} = \phi$ and $F_Y(y) = 0$.
2. Let $y > 1$. Then $\{x: F_X(x) \le y\} = [-\infty, \infty]$ and $F_Y(y) = 1$.
3. Let $0 \le y \le 1$. Then $\{x: F_X(x) \le y\} = \{x: x \le F_X^{-1}(y)\}$

and

$$F_Y(y) = \int_{-\infty}^{F_X^{-1}(y)} f_X(x)dx = F_X(F_X^{-1}(y)) = y.$$

Hence

$$F_Y(y) = \begin{cases} 0, & y < 0 \\ y, & 0 \le y \le 1 \\ 1, & y > 1. \end{cases} \tag{3.2-13}$$

Equation 3.2-13 says that whatever probability law X obeys, $Y \triangleq F_X(X)$ will be a *uniform* r.v. Conversely, given a uniform r.v. Y, the transformation $X \triangleq F_X^{-1}(Y)$ will generate a r.v. with PDF $F_X(x)$ (Figure 3.2-5). This technique is sometimes used in *simulation* to generate r.v.'s with specified distributions from a uniform r.v.

Example 3.2-6

(Quantizing.) In analog-to-digital conversion, an analog waveform is sampled, quantized, and coded (Figure 3.2-6).

A quantizer is a function that assigns to each sample x_i, $i = 1, 2, \ldots$ a value from a set $Q \triangleq \{y_{-N}, \ldots, y_0, \ldots, y_N\}$ of $2N + 1$ predetermined values [3-2]. Thus an uncountably infinite set of values (the analog input x_i) is reduced to a finite set (the digital output y_i).

---

†In other words $x_2 > x_1$ implies $F_X(x_2) > F_X(x_1)$.

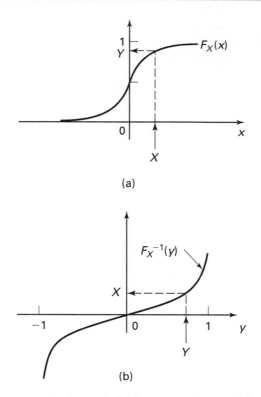

(a)

(b)

Figure 3.2-5   Generating an r.v. with PDF $F_X(x)$ from a uniform r.v. (a) Creating a uniform r.v. Y from an r.v. X with PDF $F_X(x)$; (b) creating an r.v. with PDF $F_X(x)$ from a uniform r.v. Y.

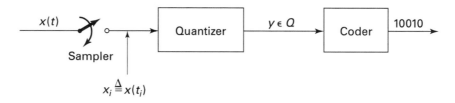

Figure 3.2-6   An analog-to-digital converter.

Note that a practical quantizer is also a limiter, that is, for x greater than some x_0 or less than some x_1, $y = y_N$ or y_{-N}, respectively.

A common quantizer is the uniform quantizer, which is a staircase function of uniform step size a, that is,

$$g(x) = ia \qquad (i-1)a < x \le ia, \qquad i \text{ an integer.} \tag{3.2-14}$$

Thus the quantizer assigns to each x the closest value of ia above continuous sample value and is shown by the staircase function in Figure 3.2-7.

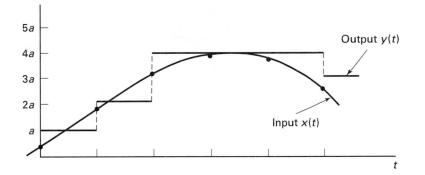

Figure 3.2-7   Quantizer output (staircase function) versus input (continuous line).

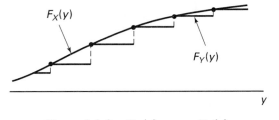

Figure 3.2-8 $F_X(y)$ versus $F_Y(y)$.

If X is an r.v. denoting the sampled value of the input and Y denotes the quantizer output, then with $a = 1$:

$$P[Y = i] = P[i - 1 < X \le i]$$
$$= F_X(i) - F_X(i - 1).$$

Also

$$F_Y(y) = \sum_i P[Y = i]u(y - i)$$

$$= \sum_i [F_X(i) - F_X(i - 1)]u(y - i). \tag{3.2-15}$$

If limiting is ignored, then when $y = n$ (an integer), $F_Y(n) = F_X(n)$, otherwise $F_Y(y) < F_X(y)$ (Figure 3.2-8).

Example 3.2-7 _____

(Sine wave.) A classic problem is to determine the pdf of $Y = \sin X$ where X is a uniform r.v. with pdf

$$f_X(x) = \begin{cases} \dfrac{1}{2\pi}, & -\pi < x \le \pi \\ 0, & \text{otherwise.} \end{cases} \tag{3.2-16}$$

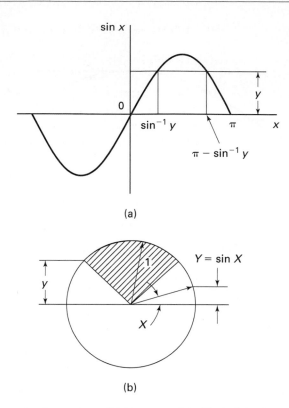

(a)

(b)

Figure 3.2-9 (a) The roots of $y = \sin x$; (b) the event $\{\sin X \le y\}$ is viewed as a unit vector that can rotate to any angle in the clear zone only.

From Figure 3.2-9 we see that for $0 \le y \le 1$, the event $\{Y \le y\}$ satisfies

$$\{Y \le y\} = \{\sin X \le y\}$$
$$= \{\pi - \sin^{-1} y < X \le \pi\} \cup \{-\pi < X \le \sin^{-1} y\}.$$

Since the two events on the last line are disjoint, we obtain

$$F_Y(y) = F_X(\pi) - F_X(\pi - \sin^{-1} y) + F_X(\sin^{-1} y) - F_X(-\pi). \qquad (3.2\text{-}17)$$

Hence

$$f_Y(y) = \frac{dF_Y(y)}{dy} = f_X(\pi - \sin^{-1} y)\frac{1}{\sqrt{1 - y^2}}$$

$$+ f_X(\sin^{-1} y)\frac{1}{\sqrt{1 - y^2}} \qquad (3.2\text{-}18)$$

$$= \frac{1}{\pi}\frac{1}{\sqrt{1 - y^2}} \qquad 0 \le y < 1. \qquad (3.2\text{-}19)$$

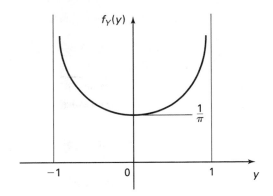

Figure 3.2-10   The probability density function of $Y = \sin X$.

If this calculation is repeated for $-1 < y \leq 0$ and for $|y| > 1$, we obtain the complete solution (Figure 3.2-10):

$$f_Y(y) = \begin{cases} \dfrac{1}{\pi} \dfrac{1}{\sqrt{1 - y^2}}, & |y| < 1, \\ 0, & \text{otherwise.} \end{cases} \qquad (3.2\text{-}20)$$

The details are left as an exercise.

We shall now derive a formula that will enable us, by "turning the crank," to solve many problems of the type $Y = g(X)$. For some problems, however, the preceding indirect approach may be easier. For lack of a better term we shall call this new approach the *direct* method.

General Formula of Determining the pdf of $Y = g(X)$

We are given the continuous r.v. X with pdf $f_X(x)$ and the differentiable function $g(x)$ of the real variable x. What is the pdf of $Y \triangleq g(X)$?

Solution   The event $\{y < Y \leq y + dy\}$ can be written as a union of disjoint elementary events $\{E_i\}$ in the Borel field generated under X. If the equation $y = g(x)$ has n real roots[†] $x_1, \ldots, x_n$, then the disjoint events have the form $E_i = \{x_i - |dx_i| < X < x_i\}$ if $g'(x_i)$ is negative or $E_i = \{x_i < X < x_i + |dx_i|\}$ if $g'(x_i)$ is positive.[‡] (See Figure 3.2-11.) In either case, it follows from the definition of the pdf that $P[E_i] = f_X(x_i)|dx_i|$. Hence

$$P[y < Y \leq y + dy] = f_Y(y)|dy|$$

$$= \sum_{i=1}^{n} f_X(x_i)|dx_i| \qquad (3.2\text{-}21)$$

[†]By roots we mean the set of points x_i such that $y - g(x_i) = 0 \; i = 1, \ldots, n$.
[‡]The prime indicates derivatives with respect to x.

or, equivalently, if we divide through by $|dy|$

$$f_Y(y) = \sum_{i=1}^{n} f_X(x_i) \left| \frac{dx_i}{dy} \right| = \sum_{i=1}^{n} f_X(x_i) \left| \frac{dy}{dx_i} \right|^{-1}.$$

At the roots of $y = g(x)$, $dy/dx_i = g'(x_i)$, and we obtain the important formula

$$f_Y(y) = \sum_{i=1}^{n} f_X(x_i)/|g'(x_i)| \qquad x_i = x_i(y), \qquad g'(x_i) \neq 0. \qquad (3.2\text{-}22)$$

Equation 3.2-22 is a fundamental equation that is very useful in solving problems where the transmittance $g(x)$ has several roots. If, for a given y, the equation $y - g(x) = 0$ has no real roots, then $f_Y(y) = 0.$[†] Figure 3.2-11 illustrates the case when $n = 2$.

Example 3.2-8

To illustrate the use of Equation 3.2-22 we solve Example 3.2-7 by the formula. Thus we seek the pdf of $Y = \sin X$ with $f_X(x) = 1/2\pi$ for $-\pi < x \leq \pi$. The function $g(x)$ is $g(x) = \sin x$. The roots of $y - \sin x = 0$ for $y > 0$ are $x_1 = \sin^{-1} y$, $x_2 = \pi - \sin^{-1} y$. Also

$$\frac{dg}{dx} = \cos x$$

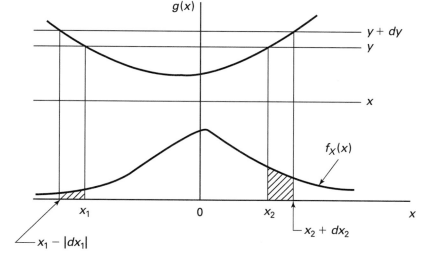

Figure 3.2-11   The event $\{y < Y \leq y + dy\}$ is the union of two disjoint events on the probability space of X.

[†] The r.v. X, being real, cannot take on values that are imaginary with nonzero probability.

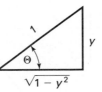

Figure 3.2-12 Evaluating $\cos(\sin^{-1} y)$.

which must be evaluated at the two roots x_1 and x_2. At $x_1 = \sin^{-1} y$ we get $dg/dx|_{x=x_1} = \cos(\sin^{-1} y)$. Likewise when $x_2 = \pi - \sin^{-1} y$, we get

$$\frac{dg}{dx}\bigg|_{x=x_2} = \cos(\pi - \sin^{-1} y) = \cos \pi \cos(\sin^{-1}) + \sin \pi \sin(\sin^{-1} y)$$

$$= -\cos(\sin^{-1} y).$$

The quantity $\cos(\sin^{-1} y)$ can be further evaluated with the help of Figure 3.2-12. There we see that $\theta = \sin^{-1} y$ and $\cos \theta = \sqrt{1 - y^2} = \cos(\sin^{-1} y)$. Hence

$$\left|\frac{dg}{dx}\right|_{x_1} = \left|\frac{dg}{dx}\right|_{x_2} = \sqrt{1 - y^2}.$$

Finally, $f_X(\sin^{-1} y) = f_X(\pi - \sin^{-1} y) = 1/2\pi$. Using these results in Equation 3.2-22 enables us to write

$$f_Y(y) = \frac{1}{\pi} \frac{1}{\sqrt{1 - y^2}} \qquad 0 \le y < 1,$$

which is the same result as in Equation 3.2-19. The solution for all y is as given in Equation 3.2-20.

Example 3.2-9

(Nonlinear devices.) A number of nonlinear zero-memory devices can be modeled by a transmittance function $g(x) = x^n$. Let $Y = X^n$. The pdf of Y depends on whether n is even or odd. We solve the case of n odd, leaving n even as an exercise. For n odd and $y > 0$, the only real root to $y = x^n = 0$ is $x_r = y^{1/n}$. Also

$$\frac{dg}{dx} = nx^{n-1} = ny^{(n-1)/n}.$$

For $y < 0$, the only real root is $x_r = -|y|^{1/n}$. See Figure 3.2-13. Also

$$\frac{dg}{dx} = n|y|^{(n-1)/n}.$$

Hence

$$f_Y(y) = \frac{1}{n} = y^{(1-n)/n} \cdot f_X(y^{1/n}), \qquad y \ge 0$$

$$= \frac{1}{n}|y|^{(1-n)/n} \cdot f_X(-|y|^{1/n}), \qquad y < 0.$$

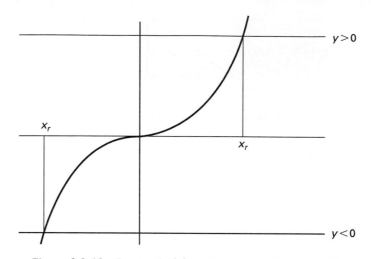

Figure 3.2-13   Roots of $g(x) = x^n - y = 0$ when n is odd.

In problems in which $g(x)$ assumes a constant value, say $g(x) = c$, over some interval Equation 3.2-22 cannot be used to compute $f_Y(y)$ because $g'(x) = 0$ over the interval.

Example 3.2-10

(Linear amplifier with cutoff.) Consider a nonlinear device with transmittance as shown in Figure 3.2-14.

The transmittance $g(x)$ is given by

$$g(x) = 0, \quad |x| \geq 1$$

$$g(x) = x, \quad -1 < x < 1.$$

Thus $g'(x) = 0$ for $|x| \geq 1$, and $g'(x) = 1$ for $-1 < x < 1$. For $y \geq 1$ and $y \leq -1$ there are no real roots to $y - g(x) = 0$. Hence $f_Y(y) = 0$ in this range. For $-1 < y < 1$, the

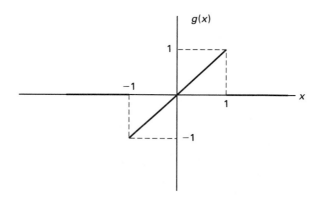

Figure 3.2-14   A linear amplifier with shutoff.

only root to $y - g(x) = y - x = 0$ is $x = y$. Hence in this range Equation 3.2-22 applies with $|g'(x)| = 1$ and $f_Y(y) = f_X(y)$. We note that $P[Y = 0] = P[X \geq 1] + P[X \leq -1]$. If $X: N(0, 1)$, $P[X \geq 1] = 1/2 - \mathrm{erf}(1) = P[X \leq -1]$ and $P[Y = 0] = 1 - 2\mathrm{erf}(1) = 0.317$. Thus the complete solution is

$$f_Y(y) = 0, \quad |y| \geq 1$$

$$f_Y(y) = \frac{1}{\sqrt{2\pi}} \exp\left[-\frac{1}{2}y^2\right], \quad -1 < y < 1,$$

$$P[Y = 0] = 0.317.$$

We might want to relate $P[Y = 0]$ to a term in the pdf. Note that if we write

$$P[Y = 0] = \lim_{\varepsilon \to 0} \int_{0-\varepsilon}^{0+\varepsilon} f_Y(y)dy$$

and use the existing $f_Y(y)$ valid in this range, that is, $(2\pi)^{-1/2} \exp\{-\frac{1}{2}y^2\}\exp\{-\frac{1}{2}y^2\}$, we get $P[y = 0] = 0$, because the integral of a continuous function over an infinitesimal interval yields zero. Hence the pdf must be modified to give the correct answer at $y = 0$ If, however, we write

$$f_Y(y) = 2(\pi)^{-1/2} \exp\left\{-\frac{1}{2}y^2\right\} + 0.317\delta(y),$$

then we get the correct answer, that is,

$$P[Y = 0] = \lim_{\varepsilon \to 0} \int_{0-\varepsilon}^{0+\varepsilon} \{(2\pi)^{-1/2} \exp\left\{-\frac{1}{2}y^2\right\} + 0.317\delta(y)\}dy = 0.317.$$

Thus a convenient way of describing the pdf of Y is by writing

$$f_Y(y) = \begin{cases} 0, & |y| \geq 1 \\ (2\pi)^{-1/2} \exp\left\{-\frac{1}{2}y^2\right\} + 0.317\delta(y), & -1 < y < 1. \end{cases}$$

Example 3.2-11 ————————————————————————
(Infinite roots.) Here we consider the periodic extension of the transmittance shown in Example 3.2-10. The extended $g(x)$ is shown in Figure 3.2-15.

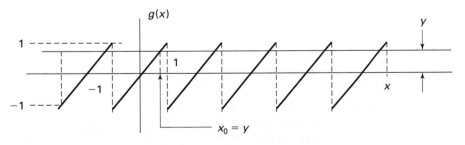

Figure 3.2-15   Periodic transmittance.

The transmittance in this case is described by

$$g(x) = \sum_{n=-\infty}^{\infty} (x - 2n) \text{ rect}\left[\frac{x-2n}{2}\right].$$

As in the previous example $f_Y(y) = 0$ for $|y| \geq 1$ because there are no real roots to the equation $y - g(x) = 0$ in this range. On the other hand, when $-1 < y < 1$, there are an infinite number of roots to $y - g(x) = 0$ and these are given by $x_n = y + 2n$ for $n = \ldots, -2, -1, 0, 1, 2, \ldots$. At each root $|g'(x_n)| = 1$. Hence, from Equation 3.2-22 we obtain $f_Y(y) = \sum_{n=-\infty}^{\infty} f_X(y + 2n) \text{ rect}\left[\frac{y}{2}\right]$. In the case that $X: N(0,1)$ this specializes to

$$f_Y(y) = (2\pi)^{-1/2} \sum_{n=-\infty}^{\infty} \exp\left\{-\frac{1}{2}(y + 2n)^2\right\} \times \text{rect}\left[\frac{y}{2}\right].$$

While this result is correct, it seems hard to believe that the sum of infinite positive terms yields a function whose area is restricted to one. To show that $f_Y(y)$ does indeed integrate to one, we proceed as follows:

$$\int_{-\infty}^{\infty} f_Y(y)dy = \frac{1}{\sqrt{2\pi}} \sum_{n=-\infty}^{\infty} \int_{-1}^{1} \exp\left\{-\frac{1}{2}(y + 2n)^2\right\} dy$$

$$= \frac{1}{\sqrt{2\pi}} \sum_{n=-\infty}^{\infty} \int_{-1+2n}^{1+2n} \exp\left\{-\frac{1}{2}y^2\right\} dy$$

$$= \sum_{n=-\infty}^{\infty} \{\text{erf}(1 + 2n) - \text{erf}(-1 + 2n)\}.$$

If this last sum is written out, the reader will quickly find that all the terms cancel except the first $(n = -\infty)$ and the last $(n = \infty)$. This leaves that

$$\int_{-\infty}^{\infty} f_Y(y)dy = \text{erf}(\infty) - \text{erf}(-\infty) = 2 \times \text{erf}(\infty) = 1.$$

3.3 SOLVING PROBLEMS OF THE TYPE $Z = g(X, Y)$

In many problems in science and engineering a random variable Z is functionally related to two (or more) random variables X, Y. Some examples are

1. The signal Z at the input of an amplifier consists of a signal X to which is added independent random noise Y. Thus $Z = X + Y$. If X is also a r.v., what is the pdf of Z? (See Figure 3.3-1.)
2. A two-engine airplane is capable of flight as long as at least one of its two engines is working. If the time-to-failures of the starboard and port engines are X and Y, respectively, then clearly the time-to-crash of the airplane is $Z \overset{\Delta}{=} \max(X, Y)$. What is the pdf of Z?

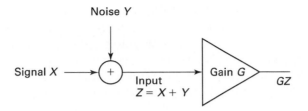

Figure 3.3-1 The signal plus independent additive noise problem.

3. Many signal processing systems use electronic multipliers to multiply two signals together (modulators, demodulators, correlators, and so forth). If X is the signal on one input and Y is the signal on the other input, what is the pdf of the output $Z \triangleq XY$?

4. In the famous "random-walk" problem that applies to a number of important physical problems, a particle undergoes random independent displacements X and Y in the x and y directions, respectively. What is the pdf of the total displacement $Z \triangleq [X^2 + Y^2]^{1/2}$? (See Figure 3.3-2.)

Problems of the type $Z = g(X, Y)$ are not fundamentally different from the type of problem we discussed in Section 3.2. Recall that for $Y = g(X)$ the basic problem was to find the point set C_y such that the events $\{\zeta : Y(\zeta) \leq y\}$ and $\{\zeta : X(\zeta) \in C_y\}$ were equal. Essentially, the same problem occurs here as well: Find the point set C_z in the (x, y) plane such that the events $\{\zeta : Z(\zeta) \leq z\}$ and $\{\zeta : X(\zeta), Y(\zeta) \in C_z\}$ are equal, this being indicated in our usual shorthand notation by

$$\{Z \leq z\} = \{(X, Y) \in C_z\} \tag{3.3-1}$$

and

$$F_Z(z) = \iint_{(x,y) \in C_z} f_{XY}(x, y)dx\, dy. \tag{3.3-2}$$

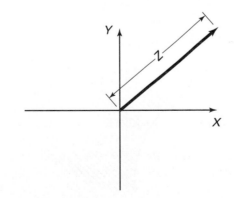

Figure 3.3-2 Displacement in the random-walk problem.

The point set C_z is determined from $g(x, y) \leq z$. Clearly in problems of the type $Z = g(X, Y)$ we deal with joint densities or distributions and double integrals (or summations) instead of single ones. Thus, in general, the computation of $f_Z(z)$ is more complicated than the computation of $f_Y(y)$ in $Y = g(X)$. However, we have access to two great labor-saving devices, which we shall learn about later: (1) We can solve many $Z = g(X, Y)$-type problems by a "turn-the-crank" type formula through the use of *auxiliary* variables (Section 3.4); and (2) we can solve problems of the type $Z = X + Y$ through the use of *characteristic functions* (Chapter 4). However, use of these shortcut methods at this stage would obscure the underlying principles.

Let us now solve the problems mentioned earlier from first principles.

Example 3.3-1

(Multiplier.) To find C_z in Equation 3.3-2 for the PDF of $Z = XY$, we need to determine the region where $g(x, y) \overset{\Delta}{=} xy \leq z$. This region is shown in Figure 3.3-3 for $z > 0$.

Thus, reasoning from the diagram, we compute

$$F_Z(z) = \int_0^\infty \left(\int_{-\infty}^{z/y} f_{XY}(x, y) dx \right) dy + \int_{-\infty}^0 \left(\int_{z/y}^\infty f_{XY}(x, y) dx \right) dy \quad \text{for } z \geq 0. \quad (3.3\text{-}3)$$

Now define the indefinite integral $G_{XY}(x, y)$ by

$$G_{XY}(x, y) \overset{\Delta}{=} \int f_{XY}(x, y) dx. \quad (3.3\text{-}4)$$

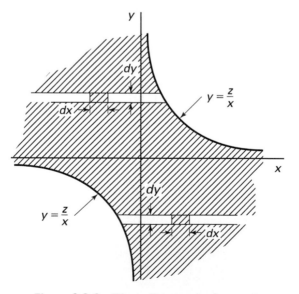

Figure 3.3-3   The region $xy \leq z$ for $z > 0$.

Then

$$F_Z(z) = \int_0^\infty [G_{XY}(z/y, y) - G_{XY}(-\infty, y)]dy$$

$$+ \int_{-\infty}^0 [G_{XY}(\infty, y) - G_{XY}(z/y, y)]dy$$

and

$$f_Z(z) = \frac{dF_Z(z)}{dz} = \int_{-\infty}^\infty \frac{1}{|y|} f_{XY}(z/y, y)dy. \tag{3.3-5}$$

We leave it as an exercise to show that Equation 3.3-5 is valid for $z < 0$ as well.

As a special case, assume X and Y are independent, identically distributed random variables (i.i.d. r.v.) with

$$f_X(x) = f_Y(x) \triangleq \frac{\alpha/\pi}{\alpha^2 + x^2}. \tag{3.3-6}$$

This is known as the Cauchy[†] probability law. Because of independence

$$f_{XY}(x, y) = f_X(x)f_Y(y)$$

and because of the evenness of the integrand in Equation 3.3-5, we obtain,[‡] after a change of variable,

$$f_Z(z) = \left(\frac{\alpha}{\pi}\right)^2 \int_0^\infty \frac{1}{z^2 + \alpha^2 x} \cdot \frac{1}{\alpha^2 + x} dx$$

$$= \left(\frac{\alpha}{\pi}\right)^2 \frac{1}{z^2 - \alpha^4} \ln \frac{z^2}{\alpha^4}. \tag{3.3-7}$$

See Figure 3.3-4 for a MATLAB plot of the $f_Z(z)$ for $\alpha = 1$.

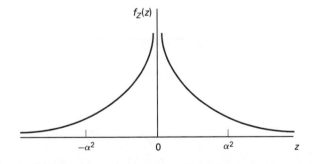

Figure 3.3-4   The pdf $f_Z(z)$ of $Z = XY$ when X and Y are i.i.d. r.v. and Cauchy.

[†]Auguste Louis Cauchy (1789–1857). French mathematician who wrote copiously on astronomy, optics, hydrodynamics, function theory, and the like.

[‡]See B. O. Peirce and R. M. Foster, *A Short Table of Integrals*, 4th ed. (Boston: Ginn & Company, 1956), p. 8.

Example 3.3-2

(Parallel operation.) We wish to compute the pdf of $Z = \max(X, Y)$ if X and Y are independent r.v.'s. Then

$$F_Z(z) = P[\max(X, Y) \leq z].$$

But the event $\{\max(X, Y) \leq z\}$ is equal to $\{X \leq z, Y \leq z\}$. Hence

$$P[Z \leq z] = P[X \leq z, Y \leq z] = F_X(z)F_Y(z) \tag{3.3-8}$$

and

$$f_Z(z) = f_Y(z)F_X(z) + f_X(z)F_Y(z). \tag{3.3-9}$$

Again as a special case, let $f_X(x) = f_Y(x)$ be the uniform $[0, 1]$ pdf. Then

$$f_Z(z) = 2z[u(z) - u(z - 1)]. \tag{3.3-10}$$

(See Figure 3.3-5.) The computation of $Z = \min(X, Y)$ is left as a homework problem.

Example 3.3-3

Let X, Y be independent, identically distributed (i.i.d) random variables with $f_X(x) = e^{-x}u(x)$. Let $Z = \max(X, Y)$. Compute $f_Z(z)$ and $P[Z \leq 1]$.

Solution   From $P[Z \leq z] = P[X \leq z, Y \leq z] = P[X \leq z]P[Y \leq z]$ we obtain

$$F_Z(z) = F_X(z)F_Y(z) = (1 - e^{-z})^2 u(z)$$

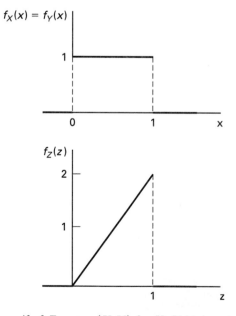

Figure 3.3-5   The pdf of $Z = \max(X, Y)$ for X, Y i.i.d. and uniform in $[0, 1]$.

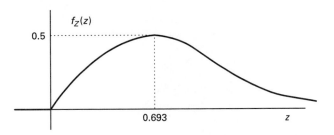

Figure 3.3-6 The pdf of the maximum of two independent exponential random variables.

and

$$f_Z(z) = \frac{dF_Z(z)}{dz} = 2e^{-z}(1 - e^{-z})u(z).$$

The pdf is shown in Figure 3.3-6. Finally, $F_Z(1) = (1 - e^{-1})^2 u(1) \approx 0.4$.

The sum of two independent random variables. The situation modeled by $Z = X + Y$ $\left(\text{and its extension } Z = \sum_{i=1}^{N} X_i\right)$ occurs so frequently in engineering and science that the computation of $f_Z(z)$ is perhaps the most important of all problems of the type $Z = g(X, Y)$.

As in other problems of this type, we must find the set of points C_z such that the event $\{Z \le z\}$ that, by definition, is equal to the event $\{X + Y \le z\}$, is also equal to $\{(X, Y) \in C_z\}$. The set of points C_z is the set of all points such that $g(x, y) \overset{\Delta}{=} x + y \le z$ and therefore represents the shaded region to the left of the line in Figure 3.3-7; any point in the shaded region satisfies $x + y \le z$.

Using Equation 3.3-2, specialized for this case, we obtain

$$F_Z(z) = \iint_{x+y \le z} f_{XY}(x, y) dx\, dy$$

$$= \int_{-\infty}^{\infty} \left(\int_{-\infty}^{z-y} f_{XY}(x, y) dx\right) dy$$

$$= \int_{-\infty}^{\infty} [G_{XY}(z - y, y) - G_{XY}(-\infty, y)] dy, \tag{3.3-11}$$

where $G_{XY}(x, y)$ is the indefinite integral

$$G_{XY}(x, y) \overset{\Delta}{=} \int f_{XY}(x, y) dx. \tag{3.3-12}$$

The pdf is obtained by differentiation $F_Z(z)$. Thus

$$f_Z(z) = \frac{dF_Z(z)}{dz} = \int_{-\infty}^{\infty} \frac{d}{dz} [G_{XY}(z - y, y)] dy$$

$$= \int_{-\infty}^{\infty} f_{XY}(z - y, y) dy. \tag{3.3-13}$$

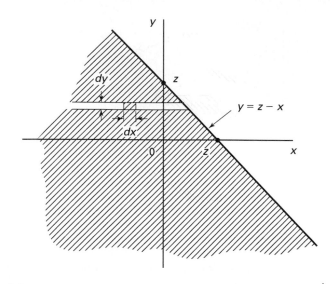

Figure 3.3-7   The region C_z (shaded) for computing the pdf of $Z \triangleq X + Y$.

Equation 3.3-13 is an important result (compare with Equation 3.3-5 for $Z = XY$). In many instances X and Y are independent r.v.'s so that $f_{XY}(x, y) = f_X(x)f_Y(y)$. Then Equation 3.3-13 takes the special form

$$f_Z(z) = \int_{-\infty}^{\infty} f_X(z - y)f_Y(y)dy \qquad (3.3\text{-}14)$$

which is known as the *convolution integral* or, more specifically, the *convolution of f_X with f_Y*.[†] It is a simple matter to prove that Equation 3.3-14 can be rewritten as

$$f_Z(z) = \int_{-\infty}^{\infty} f_X(x)f_Y(z - x)dx. \qquad (3.3\text{-}15)$$

Equations 3.3-14 and 3.3-15 can easily be extended to computing the pdf of $Z = aX + bY$. To be specific, let $a > 0$, $b > 0$. Then the region $g(x, y) \triangleq ax + by \leq z$ is to the left of the line $y = z/b - ax/b$ (Figure 3.3-8).
        Hence

$$F_Z(z) = \iint_{g(x,y) \leq z} f_{XY}(x, y)dx \, dy$$

$$= \int_{-\infty}^{\infty} f_X \left(\int_{-\infty}^{z/a - by/a} f(x, y)dx \right) dy. \qquad (3.3\text{-}16)$$

[†]A very common notation for the convolution integral as in Equation 3.3-15 is $f_Z = f_X * f_Y$.

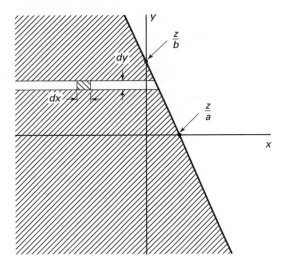

Figure 3.3-8   The region of integration for computing the pdf of $Z = aX + bY$ shown for $a > 0$, $b > 0$.

As usual, to obtain $f_Z(z)$ we differentiate with respect to z; this furnishes

$$f_Z(z) = \frac{1}{a} \int_{-\infty}^{\infty} f_X\left(\frac{z}{a} - \frac{b}{a}\right) f_Y(y) dy, \tag{3.3-17}$$

where we assumed that X and Y are independent r.v.'s. Equivalently, we can compute $f_Z(z)$ by writing

$$V \triangleq aX$$
$$W \triangleq bY$$
$$Z \triangleq V + W.$$

Then again, assuming $a > 0$, $b > 0$ and X, Y independent, we obtain from Equation 3.3-14

$$f_Z(z) = \int_{-\infty}^{\infty} f_V(z - w) f_W(w) dw,$$

where, from Equation 3.2-2,

$$f_V(v) = \frac{1}{a} f_X\left(\frac{v}{a}\right),$$

and

$$f_W(w) = \frac{1}{b} f_Y\left(\frac{w}{b}\right).$$

Thus

$$f_Z(z) = \frac{1}{ab} \int_{-\infty}^{\infty} f_X\left(\frac{z - w}{a}\right) f_Y\left(\frac{w}{b}\right) dw. \tag{3.3-18}$$

Although Equation 3.3-18 doesn't "look" like Equation 3.3-17, in fact it is identical to it. We need only make the change of variable $y \triangleq w/b$ in Equation 3.3-18 to obtain Equation 3.3-17.

Example 3.3-4 _____

Let X and Y be independent r.v.'s with $f_X(x) = e^{-x}u(x)$ and $f_Y(y) = \frac{1}{2}[u(y+1) - u(y-1)]$ and let $Z \triangleq X + Y$. What is the pdf of Z?

Solution   A big help in solving convolution-type problems is to keep track of what is going on graphically. Thus in Figure 3.3-9(a) is shown $f_X(y)$ and $f_Y(y)$; in Figure 3.3-9(b) is shown $f_X(z-y)$. Note that $f_X(z-y)$ is the *reverse and shifted* image of $f_X(y)$. How do we know that the point at the leading edge of the reverse/shifted image is $y = z$? Consider

$$f_X(z-y) = e^{-(z-y)}u(z-y).$$

But $u(z-y) = 0$ for $y > z$. Therefore the reverse/shifted function is nonzero for $(-\infty, z]$ and the leading edge of $f_X(z-y)$ is at $y = z$.

Since f_X and f_Y are discontinuous functions, we do not expect $f_Z(z)$ to be described by the same expression for all values of z. This means that we must do a careful step-by-step evaluation of Equation 3.3-14 for different regions of z-values.

(a)   *Region 1.* $z < -1$. For $z < -1$ the situation is as shown in Figure 3.3-10(a). Since there is no overlap, Equation 3.3-14 yields zero. Thus $f_Z(z) = 0$ for $z < -1$.

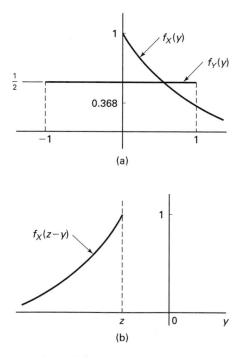

Figure 3.3-9   (a) The pdf's $f_X(y)$, $f_Y(y)$; (b) the reverse/shifted pdf $f_X(z-y)$.

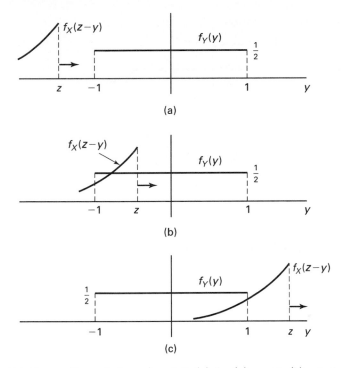

Figure 3.3-10 Relative positions $f_X(z - y)$ and $f_Y(y)$ for (a) $z < 1$; (b) $-1 \le z < 1$; (c) $z > 1$.

(b) *Region 2.* $-1 \le z < 1$. In this region the situation is as in Figure 3.3-10(b). Thus Equation 3.3-14 yields

$$f_Z(z) = \frac{1}{2} \int_{-1}^{z} e^{-(z-y)} dy$$

$$= \frac{1}{2}[1 - e^{-(z+1)}]. \qquad (3.3\text{-}19)$$

(c) *Region 3.* $z \ge 1$. In this region the situation is as in Figure 3.3-10(c). From Equation 3.3-14 we obtain

$$f_Z(z) = \frac{1}{2} \int_{-1}^{1} e^{-(z-y)} dy$$

$$= \frac{1}{2}[e^{-(z-1)} - e^{-(z+1)}]. \qquad (3.3\text{-}20)$$

Before collecting these results to form a graph we make one final important observation: Since no delta functions were involved in the computation, $f_Z(z)$ must be a *continuous* function of z. Hence, as a check on the solution, the $f_Z(z)$ values at the boundaries of the regions must match. For example, at the junction $z = 1$ between region 2 and region 3

$$\tfrac{1}{2}[1 - e^{-(z+1)}]_{z=1} \stackrel{?}{=} \tfrac{1}{2}[e^{-(z-1)} - e^{-(z+1)}]_{z=1}. \qquad (3.3\text{-}21)$$

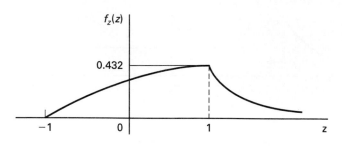

Figure 3.3-11   The pdf $f_Z(z)$ from Example 3.3-4.

Obviously the right and left sides of Equation 3.3-21 agree so we have some confidence in our solution (Figure 3.3-11).

Example 3.3-5

(Square-law detector.) Let X and Y be independent, uniform r.v.'s in $(-1, 1)$. Compute the pdf of $V \triangleq (X + Y)^2$.

Solution   We solve this problem in two steps. First, we compute the pdf of $Z \triangleq X + Y$; then we compute the pdf of $V = Z^2$. With the notation

$$\text{rect}\left(\frac{x}{D}\right) \triangleq u\left(x + \frac{D}{2}\right) - u\left(x - \frac{D}{2}\right)$$

we obtain

$$f_X(x) = \frac{1}{2}\text{rect}\left(\frac{x}{2}\right)$$

$$f_Y(y) = \frac{1}{2}\text{rect}\left(\frac{y}{2}\right)$$

$$f_X(z - y) = \frac{1}{2}\text{rect}\left(\frac{z - y}{2}\right).$$

From Equation 3.3-14 we get

$$f_Z(z) = \frac{1}{4}\int_{-\infty}^{\infty} \text{rect}\left(\frac{y}{2}\right)\text{rect}\left(\frac{z - y}{2}\right) dy. \tag{3.3-22}$$

The evaluation of Equation 3.3-22 is best done by keeping track graphically of where the "moving," that is, z-dependent function $\text{rect}((z-y)/2)$ is centered vis-à-vis the "stationary," that is, z-independent function $\text{rect}(y/2)$. The term *moving* is used because as z is varied, the function $f_X((z - y)/2)$ has the appearance of moving past $f_Y(y)$. The situation for four different values of z is shown in Figure 3.3-12.

The evaluation of $f_Z(z)$ for the four distinct regions is as follows:

(a) $z < -2$. In this region, there is no overlap so

$$f_Z(z) = 0.$$

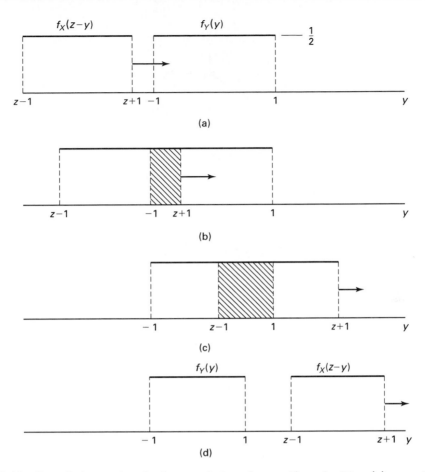

Figure 3.3-12   Four distinct regions in the convolution of two uniform densities: (a) $z < -2$; (b) $-2 \le z < 0$; (c) $0 \le z < 2$; (d) $z \ge 2$.

(b) $-2 \le z < 0$. In this region there is overlap in the interval $(-1, z+1)$ so

$$f_Z(z) = \frac{1}{4} \int_{-1}^{z+1} dy = \frac{1}{4}(z+2).$$

(c) $0 \le z < 2$. In this region there is overlap in the interval $(z-1, 1)$ so

$$f_Z(z) = \frac{1}{4} \int_{z-1}^{1} dy = \frac{1}{4}(2-z).$$

(d) $2 \le z$. In this region there is no overlap so

$$f_Z(z) = 0.$$

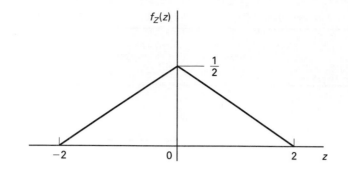

Figure 3.3-13 The pdf of $Z = X + Y$ for X, Y i.i.d. r.v.'s uniform in $(-1, 1)$.

If we put all these results together, we obtain

$$f_Z(z) = \frac{1}{4}(2 - |z|)\text{rect}\left(\frac{z}{4}\right),$$ (3.3-23)

which is graphed in Figure 3.3-13.

To complete the solution to this problem, we still need the pdf of $V = Z^2$. We compute $f_V(v)$ using Equation 3.2-22 with $g(z) = z^2$. For $v > 0$, the equation $v - z^2 = 0$ has two real roots, that is, $z_1 = \sqrt{v}$, $z_2 = -\sqrt{v}$; for $v < 0$ there are no real roots. Hence, using Equation 3.3-23 in

$$f_V(v) = \sum_{i=1}^{2} f_Z(z_i)/(2|z|)$$

yields

$$f_V(v) = \begin{cases} \dfrac{1}{4}\left(\dfrac{2}{\sqrt{v}} - 1\right) & 0 < v \le 4 \\ 0, & \text{otherwise.} \end{cases}$$ (3.3-24)

which is shown in Figure 3.3-14.

The pdf of the sum of discrete random variables is also computed by convolution, albeit a discrete convolution. For instance, let X and Y be two r.v.'s that take on values $x_1, \ldots, x_k, \ldots$ and $y_1, \ldots, y_j, \ldots$, respectively. Then $Z \triangleq X + Y$ is obviously discrete as well and its PMF is given by

$$P_Z(z_n) = \sum_{x_k + y_j = z_n} P[X = x_k, Y = y_j].$$ (3.3-25)

If X and Y are independent, Equation 3.3-25 becomes

$$P_Z(z_n) = \sum_{x_k + y_j = z_n} P_X(x_k)P_Y(y_j)$$

$$= \sum_{x_k} P_X(x_k)P_Y(z_n - x_k).$$ (3.3-26a)

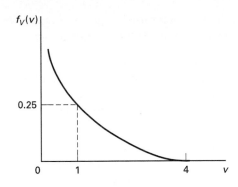

Figure 3.3-14 The pdf of V in Example 3.3-5.

If the z_n's and x_k's are equally spaced[†] then Equation 3.3-26a is recognized as a discrete convolution, in which case it can be written as

$$P_Z(n) = \sum_k P_X(k) P_Y(n - k). \tag{3.3-26b}$$

An illustration of the use of Equation 3.3-26b is given below.

Example 3.3-6

(Sum of Poisson random variables.) Let X and Y be two independent Poisson r.v.'s with PMF's $P_X(k) = \frac{1}{k!} e^{-a} a^k$ and $P_Y(i) = \frac{1}{i!} e^{-b} b^i$, where a and b are the Poisson parameters for X and Y, respectively. Let $Z \triangleq X + Y$. Then $P_Z(n)$ is given by

$$P_Z n = \sum_{k=0}^{n} P_X(k) P_Y(n - k)$$

$$= \sum_{k=0}^{n} \frac{1}{k!} \frac{1}{(n-k)!} e^{-(a+b)} a^k b^{n-k}. \tag{3.3-27}$$

Recall the binomial theorem:

$$\sum_{k=0}^{n} \binom{n}{k} a^k b^{n-k} = (a + b)^n. \tag{3.3-28}$$

Then

$$P_Z(n) = \frac{1}{n!} e^{-(a+b)} \sum_{k=0}^{n} \binom{n}{k} a^k b^{n-k}$$

$$= \frac{(a + b)^n}{n!} e^{-(a+b)}, \qquad n \geq 0, \tag{3.3-29}$$

[†]For example, let $z_n = n\Delta$, $x_k = k\Delta$, Δ a constant.

which is the Poisson law with parameter $a+b$. Thus we obtain the important result that the sum of two independent Poisson r.v.'s with parameters a, b is a Poisson r.v. with parameter $(a+b)$.

Example 3.3-7

(Sums of binomial random variables.) A more challenging example than the previous one involves computing the sums of independent, identically distributed (i.i.d) binomial r.v.'s X and Y. Let $Z = X + Y$; then $P_Z(m)$ is given by

$$P_Z(m) = \sum_{k=-\infty}^{\infty} P_X(k)P_Y(m-k),$$

where

$$P_X(k) = P_Y(k) = \begin{cases} 0 & k < 0 \\ \binom{n}{k} p^k q^{n-k} & 0 \le k \le n \\ 0 & k > n. \end{cases}$$

Thus

$$P_Z(m) = \sum_{k=0}^{n} \binom{n}{k} p^k q^{n-k} \binom{n}{m-k} p^{m-k} q^{n-(m-k)}$$

$$= p^m q^{2n-m} \sum_{k=0}^{n} \binom{n}{k} \binom{n}{m-k}.$$

Now observe that the lower limit on the sum must either be $k = 0$ or $k \ge m - n$ whichever is greater (the reader should check this for him or herself). Also the upper limit on the sum must be m or n, whichever is less (why?). Hence

$$P_Z(m) = p^m q^{2n-m} \sum_{k=\max(0,m-n)}^{\min(n,m)} \binom{n}{k} \binom{n}{m-k}.$$

Somewhat amazingly

$$\sum_{k=\max(0,m-n)}^{\min(n,m)} \binom{n}{k} \binom{n}{m-k} = \binom{2n}{m} \tag{3.3-30}$$

so that

$$P_Z(m) = \binom{2n}{m} p^m q^{2n-m} \triangleq b(m; 2n, p). \tag{3.3-31}$$

Thus the sum of i.i.d. binomial r.v.'s with PMF's $b(k; n, p)$ is a binomial r.v. with PMF $b(k; 2n, p)$.

To show that Equation 3.3-30 is true we first ask the reader to show that the left-hand side (LHS) has the same value whether $m > n$ (in which case the sum goes from $k = m - n$

to $k = n$) or whether $m \geq n$ (in which case the sum goes from $k = 0$ to $k = m$). Indeed an expansion of the LHS (say for $m \leq n$) yields

$$\binom{n}{0}\binom{n}{m} + \binom{n}{1}\binom{n}{m-1} + \cdots + \binom{n}{m}\binom{n}{0}. \tag{3.3-32}$$

Finally we ask what is the number of subpopulations of size m that can be formed from a population of size $2n$. This is clearly C_m^{2n}. Another way to compute this is to break the population of size $2n$ into two populations of size n each. Call these two populations of size n each, A and B, respectively. Then $C_k^n C_{m-k}^n$ is the number of ways of choosing k subpopulations from A and $m - k$ from B. Then clearly

$$\sum_{k=0}^{m} C_k^n C_{m-k}^n = C_m^{2n} \tag{3.3-33}$$

and since, as we said earlier

$$\sum_{k=m-n}^{n} C_k^n C_{m-k}^n = \sum_{k=0}^{m} C_k^n C_{m-k}^n$$

the result in Equation 3.3-30 is equally valid when k goes from $m - n$ to n.

We mentioned earlier in Section 3.2 that although the formula in Equation 3.2-22 (and its extensions to be discussed in Section 3.4) is very handy for solving problems of this type, the indirect approach is sometimes easier. We illustrate with the following example.

Example 3.3-8
Let X and Y be i.i.d. r.v.'s with $X = N(0, \sigma^2)$. What is the pdf of $Z \triangleq X^2 + Y^2$?

Solution   We begin with the fundamental result given in Equation 3.3-2:

$$F_Z(z) = \iint_{(x,y) \in C_z} f_{XY}(x, y)dx\, dy \quad \text{for} \quad z \geq 0$$

$$= \frac{1}{2\pi\sigma^2} \iint_{x^2 + y^2 \leq z} e^{-(1/2\sigma^2)(x^2 + y^2)} dx\, dy. \tag{3.3-34}$$

The region C_z consists of the shaded region in Figure 3.3-15.
    Equation 3.3-34 is easily evaluated using polar coordinates. Let

$$x = r \cos \theta \qquad y = r \sin \theta$$

$$dx\, dy \to r dr d\theta.$$

Then $x^2 + y^2 \leq z \to r \leq \sqrt{z}$ and Equation 3.3-34 is transformed into

$$F_Z(z) = \frac{1}{2\pi\sigma^2} \int_0^{2\pi} d\theta \int_0^{\sqrt{z}} r \exp\left[-\frac{1}{2\sigma^2}r^2\right] dr$$

$$= [1 - e^{-z/2\sigma^2}]u(z) \tag{3.3-35}$$

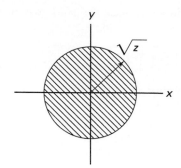

Figure 3.3-15   The region C_z for the event $\{X^2 + Y^2 \leq z\}$ for $z \geq 0$.

and

$$f_Z(z) = \frac{dF_Z(z)}{dz} = \frac{1}{2\sigma^2} e^{-z/2\sigma^2} u(z). \tag{3.3-36}$$

Thus $Z = X^2 + Y^2$ is an *exponential* r.v. if X and Y are i.i.d. zero-mean Gaussian. Equivalently Z is chi-square with two degrees of freedom.

Example 3.3-9

If the previous example is modified to finding the pdf of $Z \triangleq (X^2 + Y^2)^{1/2}$, a radically different pdf results. Again we use Equation 3.3-2 except that now C_z consists of the shaded region in Figure 3.3-16.

Thus

$$F_Z(z) = \frac{1}{2\pi\sigma^2} \int_0^{2\pi} d\theta \int_0^z r \exp\left[-\frac{1}{2\sigma^2} r^2\right] dr$$

$$= (1 - e^{-z^2/2\sigma^2})u(z). \tag{3.3-37}$$

and

$$f_Z(z) = \frac{z}{\sigma^2} = e^{-z^2/2\sigma^2} u(z) \tag{3.3-38}$$

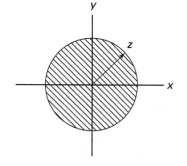

Figure 3.3-16   The region C_z for the event $\{(X^2 + Y^2)^{1/2} \leq z\}$.

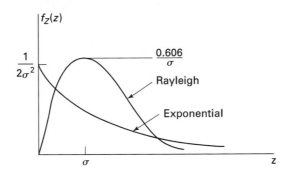

Figure 3.3-17 Rayleigh and exponential pdf's.

which is the Rayleigh density function. It is also known as the χ ("chi") distribution with two degrees of freedom. The exponential and Rayleigh pdf's are compared in Figure 3.3-17.

Stephen O. Rice [3-3], who in the 1940s did pioneering work in the analysis of electrical noise, showed that narrow-band noise signals at center frequency ω can be represented by the wave

$$Z(t) = X \cos \omega t + Y \sin \omega t \tag{3.3-39}$$

where t is time, ω is the radian frequency in radians per second and, at any particular instant t, X and Y are i.i.d. r.v.'s as $N(0, \sigma^2)$. The envelope $Z \triangleq (X^2 + Y^2)^{1/2}$ has, therefore, a Rayleigh distribution with parameter σ.

The next example generalizes the results of Example 3.3-8 and is a result of considerable interest in communication theory.

*Example 3.3-10†

(The Rician density.)‡ S. O. Rice considered a version of the following problem: Let $X: N(P, \sigma^2)$ and $Y: N(0, \sigma^2)$ be independent Gaussian r.v.'s. What is the pdf of $Z = (X^2 + Y^2)^{1/2}$? Note that with the parameter $P = 0$, we obtain the solution of Example 3.3-9.

We write

$$F_Z(z) = \begin{cases} \dfrac{1}{2\pi\sigma^2} \displaystyle\iint_{(x^2+y^2)^{1/2} \leq z} \exp\left[-\frac{1}{2}\left(\left[\frac{x-P}{\sigma}\right]^2 + \left(\frac{y}{\sigma}\right)^2\right)\right] dx\,dy, & z > 0 \\ 0, & z < 0 \end{cases} \tag{3.3-40}$$

The usual Cartesian-to-polar transformation $x = r\cos\theta$, $y = r\sin\theta$, $r = x^2 + y^2$, $\theta = \tan^{-1}(y/x)$ yields

$$F_Z(z) = \frac{\exp\left[-\frac{1}{2}\left(\frac{P}{\sigma}\right)^2\right]}{2\pi\sigma^2} \int_0^z e^{-\frac{1}{2}(r/\sigma)^2} \left(\int_0^{2\pi} e^{rP\cos\theta/\sigma^2}\,d\theta\right) r\,dr \cdot u(z). \tag{3.3-41}$$

---

†Starred examples are somewhat more involved and can be omitted on a first reading.

‡Sometimes called the Rice-Nakagami pdf in recognition of the work of Nakagami around the time of World War II.

The function

$$I_o(x) \triangleq \frac{1}{2\pi} \int_0^{2\pi} e^{x \cos \theta} d\theta$$

is called the *zero-order modified Bessel function of the first kind* and is monotonically increasing like e^x. With this notation, the cumbersome Equation 3.3-41 can be rewritten as

$$F_Z(z) = \frac{\exp\left[-\frac{1}{2}\left(\frac{P}{\sigma}\right)^2\right]}{\sigma^2} \int_0^z r I_o\left(\frac{rP}{\sigma^2}\right) e^{-\frac{1}{2}[(z-P)/\sigma]^2} dr \cdot u(z), \qquad (3.3\text{-}42)$$

where $u(z)$ ensures that the above is valid for $z > 0$. To obtain $f_Z(z)$ we differentiate with respect to z. This produces

$$f_Z(z) = \frac{z}{\sigma^2} \exp\left[-\frac{1}{2}\left(\frac{P^2 + z^2}{\sigma^2}\right)\right] I_o\left(\frac{zP}{\sigma^2}\right) \cdot u(z). \qquad (3.3\text{-}43)$$

The pdf given in Equation 3.3-43 is called the *Rician* probability density function. Since $I_o(0) = 1$, we obtain the Rayleigh law when $P = 0$. When $zP \gg \sigma^2$, that is, the argument of $I_o(\cdot)$ is large, we use the approximation

$$I_o(x) \approx \frac{e^x}{(2\pi x)^{1/2}}$$

to obtain

$$f_Z(z) \approx \frac{1}{\sqrt{2\pi\sigma^2}} \left(\frac{z}{P}\right)^{1/2} e^{-\frac{1}{2}[(z-P)/\sigma]^2}$$

which is almost Gaussian [except for the factor $(z/P)^{1/2}$]. This is the pdf of the *envelope* of the sum of a strong sine wave and weak narrow-band Gaussian noise, a situation that occurs not infrequently in electrical communications.

3.4 SOLVING PROBLEMS OF THE TYPE $V = g(X, Y)$, $W = h(X, Y)$

The problem of two functions of two random variables is essentially an extension of the earlier cases except that the algebra is somewhat more involved.

Fundamental Problem

We are given two r.v.'s X, Y with joint pdf $f_{XY}(x, y)$ and two differentiable functions $g(x, y)$ and $h(x, y)$. Two new random variables are constructed according to $V = g(X, Y)$, $W = h(X, Y)$. How do we compute the joint PDF $F_{VW}(v, w)$ (or joint pdf $f_{VW}(v, w)$) of V and W?

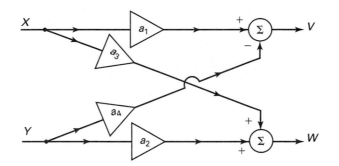

Figure 3.4-1   A two-variable-to-two-variable matrixer.

Illustrations.   1. The circuit shown in Figure 3.4-1 occurs in communication systems such as in the generation of stereo baseband systems [3-2]. The $\{a_i\}$ are gains. When $a_1 = a_2 = \cos\theta$ and $a_3 = a_4 = \sin\theta$, the circuit is known as θ-*rotational transformer*. In another application if X and Y are used to represent, for example, the left and right pick-up signals in stereo broadcasting, then V and W represent the difference and sum signals if all the a_i's are set to unity. The sum and difference signals are then used to generate the signal to be transmitted. Suppose for the moment that there are no source signals and that X and Y therefore represent only Gaussian noise. What is the pdf of V and W?

2. The error in the landing location of a spacecraft from a prescribed point is denoted by (X, Y) in Cartesian coordinates. We wish to specify the error in polar coordinates $V \triangleq (X^2 + Y^2)^{1/2}$ $W = \tan^{-1}(Y/X)$. Given the joint pdf $f_{XY}(x, y)$ of landing error coordinates in Cartesian coordinates, how do we compute the pdf of the landing error in polar coordinates?

The solution to the problem at hand is, as before, to find a point set C_{vw} such that the two events $\{V \le v, W \le w\}$ and $\{(X, Y) \in C_{vw}\}$ are equal.[†] Thus the fundamental relation is

$$P[V \le v, W \le w] \triangleq F_{VW}(v, w)$$

$$= \iint_{(x,y) \in C_{vw}} f_{XY}(x, y)dx\,dy. \tag{3.4-1}$$

The region C_{vw} is given by the points x, y that satisfy

$$C_{vw} = \{(x, y) : g(x, y) \le v, h(x, y) \le w\}. \tag{3.4-2}$$

We illustrate the application of Equation 3.4-1 with an example.

Example 3.4-1 _____

We are given $V \triangleq X + Y$ and $W \triangleq X - Y$ and wish to calculate the pdf $f_{VW}(v, w)$. The point set C_{vw} is described by the combined constraints $g(x, y) \triangleq x + y \le v$ and $h(x, y) \triangleq x - y \le w$; it is shown in Figure 3.4-2 for $v > 0$, $w > 0$.

_____

[†]In more elaborate notation we would write $\{\zeta : V(\zeta) \le v \text{ and } W(\zeta) \le w\} = \{\zeta : (X(\zeta), Y(\zeta)) \in C_{vw}\}$.

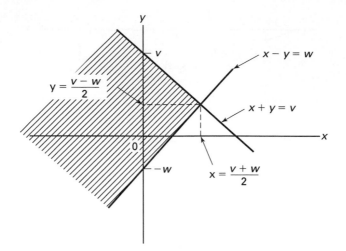

Figure 3.4-2   Point set C_{vw} (shaded region) for Example 3.4-1.

The integration over the shaded region yields

$$F_{VW}(v,w) = \int_{-\infty}^{(v+w)/2} \left(\int_{x-w}^{v-x} f_{XY}(x,y)dy \right) dx. \tag{3.4-3}$$

To obtain the joint density $f_{VW}(u,w)$ we use Equation 2.6-22. Hence (after considerable algebra)

$$f_{VW}(v,w) = \frac{\partial^2 F_{VW}(v,w)}{\partial v\, \partial w}$$

$$= \frac{1}{2} f_{XY}\left(\frac{v+w}{2}, \frac{v-w}{2} \right). \tag{3.4-4}$$

Thus even this simple problem, involving only linear functions, requires a considerable amount of work and care to obtain a solution. (We leave the details as an exercise—see Problem 3.31.) For this reason, problems of the type discussed in this section and their extensions, that is, $Y_1 = g_1(X_1, \ldots, X_n)$, $Y_2 = g_2(X_1, \ldots, X_n), \ldots, Y_n = g_n(X_1, \ldots, X_n)$, are generally solved by the technique discussed next.

Obtaining f_{VW} Directly from f_{XY}

Instead of attempting to find $f_{VW}(v,w)$ through Equation 3.4-1, we shall instead take a different approach. Consider the elementary event

$$\{v < V \le v + dv, w < W \le w + dw\}$$

and the one-to-one[†] differentiable functions $v = g(x, y)$, $w = h(x, y)$. The inverse mappings exist and are given by $x = \phi(v, w)$, $y = \psi(u, w)$. Later we shall consider the more general case where, possibly, more than one pair of (x_i, y_i) produce a given (v, w).

The probability $P[v < V \leq v + dv, w < W \leq w + dw]$ is the probability that V and W lie in an infinitesimal rectangle of area $dv\, dw$ with vertices at (v, w), $(v + dv, w)$, $(v, w + dw)$, and $(v + dv, w + dw)$. The *image* of this square in the x, y coordinate system is (for example, see Marsden and Weinstein [3-4, p. 769]) an infinitesimal parallelogram with vertices at

$$P_1 = (x, y),$$

$$P_2 = \left(x + \frac{\partial \phi}{\partial v} dv, y + \frac{\partial \psi}{\partial v} dv \right),$$

$$P_3 = \left(x + \frac{\partial \phi}{\partial w} dw, y + \frac{\partial \psi}{\partial w} dw \right),$$

$$P_4 = \left(x + \frac{\partial \phi}{\partial v} dv + \frac{\partial \phi}{\partial w} dw, y + \frac{\partial \psi}{\partial v} dv + \frac{\partial \psi}{\partial w} dw \right).$$

This mapping is shown in Figure 3.4-3.

With $\mathscr{R}$ denoting the rectangular region shown in Figure 3.4-3(a) and $\mathscr{S}$ denoting the parallelogram in Figure 3.4-3(b) and $A(\mathscr{R})$ and $A(\mathscr{S})$ denoting the areas of $\mathscr{R}$ and $\mathscr{S}$ respectively, we obtain

$$P[v < V \leq v + dv, w < W \leq w + dw] = \iint_{\mathscr{R}} f_{VW}(\xi, \eta) d\xi\, d\eta \qquad (3.4\text{-}5)$$

$$= f_{VW}(v, w) A(\mathscr{R}) \qquad (3.4\text{-}6)$$

$$= \iint_{\mathscr{S}} f_{XY}(\xi, \eta) d\xi\, d\eta \qquad (3.4\text{-}7)$$

$$= f_{XY}(x, y) A(\mathscr{S}). \qquad (3.4\text{-}8)$$

Equation 3.4-5 follows from the fundamental relation given in Equation 3.4-1; Equation 3.4-6 follows from the interpretation of the pdf given in Equation 2.4-6; Equation 3.4-7 follows by definition of the point set ℓ, that is, ℓ is the set of points that makes the events $\{(V, W) \in \mathscr{R}\}$ and $\{(X, Y) \in \ell\}$ equal and Equation 3.4-8 again follows from the interpretation of pdf.

From Equation 3.4-6 and 3.4-8 we find that

$$f_{VW}(v, w) = \frac{A(\mathscr{S})}{A(\mathscr{R})} f_{XY}(x, y), \qquad (3.4\text{-}9)$$

where $x = \phi(v, w)$ and $y = \phi(v, w)$.

Essentially then, all that remains is to compute the ratio of the two areas. This is done in Appendix C. There we show that the ratio $A(\mathscr{S})/A(\mathscr{R})$ is the magnitude of a quantity called the Jacobian of the transformation $x = \phi(v, w)$, $y = \varphi(v, w)$ and given the symbol $\tilde{J}$. If there

---

[†]Every point (x, y) maps into a unique (v, w) and vice versa.

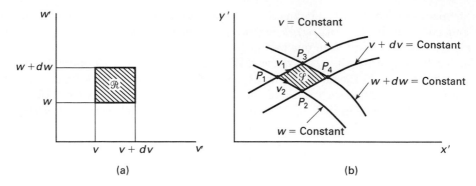

Figure 3.4-3 An infinitesimal rectangle in the v, w system (a) maps into an infinitesimal parallelogram (b) in the x, y system.

is more than one solution to the equations $v = g(x, y)$, $w = h(x, y)$, say, $x_1 = \phi_1(v, w)$, $y_1 = \varphi_1(v, w)$, $x_2 = \phi_2(v, w)$, $y_2 = \varphi_2(v, w), \ldots, x_n = \phi_n(v, w)$, $y_n = \varphi_n(v, w)$, then $\mathcal{R}$ maps into multiple, disjoint infinitesimal regions $\mathcal{S}_1, \mathcal{S}_2, \ldots, \mathcal{S}_n$ and $A(\mathcal{S}_i)/A(\mathcal{R}) = |\tilde{J}_i|$, $i = 1, \ldots, n$. The $|\tilde{J}_i|$ are often written as the magnitude of determinants, that is,

$$|\tilde{J}_i| = \text{mag} \begin{vmatrix} \partial\phi_i/\partial v & \partial\phi_i/\partial w \\ \partial\varphi_i/\partial v & \partial\varphi_i/\partial w \end{vmatrix} = |\partial\phi_i/\partial v \times \partial\varphi_i/\partial w - \partial\varphi_i/\partial v \times \partial\phi_i/\partial w|. \quad (3.4\text{-}10)$$

The end result is the important formula

$$f_{VW}(v, w) = \sum_{i=1}^{n} f_{XY}(x_i, y_i)|\tilde{J}_i|. \quad (3.4\text{-}11)$$

It is shown in the Appendix that $|\tilde{J}_i^{-1}| = |J_i| \triangleq |\partial g/\partial x \times \partial h/\partial y - \partial g/\partial y \times \partial h/\partial x|$. Then we get the equally important formula

$$f_{VW}(v, w) = \sum_{i=1}^{n} f_{XY}(x_i, y_i)/|J_i|. \quad (3.4\text{-}12)$$

Example 3.4-2
We are given two functions

$$v \triangleq g(x, y) = 3x + 5y$$

$$w \triangleq h(x, y) = x + 2y \quad (3.4\text{-}13)$$

and the joint pdf f_{XY} of two r.v.'s X, Y. What is the joint pdf of two new random variables $V = g(X, Y)$, $W = h(X, Y)$?

Solution   The inverse mappings are computed from Equation 3.4-13 to be

$$x = \phi(v, w) = 2v - 5w$$

$$y = \Phi(v, w) = -v + 3w.$$

Then

$$\frac{\partial \phi}{\partial v} = 2, \frac{\partial \phi}{\partial w} = -5, \frac{\partial \Phi}{\partial v} = -1, \frac{\partial \Phi}{\partial w} = 3$$

and

$$|\tilde{J}| = \text{mag} \begin{vmatrix} 2 & -5 \\ -1 & 3 \end{vmatrix} = 1.$$

Assume $f_{XY}(x,y) = (2\pi)^{-1} \exp[-\frac{1}{2}(x^2 + y^2)]$. Then, from Equation 3.4-11

$$f_{VW}(v,w) = \frac{1}{2\pi} \exp\left[-\frac{1}{2}[(2v - 5w)^2 + (-v + 3w)^2]\right]$$

$$= \frac{1}{2\pi} \exp\left[-\frac{1}{2}(5v^2 - 26vw + 34w^2)\right].$$

Thus the transformation converts uncorrelated Gaussian r.v.'s into correlated Gaussian r.v.'s.

3.5 ADDITIONAL EXAMPLES

To enable the reader to become familiar with the methods discussed in Section 3.4 we work out a number of examples.

Example 3.5-1
Consider the r.v.'s

$$V \triangleq g(X,Y) = \sqrt{X^2 + Y^2} \tag{3.5-1}$$

$$w = h(X,Y) = \begin{cases} \tan^{-1}\left(\dfrac{Y}{X}\right), & X > 0 \\ \tan^{-1}\left(\dfrac{Y}{X}\right) + \pi, & X < 0 \end{cases} \tag{3.5-2}$$

The r.v. V is called the magnitude or envelope while W is called the phase. Equation 3.5-2 has been written in this form because we seek a solution for w over a 2π interval while the function $\tan^{-1}(y/x)$ has range $(-\pi/2, \pi/2)$ (i.e., its principle value).

To find the roots of

$$v = \sqrt{x^2 + y^2} \tag{3.5-3a}$$

$$w = \begin{cases} \tan^{-1}\left(\dfrac{y}{x}\right), & x \geq 0 \\ \tan^{-1}\left(\dfrac{y}{x}\right) + \pi, & x < 0 \end{cases} \tag{3.5-3b}$$

we observe that for $x \geq 0$, $\frac{-\pi}{2} \leq w \leq \frac{\pi}{2}$ and $\cos w \geq 0$. Similarly, for $x < 0$, $\frac{\pi}{2} < w < \frac{3\pi}{2}$ and $\cos w < 0$. Hence the only solution to Equations 3.5-3 is

$$x = v \cos w \triangleq \phi(v, w)$$

$$y = v \sin w \triangleq \psi(v, w).$$

The Jacobian $\tilde{J}$ is given by

$$\tilde{J} = \frac{\partial(\phi, \psi)}{(v, w)} = \begin{vmatrix} \cos w & -v \sin w \\ \sin w & v \cos w \end{vmatrix} = v.$$

Hence the solution is, from Equation 3.4-11,

$$f_{VW}(v, w) = v f_{XY}(v \cos w, v \sin w). \tag{3.5-4}$$

Suppose that X and Y are i.i.d. with probability law $N(0, \sigma^2)$, that is,

$$f_{XY}(x, y) = \frac{1}{2\pi\sigma^2} e^{-[(x^2+y^2)/2\sigma^2]}.$$

Then from Equation 3.5-4

$$f_{VW}(v, w) = \begin{cases} \left(\dfrac{v}{\sigma^2} e^{-v^2/2\sigma^2} \right) \dfrac{1}{2\pi}, & v > 0, -\dfrac{\pi}{2} \le w < \dfrac{3\pi}{2} \\ 0, & \text{otherwise} \end{cases} \tag{3.5-5}$$

$$= f_V(v) f_W(w).$$

Thus V and W are independent random variables. The envelope V has a Rayleigh pdf and the phase W is uniform over a 2π interval.

Example 3.5-2

Consider now a modification of the previous problem. Let $V \triangleq \sqrt{X^2 + Y^2}$ and $W \triangleq Y/X$. Then with $g(x, y) = \sqrt{x^2 + y^2}$ and $h(x, y) = y/x$, the equations

$$v - g(x, y) = 0$$
$$w - h(x, y) = 0$$

have two solutions:

$$x_1 = v(1 + w^2)^{-1/2}, \qquad y_1 = w x_1$$
$$x_2 = -v(1 + w^2)^{-1/2}, \qquad y_2 = w x_2$$

for $-\infty < w < \infty$ and $v > 0$, and no real solutions for $v < 0$.

A direct evaluation yields $|J_1| = |J_2| = (1 + w^2)/v$. Hence

$$f_{VW}(v, w) = \frac{v}{1 + w^2} [f_{XY}(x_1, y_1) + f_{XY}(x_2, y_2)].$$

With $f_{XY}(x, y)$ given by

$$f_{XY}(x, y) = \frac{1}{2\pi\sigma^2} = \exp[-(x^2 + y^2)/2\sigma^2],$$

we obtain

$$f_{VW}(v, w) = \frac{v}{\sigma^2}e^{-v^2/2\sigma^2}u(v) \cdot \frac{1/\pi}{1 + w^2}$$

$$= f_V(v)f_W(w).$$

Thus the r.v.'s V, W are independent. V is a Rayleigh r.v. as in Example 3.5-1, but W is Cauchy.

Example 3.5-3

Let θ be a prescribed angle and consider the rotational transformation

$$V \triangleq X \cos\theta + Y \sin\theta$$

$$W \triangleq X \sin\theta - Y \cos\theta \qquad (3.5\text{-}6)$$

with

$$f_{XY}(x, y) = \frac{1}{2\pi\sigma^2}e^{-[(x^2+y^2)/2\sigma^2]}.$$

The only solution to

$$v = x \cos\theta + y \sin\theta$$

$$w = x \sin\theta - y \cos\theta$$

is

$$x = v \cos\theta + w \sin\theta$$

$$y = v \sin\theta - w \cos\theta.$$

The Jacobian $\tilde{J}$ is

$$\begin{vmatrix} \dfrac{\partial x}{\partial v} & \dfrac{\partial x}{\partial w} \\ \dfrac{\partial y}{\partial v} & \dfrac{\partial y}{\partial w} \end{vmatrix} = \begin{vmatrix} \cos\theta & \sin\theta \\ \sin\theta & -\cos\theta \end{vmatrix} = -1.$$

Hence

$$f_{VW}(v, w) = \frac{1}{2\pi\sigma^2} e^{-[(v^2+w^2)/2\sigma^2]}.$$

Thus under the rotational transformation $V = g(X, Y)$, $W = h(X, Y)$ given in Equation 3.5-6, V and W are i.i.d. Gaussian r.v.'s just like X and Y. If X and Y are Gaussian but not independent r.v.'s, it is still possible to find a transformation[†] so that V, W will be independent Gaussians if the joint pdf of X, Y is Gaussian.

Example 3.5-4

Consider again the problem of solving for the pdf of $Z = \sqrt{X^2 + Y^2}$ as in Example 3.3-9. This time we shall use Equation 3.4-11 to compute $f_Z(z)$. First we note that $Z = \sqrt{X^2 + Y^2}$ is one function of two r.v.'s while Equation 3.4-11 applies to two functions of two r.v.'s. To

---

[†]Chapter 5.

convert from one kind of problem to the other we introduce an *auxiliary r.v.* $W \triangleq X$. The introduction of the r.v. W will enable us to use Equation 3.4-11 directly. Hence

$$Z \triangleq g(X, Y) = \sqrt{X^2 + Y^2}$$

$$W \triangleq h(X, Y) = X.$$

The equations

$$z - g(x, y) = 0$$

$$w - h(x, y) = 0$$

have two real roots, for $|w| < z$, namely

$$x_1 = w \qquad\qquad x_2 = w$$

$$y_1 = \sqrt{z^2 - w^2} \qquad y_2 = -\sqrt{z^2 - w^2}.$$

At both roots, $|\tilde{J}|$ has the same value:

$$|\tilde{J}_1| = |\tilde{J}_2| = \frac{z}{\sqrt{z^2 - w^2}}.$$

Hence a direct application of Equation 3.4-11 yields

$$f_{ZW}(z, w) \frac{z}{\sqrt{z^2 - w^2}} [f_{XY}(x_1, y_1) + f_{XY}(x_2, y_2)].$$

Now assume that

$$f_{XY}(x, y) = \frac{1}{2\pi\sigma^2} e^{-[(x^2 + y^2)/2\sigma^2]}.$$

Then, since in this case $f_{XY}(x, y) = f_{XY}(x, -y)$, we obtain

$$f_{ZW}(z, w) = \begin{cases} \dfrac{1}{\pi\sigma^2} \dfrac{z}{\sqrt{z^2 - w^2}} e^{-z^2/2\sigma^2}, & z > 0, |w| < z \\ 0, & \text{otherwise.} \end{cases}$$

However, we don't really want $f_{ZW}(z, w)$, but only the marginal pdf $f_Z(z)$. To obtain this we use Equation 2.6-34 with x replaced by z and y replaced by w. This gives

$$f_Z(z) = \int_{-\infty}^{\infty} f_{ZW}(z, w) dw$$

$$= \frac{z}{\sigma^2} e^{-z^2/2\sigma^2} \left[\frac{2}{\pi} \int_0^z \frac{dw}{\sqrt{z^2 - w^2}} \right] u(z).$$

The term in parentheses has value unity. To see this consider the little triangle in Figure 3.5-1 and let $w \triangleq z \sin\theta$. Then $dw = z \cos\theta \, d\theta$ and $[z^2 - w^2]^{1/2} = z \cos\theta$ and the term in parentheses becomes

$$\frac{2}{\pi} \int_0^z \frac{dw}{\sqrt{z^2 - w^2}} = \frac{2}{\pi} \int_0^{\pi/2} d\theta = 1.$$

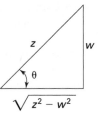

Figure 3.5-1   Trigonometric transformation $w = z \sin \theta$.

Hence

$$f_Z(z) = \frac{z}{\sigma^2} e^{-z^2/2\sigma^2} u(z)$$

which is the same result as obtained in Equation 3.3-38 except obtained by a different method.

Example 3.5-5

Finally, let us return to the problem considered in Example 3.4-1:

$$V \triangleq X + Y$$

$$W \triangleq X - Y.$$

The only root to

$$v - (x + y) = 0$$

$$w - (x - y) = 0$$

is

$$x = \frac{v + w}{2}$$

$$y = \frac{v - w}{2}$$

and $|\tilde{J}| = \frac{1}{2}$. Hence

$$f_{VW}(v, w) = \frac{1}{2} f_{XY}\left(\frac{v + w}{2}, \frac{v - w}{2}\right).$$

We leave it as an exercise to the reader to show that

$$f_V(v) = \int_{-\infty}^{\infty} f_{VW}(t, v - t)dt.$$

This important result was derived in Section 3.3 (Equation 3.3-13) by different means.

3.6 SUMMARY

The material in this chapter discussed functions of random variables, a subject of great significance in applied science and engineering and fundamental to the study of random processes. The basic problem dealt with computing the probability law of an "output" random variable Y produced by a system transformation $g(\cdot)$ operating on an "input" random variable X (i.e., $Y = g(X)$). The problem can be extended to two input random variables X, Y being operated upon by system transformation $g(\cdot)$ and $h(\cdot)$ to produce two output random variables $V = g(X,Y)$ and $W = h(X,Y)$. Then the problem is to compute the joint pdf of V, W from the joint pdf of X, Y.

We showed how most problems involving functions of r.v.'s could be computed in at least two ways:

1. the so-called indirect approach; and
2. directly through the use of a "turn-the-crank" direct method.

A number of important problems involving transformations of random variables were worked out including computing the pdf of the sum of two random variables.

PROBLEMS

3.1. Let X have PDF $F_X(x)$ and consider $Y = aX + b$ where $a < 0$. Show that if X is not necessarily continuous Equation 3.2-3 should be modified to

$$F_Y(y) = 1 - F_X\left(\frac{y-b}{a}\right) + P\left[X = \frac{y-b}{a}\right].$$

3.2. Let Y be a function of the r.v. X as follows:

$$Y \triangleq \begin{cases} X, & X \geq 0 \\ X^2, & X \leq 0. \end{cases}$$

Compute $f_Y(y)$ in terms of $f_X(x)$. Let $f_X(x)$ be given by

$$f_X(x) = \frac{1}{\sqrt{2\pi}}e^{-\frac{1}{2}x^2}.$$

3.3. Let X have pdf

$$f_X(x) = \alpha e^{-ax}u(x).$$

Compute the pdf of (a) $Y = X^3$; (b) $Y = 2X + 3$.

3.4. In medical imaging such as computer tomography the relation between detector readings y and body absorptivity x follows a $y = e^x$ law. Let X be $N(\mu, \sigma^2)$, compute the pdf of Y.

3.5. In homomorphic image processing, images are enhanced by applying non-linear transformations to the image functions. Assume that the image function is modeled as r.v. X and the enhanced image Y is $Y = \ln X$. Note that X cannot assume negative values. Compute the pdf of Y if X has an exponential density $f_X(x) = \frac{1}{3}e^{-\frac{1}{3}x}u(x)$.

3.6. Assume that $X: N(0, 1)$ and let Y be defined by

$$Y = \begin{cases} \sqrt{X}, & X \geq 0 \\ 0, & X < 0. \end{cases}$$

Compute the pdf of Y.

3.7. (a) Let $X: N(0, 1)$ and let $Y \triangleq g(X)$ where the function $g(\cdot)$ is shown in Figure P3.7. Use the indirect approach to compute $F_Y(y)$ and $f_Y(y)$ from $f_X(x)$. (b) Compute $f_Y(y)$ from Equation 3.2-22. Why can't Equation 3.2-22 be used to compute $f_Y(y)$ at $y = 0, 1$?

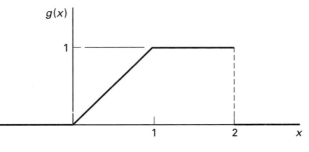

Figure P3.7

3.8. Let X be a uniform r.v. on $[0, 2]$. Compute the pdf of Y if $Y = g(X)$ and $g(\cdot)$ is as shown in Figure P3.8.

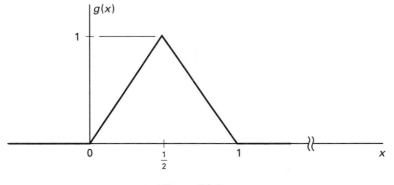

Figure P3.8

3.9. Let X be a uniform r.v. on $[0, 2]$ compute the pdf of Y if $Y = g(X)$ and $g(\cdot)$ is as shown in Figure P3.9.

3.10. Let X be a Gaussian r.v. with pdf

$$f_X(x) = \frac{1}{\sqrt{2\pi}} e^{\frac{-x^2}{2}}, \qquad -\infty < x < +\infty.$$

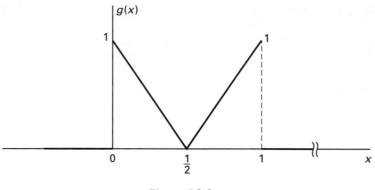

Figure P3.9

Let $Y = g(X)$ where $g(\cdot)$ is the nonlinear function given as

$$g(x) \triangleq \begin{cases} -1, & x < -1 \\ x, & -1 \le x \le 1, \\ 1, & x > 1. \end{cases}$$

It is called a *saturable limiter* function.

a) Sketch $g(x)$; b) Find $F_Y(y)$; c) Find and sketch $f_Y(y)$.

3.11. Compute the pdf of $Y = a/X(a > 0)$. Show that if X is Cauchy with parameter α, Y is Cauchy with parameter a/α.

3.12. Let $Y = \sec X$. Compute $f_Y(y)$ in terms of $f_X(x)$. What is $f_Y(y)$ when $f_X(x)$ is uniform in $(-\pi, \pi]$?

3.13. Let X and Y be independent, identically distributed r.v.'s with

$$f_X(x) = f_Y(x) = \alpha e^{-ax} u(x).$$

Compute the pdf of $Z \triangleq X - Y$.

3.14. Repeat Example 3.2-10 for $f_X(x) = e^{-x} u(x)$.

3.15. Repeat Example 3.2-11 for $f_X(x) = e^{-x} u(x)$.

3.16. The objective is to generate numbers from the pdf shown in Figure P3.16.

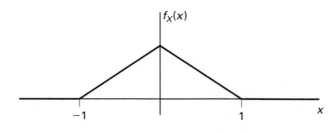

Figure P3.16

All that is available is a random number generator that generates numbers uniformly distributed in $(0,1)$. Explain what procedure you would use to meet the objective.

3.17. It is desired to generate zero-mean Gaussian numbers. All that is available is a random number generator that generates numbers uniformly distributed on $(0,1)$. It has been suggested Gaussian numbers might be generated by adding 12 uniformly distributed numbers and subtracting 6 from the sum. Write a program in which you use the procedure to generate 10,000 numbers and plot a histogram of your result. A histogram is a bar graph that has bins along the x-axis and number of points in the bin along the y-axis. Choose 200 bins of width 0.1 to span the range from -10 to 10. In what region of the histogram does the data look most Gaussian? Where does it look least Gaussian? Do you have any idea why this approach works?

3.18. In Problem 3.13 compute the pdf of $|Z|$.

3.19. Let X and Y be independent, continuous r.v.'s Let $Z = \min(X, Y)$. Compute $F_Z(z)$ and $f_Z(z)$. Sketch the result if X and Y are uniform r.v.'s in $(0,1)$. Repeat for $f_X(x) = f_Y(x) = \alpha \exp[-\alpha x] \cdot u(x)$.

3.20. Consider n independent, identically distributed r.v.'s $X_1, X_2, \ldots, X_n$ with PDF $F_{X_i}(x) \triangleq F(x)$. Let $Z \triangleq \max[X_1, X_2, \ldots, X_n]$. Compute the PDF of Z in terms of $F(x)$.

3.21. Consider n independent, identically, distributed r.v.'s $X_1, X_2, \ldots, X_n$ with PDF $F_{X_i}(x) \triangleq F(x)$. Let $Z \triangleq \min[X_1, X_2, \ldots, X_n]$. Compute the PDF of Z in terms of $F(x)$.

3.22. Let $X_1, X_2, \ldots, X_n$ be n i.i.d exponential random variables with $f_{X_i}(x) = e^{-x}u(x)$. Compute an explicit expression for the pdf of $Z_n = \max(X_1, X_2, \ldots, X_n)$. Sketch the pdf for $n = 3$.

3.23. Let $X_1, X_2, \ldots, X_n$ be n i.i.d exponential random variables with $f_{X_i}(x) = e^{-x}u(x)$. Compute an explicit expression for the pdf of $Z_n = \min(X_1, X_2, \ldots, X_n)$. Sketch the pdf for $n = 3$.

3.24. Let X, Y be i.i.d random variables with X uniformly distributed in $(-1,1)$. Compute and sketch the pdf of Z for the system shown in Figure P3.24. The square-root operation is valid only for positive numbers. Otherwise the output of the $\sqrt{}$ is zero.

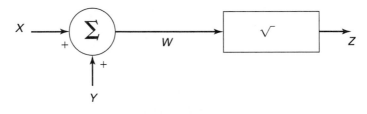

Figure P3.24

3.25. The length of time Z, an airplane, can fly is given by $Z = \alpha X$ where X is the amount of fuel in its tank and $\alpha > 0$ is a constant of proportionality. Suppose a plane has two independent fuel tanks so that when one gets empty the other switches on automatically. Because of lax maintenance a plane takes off with neither of its fuel tanks checked. Let X_1 be the fuel in the first tank and X_2 the fuel in the second tank. Let X_1 and X_2 be modeled as independent, identically distributed r.v.'s with $f_{X_1}(x) = f_{X_2}(x) = \frac{1}{b}[u(x) - u(x - b)]$. Compute the pdf of Z, the maximum flying time of the plane. If $b = 100$, say in liters, and $\alpha = $ one hour/10 liters, what is the probability that the plane will fly *at least* five hours?

3.26. Let X and Y be two independent Poisson r.v.'s with

$$P_X(k) = \frac{1}{k!}e^{-2}2^k$$

$$P_Y(j) = \frac{1}{j!}e^{-3}3^j.$$

Compute $P[Z \leq 5]$ where $Z = X + Y$. $\left[\text{Hint:} \sum_{j=0}^{n} \binom{n}{j} a^j b^{n-j} = (a + b)^n.\right]$

3.27. Let X and Y be independent, uniform r.v.'s with $f_X(x) = \frac{1}{2}$, $|x| < 1$ and zero otherwise and $f_Y(y) = \frac{1}{4}$, $|y| < 2$ and zero otherwise. Compute (a) the pdf of $Z \triangleq X + Y$; (b) the pdf of $Z \triangleq 2X - Y$.

3.28. Compute the joint pdf of

$$Z \triangleq X^2 + Y^2$$
$$W \triangleq X$$

when

$$f_{XY}(x, y) = \frac{1}{2\pi\sigma^2}e^{-[(x^2+y^2)/2\sigma^2]}.$$

Compute $f_Z(z)$ from your results.

3.29. Consider the transformation

$$Z = aX + bY$$
$$W = dX + dY.$$

Let

$$f_{XY}(x, y) = \frac{1}{2\pi\sigma^2\sqrt{1 - \rho^2}}e^{-Q(x,y)}$$

where

$$Q(x, y) = \frac{1}{2\sigma^2(1 - \rho^2)}[x^2 - 2\rho xy + y^2].$$

What combination of coefficients a, b, c, d will enable Z, W to be independent Gaussian r.v.'s?

3.30. Let

$$f_{XY}(x, y) = \frac{1}{2\pi\sqrt{1 - \rho^2}}\exp\left[-\left(\frac{x^2 - 2\rho xy + y^2}{2(1 - \rho^2)}\right)\right].$$

Compute the joint pdf $f_{VW}(v, w)$ of

$$V = \tfrac{1}{2}(X^2 + Y^2)$$
$$W = \tfrac{1}{2}(X^2 - Y^2).$$

3.31. Derive Equation 3.4-4 by the indirect method, that is, use Equations 3.4-3 and 2.6-22.

3.32. Consider the transformation

$$Z = X \cos\theta + Y \sin\theta$$
$$W = X \sin\theta - Y \cos\theta.$$

Compute the joint pdf $f_{ZW}(z, w)$ in terms of $f_{XY}(x, y)$ if

$$f_{XY}(x, y) = \frac{1}{2\pi} e^{-\frac{1}{2}(x^2 + y^2)} \quad -\infty < x < \infty, -\infty < y < \infty.$$

3.33. Let $f_{XY}(x, y) = A(x^2 + y^2)$ for $0 \le x \le 1$, $|y| \le 1$ and zero otherwise. Compute the PDF $F_{XY}(x, y)$ for all x, y. Determine the value of A.

3.34. Consider the input-output view mentioned in Section 3.1. Let the underlying experiment be observations on a r.v. X, which is the input to a system that generates an output $Y = g(X)$.

    (a) What is the range of Y?
    (b) What are reasonable probability spaces for X and Y?
    (c) What subset of R consists of the event $\{Y \le y\}$?
    (d) What is the inverse image under Y of the event $(-\infty, y]$ if $Y = 2X + 3$?

3.35. In the circuit shown (Figure P3.35) it is attempted to deliver the signal X from points a to b. The two links L1 and L2 operate independently, with times-to-failure T_1, T_2, respectively, which are exponentially and identically distributed. Denote the output by Y and compute $F_Y(y, t)$, meaning the PDF of Y at time t. Show that for any fixed t $F_Y(\infty, t) = 1$.

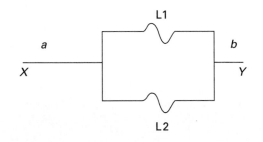

Figure P3.35

REFERENCES

3-1. W. F. Davenport, *Probability and Random Processes: An Introduction for Applied Scientists and Engineers*. New York: McGraw-Hill, 1970.
3-2. H. Stark, F. B. Tuteur, and J. B. Anderson, *Modern Electrical Communications*. 2nd edition, Upper Saddle River, N.J.: Prentice-Hall, 1988.
3-3. S. O. Rice, "Mathematical Analysis of Random Noise," *Bell System Technical Journal*, Vols. 23, 24, 1944, 1945.
3-4. J. Marsden and A. Weinstein, *Calculus*. Menlo Park, Calif.: Benjamin/Cummings, 1980.

ADDITIONAL READING

Cooper, G. R. and C. D. McGillem, *Probabilistic Methods of Signal and System Analysis*, 3rd edition. New York: Holt, Rinehart and Winston, 1999.
Deebles, P. Z. Jr., *Probability, Random Variables, and Random Signal Principles*, 4th edition. New York: McGraw-Hill, 2001.
Garcia, L.-G., *Probability and Random Processes for Electrical Engineering*, 2nd edition. Reading, MA: Addison-Wesley, 1994.
Helstrom, C. W., *Probability and Stochastic Processes for Engineers*. 2nd edition. New York: Macmillan, 1991.
Papoulis, A., *Probability, Random Variables, and Stochastic Processes*. 3rd edition. New York: McGraw-Hill, 1991.
Papoulis, A., *Probability & Statistics*. Englewood Cliffs, NJ: Prentice Hall, 1990.
Scheaffer, R. L., *Introduction to Probability and Its Applications*. Belmont, CA: Duxbury, 1990.
Viniotis, Y., *Probability and Random Processes for Electrical Engineers*. New York: McGraw-Hill, 1998.
Yates, R. D. and D. J. Goodman, *Probability and Stochastic Processes*. New York: Wiley, 1999.
Ziemer, R. E., *Elements of Engineering Probability & Statistics*. Upper Saddle River, NJ: Prentice Hall, 1997.

4

Expectation and Introduction to Estimation

4.1 EXPECTED VALUE OF A RANDOM VARIABLE

It is often desirable to summarize certain properties of a random variable (r.v.) and its probability law by a few numbers. Such numbers are furnished to us by the various averages, or *expectations* of a r.v.; the term *moments* is often used to describe a broad class of averages, and we shall use it later.

We are all familiar with the notion of the average of a set of numbers, for example, the average class grade for an exam, the average height and weight of children at age five, the average lifetime of men versus women, and the like. Basically, we compute the average of a set of numbers $x_1, x_2, \ldots, x_N$ as follows:

$$\mu_s = \frac{1}{N} \sum_{i=1}^{N} x_i, \qquad (4.1\text{-}1)$$

where the subscript s is a reminder that μ_s is the average of a set.

The average μ_s of a set of numbers $x_1, x_2, \ldots, x_N$ can be viewed as the "center of gravity" of the set. More precisely the average is the number that is simultaneously closest to all the numbers in the set in the sense that the sum of the distances from it to all the points in the set is smallest. To demonstrate this we need only ask what number z minimizes the summed distance D or summed distance-square D^2 to all the points. Thus with

$$D^2 \triangleq \sum_{i=1}^{N} (z - x_i)^2$$

the minimum occurs when $dD^2/dz = 0$ or

$$\frac{dD^2}{dz} = 2Nz - 2\sum_{i=1}^{N} x_i = 0$$

which implies that

$$z = \mu_s = \frac{1}{N}\sum_{i=1}^{N} x_i.$$

Note that each number in Equation 4.1-1 is given the same weight (i.e., each x_i is multiplied by the same factor $1/N$). If, for some reason we wish to give some numbers more weight than others when computing the average, we then obtain a *weighted* average. However, we won't pursue the idea of a weighted average any further in this chapter.

Although the average as given in Equation 4.1-1 gives us the "most likely" value or the "center of gravity" of the set, it does not tell us how much the numbers spread or deviate from the average. For example, the sets of numbers $S_1 = \{0.9, 0.98, 0.95, 1.1, 1.02, 1.05\}$ and $S_2 = \{0.2, -3, 1.8, 2, 4, 1\}$ have the same average but the spread of the numbers in S_2 is much greater than that of S_1. An average that summarizes this spread is the *standard deviation of the set*, σ_s, computed from

$$\sigma_s = \left[\frac{1}{N}\sum_{i=1}^{N}(x_i - \mu_s)^2\right]^{1/2}. \tag{4.1-2}$$

Equations 4.1-1 and 4.1-2, important as they are, fall far short of disclosing the usefulness of averages. To exploit the full range of applications of averages we must develop a calculus of averages from probability theory.

Consider a probability space $(\Omega, \mathscr{F}, P)$ associated with an experiment $\mathscr{H}$ and a discrete r.v. X. Associated with each outcome ζ_i of $\mathscr{H}$, there is a value $X(\zeta_i) \triangleq x_i$, which the r.v. X takes on. Let $x_1, x_2, \ldots, x_M$ be the M distinct values that X can take. Now assume that $\mathscr{H}$ is repeated N times and let $x^{(k)}$ be the observed outcome at the kth trial. Note that $x^{(k)}$ must assume one of the numbers $x_1, \ldots, x_M$. Suppose that in the N trials x_1 occurs n_1 tries, x_2 occurs n_2 times, and so forth. Then for N large, we can estimate the average value $\overline{X}$ of X from the formula

$$\overline{X} \simeq \frac{1}{N}\sum_{k=1}^{N} x^{(k)} \tag{4.1-3}$$

$$= \frac{1}{N}\sum_{i=1}^{N} n_i x_i = \sum_{i=1}^{M} x_i \left(\frac{n_i}{N}\right)$$

$$= \sum_{i=1}^{M} x_i P[X = x_i]. \tag{4.1-4}$$

Equation 4.1-4, which follows from the frequency definition of probability, leads us to our first definition.

Definition 4.1-1   The *expected* or *average* value of a discrete r.v. X taking on values x_i with PMF $P_X(x_i) \triangleq P[X = x_i]i = 1, 2, \ldots$, is defined by

$$E[X] \triangleq \sum_i x_i P_X(x_i) \quad \blacksquare \tag{4.1-5}$$

As given, the expectation is computed in the probability space generated by the random variable. We can also compute the expectation by summing over all points of the discrete sample space, that is, $E[X] = \sum_\Omega X(\zeta_i)P[\{\zeta_i\}]$, where the ζ_i are the discrete points in the sample description space, and the $\{\zeta_i\}$ are the elementary, discrete outcomes in Ω, the sample description space.

A definition that applies to both continuous and discrete r.v.'s is the following:

Definition 4.1-2   The expected value, if it exists,[†] of a real r.v. X with pdf $f_X(x)$ is defined by

$$E[X] = \int_{-\infty}^{\infty} x f_X(x)\, dx. \quad \blacksquare \tag{4.1-6}$$

Here, as well as in Definition 4.1-2, the expectation can be computed in the original probability space, if the sample description space is not discrete but continuous, for example, an uncountable infinite set of outcomes such as the real line. Then $E[X] = \int_\Omega X(\zeta)P[\{d\zeta\}]$, where $P[\{d\zeta\}]$ is the probability of the infinitesimal event $\{\zeta < \zeta' \le \zeta + d\zeta\}$.

The symbols $E[X]$, $\overline{X}$, μ_X, or simply μ are often used interchangeably for the expected value of X. Consider now a function of a r.v. $Y = g(X)$. The expected value of Y is, from Equation 4.1-6,

$$E[Y] = \int_{-\infty}^{\infty} y f_Y(y) dy. \tag{4.1-7}$$

However, Equation 4.1-7 requires computing $f_Y(y)$ from $f_X(x)$. If all we want is $E[Y]$, is there a way to compute it without first computing $f_Y(y)$? The answer is given by Theorem 4.1-1 which follows.

Theorem 4.1-1   The expected value of $Y = g(X)$ can be computed from

$$E[Y] = \int_{-\infty}^{\infty} g(x) f_X(x)\, dx \tag{4.1-8}$$

where $g(\cdot)$ is a measurable (Borel) function.[‡] Equation 4.1-8 is one of the most important results in the theory of probability. Unfortunately, a rigorous proof of Equation 4.1-8 requires some knowledge of Lebesgue integration. We offer instead an informal argument to argue that Equation 4.1-8 is valid; readers interested in more rigorous arguments can consult William Feller [4-1, p. 5] or Wilbur Davenport [4-2, p. 223]. $\blacksquare$

[†]The expected value will exist if the integral is absolutely convergent, that is, if $\int_{-\infty}^{\infty} |x| f(x)\, dx < \infty$.
[‡]See definition of a measurable function in Section 3.1.

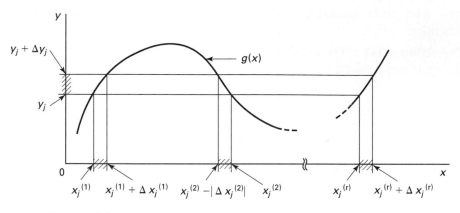

Figure 4.1-1   Equivalence between the events given in Equation 4.1-9.

On the Validity of Equation 4.1-8

Recall from Section 3.2 that if $Y = g(X)$ then for any y_j (Figure 4.1-1)

$$\{y_j < Y \le y_j + \Delta y_j\} = \bigcup_{k=1}^{r_j} \{x_j^{(k)} < X \le x_j^{(k)} + \Delta x_j^{(k)}\}, \tag{4.1-9}$$

where r_j is the number of real roots of the equation $y_j - g(x) = 0$, that is,

$$y_j = g(x_j^{(1)}) = \ldots = g(x_j^{(r_j)}). \tag{4.1-10}$$

The equal sign in Equation 4.1-9 means that the underlying event is the same for both mappings X and Y. Hence the probabilities of the events on either side of the equal sign are equal. The events on the right side of Equation 4.1-9 are disjoint and therefore the probability of the union is the sum of the probabilities of the individual events. Now partition the y-axis into many fine subintervals $y_1, y_2, \ldots, y_j, \ldots$. Then, approximating Equation 4.1-7 with a Riemann[†] sum and using Equation 2.4-6, we can write

$$E[Y] = \int_{-\infty}^{\infty} y f_Y(y)\,dy$$

$$\simeq \sum_{j=1}^{m} y_j P[y_j < Y \le y_j + \Delta y_j]$$

$$= \sum_{j=1}^{m} \sum_{k=1}^{r_j} g(x_j^{(k)}) P[x_j^{(k)} < X \le x_j^{(k)} + \Delta x_j^{(k)}]. \tag{4.1-11}$$

[†]Bernhard Riemann (1826–1866). German mathematician who made numerous contributions to the theory of integration.

The last line of Equation 4.1-11 is obtained with the help of Equations 4.1-9 and 4.1-10. But the points $x_j^{(k)}$ are distinct, so that the cumbersome double indices j and k can be replaced with a single subscript index, say, i, The Equation 4.1-11 becomes

$$E[Y] \simeq \sum_{i=1}^{n} g(x_i) P[x_i < X \le x_i + \Delta x_i],$$

and as $\Delta y, \Delta x \to 0$ we obtain the exact result that

$$E[Y] = \int_{-\infty}^{\infty} g(x) f_X(x)\, dx. \tag{4.1-12}$$

Equation 4.1-12 follows from the Riemann sum approximation and Equation 2.4-6; the x_i have been arranged in increasing order $x_1 < x_2 < x_3 < \ldots$ As Athanasios Papoulis [4-3, p. 141] points out, with the use of Lebesgue integration one can almost immediately and rigorously establish Equation 4.1-12.

In the special case where X is a discrete r.v., then

$$E[Y] = \sum_i g(x_i) P_X(x_i). \tag{4.1-13}$$

This result follows immediately from Equation 4.1-12, since the pdf of a discrete r.v. involves delta functions that have the property given in Equation B.2-1 in Appendix B. ■

Example 4.1-1 _____
Let $X: N(\mu, \sigma^2)$. The expected value of X is

$$E[X] = \int_{-\infty}^{\infty} x \left(\frac{1}{\sqrt{2\pi\sigma^2}} \exp\left(-\frac{1}{2} \left(\frac{x-\mu}{\sigma} \right)^2 \right) \right) dx.$$

Let $z \triangleq (x - \mu)/\sigma$. Then

$$E[X] = \frac{\sigma}{\sqrt{2\pi}} \int_{-\infty}^{\infty} z e^{-\frac{1}{2}z^2}\, dz + \mu \left(\frac{1}{\sqrt{2\pi}} \int_{-\infty}^{\infty} e^{-\frac{1}{2}z^2}\, dz \right).$$

The first term is zero because the integrand is odd, and the second term is μ because the term in parentheses is $P[Z \le \infty]$ for $Z: N(0,1)$. Hence

$$E[X] = \mu \qquad \text{for } X: N(\mu, \sigma^2).$$

Thus the parameter μ in $N(\mu, \sigma^2)$ is indeed the expected value of X (the mean) as claimed in Section 2.4.

Example 4.1-2 _____
Let X be a Poisson r.v. with parameter $a > 8$. Then

$$E[X] = \sum_{k=0}^{\infty} k \frac{e^{-a}}{k!} a^k$$

$$= a \sum_{k=0}^{\infty} \frac{e^{-a}}{(k-1)!} a^{k-1}$$

$$= a \sum_{k=1}^{\infty} \frac{e^{-a}}{(k-1)!} a^{k-1}$$

$$= a \sum_{i=0}^{\infty} \frac{e^{-a}}{i!} a^{i}$$

$$= a. \tag{4.1-14}$$

Thus the expected value of Poisson r.v. is the parameter a.

Example 4.1-3

Let $Y = X^2$ with $X: N(0, \sigma^2)$. Then using Equation 4.1-12 we write

$$E[Y] = \frac{1}{\sqrt{2\pi\sigma^2}} \int_{-\infty}^{\infty} x^2 \exp\left(-\frac{1}{2}\left(\frac{x}{\sigma}\right)^2\right) dx.$$

With $z \triangleq x/\sigma$ this becomes (for $\sigma > 0$)

$$E[Y] = \sigma^2 \left(\frac{1}{\sqrt{2\pi}} \int_{-\infty}^{\infty} z^2 e^{-\frac{1}{2}z^2} dz\right).$$

The term in brackets can be reduced to

$$\frac{1}{\sqrt{2\pi}} \int_{-\infty}^{\infty} e^{-\frac{1}{2}z^2} dz = 1$$

with the help of integration by parts. Thus for $Y = X^2$ and $X: N(0, \sigma^2)$

$$E[Y] = \sigma^2. \tag{4.1-15}$$

More generally, for $X: N(\mu, \sigma^2)$, $E[X^2] = \mu^2 + \sigma^2$ and $E[(X - \mu)^2] = \sigma^2$. We leave the proof of these important results as exercises for the reader. The parameter σ^2 is called the variance, and μ is the mean. In practice these are often estimated by making many observations on X and using Equation 4.1-1 to estimate the mean and Equation 4.1-2 to estimate σ.

Example 4.1-4

The Cauchy pdf with parameters $\alpha(-\infty < \alpha < \infty)$ and $\beta(\beta > 0)$ is given by

$$f_X(x) = \frac{1}{\pi\beta\left(1 + \left(\dfrac{x-\alpha}{\beta}\right)^2\right)} \qquad -\infty < x < \infty. \tag{4.1-16}$$

Let X be Cauchy with $\beta = 1$, $\alpha = 0$. Then

$$E[X] = \int_{-\infty}^{\infty} x \left(\frac{1}{\pi(x^2 + 1)} \right) dx$$

is an improper integral and doesn't converge in the ordinary sense. However, if we evaluate the integral in the Cauchy principal value sense, that is,

$$E[X] = \lim_{x_0 \to \infty} \left[\int_{-x_0}^{x_0} x \left(\frac{1}{\pi(x^2 + 1)} \right) dx \right], \tag{4.1-17}$$

then $E[X] = 0$. Note, however, that with $Y \triangleq X^2$, $E[Y]$ doesn't exist in any sense because

$$E[Y] = \int_{-\infty}^{\infty} x^2 \left[\frac{1}{\pi(x^2 + 1)} \right] dx = \infty \tag{4.1-18}$$

and thus fails to converge in any sense. Thus the finite variance of a Cauchy r.v. does not exist.

From the linearity of the expectation operator[†] we can easily obtain the important result that for any X

$$E\left[\sum_{i=1}^{N} g_i(X) \right] = \sum_{i=1}^{N} E[g_i(X)] \tag{4.1-19}$$

provided that these exist. The demonstration of Equation 4.1-19 is left as an exercise.

For a function of two random variables, that is, $Z = g(X, Y)$, the expected value of Z can be computed from

$$E[Z] = \int_{-\infty}^{\infty} z f_Z(z) dz$$
$$= \int_{-\infty}^{\infty} \int_{-\infty}^{\infty} g(x, y) f_{XY}(x, y) \, dx \, dy. \tag{4.1-20}$$

To prove that Equation 4.1-20 can be used to compute $E[Z]$ requires an argument similar to the one we used in establishing Equation 4.1-8. Indeed one would start with an equation very similar to Equation 4.1-9, e.g.,

$$\{z_j < Z \le z_j + \Delta z\} = \bigcup_{k=1}^{N_j} \{(X, Y) \in D_k\},$$

where the D_k are very small disjoint regions containing the points $(x_k^{(j)}, y_k^{(j)})$ such that $g(x_k^{(j)}, y_k^{(j)}) = z_j$. Taking probabilities of both sides and recalling that the D_k are disjoint,

---

[†]An operator $\mathscr{A}$ is said to be linear if for any two functions g_1 and g_2 in its domain, $\mathscr{A}[a_1 g_1 + a_2 g_2] = a_1 \mathscr{A} g_1 + a_2 \mathscr{A} g_2$. where a_1 and a_2 are arbitrary coefficients.

yields

$$f_Z(z_j)\Delta z_j \simeq \sum_{k=1}^{N_j} f(x_k^{(j)}, y_k^{(j)})\Delta a_k^{(j)},$$

where $\Delta a_k^{(j)}$ is an infinitesimal area.

Now multiply both sides by z_j and recall that $z_j = g(x_k^{(j)}, y_k^{(j)})$. Then

$$z_j f_Z(z_j)\Delta z_j \simeq \sum_{k=1}^{N_i} g(x_k^{(j)}, y_k^{(j)}) f_{XY}(x_k^{(j)}, y_k^{(j)})\Delta a_k^{(j)}$$

and, as $j \to \infty$, $\Delta z_j \to 0$, $\Delta a_k^{(j)} \to da = dx\,dy$,

$$\int_{-\infty}^{\infty} z f_Z(z)dz = \int_{-\infty}^{\infty}\int_{-\infty}^{\infty} g(x,y)f_{XY}(x,y)\,dx\,dy.$$

Carl Helstrom[†] furnishes an alternative proof that is of interest. As before let $Z = g(X,Y)$ and write

$$E[Z] = \int_{-\infty}^{\infty} z f_Z(z)dz$$

$$= \int_{-\infty}^{\infty}\int_{-\infty}^{\infty} z f_{Z|Y}(z|y)f_Y(y)dy\,dz.$$

The second line follows from the definition of a marginal pdf. Now recall that if $Z = g(X)$ then

$$\int_{-\infty}^{\infty} z f_Z(z)dz = \int_{-\infty}^{\infty} g(x)f_X(x)\,dx.$$

We can use this result in the present problem as follows. If we hold Y, fixed at $Y = y$, then $g(X,y)$ depends only on X, and the conditional expectations of z with $Y = y$ is

$$\int_{-\infty}^{\infty} z f_{Z|Y}(z|y)dz = \int_{-\infty}^{\infty} g(x,y)f_{X|Y}(x|y)\,dx.$$

Using this result in the above yields

$$E[Z] = \int_{-\infty}^{\infty} z f_Z(z)dz$$

$$= \int_{-\infty}^{\infty}\left\{\int_{-\infty}^{\infty} z f_{Z|Y}(z|y)dz\right\} f_Y(y)dy$$

$$= \int_{-\infty}^{\infty}\int_{-\infty}^{\infty} g(x,y)f_{X|Y}(x,y)f_Y(y)\,dx\,dy$$

$$= \int_{-\infty}^{\infty}\int_{-\infty}^{\infty} g(x,y)f_{XY}(x,y)\,dx\,dy.$$

[†]Carl W. Helstrom, *Probability and Stochastic Processes for Engineers*, 2nd edition, (New York, Macmillan, 1991).

Example 4.1-5

Let $g(x, y) = xy$. Compute $E[Z]$ if $Z = g(X, Y)$ and

$$f_{XY}(x, y) = \frac{1}{2\pi\sigma^2} \exp\left[-\frac{1}{2\sigma^2}\left((x-a)^2 + (y-b)^2\right)\right].$$

Solution Direct substitution into Equation 4.1-20 and recognizing that the resulting double integral factors into the product of two single integrals enables us to write

$$E[Z] = \frac{1}{\sqrt{2\pi}\sigma} \int_{-\infty}^{\infty} x \exp\left[-\frac{1}{2\sigma^2}(x-a)^2\right] dx$$

$$\times \frac{1}{\sqrt{2\pi}\sigma} \int_{-\infty}^{\infty} y \exp\left[-\frac{1}{2\sigma^2}(y-b)^2\right] dy$$

$$= ab.$$

Equation 4.1-20 can be used to compute $E[X]$ or $E[Y]$. Thus with $Z = g(X, Y) = X$, we obtain

$$E[X] = \int_{-\infty}^{\infty} \int_{-\infty}^{\infty} x f_{XY}(x, y)\, dx\, dy$$

$$= \int_{-\infty}^{\infty} \left[\int_{-\infty}^{\infty} f_{XY}(x, y)dy\right] x\, dx. \tag{4.1-21}$$

By Equation 2.6-34, the integral in brackets is the marginal pdf $f_X(x)$. Hence Equation 4.1-20 is completely consistent with the definition

$$E[X] \triangleq \int_{-\infty}^{\infty} x f_X(x)\, dx.$$

With the help of marginal densities we can conclude that

$$E[X + Y] = \int_{-\infty}^{\infty} \int_{-\infty}^{\infty} (x + y) f_{XY}(x, y)\, dx\, dy$$

$$= \int_{-\infty}^{\infty} x \left(\int_{-\infty}^{\infty} f_{XY}(x, y)dy\right) dx + \int_{-\infty}^{\infty} y \left(\int_{-\infty}^{\infty} f_{XY}(x, y)\, dx\right) dy$$

$$= E[X] + E[Y]. \tag{4.1-22}$$

Equation 4.1-21 can be extended to N random variables $X_1, X_2, \ldots, X_N$. Thus

$$E\left[\sum_{i=1}^{N} X_i\right] = \sum_{i=1}^{N} E[X_i]. \tag{4.1-23}$$

Note that *independence is not required*.

Example 4.1-6 _____

Let X, Y be jointly normal, independent r.v.'s with pdf

$$f_{XY}(x, y) = \frac{1}{2\pi\sigma_1\sigma_2} \exp\left\{-\frac{1}{2}\left[\left(\frac{x-\mu_1}{\sigma_1}\right)^2 + \left(\frac{y-\mu_2}{\sigma_2}\right)^2\right]\right\}.$$

It is clear that X and Y are independent since $f_{XY}(x,y) = f_X(x)f_Y(y)$. The marginal pdf's are obtained using Equations 2.6-33 and 2.6-34:

$$f_X(x) = \frac{1}{\sqrt{2\pi\sigma_1^2}} \exp\left[-\frac{1}{2}\left(\frac{x-\mu_1}{\sigma_1}\right)^2\right]$$

$$f_Y(y) = \frac{1}{\sqrt{2\pi\sigma_2^2}} \exp\left[-\frac{1}{2}\left(\frac{y-\mu_2}{\sigma_2}\right)^2\right].$$

Thus Equation 4.1-22 yields

$$E[X + Y] = \mu_1 + \mu_2.$$

Example 4.1-7 _____

(Chi-square law.) In a number of problems in engineering and science, signals add incoherently, meaning that the power in the sum of the signals is merely the sum of the powers. This occurs, for example, in optics when a surface is illuminated by light sources of different wavelengths. Then the power measured on the surface is just the sum of the powers contributed by each of the sources. In electric circuits, when the sources are sinusoidal at different frequencies, the power dissipated in any resistor is the sum of the powers contributed by each of the sources. Suppose the individual source signals, at a given instant of time, are modeled as normal r.v.'s. In particular let $X_1, X_2, \ldots, X_n$ be i.i.d with $X_i: N(0,1)$ for $i = 1, 2, \ldots, n$ and let $Y_i = X_i^2$. We know from Example 3.2-2 in Chapter 3 that the pdf of Y_i is given by

$$f_{Y_i}(y) = \frac{1}{\sqrt{2\pi y}}e^{-y/2}u(y).$$

Consider now the sums $Z_2 = Y_1 + Y_2$, $Z_3 = Y_1 + Y_2 + Y_3, \ldots, Z_n = \Sigma_{i=1}^{n}Y_i$. The pdf of Z_2 is easily computed by convolution as

$$f_{Z_2}(z) = \int_{-\infty}^{\infty} \frac{1}{\sqrt{2\pi x}}e^{-x/2}u(x) \times \frac{1}{\sqrt{2\pi(z-x)}}e^{-\frac{1}{2}(z-x)}u(z-x)\,dx$$

$$= \frac{1}{\pi}e^{-z/2}\int_0^{\sqrt{z}}\frac{dy}{\sqrt{z-y^2}}u(z)$$

$$= \frac{1}{2}e^{-z/2}u(z) \text{ (exponential pdf)}.$$

To get from line 1 to line 2 we let $x = y^2$. To get from line 2 to line 3, we used that the integral is an elementary trigonometric function integral in disguise. To get the pdf of Z_3 we convolve the pdf of Z_2 with that of Y_3. The result is

$$f_{Z_3} = \frac{1}{2} \int_{-\infty}^{\infty} e^{-x/2} u(x) \times \frac{1}{\sqrt{2\pi(z-x)}} e^{-\frac{1}{2}(z-x)} u(z-x)\, dx$$

$$= \frac{1}{\sqrt{2\pi}} z^{\frac{1}{2}} e^{-\frac{1}{2}z} u(z).$$

We leave the intermediate steps which involve only elementary transformations to the reader. Proceeding in this way, or using mathematical induction, we find that

$$f_{Z_n}(z) = \frac{1}{2^{n/2}\,\Gamma(n/2)} z^{\frac{n-2}{2}} e^{-z/2} u(z).$$

This pdf was introduced in Chapter 2 as the Chi-square pdf. More precisely, it is known as the *Chi-square distribution with n degrees-of-freedom*. For $n > 2$, the pdf has value zero at $z = 0$, reaches a peak, and then exhibits monotonically decreasing tails. For large values of n, it resembles a Gaussian pdf with mean in the vicinity of n. However the Chi-square can never be truly Gaussian because the Chi-square random variable never takes on negative values. The character of the Chi-square pdf if shown in Figure 4.1-2 for different values of large n.

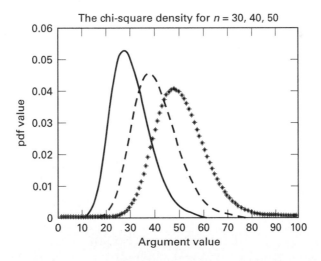

Figure 4.1-2   The Chi-square pdf for three large values for the parameter n: $n = 30$ (solid); $n = 40$ (dashed); $n = 50$ (stars). For large values of n, the Chi-square pdf can be approximated by a normal $N(n, 2n)$ for computing probabilities not too far from the mean. For example, for $n = 30$, $P[\mu - \sigma < X < \mu + \sigma] = 0.6827$ assuming $X : N(30, 60)$. The value computed, using single-precision arithmetic, using the Chi-square pdf, yields 0.6892.

The mean and variance of the Chi-square r.v are readily computed from the definition $Z_n \triangleq \Sigma_{i=1}^n X_i^2$. Thus $E\{Z_n\} = E\{\Sigma_{i=1}^n X_i^2\} = \Sigma_{i=1}^n E\{X_i^2\} = n$. Also $Var\{Z_n\} = E\{(Z_n - n)^2\}$. After simplifying, we obtain $Var(Z_n) = E\{Z_n^2\} - n^2$. We leave it to the reader to show that $E\{Z_n^2\} = 2n + n^2$ and, hence, that $Var\{Z_n\} = 2n$.

Example 4.1-8

At the famous University of Politicalcorrectness (U of P), the administration requires that each professor be equipped with an electronic Rolodex which contains the names of every student in the class. When the professor wishes to call on a student, she merely hits the "call" button on the Rolodex, and a student's name is selected randomly by an electronic circuit inside the Rolodex. By using this device the professor becomes immune to charges of bias in the selection of students she calls on to answer her questions. Find an expression for the *average number* of "calls" required so each student is called upon at least once.

Solution   The use of the electronic Rolodex implies that some students may not be called at all during the entire semester and other students may be called twice or three times in a row. It will depend on how big the class is. Nevertheless the average is well defined because extremely long bad runs, that is, where one or more students are not called on, are very rare. The careful reader may have observed that this is an occupancy problem if we associate "calls" with *balls* and students with *cells*. Let $E(r, n)$ denote the event that all cells have been *filled for the first time* after the rth ball has been placed among the n cells. The only way that $E(r, n)$ can happen is that exactly one cell remains unfilled by the distribution of the first $r - 1$ balls, and that the rth ball is assigned to that cell. Translated to the class situation, this means that after $r - 1$ "calls," all but one student will have been called and that this student is called on the rth "call". Let C be the event that a given cell receives a ball on any one try, for example, the rth ball going into the unoccupied cell. Then $E(r, n) = C \cap D_1(r - 1, n)$, where $D_1(r - 1, n)$ is the event that exactly one out of n cells remains empty after $r - 1$ balls have been distributed. Since these two events are independent, we find that

$$P[E(r, n)] = P_1[r - 1, n] \times P[C],$$

where $P_1(r - 1, n)$ is given by Equation 1.8-10 with $m = 1$, and $P[C] = 1/n$. The final result is

$$P[E(r, n)] = \sum_{i=0}^n \binom{n}{i} (-1)^i \left(1 - \frac{i + 1}{n}\right)^r \qquad (4.1\text{-}24)$$

and $E\{r\}$ is given by

$$E\{r\} = \sum_{r=1}^\infty r \times P[E(r, n)]. \qquad (4.1\text{-}25)$$

Example 4.1-9

Write a MATLAB program for computing the probability that all the students in Example 4.1-8 are called upon at least once in r calls from the electronic rolodex. Assume there are 20 students in the class.

Solution   The appropriate equation to be coded is Equation 4.1-24. The result is shown in Figure 4.1-3.

```
function [tries,prob]=occupancy(balls,cells)
tries=1:balls; % identifies a vector ''tries''
prob=zeros(1,balls); % identifies a vector ''prob''
c=zeros(1,cells); % identifies a vector ''c''
d=zeros(1,cells); % identifies a vector ''d''

term=zeros(1,cells); % identifies a vector ''term''
% next follows the realization of Equation (4.1-24)
for m=1:balls

    for k=1:cels
        c(k)=(-1)^k*prod(1:cells)/(prod(1:k)*prod(1:cells-k));
        d(k)=(1-(k/cells))^m;
        term(k)=c(k)*d(k);
    end
prob(m)=1+sum(term);
end
  plot(tries,prob)
  title(['Probability of all ' num2str(cells) ' students in the class
being called in r tries'])
xlabel('number of tries')
ylabel(['Probability of all ' num2str(cells) ' students being called
'])
```

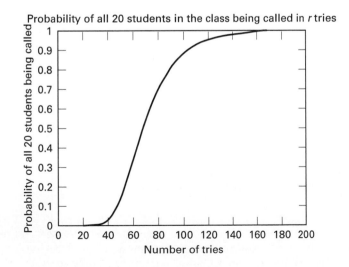

Figure 4.1-3   MATLAB result for Example 4.1-9.

Example 4.1-10

Write a MATLAB program for computing the average number of calls required for each student to be called at least once. Assume a maximum of 50 students and make sure the number of calls is large ($n \geq 400$).

Solution   The appropriate equation to be coded is Equation 4.1-25. The result is shown in Figure 4.1-4.

```
function [cellvec,avevec]=avertries(ballimit,cellimit)
cellvec = 1:cellimit;
termvec = zeros(1,ballimit);
avevec = zeros(1,cellimit);
brterm=zeros(1,ballimit);
srterm=zeros(1,ballimit);
for n=1:cellimit;
   c = zeros(1,n);
   d = zeros(1,n);
   termvec = zeros(1,n);
   for r=1:ballimit
      for l=1:n-1
         c(l) = ((-1)^l)*prod(1:n-1)/(prod(1:l)*prod(1:n-1-1));
         d(l) = (1-((l+1)/n))^(r-1);
         termvec(l) = c(l)*d(l);
      end
      brterm(r)=r*sum(termvec);
      lrterm(r)=r*((1-(1/n)))^(r-1);
```

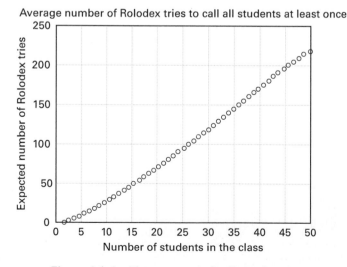

Figure 4.1-4   MATLAB result for Example 4.1-10.

```
    end
    avevec(n)=sum(brterm)+sum(lrterm);
end
plot(cellvec,avec,'o')
title('Average number of Rolodex tries to call all students at least
once')
xlabel('number of students in the class')
ylabel('Expected number of Rolodex tries to reach all students at least
once')
grid
```

Example 4.1-11 _____

(Geometric distribution.) The random variable X is said to have a *geometric distribution* if its probability mass function is given by

$$P_X(n) = (1-a)a^n u(n),$$

where $n = 0, 1, \ldots$ and $0 < a < 1$. Clearly $\Sigma_{n=0}^{\infty} P_X(n) = 1$, a result easily obtained from $\Sigma_{n=0}^{\infty} a^n = (1-a)^{-1}$ for $0 < a < 1$. The expected value is found from

$$E[X] = \mu = (1-a)\sum_{n=0}^{\infty} na^n = (1-a) \times a \times \frac{d}{da}\{(1-a)^{-1}\} = \frac{a}{1-a}.$$

Solving for a, we obtain

$$a = \frac{\mu}{1+\mu}.$$

Thus we can rewrite the geometric PMF as

$$P_X(n) = \frac{1}{1+\mu}\left(\frac{\mu}{1+\mu}\right)^n u(n).$$

_____

4.2 CONDITIONAL EXPECTATIONS

In many practical situations we want to know the average of a subset of the population: the average of the *passing grades* of an exam; the average lifespan of people who are still alive at age 70; the average height of fighter pilots (many air forces have both an upper and lower limit on the acceptable height of a pilot); the average blood pressure of long-distance runners, and so forth. Problems of this type fall within the realm of *conditional expectations*.

In conditional expectations we compute the average of a subset of a population that shares some property due to the outcome of an event. For example in the case of the average of passing grades, the subset is those exams that received passing grades. What all these exams share is that their grade is, say, ≥ 65. The event that has occurred is that they received passing grades.

Definition 4.2-1   The conditional expectation of X given that the event B has occurred is

$$E[X|B] \triangleq \int_{-\infty}^{\infty} x f_{X|B}(x|B)\, dx. \tag{4.2-1}$$

If X is a discrete, then Equation 4.2-1 can be replaced with

$$E[X|B] \triangleq \sum_{i} x_i P_{X|B}(x_i|B). \quad \blacksquare \tag{4.2-2}$$

To give the reader a feel for the notion of conditional expectation, consider the following exam scores in a course on Probability Theory: 28, 35, 44, 66, 68, 75, 77, 80, 85, 87, 90, 100, 100. Assume that the the passing grade is 65. Then the average score is 71.9; however the *average passing score* is 82.8. A closely related example is worked out as follows.

Example 4.2-1

Consider a continuous r.v. X and the event $B \triangleq \{X \geq a\}$. From Equations 2.6-1 and 2.6-2 and a little bit of work we obtain

$$F_{X|B}(x|X \geq a) = \begin{cases} 0, & x < a \\ \dfrac{F_X(x) - F_X(a)}{1 - F_X(a)}, & x \geq a. \end{cases} \tag{4.2-3}$$

Hence

$$f_{X|B}(x|X \geq a) = \begin{cases} 0, & x < a \\ \dfrac{f_X(x)}{1 - F_X(a)}, & x \geq a. \end{cases} \tag{4.2-4}$$

and

$$E[X|X \geq a] = \frac{\displaystyle\int_a^{\infty} x f_X(x)\, dx}{\displaystyle\int_a^{\infty} f_X(x)\, dx}. \tag{4.2-5}$$

Assume that X is a uniform r.v. in $[0, 100]$. Then

$$E[X] = \frac{1}{100} \int_0^{100} x\, dx = 50,$$

but using Equation 4.2-5 with $a = 65$

$$E[X|X \geq 65] = 82.5.$$

Conditional expectations often occur when dealing with random variables that are related in some way. For example let Y denote the lifetime of a person chosen at random, and let X be a binary r.v. that denotes whether the person smokes or not, that is, $X = 0$ if a nonsmoker, $X = 1$ if a smoker. Then clearly $E[Y|X = 0]$ is expected to be larger[†]

[†]Statistical evidence indicates that each cigarette smoked reduces longevity by about eight minutes. Hence smoking one pack a day for a whole year reduces the expected longevity of the smoker by 40 days!

than $E[Y|X = 1]$. Or let X be the intensity of the incident illumination and let Y be the instantaneous photocurrent generated by a photodetector. Clearly the expected value of Y will be larger for stronger illumination and smaller for weaker illumination. We define some important concepts as follows.

Definition 4.2-2 Let X and Y be discrete r.v.'s with joint PMF $P_{X,Y}(x_i, y_j)$. Then the conditional expectation of Y given $X = x_i$ denoted by $E[Y|X = x_i]$ is

$$E[Y|X = x_i] \triangleq \sum_j y_j P_{Y|X}(y_j|x_i) \qquad (4.2\text{-}6)$$

Here $P_{Y|X}(y_j|x_i)$ is the conditional probability that $\{Y = y_j\}$ occurs given that $\{X = x_i\}$ has occurred and is of course given by $P_{X,Y}(x_i, y_j)/P_X(x_i)$. ■

We can derive an interesting and useful formula for $E[Y]$ in terms of the conditional expectation of Y given $X = x$. The reasoning is much the same as that which we used in computing the average or total probability of an event in terms of its conditional probabilities (see Equations 1.6-4 or 2.6-4). Thus

$$E[Y] = \sum_j y_j P_Y(y_j) \qquad (4.2\text{-}7)$$

$$= \sum_j y_j \sum_i P_{X,Y}(x_i, y_j)$$

$$= \sum_i \left[\sum_j y_j P_{Y|X}(y_j|x_i) \right] P_X(x_i)$$

$$= \sum_i E[Y|X = x_i] P_X(x_i). \qquad (4.2\text{-}8)$$

Equation 4.2-8 is a very neat result and says that we can compute $E[Y]$ by averaging the conditional expectation of Y given X with respect to X.[†] Thus in the smoking-longevity example discussed earlier, suppose $E[Y|X = 0] = 79.2$ years and $E[Y|X = 1] = 69.4$ years and $P_X(0) = 0.75$ and $P_X(1) = 0.25$. Then

$$E[Y] = 79.2 \times 0.75 + 69.4 \times 0.25 = 76.75$$

is the expected lifetime of the general population.

[†]Notice that this statement implies that the conditional expectation of Y given X is a r.v. We shall elaborate on this important concept shortly. For the moment we assume that X assumes the fixed value x_i (or x).

A result similar to Equation 4.2-8 holds for the continuous case as well. It is derived using Equation 2.6-49 from Chapter 2, that is,

$$f_{Y|X}(y|x) = \frac{f_{XY}(x,y)}{f_X(x)} \qquad f_X(x) \neq 0. \tag{4.2-9}$$

The definition of conditional expectation for a continuous r.v. follows.

Definition 4.2-3   Let X and Y be continuous r.v.'s with joint pdf $f_{XY}(x,y)$. Let the conditional pdf of Y given that $X = x$ be denoted as in Equation 4.2-9. Then the conditional expectation of Y given that $X = x$ is given by

$$E[Y|X = x] \triangleq \int_{-\infty}^{\infty} y f_{Y|X}(y|x) dy. \quad \blacksquare \tag{4.2-10}$$

Since

$$E[Y] = \int_{-\infty}^{\infty} \int_{-\infty}^{\infty} y f_{XY}(x,y)\, dx\, dy \tag{4.2-11}$$

it follows from Equations 4.2-9 and 4.2-10 that

$$E[Y] = \int_{-\infty}^{\infty} f_X(x) \left[\int_{-\infty}^{\infty} y f_{Y|X}(y|x) dy \right] dx$$

$$= \int_{-\infty}^{\infty} E[Y|X = x] f_X(x)\, dx. \tag{4.2-12}$$

Equation 4.2-12 is the continuous r.v. equivalent of Equation 4.2-8. It can be used to good advantage (over the direct method) for computing $E[Y]$. We illustrate this point with an example from optical communications.

Example 4.2-2 _____

In the photoelectric detector shown in Figure 4.2-1, the number of photoelectrons Y produced in time τ depends on the (normalized) incident energy X. If X were constant, say $X = x$, Y would be a Poisson r.v. [4-4] with parameter x, but as real light sources—except for gain-stabilized lasers—do not emit constant energy signals, X must be treated as a r.v. In certain situations the pdf of X is accurately modeled by

$$f_X(x) = \begin{cases} \dfrac{1}{\mu_X} \exp\left(-\dfrac{x}{\mu_X} \right), & x \geq 0 \\ 0, & x < 0, \end{cases} \tag{4.2-13}$$

where μ_X is a parameter that equals $E[X]$. We shall now compute $E[Y]$ by Equation 4.2-12 and by the direct method.

Solution   Since for $X = x$, Y is Poisson, we can write

$$P[Y = k|X = x] = \frac{x^k}{k!} e^{-x} \qquad k = 0, 1, 2, \ldots$$

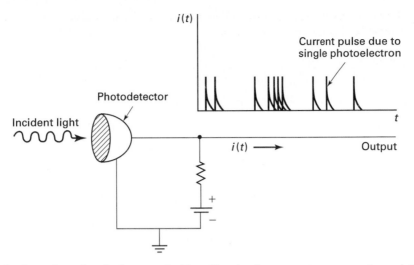

Figure 4.2-1 In a photoelectric detector, incident illumination generates a current consisting of photo-generated electrons.

and, from Example 4.1-2,

$$E[Y|X = x] = x.$$

Finally, using Equation 4.2-12 with the appropriate substitutions, that is,

$$[E[Y] = \int_0^\infty x \left[\frac{1}{\mu_X} \exp\left(-\frac{x}{\mu_X} \right) \right] dx$$

we easily obtain, by integration by parts,

$$E[Y] = \mu_X.$$

In contrast to the simplicity with which we obtained this result, consider the direct approach, that is,

$$E[Y] = \sum_{k=0}^\infty k P_Y(k). \tag{4.2-14}$$

To compute $P_Y(k)$ we use the Poisson transform (Equation 2.6-9) with $f_X(x)$, as given by Equation 4.2-13. This furnishes (see Equation 2.6-17)

$$P_Y(k) = \frac{\mu_X^k}{(1 + \mu_X)^{k+1}}. \tag{4.2-15}$$

Finally, using Equation 4.2-15 in 4.2-14 yields

$$E[Y] = \sum_{k=0}^\infty k \frac{\mu_X^k}{(1 + \mu_X)^{k+1}}.$$

It is known that this series sums to μ_X. Alternatively one can evaluate the sum indirectly using some clever tricks involving derivatives.

Example 4.2-3 _____

Let X and Y be two zero-mean r.v.'s with joint density

$$f_{XY}(x,y) = \frac{1}{2\pi\sigma^2\sqrt{1-\rho^2}}\exp\left(-\frac{x^2+y^2-2\rho xy}{\sigma^2(1-\rho^2)}\right) \quad |\rho| < 1. \tag{4.2-16}$$

We shall soon find out (Section 4.3) that the pdf in Equation 4.2-16 is a special case of the general joint Gaussian law for two r.v.'s. First we see that when $\rho \neq 0$, $f_{XY}(x,y) \neq f_X(x)f_Y(y)$; hence X and Y are not independent when $\rho \neq 0$. When $\rho = 0$, we can indeed write $f_{XY}(x,y) = f_X(x)f_Y(y)$ so that $\rho = 0$ implies independence. For the present, however, our unfamiliarity with the meaning of ρ (ρ is called the normalized *covariance* or *correlation coefficient*) is not important. From Equations 2.6-34 and 2.6-35 it is easy to show that X and Y are zero-mean Gaussian r.v.'s, that is,

$$f_X(x) = f_Y(x) = \frac{1}{\sqrt{2\pi\sigma^2}}e^{-x^2/2\sigma^2}.$$

However, the conditional expectation of Y given $X = x$ is not zero even though Y is a zero-mean r.v.! In fact from Equation 4.2-9:

$$f_{Y|X}(y|x) = \frac{1}{\sqrt{2\pi\sigma^2(1-\rho^2)}}\exp\left(-\frac{(y-\rho x)^2}{2\sigma^2(1-\rho^2)}\right). \tag{4.2-17}$$

Hence $f_{Y|X}(y|x)$ is Gaussian with mean ρx. Thus

$$E[Y|X=x] = \int_{-\infty}^{\infty} y f_{Y|X}(y|x)\,dx$$

$$= \rho x. \tag{4.2-18}$$

When ρ is close to unity, $E[Y|X=x] \simeq x$, which implies that Y tracks X quite closely (exactly if $\rho = 1$), and if we wish to predict Y, say, with Y_P upon observing $X = x$, a good bet is to choose our *predicted value* $Y_P = x$. On the other hand, when $\rho = 0$, observing X doesn't help us to predict Y. Thus we see that in the Gaussian case at least and somewhat more generally, ρ is related to the predictability of one r.v. from observing another. A cautionary note should be sounded, however: The fact that one r.v. doesn't help us to linearly predict another doesn't generally mean that the two r.v.'s are independent.

Example 4.2-4 _____

(Expectations conditioned on sums of random variables.) Consider the two independent r.v.'s X_1 and X_2. We wish to compute $E[X_1|X_1+X_2 = y]$. It is first necessary to determine the conditional probability $P[X_1 = x_1|X_1 + X_2 = y]$. This conditional probability can be written as

$$P[X_1 = x_1 | X_1 + X_2 = y] = \frac{P[X_1 = x_1, X_1 + X_2 = y]}{P[X_1 + X_2 = y]}$$

$$= \frac{P[X_1 = x_1, X_2 = y - x_1]}{P[X_1 + X_2 = y]} \qquad (4.2\text{-}19)$$

$$= \frac{P[X_1 = x_1] P[X_2 = y - x_1]}{P[X_1 + X_2 = y]}.$$

Let X_1 and X_2 be i.i.d Poisson with parameter θ. Then, from Equation 4.2-19 and Example 3.3-6 we get

$$P[X_1 = x_1 | X_1 + X_2 = y]$$

$$= \frac{(e^{-\theta_1} \theta_1^{x_1}/x_1!) \times (e^{-\theta_2} \theta_2^{y-x_1}/(y - x_1)!)}{e^{-(\theta_1+\theta_2)}(\theta_1 + \theta_2)^y/y!} \qquad (4.2\text{-}20)$$

$$= \binom{y}{x_1} \theta_1^{x_1} \theta_2^{y-x_1} \times (\theta_1 + \theta_2)^{-y}.$$

Now recall that $E[X_1 | X_1 + X_2 = y] \triangleq \sum_{x_1=0}^{y} x_1 P[X_1 = x_1 | X_1 + X_2 = y]$ and the binomial expansion formula is given by $\sum_{k=0}^{n} \binom{n}{k} \theta_1^k \theta_2^{n-k} = (\theta_1 + \theta_2)^n$. Then using Equation 4.2-20 finally yields

$$E[X_1 | X_1 + X_2 = y] = y \times \left(\frac{\theta_1}{\theta_1 + \theta_2} \right). \qquad (4.2\text{-}21)$$

Example 4.2-5 _____
(Continuation of Example 4.2-4.) Let X_1, X_2, X_3 denote multinomial random variables. Then

$$P_X[X_1 = x_1, X_2 = x_2, X_3 = x_3] = \frac{n!}{x_1! x_2! x_3!} p_1^{x_1} p_2^{x_2} p_3^{x_3}, \qquad (4.2\text{-}22)$$

where $n = x_1 + x_2 + x_3$ and $p_1 + p_2 + p_3 = 1$. We wish to compute $E[X_1 | X_1 + X_2 = y]$.

Solution As in the previous example, we need to compute $P[X_1 = x_1 | X_1 + X_2 = y]$. We write

$$P[X_1 = x_1 | X_1 + X_2 = y] = \frac{P[X_1 = x_1, X_1 + X_2 = y]}{P[X_1 + X_2 = y]}.$$

Note that for the multinomial, the event $\{\zeta : X_1(\zeta) + X_2(\zeta) = y\} \cap \{\zeta : X_1(\zeta) = x_1\}$ is identical to the event $\{\zeta : X_1(\zeta) = x_1, X_2(\zeta) = y - x_1, X_3(\zeta) = n - y\}$. Hence

$$P[X_1 = x_1 | X_1 + X_2 = y] = \frac{P[X_1 = x_1, X_2 = y - x_1, X_3 = n - y]}{P[X_3 = n - y]} \qquad (4.2\text{-}23)$$

$$= \frac{n!}{x_1!(y - x_1)!(n - y)!} p_1^{x_1} p_2^{y-x_1} p_3^{n-y}.$$

$$\div \frac{n!}{(n-y)!y!} p_3^{n-y}(1-p_3)^y$$

$$= \binom{y}{x_1} p_1^{x_1} p_2^{y-x_1} (p_1+p_2)^{-y}. \tag{4.2-24}$$

Finally, using

$$E[X_1|X_1+X_2=y] = \sum_{x_1} x_1 P[X_1=x_1|X_1+X_2=y]$$

we obtain that

$$E[X_1|X_1+X_2=y] = y\frac{p_1}{p_1+p_2}. \tag{4.2-25}$$

We leave it to the reader to compute that

$$E[X_2|X_1+X_2=y] = y\frac{p_2}{p_1+p_2}. \tag{4.2-26}$$

These kinds of problems occur in the estimation procedure known as the expectation-maximization algorithm, discussed in detail in Chapter 9.

Conditional Expectation as a Random Variable

Consider, for the sake of being specific, a function $Y = g(X)$ of a discrete r.v. X. Then its expected value is

$$E[Y] = \sum_i g(x_i) P_X(x_i)$$

$$= E[g(X)].$$

This suggests that we could write Equation 4.2-8 in similar notation, that is,

$$E[Y] = \sum_i E[Y|X=x_i] P_X(x_i)$$

$$= E[E[Y|X]]. \tag{4.2-27}$$

It is important to note that the object $E[Y|X=x_i]$ is a number, as is $g(x_i)$, but the object $E[Y|X]$ is a function of the r.v. X *and therefore is itself a r.v.* Given a probability space $\mathscr{P} = (\Omega, \mathscr{F}, P)$ and a r.v. X defined on $\mathscr{P}$, for each outcome $\zeta \in \Omega$ we generate the real number $E[Y|X(\zeta)]$. Thus $E[Y|X]$ is a r.v. that assumes the value $E[Y|X(\zeta)]$ when ζ is the outcome of the underlying experiment. As always, the functional dependence of X on ζ is suppressed, and we specify X rather than the underlying probability space $\mathscr{P}$. The following example illustrates the use of the conditional expectation as a r.v.

Example 4.2-6

Consider a communication system in which the message delay (in milliseconds) is Y and the channel choice is X. Let $X = 1$ for a satellite channel, $X = 2$ for a coaxial cable channel,

$X = 3$ for a microwave surface link, and $X = 4$ for a fiber-optical link. A channel is chosen based on availability, which is a random phenomenon. Suppose $P_X(k) = 1/4 \, k = 1, \ldots, 4$. Assume that it is known that $E[Y|X = 1] = 500$, $E[Y|X = 2] = 300$, $E[Y|X = 3] = 200$, and $E[Y|X = 4] = 100$. Then the r.v. $g(X) \triangleq E[Y|X]$ is defined by

$$g(X) = \begin{cases} 500, & \text{for } X = 1 \quad P_X(1) = \frac{1}{4} \\ 300, & \text{for } X = 2 \quad P_X(2) = \frac{1}{4} \\ 200, & \text{for } X = 3 \quad P_X(3) = \frac{1}{4} \\ 100, & \text{for } X = 4 \quad P_X(4) = \frac{1}{4} \end{cases}$$

and $E[Y] = E[g(X)] = 500 \times \frac{1}{4} + 300 \times \frac{1}{4} + 200 \times \frac{1}{4} + 100 \times \frac{1}{4} = 275$.

The notion of $E[Y|X]$ being a r.v. is equally valid for discrete, continuous, or mixed r.v.'s X. For example, Equation 4.2-12

$$E[Y] = \int_{-\infty}^{\infty} E[Y|X = x] f_X(x) \, dx$$

can also be written as $E[Y] = E\{E[Y|X]\}$, where $E[Y|X]$ in this case is a function of the continuous r.v. X. The inner expectation is with respect to Y and the outer with respect to X.

The foregoing can be extended to more complex situations. For example, the object $E[Z|X, Y]$ is a function of the r.v.'s X and Y and therefore is a function of two r.v.s. For a particular outcome $\zeta \in \Omega$, it assumes the value $E[Z|X(\zeta), Y(\zeta)]$. To compute $E[Z]$ we would write $E[Z] = E[E[Z|X, Y]]$

$$E[Z] = E[E[Z|X, Y]] \tag{4.2-28a}$$

$$= \int_{-\infty}^{\infty} \int_{-\infty}^{\infty} \int_{-\infty}^{\infty} z f_{Z|X,Y}(z|x, y) f_{XY}(x, y) \, dx \, dy \, dz. \tag{4.2-28b}$$

In Equation 4.2-28a the inner expectation is with respect to Z and the outer is with respect to X and Y.

We conclude this section by summarizing some properties of conditional expectations.

Property (i). $E[Y] = E[E[Y|X]]$.

Proof   See arguments leading up to Equation 4.2-8 for the discrete case and Equation 4.2-12 for the continuous case. The inner expectation is with respect to Y, the outer with respect to X. ■

Property (ii). If X and Y are independent, then $E[Y|X] = E[Y]$.

Proof

$$E[Y|X = x] = \int_{-\infty}^{\infty} y f_{Y|X}(y|x) dy.$$

But $f_{XY}(x,y) = f_{Y|X}(y|x)f_X(x) = f_Y(y)f_X(x)$ if X and Y are independent. Hence $f_{Y|X}(y|x) = f_Y(y)$ and

$$E[Y|X = x] = \int_{-\infty}^{\infty} y f_Y(y) dy = E[Y]$$

for each x. Thus

$$E[Y|X] = \int_{-\infty}^{\infty} y f_Y(y) dy = E[Y].$$

An analogous proof holds for the discrete case. ∎

Property (iii). $E[Z|X] = E[E[Z|X,Y]|X]$.

   Proof

$$E[Z|X = x] = \int_{-\infty}^{\infty} z f_{Z|X}(z|x) dz$$

$$= \int_{-\infty}^{\infty} \int_{-\infty}^{\infty} z f_{Z|X,Y}(z|x,y) f_{Y|X}(y|x) dz\, dy$$

$$= \int_{-\infty}^{\infty} dy\, f_{Y|X}(y|x) \int_{-\infty}^{\infty} z f_{Z|X,Y}(z|x,y) dz$$

$$= E\left[E[Z|X,Y]|X = x\right],$$

where the inner expectation is with respect to Z and the outer with respect to Y. Since this is true for all x, we have $E[Z|X] = E[E[Z|X,Y]|X]$. The mean $\mu_Y = E[Y]$ is an estimate of the random variable Y. The mean-square error in this estimate is $\varepsilon^2 = E[(Y - \mu_Y)^2]$. In fact this estimate is optimal in that any constant other than μ_Y would lead to an increased ε^2. ∎

4.3 MOMENTS

Although the expectation is an important "summary" number for the behavior of a random variable, it is far from adequate in describing the complete behavior of the r.v. Indeed, we saw in Section 4.1 that two sets of numbers could have the same sample mean but the sample deviations could be quite different. Likewise, for two r.v.'s: Their expectations could be the same but their standard deviations could be very different. Summary numbers like μ_X, σ_X^2, $E[X^2]$, and others are called *moments*. Generally, a r.v. will have many non-zero higher-order moments and, under certain conditions (Section 4.5), it is possible to completely describe the behavior of the r.v., that is, reconstruct its pdf from knowledge of all the moments. In the following definitions we shall assume that the moments exist.

Definition 4.3-1   The rth moment of X is defined as

$$\xi_r \triangleq E[X^r] = \int_{-\infty}^{\infty} x^r f_X(x)\, dx, \qquad \text{where } r = 0, 1, 2, 3, \ldots . \qquad (4.3\text{-}1)$$

If X is a discrete r.v., the rth moment can be computed from the PMF as

$$\xi_r \triangleq \sum_i x_i^r P_X(x_i).$$

We note that $\xi_0 = 1$ $\xi_1 = \mu$ (the mean).   ■

Definition 4.3-2   The rth *central moment* of X is defined as

$$m_r \triangleq E[(X - \mu)^r] \qquad \text{where } r = 0, 1, 2, 3, \ldots. \tag{4.3-2a}$$

For a discrete r.v. we can compute m_r from

$$m_r \triangleq \sum_i (x_i - \mu)^r P_X(x_i).   ■ \tag{4.3-2b}$$

The most frequently used central moment is m_2. It is called the variance and is denoted by σ^2 and also sometimes as $\text{Var}[X]$. Note that $m_0 = 1$, $m_1 = 0$, $m_2 = \sigma^2$. A very important formula that connects the variance to $E[X^2]$ and μ is obtained as follows:

$$\sigma^2 = E\left[[X - \mu]^2\right] = E[X^2] - E[2\mu X] + E[\mu^2].$$

But for any constant c, $E[cX] = cE[X]$ and $E[c^2] = c^2$. Thus

$$\sigma^2 = E[X^2] - 2\mu E[X] + \mu^2$$
$$= E[X^2] - \mu^2 \tag{4.3-3}$$

since $E[X] \triangleq \mu$. In order to save symbology, an overbar is often used to denote expectation. Thus $\overline{X^r} \triangleq E[X^r]$, and so forth, for other moments. Using this notation, Equation 4.3-3 appears as

$$\sigma^2 = \overline{X^2} - \mu^2 \tag{4.3-4a}$$

or, equivalently,

$$\overline{X^2} = \sigma^2 + \mu^2. \tag{4.3-4b}$$

Equations 4.3-4a relate the second central moment m_2 to ξ_2 and μ. We can generalize this result as follows. Observe that

$$(X - \mu)^r = \sum_{i=0}^r \binom{r}{i} (-1)^i \mu^i X^{r-i}. \tag{4.3-5a}$$

By taking the expectation of both sides of Equation 4.3-5a and recalling the linearity of the expectation operator, we obtain

$$m_r = \sum_{i=0}^r \binom{r}{i} (-1)^i \mu^i \xi_{r-i}. \tag{4.3-5b}$$

Example 4.3-1
Let us compute ξ_2 for X, a binomial r.v. By definition

$$P_X(k) = \binom{n}{k} p^k q^{n-k}$$

and

$$\xi_2 = \sum_{k=0}^{n} k^2 \binom{n}{k} p^k q^{n-k}$$

$$= p^2 n(n-1) + np$$

$$= n^2 p^2 + npq. \tag{4.3-6}$$

In going from line 2 to line 3 several steps of algebra were used whose duplication we leave as an exercise. In going from line 3 to line 4, we rearranged terms and used the fact that $q \overset{\Delta}{=} 1 - p$. The expected value of X is

$$\xi_1 = \sum_{k=0}^{n} k \frac{n!}{k!(n-k)!} p^k q^{n-k}$$

$$= np = \mu. \tag{4.3-7}$$

Using this result in Equation 4.3-6 and recalling Equation 4.3-4 allow us to conclude that for a binomial r.v. with PMF $b(k; n, p)$

$$\sigma^2 = npq. \tag{4.3-8}$$

For any given n, maximum variance is obtained when $p = q = 0.5$ (Figure 4.3-1).

Example 4.3-2
Let us compute m_2 for $X: N(0, \sigma^2)$. Since $\mu = 0$, $m_2 = \xi_2$ and

$$m_2 = \frac{1}{\sqrt{2\pi\sigma^2}} \int_{-\infty}^{\infty} x^2 e^{-\frac{1}{2}(x/\sigma)^2} \, dx.$$

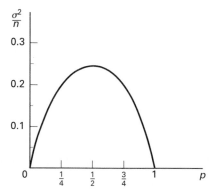

Figure 4.3-1   Variance of a binomial r.v. versus p.

But this integral was already evaluated in Example 4.1-3, where we found $\overline{X^2} = \sigma^2$. Thus the variance of a Gaussian r.v. is indeed the parameter σ^2 regardless of whether X is zero-mean or not.

An interesting and somewhat more difficult example that illustrates a useful application of moments is given next.

Example 4.3-3 _____

The maximum entropy (ME) principle states that if we don't know the pdf $f_X(x)$ of X but would like to estimate it with a function, say $p(x)$, a very good choice is the function $p(x)$ which maximizes the *entropy*, defined by [4-5],

$$H[X] \overset{\Delta}{=} - \int_{-\infty}^{\infty} p(x) \ln p(x) \, dx \tag{4.3-9}$$

and which satisfies the constraints

$$p(x) \geq 0 \tag{4.3-10a}$$

$$\int_{-\infty}^{\infty} p(x) \, dx = 1 \tag{4.3-10b}$$

$$\int_{-\infty}^{\infty} x p(x) \, dx = \mu \tag{4.3-10c}$$

$$\int_{-\infty}^{\infty} x^2 p(x) \, dx = \xi_2, \text{ and so forth.} \tag{4.3-10d}$$

Suppose we know from measurements or otherwise only μ in Equation 4.3-10c and that $x \geq 0$. Thus we wish to find $p(x)$ that maximizes $H[X]$ of Equation 4.3-9 subject to the first three constraints of Equations 4.3-10a. According to the method of Lagrange multipliers [4-6], the solution is obtained by maximizing the expression

$$- \int_{0}^{\infty} p(x) \ln p(x) \, dx + \lambda_1 \int_{0}^{\infty} p(x) \, dx + \lambda_2 \int_{0}^{\infty} x p(x) \, dx$$

by differentiation with respect to $p(x)$. The constants λ_1 and λ_2 are Lagrange multipliers and must be determined. After differentiating we obtain

$$\ln p(x) = -(1 + \lambda_1) - \lambda_2 x$$

or

$$p(x) = e^{-(1 + \lambda_1 + \lambda_2 x)}. \tag{4.3-11}$$

When this result is substituted in Equations 4.3-10b and 4.3-10c we find that

$$e^{-(1+\lambda_1)} = \frac{1}{\mu}, \quad \mu > 0$$

and
$$\lambda_2 = \frac{1}{\mu}.$$

Hence our ME estimate of $f_X(x)$ is

$$p(x) = \begin{cases} \dfrac{1}{\mu} e^{-x/\mu}, & x \geq 0 \\ 0, & x < 0. \end{cases} \tag{4.3-12}$$

The problem of obtaining the ME estimate of $f_X(x)$ when both μ and σ^2 are known is left as an exercise. In this case $p(x)$ is the Normal distribution with mean μ and variance σ^2.

Less useful than ξ_r or m_r are the absolute moments and generalized moments about some arbitrary point, say a, defined by, respectively,

$$E[|X|^r] \overset{\Delta}{=} \int_{-\infty}^{\infty} |x|^r f_X(x)\, dx \quad \text{(absolute moment)}$$

$$E[(X-a)^r] \overset{\Delta}{=} \int_{-\infty}^{\infty} (x-a)^r f_X(x)\, dx \quad \text{(generalized moment)}.$$

Note that if we set $a = \mu$, the generalized moments about a are then the central moments. If $a = 0$, the generalized moments are simply the moments ξ_r.

Joint Moments

Let us now turn to a topic first touched upon in Example 4.2-3. Suppose we are given two random variables X and Y and wish to have a measure of how good a linear prediction we can make of the value of, say, Y upon observing what value X has. At one extreme if X and Y are independent, observing X tells us nothing about Y. At the other extreme if, say, $Y = aX + b$, then observing the value of X immediately tells us the value of Y. However, in many situations in the real world, two random variables are neither completely independent nor linearly dependent. Given this state of affairs, it then becomes important to have a measure of how much can be said about one random variable from observing another. The quantities called *joint moments* offer us such a measure. Not all joint moments, to be sure, are equally important in this task; especially important are certain *second-order joint moments* (to be defined shortly). However, as we shall see later, in various applications other joint moments are important as well and so we shall deal with the general case below.

Definition 4.3-3   The ijth joint moment of X and Y is given by

$$\xi_{ij} \overset{\Delta}{=} E[X^i Y^j]$$

$$= \int_{-\infty}^{\infty} \int_{-\infty}^{\infty} x^i y^j f_{XY}(x, y)\, dx\, dy. \tag{4.3-13}$$

If X and Y are discrete, we can compute ξ_{ij} from the PMF as

$$\xi_{ij} \overset{\Delta}{=} \sum_l \sum_m x_l^i y_m^j P_{X,Y}(x_l, y_m). \quad \blacksquare \tag{4.3-14}$$

Definition 4.3-4   The ijth joint *central moment* of X and Y is defined by

$$m_{ij} \triangleq E[(X - \overline{X})^i (Y - \overline{Y})^j], \qquad (4.3\text{-}15)$$

where, in the notation introduced earlier, $\overline{X} \triangleq E[X]$, and so forth, for $\overline{Y}$. The order of the moment is $i + j$. Thus all of the following are second-order moments

$$\xi_{02} = E[Y^2] \qquad m_{02} = E[(Y - \overline{Y})^2]$$
$$\xi_{20} = E[X^2] \qquad m_{20} = E[(X - \overline{X})^2]$$
$$\xi_{11} = E[XY] \qquad m_{11} = E[(X - \overline{X})(Y - \overline{Y})]$$
$$= E[XY] - \overline{X}\,\overline{Y}$$
$$\triangleq \text{Cov}[X, Y]. \quad \blacksquare$$

As measures of predictability and in some cases statistical dependence, the most important joint moments are ξ_{11} and m_{11}; they are known as the *correlation* and *covariance* of X and Y, respectively. The *correlation coefficient*[†] defined by

$$\rho \triangleq \frac{m_{11}}{\sqrt{m_{20} m_{02}}} \qquad (4.3\text{-}16)$$

was already introduced in Section 4.2 (Equation 4.2-16). It satisfies $|\rho| \leq 1$. To show this consider the nonnegative expression

$$E[(\lambda(X - \overline{X}) - (Y - \overline{Y}))^2] \geq 0,$$

where λ is any real constant. To verify that the left side is indeed nonnegative, we merely rewrite it in the form

$$Q(\lambda) \triangleq \int_{-\infty}^{\infty} \int_{-\infty}^{\infty} [\lambda(x - \overline{X}) - (y - \overline{Y})]^2 f_{XY}(x, y)\, dx\, dy \geq 0,$$

where the $\geq$ follows from the fact that the integral of a nonnegative quantity cannot be negative.

The previous equation is a quadratic in λ. Indeed, after expanding we obtain

$$Q(\lambda) = \lambda^2 m_{20} + m_{02} - 2\lambda m_{11} \geq 0.$$

Thus $Q(\lambda)$ can have at most one real root. Hence its discriminant must satisfy

$$\left(\frac{m_{11}}{m_{20}}\right)^2 - \frac{m_{02}}{m_{20}} \leq 0$$

[†]Note that it would be more properly termed the *covariance coefficient* or *normalized covariance*.

or

$$m_{11}^2 \leq m_{02}m_{20} \tag{4.3-17}$$

whence the condition $|\rho| \leq 1$ follows.

When $m_{11}^2 = m_{02}m_{20}$, that is, $|\rho| = 1$ then it is easy to establish that

$$E\left[\left(\frac{m_{11}}{m_{20}}(X - \overline{X}) - (Y - \overline{Y})\right)^2\right] = 0$$

or, equivalent, that

$$\int_{-\infty}^{\infty}\int_{-\infty}^{\infty} \left(\frac{m_{11}}{m_{20}}(x - \overline{X}) - (y - \overline{Y})\right)^2 f_{XY}(x,y)\, dx\, dy = 0. \tag{4.3-18}$$

Since $f_{XY}(x,y)$ is never negative, Equation 4.3-18 implies that the term in parentheses is zero everywhere.[†] Thus we have from Equation 4.3-18 that when $|\rho| = 1$

$$Y = \frac{m_{11}}{m_{20}}(X - \overline{X}) + \overline{Y} \tag{4.3-19}$$

that is, Y is a linear function of X. When $\text{Cov}[X,Y] = 0$, then $\rho = 0$ and X and Y are *said to be uncorrelated*.

Properties of Uncorrelated Random Variables

(A) If X and Y are uncorrelated, then

$$\sigma_{X+Y}^2 = \sigma_X^2 + \sigma_Y^2, \tag{4.3-20}$$

where

$$\sigma_{X+Y}^2 \triangleq E[(X + Y)^2] - (E[X + Y])^2.$$

(B) If X and Y are independent, they are uncorrelated. Proof of (A): We leave this as an exercise to the reader; proof of (B): Since $\text{Cov}[X,Y] = E[XY] - E[X]E[Y]$, we must show that $E[XY] = E[X]E[Y]$. But

$$E[XY] = \int_{-\infty}^{\infty}\int_{-\infty}^{\infty} xy f_{XY}(x,y)\, dx\, dy$$

$$= \int_{-\infty}^{\infty} x f_X(x)\, dx \int_{-\infty}^{\infty} y f_Y(y)\, dy \quad \text{(by independence assumption)}$$

$$= E[X]E[Y]. \quad \blacksquare$$

[†]Except possibly over a set of points of zero probability. To be more precise we should exchange the word "everywhere" in the text to "almost everywhere," often abbreviated a.e.

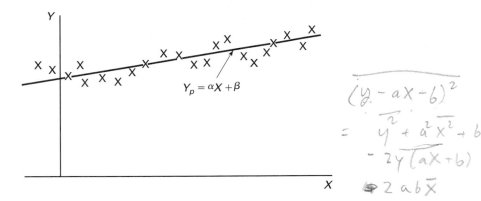

Figure 4.3-2   Pairwise observations on (X, Y) constitute a scatter diagram. The relationship between X and Y is approximated with a straight line.

Example 4.3-4

(Linear prediction.) Suppose we wish to predict the values of a r.v. Y by observing the values of another r.v. X. In particular, the available data (Figure 4.3-2) suggest that a good prediction model for Y is the linear function

$$Y_P \triangleq \alpha X + \beta. \qquad (4.3\text{-}21)$$

Now although Y may be related to X, the values it takes on may be influenced by other sources that do not affect X. Thus, in general, $|\rho| \neq 1$ and we expect that there will be an error between the predicted value of Y, that is, Y_P and the value that Y actually assumes. Our task becomes then to adjust the coefficients α and β in order to minimize the mean-square error

$$\varepsilon^2 \triangleq E[(Y - Y_P)^2]. \qquad (4.3\text{-}22)$$

This problem is a simple version of *optimum linear prediction*. In statistics it is called *linear regression*.

Solution   Upon expanding Equation 4.3-22 we obtain

$$\varepsilon^2 = \overline{Y^2} - 2a\overline{XY} - 2\beta\overline{Y} + 2\alpha\beta\overline{X} + \alpha^2\overline{X^2} + \beta^2.$$

To minimize ε with respect to α and β we solve for α and β that satisfy

$$\frac{\partial \varepsilon^2}{\partial \alpha} = 0 \qquad \frac{\partial \varepsilon^2}{\partial \beta} = 0. \qquad (4.3\text{-}23)$$

This yields the best α and β, which we denote by α_0, β_0 in the sense that they minimize ε. A little algebra establishes that

$$\alpha_0 = \frac{\text{Cov}[X, Y]}{\sigma_X^2} = \frac{\rho\sigma_Y}{\sigma_X} \qquad (4.3\text{-}24a)$$

and

$$\beta_0 = \overline{Y} - \frac{Cov[X,Y]}{\sigma_X^2}\overline{X}$$

$$= \overline{Y} - \rho\frac{\sigma_Y}{\sigma_X}\overline{X}. \tag{4.3-24b}$$

Thus the best linear predictor is given by

$$Y_P - \overline{Y} = \rho\frac{\sigma_Y}{\sigma_X}(X - \overline{X}) \tag{4.3-25}$$

and passes through the point $(\overline{X}, \overline{Y})$. If we use α_0, β_0 in Equation 4.3-22 we obtain the smallest mean-square error ε_{min}^2, which is (Problem 4.22)

$$\varepsilon_{min}^2 = \sigma_Y^2(1 - \rho^2). \tag{4.3-26}$$

Something rather strange happens when $\rho = 0$. From Equation 4.3-25 we see that for $\rho = 0$ $Y_P = \overline{Y}$ regardless of X! This means that observing X has no bearing on our prediction of Y, and the best predictor is merely $Y_P = \overline{Y}$. We encountered somewhat the same situation in Example 4.2-3. Thus associating the correlation coefficient with ability to predict seems justified in problems involving linear prediction and the joint Gaussian pdf. In some fields, a lack of correlation between two r.v.'s is taken to be *prima facie* evidence that they are unrelated, that is, independent. No doubt this conclusion arises in part from the fact that if two r.v.'s, say, X and Y, are indeed independent, they will be uncorrelated. As stated earlier, the opposite is generally not true. An example follows.

Example 4.3-5

Consider two r.v.'s X and Y with joint PMF $P_{X,Y}(x_i, y_j)$ as shown.

Values of $P_{X,Y}(x_i, y_j)$

	$x_1 = -1$	$x_2 = 0$	$x_3 = +1$
$y_1 = 0$	0	$\frac{1}{3}$	0
$y_2 = 1$	$\frac{1}{3}$	0	$\frac{1}{3}$

X and Y are not independent, since $P_{X,Y}(0,1) = 0 \neq P_X(0)P_Y(1) = \frac{2}{9}$. Furthermore, $\overline{X} = 0$ so that $Cov(X,Y) = \overline{XY} - \overline{X}\,\overline{Y} = \overline{XY}$. We readily compute

$$\overline{XY} = (-1)(1)\tfrac{1}{3} + (1)(1)\tfrac{1}{3} = 0.$$

Hence X and Y are uncorrelated but not independent.

There is an important special case for which $\rho = 0$ always implies independence. We now discuss this case.

Jointly Gaussian Random Variables

We say that two r.v.'s are jointly Gaussian (or jointly normal) if their joint pdf is

$$f_{XY}(x, y) = \frac{1}{2\pi\sigma_X\sigma_Y\sqrt{1 - \rho^2}} \times \exp\left(\frac{-1}{2(1 - \rho^2)}\left\{\left(\frac{x - \overline{X}}{\sigma_X}\right)^2 \right.\right.$$

$$\left.\left. -2\rho\frac{(x - \overline{X})(y - \overline{Y})}{\sigma_X\sigma_Y} + \left(\frac{y - \overline{Y}}{\sigma_Y}\right)^2\right\}\right). \tag{4.3-27}$$

Five parameters are involved: σ_X, σ_Y, $\overline{X}$, $\overline{Y}$, and ρ. If $\rho = 0$ we observe that

$$f_{XY}(x, y) = f_X(x)f_Y(y),$$

where

$$f_X(x) = \frac{1}{\sqrt{2\pi\sigma_X^2}}\exp\left(-\frac{1}{2}\left(\frac{x - \overline{X}}{\sigma_X}\right)^2\right) \tag{4.3-28}$$

and

$$f_Y(y) = \frac{1}{\sqrt{2\pi\sigma_Y^2}}\exp\left(-\frac{1}{2}\left(\frac{y - \overline{Y}}{\sigma_Y}\right)^2\right). \tag{4.3-29}$$

Thus two jointly Gaussian r.v.'s that are uncorrelated (that is, $\rho = 0$) are also independent. The marginal densities $f_X(x)$ and $f_Y(y)$ for jointly normal r.v.'s are always normal regardless of what ρ is. However, the converse does not hold; that is, if $f_X(x)$ and $f_Y(y)$ are Gaussian, one cannot conclude that X and Y are jointly Gaussian.

To see this we borrow from a popular x-ray imaging technique called *computerized tomography* (CT) useful for detecting cancer and other abnormalities in the body. Suppose we have an object with x-ray absorptivity function $f(x, y) \geq 0$. This function is like a joint pdf in that it is real, never negative, and easily normalized to a unit volume—however this last feature is not important. Thus we can establish a one-to-one relationship between a joint pdf $f_{XY}(x, y)$ and the x-ray absorptivity $f(x, y)$. In CT, x-rays are passed through the object along different lines, for some fixed angle, and the integrals of the absorptivity are measured and recorded. Each integral is called a projection and the set of all projections for given angle θ is called the profile function at θ. Thus the projection for a line at angle θ and displacement s from the center is given by [Figure. 4.3-3(a)]

$$f_\theta(s) = \int_{L(s,\theta)} f(x, y)dl$$

where $L(s, \theta)$ are the points along a line displaced from the center by s at angle θ and dl is a differential length along $L(s, \theta)$. If we let s vary from its smallest to largest value, we obtain the profile function for that angle. By collecting all the profiles for all the angles and using a sophisticated algorithm called filtered-convolution back-projection, it is possible to get a high-quality x-ray image of the body. Suppose we measure the profiles at 0 degrees

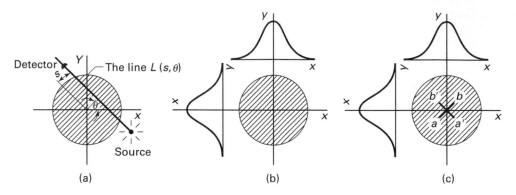

Figure 4.3-3   Using the computerized tomography paradigm to show that Gaussian marginal pdf's do not imply a joint Gaussian distribution. (a) A projection is the line integral at displacement s and angle θ. The set of all projections for a given angle is the profile function for that angle. (b) A joint Gaussian x-ray object produces Gaussian-shaped profile functions in the horizontal and vertical directions; (c) by adding a constant absorptivity along a–b and subtracting an absorptivity along a'–b', the profile functions remain the same but the underlying absorptivity is not Gaussian anymore.

and 90 degrees as shown in Figure 4.3-3(b). Then we obtain

$$f_1(x) = \int_{-\infty}^{\infty} f(x,y) dy \qquad \text{(horizontal profile)}$$

$$f_2(y) = \int_{-\infty}^{\infty} f(x,y)\, dx \qquad \text{(vertical profile)}$$

If $f(x,y)$ is Gaussian, then we already know that $f_1(x)$ and $f_2(y)$ will be Gaussian because f_1 and f_2 are analogous to marginal pdfs. Now is it possible to modify $f(x,y)$ from Gaussian to non-Gaussian without observing a change in the Gaussian profile? If yes, we have demonstrated our assertion that Gaussian marginals do not necessarily imply a joint Gaussian pdf. In Figure 4.3-3(c) we increase the absorptivity of the object by an amount P along the 45 degree strip running from a to b and decrease the absorptivity by the same amount P along the 135-degree strip running from a' to b'. Then since the profile integrals add and subtract P in both horizontal and vertical directions, the net change in $f_1(x)$ and $f_2(y)$ is zero. This proves our assertion. We assume that P is not so large that when subtracted from $f(x,y)$ along a'–b', the result is negative. The reason we must make this assumption is that pdf's and x-ray absorptivities can never be negative.

To illustrate a joint normal distribution consider the following somewhat idealized situation. Let X and Y denote the height of the husband and wife, respectively, of a married pair picked at random from the population of married people. It is often assumed that X and Y are individually Gaussian although this is obviously only an approximation since heights are bounded from below by zero and from above by physiological constraints. Conventional wisdom has it that in our society tall people prefer tall mates and short people prefer short mates. If this is indeed true, then X and Y are *positively correlated* that is, $\rho > 0$. On the other hand, in certain societies it may be fashionable for tall men to marry short women and for tall women to marry short men. Again we can expect X and Y to be correlated albeit negatively this time, that is, $\rho < 0$. Finally, if all marriages are the result of a lottery, we would expect ρ to be zero or very small.

Contours of Constant Density of the Joint Gaussian pdf

It is of interest to determine the locus of points in the xy plane when $f_{XY}(x, y)$ is set constant. Clearly $f_{XY}(x, y)$ will be constant if the exponent is set to a constant, say, c^2:

$$\left(\frac{x - \overline{X}}{\sigma_X}\right)^2 - 2\rho\frac{(x - \overline{X})(y - \overline{Y})}{\sigma_X \sigma_Y} + \left(\frac{y - \overline{Y}}{\sigma_Y}\right)^2 = c^2.$$

This is the equation of an ellipse centered at $x = \overline{X}$, $y = \overline{Y}$. For simplicity we set $\overline{X} = \overline{Y} = 0$. When $\rho = 0$, the major and minor diameters of the ellipse are parallel to the x- and y-axes, a condition we know to associate with independence of X and Y. If $\rho = 0$ and $\sigma_X = \sigma_Y$, the ellipse degenerates into a circle. Several cases are shown in Figure 4.3-4.

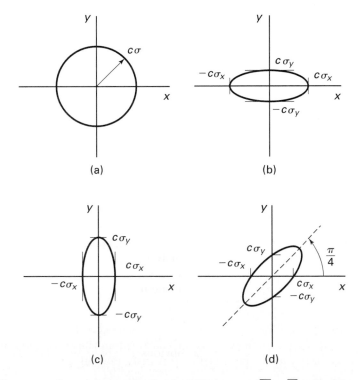

Figure 4.3-4   Contours of constant density for the joint normal $(\overline{X} = \overline{Y} = 0)$: (a) $\sigma_X = \sigma_Y$, $\rho = 0$; (b) $\sigma_X > \sigma_Y$, $\rho = 0$; (c) $\sigma_X < \sigma_Y$, $\rho = 0$; (d) $\sigma_X = \sigma_Y$; $\rho > 0$.

Surprisingly the marginal densities $f_X(x)$ and $f_Y(y)$ computed from the joint pdf of Equation 4.3-27 do not depend on the parameter ρ. To see this we compute

$$f_X(x) = \int_{-\infty}^{\infty} f_{XY}(x, y) dy$$

with $\overline{X} = \overline{Y} = 0$ for simplicity. The integration, while somewhat messy, is easily done by following these three steps:

1. Factor out of the integral all terms that do not depend on y;
2. Complete the squares in the exponent of e (see "completing the square" in Appendix A; and
3. Recall that for $b > 0$ and real y

$$\frac{1}{\sqrt{2\pi b^2}} \int_{-\infty}^{\infty} \exp\left[-\frac{1}{2}\left(\frac{y - a}{b}\right)^2\right] dy = 1.$$

Indeed after step 2 we obtain

$$f_X(x) = \frac{1}{\sqrt{2\pi}\sigma_X} \exp\left[-\frac{1}{2}\left(\frac{x}{\sigma_X}\right)^2\right]$$

$$\times \left\{\frac{1}{\sqrt{2\pi\sigma_Y^2(1 - \rho^2)}} \int_{-\infty}^{\infty} \exp\left[\frac{-(y - \rho x\sigma_Y/\sigma_X)^2}{2\sigma_Y^2(1 - \rho^2)}\right] dy\right\}. \qquad (4.3\text{-}30)$$

But the term in curly brackets is unity. Hence

$$f_X(x) = \frac{1}{\sqrt{2\pi}\sigma_X} \exp\left[-\frac{1}{2}\left(\frac{x}{\sigma_X}\right)^2\right]. \qquad (4.3\text{-}31)$$

A similar calculation for $f_Y(y)$ would furnish

$$f_Y(y) = \frac{1}{\sqrt{2\pi}\sigma_Y} \exp\left[-\frac{1}{2}\left(\frac{y}{\sigma_Y}\right)^2\right]. \qquad (4.3\text{-}32)$$

As we stated earlier, if $\rho = 0$, then X and Y are independent. On the other hand as $\rho \to \pm 1$, X and Y become linearly dependent. For simplicity let $\sigma_X = \sigma_Y \overset{\Delta}{=} \sigma$ and $\overline{X} = \overline{Y} = 0$; then the contour of constant density becomes

$$x^2 - 2\rho xy + y^2 = c^2\sigma^2$$

which is a 45-degree tilted ellipse (with respect to the x-axis) for $\rho > 0$ and a 135-degree tilted ellipse for $\rho < 0$. We can generate a coordinate system that is rotated 45-degrees from the $x - y$ system by introducing the coordinate transformation

$$v = \frac{x + y}{\sqrt{2}} \quad w = \frac{x - y}{\sqrt{2}}.$$

Then the contour of constant density becomes

$$v^2[1-\rho] + w^2[1+\rho] = \sigma^2 c^2$$

which is an ellipse with major and minor diameters parallel to the v-w-axes. If $\rho > 0$, the major diameter is parallel to the v-axis; if $\rho < 0$, the major diameter is parallel to the w-axis. As $\rho \to \pm 1$, the lengths of the major diameters become infinitely long and all of the pdf concentrates along the line $y = x(\rho \to 1)$ or $y = -x(\rho \to -1)$.

Finally by introducing two new r.v.'s

$$V \triangleq (X+Y)/\sqrt{2}$$

$$W \triangleq (X-Y)/\sqrt{2}$$

it is not difficult to show that as $\rho \to 1$

$$f_{XY}(x,y) \to \frac{1}{\sqrt{2\pi}\sigma} \exp\left[-\frac{1}{2}\left(\frac{x}{\sigma}\right)^2\right] \times \delta(y-x)$$

or, equivalently,

$$f_{XY}(x,y) \to \frac{1}{\sqrt{2\pi}\sigma} \exp\left[-\frac{1}{2}\left(\frac{y}{\sigma}\right)^2\right] \times \delta(y-x).$$

This degeneration of the joint Gaussian into a pdf of only one variable along the line $y = x$ is due to the fact that as $\rho \to 1$, X and Y become equal. We leave the details as an exercise to the student.

A computer rendition of the joint Gaussian pdf and its contours of constant density is shown in Figure 4.3-5 for $\overline{X} = \overline{Y} = 0$, $\sigma_X = \sigma_X = 2$, and $\rho = 0.9$.

4.4 CHEBYSHEV AND SCHWARZ INEQUALITIES

The Chebyshev[†] inequality furnishes a bound on the probability of how much a r.v. X can deviate from its mean value $\overline{X}$.

Theorem 4.4-1 Let X be an arbitrary r.v. with mean $\overline{X}$ and finite variance σ^2. Then for any $\delta > 0$

$$P[|X - \overline{X}| \geq \delta] \leq \frac{\sigma^2}{\delta^2}. \tag{4.4-1}$$

Proof Equation 4.4-1 follows directly from the following observation:

$$\sigma^2 \triangleq \int_{-\infty}^{\infty} (x-\overline{X})^2 f_X(x)\, dx \geq \int_{|x-\overline{X}|\geq\delta} (x-\overline{X})^2 f_X(x)\, dx$$

$$\geq \delta^2 \int_{|x-\overline{X}|\geq\delta} f_X(x)\, dx$$

$$= \delta^2 P[|X - \overline{X}| \geq \delta]$$

[†]Pafnuti L. Chebyshev (1821–1894), Russian mathematician.

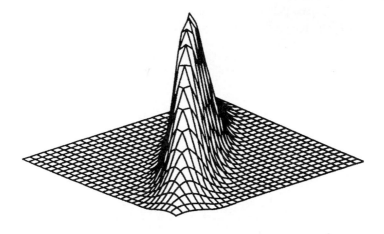

(a)

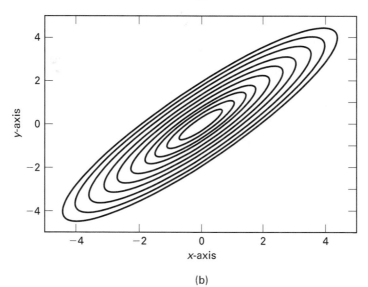

(b)

Figure 4.3-5   (a) Gaussian pdf with $\overline{X} = \overline{Y} = 0$, $\sigma_X = \sigma_Y = 2$, and $\rho = 0.9$; (b) Contours of constant density.

whence Equation 4.4-1 follows. Since

$$\{|X - \overline{X}| \geq \delta\} \cup \{|X - \overline{X}| < \delta\} = \Omega \quad (\Omega \text{ being the certain event}),$$

it immediately follows that

$$P[|X - \overline{X}| < \delta] \geq 1 - \frac{\sigma^2}{\delta^2}. \tag{4.4-2}$$

Sometimes it is convenient to express δ in terms of σ, that is, $\delta \triangleq k\sigma$, where k is a constant. Then Equations 4.4-1 and 4.4-2 become, respectively,

$$P[|X - \overline{X}| \geq k\sigma] \leq \frac{1}{k^2} \tag{4.4-3}$$

$$P[|X - \overline{X}| < k\sigma] \geq 1 - \frac{1}{k^2}. \quad \blacksquare \tag{4.4-4}$$

Example 4.4-1 ~Problem 3~
(Deviation from the mean for a Normal r.v.) Let $X \colon N(\overline{X}, \sigma^2)$. How do $P[|X - \overline{X}| < k\sigma]$ and $P[|X - \overline{X}| \geq k\sigma]$ compare with the Chebyshev bound (CB)?

Solution   Using Equations 2.4-14d and 2.4-14e, it is easy to show that $P[|X - \overline{X}| < k\sigma] = 2\mathrm{erf}(k)$ and $P[|X - \overline{X}| \geq k\sigma] = 1 - 2\mathrm{erf}(k)$, where erf(k) is defined in Equation 2.4-12. Using Table 2.4-1 and Equations 4.4-3 and 4.4-4 we obtain Table 4.4-1 for comparison
From Table 4.4-1 we see that the Chebyshev bound is not very good; however, it must be recalled that the bound applies to *any* r.v. X as long as σ^2 exists.
There are a number of extensions of the Chebyshev inequality (see Wilbur Davenport, [4-2, p. 256]). We consider such an extension in what follows.

Table 4.4-1

k	$P[|X - \overline{X}| < k\sigma]$	CB	$P[|X - \overline{X}| > k\sigma]$	CB
0	0	0	1	1
0.5	0.383	0	0.617	1
1.0	0.683	0	0.317	1
1.5	0.866	0.556	0.134	0.444
2.0	0.955	0.750	0.045	0.250
2.5	0.988	0.840	0.012	0.160
3.0	0.997	0.889	0.003	0.111

Random Variables with Nonnegative Values

Consider a r.v. X for which $f_X(x) = 0$ for $x < 0$. Then X is said to take on only nonnegative values. For such a r.v., the following inequality applies:

$$P[X \geq \delta] \leq \frac{E[X]}{\delta}. \tag{4.4-5}$$

In contrast to the Chebyshev bound, this bound involves only the mean of X.

Proof of Equation 4.4-5

$$E[X] = \int_0^\infty x f_X(x)\, dx \geq \int_\delta^\infty x f_X(x)\, dx \geq \delta \int_\delta^\infty f_X(x)\, dx$$

$$\geq \delta P[X \geq \delta]$$

whence Equation 4.4-5 follows. Equation 4.4-5 puts a bound on what fraction of a population can exceed δ. ∎

problem 4

Example 4.4-2

Assume that in the manufacturing of very low grade electrical 1000-ohm resistors the average resistance, as determined by measurements, is indeed 1000 ohms but there is a large variation about this value. If all resistors over 1500 ohms are to be discarded, what is the maximum fraction of resistors to meet such a fate?

Solution   With $\overline{X} = 1000$, and $\delta = 1500$, we obtain

$$P[X \geq 1500] \leq \frac{1000}{1500} = 0.67.$$

Thus, if nothing else, the manufacturer has the assurance that the percentage of discarded resistors cannot exceed 67 percent of the total.

The Schwarz Inequality

We have already encountered the probabilistic form of the Schwarz[†] inequality in Equation 4.3-17 repeated here as

$$\mathrm{Cov}^2(X, Y) \leq E[(X - \overline{X})^2] E[(Y - \overline{Y})^2]$$

with equality if and only if Y is a linear function of X. In later work we shall need another version of the Schwarz inequality that is commonly used in obtaining results in signal processing and stochastic processes. Consider two ordinary, that is, nonrandom functions h and g not necessarily real and define the *norm* of h, if it exists, by

$$\|h\| \triangleq \left(\int_{-\infty}^\infty |h(x)|^2\, dx \right)^{1/2}, \tag{4.4-6}$$

and so forth, for the norm of g. The *scalar* or *inner* product of h with g is denoted by (h, g) and is defined by

$$(h, g) \triangleq \int_{-\infty}^\infty h(x) g^*(x)\, dx$$

$$= (g, h)^*. \tag{4.4-7}$$

[†]H. Amandus Schwarz (1843–1921), German mathematician.

Then another form of the Schwarz inequality is

$$|(h,g)| \le \|h\|\,\|g\| \tag{4.4-8}$$

with equality if and only if h is proportional to g, that is, $h(x) = cg(x)$. The proof of Equation 4.4-8 is easily obtained by considering the norm of $\lambda h(x) + g(x)$ as a function of the variable λ,

$$\|\lambda h(x) + g(x)\|^2 = |\lambda|^2\|h\|^2 + \lambda(h,g) + \lambda^*(h,g)^* + \|g\|^2 \ge 0 \tag{4.4-9}$$

with

$$\lambda = -\frac{(h,g)^*}{\|h\|^2} \tag{4.4-10}$$

whence Equation 4.4-8. In the special case where h and g are real functions of real random variables, that is, $h(X)$, $g(X)$, then Equation 4.4-8 still is valid provided that the definitions or norm and inner product are modified as follows:

$$\|h\|^2 \triangleq \int_{-\infty}^{\infty} h^2(x) f_X(x)\, dx = E[h^2(X)] \tag{4.4-11}$$

$$(h,g) \triangleq \int_{-\infty}^{\infty} h(x) g(x) f_X(x)\, dx = E[h(X)g(X)] \tag{4.4-12}$$

whence we obtain

$$|E[h(X)g(X)]| \le (E[h^2(X)])^{1/2} (E[g^2(X)])^{1/2}. \tag{4.4-13}$$

Example 4.4-3 ⎯⎯⎯
(The [weak] Law of Large Numbers [LLN].) Let $X_1, \ldots, X_n$ be independent and identically distributed random variables with mean μ_X and variance σ_X^2. Assume that we don't know the value of μ_X (or σ_X) and thus consider the *sample mean estimator*[†]

$$\hat{\mu}_n \triangleq \frac{1}{n} \sum_{i=1}^{n} X_i$$

as an estimator for μ_X. We can use the Chebyshev inequality to show that $\hat{\mu}_n$ is asymptotically a perfect estimator for of μ_X. First we compute

$$E[\hat{\mu}_n] = \frac{1}{n} \sum_{i=1}^{n} E[X_i]$$

$$= \frac{1}{n} n \mu_X$$

$$= \mu_X.$$

⎯⎯⎯⎯⎯⎯⎯⎯⎯⎯⎯

[†]An estimator is a function of the observations $X_1, X_2, \ldots, X_n$ that estimates a parameter of the distribution. Estimators are random variables. When an estimator takes on a particular value, that is, a realization, that number is sometimes called the *estimate*.

Next we compute

$$\text{Var}[\hat{\mu}_n] = \frac{1}{n^2} \text{Var}\left[\sum_{i=1}^{n} X_i\right]$$

$$= \left(\frac{1}{n^2}\right) n\sigma_X^2$$

$$= \frac{1}{n}\sigma_X^2.$$

Thus by the Chebyshev inequality (Equation 4.4-1) we have

$$P[|\hat{\mu}_n - \mu_X| \geq \delta|] \leq \sigma_X^2/n\delta^2.$$

Clearly for any fixed $\delta > 0$, the right side can be made arbitrarily small by choosing n large enough. Thus

$$\lim_{n\to\infty} P[|\hat{\mu}_n - \mu_X| \geq \delta] = 0$$

for every $\delta > 0$.

The law of large numbers is the theoretical basis for estimating μ_X from measurements. When an experimenter takes the sample mean of n measurements, he is relying on the LLN in order to use the sample mean as an estimate of the unknown theoretical expectation μ_X.

Sometimes inequalities can be derived from properties of the pdf. We illustrate with the following example due to Yongyi Yang.

Example 4.4-4
Let the pdf of the real r.v. X satisfy $f_X(x) = f_X(-x)$. Show that $\sigma_X \geq E[|X|]$ with equality if $\text{Var}|X| = 0$.

Solution   Let $Y \triangleq |X|$. Then $E[Y^2] = E[X^2] = \mu_X^2 + \sigma_X^2 = \sigma_X^2$ since $\mu_X = 0$. Also $E[Y^2] = \mu_Y^2 + \sigma_Y^2 = E^2[|X|] + \sigma_Y^2 = \sigma_X^2$. But $\sigma_Y^2 \geq 0$. Hence $E^2[|X|] \leq \sigma_X^2$ with equality if $\sigma_Y^2 = 0$. Such a case arises when the pdf of X has the form $f_X(x) = \frac{1}{2}[\delta(x - a) + \delta(x + a)]$, where a is some positive number. Then $Y = a$, $\sigma_Y = 0$ and $E[|X|] = \sigma_X$.

Another important inequality is furnished by the *Chernoff bound*. We discuss this bound in Section 4.6 after introducing the moment-generating function $\theta(t)$ in the next section.

4.5 MOMENT-GENERATING FUNCTIONS

The moment-generating function, if it exists, of a random variable X is defined by[†]

$$\theta(t) \overset{\Delta}{=} E[e^{tX}] \tag{4.5-1}$$

$$= \int_{-\infty}^{\infty} e^{tx} f_X(x)\,dx, \tag{4.5-2}$$

where t is a complex variable.

For discrete r.v.'s, we can define $\theta(t)$ using the PMF as

$$\theta(t) = \sum_i e^{tx_i} P_X(x_i). \tag{4.5-3}$$

From Equation 4.5-2 we see that except for a sign reversal in the exponent, the moment-generating function is the two-sided Laplace transform of the pdf for which there is a known inversion formula. Thus, in general, knowing $\theta(t)$ is equivalent to knowing $f_X(x)$ and vice versa.

The main reasons for introducing $\theta(t)$ are (1) it enables a convenient computation of the moments of X; (2) it can be used to estimate $f_X(x)$ from experimental measurements of the moments; (3) it can be used to solve problems involving the computation of the sums of random variables; and (4) it is an important analytical instrument that can be used to demonstrate basic results such as the *Central Limit Theorem*.[‡]

Proceeding formally, if we expand e^{tX} and take expectations, then

$$E[e^{tX}] = E\left[1 + tX + \frac{(tX)^2}{2!} + \ldots + \frac{(tX)^n}{n!} + \ldots\right]$$

$$= 1 + t\mu + \frac{t^2}{2!}\xi_2 + \ldots + \frac{t^n}{n!}\xi_n + \ldots. \tag{4.5-4}$$

Since the moments ξ_i may not exist, for example, none of the moments above the first exist for the Cauchy pdf, $\theta(t)$, may not exist. However, if $\theta(t)$ does exist, computing any moment is easily obtained by differentiation. Indeed, if we allow the notation

$$\theta^{(k)}(0) \overset{\Delta}{=} \left.\frac{d^k}{dt^k}(\theta(t))\right|_{t=0,}$$

then

$$\xi_k = \theta^{(k)}(0) \qquad k = 0, 1, \ldots. \tag{4.5-5}$$

Example 4.5-1

Let $X: N(\mu, \sigma^2)$. Its moment-generating function is

$$\theta(t) = \frac{1}{\sqrt{2\pi\sigma^2}} \int_{-\infty}^{\infty} \exp\left(-\frac{1}{2}\left(\frac{x-\mu}{\sigma}\right)^2\right) e^{tx}\,dx. \tag{4.5-6}$$

[†]The terminology varies (see Feller [4-1], p. 411).
[‡]To be discussed in Section 4.7.

Using the trick known as "completing the square"[†] in the exponent, we can write Equation 4.5-6 as

$$\theta(t) = \exp[\mu t + \sigma^2 t^2 / 2]$$

$$\times \frac{1}{\sqrt{2\pi\sigma^2}} \int_{-\infty}^{\infty} \exp\left\{ -\frac{1}{2\sigma^2} (x - (\mu + \sigma^2 t))^2 \right\} \, dx.$$

But the factor on the second line is unity since it is the integral of a Gaussian pdf with mean $\mu + \sigma^2 t$ and variance σ^2. Hence

$$\theta(t) = \exp(\mu t + \sigma^2 t^2 / 2), \tag{4.5-7}$$

from which we obtain

$$\theta^{(1)}(0) = \mu$$

$$\theta^{(2)}(0) = \mu^2 + \sigma^2.$$

Example 4.5-2

Let X be a binomial r.v. with parameters n (number of tries), p (probability of a success per trial), and $q = 1 - p$. Then

$$\theta(t) = \sum_{k=0}^{n} e^{tk} \binom{n}{k} p^k q^{n-k}$$

$$= \sum_{k=0}^{n} \binom{n}{k} [e^t p]^k q^{n-k}$$

$$= (pe^t + q)^n. \tag{4.5-8}$$

We obtain

$$\theta^{(1)}(0) \triangleq np = \mu$$

$$\theta^{(2)}(0) = \{ np e^t (pe^t + q)^{n-1} + n(n-1) p^2 e^{2t} (pe^t + q)^{n-2} \}_{t=0} \tag{4.5-9}$$

$$= npq + \mu^2.$$

Hence

$$\mathrm{Var}[X] = npq. \tag{4.5-10}$$

Example 4.5-3

(Geometric distribution.) Let X follow the geometric distribution. Then the PMF is $P_X(n) = a^n (1-a) u(n), n = 0, 1, 2, \ldots$ and $0 < a < 1$. The moment-generating function is computed as

---

[†]See "Completing the squares" in Appendix A.

$$\theta(t) = \sum_{n=0}^{\infty} (1-a)a^n e^{tn}$$

$$= (1-a)\sum_{n=0}^{\infty}(ae^t)^n = \frac{1-a}{1-ae^t}.$$

Then the mean is computed from $\theta'(0) = (1-a)(1-ae^t)^{-2}ae^{-t}|_{t=0} = a/(1-a)$.

We make the observation that if all the moments exist and are known, then $\theta(t)$ is known as well (see Equations 4.5-4 and 4.5-2). Since $\theta(t)$ is related to $f_X(x)$ through the Laplace transform, we can, in principle at least, determine $f_X(x)$ from its moments if they exist.[†] In practice, if X is the r.v. whose pdf is desired and X_i represents our ith observation of X, then we can *estimate* the rth moment of X, ξ_r, from

$$\hat{\Theta}_r = \frac{1}{n}\sum_{i=1}^{n} X_i^r, \tag{4.5-11}$$

where $\hat{\Theta}_r$ is called a *moment estimator* and is a random variable, and n is the number of observations. Even though $\hat{\Theta}_r$ is an r.v., its variance becomes small as n becomes large and for n large enough we can have confidence that $\hat{\Theta}_r$ is reasonably close to ξ_r (not a r.v.).

The moment-generating function $\theta_{XY}(t_1, t_2)$ of two r.v.'s X and Y is defined by

$$\theta_{XY}(t_1, t_2) \triangleq E[e^{(t_1 X + t_2 Y)}],$$

$$= \int_{-\infty}^{\infty}\int_{-\infty}^{\infty} \exp(t_1 x + t_2 y) f_{XY}(x,y)\, dx\, dy. \tag{4.5-12}$$

Proceeding as we did in Equation 4.5-4, we can easily establish with the help of a power series expansion that

$$\theta_{XY}(t_1, t_2) = \sum_{i=0}^{\infty}\sum_{j=0}^{\infty} \frac{t_1^i t_2^j}{i!j!}\xi_{ij}, \tag{4.5-13}$$

where ξ_{ij} is defined in Equation 4.3-13. Using the notation

$$\theta_{XY}^{(l,n)}(0,0) \triangleq \left.\frac{\partial^{l+n}\theta_{XY}(t_1, t_2)}{\partial t_1^l \partial t_2^n}\right|_{t_1=t_2=0}$$

we can easily show from Equation 4.5-12 or 4.5-13 that

$$\xi_{ln} = \theta_{XY}^{(l,n)}(0,0). \tag{4.5-14}$$

[†]For some distributions not all moments exist. For example, as stated earlier for the Cauchy distribution, all moments above the first do not exist.

In particular

$$\theta_{XY}^{(1,0)}(0,0) = \overline{X}, \qquad \theta_{XY}^{(0,1)}(0,0) = \overline{Y} \qquad (4.5\text{-}15)$$

$$\theta_{XY}^{(2,0)}(0,0) = \overline{X^2}, \qquad \theta_{XY}^{(0,2)}(0,0) = \overline{Y^2} \qquad (4.5\text{-}16)$$

$$\theta_{XY}^{(1,1)}(0,0) = \overline{XY} = \text{Cov}[X,Y] + \overline{X}\,\overline{Y}. \qquad (4.5\text{-}17)$$

One could similarly define a joint moment-generating function for N random variables $X_1, \ldots, X_N$ by

$$\theta(t_1, t_2, \ldots, t_N) = E\left[\exp \sum_{i=1}^{N} t_i X_i\right]$$

$$= \sum_{k_1=0}^{\infty} \sum_{k_2=0}^{\infty} \cdots \sum_{k_N=0}^{\infty} \frac{t_1^{k_1}}{k_1!} \cdots \frac{t_N^{k_N}}{k_N!} \overline{X_1^{k_1} X_2^{k_2} \ldots X_N^{k_N}} \qquad (4.5\text{-}18)$$

from which the moments can be computed using straightforward extensions of the earlier cases.

4.6 CHERNOFF BOUND

The Chernoff bound furnishes an upper bound on the tail probability $P[X \geq a]$ where a is some prescribed constant. First note that $u(x - a) \leq e^{t(x-a)}$ for any $t > 0$. Assume that X is a continuous r.v. Then

$$P[X \geq a] = \int_a^{\infty} f_X(x)\,dx$$

$$= \int_{-\infty}^{\infty} f_X(x)u(x - a)\,dx \qquad (4.6\text{-}1)$$

and, by the observation made above, it follows that

$$P[X \geq a] \leq \int_{-\infty}^{\infty} f_X(x)e^{t(x-a)}\,dx \qquad (4.6\text{-}2)$$

and this must hold for any $t > 0$. But, from Equation 4.5-2,

$$\int_{-\infty}^{\infty} f_X(x)e^{t(x-a)}\,dx = e^{-at}\theta_X(t), \qquad (4.6\text{-}3)$$

where the subscript has been added to emphasize that the moment-generating function is associated with X. Combining Equations 4.6-3 and 4.6-2, we obtain

$$P[X \geq a] \leq e^{-at}\theta_X(t). \qquad (4.6\text{-}4)$$

The tightest bound, which occurs when the right-hand side is minimized with respect to t, is called the Chernoff bound. We illustrate with some examples.

Example 4.6-1 _____

Let $X : N(\mu, \sigma^2)$ and consider the Chernoff bound on $P(X \geq a)$ where $a > \mu$. From Equations 4.5-7 and 4.6-3 we obtain

$$P[X \geq a] \leq e^{-(a-\mu)t + \sigma^2 t^2/2}.$$

The minimum of the right-hand side is obtained by differentiating with respect to t and occurs when $t = (a - \mu)/\sigma^2$. Hence the Chernoff bound is

$$P[X \geq a] \leq e^{-(a-\mu)^2/2\sigma^2}. \tag{4.6-5}$$

The Chernoff bound can be derived for discrete random variables also. For example, assume that an r.v. X takes values $X = i$, $i = 0, 1, 2, \ldots$ with probabilities $P[X = i] \triangleq P_X(i)$. For any integers n, k, define

$$u(n - k) = \begin{cases} 1 & u(n \geq k) \\ 0, & \text{otherwise.} \end{cases}$$

If follows, therefore, that

$$P[X \geq k] = \sum_{n=k}^{\infty} P_X(n)$$

$$= \sum_{n=0}^{\infty} P_X(n) u(n - k)$$

$$= \sum_{n=0}^{\infty} P_X(n) e^{t(n-k)} \qquad \text{for } t \geq 0.$$

The last line follows from the fact that

$$e^{t(n-k)} \geq u(n - k) \qquad \text{for } t \geq 0.$$

We note that

$$\sum_{n=0}^{\infty} P_X(n) e^{t(n-k)} = e^{-tk} \sum_{n=0}^{\infty} P_X(n) e^{tn}$$

$$= e^{-tk} \theta_X(t) \qquad \text{(by Equation 4.5-3).}$$

Hence we establish the important result

$$P[X \geq k] \leq e^{-tk} \theta_X(t). \tag{4.6-6}$$

As before, the Chernoff bound is determined by minimizing the right-hand side of Equation 4.6-6. We illustrate with an example.

Example 4.6-2 _____

Let X be a Poisson r.v. with parameter $a > 0$. Compute the Chernoff bound for $P[X \geq k]$ where $k > a$. From homework Problem 4.26 we find that

$$\theta_X(t) = e^{a[e^t - 1]}$$

and

$$e^{-tk}\theta_X(t) = e^{-a}e^{[ae^t - kt]}.$$

By setting

$$\frac{d}{dt}[e^{-tk}\theta_X(t)] = 0$$

we find that the minimum is reached when $t = t_m$ where

$$t_m = \ln\frac{k}{a}.$$

Thus with $a = 2$ and $k = 5$, we find

$$P[X \geq 5] \leq e^{-2}\exp[5 - 5\ln(5/2)]$$

$$\leq 0.2.$$

_____

4.7 CHARACTERISTIC FUNCTIONS

If in Equation 4.5-1 we replace the parameter t by $j\omega$ where $j \triangleq \sqrt{-1}$, we obtain the *characteristic* function of X defined by

$$\Phi_X(\omega) \triangleq E[e^{j\omega X}]$$

$$= \int_{-\infty}^{\infty} f_X(x)e^{j\omega x}\, dx, \tag{4.7-1}$$

which except for a minus sign difference in the exponent, we recognize as the Fourier transform of $f_X(x)$. For discrete r.v.'s we can define $\Phi_X(\omega)$ in terms of the PMF by

$$\Phi_X(\omega) = \sum_i e^{j\omega x_i}P_X(x_i). \tag{4.7-2}$$

For our purposes, the characteristic function has all the properties of the moment-generating function. The Fourier transform is widely used in statistical communication theory, and since the inversion of Equation 4.7-1 is often easy to achieve, either by direct integration or through the availability of extensive tables of Fourier transforms (for example, Reference 4-7), the characteristic function is widely used to solve problems involving the computation of the sums of independent r.v.'s. We have seen that the pdf of the sum of independent random variables involves the convolution of their pdf's. Thus if $Z = X_1 + \ldots + X_N$, where $X_i, i = 1, \ldots, N$ are independent r.v.'s the pdf of Z is furnished by

$$f_Z(z) = f_{X_1}(z) * f_{X_2}(z) * \ldots * f_{X_N}(z) \tag{4.7-3}$$

that is, the repeated convolution product.

The actual evaluation of Equation 4.7-3 can be very tedious. However, we know from our studies of Fourier transforms that the *Fourier transform of a convolution product is the product of the individual transforms*. We illustrate the use of characteristic functions in the following examples.

Example 4.7-1

Let $Z \overset{\Delta}{=} X_1 + X_2$ with $f_{X_1}(x)$, $f_{X_2}(x)$ and $f_Z(z)$ denoting the pdf's of X_1, X_2, and Z, respectively. Show that $\Phi_Z(\omega) = \Phi_{X_1}(\omega)\Phi_{X_2}(\omega)$.

Solution   From the main result of Section 3.3 (Equation 3.3-15) we have

$$f_Z(z) = \int_{-\infty}^{\infty} f_{X_1}(x)f_{X_2}(z - x)\,dx$$

and

$$\Phi_Z(\omega) = \int_{-\infty}^{\infty} e^{j\omega z} \left[\int_{-\infty}^{\infty} f_{X_1}(x)f_{X_2}(z - x)\,dx \right] dz$$

$$= \int_{-\infty}^{\infty} f_{X_1}(x) \int_{-\infty}^{\infty} f_{X_2}(z - x)e^{j\omega z}\,dx\,dz.$$

With a change of variable $\alpha \overset{\Delta}{=} z - x$ we obtain

$$\Phi_Z(\omega) = \Phi_{X_1}(\omega)\Phi_{X_2}(\omega).$$

This result is easily extended, by induction, to N r.v.'s. Thus if $Z_1 = X_1 + \cdots + X_N$, then

$$\Phi_Z = \Phi_{X_1}(\omega)\Phi_{X_2}(\omega)\ldots\Phi_{X_N}(\omega).$$

Example 4.7-2

Let $X_i, i = 1, \ldots, N$ be a sequence of *independent, identically distributed* (i.i.d.) r.v.'s with $X: N(0, 1)$. Compute the pdf of

$$Z = \sum_{i=1}^{N} X_i.$$

Solution   The pdf of Z can be computed by Equation 4.7-3. On the other hand with $\Phi_{X_i}(\omega)$ denoting the characteristic function of X_i we have

$$\Phi_Z(\omega) = \Phi_{X_1}(\omega) \times \ldots \times \Phi_{X_N}(\omega). \tag{4.7-4}$$

However, since the X_i's are i.i.d. $N(0,1)$, the characteristic function of all the X_i's are the same, and we define $\Phi_X(\omega) \overset{\Delta}{=} \Phi_{X_2}(\omega) = \ldots = \Phi_{X_N}(\omega)$. Thus

$$\Phi_X(\omega) = \int_{-\infty}^{\infty} \frac{1}{\sqrt{2\pi}} e^{-\frac{1}{2}x^2} e^{j\omega x}\, dx. \tag{4.7-5}$$

By completing the squares in the exponent we obtain

$$\Phi_X(\omega) = \frac{1}{\sqrt{2\pi}} \int_{-\infty}^{\infty} e^{-\frac{1}{2}[x^2 - 2j\omega x + (j\omega)^2 - (j\omega)^2]}\, dx$$

$$= e^{-\frac{\omega^2}{2}} \int_{-\infty}^{\infty} \frac{1}{\sqrt{2\pi}} e^{-\frac{1}{2}(x - j\omega)^2}\, dx.$$

But the integral can be regarded as the area under a "Gaussian pdf" with "mean" $j\omega$. Hence its value is unity and we obtain

$$\Phi_X(\omega) = e^{-\frac{\omega^2}{2}}$$

and

$$\Phi_Z(\omega) = [\Phi_X(\omega)]^n = e^{-\frac{1}{2}n\omega^2}. \tag{4.7-6}$$

From the form of $\Phi_Z(\omega)$ we deduce that $f_Z(z)$ must also be Gaussian. To obtain $f_Z(z)$ we use the Fourier inversion formula:

$$f_Z(z) = \frac{1}{2\pi} \int_{-\infty}^{\infty} \Phi_Z(\omega) e^{-j\omega z}\, d\omega. \tag{4.7-7}$$

Inserting Equation 4.7-6 into Equation 4.7-7 and manipulating terms enables us to obtain

$$f_Z(z) = \frac{1}{\sqrt{2\pi n}} e^{-\frac{1}{2}(z^2/n)}.$$

Hence $f_Z(z)$ is indeed Gaussian. The variance of Z is n, and its mean is zero.

Example 4.7-3 _____
Consider two i.i.d. r.v.'s X and Y with

$$f_X(x) = f_Y(x) = \frac{1}{a}\mathrm{rect}\left(\frac{x}{a}\right).$$

Compute the pdf of $Z = X + Y$ using characteristic functions.

Solution   We can, of course, compute $f_Z(z)$ by convolving $f_X(x)$ and $f_Y(y)$. However, using characteristic functions, we obtain $f_Z(z)$ from

$$f_Z(z) = \frac{1}{2\pi} \int_{-\infty}^{\infty} \Phi_X(\omega)\Phi_Y(\omega) e^{-j\omega z}\, d\omega,$$

where

$$\Phi_X(\omega)\Phi_Y(\omega) = \Phi_Z(\omega).$$

Since the pdf's of X and Y are the same, we can write

$$\Phi(\omega) \overset{\Delta}{=} \Phi_X(\omega) = \Phi_Y(\omega)$$

$$= \frac{1}{a} \int_{-a/2}^{a/2} e^{j\omega x}\, dx$$

$$= \frac{\sin(a\omega/2)}{a\omega/2}.$$

Hence

$$\Phi_Z(\omega) = \left(\frac{\sin(a\omega/2)}{a\omega/2} \right)^2 \tag{4.7-8}$$

and

$$f_Z(z) = \frac{1}{2\pi} \int_{-\infty}^{\infty} \Phi_Z(\omega) e^{-j\omega z} d\omega$$

$$= \frac{1}{a}\left(1 - \frac{|z|}{a} \right) \mathrm{rect}\left(\frac{z}{2a} \right), \tag{4.7-9}$$

which is shown in Figure 4.7-1. The easiest way to obtain the result in Equation 4.7-9 is to look up the Fourier transform (or its inverse) of Equation 4.7-8 in a table of elementary Fourier transforms.

As in the case of moment-generating functions, we can compute the moments from the characteristic functions by differentiation, provided that these exist. If we expand $\exp(j\omega X)$ into a power series and take the expectation, we obtain

$$\Phi_X(\omega) = E[e^{j\omega X}]$$

$$= \sum_{n=0}^{\infty} \frac{(j\omega)^n}{n!} \xi_n. \tag{4.7-10}$$

Figure 4.7-1   The pdf of $Z = X + Y$ when X and Y are i.i.d. uniformly distributed in $(-a/2, a/2)$.

From Equation 4.7-10 it is easily established that

$$\xi_n = \frac{1}{j^n} \Phi_X^{(n)}(0) \tag{4.7-11}$$

where we have used the notation

$$\Phi_N^{(n)}(0) \triangleq \frac{d^n}{d\omega^n} \Phi_X(\omega)\bigg|_{\omega=0} .$$

Example 4.7-4

Compute the first few moments of $Y = \sin \Theta$ if Θ is uniformly distributed on $[0, 2\pi]$.

Solution   We use the result in Equation 4.1-8; that is, if $Y = g(X)$, then

$$\overline{Y} = \int_{-\infty}^{\infty} y f_Y(y) dy = \int_{-\infty}^{\infty} g(x) f_X(x) \, dx.$$

Hence

$$E[e^{j\omega Y}] = \int_{-\infty}^{\infty} e^{j\omega y} f_Y(y) dy$$

$$= \frac{1}{2\pi} \int_0^{2\pi} e^{j\omega \sin \theta} d\theta$$

$$= J_0(\omega),$$

where $J_0(\omega)$ is the Bessel function of the first kind of order zero. A power series expansion of $J_0(\omega)$ gives

$$J_0(\omega) = 1 - \left(\frac{\omega}{2}\right)^2 + \frac{1}{2!2!}\left(\frac{\omega}{2}\right)^4 - \dots .$$

Hence all the odd-order moments are zero. From Equation 4.7-11 we compute

$$\overline{Y^2} = \xi_2 \triangleq (-1)\Phi_X^{(2)}(0) = \tfrac{1}{2}$$

$$\overline{Y^4} = \xi_4 = (+1)\Phi_X^{(4)}(0) = \tfrac{3}{8}.$$

Example 4.7-5

Let X and Y be i.i.d. binomial r.v.'s with parameters n and p, that is,

$$P_X(k) = P_Y(k) = \binom{n}{k} p^k q^{n-k}.$$

Compute the PMF of $Z = X + Y$.

Solution   Since X and Y take on nonnegative integer values, so must Z. We can solve this problem by (1) convolution of the pdf's, which involves delta functions; (2) *discrete* *convolution* of the PMF's; and (3) by characteristic functions. The discrete convolution for this case is

$$P[Z = k] = \sum_i P_X(i) P_Y(k - i)$$

$$= \sum_i \binom{n}{i} p^i q^{n-i} \binom{n}{k-i} p^{k-i} q^{n-(k-i)}$$

$$= p^k q^{2n-k} \sum_i \binom{n}{i} \binom{n}{k-i}, \quad \text{for } k = 0, 1, \ldots, 2n.$$

The trouble is that we may not immediately recognize the closed form of the sum of products of binomial coefficients.[†] The computation of the PMF of Z by characteristic functions is very simple. First observe that

$$\Phi_X(\omega) = \Phi_Y(\omega) = \sum_{k=0}^n e^{j\omega k} \binom{n}{k} p^k q^{n-k}$$

$$= (p e^{j\omega} + q)^n.$$

Thus, by virtue of the independence of X and Y, we obtain

$$\Phi_Z(\omega) = E[\exp j\omega(X + Y)]$$

$$= E[\exp(j\omega X)] E[\exp(j\omega Y)]$$

$$= (p e^{j\omega} + q)^{2n}.$$

Thus Z is binomial with parameters $2n$ and p, that is,

$$P_Z(k) = \binom{2n}{k} p^k q^{2n-k}, \quad \text{for } k = 0, \ldots, 2n.$$

As a by-product of the computation of $P_Z(k)$ by c.f.'s we obtain the result that

$$\binom{2n}{k} = \sum_{i=0}^n \binom{n}{i} \binom{n}{k-i}.$$

An extension of this result is the following: If $X_1, X_2, \ldots, X_N$ are i.i.d. binomial with parameters n, p then $Z = \sum_{i=1}^N X_i$ is binomial with parameters Nn, p. Regardless of how large N gets, Z remains a discrete r.v. with a binomial PMF.[‡]

Example 4.7-6

A lottery game called "three players for six hits" is played as follows. A bettor bets the bank that three baseball players of the bettor's choosing will get a combined total of six

---

[†] Recall that we ran into this problem in Example 3.3-7 in Chapter 3.
[‡] Recall this statement for future reference in connection with the *Central Limit Theorem*.

hits or more in the games in which they play. Many combinations can lead to a win; for example, player A can go hitless in his game, but player B can collect three hits in his game, and player C can collect three hits in his game. The players can be on the same team or on different teams. The bet is at even odds and the bettor receives back $2 on a bet of $1 in case of a win. Is this a "fair" game, that is, is the probability of a win close to one-half?

Solution   Let X_1, X_2, X_3 denote the number of hits by players A, B, C, respectively. Clearly X_1, X_2, X_3 are individually binomial. The total number of hits is $Y = \sum_{i=1}^{3} X_i$. We wish to compute $P[Y \geq 6]$. To simplify the problem, assume that each player bats five times per game, and their batting averages are the same, say 300 (for those unfamiliar with baseball nomenclature, this means that the probability of getting a hit while batting is 0.3). Then from the results of Example 4.7-5, we find Y is binomial with parameters $n = 15$, $p = 0.3$. Thus

$$P[Y \geq 6] = \sum_{i=0}^{i=15} \binom{15}{i} (0.3)^i (0.7)^{15-i}$$

$$\approx \operatorname{erf}(6.76) - \operatorname{erf}(0.56)$$

$$\approx 0.29.$$

In arriving at this result, we used the normal approximation to the binomial as suggested in Section 1.11. The bettor has less than a *one-third chance* of winning. Despite the poor odds, the game can be modified to be fairer to the bettor. Define the random variables G as the *gain* to the bettor and define a fair game as one in which the expected gain is zero. Then if the bettor were to receive winnings of $2.45 per play instead of $1, we would find that $E\{G\} = \$2.45 \times 0.29 - \$1 \times 0.71 \approx 0$. Of course if $E\{G\} > 0$, then in a sense, the game favors the bettor. Some people play the state lottery using this criterion.

Joint Characteristic Functions

As in the case of joint moment-generating functions we can define the joint characteristic function by

$$\Phi_{X_1 \ldots X_N}(\omega_1, \omega_2, \ldots, \omega_N) = E\left[\exp\left(j \sum_{i=1}^{N} \omega_i X_i\right)\right]. \tag{4.7-12}$$

By the Fourier inversion property, the joint pdf is the inverse Fourier transform (with a sign reversal) of $\Phi_{X_1 \ldots X_N}(\omega_1, \ldots, \omega_N)$. Thus

$$f_{X_1 \ldots X_N}(x_1, \ldots, x_N) = \frac{1}{(2\pi)^n} \int_{-\infty}^{\infty} \cdots \int_{-\infty}^{\infty} \Phi_{X_1 \ldots X_N}(\omega_1, \ldots, \omega_N)$$

$$\times \exp\left(-j \sum_{i=1}^{N} \omega_i x_i\right) d\omega_1 \, d\omega_2 \ldots d\omega_N. \tag{4.7-13}$$

We can obtain the moments by differentiation. For instance, with X, Y denoting any two random variables ($N = 2$) we have

$$\xi_{rk} \overset{\Delta}{=} E[X^r Y^k] = (-j)^{r+k} \Phi_{XY}^{(r,k)}(0,0), \tag{4.7-14}$$

where

$$\Phi_{XY}^{(rk)}(0,0) \overset{\Delta}{=} \frac{\partial^{r+k} \Phi_{XY}(\omega_1, \omega_2)}{\partial \omega_1^r \partial \omega_2^k}\bigg|_{\omega_1 = \omega_2 = 0}. \tag{4.7-15}$$

Finally for discrete r.v.'s we can define the joint characteristic function in terms of the joint PMF. For instance for two r.v.'s, X and Y, we obtain

$$\Phi_{XY}(\omega_1, \omega_2) \overset{\Delta}{=} \sum_i \sum_j e^{j(\omega_1 x_i + \omega_2 y_j)} P_{XY}(x_i, y_j). \tag{4.7-16}$$

Example 4.7-7

Compute the joint characteristic function of X and Y if

$$f_{XY} = \frac{1}{2\pi} \exp\left[-\frac{1}{2}(x^2 + y^2)\right]$$

Solution   Applying the definition in Equation 4.7-12 we get

$$\Phi_{XY}(\omega_1, \omega_2) = \frac{1}{2\pi} \int_{-\infty}^{\infty} \int_{-\infty}^{\infty} e^{-\frac{1}{2}(x^2+y^2)} e^{j\omega_1 x + j\omega_2 y} \, dx \, dy.$$

Completing the squares in both x *and* y, we get

$$\Phi_{XY}(\omega_1, \omega_2) = e^{-\frac{1}{2}(\omega_1^2 + \omega_2^2)} \int_{-\infty}^{\infty} e^{-\frac{1}{2}[x^2 - 2j\omega_1 x + (j\omega_1)^2]} \frac{dx}{\sqrt{2\pi}}$$

$$\times \int_{-\infty}^{\infty} e^{-\frac{1}{2}[y^2 - j\omega_2 x + (j\omega_2)^2]} \frac{dy}{\sqrt{2\pi}}$$

$$= e^{-\frac{1}{2}(\omega_1^2 + \omega_2^2)} \int_{\infty}^{\infty} e^{\frac{1}{2}(x - j\omega_1)^2} \frac{dx}{\sqrt{2\pi}} \int_{-\infty}^{\infty} e^{\frac{1}{2}(y - j\omega_2)^2} \frac{dy}{\sqrt{2\pi}}$$

$$= e^{-\frac{1}{2}(\omega_1^2 + \omega_2^2)},$$

since the integrals are the areas under unit-variance Gaussian curves.

Example 4.7-8

Compute the joint characteristic function of the discrete r.v.'s X and Y if the joint PMF is

$$P_{XY}(k, l) = \begin{cases} \frac{1}{3}, & k = l = 0 \\ \frac{1}{6}, & k = \pm 1, l = 0 \\ \frac{1}{6}, & k = l = \pm 1 \\ 0, & \text{else.} \end{cases}$$

Solution   Using Equation 4.7-16 we obtain

$$\Phi_{XY}(\omega_1, \omega_2) = \sum_{k=-1}^{1} \sum_{l=-1}^{1} e^{j(\omega_1 k + \omega_2 l)} P_{XY}(k, l)$$

$$+ \frac{1}{3} + \frac{1}{3} \cos \omega_1 + \frac{1}{3} \cos(\omega_1 + \omega_2).$$

From Equations 4.7-14 and 4.7-15 we obtain, since $\overline{X} = \overline{Y} = 0$,

$$\sigma_X^2 \triangleq \xi_{20} = -(-j)^2 [\cos \omega_1 + \cos(\omega_1 + \omega_2)] \left. \frac{1}{3} \right|_{\omega_1 = \omega_2 = 0}$$

$$= \frac{2}{3};$$

$$\sigma_Y^2 \triangleq \xi_{02} = -(-j)^2 \frac{1}{3} \cos(\omega_1 + \omega_2) \Big|_{\omega_1 = \omega_2 = 0}$$

$$= \frac{1}{3};$$

$$\xi_{11} = -(-j)^2 \frac{1}{3} \cos(\omega_1 + \omega_2) \Big|_{\omega_1 = \omega_0 = 0}$$

$$= \frac{1}{3}.$$

Hence the correlation coefficient ρ is computed to be

$$\rho = \frac{\xi_{11}}{\sigma_X \sigma_Y} = \frac{\frac{1}{3}}{\sqrt{\frac{2}{3}} \sqrt{\frac{1}{3}}} = \frac{1}{\sqrt{2}} = 0.707.$$

Example 4.7-9

As another example we compute the joint characteristic function of X and Y with

$$f_{XY}(x, y) = \frac{1}{2\pi \sqrt{1 - \rho^2}} \exp\left(-\frac{x^2 + y^2 - 2\rho xy}{2(1 - \rho^2)}\right).$$

To solve this problem we use two facts:

(1) A zero-mean Gaussian r.v. Z with variance σ_Z^2 has characteristic function

$$E[e^{j\omega Z}] = \exp\left[-\frac{1}{2}\sigma_Z^2 \omega^2\right] \qquad (4.7\text{-}17)$$

and, in particular, with $\omega = 1$,

$$E[e^{jZ}] = \exp\left[-\frac{1}{2}\sigma_Z^2\right]. \qquad (4.7\text{-}18)$$

Proof of fact (1) Use the definition of the characteristic function with $f_Z(z) = (2\pi\sigma_Z^2)^{-1/2} \exp\left(-\dfrac{1}{2}\dfrac{z^2}{\sigma_Z^2}\right)$ and apply the complete-the-square technique described in Appendix A.

(2) If X and Y are zero-mean jointly Gaussian r.v.'s, then for any real ω_1, ω_2, the r.v.'s

$$Z \overset{\Delta}{=} \omega_1 X + \omega_2 Y$$

$$W \overset{\Delta}{=} X$$

are jointly Gaussian and, as a direct by-product, the marginal density of Z is Gaussian.

Proof of fact (2) Simply use Equation 3.4-11 or 3.4-12 to compute $f_{ZW}(z, w)$. One easily finds that Z, W are jointly Gaussian and that, therefore, the marginal pdf of Z alone is Gaussian with $\overline{Z} = 0$. The variance of Z is computed as

$$\text{Var}(Z) = E[(\omega_1 X + \omega_2 Y)^2]$$
$$= \omega_1^2 \text{Var}[X] + \omega_2^2 \text{Var}[Y] + 2\omega_1\omega_2\overline{XY}.$$

With $\sigma_X^2 = \sigma_Y^2 \overset{\Delta}{=} 1$, we obtain $\sigma_Z^2 = \omega_1^2 + \omega_2^2 + 2\omega_1\omega_2\rho$.

Finally recalling that $Z = \omega_1 X + \omega_2 Y$ and using Equation 4.7-18, we write

$$E[e^{j(\omega_1 X + \omega_2 Y)}] = e^{-\frac{1}{2}(\omega_1^2 + \omega_2^2 + \omega_1\omega_2\rho)}. \tag{4.7-19}$$

Equation 4.7-19 is the joint characteristic function of two zero-mean, unity variance correlated Gaussian r.v.'s. When $\rho = 0$, the r.v.'s become uncorrelated and therefore independent and we obtain the result in Example 4.7-7.

The extension to more than two discrete r.v.'s is straightforward, although the notation becomes a little clumsy, unless matrices are introduced.

The Central Limit Theorem

It is sometimes said that the sum of a large number of random variables tends toward the Normal. Under what conditions is this true? The *Central Limit Theorem* deals with this important point.

Basically the central limit theorem[†] says that the normalized sum of a large number of mutually independent random variables $X_1, \ldots, X_n$ with zero means and finite variances $\sigma_1^2, \ldots, \sigma_n^2$ tends to the Normal probability distribution function provided that the individual variances σ_k^2, $k = 1, \ldots, n$ are small compared to $\sum_{i=1}^n \sigma_i^2$. The constraints on the variances are known as the Lindeberg conditions and are discussed in detail by Feller [4-1, p. 262]. We state a general form of the Central Limit Theorem in the following and furnish a proof for a special case.

[†]First proved by Abraham De Moivre in 1733 for the special case of Bernoulli r.v.'s. A more general proof was furnished by J. W. Lindeberg in *Mathematische Zeitschrift*, vol. 15 (1922), pp. 211–225.

Theorem 4.7-1   Let $X_1, \ldots, X_n$ be n mutually independent (scalar) r.v.'s with PDF's $F_{X_1}(x_1)$, $F_{X_2}(x_2), \ldots, F_{X_n}(x_n)$, respectively, such that

$$\overline{X}_k = 0, \qquad \mathrm{Var}[X_k] = \sigma_k^2$$

and let

$$s_n^2 \triangleq \sigma_1^2 + \ldots + \sigma_n^2.$$

If for a given $\varepsilon > 0$ and n sufficiently large the σ_k satisfy

$$\sigma_k < \varepsilon s_n \qquad k = 1, \ldots, n,$$

then the normalized sum

$$Z_n \triangleq (X_1 + \ldots + X_n)/s_n$$

converges to the standard Normal PDF, denoted $1/2 + \mathrm{erf}(z)$, that is, $\lim_{n \to \infty} F_{Z_n}(z) = 1/2 + \mathrm{erf}(z)$.   ■

A discussion of convergence in distribution is given later in this section.

We now prove a special case of the foregoing.

Theorem 4.7-2   Let $X_1, X_2, \ldots, X_n$ be i.i.d. r.v.'s with $\overline{X}_i = 0$, and $\mathrm{Var}[X_i] = 1$, $i = 1, \ldots, n$. Then

$$Z_n \triangleq \frac{1}{\sqrt{n}} \sum_{i=1}^{n} X_i$$

tends to the normal in the sense that its characteristic function Φ_{Z_n} satisfies

$$\lim_{n \to \infty} \Phi_{Z_n}(\omega) = e^{-\frac{1}{2}\omega^2},$$

which is the characteristic function of the $N(0, 1)$ r.v.

Proof   Let $W_i \triangleq X_i/\sqrt{n}$. Also, let $\Phi_{X_i}(\omega)$ and $f_{X_i}(x)$ be the characteristic function and pdf of X, respectively. Then

$$\Phi_{W_i} \triangleq E[e^{j\omega W_i}]$$

$$= E[e^{j(\omega/\sqrt{n})X_i}]$$

$$= \Phi_{X_i}\left(\frac{\omega}{\sqrt{n}}\right).   ■$$

Since $\Phi_{X_i}(\omega)$ and $\Phi_{W_i}(\omega)$ do not depend on i, we write $\Phi_{X_i}(\omega) \triangleq \Phi_X(\omega)$ and $\Phi_{W_i}(\omega) \triangleq \Phi_W(\omega)$. From calculus we know that any function $\Phi(\omega)$ whose derivative exists in a neighborhood about ω_0 can be represented by a Taylor series

$$\Phi(\omega) = \sum_{l=0}^{\infty} \frac{1}{l!} \Phi^{(l)}(\omega_0)(\omega - \omega_0)^l,$$

where $\Phi^{(l)}(\omega_0)$ is the lth derivative of $\Phi(\omega)$ at ω_0. Moreover if the derivatives are continuous in the interval $[\omega_0, \omega]$, $\Phi(\omega)$ can be expressed as a finite Taylor series plus a remainder $A_L(\omega)$, that is,

$$\Phi(\omega) = \sum_{l=0}^{L-1} \frac{1}{l!} \Phi^l(\omega_0)(\omega - \omega_0)^l + A_L(\omega),$$

where

$$A_L(\omega) \triangleq \frac{1}{L!} \Phi^L(\xi)(\omega - \omega_0)^L$$

and ξ is some point in the interval $[\omega_0, \omega]$. Let us apply this result to $\Phi_W(\omega)$ with $\omega_0 = 0$. Then

$$\Phi_W(\omega) = \int_{-\infty}^{\infty} e^{j\omega x/\sqrt{n}} f_X(x)\, dx$$

$$\Phi_W^{(0)}(0) = 1$$

$$\Phi_W^{(1)}(0) = \int_{-\infty}^{\infty} j\, \frac{x}{\sqrt{n}} e^{j\omega x/\sqrt{n}} f_X(x)\, dx \Big|_{\omega=0} = 0$$

$$\Phi_W^{(2)}(0) = \int_{-\infty}^{\infty} \left(\frac{jx}{\sqrt{n}} \right)^2 e^{j\omega x/\sqrt{n}} f_X(x)\, dx \Big|_{\omega=0} = -\frac{1}{n}.$$

Hence

$$\Phi_W(\omega) = 1 - \frac{1}{2n}\omega^2 + \frac{R_2'(\omega)}{n\sqrt{n}},$$

where

$$R_2'(\omega) \triangleq -j\omega^3 \int_{-\infty}^{\infty} x^3 e^{j\xi x/\sqrt{n}} f_X(x)\, dx/6.$$

Since $Z_n = \sum_{i=1}^{n} W_i$, we obtain

$$\Phi_{Z_n}(\omega) = [\Phi_W(\omega)]^n,$$

or

$$\ln \Phi_{Z_n}(\omega) = n \ln \Phi_W(\omega).$$

Now recall that for any h such that $|h| < 1$,

$$\ln(1 + h) = h - \frac{h^2}{2} + \frac{h^3}{3} - \cdots.$$

For any *fixed* ω, we can choose an n large enough so that (let $R_2' \triangleq R_2'(\omega)$)

$$\left| -\frac{\omega^2}{2n} + \frac{R_2'}{n\sqrt{n}} \right| < 1.$$

Assuming this to have been done, we can write

$$\ln \Phi_{Z_n}(\omega) = n \ln \left[1 - \frac{\omega^2}{2n} + \frac{R_2'}{n\sqrt{n}} \right]$$

$$\simeq n \left[-\frac{\omega^2}{2n} + \frac{R_2'}{n\sqrt{n}} - \left(-\frac{\omega^2}{2n} + \frac{R_2'}{n\sqrt{n}} \right)^2 + \left(-\frac{\omega^2}{2n} + \frac{R_2'}{n\sqrt{n}} \right)^3 + \cdots \right]$$

$$= -\frac{\omega^2}{2} + \text{ terms involving factors of } n^{-1/2}, n^{-1}, n^{-3/2}, \ldots.$$

Hence

$$\lim_{n\to\infty} [\ln \Phi_{Z_n}(\omega)] = -\frac{\omega^2}{2}$$

or, equivalently,

$$\lim_{n\to\infty} \Phi_{Z_n}(\omega) = e^{-\omega^2/2},$$

which is the characteristic function of the $N(0,1)$ r.v. Note that to argue that $\lim_{n\to\infty} f_{Z_n}(z)$ is the normal pdf we should have to argue that

$$\lim_{n\to\infty} \Phi_{Z_n}(\omega) \overset{\Delta}{=} \lim_{n\to\infty} \left(\int_{-\infty}^{\infty} f_{Z_n}(z) e^{j\omega z} dz \right)$$

$$\overset{?}{=} \int_{-\infty}^{\infty} \left(\lim_{n\to\infty} f_{Z_n}(z) \right) e^{j\omega z} dz.$$

However, the operations of limiting and integrating are not always interchangeable. Hence we cannot say that the pdf of Z_n converges to $N(0,1)$. Indeed we already know from Example 4.7-5 that the sum of n i.i.d. binomial r.v.'s is binomial regardless of how large n is; moreover, the binomial PMF or pdf is a discontinuous function while the Gaussian is continuous and no matter how large n is, this fact cannot be altered. However, the integrals of the binomial pdf, for large n, behave like integrals of the Gaussian pdf. This is why the distribution function of Z_n tends to a Gaussian distribution function but not necessarily to a Gaussian pdf.

The astute reader will have noticed that in the prior development we showed the normal convergence of the characteristic function but not as yet the normal convergence of the PDF. To prove the latter true we can use a continuity theorem (see Feller [4-1, p. 508]) which states the following: Consider a sequence of r.v.'s $Z_i, i = 1, \ldots, n$ with characteristic functions and PDF's $\Phi_i(\omega)$ and $F_i(z)$, $i = 1, \ldots, n$ respectively; with $\Phi(\omega) \overset{\Delta}{=} \lim_{n\to\infty} \Phi_n(\omega)$ and $\Phi(\omega)$ continuous at $\omega = 0$ then $F(z) = \lim_{n\to\infty} F_n(z)$.

Example 4.7-10 _____

(Application of the Central Limit Theorem [CLT].) Let $X_i, i = 1, \ldots, n$ be a sequence of i.i.d. r.v.'s with $E[X_i] = \mu_X$ and $\text{Var}[X_i] = \sigma_X^2$. Let $Y \overset{\Delta}{=} \sum_{i=1}^{n} X_i$. We wish to compute $P[a \leq Y \leq b]$ using the CLT. Clearly with $Z \overset{\Delta}{=} (Y - E[Y])/\sigma_Y$, and $\sigma_Y > 0$

$$P[a \leq Y \leq b] = P[a' \leq Z \leq b'],$$

where

$$a' \triangleq \frac{a - E[Y]}{\sigma_Y}$$

$$b' = \frac{b - E[Y]}{\sigma_Y}$$

and

$$\sigma_Y = \sqrt{n}\sigma_X.$$

Note that Z is a zero-mean, unity variance r.v. involving the sum of a large number (n assumed large) of i.i.d. r.v.'s. Indeed it is easy to show that

$$Z = \frac{1}{\sqrt{n}} \sum_{i=1}^{n} \left(\frac{X_i - \mu_X}{\sigma_X} \right).$$

Hence

$$P[a' \leq Z \leq b'] \simeq \frac{1}{\sqrt{2\pi}} \int_{a'}^{b'} e^{-\frac{1}{2}z^2} dz.$$

Although the CLT might be more appropriately called the "Normal convergence theorem," the word *central* in Central Limit Theorem is useful as a reminder that PDF's converge to the normal PDF around the center, that is, around the mean. Although all PDF's converge together at $\pm\infty$, it is in fact in the tails that the CLT frequently gives the poorest estimates of the correct probabilities, if these are small. An illustration of this phenomenon is given in Problem 4.38.

In a type of computer-based engineering analysis called *Monte-Carlo simulation* it is often necessary to have access to random numbers. There are several random number generators available in software that generate numbers that appear random but in fact are not: They are generated using an algorithm that is completely deterministic and therefore they can be duplicated by anyone who has a copy of the algorithm. The numbers, called pseudo-random numbers, are often adequate for situations where not too many random numbers are needed. For situations where a very large number of random numbers are needed, for example, modelling atomic processes, it turns out that it is very difficult to find an adequate random number generator. Most will eventually display number sequences that repeat, that is, are periodic, are highly correlated, or show other biases. Note that the alternative, that is, using a naturally random process such as the emission of photons from x-ray sources or the liberation of photoelectrons from valence bands in photodetectors also suffers from a major problem: We cannot be certain what underlying probability law is truly at work. And even if we knew what law was at work, the very act of counting photons or photoelectrons might bias the distribution of random numbers.

In any case, if we assume that for our purposes the uniform random number generators (URNG) commonly available with most PC software packages are adequate in that they create unbiased realizations of a uniform r.v. X, the next question is how can we convert uniform random numbers, that is, those that are assumed to obey the uniform pdf in $(0, 1)$

to Gaussian random numbers? For this purpose we can use the CLT as follows. Let X_i represent the ith random number generated by the URNG. Then

$$Z = X_1 + \ldots + X_n$$

will be approximately Gaussian for a reasonably large n (say > 10). Note that the pdf of Z is the n-repeated convolution of a unit pulse which starts to look like a Gaussian very quickly everywhere except in the tails. The reasons there is a problem in the tails is that Z is confined to the range $0 \leq Z \leq n$ while if Z were a true Gaussian r.v., then $-\infty < Z < \infty$.

4.8 ESTIMATORS FOR THE MEAN AND VARIANCE OF THE NORMAL LAW

In the previous section we saw how important the Normal (Gaussian) distribution is to probability. Unfortunately the parameters μ and σ^2 are usually not known beforehand and must be estimated. The sample mean estimator $\hat{\mu}_n$, defined by

$$\hat{\mu}_n \triangleq \frac{1}{n} \sum_{i=1}^{n} X_i \tag{4.8-1}$$

where the X_i, $i = 1, \ldots, n$ are independent observations of a Normal random variable, is often used to estimate the mean μ. The sample variance estimator $\hat{\sigma}_n^2$, defined by

$$\hat{\sigma}_n^2 \triangleq \frac{1}{n-1} \sum_{i=1}^{n} (X_i - \hat{\mu}_n)^2, \tag{4.8-2}$$

is often used to estimate the variance σ^2. How good these estimators are depends on what we mean by "good." Both estimators have the following satisfying properties:

(1) $E\{\hat{u}_n\} = \mu$, and $E\{\hat{\sigma}_n^2\} = \sigma^2$;
(2) $E_{n\to\infty}\{(\hat{\mu}_n - \mu)^2\} = 0$, and $E_{n\to\infty}\{(\hat{\sigma}_n^2 - \sigma^2)^2\} = 0.$ \hfill (4.8-3)

The proof of these results requires some work but involves only straightforward computations. The first property is called *unbiasedness*; the second property is called *square-error consistency*. Not all estimators have these desirable properties, although some have other desirable properties. A more detailed discussion on the properties of estimators is furnished in Chapter 5.

Given realizations of $\hat{\mu}_n$ and $\hat{\sigma}_n^2$, we would like to say something about how confident we are in them. Alternatively, having defined a required level of confidence, we might ask how many observations, n, are required to meet this level of confidence.

In what follows we briefly discuss a confidence technique known as *confidence interval estimation*.

Confidence Intervals for the Mean

Recall from Chapter 2, Equation 2.4-22, that a random variable X has a student-t distribution if its pdf given by

$$f_X(x) = K_{st}\left(1 + \frac{x^2}{n}\right)^{-\left(\frac{n+1}{2}\right)}, \tag{4.8-3}$$

where $K_{st} = \frac{\Gamma[(n+1)/2]}{\Gamma(n/2)\sqrt{\pi n}}$. It is common in the literature to use the symbol t or $t(n)$ to denote a random variable that has a Student-t distribution.[†] We let $f_t(z, n)$ be used to denote its pdf. Here the parameter n is included in the argument of the pdf because of its importance in what follows.

The r.v.

$$t \triangleq \frac{\hat{\mu}_n - \mu}{\left[\sum\limits_{i=1}^{n}(X_i - \hat{\mu}_n)^2/n(n-1)\right]^{1/2}} \tag{4.8-4}$$

can be shown to be $t(n-1)$ (see Chapter 7 in [4-9]). One can therefore find numbers t_ζ and $-t_\zeta$ such that

$$P[-t_\zeta \leq t \leq t_\zeta] = \int_{-t_\zeta}^{t_\zeta} f_t(z, n-1)dz = 1 - 2\zeta.$$

To simplify the notation, define

$$A(n) \triangleq \left[\frac{\sum\limits_{i=1}^{n}(X_i - \hat{\mu}_n)^2}{n(n-1)}\right]^{1/2}.$$

Then the event $\{-t_\zeta \leq t \leq t_\zeta\}$ is identical with the event

$$\left\{-t_\zeta \leq \frac{\hat{\mu}_n - \mu}{A(n)} \leq t_\zeta\right\} \tag{4.8-5}$$

or, equivalently,

$$\{\hat{\mu}_n - t_\zeta A(n) \leq \mu \leq \hat{\mu}_n + t_\zeta A(n)\}. \tag{4.8-6}$$

The numbers t_ζ are called the ζ percent level of t and locate points which cut off ζ percent of the area under $f_t(z, n-1)$; this is shown in Figure 4.8-1

Usually tables of the cumulative t distribution are furnished with the degrees of freedom n as a parameter. Thus

$$F_t(t_\zeta, n) = \int_{-\infty}^{t_\zeta} f_t(z, n)dz \tag{4.8-7}$$

[†]In using t or $t(n)$ as a symbol for a random variable we depart from our usual notation of using capital letters for r.v.'s.

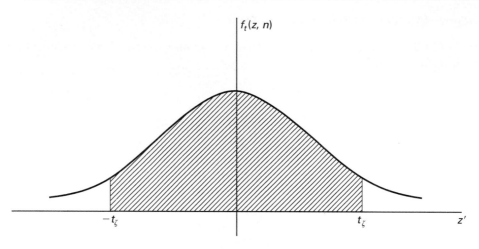

Figure 4.8-1   The numbers t_ζ. The area between $-t_\zeta$ and t_ζ is $1 - 2\zeta$.

is furnished for various n. Because of the symmetry of $f_t(z, n)$, we obtain

$$P[-t_\zeta \le t \le t_\zeta] = 2F_t(t_\zeta, n) - 1$$

or

$$F_t(t_\zeta, n) = \tfrac{1}{2}(1 + P[-t_\zeta \le t \le t_\zeta]). \qquad (4.8\text{-}8)$$

Having computed $f_t(t_\zeta, n)$ from the level of confidence, we can find t_ζ from the table for the appropriate degrees of freedom. The confidence interval is

$$[\hat\mu_n - t_\zeta A(n), \hat\mu_n + t_\zeta A(n)]$$

and its length is $2t_\zeta A(n)$.

Example 4.8-1

Twenty-one independent observations ($n = 21$) are made on a Gaussian r.v. X. Call these observations $X_1, X_2, \ldots, X_{21}$. Based on the data the realization of $\hat\mu_n$ is 3.5 and the realization of

$$\left[\frac{\displaystyle\sum_{i=1}^{n}(X_i - \hat\mu_n)^2}{n(n - 1)} \right]^{1/2}$$

is 0.45. A 90 percent confidence interval on μ is desired.

Solution   Since $P[-t_{.05} \le t \le t_{.05}] = 0.9$, we obtain from Equation 4.8-8 $F_t(t_{.05}, 20) = 0.95$. From Table 4.8-1, for $n = 20$, we obtain $t_{.05} = 1.725$. The corresponding confidence interval is

$$[3.5 - 1.725(0.45), 3.5 + 1.725(0.45)]$$

$$= [2.72, 4.28].$$

Table 4.8-1 Student-t Distribution

$$F_t(t_\zeta, n) = \int_{-\infty}^{t_\zeta} f_t(z, n)dz$$

F n	0.60	0.75	0.90	0.95	0.975	0.99	0.995	0.9995
1	0.325	1.000	3.078	6.314	12.706	31.821	63.657	636.619
2	0.289	0.816	1.886	2.920	4.303	6.965	9.925	31.598
3	0.277	0.765	1.638	2.353	3.182	4.541	5.841	12.924
4	0.271	0.741	1.533	2.132	2.776	3.747	4.604	8.610
5	0.267	0.727	1.476	2.015	2.571	3.365	4.032	6.869
6	0.265	0.718	1.440	1.943	2.447	3.143	3.707	5.959
7	0.263	0.711	1.415	1.895	2.365	2.998	3.499	5.408
8	0.262	0.706	1.397	1.860	2.306	2.896	3.355	5.041
9	0.261	0.703	1.383	1.833	2.262	2.821	3.250	4.781
10	0.260	0.700	1.372	1.812	2.228	2.764	3.169	4.587
11	0.260	0.697	1.363	1.796	2.201	2.718	3.106	4.437
12	0.259	0.695	1.356	1.782	2.179	2.681	3.055	4.318
13	0.259	0.694	1.350	1.771	2.160	2.650	3.012	4.221
14	0.258	0.692	1.345	1.761	2.145	2.624	2.977	4.140
15	0.258	0.691	1.341	1.753	2.131	2.602	2.947	4.073
16	0.258	0.690	1.337	1.746	2.120	2.583	2.921	4.015
17	0.257	0.689	1.333	1.740	2.110	2.567	2.898	3.965
18	0.257	0.688	1.330	1.734	2.101	2.552	2.878	3.922
19	0.257	0.688	1.328	1.729	2.093	2.539	2.861	3.883
20	0.257	0.687	1.325	1.725	2.086	2.528	2.845	3.850
21	0.257	0.686	1.323	1.721	2.080	2.518	2.831	3.819
22	0.256	0.686	1.321	1.717	2.074	2.508	2.819	3.792
23	0.256	0.685	1.319	1.714	2.069	2.500	2.807	3.767
24	0.256	0.685	1.318	1.711	2.064	2.492	2.797	3.745
25	0.256	0.684	1.316	1.708	2.060	2.485	2.787	3.725
26	0.256	0.684	1.315	1.706	2.056	2.479	2.779	3.707
27	0.256	0.684	1.314	1.703	2.052	2.473	2.771	3.690
28	0.256	0.683	1.313	1.701	2.048	2.467	2.763	3.674
29	0.256	0.683	1.311	1.699	2.045	2.462	2.756	3.659
30	0.256	0.683	1.310	1.697	2.042	2.457	2.750	3.646
40	0.255	0.681	1.303	1.684	2.021	2.423	2.704	3.551
60	0.254	0.679	1.296	1.671	2.000	2.390	2.660	3.460
120	0.254	0.677	1.289	1.658	1.980	2.358	2.617	3.373
∞	0.253	0.674	1.282	1.645	1.960	2.326	2.576	3.291

Adapted from W.H. Beyer, Ed., in *CRC Handbook of Tables for Probability and Statistics*, 2d ed., The Chemical Rubber Co., Cleveland, 1968; p. 283. With permission of CRC Press, Inc.

Confidence Interval for the Variance

We make n observations on a Normal random variable X with variance σ^2 (unknown) and mean μ (unknown). We already know that the r.v

$$U_i \triangleq \frac{X_i - \mu}{\sigma}$$

is $N(0,1)$ and hence $Z_n \triangleq \sum_{i=1}^n U_i^2$ is Chi-square with n degrees of freedom. However a careful examination of the r.v

$$W_n \triangleq \sum_{i=1}^n \left(\frac{X_i - \hat{\mu}_n}{\sigma}\right)^2$$

reveals that W_n is Chi-square with $n - 1$ degrees of freedom (Problem 4.48). Hence we replace W_n by the symbol Z_{n-1}, that is, a Chi-square r.v with $n-1$ degrees of freedom and consider the computation of

$$P[a \leq Z_{n-1} \leq b] = \int_a^b f_{\chi^2}(z, n-1)dz,$$

where $f_{\chi^2}(z, n-1)$ is the Chi-square with $n-1$ degrees of freedom and $0 < a < b$. But

$$P[a \leq Z_{n-1} \leq b] = P\left[\frac{1}{b}\sum_{i=1}^n (X_i - \hat{\mu}_n)^2 \leq \sigma^2 \leq \frac{1}{a}\sum_{i=1}^n (X_i - \hat{\mu}_n)^2\right]$$

which relates the numbers ab to the end points of the confidence interval. To find numbers ab such that $P[a \leq Z_{n-1} \leq b] = \gamma$ (the level of confidence) is not difficult, but to find numbers a^*, b^* such that the length, L, of the confidence interval is the *smallest* subject to the given confidence level γ is more difficult because of the asymmetry of the Chi-square pdf. The length of the confidence interval is $(1/a - 1/b) \sum_{i=1}^n (X_i - \hat{\mu}_n)^2$. For moderate and large values of n, a *near-shortest* confidence interval is obtained by the procedure in Example 4.8-2.

Example 4.8-2

Sixteen observations are made on a normal r.v. A 95% confidence interval on the estimation of σ^2 is required. Find the values of a, b that yield an almost-shortest confidence interval.

Solution   Assume that the available tables contain $P[Z_{15} \leq z] = \int_0^z f_{\chi^2}(x, 15)\,dx$. We are going to choose a, b such that equal areas are removed from the lower and upper tails of the pdf (see Figure 4.8-2). For a 95% confidence level, 5% of the total area under the pdf is to be omitted, or 2.5% from each tail. Thus to get a we look up Table 4.8-2 to get $P[Z_{15} \leq a] = 0.025$ or $a = 6.26$. To get b we evaluate $1 - P[Z_{15} \leq b] = 0.025$ or $b = 27.5$. The length, L, of the confidence interval is then $L = 0.123 \sum_{i=1}^{16} (X_i - \hat{\mu}_n)^2$ (a random variable). If we repeated the experiment with 16 fresh samples one hundred times, we could expect the intervals, so computed, to cover the true variance about 95% of the time.

Table 4.8-2 Chi-Square Distribution

$$F_{\chi^2}(u,n) = \int_0^u \frac{x^{(n-2)/2} e^{-x/2}\,dx}{2^{n/2}[(n-2)/2]!}$$

n\F	.005	.010	.025	.050	.100	.250	.500	.750	.900	.950	.975	.990	.995
1	$.0^4393$	$.0^3157$	$.0^3982$	$.0^2393$	.0158	.102	.455	1.32	2.71	3.84	5.02	6.63	7.88
2	.0100	.0201	.0506	.103	.211	.575	1.39	2.77	4.61	5.99	7.38	9.21	10.6
3	.0717	.115	.216	.352	.584	1.21	2.37	4.11	6.25	7.81	9.35	11.3	12.8
4	.207	.297	.484	.711	1.06	1.92	3.36	5.39	7.78	9.49	11.1	13.3	14.9
5	.412	.554	.831	1.15	1.61	2.67	4.35	6.63	9.24	11.1	12.8	15.1	16.7
6	.676	.872	1.24	1.64	2.20	3.45	5.35	7.84	10.6	12.6	14.4	16.8	18.5
7	.989	1.24	1.69	2.17	2.83	4.25	6.35	9.04	12.0	14.1	16.0	18.5	20.3
8	1.34	1.65	2.18	2.73	3.49	5.07	7.34	10.2	13.4	15.5	17.5	20.1	22.0
9	1.73	2.09	2.70	3.33	4.17	5.90	8.34	11.4	14.7	16.9	19.0	21.7	23.6
10	2.16	2.56	3.25	3.94	4.87	6.74	9.34	12.5	16.0	18.3	20.5	23.2	25.2
11	2.60	3.05	3.82	4.57	5.58	7.58	10.3	13.7	17.3	19.7	21.9	24.7	26.8
12	3.07	3.57	4.40	5.23	6.30	8.44	11.3	14.8	18.5	21.0	23.3	26.2	28.3
13	3.57	4.11	5.01	5.89	7.04	9.30	12.3	16.0	19.8	22.4	24.7	27.7	29.8
14	4.07	4.66	5.63	6.57	7.79	10.2	13.3	17.1	21.1	23.7	26.1	29.1	31.3
15	4.60	5.23	6.26	7.26	8.55	11.0	14.3	18.2	22.3	25.0	27.5	30.6	32.8
16	5.14	5.81	6.91	7.96	9.31	11.9	15.3	19.4	23.5	26.3	28.8	32.0	34.3
17	5.70	6.41	7.56	8.67	10.1	12.8	16.3	20.5	24.8	27.6	30.2	33.4	35.7
18	6.26	7.01	8.23	9.39	10.9	13.7	17.3	21.6	26.0	28.9	31.5	34.8	37.2
19	6.84	7.63	8.91	10.1	11.7	14.6	18.3	22.7	27.2	30.1	32.9	36.2	38.6
20	7.43	8.26	9.59	10.9	12.4	15.5	19.3	23.8	28.4	31.4	34.2	37.6	40.0
21	8.03	8.90	10.3	11.6	13.2	16.3	20.3	24.9	29.6	32.7	35.5	38.9	41.4
22	8.64	9.54	11.0	12.3	14.0	17.2	21.3	26.0	30.8	33.9	36.8	40.3	42.8
23	9.26	10.2	11.7	13.1	14.8	18.1	22.3	27.1	32.0	35.2	38.1	41.6	44.2
24	9.89	10.9	12.4	13.8	15.7	19.0	23.3	28.2	33.2	36.4	39.4	43.0	45.6
25	10.5	11.5	13.1	14.6	16.5	19.9	24.3	29.3	34.4	37.7	40.6	44.3	46.9
26	11.2	12.2	13.8	15.4	17.3	20.8	25.3	30.4	35.6	38.9	41.9	45.6	48.3
27	11.8	12.9	14.6	16.2	18.1	21.7	26.3	31.5	36.7	40.1	43.2	47.0	49.6
28	12.5	13.6	15.3	16.9	18.9	22.7	27.3	32.6	37.9	41.3	44.5	48.3	51.0
29	13.1	14.3	16.0	17.7	19.8	23.6	28.3	33.7	39.1	42.6	45.7	49.6	52.3
30	13.8	15.0	16.8	18.5	20.6	24.5	29.3	34.8	40.3	43.8	47.0	50.9	53.7

This table is abridged from "Tables of Percentage Points of the Incomplete Beta Function and of the Chi-square Distribution," *Biometrika* Vol. 32 (1941). It is here published with the kind permission of the author, Catherine M. Thompson, and the editor of *Biometrika*.

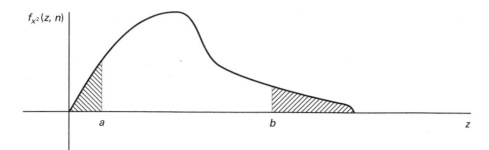

Figure 4.8-2   The points a, b are chosen so that equal amounts of area are removed from the tails. The area of the remainder should equal γ, the confidence level.

4.9 SUMMARY

In this chapter we discussed the various averages of one or more random variables and the implication of those averages. We began by defining the average or expected value of a random variable X and then showed that the expected value of $Y = g(X)$ could be computed directly from the pdf or PMF of X. We briefly discussed the important notion of conditional expectation and showed how the expected value of a r.v. could be advantageously computed by averaging over its conditional expectation. We then argued that a single summary number such as the average value, $\overline{X}$, of X was insufficient for describing the behavior of X. This led to the introduction of moments, that is, the average of powers of X. We illustrated how moments can be used to estimate pdf's by the maximum entropy principle and introduced the concept of *joint moments*. We showed how the covariance of two random variables could be interpreted as a measure of how well we can predict one r.v. from observing another using a linear predictor model. By giving a counterexample, we demonstrated that uncorrelatedness does not imply independence of two r.v.'s, the latter being a stronger condition. The joint Gaussian pdf for two r.v.'s was discussed, and it was shown that in the Gaussian case, independence and uncorrelatedness are equivalent. We then introduced the reader to some important bounds and inequalities known as the Chebyshev and Schwarz inequalities and the Chernoff bound and illustrated how these are used in problems in probability.

The second half of the chapter dealt mostly with moment-generating (m.g.f.) and characteristic functions (c.f.) and the Central Limit theorem. We showed how the m.g.f. and c.f. are essentially the Laplace and Fourier transforms, respectively, of the pdf of a random variable and how we could compute all the moments, provided that these exist, from either of these functions. Several properties of these important functions were explored. We illustrated how the c.f. could be used to solve problems involving the computation of the pdf's of the sums of random variables.

We then discussed the Central Limit theorem, one of the most important results in probability theory, and the basis for the ubiquitous Normal behavior of many random phenomena. The CLT states that under relatively loose mathematical constraints, the probability distribution function of the sum of independent random variables tends toward the Normal PDF.

We ended the chapter by reminding the reader that such parameters as the mean and variance are often not known *a priori* and must be estimated from observations on random variables. The process of estimating these and other parameters is within the realm of a branch of applied probability called *statistics*. We furnished examples of estimators for the mean and variance and used the Student-t and Chi-square distribution to establish confidence intervals for these estimators. In Chapter 5, we will resume our discussion of estimators and provide a more theoretical framework, so-called *maximum likelihood*, for determining estimators with desirable properties.

PROBLEMS

4.1. Compute the average and standard deviation of the following set: 3.02, 5.61, −2.37, 4.94, −6.25, −1.05, −3.25, 5.81, 2.27, 0.54, 6.11, −2.56.

4.2. Compute $E[X]$ when X is a Bernoulli r.v., that is,

$$X = \begin{cases} 1, & P_X(1) = p > 0 \\ 0, & P_X(0) = 1 - p > 0. \end{cases}$$

4.3. Compute $E[X]$ when X is a binomial r.v., that is,

$$P_X(k) = \binom{n}{k} p^k (1-p)^{n-k} \qquad k = 0, \ldots, n, \qquad 0 < p < 1.$$

4.4. Let X be a uniform r.v., that is,

$$f_X(x) = \begin{cases} (b-a)^{-1} & 0 < a < x < b \\ 0, & \text{otherwise.} \end{cases}$$

Compute $E[X]$.

4.5. In Problem 4.4, let $Y \triangleq X^2$. Compute the pdf of Y and $E[Y]$ by Equation 4.1-7. Then compute $E[Y]$ by Equation 4.1-8.

4.6. Let X be a Poisson r.v. with parameter a. Compute $E[Y]$ when $Y \triangleq X^2 + b$.

4.7. Prove that if $E[X]$ exists and X is a continuous r.v., then $|E[X]| \leq E[|X|]$. Repeat for X discrete.

4.8. Show that if $E[g_i(X)]$ exists for $i = 1, \ldots, N$, then

$$E\left[\sum_{i=1}^{N} g_i(X)\right] = \sum_{i=1}^{N} E[g_i(X)].$$

4.9. A random sample of 20 households shows the following numbers of children per household: 3, 2, 0, 1, 0, 0, 3, 2, 5, 0, 1, 1, 2, 2, 1, 0, 0, 0, 6, 3. (a) For this set what is the average number of children per household? (b) What is the average number of children in households given that there is at least one child?

4.10. Let $B \triangleq \{a < X \leq b\}$. Derive a general expression for $E[X|B]$ if X is a continuous r.v. Let $X: N(0,1)$ with $B = \{-1 < X \leq 2\}$. Compute $E[X|B]$.

4.11. A particular color TV model is manufactured in three different plants, say, A, B, and C of the same company. Because the workers at A, B, and C are not equally experienced, the quality of the sets differs from plant to plant. The pdf's of the time-to-failure, X, in years are

$$f_X(x) = \frac{1}{5}\exp(-x/5)u(x) \text{ for } A$$

$$f_X(x) = \frac{1}{6.5}\exp(-x/6.5)u(x) \text{ for } B$$

$$f_X(x) = \frac{1}{10}\exp(-x/10)u(x) \text{ for } C,$$

where $u(x)$ is the unit step. Plant A produces three times as many sets as B, which produces twice as many as C. The sets are all sent to a central warehouse, intermingled, and shipped to retail stores all around the country. What is the expected lifetime of a set purchased at random?

4.12. A source transmits a signal Θ with pdf

$$f_\Theta(\theta) = \begin{cases} (2\pi)^{-1} & 0 < \theta \le 2\pi \\ 0, & \text{otherwise.} \end{cases}$$

Because of additive Gaussian noise, the pdf of the received signal Y when $\Theta = \theta$ is

$$f_{Y|\Theta}(y|\theta) = \frac{1}{\sqrt{2\pi\sigma^2}}\exp\left[-\frac{1}{2}\left(\frac{y-\theta}{\sigma}\right)^2\right].$$

Compute $E[Y]$.

4.13. Compute the variance of X if X is (a) Bernoulli; (b) binomial; (c) Poisson; (d) Gaussian; (e) Rayleigh.

4.14. Let X and Y be independent r.v.'s, each $N(0,1)$. Find the mean and variance of $Z \triangleq \sqrt{X^2 + Y^2}$.

4.15. (Papoulis [4-3]). Let $Y = h(X)$. We wish to compute approximation to $E[h(X)]$ and $E[h^2(X)]$. Assume that $h(x)$ admits to a power series expansions, that is, all derivatives exist. Assume further that all derivatives above the second are small enough to be omitted. Show that ($\overline{X} = \mu$, $\text{Var}(X) = \sigma^2$).

    (a) $E[h(X)] \simeq h(\mu) + h''(\mu)\sigma^2/2$;
    (b) $E[h^2(X)] \simeq h^2(\mu) + ([h'(\mu)]^2 + h(\mu)h''(\mu))\sigma^2$.

4.16. Let $f_{XY}(x,y)$ be given by

$$f_{XY}(x,y) = \frac{1}{2\pi\sigma^2\sqrt{1-\rho^2}}\exp\left(-\frac{x^2+y^2-2\rho xy}{2\sigma^2(1-\rho^2)}\right),$$

where $|\rho| < 1$. Show that $E[Y] = 0$ but $E[Y|X = x] = \rho x$. What does this result say about predicting the value of Y upon observing the value of X?

4.17. Show that in the joint Gaussian pdf with $\overline{X} = \overline{Y} = 0$ and $\sigma_X = \sigma_Y \triangleq \sigma$, when

$$\rho \to 1, \qquad f_{XY}(x, y) \to \frac{1}{\sqrt{2\pi}\sigma} \exp\left[-\frac{1}{2}\left(\frac{x}{\sigma}\right)^2\right]\delta(y - x).$$

4.18. Consider a probability space $\mathscr{P} = (\Omega, \mathscr{F}, P)$. Let $\Omega = \{\zeta_1, \ldots, \zeta_5\} = \{-1, -\frac{1}{2}, 0, \frac{1}{2}, 1\}$ with $P[\{\zeta_i\}] = \frac{1}{5}$, $i = 1, \ldots, 5$. Define two r.v.'s on $\mathscr{P}$ as follows:

$$X(\zeta) \triangleq \zeta \qquad \text{and} \qquad Y(\zeta) \triangleq \zeta^2.$$

  (a) Show that X and Y are dependent r.v.'s.
  (b) Show that X and Y are uncorrelated.

4.19. We wish to estimate the pdf of X with a function $p(x)$ that maximizes the entropy

$$H[X] \triangleq -\int_{-\infty}^{\infty} p(x) \ln p(x)\, dx.$$

It is known from measurements that $E[X] = \mu$ and $\text{Var}[X] = \sigma^2$. Find the maximum entropy estimate of the pdf of X.

4.20. Let $X: N(0, \sigma^2)$. Show that

$$\xi_n \triangleq E[X^n] = 1 \cdot 3 \ldots (n-1)\sigma^n \qquad n \text{ even}$$

$$\xi_n = 0 \qquad n \text{ odd}.$$

4.21. With $\overline{X} \triangleq E[X]$ and $\overline{Y} \triangleq E[Y]$, show that if $m_{11} = \sqrt{m_{20} m_{02}}$, then

$$E\left[\left(\frac{m_{11}}{m_{20}}(X - \overline{X}) - (Y - \overline{Y})\right)^2\right] = 0.$$

Use this result to show that when $|\rho| = 1$, then Y is a linear function of X, that is, $Y = \alpha X + \beta$. Relate α, β to the moments of X and Y.

4.22. Show that in the optimum linear predictor in Example 4.3-4 the smallest mean-square error is

$$\varepsilon_{\min}^2 = \sigma_Y^2 (1 - \rho^2).$$

Explain why $\varepsilon_{\min}^2 = 0$ when $|\rho| = 1$.

4.23. Let $E[X_i] = \mu$, $\text{Var}[X_i] = \sigma^2$. We wish to estimate μ with the *sample mean*

$$\hat{\mu}_N \triangleq \frac{1}{N}\sum_{i=1}^{N} X_i.$$

Compute the mean and variance of $\hat{\mu}_N$ assuming the X_i for $i = 1, \ldots, N$ are independent.

4.24. In Problem 4.23, how large should N be so that

$$P[|\hat{\mu}_N - \mu| > 0.1\sigma] \leq 0.01.$$

4.25. Let X be a uniform r.v. in $(-\frac{1}{2}, \frac{1}{2})$. Compute (a) its moment-generating function; and (b) its mean by Equation 4.5-5. [*Hint*: $\sinh z \triangleq (e^z - e^{-z})/2$. Use limits when computing the mean.]

4.26. Let X be a Poisson r.v. Compute its (a) moment-generating function; and (b) its mean by Equation 4.5-5.

4.27. The *negative binomial distribution* with parameters N, Q, P where $Q - P = 1, P > 0$ and $N \geq 1$ is defined by

$$P[X = k] = \binom{N + k - 1}{N - 1} \left(\frac{P}{Q}\right)^k \left(1 - \frac{P}{Q}\right)^N \qquad (k = 0, 1, 2, \ldots).$$

It is sometimes used as an alternative to the Poisson distribution when one cannot guarantee that individual events occur independently (the "strict" randomness requirement for the Poisson distribution). Show that the moment-generating function is

$$\theta_X(t) = (Q - Pe^t)^{-N}.$$

[*Hint*: Either compute or look up the expansion formula for $(Q - Pe^t)^{-N}$, for example, see *Discrete Distributions* by N. L. Johnson and S. Kotz, John Wiley and Sons, 1969.]

4.28. Compute the Chernoff bound on $P[X \geq a]$ where X is a r.v. that satisfies the exponential law $f_X(x) = \lambda e^{-\lambda x} u(x)$.

4.29. Let $N = 1$ in Problem 4.27. (a) Compute the Chernoff bound on $P[X \geq k]$; (b) generalize the result for arbitrary N.

4.30. Let X have a Cauchy pdf

$$f_X(x) = \frac{\alpha}{\pi(\alpha^2 + x^2)}.$$

Compute the characteristic function (cf) $\Phi_X(\omega)$ of X.

4.31. Let

$$Y = \frac{1}{N} \sum_{i=1}^{N} X_i$$

where the X_i are independent, identically distributed Cauchy r.v.'s with

$$f_{X_i}(x) = \frac{1}{\pi[1 + (x - \mu)^2]} \qquad i = 1, \ldots, N.$$

Show that the pdf of Y is

$$f_Y(x) = \frac{1}{\pi[1 + (x - \mu)^2]},$$

that is, is identical to the pdf of the X_i's and independent of N. (*Hint*: Use the c.f. approach.)

4.32. Let X be uniform over $(-a, a)$. Let Y be independent of X and uniform over $([n - 2]a, na)$, $n = 1, 2, \ldots$. Compute the expected value of $Z = X + Y$ for each n. From this result sketch the pdf of Z. What is the only effect of n?

4.33. Consider the recursion known as a *first-order moving average* given by

$$X_n = Z_n - aZ_{n-1} \qquad |a| < 1,$$

where X_n, Z_n, Z_{n-1} are all r.v.'s for $n = \ldots, -1, 0, 1, \ldots$. Assume $E[Z_n] = 0$ all n; $E[Z_n Z_j] = 0$ all $n \neq j$; and $E[Z_n^2] = \sigma^2$ all n. Compute $R_n(k) \triangleq E[X_n X_{n-k}]$ for $k = 0, \pm 1, \pm 2, \ldots$.

4.34. Consider the recursion known as a *first-order autoregression*

$$X_n = bX_{n-1} + Z_n \qquad |b| < 1.$$

The following is assumed true: $E[Z_n] = 0$, $E[Z_n^2] = \sigma^2$ all n; $E[Z_n Z_j] = 0$ all $n \neq j$. Also $E[Z_n X_{n-j}] = 0$ for $j = 1, 2, \ldots$. Compute $R_n(k) = E[X_n X_{n-k}]$ for $k = \pm 1$, $\pm 2, \ldots$. Assume $E[X_n^2] \triangleq K$ independent of n.

4.35. Let $Z \triangleq aX + bY$, $W \triangleq cX + dY$. Compute the joint c.f. of Z and W in terms of the joint c.f. of X and Y.

4.36. Let X and Y be two independent Poisson r.v.'s with

$$P_X(k) = \frac{1}{k!}e^{-2}2^k$$

$$P_Y(k) = \frac{1}{k!}e^{-3}3^k.$$

Compute the PMF of $Z = X + Y$ using moment-generating or characteristic functions.

4.37. Your company manufactures toaster ovens. Let the probability that a toaster oven has a dent or scratch be $p = 0.05$. Assume different ovens get dented or scratched independently. In one week the company makes 2000 of these ovens. What is the approximate probability that in this week more than 109 ovens are dented or scratched?

4.38. Let X_i for $i = 1, \ldots, n$ be a sequence of i.i.d. Bernoulli r.v.'s with $P_X(1) = p$ and $P_X(0) = q = 1 - p$. Let the event of a $\{1\}$ be a success and the event of a $\{0\}$ be a failure.

    (a) Show that

$$Z_n \triangleq \frac{1}{\sqrt{n}} \sum_{i=1}^{n} W_i,$$

        where $W_i \triangleq (X_i - p)/\sqrt{pq}$, is a zero-mean, unity variance r.v., involving the sum of a large number of independent r.v.'s, if n is large.

    (b) For $n = 2000$ and $k = 110, 130, 150$ compute $P[k$ successes in n tries$]$ using (i) the exact binomial expression; (ii) the Poisson approximation to the binomial; and (iii) the CLT approximations. Do this by writing three MATLAB miniprograms. Verify that as the correct probabilities decrease, the error in the CLT approximation increases.

4.39. Assume that we have uniform random number generator (URNG) that is well modeled by a sequence of i.i.d uniform random variables X_i, $1 = 1, \ldots, n$ where X_i is the ith output of the URNG. Assume that

$$f_{X_i}(X_i) = \frac{1}{a} \text{rect}\left(\frac{x_i - a/2}{a}\right).$$

(a) Show that with $Z_n = X_1 + \ldots + X_n$, $E[Z_n] = na/2$. (b) Show that $\text{var}(Z_n) = na^2/12$. (c) Write a Matlab.m program that computes the plots $f_{Z_n}(z)$ for $n = 2, 3, 10, 20$. (d) Write a Matlab.m program that plots Gaussian pdfs $N\left(\frac{na}{2}, \frac{na^2}{12}\right)$ for $n = 2, 3, 10, 20$ and compare $f_{Z_n}(z)$ with $N\left(\frac{na}{2}, \frac{na^2}{12}\right)$ for each n. (e) For each n compute $P[\mu_n - k\sigma_n \leq Z_n \leq \mu_n + k\sigma_n]$, where $\mu_n = na/2$, $\sigma_n^2 = na^2/12$ for a few values of k e.g. $k = 0.1, 0.5, 1, 2, 3$. Do this using both $f_{Z_n}(z)$ and $N\left(\frac{na}{2}, \frac{na^2}{12}\right)$. Choose any reasonable value of a, for example, $a = 1$.

4.40. Let $f_X(x)$ be the pdf of a real, continuous random variable X. Show that if $f_X(x) = f_X(-x)$, then $E\{X\} = 0$.

4.41. Compute the variance of the Chi-square random variable $Z_n = \sum_{i=1}^{n} X_i^2$, where $X_i : N(0, 1)$ and the X_i, for $i = 1, 2, \cdots, n$, are mutually independent. (*Hint*: Use the result in Problem 4.43.)

4.42. Let X be a Normal r.v. with $X : N(\mu, \sigma^2)$. Show that $E\{(X - \mu)^{2k+1}\} = 0$, while $E\{(X - \mu)^{2k}\} = [(2k)!/2^k k!]\sigma^{2k}$.

4.43. (a) Write a Matlab.m file that will compute the pdf for a Chi-square r.v, Z_n and display it as a graph for $n = 30, 40, 50$. (b) Add to your program the capability to compute $P[\mu - \sigma \leq Z_n \leq \mu + \sigma]$. Compare your result with a Gaussian approximation $P[\mu - \sigma \leq X \leq \mu + \sigma]$, where $X : N(n, 2n)$.

4.44. Let $X_i, i = 1, \ldots, 4$ be four zero-mean Gaussian random variables. Use the joint characteristic function to show that

$$E\{X_1 X_2 X_3 X_4\} = E\{X_1 X_2\}E\{X_3 X_4\} + E\{X_1 X_3\}E\{X_2 X_4\}$$
$$+ E\{X_2 X_3\}E\{X_1 X_4\}.$$

4.45. Compute the moment-generating and characteristic functions for the Chi-square r.v.

4.46. Is the converse statement of Problem 4.40 true? That is, if $E\{X\} = 0$, does that imply that $f_X(x) = f_X(-x)$?

4.47. Let X be uniform on (a, b) where $0 < a < b$. With $r \triangleq b/a$, show the mean-to-standard deviation ratio, μ/σ, approaches the value 1.73 as $r \to \infty$.

4.48. Show that $W_n \triangleq \sum_{i=1}^{n}[(X_i - \frac{1}{n}\sum_{j=1}^{n} X_j)/\sigma]^2$ is Chi-square with $n - 1$ degrees of freedom.

4.49. Derive the inequality $\sigma_X P[|X| \geq \sigma_X] \leq E\{|X|\} \leq \sigma_X$ that holds true if $f_X(x) = f_X(-x)$.

REFERENCES

4-1. W. Feller, *An Introduction to Probability Theory and Its Applications*, Vol. 2, 2nd edition. New York: John Wiley, 1971.

4-2. W. B. Davenport, Jr., *Probability and Random Processes*. New York: McGraw-Hill, p. 99.

4-3. A. Papoulis, *Probability, Random Variables, and Stochastic Processes* 3rd edition, New York: McGraw-Hill, 1991.

4-4. B. Saleh, *Photoelectron Statistics*. New York: Springer-Verlag, 1978, Chapter 5.

4-5. R. G. Gallagher, *Information Theory and Reliable Communications*. New York: John Wiley, 1968.

4-6. P. M. Morse and H. Feshbach, *Methods of Theoretical Physics*, Part 1. New York: McGraw-Hill, 1953, p. 279.

4-7. G. A. Korn and T. S. Korn, *Mathematical Handbook for Scientists and Engineers*. New York: McGraw-Hill, 1961.

4-8. W. Feller, *An Introduction to Probability Theory and Its Applications*, Vol. 1, 2nd edition. New York: John Wiley, 1957.

4-9. A. M. Mood and F. A. Grayball, *Introduction to the Theory of Statistics*. New York: McGraw-Hill, 1963.

ADDITIONAL READING

Cooper, G. R. and C. D. McGillem, *Probabilistic Methods of Signal and System Analysis*, 3rd edition. New York: Holt, Rinehart and Winston, 1999.

Deebles, P. Z. Jr., *Probability, Random Variables, and Random Signal Principles*, 4th edition. New York: McGraw-Hill, 2001.

Garcia, L.-G., *Probability and Random Processes for Electrical Engineering*, 2nd edition. Reading, MA: Addison-Wesley, 1994.

Helstrom, C. W., *Probability and Stochastic Processes for Engineers*. 2nd edition. New York: Macmillan, 1991.

Johnson, R. A., *Miller & Freund's Probability and Statistics for Engineers*, 5th edition. Englewood Cliffs, NJ: Prentice Hall, 1994.

Papoulis, A., *Probability, Random Variables, and Stochastic Processes*. 3rd edition. New York: McGraw-Hill, 1991.

Papoulis, A., *Probability & Statistics*. Englewood Cliffs, NJ: Prentice Hall, 1990.

Scheaffer, R. L., *Introduction to Probability and Its Applications*. Belmont, CA: Duxbury, 1990.

Viniotis, Y., *Probability and Random Processes for Electrical Engineers*. New York: McGraw-Hill, 1998.

Yates, R. D. and D. J. Goodman, *Probability and Stochastic Processes*. New York: Wiley, 1999.

Ziemer, R. E., *Elements of Engineering Probability & Statistics*. Upper Saddle River, NJ: Prentice Hall, 1997.

5

Random Vectors and Parameter Estimation

5.1 JOINT DISTRIBUTION AND DENSITIES

In many practical problems involving random phenomena we make observations that are essentially of a vector nature. We illustrate with three examples.

Example 5.1-1
A seismic waveform $X(t)$ is received at a geophysical recording station and is sampled at the instants $t_1, t_2, \ldots, t_n$. We thus obtain a vector $\mathbf{X} = (X_1, \ldots, X_n)^T$ where $X_i \triangleq X(t_i)$ and T denotes transpose.[†] For political and military reasons it is important to determine whether the waveform was radiated from an earthquake or an underground explosion. Assume that an expert computer system has available a lot of stored data regarding both earthquakes and underground explosions. The vector $\mathbf{X}$ is compared to the stored data. What is the probability that $X(t)$ is correctly identified?

Example 5.1-2
To evaluate the health of grade-school children, the Health Department of a certain region measures the height, weight, blood pressure, red-blood cell count, white-blood cell count, pulmonary capacity, heart rate, blood-lead level, and vision acuity of each child. The resulting vector $\mathbf{X}$ is taken as a summary of the health of each child. What is the probability that a child chosen at random is healthy?

[†]All vectors will be assumed to be column vectors unless otherwise stated.

244

Example 5.1-3

A computer system equipped with a TV camera is designed to recognize black-lung disease from x-rays. It does this by counting the number of radio-opacities in six lung zones (three in each lung) and estimating the average size of the opacities in each zone. The result is a 12-component vector $\mathbf{X}$ from which a decision is made. What is the best computer decision?

The three previous examples are illustrative of many problems encountered in engineering and science that involve sets of random variables that are grouped for some purpose. Such sets of random variables are conveniently studied by vector methods. For this reason we treat these grouped random variables as a single object called a *random vector*. As in earlier chapters, capital letters at the lower end of the alphabet will denote random variables; bold capital letters will denote random vectors and matrices and lower-case bold letters are deterministic vectors, for example, the values that random vectors assume.

Let $\mathbf{X}$ be a random vector with probability distribution function (PDF) $F_{\mathbf{X}}(\mathbf{x})$. Then by definition[†]

$$F_{\mathbf{X}}(\mathbf{x}) \triangleq P[X_1 \leq x_1, \ldots, X_n \leq x_n]. \tag{5.1-1}$$

By defining $\{\mathbf{X} \leq \mathbf{x}\} \triangleq \{X_1 \leq x_1, \ldots, X_n \leq x_n\}$, we can rewrite Equation 5.1-1 as

$$F_{\mathbf{X}}(\mathbf{x}) \triangleq P[\mathbf{X} \leq \mathbf{x}]. \tag{5.1-2}$$

We associate the events $\{\mathbf{X} \leq \infty\}$ and $\{\mathbf{X} \leq -\infty\}$ with the certain and impossible events respectively. Hence

$$F_{\mathbf{X}}(\infty) = 1 \tag{5.1-3a}$$

$$F_{\mathbf{X}}(-\infty) = 0. \tag{5.1-3b}$$

If the nth-mixed partial of $F_{\mathbf{X}}(\mathbf{x})$ exists we can define a *probability density function* (pdf) as

$$f_{\mathbf{X}}(\mathbf{x}) \triangleq \frac{\partial^n F_{\mathbf{X}}(\mathbf{x})}{\partial x_1 \ldots \partial x_n}. \tag{5.1-4}$$

The reader will observe that these definitions are completely analogous to the scalar definitions given in Chapter 2. We could have defined

$$f_{\mathbf{X}}(\mathbf{x}) \triangleq \lim_{\substack{\Delta x_1 \to 0 \\ \vdots \\ \Delta x_n \to 0}} \frac{P[x_1 < X_1 \leq x_1 + \Delta x_1, \ldots, x_n < X_n \leq x_n + \Delta x_n]}{\Delta x_1 \ldots \Delta x_n} \tag{5.1-5}$$

and arrived at Equation 5.1-4. For example, for $n = 2$

$$P[x_1 < X_1 \leq x_1 + \Delta x_1, x_2 < X_2 \leq x_2 + \Delta x_2]$$
$$= F_{\mathbf{X}}(x_1 + \Delta x_1, x_2 + \Delta x_2) - F_{\mathbf{X}}(x_1, x_2 + \Delta x_2) - F_{\mathbf{X}}(x_1 + \Delta x_1, x_2) + F_{\mathbf{X}}(x_1, x_2).$$

[†]We remind the reader that the event $\{X_1 \leq x_1, \ldots, X_n \leq x_n\}$ is an *intersection* of the n events $\{X_i \leq x_i\}$ for $i = 1, \ldots n$.

Thus (still for $n = 2$)

$$f_{\mathbf{X}}(\mathbf{x}) = \lim_{\substack{\Delta x_1 \to 0 \\ \Delta x_2 \to 0}} \frac{1}{\Delta x_1 \Delta x_2} [F_{\mathbf{X}}(x_1 + \Delta x_1, x_2 + \Delta x_2) - F_{\mathbf{X}}(x_1 + \Delta x_1, x_2)$$

$$- F_{\mathbf{X}}(x_1, x_2 + \Delta x_2) + F_{\mathbf{X}}(x_1, x_2)]$$

which is by definition the second mixed partial derivative, and thus

$$f_{\mathbf{X}}(x_1, x_2) = \frac{\partial^2 F_{\mathbf{X}}(x_1, x_2)}{\partial x_1 \partial x_2}.$$

From Equation 5.1-5 we make the useful observation that

$$f_{\mathbf{X}}(\mathbf{x}) \Delta x_1 \ldots \Delta x_n \simeq P[x_1 < X_1 \leq x_1 + \Delta x_1, \ldots, x_n < X_n \leq x_n + \Delta x_n] \qquad (5.1\text{-}6)$$

if the increments are small. If we integrate Equation 5.1-4, we obtain the PDF as

$$F_{\mathbf{X}}(\mathbf{x}) = \int_{-\infty}^{x_1} \ldots \int_{-\infty}^{x_n} f_{\mathbf{X}}(\mathbf{x}') \, dx_1' \ldots dx_n',$$

which we can write in compact notation as

$$F_{\mathbf{X}}(\mathbf{x}) = \int_{-\infty}^{\mathbf{x}} f_{\mathbf{X}}(\mathbf{x}') d\mathbf{x}'.$$

More generally, for any event $B \subset R^N$ (R^N being Euclidean N-space) consisting of the countable union and intersection of parallelepipeds

$$P[B] = \int_{\mathbf{x} \in B} f_{\mathbf{X}}(\mathbf{x}) d\mathbf{x}. \qquad (5.1\text{-}7)$$

(Compare with Equation 2.5-3.) The argument behind the validity of Equation 5.1-7 follows very closely the argument furnished in the one-dimensional case (Section 2.5). Davenport [5-1, p. 149] discusses the validity of Equation 5.1-7 for the case $n = 2$. For $n > 2$ one can proceed by induction.

The conditional distribution function of $\mathbf{X}$ given the event B is defined by

$$F_{\mathbf{X}|B}(\mathbf{x}|B) \stackrel{\Delta}{=} P[\mathbf{X} \leq \mathbf{x}|B]$$

$$= \frac{P[\mathbf{X} \leq \mathbf{x}, B]}{P[B]} \quad (P[B] \neq 0).$$

These and subsequent results closely parallel the one-dimensional case. Consider next the n disjoint and exhaustive events $\{B_i, i = 1, \ldots, n\}$ with $P[B_i] > 0$. Then $\cup_{i=1}^{n} B_i = \Omega$ and $B_i B_j = \phi$ all $i \neq j$. We leave it as an exercise to the reader to show that

$$F_{\mathbf{X}}(\mathbf{x}) = \sum_{i=1}^{n} F_{\mathbf{X}|B_i}(\mathbf{x}|B_i) P[B_i]. \qquad (5.1\text{-}8)$$

The unconditional PDF on the left is sometimes called a *mixture* distribution function. The conditional pdf of $\mathbf{X}$ given the event B is an nth mixed partial derivative of $F_{\mathbf{X}|B}(\mathbf{x}|B)$ if it exists. Thus

$$f_{\mathbf{X}|B}(\mathbf{x}|B) \triangleq \frac{\partial^n F_{\mathbf{X}|B}(\mathbf{x}|B)}{\partial x_1 \ldots \partial x_n}. \tag{5.1-9}$$

It follows from Equations 5.1-8 and 5.1-9 that

$$f_{\mathbf{X}}(\mathbf{x}) = \sum_{i=1}^{n} f_{\mathbf{X}|B}(\mathbf{x}|B_i) P[B_i]. \tag{5.1-10}$$

Because $f_{\mathbf{X}}(\mathbf{x})$ is a mixture, that is, a linear combination of conditional pdf's it is sometimes called a *mixture* pdf.[†]

The joint distribution of two random vectors $\mathbf{X} = (X_1, \ldots, X_n)^T$ and $\mathbf{Y} = (Y_1, \ldots, Y_M)^T$ is

$$F_{\mathbf{XY}}(\mathbf{x}, \mathbf{y}) = P[\mathbf{X} \leq \mathbf{x}, \mathbf{Y} \leq \mathbf{y}]. \tag{5.1-11a}$$

The *joint density* of $\mathbf{X}$ and $\mathbf{Y}$, if it exists, is given by

$$f_{\mathbf{XY}}(\mathbf{x}, \mathbf{y}) = \frac{\partial^{(n+m)} F_{\mathbf{XY}}(\mathbf{x}, \mathbf{y})}{\partial x_1 \ldots \partial x_n \, \partial y_1 \ldots \partial y_m}. \tag{5.1-11b}$$

The *marginal density* of $\mathbf{X}$ alone, $f_{\mathbf{X}}(\mathbf{x})$, can be obtained from $f_{\mathbf{XY}}(\mathbf{x}, \mathbf{y})$ by integration, that is,

$$f_{\mathbf{X}}(\mathbf{x}) = \int_{-\infty}^{\infty} \cdots \int_{-\infty}^{\infty} f_{\mathbf{XY}}(\mathbf{x}, \mathbf{y}) \, dy_1 \ldots dy_m.$$

Similarly, the marginal pdf of the vector $\mathbf{X}' \triangleq (X_1, \ldots, X_{n-1})^T$ is obtained from the pdf of $\mathbf{X}$ by

$$f_{\mathbf{X}'}(\mathbf{x}') \triangleq \int_{-\infty}^{\infty} f_{\mathbf{X}}(\mathbf{x}) \, dx_n \quad \text{where } \mathbf{x}' \triangleq (x_1, \ldots, x_{n-1})^T. \tag{5.1-12}$$

Obviously, Equation 5.1-12 can be extended to other kinds of marginal pdf's as well by merely integrating over the appropriate variable.

Example 5.1-4

Let $\mathbf{X} = (X_1, X_2, X_3)^T$ denote the position of a particle inside a sphere of radius a centered about the origin. Assume that at the instant of observation, the particle is equally likely to be anywhere in the sphere, that is,

$$f_{\mathbf{X}}(\mathbf{x}) = \begin{cases} \dfrac{3}{4\pi a^3}, & \sqrt{x_1^2 + x_2^2 + x_3^2} < a \\ 0, & \text{otherwise.} \end{cases}$$

Compute the probability that the particle lies within a subsphere of radius $2a/3$ contained within the larger sphere.

[†]This usage is prevalent in statistical pattern recognition.

Solution   Let E denote the event that the particle lies within the subsphere (centered at the origin for simplicity) and let

$$\mathcal{R} \triangleq \{x_1, x_2, x_3\colon \sqrt{x_1^2 + x_2^2 + x_3^2} < 2a/3\}.$$

Then the evaluation of

$$P[E] = \iiint\limits_{\mathcal{R}} f_{\mathbf{x}}(x_1, x_2, x_3) \, dx_1 \, dx_2 \, dx_3$$

is best done using spherical coordinates, that is,

$$P[E] = \frac{3}{4\pi a^3} \int_{r=0}^{2a/3} \int_{\phi=0}^{\pi} \int_{\theta=0}^{2\pi} r^2 \sin \phi \, dr \, d\phi \, d\theta.$$

Note that in this simple case the answer can be obtained directly by noting the ratio of volumes, that is, $(2a/3)^3 \div a^3 = 8/27 \simeq 0.3$.

5.2 MULTIPLE TRANSFORMATION OF RANDOM VARIABLES

The material in this section is a direct extension of Section 3.4 in Chapter 3. Consider a sample description space Ω with point ζ and a set of n real random variables $X_1, X_2, \cdots, X_n$ from Ω to the real line R. For each $\zeta \in \Omega$ we generate the n-component vector of numbers $\mathbf{X}(\zeta) \triangleq (X_1(\zeta), X_2(\zeta), \ldots, X_n(\zeta)) \in R^n$. Then $\mathbf{X} \triangleq (X_1, X_2, \ldots, X_n)$ is said to be an n-dimensional real random vector. The definition is easily extended to a complex random vector. Now consider the n real functions

$$
\begin{aligned}
y_1 &= g_1(x_1, x_2, \cdots, x_n) \\
y_2 &= g_2(x_1, x_2, \cdots, x_n) \\
&\;\;\vdots \\
y_n &= g_n(x_1, x_2, \cdots, x_n),
\end{aligned}
\tag{5.2-1}
$$

where the $g_i, i = 1, \ldots, n$ are *functionally independent*, meaning that there exists no function $H(y_1, y_2, \ldots, y_n)$ that is identically zero. For example, the three linear functions

$$
\begin{aligned}
y_1 &= x_1 - 2x_2 + x_3 \\
y_2 &= 3x_1 + 2x_2 + 2x_3 \\
y_3 &= 5x_1 - 2x_2 + 4x_3
\end{aligned}
\tag{5.2-2}
$$

are not functionally independent because $H(y_1, y_2, \ldots, y_n) = 2y_1 + y_2 - y_3 = 0$ for all values of x_1, x_2, x_3. We create the vector of n r.v.'s $\mathbf{Y} \triangleq (Y_1, Y_2, \ldots, Y_n)$ according to

$$
\begin{aligned}
Y_1 &= g_1(X_1, X_2, \cdots, X_n) \\
Y_2 &= g_2(X_1, X_2, \cdots, X_n) \\
&\;\;\vdots \\
Y_n &= g_n(X_1, X_2, \cdots, X_n).
\end{aligned}
\tag{5.2-3}
$$

In this way we have generated n functions of n random variables. In order to save on notation, we let $\mathbf{x} \triangleq (x_1, x_2, \ldots, x_n)$, $\mathbf{y} \triangleq (y_1, y_2, \ldots, y_n)$ and ask: Given the joint pdf $f_{\mathbf{X}}(\mathbf{x})$, how do we compute the joint pdf of the $Y_i, i = 1, \ldots, n$, that is $f_{\mathbf{Y}}(\mathbf{y})$? Note that if we start out with fewer r.v.'s Y_i, say, $i = 1, \ldots, m$ than the number of X_i, say $i = 1, \ldots, n$ with $m < n$, we can add more Y_i by introducing auxiliary functions as we did in Example 3.5-4.

We assume that we can solve the set of Equations 5.2-1 uniquely for the $x_i, i = 1, \ldots, n$ as

$$
\begin{aligned}
x_1 &= \phi_1(y_1, y_2, \cdots, y_n) \\
x_2 &= \phi_2(y_1, y_2, \cdots, y_n) \\
&\;\;\vdots \\
x_n &= \phi_n(y_1, y_2, \cdots, y_n).
\end{aligned}
\tag{5.2-4}
$$

Now consider the elementary event $A \triangleq \{\zeta \colon y_i \le Y_i \le y_i + dy_i, \; i = 1, \ldots, n\}$. Here the Y_i are restricted to take on values in the infinitesimal *rectangular* parallelepiped that we denote by $\mathscr{P}_y$. Following the procedure in Equations 3.4-5 to 3.4-8, we write

$$
P[A] = \int_{\mathscr{P}_y} f_{\mathbf{Y}}(\mathbf{y})\, dy = f_{\mathbf{Y}}(\mathbf{y}) \mathbf{V}_y = \int_{\mathscr{P}_x} f_{\mathbf{X}}(\mathbf{x})\, dx = f_{\mathbf{X}}(\mathbf{x}) \mathbf{V}_x,
\tag{5.2-5}
$$

where $\mathscr{P}_x$ is an infinitesimal parallelepiped (not necessarily rectangular), $\mathbf{V}_y$ is the volume of $\mathscr{P}_y$, and $\mathbf{V}_x$ is the volume of $\mathscr{P}_x$. From Equation 5.2-5 we obtain

$$
f_{\mathbf{Y}}(\mathbf{y}) = f_{\mathbf{X}}(\mathbf{x}) \frac{\mathbf{V}_x}{\mathbf{V}_y}.
\tag{5.2-6}
$$

The ratio of infinitesimal volumes is shown in Appendix C to be the magnitude of the determinant, $\tilde{J}$, given by

$$
\tilde{J} =
\begin{vmatrix}
\dfrac{\partial \phi_1}{\partial y_1} & \cdots & \dfrac{\partial \phi_1}{\partial y_1} \\
\vdots & & \vdots \\
\dfrac{\partial \phi_n}{\partial y_1} & \cdots & \dfrac{\partial \phi_n}{\partial y_n}
\end{vmatrix}
\tag{5.2-7}
$$

$$
=
\begin{vmatrix}
\dfrac{\partial g_1}{\partial x_1} & \cdots & \dfrac{\partial g_1}{\partial x_n} \\
\vdots & & \vdots \\
\dfrac{\partial g_n}{\partial x_1} & \cdots & \dfrac{\partial g_n}{\partial x_n}
\end{vmatrix}^{-1}
= J^{-1}.
\tag{5.2-8}
$$

Hence

$$
f_{\mathbf{Y}}(\mathbf{y}) = f_{\mathbf{X}}(\mathbf{x}) |\tilde{J}| = f_{\mathbf{X}}(\mathbf{x}) / |J|.
\tag{5.2-9}
$$

In general, the infinitesimal rectangular parallelepiped in the $\mathbf{y}$ system maps into n disjoint, infinitesimal parallelepipeds in the $\mathbf{x}$ system. Then the event A, as defined above, is the union of the events $E_i, i = l, \ldots, r$, where $E_i = \{\mathbf{X} \in \mathscr{P}_x^{(i)}\}$ and $\mathscr{P}_x^{(i)}$ is one of the r parallelepipeds in the $\mathbf{x}$ system with volume $V_x^{(i)}$. Since the regions and, therefore, the events are disjoint, the elementary probabilities $P[E_i]$ add, and we obtain the main result of this section, that is

$$f_\mathbf{Y}(\mathbf{y}) = \sum_{i=1}^{r} f_\mathbf{X}(\mathbf{x}^{(i)}) |\tilde{J}_i| \tag{5.2-10}$$

$$= \sum_{i=1}^{r} f_\mathbf{X}(\mathbf{x}^{(i)}) / |J_i|. \tag{5.2-11}$$

In Equations 5.2-10 and 5.2-11 $|\tilde{J}_i| \triangleq V_x^{(i)}/V_y$ and $|\tilde{J}_i| = |J_i|^{-1}$.

Example 5.2-1
We are given three transmittances

$$g_1(\mathbf{x}) = x_1^2 - x_2^2$$

$$g_2(\mathbf{x}) = x_1^2 + x_2^2$$

$$g_3(\mathbf{x}) = x_3.$$

There are four solutions (roots) to the system

$$y_1 = x_1^2 - x_2^2$$

$$y_2 = x_1^2 + x_2^2$$

$$y_3 = x_3.$$

They are

$$x_1^{(1)} = ((y_1 + y_2)/2)^{1/2} \qquad x_1^{(2)} = ((y_1 + y_2)/2)^{1/2}$$

$$x_2^{(1)} = ((y_2 - y_1)/2)^{1/2} \qquad x_2^{(2)} = -((y_2 - y_1)/2)^{1/2}$$

$$x_3^{(1)} = y_3 \qquad x_3^{(2)} = y_3$$

$$x_1^{(3)} = -((y_1 + y_2)/2)^{1/2} \qquad x_1^{(4)} = -((y_1 + y_2)/2)^{1/2}$$

$$x_2^{(3)} = ((y_2 - y_1)/2)^{1/2} \qquad x_2^{(4)} = -((y_2 - y_1)/2)^{1/2}$$

$$x_3^{(3)} = y_3 \qquad x_3^{(4)} = y_3.$$

$$\tag{5.2-12}$$

For the roots to be real, $y_1 > 0$ and $y_2 > y_1$. In this case the single rectangular parallelepiped in the $\mathbf{y}$ system maps into four disjoint, infinitesimal parallelepipeds in the $\mathbf{x}$ system.

Example 5.2-2 _____

For the transformation considered in Example 5.2-1 compute $f_{\mathbf{Y}}(\mathbf{y})$ if

$$f_{\mathbf{X}}(\mathbf{x}) = (2\pi)^{-3/2} \exp\left[-\frac{1}{2}(x_1^2 + x_2^2 + x_3^2)\right].$$

Solution We must compute the Jacobian $|J|$ at each of the four roots. The Jacobian is computed as

$$J = \begin{vmatrix} 2x_1 & 2x_1 & 0 \\ -2x_2 & 2x_1 & 0 \\ 0 & 0 & 1 \end{vmatrix} = 8x_1 x_2.$$

For example at the first root we compute

$$J_1 = 0.5(y_2^2 - y_1^2)^{1/2}.$$

A direct calculation shows that $|J_1| = |J_2| = |J_3| = |J_4|$. Finally labeling the four solutions in Equation 5.2-12 as $\mathbf{x}_1, \mathbf{x}_2, \mathbf{x}_3, \mathbf{x}_4$, we obtain

$$f_{\mathbf{Y}}(\mathbf{y}) = \frac{2}{(y_2^2 - y_1^2)^{1/2}} \sum_{i=1}^{4} f_{\mathbf{X}}(\mathbf{x}_i) = \frac{8(2\pi)^{-3/2}}{(y_2^2 - y_1^2)^{1/2}} \exp\left[-\frac{1}{2}(y_2 + y_3^2)\right] \times u(y_2)u(y_2 - y_1).$$

_____

Although a random vector is completely characterized by its distribution or density functions, the latter are often hard to come by except for some notable exceptions. By far the two most important exceptions are (1) when $F_{\mathbf{X}}(\mathbf{x}) = F_{X_1}(x_1) \ldots F_{X_n}(x_n)$, that is, the n components of $\mathbf{X}$ are independent, and (2) when $\mathbf{X}$ obeys the multidimensional Gaussian law. Case (1) is easily handled, since it is essentially an extension of the scalar case. Case (2) will be discussed in Section 5.6. But what to do when neither case (1) nor (2) applies? Estimating multidimensional distributions involving dependent variables is often not practical and even if available might be too complex to be of any real use. Therefore, when we deal with vector r.v.'s, we often settle for a less complete but more computable characterization based on moments. For most engineering applications the most important moments are the expectation vector (the first moment) and the covariance matrix (a second moment). These quantities and their use are discussed in the next several sections.

5.3 EXPECTATION VECTORS AND COVARIANCE MATRICES†

Definition 5.3-1 The expectation of the (column) vector $\mathbf{X} = (X_1, \ldots, X_n)^T$ is a vector $\boldsymbol{\mu}$ (or $\bar{\mathbf{X}}$) whose elements $\mu_1, \ldots, \mu_n$ are given by

$$\mu_i \triangleq \int_{-\infty}^{\infty} \ldots \int_{-\infty}^{\infty} x_i f_{\mathbf{X}}(x_1, \ldots, x_n) \, dx_1 \ldots dx_n. \tag{5.3-1}$$

_____

†This section requires some familiarity with matrix theory.

Equivalently with

$$f_{X_i}(x_i) \triangleq \int_{-\infty}^{\infty} \cdots \int_{-\infty}^{\infty} f_{\mathbf{X}}(\mathbf{x})\, dx_1 \ldots dx_{i-1}\, dx_{i+1} \ldots dx_n$$

the marginal pdf of X_i, we can write

$$\mu_i = \int_{-\infty}^{\infty} x_i f_{X_i}(x_i)\, dx_i \qquad i = 1, \ldots, n. \quad \blacksquare$$

Definition 5.3-2   The covariance matrix $\mathbf{K}$ associated with a real random vector $\mathbf{X}$ is the expected value of the outer vector product $(\mathbf{X} - \boldsymbol{\mu})(\mathbf{X} - \boldsymbol{\mu})^T$, that is,

$$\mathbf{K} \triangleq E[(\mathbf{X} - \boldsymbol{\mu})(\mathbf{X} - \boldsymbol{\mu})^T]. \tag{5.3-2}$$

Define

$$
\begin{aligned}
K_{ij} &\triangleq E[(X_i - \mu_i)(X_j - \mu_j)] \\
&= E[(X_j - \mu_j)(X_i - \mu_i)] \\
&= K_{ji} \qquad i, j = 1, \ldots, n.
\end{aligned} \tag{5.3-3}
$$

In particular with $\sigma_i^2 \triangleq K_{ii}$, we can write $\mathbf{K}$ in expanded form as

$$
\mathbf{K} =
\begin{bmatrix}
\sigma_1^2 & & \cdots & & K_{1n} \\
& \ddots & & & \\
\vdots & & \sigma_i^2 & & \vdots \\
& & & \ddots & \\
K_{n1} & & \cdots & & \sigma_n^2
\end{bmatrix} . \quad \blacksquare \tag{5.3-4}
$$

If $\mathbf{X}$ is real, all the elements of $\mathbf{K}$ are real. Also since $K_{ij} = K_{ji}$, covariance matrices fall within the class of matrices called *real symmetric* (r.s.). Such matrices fall within the larger class of Hermitian matrices.[†] Real symmetric matrices have many interesting properties, several of which we shall discuss in the next section.

The diagonal elements σ_i^2 are the variances associated with the individual r.v.'s X_i for $i = 1, \ldots, n$. The covariance matrix $\mathbf{K}$ is closely related to the correlation matrix $\mathbf{R}$ defined by

$$\mathbf{R} \triangleq E[\mathbf{X}\mathbf{X}^T]. \tag{5.3-5}$$

Indeed expanding Equation 5.3-2 yields

$$\mathbf{K} = \mathbf{R} - \boldsymbol{\mu}\boldsymbol{\mu}^T$$

[†]The class of $n \times n$ matrices for which $K_{ij} = K_{ji}^*$. For a thorough discussion of the properties of such matrices see [5-2]. When $\mathbf{X}$ is complex, the covariance is generally not r.s. but is Hermitian.

or

$$\mathbf{R} = \mathbf{K} + \boldsymbol{\mu}\boldsymbol{\mu}^T. \tag{5.3-6}$$

The correlation matrix $\mathbf{R}$ is also real symmetric; it is sometimes called the *autocorrelation* matrix.

Random vectors are often classified according to whether they are uncorrelated, orthogonal, or independent.

Definition 5.3-3   Consider two real n-dimensional random vectors $\mathbf{X}$ and $\mathbf{Y}$ with respective mean vectors $\boldsymbol{\mu}_{\mathbf{X}}$ and $\boldsymbol{\mu}_{\mathbf{Y}}$. Then if the expected value of their *outer product* satisfies

$$E\{\mathbf{X}\mathbf{Y}^T\} = \boldsymbol{\mu}_{\mathbf{X}}\boldsymbol{\mu}_{\mathbf{Y}}^T, \tag{5.3-7}$$

$\mathbf{X}$ and $\mathbf{Y}$ are sait to be *uncorrelated*. If

$$E\{\mathbf{X}\mathbf{Y}^T\} = \mathbf{0} \quad \text{(an } n \times n \text{ matrix of all zeros)}, \tag{5.3-8}$$

$\mathbf{X}$ and $\mathbf{Y}$ are sait to be *orthogonal*. Note that in the orthogonal case $E\{X_i Y_j\} = 0$ for all $0 \leq i, j \leq n$. Thus the expected value of the *inner product* is zero i.e., $\mathbf{E}[\mathbf{X}^{\mathbf{T}}\mathbf{Y}] = 0$ which reminds us of the meaning of orthogonality for two ordinary (nonrandom) vectors, that is, $\mathbf{x}^T\mathbf{y} = 0$. Finally if

$$f_{\mathbf{X}\mathbf{Y}}(\mathbf{x}, \mathbf{y}) = f_{\mathbf{X}}(\mathbf{x})f_{\mathbf{Y}}(\mathbf{y}), \tag{5.3-9}$$

$\mathbf{X}$ and $\mathbf{Y}$ are said to be *independent*. ∎

Independence always implies uncorrelatedness but the converse is not generally true. An exception is the multidimensional Gaussian pdf discussed in Section 5.6. It is often difficult, in practice, to show that two random vectors are independent. However, statistical tests exist to determine, within prescribed confidence levels, the extent to which they are correlated.

Example 5.3-1 _____

(Dependent random variables with an almost-diagonal covariance matrix.) Consider two r.v.'s X_1 and X_2 with joint pdf $f_{X_1 X_2}(x_1, x_1) = x_1 + x_2$ for $0 < x_1 \leq 1$, $0 < x_2 \leq 1$, and zero elsewhere. We find that while X_1 and X_2 are not independent, they are essentially uncorrelated. To demonstrate this we shall compute $E[(X_1 - \mu_1)(X_2 - \mu_2)]$ as

$$K_{12} = K_{21} = R_{21} - \mu_2\mu_1.$$

We first compute

$$\mu_1 = \mu_2 = \iint_S x(x + y)\, dx\, dy = 0.583,$$

where $S = \{(x_1, x_2) \colon 0 < x_1 \leq 1,\ 0 < x_2 \leq 1\}$.

Next we compute the correlation products

$$R_{12} = R_{21} \triangleq \iint_S xy(x + y)\, dx\, dy = 0.333.$$

Hence $K_{12} = K_{21} = 0.333 - (0.583)^2 = -0.007$. Also we compute

$$\sigma_1^2 = \sigma_2^2 = \int_0^1 x^2(x + 1/2 \cdot x)\, dx - (0.583)^2 = 0.077.$$

Hence the correlation coefficient (normalized covariance) is computed to be $\rho = K_{12}/\sigma_1\sigma_2 = -0.091$. For the purpose of predicting X_2 by observing X_1, or vice versa, one may consider these r.v.'s as being uncorrelated. Indeed the prediction error ε in Equation 4.3-22 from Example 4.3-4 is 0.076. Were X_1, X_2 truly uncorrelated, the prediction error would have been 0.077! The covariance matrix $\mathbf{K}$ for this case is

$$\mathbf{K} = \begin{bmatrix} 0.077 & -0.007 \\ -0.007 & 0.077 \end{bmatrix} = 0.077 \begin{bmatrix} 1 & -0.09 \\ -0.09 & 1 \end{bmatrix}.$$

5.4 PROPERTIES OF COVARIANCE MATRICES

Since covariance matrices are r.s., we study some of the properties of such matrices. Let $\mathbf{M}$ be any $n \times n$ r.s. matrix. The quadratic form associated with $\mathbf{M}$ is the scalar $q(\mathbf{z})$ defined by

$$q(\mathbf{z}) \triangleq \mathbf{z}^T \mathbf{M} \mathbf{z}, \tag{5.4-1}$$

where $\mathbf{z}$ is any column vector. $\mathbf{M}$ is said to be *positive semidefinite* (p.s.d.) if

$$\mathbf{z}^T \mathbf{M} \mathbf{z} \geq 0$$

for all $\mathbf{z}$. If $\mathbf{z}^T \mathbf{M} \mathbf{z} > 0$ for all $\mathbf{z} \neq 0$, $\mathbf{M}$ is said to be *positive definite* (p.d.). A covariance matrix $\mathbf{K}$ is always (at least) p.s.d. since for any vector $\mathbf{z} \triangleq (z_1, \ldots, z_n)^T$

$$0 \leq E\{[\mathbf{z}^T(\mathbf{X} - \boldsymbol{\mu})]^2\}$$
$$= \mathbf{z}^T E[(\mathbf{X} - \boldsymbol{\mu})(\mathbf{X} - \boldsymbol{\mu})^T]\mathbf{z}$$
$$= \mathbf{z}^T \mathbf{K} \mathbf{z}, \quad \mathbf{K} \triangleq E[(\mathbf{X} - \boldsymbol{\mu})(\mathbf{X} - \boldsymbol{\mu})^T]. \tag{5.4-2}$$

We shall show later that when $\mathbf{K}$ is full-rank, then $\mathbf{K}$ is p.d.

We now state some definitions and theorems (most without proof) from linear algebra [5-2, Chapter 4] that we shall need for developing useful operations on covariance matrices.

Definition 5.4-1   The *eigenvalues* of an $n \times n$ matrix $\mathbf{M}$ are those numbers λ for which the characteristic equation $\mathbf{M}\boldsymbol{\phi} = \lambda\boldsymbol{\phi}$ has a solution $\boldsymbol{\phi} \neq \mathbf{0}$. The column vector $\boldsymbol{\phi} = (\phi_1, \phi_2, \ldots, \phi_n)^T$ is called an *eigenvector*.

Eigenvectors are often normalized so that $\boldsymbol{\phi}^T\boldsymbol{\phi} \triangleq ||\boldsymbol{\phi}||^2 = 1$. ∎

Theorem 5.4-1   The number λ is an eigenvalue of the square matrix $\mathbf{M}$ if and only if $\det(\mathbf{M} - \lambda\mathbf{I}) = 0$.[†] ∎

[†]det is short for determinant and $\mathbf{I}$ is the identity matrix.

Example 5.4-1

Consider the matrix

$$\mathbf{M} = \begin{bmatrix} 4 & 2 \\ 2 & 4 \end{bmatrix}.$$

The eigenvalues are obtained with the help of Theorem 5.4-1, that is,

$$\det \begin{bmatrix} 4 - \lambda & 2 \\ 2 & 4 - \lambda \end{bmatrix} = (4 - \lambda)^2 - 4 = 0,$$

whence

$$\lambda_1 = 6, \qquad \lambda_2 = 2.$$

The (normalized) eigenvector associated with $\lambda_1 = 6$ is obtained from

$$(\mathbf{M} - 6\mathbf{I})\boldsymbol{\phi} = 0,$$

which, written out as a system of equations, yields

$$\left.\begin{array}{r} -2\phi_1 + 2\phi_2 = 0 \\ -2\phi_1 + 2\phi_2 = 0 \end{array}\right\} \Rightarrow \boldsymbol{\phi}_1 = \frac{1}{\sqrt{2}}(1, 1)^T.$$

The double arrow $\Rightarrow$ means "implies that." The eigenvector associated with $\lambda_2 = 2$, following the same procedure as above, is found from

$$\left.\begin{array}{r} 2\phi_1 + 2\phi_2 = 0 \\ 2\phi_1 + 2\phi_2 = 0 \end{array}\right\} \Rightarrow \boldsymbol{\phi}_2 = \frac{1}{\sqrt{2}}(1, 1)^T.$$

Not all $n \times n$ matrices have n distinct eigenvalues or n eigenvectors. Sometimes a matrix can have fewer than n distinct eigenvalues but still have n distinct eigenvectors.

Definition 5.4-2   Two $n \times n$ matrices $\mathbf{A}$ and $\mathbf{B}$ are called *similar* if there exists an $n \times n$ matrix $\mathbf{T}$ with $\det \mathbf{T} \neq 0$ such that

$$\mathbf{T}^{-1}\mathbf{A}\mathbf{T} = \mathbf{B}. \quad \blacksquare \qquad\qquad (5.4\text{-}3)$$

Theorem 5.4-2   An $n \times n$ matrix $\mathbf{M}$ is similar to a diagonal matrix if and only if $\mathbf{M}$ has n linearly independent eigenvectors. $\blacksquare$

Theorem 5.4-3   Let $\mathbf{M}$ be a r.s. matrix with eigenvalues $\lambda_1, \ldots, \lambda_n$. Then $\mathbf{M}$ has n mutually orthogonal unit eigenvectors $\boldsymbol{\phi}_1, \ldots, \boldsymbol{\phi}_n$. $\blacksquare$

Discussion.   Since $\mathbf{M}$ has n mutually orthogonal (and therefore independent) unit eigenvectors, it is similar to some diagonal matrix $\boldsymbol{\Lambda}$ under a suitable transformation $\mathbf{T}$. What are $\boldsymbol{\Lambda}$ and $\mathbf{T}$? The answer is furnished by the following important theorem.

Theorem 5.4-4   Let $\mathbf{M}$ be a real symmetric matrix with eigenvalues $\lambda_1, \ldots, \lambda_n$. Then $\mathbf{M}$ is similar to the diagonal matrix $\boldsymbol{\Lambda}$ given by

$$\boldsymbol{\Lambda} \triangleq \begin{bmatrix} \lambda_1 & & \mathbf{0} \\ & \ddots & \\ \mathbf{0} & & \lambda_n \end{bmatrix}$$

under the transformation

$$\mathbf{U}^{-1}\mathbf{M}\mathbf{U} = \boldsymbol{\Lambda} \tag{5.4-4}$$

where $\mathbf{U}$ is a matrix whose columns are the ordered[†] orthogonal unit eigenvectors $\boldsymbol{\phi}_i$, $i = 1, \ldots, n$ of $\mathbf{M}$. Thus

$$\mathbf{U} = (\boldsymbol{\phi}_1, \ldots, \boldsymbol{\phi}_n). \tag{5.4-5}$$

Moreover, it can be shown that $\mathbf{U}^T\mathbf{U} = \mathbf{I}$ (and that $\mathbf{U}^T = \mathbf{U}^{-1}$) so that Equation 5.4-4 can be written as

$$\mathbf{U}^T\mathbf{M}\mathbf{U} = \boldsymbol{\Lambda}. \quad \blacksquare \tag{5.4-6}$$

Discussion.   Matrices such as $\mathbf{U}$, which satisfy $\mathbf{U}^T\mathbf{U} = \mathbf{I}$, are called *unitary*. They have the property of *distance preservation* in the following sense: Consider a vector $\mathbf{x} = (x_1, \ldots, x_n)^T$. The Euclidean distance of $\mathbf{x}$ from the origin is

$$||\mathbf{x}|| \triangleq (\mathbf{x}^T\mathbf{x})^{1/2},$$

where $||\mathbf{x}||$ is called the *norm* of $\mathbf{x}$. Now consider the transformation $\mathbf{y} = \mathbf{U}\mathbf{x}$ where $\mathbf{U}$ is unitary. Then

$$||\mathbf{y}||^2 = \mathbf{y}^T\mathbf{y} = \mathbf{x}^T\mathbf{U}^T\mathbf{U}\mathbf{x} = ||\mathbf{x}||^2.$$

Thus the new vector $\mathbf{y}$ has the same distance from the origin as the old vector $\mathbf{x}$ under the transformation $\mathbf{y} = \mathbf{U}\mathbf{x}$.

Since a covariance matrix $\mathbf{K}$ is real symmetric, it can be easily diagonalized according to Equation 5.4-6 once $\mathbf{U}$ is known. The columns of $\mathbf{U}$ are just the eigenvectors of $\mathbf{K}$ and these can be obtained once the eigenvalues are known. The diagonalization of covariance matrices is a very important procedure in applied probability theory. It is used to transform correlated r.v.'s into uncorrelated r.v.'s; in statistical pattern recognition (for example, computing the Fisher discriminant—defined in Section 5.5); in solving problems involving the multidimensional normal pdf; and numerous other places.

Example 5.4-2 _____

(Decorrelation of random vectors.) A random vector $\mathbf{X} = (X_1, X_2, X_3)^T$ has covariance matrix

$$\mathbf{K_X} = \begin{bmatrix} 2 & -1 & 1 \\ -1 & 2 & 0 \\ 1 & 0 & 2 \end{bmatrix}.$$

_____

[†]That is, $\boldsymbol{\phi}_i$ goes with λ_i for $i = 1, \ldots, n$.

Design a nontrivial transformer (a circuit that consists of adders and multipliers) that will generate from $\mathbf{X}$ a new random vector $\mathbf{Y}$ whose components are uncorrelated.

Solution   First we compute the eigenvalues by solving the matrix equation $\mathbf{K} - \lambda\mathbf{I} = 0$. This yields $\lambda_1 = 2$, $\lambda_2 = 2 + \sqrt{2}$, $\lambda_3 = 2 - \sqrt{2}$. Next we compute the three orthogonal eigenvectors by solving the equation $(\mathbf{K} - \lambda_i\mathbf{I})\boldsymbol{\phi}_i = 0$, $i = 1, 2, 3$. This yields

$$\boldsymbol{\phi}_1 = \left(0, \frac{1}{\sqrt{2}}, \frac{1}{\sqrt{2}}\right)^T,$$

$$\boldsymbol{\phi}_2 = \left(\frac{1}{\sqrt{2}}, -\frac{1}{2}, \frac{1}{2}\right)^T,$$

$$\boldsymbol{\phi}_3 = \left(\frac{1}{\sqrt{2}}, \frac{1}{2}, -\frac{1}{2}\right)^T.$$

Now we create the eigenvector matrix $\mathbf{U} = (\boldsymbol{\phi}_1\boldsymbol{\phi}_2\boldsymbol{\phi}_3)$ that, upon transposing, becomes an appropriate transformer to make the components of $\mathbf{Y}$ uncorrelated. With

$$\mathbf{A} = \mathbf{U}^T = \begin{bmatrix} 0 & \frac{1}{\sqrt{2}} & \frac{1}{\sqrt{2}} \\ \frac{1}{\sqrt{2}} & -\frac{1}{2} & \frac{1}{2} \\ \frac{1}{\sqrt{2}} & \frac{1}{2} & -\frac{1}{2} \end{bmatrix}.$$

The transformation $\mathbf{Y} = \mathbf{AX}$ yields the components

$$Y_1 = \frac{1}{\sqrt{2}}(X_2 + X_3)$$

$$Y_2 = \frac{1}{\sqrt{2}}X_1 - \frac{1}{2}X_2 + \frac{1}{2}X_3$$

$$Y_3 = \frac{1}{\sqrt{2}}X_1 + \frac{1}{2}X_2 - \frac{1}{2}X_3.$$

The covariance of $\mathbf{Y}$ is given by

$$\mathbf{K_Y} = \begin{bmatrix} 2 & 0 & 0 \\ 0 & 2 + \sqrt{2} & 0 \\ 0 & 0 & 2 - \sqrt{2} \end{bmatrix}.$$

Actually we could go one step further; by passing the three components of $\mathbf{Y}$, separately, through three amplifiers, we can make the variance (average AC power) the same in each amplified component. This process is called *whitening* and is discussed in greater detail in the next section. Clearly if Y_1 is passed through an amplifier with gain proportional to $\frac{1}{\sqrt{\lambda_1}}$, Y_2 is passed through an amplifier with gain proportional to $\frac{1}{\sqrt{\lambda_2}}$, and Y_3 is passed through an amplifier with gain proportional to $\frac{1}{\sqrt{\lambda_3}}$, all three outputs will have the same power.

If $\boldsymbol{\phi}_1, \ldots, \boldsymbol{\phi}_n$ are the orthogonal unit eigenvectors of a real symmetric matrix $\mathbf{M}$, then the system of equations

$$\mathbf{M}\boldsymbol{\phi}_1 = \lambda_1 \boldsymbol{\phi}_1$$

$$\vdots$$

$$\mathbf{M}\boldsymbol{\phi}_n = \lambda_n \boldsymbol{\phi}_n$$

can be compactly written as

$$\mathbf{MU} = \mathbf{U\Lambda}. \tag{5.4-7}$$

The next theorem establishes a relation between the eigenvalues of a r.s. matrix and its positive definite character.

Theorem 5.4-5   A real symmetric matrix $\mathbf{M}$ is positive definite if and only if all its eigenvalues are positive.

Proof   First let $\lambda_i > 0$, $i = 1, \ldots, n$. Then with the linear transformation $\mathbf{x} \stackrel{\Delta}{=} \mathbf{Uy}$ we can write for any vector $\mathbf{x}$

$$\begin{aligned} \mathbf{x}^T \mathbf{Mx} &= (\mathbf{Uy})^T \mathbf{M}(\mathbf{Uy}) \\ &= \mathbf{y}^T \mathbf{U}^T \mathbf{MUy} \\ &= \mathbf{y}^T \mathbf{\Lambda y} \\ &= \sum_{i=1}^{n} \lambda_i y_i^2 > 0 \end{aligned} \tag{5.4-8}$$

unless $\mathbf{y} = 0$. But if $\mathbf{y} = 0$, then from $\mathbf{x} = \mathbf{Ay}$, $\mathbf{x} = 0$ as well. Hence we have shown that $\mathbf{M}$ is p.d. if $\lambda_i > 0$ for all i. Conversely, we must show that if $\mathbf{M}$ is p.d., then all $\lambda_i > 0$. Thus for any $\mathbf{x} \neq 0$

$$0 < \mathbf{x}^T \mathbf{Mx}. \tag{5.4-9}$$

In particular, Equation 5.4-9 must hold for $\boldsymbol{\phi}_1, \ldots, \boldsymbol{\phi}_n$. But

$$0 < \boldsymbol{\phi}_i^T \mathbf{M} \boldsymbol{\phi}_i = \lambda_i \qquad i = 1, \ldots, n.$$

Hence $\lambda_i > 0$, $i = 1, \ldots, n$. Thus a p.d. covariance matrix $\mathbf{K}$ will have all positive eigenvalues. Also since its determinant $\det(\mathbf{K})$ is the product of its eigenvalues, $\det(\mathbf{K}) > 0$. Thus when $\mathbf{K}$ is full-rank it is p.d.   ∎

The next section deals with an important problem in a branch of applied probability called *statistical pattern recognition*.

5.5 SIMULTANEOUS DIAGONALIZATION OF TWO COVARIANCE MATRICES AND APPLICATIONS IN PATTERN RECOGNITION

Theorem 5.5-1   Let $\mathbf{P}$ and $\mathbf{Q}$ be $n \times n$ real symmetric matrices. If $\mathbf{P}$ is positive definite, then there exists a $n \times n$ matrix $\mathbf{V}$ which achieves

$$\mathbf{V}^T \mathbf{P} \mathbf{V} = \mathbf{I} \tag{5.5-1}$$

and

$$\mathbf{V}^T \mathbf{Q} \mathbf{V} = \mathbf{\Lambda} = \mathrm{diag}(\lambda_1, \ldots, \lambda_n).$$

The real numbers $\lambda_1, \ldots, \lambda_n$ satisfy the generalized eigenvalue equation

$$\mathbf{Q}\mathbf{v}_i = \lambda_i \mathbf{P}\mathbf{v}_i. \tag{5.5-2}$$

The numbers λ_i and vectors $\mathbf{v}_i$ for $i = 1, \ldots, n$ are sometimes called generalized eigenvalues and eigenvectors.

Proof   Let $\mathbf{U} = (\boldsymbol{\phi}_1, \ldots, \boldsymbol{\phi}_n)$ where $\boldsymbol{\phi}_i, \gamma_i$ for $i = 1, \ldots, n$ are the unit eigenvectors and eigenvalues of $\mathbf{P}$. Then

$$\mathbf{U}^T \mathbf{P} \mathbf{U} = \mathrm{diag}(\gamma_1, \ldots, \gamma_n) \triangleq \mathbf{M}. \tag{5.5-3}$$

Since $\mathbf{P}$ is positive definite, all the $\gamma_i > 0$ and the diagonal matrix $\mathbf{Z} \triangleq \mathrm{diag}(\gamma_1^{-1/2}, \ldots, \gamma_n^{-1/2})$ exists and is real. Now

$$\mathbf{Z}^T \mathbf{U}^T \mathbf{P} \mathbf{U} \mathbf{Z} = \mathbf{Z}^T \mathbf{M} \mathbf{Z} = \mathbf{I}. \tag{5.5-4}$$

Thus by the similarity transformation $(\mathbf{U}\mathbf{Z})^T \mathbf{P} (\mathbf{U}\mathbf{Z})$ we have not only diagonalized $\mathbf{P}$ but also reduced it to an identity matrix. This process is called *whitening*. Next we inquire whether the similarity transformation $(\mathbf{U}\mathbf{Z})^T \mathbf{Q} (\mathbf{U}\mathbf{Z})$ has produced a matrix that retains a real symmetric structure. But

$$[\mathbf{Z}^T \mathbf{U}^T \mathbf{Q} \mathbf{U} \mathbf{Z}]^T = \mathbf{Z}^T \mathbf{U}^T \mathbf{Q} \mathbf{U} \mathbf{Z} \tag{5.5-5}$$

so that $\mathbf{Z}^T \mathbf{U}^T \mathbf{Q} \mathbf{U} \mathbf{Z} \triangleq \mathbf{A}$ is r.s. Since $\mathbf{A}$ is r.s., there exists a unitary similarity transformation $\mathbf{W}^T \mathbf{A} \mathbf{W} = \mathrm{diag}(\lambda_1, \ldots, \lambda_n) \triangleq \mathbf{\Lambda}$ where $\mathbf{W}^T \mathbf{W} = \mathbf{I}$. Now we ask, does applying $\mathbf{W}$ to $\mathbf{Z}^T \mathbf{M} \mathbf{Z} = \mathbf{I}$ (Equation 5.5-4), that is,

$$\mathbf{W}^T (\mathbf{Z}^T \mathbf{M} \mathbf{Z}) \mathbf{W} \tag{5.5-6}$$

affect the whitening transformation? The answer is no since substituting Equation 5.5-4 in 5.5-6 results in $\mathbf{W}^T \mathbf{I} \mathbf{W} = \mathbf{I}$.

Comparing $\mathbf{W}^T\mathbf{A}\mathbf{W} \triangleq (\mathbf{UZW})^T\mathbf{P}(\mathbf{UZW})$ with Equation 5.5-1 we see that the matrix $\mathbf{V}$, which achieves the simultaneous diagonalization of $\mathbf{P}$ and $\mathbf{Q}$, is given by

$$\mathbf{V} \triangleq \mathbf{UZW}. \tag{5.5-7}$$

Using Equation 5.5-1 we obtain

$$\det(\mathbf{V}^T\mathbf{P}\mathbf{V}) = \det(\mathbf{I}) = 1.$$

From matrix theory we know that for any real $n \times n$ matrices $\mathbf{A}$, $\mathbf{B}$, $\mathbf{C}$ $\det(\mathbf{ABC}) = \det(\mathbf{A})\det(\mathbf{B})\det(\mathbf{C})$ and $\det(\mathbf{A}^T) = \det(\mathbf{A})$. Thus

$$\det(\mathbf{V}^T\mathbf{P}\mathbf{V}) = \det(\mathbf{V})\det(\mathbf{P})\det(\mathbf{V}) = 1.$$

Hence $\det(\mathbf{V}) \neq 0$ and $\mathbf{V}$ is invertible, that is, $\mathbf{V}^{-1}$ exists. From

$$\mathbf{V}^T\mathbf{P}\mathbf{V} = \mathbf{I}$$

we obtain

$$\mathbf{P}\mathbf{V} = (\mathbf{V}^T)^{-1}. \tag{5.5-8}$$

Thus from $\mathbf{V}^T\mathbf{Q}\mathbf{V} = \mathbf{\Lambda}$ we obtain

$$\mathbf{Q}\mathbf{V} = [\mathbf{V}^T]^{-1}\mathbf{\Lambda}$$

and, using Equation 5.5-8,

$$\mathbf{Q}\mathbf{V} = \mathbf{P}\mathbf{V}\mathbf{\Lambda}, \tag{5.5-9}$$

which the reader will recognize as compact notation for the system of generalized eigenvalue equations

$$\mathbf{Q}\mathbf{v}_i = \lambda_i\mathbf{P}\mathbf{v}_i \qquad i = 1,\ldots,n \tag{5.5-10}$$

given in Equation 5.5-2. This completes the proof of Theorem 5.5-1. ∎

We observe that Equation 5.5-10 can also be written (since $\mathbf{P}$ is p.d. it is invertible) as

$$\mathbf{P}^{-1}\mathbf{Q}\mathbf{v}_i = \lambda_i\mathbf{v}_i. \tag{5.5-11}$$

Thus the $\mathbf{v}_i$ can be viewed as the eigenvalues of the matrix $\mathbf{P}^{-1}\mathbf{Q}$. However, the normalization of $\mathbf{v}_i$ must be such as to satisfy $\mathbf{v}_i^T\mathbf{P}\mathbf{v}_i = 1$, which implies $\mathbf{V}^T\mathbf{P}\mathbf{V} = \mathbf{I}$. The procedure, then, for diagonalizing two matrices $\mathbf{P}$ and $\mathbf{Q}$ simultaneously is the following:

1. Calculate the eigenvalues of Equation 5.5-11; calculate unnormalized eigenvectors $\mathbf{v}_i'$ for $i = 1,\ldots,n$ by solving

$$(\mathbf{P}^{-1}\mathbf{Q} - \lambda\mathbf{I})\mathbf{v}_i' = 0;$$

2. Find constants K_i, $i = 1,\ldots,n$ such that $\mathbf{v}_i \triangleq K_i\mathbf{v}_i'$ satisfies $\mathbf{v}_i^T\mathbf{P}\mathbf{v}_i = 1$, $i = 1,\ldots,n$.

As a final comment we point out that an important result can be obtained from Equation 5.5-4. Since $\mathbf{P}$ is positive definite, both $\mathbf{U}$ and $\mathbf{Z}$ are invertible. Moreover, since $\mathbf{U}$ is unitary $\mathbf{U}^{-1} = \mathbf{U}^T$ and $(\mathbf{Z}^{-1})^T = \mathbf{Z}^{-1}$. Then if we premultiply

$$(\mathbf{UZ})^T\mathbf{P}(\mathbf{UZ}) = \mathbf{I}$$

by $[(\mathbf{UZ})^T]^{-1}$ and post-multiply by $(\mathbf{UZ})^{-1}$, we obtain

$$\mathbf{P} = \mathbf{CC}^T, \tag{5.5-12}$$

where $\mathbf{C} \triangleq \mathbf{UZ}^{-1}$. Thus we see that any real-symmetric, positive definite matrix $\mathbf{P}$ can be factored as in Equation 5.5-12. In terms of $\mathbf{C}$ we obtain

$$\mathbf{C}^{-1}\mathbf{P}[\mathbf{C}^{-1}]^T = \mathbf{I} \tag{5.5-13a}$$

and

$$\mathbf{C}^T\mathbf{P}^{-1}\mathbf{C} = \mathbf{I}. \tag{5.5-13b}$$

Example 5.5-1

(Simultaneous diagonalization of two covariance matrices $\mathbf{P}$ and $\mathbf{Q}$.) We are given

$$\mathbf{P} = \begin{bmatrix} 2 & 1 \\ 1 & 2 \end{bmatrix} \qquad \mathbf{Q} = \begin{bmatrix} 3 & -1 \\ -1 & 3 \end{bmatrix}.$$

We compute

$$\mathbf{P}^{-1} = \frac{1}{3}\begin{bmatrix} 2 & -1 \\ -1 & 2 \end{bmatrix} \qquad \mathbf{P}^{-1}\mathbf{Q} = \begin{bmatrix} \dfrac{7}{3} & \dfrac{-5}{3} \\ \dfrac{-5}{3} & \dfrac{7}{3} \end{bmatrix}.$$

The eigenvalues are computed from $\det(\mathbf{P}^{-1}\mathbf{Q} - \lambda\mathbf{I}) = 0$. They are $\lambda_1 = 4$, $\lambda_2 = 2/3$. Next we compute unnormalized eigenvectors from

$$(\mathbf{P}^{-1}\mathbf{Q} - \lambda\mathbf{I})\mathbf{v}_i' = 0 \qquad i = 1, 2.$$

This yields

$$\mathbf{v}_1' = K_1(1, -1)^T \qquad \mathbf{v}_2' = K_2(1, 1)^T.$$

The eigenvectors in this case are orthogonal. In general they will not be.[†] Finally, we must find the constants K_1 and K_2 such that $\mathbf{v}_i^T\mathbf{P}\mathbf{v}_i = 1$, $i = 1, 2$. Hence

$$\begin{bmatrix} 1 & -1 \end{bmatrix}\begin{bmatrix} 2 & 1 \\ 1 & 2 \end{bmatrix}\begin{bmatrix} 1 \\ -1 \end{bmatrix}K_1^2 = 1$$

[†]The reader should not conclude from this example that the product of two real symmetric matrices is always real symmetric. In general, the product of two r.s. matrices is not r.s. See Problem 5-10(b).

and

$$[1 \quad 1] \begin{bmatrix} 2 & 1 \\ 1 & 2 \end{bmatrix} \begin{bmatrix} 1 \\ 1 \end{bmatrix} K_2^2 = 1.$$

Such constants are found to be $K_1 = 1/\sqrt{2}$, $K_2 = 1/\sqrt{6}$. Thus the matrix $\mathbf{V} = (\mathbf{v}_1, \mathbf{v}_2)$ is given by

$$\mathbf{V} = \begin{bmatrix} \dfrac{1}{\sqrt{2}} & \dfrac{1}{\sqrt{6}} \\ -\dfrac{1}{\sqrt{2}} & \dfrac{1}{\sqrt{6}} \end{bmatrix}.$$

It is easily verified that $\mathbf{V}^T \mathbf{P} \mathbf{V} = \mathbf{I}$ and $\mathbf{V}^T \mathbf{Q} \mathbf{V} = \mathrm{diag}(4, 2/3)$.

Projection

Given two real nonzero vectors $\mathbf{x} = (x_1, \ldots, x_n)^T$ and $\mathbf{a} = (a_1, \ldots, a_n)^T$ we define the projection of $\mathbf{x}$ onto $\mathbf{a}$ as the inner product $\mathbf{x}^T \mathbf{a} = \mathbf{a}^T \mathbf{x}$. The angle θ between $\mathbf{x}$ and $\mathbf{a}$ is implicitly defined by

$$\cos \theta = \frac{\mathbf{a}^T \mathbf{x}}{||\mathbf{x}|| \, ||\mathbf{a}||} = \frac{a_1 x_1 + \ldots + a_n x_n}{\left[\sum_i x_i^2\right]^{1/2} \left[\sum_i a_i^2\right]^{1/2}}.$$

If $||\mathbf{a}|| = 1$, we can think of $\mathbf{a}$ as a *unit direction vector* in which case $\mathbf{x}^T \mathbf{a}$ can be regarded as the projection of $\mathbf{x}$ *along the direction* $\mathbf{a}$ (Figure 5.5-1).

From the Schwarz inequality we have $|\mathbf{a}^T \mathbf{x}| \leq ||\mathbf{a}|| \, ||\mathbf{x}||$ with equality if and only if $\mathbf{a}$ and $\mathbf{x}$ are colinear, that is, $\mathbf{a} = K\mathbf{x}$ (K a constant).

Example 5.5-2

Let $\boldsymbol{\mu}_i = E[\mathbf{X}_i] i = 1, 2$. Along what direction $\mathbf{a} = (a_1, a_2)^T$ is the projection of $\boldsymbol{\mu}_1 - \boldsymbol{\mu}_2$ a maximum?

Solution　To the astute reader, this problem may seem somewhat trivial but we shall proceed as if we didn't know the Schwarz inequality. We wish to maximize $|\mathbf{a}^T (\boldsymbol{\mu}_1 - \boldsymbol{\mu}_2)|$

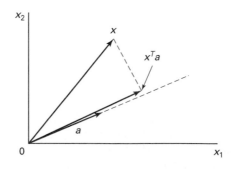

Figure 5.5-1　Projection of x along the direction a ($||\mathbf{a}|| = 1$).

subject to the constraint that $||\mathbf{a}||^2 = a_1^2 + a_2^2 = 1$. By the method of Lagrange multipliers we write

$$q(\mathbf{a}) \triangleq \mathbf{a}^T(\boldsymbol{\mu}_1 - \boldsymbol{\mu}_2) + \lambda(||\mathbf{a}||^2 - 1)$$

and set

$$\frac{\partial q(\mathbf{a})}{\partial a_i} = 0 \qquad i = 1, 2$$

subject to

$$a_1^2 + a_2^2 = 1.$$

The solution is

$$\mathbf{a} = \pm\frac{\boldsymbol{\mu}_1 - \boldsymbol{\mu}_2}{||\boldsymbol{\mu}_1 - \boldsymbol{\mu}_2||}. \tag{5.5-14}$$

Actually, the solution could have been obtained immediately from the Schwarz inequality: $|\mathbf{a}^T(\boldsymbol{\mu}_1 - \boldsymbol{\mu}_2)| \le ||\mathbf{a}|| \, ||\boldsymbol{\mu}_1 - \boldsymbol{\mu}_2||$ with equality if

$$\mathbf{a} = K(\boldsymbol{\mu}_1 - \boldsymbol{\mu}_2) \quad (K \text{ a constant}).$$

Since $||\mathbf{a}|| = 1$, we obtain

$$|K| = ||\boldsymbol{\mu}_1 - \boldsymbol{\mu}_2||^{-1},$$

whence follows Equation 5.5-14.

Maximization of Quadratic Forms

In solving problems involving random vectors we sometimes have to maximize quadratic forms. The following three theorems deal with this problem.

Theorem 5.5-2   Let $\mathbf{M}$ be a r.s. matrix with largest eigenvalue λ_1. Suppose we want to maximize the quadratic form $\mathbf{u}^T\mathbf{M}\mathbf{u}$ on the unit sphere, that is, $||\mathbf{u}|| = 1$. Then the maximum value of $\mathbf{u}^T\mathbf{M}\mathbf{u}$ subject to $||\mathbf{u}|| = 1$ is λ_1 and the maximum occurs when $\mathbf{u} = \boldsymbol{\phi}_1$, that is, the unit eigenvector associated with λ_1. ■

Theorem 5.5-3   If $\mathbf{M}$ is a r.s. matrix with largest eigenvalue λ_1, then

$$\lambda_1 = \max \frac{\mathbf{x}^T\mathbf{M}\mathbf{x}}{||\mathbf{x}||^2} \quad (\mathbf{x} \ne 0) \tag{5.5-15}$$

and the maximum is achieved whenever $\mathbf{x} = K\boldsymbol{\phi}_1$ where $\boldsymbol{\phi}_1$ is the unit eigenvector associated with λ_1 and K is any real constant. ■

Theorem 5.5-4   Let $\mathbf{M}$ be a r.s. matrix with eigenvalues $\lambda_1 \ge \lambda_2 \ge \ldots \ge \lambda_n$. Let $\boldsymbol{\phi}_1, \boldsymbol{\phi}_2, \ldots, \boldsymbol{\phi}_n$ be the eigenvectors associated with $\lambda_1, \lambda_2, \ldots, \lambda_n$, respectively. Suppose we want to maximize the quadratic form $\mathbf{u}^T\mathbf{M}\mathbf{u}$ on the unit sphere, that is, $||\mathbf{u}|| = 1$ subject to the constraints $\mathbf{u}^T\boldsymbol{\phi}_1 = \ldots = \mathbf{u}^T\boldsymbol{\phi}_{i-1} = 0$. Then the maximum value of $\mathbf{u}^T\mathbf{M}\mathbf{u}$ subject to all i constraints is λ_i and the maximum is achieved when $\mathbf{u} = \boldsymbol{\phi}_i$. ■

The proofs of these theorems can be found in books on the subject of linear algebra (for example, see pages 140–145 of the book by Franklin, [5-2]). For the reader's convenience and also because we shall use the result in the example to follow we furnish the proof of Theorem 5.5-2 below.

Proof of Theorem 5.5-2.   The complete set of eigenvectors ϕ_i, $i = 1, \ldots, n$ form a *basis* for n-dimensional Euclidean space E^n. Therefore any vector $\mathbf{u}$ can be represented as

$$\mathbf{u} = c_1\phi_1 + \ldots + c_n\phi_n,$$

where the constants c_i are determined from $c_i = \phi^T\mathbf{u}$. Now $||\mathbf{u}|| = 1$ implies

$$||\mathbf{u}|| = 1 = \left(\sum_{i=1}^{n} c_i\phi_i^T\right)\left(\sum_{j=1}^{n} c_j\phi_j\right)$$

$$= \sum_{i=1}^{n} c_i^2 \qquad \text{since } \phi_i^T\phi_j = \delta_{ij}.^\dagger$$

Furthermore

$$\mathbf{Mu} = \mathbf{M}\left(\sum_{i=1}^{n} c_i\phi_i\right)$$

$$= \sum_{i=1}^{n} c_i\mathbf{M}\phi_i = \sum_{i=1}^{n} c_i\lambda_i\phi_i.$$

Hence

$$\mathbf{u}^T\mathbf{Mu} = \left(\sum_{j=1}^{n} c_j\phi_j^T\right)\left(\sum_{i=1}^{n} c_i\lambda_i\phi_i\right)$$

$$= \sum_{i=1}^{n} \lambda_i c_i^2 \qquad \text{(again since } \phi_j^T\phi_i = \delta_{ij})$$

$$\leq \lambda_1 \sum_{i=1}^{n} c_i^2 = \lambda_1$$

with equality if and only if $c_2 = \ldots = c_n = 0$. Thus

$$\mathbf{u}^T\mathbf{Mu} \leq \lambda_1$$

with equality when $\mathbf{u} = \phi_1$ (in which case $c_1 = 1$ and all other c_i are zero).   ∎

---

†The Kronecker δ_{ij} has value 1 when $i = j$ and value zero otherwise.

The application of much of the material in this section will be illustrated in the following example from pattern recognition.

Example 5.5-3

(Automatic recognition of black-lung disease by computer [5-3].) We return to Example 5.1-3 that dealt with computer recognition of black-lung disease (BLD) from scanning x-ray films of the suspect population. We suppose that the x-rays fall into two categories: those that exhibit the BLD syndrome (say class ω_1) and those free of the BLD syndrome (say class ω_2). When an x-ray is presented to the computer, it extracts an n-component measurement vector $\mathbf{X}$ from which it must decide whether the x-ray should be assigned to class ω_1 or ω_2. We assume that the computer does not have available the conditional pdf's $f_{\mathbf{X}|\omega_i}(\mathbf{x}|\omega_i)$ $i = 1, 2$ but knows the conditional class means and covariances.[†] These are defined as

$$\boldsymbol{\mu}_i \triangleq E[\mathbf{X} \,|\, \text{given that } \mathbf{X} \text{ comes from class } \omega_i]$$

$$\triangleq E[\mathbf{X} \,|\, \omega_i] \qquad i = 1, 2 \tag{5.5-16a}$$

$$\mathbf{K}_i \triangleq E[(\mathbf{X} - \boldsymbol{\mu}_i)(\mathbf{X} - \boldsymbol{\mu}_i)^T \,|\, \text{given that } \mathbf{X} \text{ comes from class } \omega_i]$$

$$\triangleq E[(\mathbf{X} - \boldsymbol{\mu}_i)(\mathbf{X} - \boldsymbol{\mu}_i)^T \,|\, \omega_i] \qquad i = 1, 2. \tag{5.5-16b}$$

The decision whether any given value of $\mathbf{X}$ belongs to ω_1 or ω_2 will be done as follows: From $\mathbf{X}$, a single scalar r.v., Y known as a *feature* is extracted by projecting $\mathbf{X}$ along a direction $\mathbf{a}$, that is,

$$Y = \mathbf{a}^T \mathbf{X} \qquad ||\mathbf{a}|| = 1. \tag{5.5-17}$$

Next we assume that from *training data* we have available two sets of features: $\mathcal{R}_1 \triangleq \{y_1^{(1)}, \dots, y_n^{(1)}, \text{ all from } \omega_1\}$ and $\mathcal{R}_2 \triangleq \{y_1^{(2)}, \dots, y_n^{(2)}, \text{ all from } \omega_2\}$.[‡] To be specific, let the feature Y assume that value y. Our problem is then to assign to y a class membership, that is, should y be assigned to ω_1 or ω_2? We shall do this using the *N-nearest-neighbor rule* [5-4, pp. 103–105]. This is realized as follows: The N nearest neighbors of the unassigned y are found from the training sets. Being from the training sets, these nearest neighbors have a known class membership. If the majority of the N nearest neighbors belong to ω_1, then y gets assigned to ω_1; otherwise y gets assigned to ω_2. For this rule to work well, the data in $\mathcal{R}_1$ and $\mathcal{R}_2$ should cluster in different regions of the real line. To be specific, refer to Figure 5.5-2. There, a particular realization y of the r.v. Y is shown by the solid dot. We take $N = 5$. Since four of the nearest neighbors belong to ω_1 and only one belongs to ω_2, y gets assigned to ω_1. With high probability, the ω_1 classification for y is correct, since the *projected training data nicely separate.*

[†]If the conditional pdf's were known, then Bayes' decision theory could be used to decide on class membership. Bayesian theory, which can be regarded as optimum in a rather broad sense for problems of this type, is discussed in [5-4].

[‡]The y's in $\mathcal{R}_1$ and $\mathcal{R}_2$ represent particular outcomes of the r.v. Y. In practice the raw training data $\{\mathbf{X}_i^{(j)}, i = 1, \dots, n; \ j = 1, 2\}$ from which the $\{Y_i^{(j)}, i, \dots, n; \ j = 1, 2\}$ would be computed would be obtained by having expert radiologists carefully examine x-rays from *known* sufferers of BLD and *known* healthy humans. This training data would then be made available to the computer, in effect converting the computer to an "expert" system.

- Denotes y value of unknown sample
- × Denotes y value from $\mathscr{R}_1$
- ○ Denotes y value from $\mathscr{R}_2$

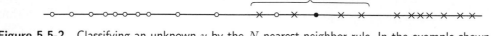

Five nearest
neighbors of
unknown y

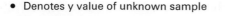

Figure 5.5-2   Classifying an unknown y by the N-nearest-neighbor rule. In the example shown, $N = 5$. Four out of five nearest neighbors belong to ω_1, hence y gets assigned to ω_1.

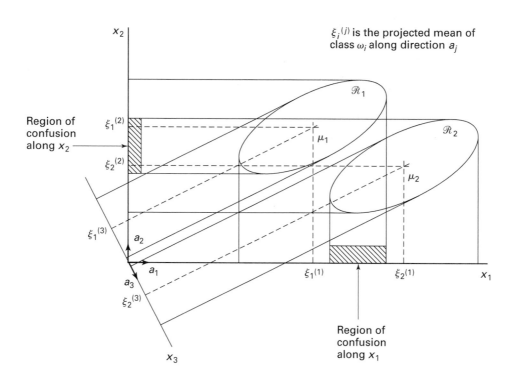

Figure 5.5-3   Projecting the data along $\mathbf{a}_1$ produces a region of confusion because the projected variances along $\mathbf{a}_1$ are too large. Projecting the data along $\mathbf{a}_2$ results in a large region of confusion because the projected difference in means $|\xi_1^{(2)} - \xi_1^{(2)}|$ is too small. Along $\mathbf{a}_3$, however, the projected difference in means is large in relation to the projected variances. Hence there is no region of confusion and $\mathbf{a}_3$ is the preferred direction.

Assuming that there exists at least one direction $\mathbf{a}$ which will cluster $\mathscr{R}_1$ and $\mathscr{R}_2$ in different regions, how do we find $\mathbf{a}$? To answer this question refer to Figure 5.5-3. There we see that $\mathbf{a}_1$ is a poor direction for projecting $\mathbf{X}$ because although the projected difference in means is relatively large, the projected variances are also large and there is a significant region of confusion (the crosshatched region). Thus $\mathbf{a}_1$ is not a good direction for projection of the data. Along $\mathbf{a}_2$, the projected variances are less but so is the difference in projected means; once again there is manifest a region of confusion. However, along $\mathbf{a}_3$ the projected difference in means is large in relation to the projected variances; the region of confusion is small or nonexistent. Thus $\mathbf{a}_3$ is a good projection direction. If we call ξ_i and σ_i^2 for $i = 1, 2$ the *projected* means and variances along an arbitrary direction $\mathbf{a}$, respectively, then an excellent indicator of how good is any particular projection direction $\mathbf{a}$ is the size of the criterion function $J(\mathbf{a})$ defined by

$$J(\mathbf{a}) \triangleq \frac{(\xi_1 - \xi_2)^2}{\sigma_1^2 + \sigma_2^2}. \tag{5.5-18}$$

If for a particular $\mathbf{a}$, $J(\mathbf{a})$ is large, the region of confusion will be small; if $J(\mathbf{a})$ is small, the converse is true. Thus having identified a measure of cluster separation, how do we now proceed to find the value of $\mathbf{a}$ that maximizes $J(\mathbf{a})$?

Solution   We are looking to maximize $J(\mathbf{a})$ in Equation 5.5-18 by a suitable choice of $\mathbf{a}$. Note that under the transformation

$$Y = \mathbf{a}^T \mathbf{X} \tag{5.5-19}$$

we obtain

$$\xi_i = E[Y \,|\, \omega_i]$$
$$= \mathbf{a}^T \boldsymbol{\mu}_i \quad i = 1, 2 \tag{5.5-20}$$

and

$$\sigma_i^2 = E[(\mathbf{a}^T (\mathbf{X} - \boldsymbol{\mu}_i))^2 \,|\, \omega_i]$$
$$= \mathbf{a}^T \mathbf{K}_i \mathbf{a} \quad i = 1, 2. \tag{5.5-21}$$

Thus

$$J(\mathbf{a}) = \frac{(\mathbf{a}^T (\boldsymbol{\mu}_1 - \boldsymbol{\mu}_2))^2}{\mathbf{a}^T (\mathbf{K}_1 + \mathbf{K}_2) \mathbf{a}} \tag{5.5-22}$$

$$= \frac{\mathbf{a}^T \mathbf{Q} \mathbf{a}}{\mathbf{a}^T \mathbf{P} \mathbf{a}}, \tag{5.5-23}$$

where

$$\mathbf{Q} \triangleq (\boldsymbol{\mu}_1 - \boldsymbol{\mu}_2)(\boldsymbol{\mu}_1 - \boldsymbol{\mu}_2)^T$$
$$\mathbf{P} \triangleq \mathbf{K}_1 + \mathbf{K}_2 \quad \text{(assumed full rank).} \tag{5.5-24}$$

Now we simultaneously diagonalize $\mathbf{Q}$ and $\mathbf{P}$ with the transformation

$$\mathbf{a} = \mathbf{V} \mathbf{b}, \tag{5.5-25}$$

where the columns $\mathbf{v}_i$ for $i = 1, \ldots, n$ of $\mathbf{V}$ are the eigenvectors of the equations (the λ_i are the eigenvalues)

$$\mathbf{P}^{-1} \mathbf{Q} \mathbf{v}_i = \lambda_i \mathbf{v}_i$$

and satisfy the generalized orthogonality $\mathbf{v}_i^T \mathbf{P} \mathbf{v}_i = \mathbf{I}$. Use of Equation 5.5-25 in Equation 5.5-23 yields

$$
J(\mathbf{a}) = \frac{\mathbf{b}^T \mathbf{V}^T \mathbf{Q} \mathbf{V} \mathbf{b}}{\mathbf{b}^T \mathbf{V}^T \mathbf{P} \mathbf{V} \mathbf{b}}
$$

$$
= \frac{\mathbf{b}^T \mathbf{\Lambda} \mathbf{b}}{||\mathbf{b}||^2}, \tag{5.5-26}
$$

where $\mathbf{V}^T \mathbf{P} \mathbf{V} = \mathbf{I}$ and $\mathbf{V}^T \mathbf{Q} \mathbf{V} = \mathbf{\Lambda} = \mathrm{diag}(\lambda_1, \dots, \lambda_n)$. Next we use Theorem 5.5-3 to find the maximum of Equation 5.5-26. The maximum is λ_1, and it occurs when $\mathbf{b} = \boldsymbol{\phi}_1$, that is, the eigenvector associated with λ_1. This eigenvector is $\mathbf{f}_1 = (1, 0, 0, \dots, 0)^T$ since $\mathbf{\Lambda}$ is diagonal. Also from $\mathbf{a} = \mathbf{V} \mathbf{b}$ we find that

$$
\mathbf{a} = \mathbf{V} \boldsymbol{\phi}_1 = \mathbf{v}_1.
$$

Hence $\mathbf{a}$ satisfies

$$
\mathbf{P}^{-1} \mathbf{Q} \mathbf{a} = \lambda_1 \mathbf{a}
$$

or

$$
\mathbf{P}^{-1}(\boldsymbol{\mu}_1 - \boldsymbol{\mu}_2)(\boldsymbol{\mu}_1 - \boldsymbol{\mu}_2)^T \mathbf{a} = \lambda_1 \mathbf{a}.
$$

Since $(\boldsymbol{\mu}_1 - \boldsymbol{\mu}_2)^T \mathbf{a}$ is a scalar, say k, we obtain

$$
\mathbf{a} = \frac{k}{\lambda_1} \mathbf{P}^{-1}(\boldsymbol{\mu}_1 - \boldsymbol{\mu}_2). \tag{5.5-27}
$$

Since we are interested only in direction, we can ignore the generalized orthogonalization implied by Equation 5.5-1, that is, that $\mathbf{a}^T \mathbf{P} \mathbf{a} = 1$. Instead we normalize $\mathbf{a}$ to unity by choosing k so that $||\mathbf{a}|| = 1$. It is easy to establish that this requires

$$
k = \lambda_1 ||\mathbf{P}^{-1}(\boldsymbol{\mu}_2 - \boldsymbol{\mu}_1)||^{-1}.
$$

Equation 5.5-27 is an important result in the field of applied probability called statistical pattern recognition. The vector $\mathbf{a}$ is called the Fisher[†] linear discriminant.

Having found the best direction for separating the two classes, we generate the scalar r.v. from

$$
Y = \mathbf{a}^T \mathbf{X}.
$$

When the dimension of $\mathbf{X}$ is large and the $a_i X_i$, $i = 1, \dots, n$ satisfy the Lindeberg conditions, then from the *Central Limit Theorem* (Section (4.7), Y can be taken to be a normal r.v. The assumption that Y is normal—often made even when the above conditions are not quite satisfied[‡]—enables an optimum partitioning of the real line into regions Q_1^* and Q_2^*

[†]Ronald Aylmer Fisher (1890–1962), British mathematician, who along with Richard von Mises and Andrei Kolmogorov developed modern probability theory.

[‡]Recall that the CLT says that Y tends to the Normal distribution.

such that (1) the probability of a correct decision is maximized and/or (2) the cost of an incorrect decision is minimized. The optimum partitioning of Gaussian classes is a well-studied problem. See, for example, [5-4].

You will have observed that Example 5.5-3 made use of many of the results of this chapter, for example, covariance matrices, their simultaneous diagonalization, maximization of quadratic forms, and others.

5.6 THE MULTIDIMENSIONAL GAUSSIAN (NORMAL) LAW

The general n-dimensional Gaussian law has a rather forbidding mathematical appearance upon first acquaintance but is, fortunately, rather easily extended from the one-dimensional Gaussian pdf. We already know that if X is a (scalar) Gaussian r.v. with mean μ and variance σ^2, its pdf is

$$f_X(x) = \frac{1}{\sqrt{2\pi}\sigma} \exp\left(-\frac{1}{2}\left(\frac{x-\mu}{\sigma}\right)^2\right).$$

Now consider the random vector $\mathbf{X} = (X_1, \ldots, X_n)^T$ with *independent* components X_i, $i = 1, \ldots, n$. Then the pdf of $\mathbf{X}$ is the product of the individual pdf's of $X_1, \ldots, X_n$, that is,

$$f_\mathbf{X}(x_1, \ldots, x_n) = \prod_{i=1}^{n} f_{X_i}(x_i)$$

$$= \frac{1}{(2\pi)^{n/2}\sigma_1 \ldots \sigma_n} \exp\left[-\frac{1}{2}\sum_{i=1}^{n}\left(\frac{x_i - \mu_i}{\sigma_i}\right)^2\right], \qquad (5.6\text{-}1)$$

where μ_i, σ_i^2 are the mean and variance, respectively, of $X_i, i = 1, \ldots, n$. Equation 5.6-1 can be written compactly as

$$f_\mathbf{X}(\mathbf{x}) = \frac{1}{(2\pi)^{n/2}[\det(\mathbf{K})]^{1/2}} \exp[-\tfrac{1}{2}(\mathbf{x} - \boldsymbol{\mu})^T \mathbf{K}^{-1}(\mathbf{x} - \boldsymbol{\mu})], \qquad (5.6\text{-}2)$$

where

$$\mathbf{K} \triangleq \begin{bmatrix} \sigma_1^2 & & 0 \\ & \ddots & \\ 0 & & \sigma_n^2 \end{bmatrix}, \qquad (5.6\text{-}3)$$

$\boldsymbol{\mu} = (\mu_1, \ldots, \mu_n)^T$, and $\det(\mathbf{K}) = \prod_{i=1}^{n} \sigma_i^2$. Note that $\mathbf{K}^{-1}$ is merely

$$\mathbf{K}^{-1} = \begin{bmatrix} \sigma_1^{-2} & & 0 \\ & \ddots & \\ 0 & & \sigma_n^{-2} \end{bmatrix}.$$

The fact that Equations 5.6-1 and 5.6-2 are identical is left as an exercise to the reader. Note that because the X_i, $i = 1, \ldots, n$ are independent, the covariance matrix $\mathbf{K}$ is diagonal, since

$$E[(X_i - \mu_i)^2] \stackrel{\Delta}{=} \sigma_i^2 \quad i = 1, \ldots, n. \tag{5.6-4}$$

$$E[(X_i - \mu_i)(X_j - \mu_j)] = 0 \quad i \neq j. \tag{5.6-5}$$

Next we ask, what happens if $\mathbf{K}$ is a positive definite covariance matrix that is not necessarily diagonal? Does Equation 5.6-2 with *arbitrary* p.d. covariance $\mathbf{K}$ still obey the requirements of a pdf? If it does, we shall call $\mathbf{X}$ a *Normal* random vector and $f_{\mathbf{X}}(\mathbf{x})$ the *multidimensional Normal* pdf. To show that $f_{\mathbf{X}}(\mathbf{x})$ is indeed a pdf we must show that

$$f_{\mathbf{X}}(\mathbf{x}) \geq 0 \tag{5.6-6a}$$

and

$$\int_{-\infty}^{\infty} f_{\mathbf{X}}(\mathbf{x})d\mathbf{x} = 1. \tag{5.6-6b}$$

(We use the vector notation $d\mathbf{x} \stackrel{\Delta}{=} dx_1\, dx_2 \ldots dx_n$ for a volume element.) We assume as always that $\mathbf{X}$ is real, that is, $X_1, \ldots, X_n$ are real r.v.'s. To show that Equation 5.6-2 with arbitrary p.d. covariance matrix $\mathbf{K}$ satisfies Equation 5.6-6a is simple and left as an exercise; to prove Equation 5.6-6b is more difficult, and follows here.

Proof of Equation 5.6-6b when $f_{\mathbf{X}}(\mathbf{x})$ is as in Equation 5.6-2 and K is an arbitrary p.d. covariance matrix. We note that with $\mathbf{z} \stackrel{\Delta}{=} \mathbf{x} - \boldsymbol{\mu}$, Equation 5.6-2 can be written as

$$f_{\mathbf{X}}(\mathbf{x}) \stackrel{\Delta}{=} \frac{1}{(2\pi)^{n/2}[\det(\mathbf{K})]^{1/2}} \phi(\mathbf{z}),$$

where

$$\phi(\mathbf{z}) \stackrel{\Delta}{=} \exp(-\tfrac{1}{2}\mathbf{z}^T\mathbf{K}^{-1}\mathbf{z}). \tag{5.6-7a}$$

With

$$\alpha \stackrel{\Delta}{=} \int_{-\infty}^{\infty} \phi(\mathbf{z})d\mathbf{z} \tag{5.6-7b}$$

we see that

$$\int_{-\infty}^{\infty} f_{\mathbf{X}}(\mathbf{x})d\mathbf{x} = \frac{\alpha}{(2\pi)^{n/2}[\det(\mathbf{K})]^{1/2}},$$

Hence we need only evaluate α to prove (or disprove) Equation 5.6-6b.

From Equations 5.5-12 and 5.5-13b we know that there exists an $n \times n$ matrix $\mathbf{C}$ such that $\mathbf{K} = \mathbf{C}\mathbf{C}^T$ and $\mathbf{C}^T\mathbf{K}^{-1}\mathbf{C} = \mathbf{I}$ (the identity matrix). Now consider the linear transformation

$$\mathbf{z} = \mathbf{C}\mathbf{y} \tag{5.6-8}$$

for use in Equations 5.6-7. To understand the effect of this transformation we note first that

$$\mathbf{z}^T\mathbf{K}^{-1}\mathbf{z} = \mathbf{y}^T\mathbf{C}^T\mathbf{K}^{-1}\mathbf{C}\mathbf{y} = ||\mathbf{y}||^2 = \sum_{i=1}^{n} y_i^2$$

so that $\phi(\mathbf{z})$ is given by

$$\phi(\mathbf{z}) = \prod_{i=1}^{n} \exp[-\tfrac{1}{2}y_i^2].$$

Next we use a result from advanced calculus (see Kenneth Miller, [5-5, p. 16]) that for a linear transformation such as in Equation 5.6-8 volume elements are related as

$$d\mathbf{z} = |\det(\mathbf{C})|d\mathbf{y},$$

where $d\mathbf{z} \overset{\Delta}{=} dz_1 \ldots dz_n$ and $d\mathbf{y} = dy_1 \ldots dy_n$. Hence Equation 5.6-7b is transformed to

$$\alpha = \int_{-\infty}^{\infty} \exp\left(-\frac{1}{2}\sum_{i=1}^{n} y_i^2\right) dy_1 \ldots dy_n |\det(\mathbf{C})|$$

$$= \left[\int_{-\infty}^{\infty} e^{-y^2/2}\, dy\right]^n |\det(\mathbf{C})|$$

$$= [2\pi]^{n/2}|\det(\mathbf{C})|.$$

But since $\mathbf{K} = \mathbf{C}\mathbf{C}^T$, $\det(\mathbf{K}) = \det(\mathbf{C})\det(\mathbf{C}^T) = [\det(\mathbf{C})]^2$ or

$$|\det(\mathbf{C})| = |\det(\mathbf{K})|^{1/2} = (\det(\mathbf{K}))^{1/2}.$$

Hence

$$\alpha = (2\pi)^{n/2}[\det(\mathbf{K})]^{1/2}$$

and

$$\frac{\alpha}{[2\pi]^{n/2}[\det(\mathbf{K})]^{1/2}} = 1,$$

which proves Equation 5.6-6b. ■

Having established that

$$f_{\mathbf{X}}(\mathbf{x}) = \frac{1}{(2\pi)^{n/2}[\det(\mathbf{K})]^{1/2}} \exp(-\tfrac{1}{2}(\mathbf{x} - \boldsymbol{\mu})^T\mathbf{K}^{-1}(\mathbf{x} - \boldsymbol{\mu})) \qquad (5.6\text{-}9)$$

indeed satisfies the requirements of a pdf and is a generalization of the univariate Normal pdf, we now ask what is the pdf of the random vector $\mathbf{Y}$ given by

$$\mathbf{Y} \overset{\Delta}{=} \mathbf{A}\mathbf{X}, \qquad (5.6\text{-}10)$$

where $\mathbf{A}$ is a nonsingular $n \times n$ transformation. The answer is furnished by the following theorem.

Theorem 5.6-1   Let $\mathbf{X}$ be an n-dimensional Normal random vector with positive definite covariance matrix $\mathbf{K}$ and mean vector $\boldsymbol{\mu}$. Let $\mathbf{A}$ be a nonsingular linear transformation in n dimensions. Then $\mathbf{Y} \triangleq \mathbf{AX}$ is an n-dimensional Normal random vector with covariance matrix $\mathbf{Q} \triangleq \mathbf{AKA}^T$ and mean vector $\beta \triangleq \mathbf{A}\boldsymbol{\mu}$.   ■

Proof   We use Equation 5.2-11, that is,

$$f_{\mathbf{Y}}(\mathbf{y}) = \sum_{i=1}^{r} \frac{f_{\mathbf{X}}(\mathbf{x}_i)}{|J_i|}, \tag{5.6-11}$$

where $\mathbf{Y}$ is some function of $\mathbf{X}$, that is, $\mathbf{Y} = \mathbf{g}(\mathbf{X}) \triangleq (g_1(\mathbf{X}), \dots, g_n(\mathbf{X}))^T$, the $\mathbf{x}_i$, $i = 1, \dots, r$ are the roots of the equation $\mathbf{g}(\mathbf{x}_i) - \mathbf{y} = \mathbf{0}$ and J_i is the Jacobian evaluated at the ith root, that is,

$$J_i = \det\left(\frac{\partial \mathbf{g}}{\partial \mathbf{x}}\right)\bigg|_{\mathbf{x}=\mathbf{x}_i} = \begin{vmatrix} \dfrac{\partial g_1}{\partial x_1} & \cdots & \dfrac{\partial g_1}{\partial x_n} \\ \vdots & & \vdots \\ \dfrac{\partial g_n}{\partial x_1} & \cdots & \dfrac{\partial g_n}{\partial x_n} \end{vmatrix}_{\mathbf{x}=\mathbf{x}_i}. \tag{5.6-12}$$

Since we are dealing with a nonsingular *linear* transformation, the only solution to

$$\mathbf{Ax} - \mathbf{y} = \mathbf{0} \quad \text{is} \quad \mathbf{x} = \mathbf{A}^{-1}\mathbf{y}. \tag{5.6-13}$$

Also

$$J_i = \det\left(\frac{\partial (\mathbf{Ax})}{\partial \mathbf{x}}\right) = \det(\mathbf{A}). \tag{5.6-14}$$

Hence

$$f_{\mathbf{Y}}(\mathbf{y}) = \frac{1}{(2\pi)^{n/2}[\det(\mathbf{K})]^{1/2}|\det(\mathbf{A})|} \exp(-\tfrac{1}{2}(\mathbf{A}^{-1}\mathbf{y} - \boldsymbol{\mu})^T \mathbf{K}^{-1}(\mathbf{A}^{-1}\mathbf{y} - \boldsymbol{\mu})). \tag{5.6-15}$$

Can this formidable expression be put in the form of Equation 5.6-9? First we note that

$$[\det(\mathbf{K})]^{1/2}|\det(\mathbf{A})| = [\det(\mathbf{AKA}^T)]^{1/2} \triangleq [\det(\mathbf{Q})]^{1/2}. \tag{5.6-16}$$

Next we note that

$$(\mathbf{A}^{-1}\mathbf{y} - \boldsymbol{\mu})^T \mathbf{K}^{-1}(\mathbf{A}^{-1}\mathbf{y} - \boldsymbol{\mu}) = (\mathbf{y} - \mathbf{A}\boldsymbol{\mu})^T(\mathbf{AKA}^T)^{-1}(\mathbf{y} - \mathbf{A}\boldsymbol{\mu}). \tag{5.6-17}$$

(We leave the algebraic manipulations as an exercise.) But $\mathbf{A}\boldsymbol{\mu} \triangleq \beta = E[\mathbf{Y}]$ and $\mathbf{AKA}^T = E[(\mathbf{Y} - \beta)(\mathbf{Y} - \beta)^T] = \mathbf{Q}$. Hence Equation 5.6-15 can be rewritten as

$$f_{\mathbf{Y}}(\mathbf{y}) = \frac{1}{(2\pi)^{n/2}[\det(\mathbf{Q})]^{1/2}} \exp[-\tfrac{1}{2}(\mathbf{y} - \beta)^T \mathbf{Q}^{-1}(\mathbf{y} - \beta)]. \quad ■ \tag{5.6-18}$$

The next question that arises quite naturally as an extension of the previous result is: Does $\mathbf{Y}$ remain a Normal random vector under more general (nontrivial) linear transformation? The answer is given by the following theorem, which is a generalization of Theorem 5.6-1.

Theorem 5.6-2   Let $\mathbf{X}$ be an n-dimensional Normal random vector with positive definite covariance matrix $\mathbf{K}$ and mean vector $\boldsymbol{\mu}$. Let $\mathbf{A}_{mn}$ be an $m \times n$ matrix of rank m. Then the random vector generated by

$$\mathbf{Y} = \mathbf{A}_{mn}\mathbf{X}$$

has an m-dimensional Normal pdf with p.d. covariance matrix $\mathbf{Q}$ and mean vector $\boldsymbol{\beta}$ given respectively by

$$\mathbf{Q} \triangleq \mathbf{A}_{mn}\mathbf{K}\mathbf{A}_{mn}^T \tag{5.6-19}$$

and

$$\boldsymbol{\beta} = \mathbf{A}_{mn}\boldsymbol{\mu}. \quad \blacksquare \tag{5.6-20}$$

The proof of this theorem is quite similar to the proof of Theorem 5.6-1; it is given by Miller in Reference [5-6, p. 22].

Some examples involving transformations of Normal random variables are given below.

Example 5.6-1 ──────────────────────────────────────

A zero-mean Normal random vector $\mathbf{X} = (X_1, X_2)^T$ has covariance matrix $\mathbf{K}$ given by

$$\mathbf{K} = \begin{bmatrix} 3 & -1 \\ -1 & 3 \end{bmatrix}.$$

Find a transformation $\mathbf{Y} = \mathbf{D}\mathbf{X}$ such that $\mathbf{Y} = (Y_1, Y_2)^T$ is a Normal random vector with uncorrelated (and therefore independent) components of unity variance.

Solution   Write

$$E[\mathbf{Y}\mathbf{Y}^T] = E[\mathbf{D}\mathbf{X}\mathbf{X}^T\mathbf{D}^T] = \mathbf{D}\mathbf{K}\mathbf{D}^T = \mathbf{I}.$$

The last equality on the right follows from the requirement that the covariance of $\mathbf{Y}$, $\mathbf{K_Y}$, satisfies

$$\mathbf{K_Y} = \begin{bmatrix} 1 & 0 \\ 0 & 1 \end{bmatrix} \triangleq \mathbf{I}.$$

From Equation 5.5-13a, the matrix $\mathbf{D}$ must be $\mathbf{D} = \mathbf{C}^{-1} = \mathbf{Z}\mathbf{U}^T$, where $\mathbf{Z}$ is the normalizing matrix

$$\mathbf{Z} = \begin{bmatrix} \lambda_1^{-1/2} & 0 \\ 0 & \lambda_2^{-1/2} \end{bmatrix} \quad (\lambda_i, i = 1, 2 \text{ are eigenvalues of } \mathbf{K})$$

and $\mathbf{U}$ is the matrix whose columns are the unit eigenvectors of $\mathbf{K}$ (recall $\mathbf{U}^{-1} = \mathbf{U}^T$). From $\det(\mathbf{K} - \lambda\mathbf{I}) = 0$ we find $\lambda_1 = 4, \lambda_2 = 2$. Hence

$$\boldsymbol{\Lambda} = \begin{bmatrix} 4 & 0 \\ 0 & 2 \end{bmatrix} \qquad \mathbf{Z} = \boldsymbol{\Lambda}^{-1/2} = \begin{bmatrix} \dfrac{1}{2} & 0 \\ 0 & \dfrac{1}{\sqrt{2}} \end{bmatrix}.$$

Next from

$$(\mathbf{K} - \lambda_1\mathbf{I})\boldsymbol{\phi}_1 = 0 \quad \text{with} \quad \|\boldsymbol{\phi}_1\| = 1$$

and

$$(\mathbf{K} - \lambda_2\mathbf{I})\boldsymbol{\phi}_2 = 0 \quad \text{with} \quad ||\boldsymbol{\phi}_2|| = 1$$

we find $\boldsymbol{\phi}_1 = (1/\sqrt{2}, -1\sqrt{2})^T$, $\boldsymbol{\phi}_2 = (1/\sqrt{2}, 1\sqrt{2})^T$. Thus

$$\mathbf{U} = (\boldsymbol{\phi}_1, \boldsymbol{\phi}_2) = \frac{1}{\sqrt{2}}\begin{bmatrix} 1 & 1 \\ -1 & 1 \end{bmatrix}$$

and

$$\mathbf{D} = \mathbf{Z}\mathbf{U}^T = \frac{1}{\sqrt{2}}\begin{bmatrix} \dfrac{1}{2} & -\dfrac{1}{2} \\ \dfrac{1}{\sqrt{2}} & \dfrac{1}{\sqrt{2}} \end{bmatrix}.$$

As a check to see if $\mathbf{D}\mathbf{K}\mathbf{D}^T$ is indeed an identity covariance matrix we compute

$$\frac{1}{2}\begin{bmatrix} \dfrac{1}{2} & -\dfrac{1}{2} \\ \dfrac{1}{\sqrt{2}} & \dfrac{1}{\sqrt{2}} \end{bmatrix}\begin{bmatrix} 3 & -1 \\ -1 & 3 \end{bmatrix}\begin{bmatrix} \dfrac{1}{2} & \dfrac{1}{\sqrt{2}} \\ -\dfrac{1}{2} & \dfrac{1}{\sqrt{2}} \end{bmatrix} = \begin{bmatrix} 1 & 0 \\ 0 & 1 \end{bmatrix}.$$

In many situations we might want to generate correlated samples of a random vector $\mathbf{X}$ whose covariance matrix $\mathbf{K}$ is not diagonal. From Example 5.6-1 we see that the transformation

$$\mathbf{X} = \mathbf{D}^{-1}\mathbf{Y}, \tag{5.6-21}$$

where $\mathbf{D} = \mathbf{Z}\mathbf{U}^T$ produces a Normal random vector whose covariance is $\mathbf{K}$. Thus one way of obtaining correlated from uncorrelated samples is to use the transformation given in Equation 5.6-21 on the computer-generated samples. This procedure is the reverse of what we did in Example 5.6-1.

Example 5.6-2 _____
Two jointly Normal r.v.'s X_1 and X_2 have joint pdf given by

$$f_{X_1X_2}(x_1, x_2) = \frac{1}{2\pi\sigma^2\sqrt{1-\rho^2}}\exp\left(\frac{-1}{2\sigma^2(1-\rho^2)}(x_1^2 - 2\rho x_1 x_2 + x_2^2)\right).$$

Let the correlation coefficient ρ be -0.5. From X_1, X_2 find two jointly Normal r.v.'s Y_1 and Y_2 such that Y_1 and Y_2 are independent. Avoid the trivial case of $Y_1 = Y_2 = 0$.

Solution   Define $\mathbf{x} \triangleq (x_1, x_2)^T$ and $\mathbf{y} = (y_1, y_2)^T$. Then with $\rho = -0.5$, the quadratic in the exponent can be written as

$$x_1^2 + x_1 x_2 + x_2^2 = \mathbf{x}^T\begin{bmatrix} a & b \\ c & d \end{bmatrix}\mathbf{x} = ax_1^2 + (b+c)x_1 x_2 + dx_2^2,$$

where the a, b, c, d are to be determined. We immediately find that $a = d = 1$ and—because of the real symmetric requirement—we find $b = c = 0.5$. We can rewrite $f_{X_1 X_2}(x_1, x_2)$ in standard form as

$$f_{X_1 X_2}(x_1, x_2) = \frac{1}{2\pi[\det(\mathbf{K})]^{1/2}} \exp\left(-\frac{1}{2}(\mathbf{x}^T \mathbf{K}^{-1} \mathbf{x})\right)$$

whence

$$\mathbf{K}^{-1} = \frac{1}{\sigma^2(1-\rho^2)} \begin{bmatrix} a & b \\ c & d \end{bmatrix} = \frac{4}{3\sigma^2} \begin{bmatrix} 1 & 0.5 \\ 0.5 & 1 \end{bmatrix}.$$

Our task is now to find a transformation that diagonalizes $\mathbf{K}^{-1}$. This will enable the joint pdf of Y_1 to Y_2 to be factored, thereby establishing that Y_1 and Y_2 are independent.

The factor $4/3\sigma^2$ affects the eigenvalues of $\mathbf{K}^{-1}$ but not the eigenvectors. To compute a set of orthonormal eigenvectors of $\mathbf{K}^{-1}$ we need only consider $\tilde{\mathbf{K}}^{-1}$ given by

$$\tilde{\mathbf{K}}^{-1} \triangleq \begin{bmatrix} 1 & 0.5 \\ 0.5 & 1 \end{bmatrix}$$

for which we obtain $\tilde{\lambda}_1 = 3/2$ $\tilde{\lambda}_2 = 1/2$. The corresponding unit eigenvectors are $\boldsymbol{\phi}_1 = 1/\sqrt{2}(1, 1)^T$ and $\boldsymbol{\phi}_2 = 1/\sqrt{2}(1, -1)^T$. Thus with

$$\tilde{\mathbf{U}} \triangleq \begin{bmatrix} 1 & 1 \\ 1 & -1 \end{bmatrix}$$

(normalization by $1/\sqrt{2}$ is not needed to obtain a diagonal covariance matrix so we dispense with these factors) we find that

$$\tilde{\mathbf{U}}^T \tilde{\mathbf{K}}^{-1} \tilde{\mathbf{U}} = \operatorname{diag}(1, 3).$$

Hence a transformation that will work is[†]

$$\mathbf{Y} = \tilde{\mathbf{U}}^T \mathbf{X},$$

that is,

$$Y_1 = X_1 + X_2$$
$$Y_2 = X_1 - X_2.$$

To find $f_{Y_1 Y_2}(y_1, y_2)$ we use Equation 3.4-12 of Chapter 3:

$$f_{Y_1 Y_2}(y_1, y_2) = \sum_{i=1}^{n} f_{X_1 X_2}(\mathbf{x}_i)/|J_i|,$$

[†]There is no requirement to whiten the covariance matrix as in Example 5.6-1. Also, diagonalizing $\mathbf{K}^{-1}$ is equivalent to diagonalizing $\mathbf{K}$.

where the $\mathbf{x}_i \triangleq (x_1^{(i)}, x_2^{(i)})^T$, $i = 1, \ldots, n$ are the n solutions to $\mathbf{y} - \tilde{\mathbf{U}}^T\mathbf{x} = 0$ and J_i is the Jacobian. The only solution, that is $n = 1$, to $\mathbf{y} - \tilde{\mathbf{U}}^T\mathbf{x} = 0$ is

$$x_1 = \frac{y_1 + y_2}{2}$$

$$x_2 = \frac{y_1 - y_2}{2}$$

and, dispensing with subscripts there being only one root,

$$J = \det\left(\frac{\partial \mathbf{g}}{\partial \mathbf{x}}\right) = \det\begin{bmatrix} 1 & 1 \\ 1 & -1 \end{bmatrix} = -2.$$

Hence

$$f_{Y_1 Y_2}(y_1, y_2) = \frac{1}{2} f_{X_1 X_2}\left(\frac{y_1 + y_2}{2}, \frac{y_1 - y_2}{2}\right)$$

$$= \frac{1}{\sqrt{2\pi\sigma^2}} \exp\left[-\frac{y_1^2}{2\sigma^2}\right] \cdot \frac{1}{\sqrt{2\pi\sigma'^2}} \exp\left[-\frac{y_2^2}{2\sigma'^2}\right],$$

where $\sigma' \triangleq \sqrt{3}\sigma$.

Examples 5.6-1 and 5.6-2 are special cases of the following theorem:

Theorem 5.6-3   Let X be a Normal, zero-mean (for convenience) random vector with positive definite covariance matrix $\mathbf{K}$. Then there exists a nonsingular $n \times n$ matrix $\mathbf{C}$ such that under the transformation

$$\mathbf{Y} = \mathbf{C}^{-1}\mathbf{X},$$

the components $Y_1, \ldots, Y_n$ of $\mathbf{Y}$ are independent.

Proof   Use the $\mathbf{C}$ from Equation 5.5-12 with $\mathbf{K}$ replacing $\mathbf{P}$, that is,

$$\mathbf{K} = \mathbf{C}\mathbf{C}^T. \quad \blacksquare$$

Example 5.6-3 _____

(The generalized Rayleigh law.) Let $\mathbf{X} = (X_1, X_2, X_3)^T$ be a Normal random vector with covariance matrix

$$\mathbf{K} = \sigma^2\mathbf{I}.$$

Compute the pdf of $R_3 \triangleq ||\mathbf{X}|| = \sqrt{X_1^2 + X_2^2 + X_3^2}$.

Solution   The probability of the event $\{R_3 \leq r\}$ is the probability distribution function $F_{R_3}(r)$ of R_3. Thus

$$F_{R_3}(r) = \frac{1}{|(2\pi)^{3/2}[\sigma^2]^{3/2}|} \iiint_{\mathscr{I}} \exp\left[-\frac{1}{2\sigma^2}(x_1^2 + x_2^2 + x_3^2)\right] dx_1\, dx_2\, dx_3$$

where $\mathscr{J} = \{(x_1, x_2, x_3): \sqrt{x_1^2 + x_2^2 + x_3^2} \leq r\}$. Now let

$$x_1 \triangleq \xi \cos \phi$$

$$x_2 \triangleq \xi \sin \phi \cos \theta$$

$$x_3 = \xi \sin \phi \sin \theta,$$

that is, a rectangular-to-spherical coordinate transformation. The Jacobian of this transformation is $\xi^2 \sin \phi$. Using this transformation in the expression for $F_{R_3}(r)$ we obtain for $r > 0$

$$F_{R_3}(r) = \frac{1}{(2\pi)^{3/2}(\sigma^2)^{3/2}} \int_0^r \int_{\theta=0}^{2\pi} \int_{\phi=0}^{\pi} \exp\left[-\frac{\xi^2}{2\sigma^2}\right] \xi^2 \sin \phi \, d\xi \, d\theta \, d\phi$$

$$= \frac{4\pi}{(2\pi)^{3/2}[\sigma^2]^{3/2}} \int_0^r \xi^2 \exp\left[-\frac{\xi^2}{2\sigma^2}\right] d\xi.$$

To obtain $f_{R_3}(r)$, we differentiate $F_{R_3}(r)$ with respect to r. This yields

$$f_{R_3}(r) = \frac{2r^2}{\Gamma(\frac{3}{2})[2\sigma^2]^{3/2}} \exp\left[-\frac{r^2}{2\sigma^2}\right] \cdot u(r), \tag{5.6-22}$$

where $u(r)$ is the unit step and $\Gamma(3/2) = \sqrt{\pi}/2$. Equation 5.6-22 is an extension of the ordinary two-dimensional Rayleigh introduced in Chapter 2. The general n-dimensional Rayleigh is the pdf associated with $R_n \triangleq \|\mathbf{X}\| = \sqrt{X_1^2 + \ldots + X_n^2}$ and is given by

$$f_{R_n}(r) = \frac{2r^{n-1}}{\Gamma\left(\frac{n}{2}\right)[2\sigma^2]^{n/2}} \exp\left[-\frac{r^2}{2\sigma^2}\right] \cdot u(r). \tag{5.6-23}$$

The proof of Equation 5.6-23 requires the use of n-dimensional spherical coordinates. Such generalized spherical coordinates are well known in the mathematical literature [5-5, p. 9]. The demonstration of Equation 5.6-23 is left as a challenging problem.

5.7 CHARACTERISTIC FUNCTIONS OF RANDOM VECTORS

In Equation 4.7-1 we defined the characteristic function (c.f.) of a random variable as

$$\Phi_X(\omega) \triangleq E[e^{j\omega X}].$$

The extension to random vectors is completely straightforward. Let $\mathbf{X} = (X_1, \ldots, X_n)^T$ be a real n-component random vector. Let $\boldsymbol{\omega} = (\omega_1, \ldots, \omega_n)^T$ be a real n-component parameter vector. The c.f. of $\mathbf{X}$ is defined as

$$\Phi_{\mathbf{X}}(\boldsymbol{\omega}) \triangleq E[e^{j\boldsymbol{\omega}^T\mathbf{X}}]. \tag{5.7-1}$$

The similarity to the scalar case is obvious. The actual evaluation of Equation 5.7-1 is done through

$$\Phi_{\mathbf{X}}(\boldsymbol{\omega}) = \int_{-\infty}^{\infty} f_{\mathbf{X}}(\mathbf{x})e^{j\boldsymbol{\omega}^T\mathbf{x}}d\mathbf{x}. \tag{5.7-2}$$

In Equation 5.7-2 we use the usual compact notation that $d\mathbf{x} = dx_1 \ldots dx_n$ and the integral sign refers to an n-fold integration. If $\mathbf{X}$ is a discrete random vector, $\Phi_{\mathbf{X}}(\boldsymbol{\omega})$ can be computed from the joint probability mass function (PMF) as

$$\Phi_{\mathbf{X}}(\boldsymbol{\omega}) = \sum_{i_1} \ldots \sum_{i_n} \exp[j(\omega_1 x_{i_1} + \ldots + \omega_n x_{i_n})] \cdot P[X_1 = x_{i_1}, \ldots, X_n = x_{i_n}] \tag{5.7-3}$$

From Equation 5.7-2 we see that $\Phi_{\mathbf{X}}(\boldsymbol{\omega})$ is, except for a sign reversal in the exponent, the n-dimensional Fourier transform of $f_{\mathbf{X}}(\mathbf{x})$. This being the case, we can recover the pdf by an inverse n-dimensional Fourier transform (again with a sign reversal). Thus

$$f_{\mathbf{X}}(\mathbf{x}) = \frac{1}{(2\pi)^n} \int_{-\infty}^{\infty} \Phi_{\mathbf{X}}(\boldsymbol{\omega})e^{-j\boldsymbol{\omega}^T\mathbf{x}}d\boldsymbol{\omega}. \tag{5.7-4}$$

The c.f. is very useful for computing joint moments. We illustrate with an example.

Example 5.7-1 _____

Let $\mathbf{X} \triangleq (X_1, X_2, X_3)^T$ and $\boldsymbol{\omega} \triangleq (\omega_1, \omega_2, \omega_3)$. Compute $E[X_1 X_2 X_3]$.

Solution   Since

$$\Phi_{\mathbf{X}}(\omega_1, \omega_2, \omega_3) = \int_{-\infty}^{\infty} \int_{-\infty}^{\infty} \int_{-\infty}^{\infty} f_{\mathbf{X}}(x_1, x_2, x_3)e^{j[\omega_1 x_1 + \omega_2 x_2 + \omega_3 x_3]} dx_1 \, dx_2 \, dx_3,$$

we obtain by partial differentiation

$$\frac{1}{j^3} \frac{\partial^3 \Phi_{\mathbf{X}}(\omega_1, \omega_2, \omega_3)}{\partial \omega_1 \partial \omega_2 \partial \omega_3} \bigg|_{\omega_1 = \omega_2 = \omega_3 = 0}$$

$$= \int_{-\infty}^{\infty} \int_{-\infty}^{\infty} \int_{-\infty}^{\infty} x_1 x_2 x_3 f_{\mathbf{X}}(x_1, x_2, x_3) \, dx_1 \, dx_2 \, dx_3$$

$$\triangleq E[X_1 X_2 X_3].$$

_____

Any moment—provided that it exists—can be computed by the method used in Example 5.7-1, that is, by partial differentiation. Thus

$$E[X_1^{k_1} \ldots X_n^{k_n}] = j^{(k_1 + \ldots + k_n)} \frac{\partial^{k_1 + \ldots + k_n} \Phi_{\mathbf{X}}(\omega_1, \ldots, \omega_n)}{\partial \omega_1^{k_1} \ldots \partial \omega_n^{k_n}} \bigg|_{\omega_1 = \ldots = \omega_n = 0} \tag{5.7-5}$$

By writing

$$E[\exp(j\boldsymbol{\omega}^T\mathbf{X})] = E\left[\exp\left(j\sum_{i=1}^{n} \omega_i X_i\right)\right] = E\left[\prod_{i=1}^{n} \exp(j\omega_i X_i)\right]$$

and expanding each term in the product into a power series, we readily obtain the rather cumbersome formula

$$\Phi_{\mathbf{X}}(\boldsymbol{\omega}) = \sum_{k_1=0}^{\infty} \cdots \sum_{k_n=0}^{\infty} E[X_1^{k_1} \ldots X_n^{k_n}] \frac{(j\omega_1)^{k_1}}{k_1!} \cdots \frac{(j\omega_n)^{k_n}}{k_n!}, \tag{5.7-6}$$

which has the advantage of explicitly revealing the relationship between the c.f. and the joint moments of the $X_i, i = 1, \ldots, n$. Of course Equation 5.7-6 has meaning only if

$$E[X_1^{k_1} \ldots X_n^{k_n}]$$

exists for all values of the nonnegative integers $k_1, \ldots, k_n$.

From Equation 5.7-2 we make the rather obvious observations that

1. $|\Phi_{\mathbf{X}}(\boldsymbol{\omega})| \le \Phi_{\mathbf{X}}(\mathbf{0}) = 1$ and
2. $\Phi_{\mathbf{X}}^*(\boldsymbol{\omega}) = \Phi_{\mathbf{X}}(-\boldsymbol{\omega})$ (* indicates conjugation).
3. All c.f.'s of *subsets* of the components of $\mathbf{X}$ can be obtained once $\Phi_{\mathbf{X}}(\boldsymbol{\omega})$ is known.

The last property is easily demonstrated with the following example. Suppose $\mathbf{X} = (X_1, X_2, X_3)^T$ has c.f.[†] $\Phi_{\mathbf{X}}(\omega_1, \omega_2, \omega_3) = E[\exp j(\omega_1 X_1 + \omega_2 X_2 + \omega_3 X_3)]$. Then

$$\Phi_{X_1 X_2}(\omega_1, \omega_2) = \Phi_{X_1 X_2 X_3}(\omega_1, \omega_2, 0)$$

$$\Phi_{X_1 X_3}(\omega_1, \omega_3) = \Phi_{X_1 X_2 X_3}(\omega_1, 0, \omega_3)$$

$$\Phi_{X_1}(\omega_1) = \Phi_{X_1 X_2 X_3}(\omega_1, 0, 0).$$

As pointed out in Chapter 4, characteristic functions are also useful in solving problems involving sums of independent random variables. Thus suppose $\mathbf{X} = (X_1, \ldots, X_n)^T$ where the X_i are *independent* r.v.'s with marginal pdf's $f_{X_i}(x_i), i = 1, \ldots, n$. The pdf of the sum

$$Z = X_1 + \ldots + X_n$$

can be obtained from

$$f_Z(z) = f_{X_1}(z) * \ldots * f_{X_n}(z). \tag{5.7-7}$$

However, the actual carrying out of the n-fold convolution in Equation 5.7-7 can be quite tedious. The computation of $f_Z(z)$ can be done more advantageously from the c.f.'s as follows. In

$$\Phi_{\mathbf{X}}(\omega_1, \ldots, \omega_n) = E[e^{j\boldsymbol{\omega}^T \mathbf{X}}]$$

choose $\boldsymbol{\omega} = (\omega, \ldots, \omega)^T$. Then

$$\Phi_{\mathbf{X}}(\omega, \ldots, \omega) = E[e^{j\omega(X_1 + \ldots + X_n)}]$$

$$= E\left[\prod_{i=1}^{n} e^{j\omega X_i}\right]$$

[†]We use $\Phi_{\mathbf{X}}(\cdot)$ and $\Phi_{X_1 X_2 X_3}(\cdot)$ interchangeably if $\mathbf{X} = (X_1, X_2, X_3)^T$.

$$= \prod_{i=1}^{n} E\left[e^{j\omega X_i}\right] = \prod_{i=1}^{n} \Phi_{X_i}(\omega)$$

$$= \Phi_Z(\omega). \tag{5.7-8}$$

In this development, line 3 follows from the fact that if $X_1, \ldots, X_n$ are n independent r.v.'s, then $Y_i = g_i(X_i)$, $i = 1, \ldots, n$ will also be n independent r.v.'s and $E[Y_1 \ldots Y_n] = E[Y_1] \ldots E[Y_n]$. We leave the demonstration of this result as an exercise. The inverse Fourier transform of Equation 5.7-8 yields the pdf $f_Z(z)$. This approach works equally well when the X_i are discrete. Then the PMF and the discrete Fourier transform can be used. We illustrate this approach to computing the pdf's of sums of random variables with an example.

Example 5.7-2

Let $\mathbf{X} = (X_1, \ldots, X_n)^T$ where the X_i, $i = 1, \ldots, n$ are i.i.d. Poisson r.v.'s with Poisson parameter λ. Let $Z = X_1 + \ldots + X_n$. Then

$$P_{X_i}(k) = \frac{\lambda^k e^{-\lambda}}{k!} \tag{5.7-9}$$

and

$$\Phi_{X_i}(\omega) = \sum_{k=0}^{\infty} \frac{\lambda^k \exp(j\omega k)}{k!} e^{-\lambda}$$

$$= e^{\lambda(\exp(j\omega)-1)}. \tag{5.7-10}$$

Hence, from the third line leading up to Equation 5.7-8 we obtain

$$\Phi_Z(\omega) = \prod_{i=1}^{n} e^{\lambda(\exp(j\omega)-1)}$$

$$= e^{n\lambda(\exp(j\omega)-1)}. \tag{5.7-11}$$

Comparing Equation 5.7-11 with Equation 5.7-10 we see by inspection that $\Phi_Z(z)$ is the c.f. of the PMF

$$P_Z(k) = \frac{\xi^k e^{-\xi}}{k!} \qquad k = 0, 1 \ldots \tag{5.7-12}$$

where $\xi \triangleq n\lambda$. Thus the PMF of the sum of n i.i.d. Poisson r.v.'s is Poisson with parameter $n\lambda$.

The Characteristic Function of the Gaussian (Normal) Law

Let $\mathbf{X}$ be a real normal random vector with nonsingular covariance matrix $\mathbf{K}$. Then from Equation 5.5-12 both $\mathbf{K}$ and $\mathbf{K}^{-1}$ can be factored as

$$\mathbf{K} = \mathbf{C}\mathbf{C}^T \tag{5.7-13}$$

$$\mathbf{K}^{-1} = \mathbf{D}\mathbf{D}^T, \qquad \mathbf{D} = [\mathbf{C}^T]^{-1}, \tag{5.7-14}$$

where $\mathbf{C}$ and $\mathbf{D}$ are nonsingular. This observation will be put to good use shortly. The c.f. of $\mathbf{X}$ is by definition

$$\Phi_{\mathbf{X}}(\boldsymbol{\omega}) = \frac{1}{(2\pi)^{n/2}[\det(\mathbf{K})]^{1/2}} \int_{-\infty}^{\infty} \exp\left(-\frac{1}{2}(\mathbf{x}-\boldsymbol{\mu})^T\mathbf{K}^{-1}(\mathbf{x}-\boldsymbol{\mu})\right) \cdot \exp(j\boldsymbol{\omega}^T\mathbf{x})\,d\mathbf{x}.$$
(5.7-15)

Now introduce the transformation

$$\mathbf{z} \triangleq \mathbf{D}^T(\mathbf{x}-\boldsymbol{\mu})$$
(5.7-16)

so that

$$\mathbf{z}^T\mathbf{z} = (\mathbf{x}-\boldsymbol{\mu})^T\mathbf{D}\mathbf{D}^T(\mathbf{x}-\boldsymbol{\mu})$$

$$= (\mathbf{x}-\boldsymbol{\mu})^T\mathbf{K}^{-1}(\mathbf{x}-\boldsymbol{\mu}).$$
(5.7-17)

The Jacobian of this transformation is $\det(\mathbf{D}^T) = \det(\mathbf{D})$. Thus under the transformation in Equation 5.7-16, Equation 5.7-15 is transformed to

$$\Phi_{\mathbf{X}}(\boldsymbol{\omega}) = \frac{\exp(j\boldsymbol{\omega}^T\boldsymbol{\mu})}{(2\pi)^{n/2}[\det(\mathbf{K})]^{1/2}|\det(\mathbf{D})|} \int_{-\infty}^{\infty} \exp\left(-\frac{1}{2}\mathbf{z}^T\mathbf{z}\right) \cdot \exp(j\boldsymbol{\omega}^T(\mathbf{D}^T)^{-1}\mathbf{z})\,d\mathbf{z}.$$
(5.7-18)

We can complete the squares in the integrand as follows:

$$\exp[-\{\tfrac{1}{2}[\mathbf{z}^T\mathbf{z} - 2j\boldsymbol{\omega}^T(\mathbf{D}^T)^{-1}\mathbf{z}]\}] = \exp(-\tfrac{1}{2}\boldsymbol{\omega}^T(\mathbf{D}^T)^{-1}(\mathbf{D})^{-1}\boldsymbol{\omega})$$

$$\cdot \exp(-\tfrac{1}{2}||\mathbf{z} - j\mathbf{D}^{-1}\boldsymbol{\omega}||^2).$$
(5.7-19)

Equations 5.7-18 and 5.7-19 will be greatly simplified if we use the following easily proven results: (A) If $\mathbf{K}^{-1} = \mathbf{D}\mathbf{D}^T$, then $\mathbf{K} = [\mathbf{D}^T]^{-1}\mathbf{D}^{-1}$; (B) $\det(\mathbf{K}^{-1}) = \det(\mathbf{D})\det(\mathbf{D}^T) = [\det(\mathbf{D})]^2 = [\det(\mathbf{K})]^{-1}$. Hence $|\det(\mathbf{D})|^{-1} = [\det(\mathbf{K})]^{1/2}$. It then follows that

$$\Phi_{\mathbf{X}}(\boldsymbol{\omega}) = \exp\left(j\boldsymbol{\omega}^T\boldsymbol{\mu} - \frac{1}{2}\boldsymbol{\omega}^T\mathbf{K}\boldsymbol{\omega}\right) \cdot \frac{1}{(2\pi)^{n/2}} \int_{-\infty}^{\infty} e^{-\frac{1}{2}||\mathbf{z}-j\mathbf{D}^{-1}\boldsymbol{\omega}||^2}\,d\mathbf{z}.$$

Finally we recognize that the n-fold integral on the right-hand side is the product of n identical integrals, each of unit variance. Hence the value of the integral is merely $(2\pi)^{n/2}$, which cancels the factor $(2\pi)^{-n/2}$ and yields as the c.f. for the normal random vector:

$$\Phi_{\mathbf{X}}(\boldsymbol{\omega}) = \exp[j\boldsymbol{\omega}^T\boldsymbol{\mu} - \tfrac{1}{2}\boldsymbol{\omega}^T\mathbf{K}\boldsymbol{\omega}],$$
(5.7-20)

where $\boldsymbol{\mu}$ is the mean vector, $\boldsymbol{\omega} = (\omega_1, \ldots, \omega_n)^T$, and $\mathbf{K}$ is the covariance. We observe in passing that $\Phi_{\mathbf{X}}(\boldsymbol{\omega})$ has a multidimensional complex Gaussian form as a function of $\boldsymbol{\omega}$. Thus the Gaussian pdf has mapped into a Gaussian c.f., a result that should not be too surprising since we already know that the one-dimensional Fourier transform maps a Gaussian function into a Gaussian function.

5.8 PARAMETER ESTIMATION

The problem of estimating the parameters μ and σ^2 (as in Section 4.8) or $\boldsymbol{\mu}$ and $\mathbf{K}$ is a problem of parameter estimation. We illustrate what is meant by parameter estimation with two examples.

Example 5.8-1

Suppose θ is an unknown scalar that we wish to estimate. We make n measurements called observations and collect n samples (sometimes called realizations) $x_1, \ldots, x_n$ where

$$x_i = \theta + \varepsilon_i \quad i = 1, \ldots, n. \tag{5.8-1}$$

In Equation 5.8-1 x_i is the value of the measurement and ε_i is the value of the measurement noise on the ith observations. After completing the n measurements and collecting all the data, a reasonable estimate $\hat{\theta}$ of θ is furnished by

$$\hat{\theta} = \frac{1}{n} \sum_{i=1}^{n} x_i. \tag{5.8-2}$$

Example 5.8-2

We observe the values $x_1^{(1)}, \ldots, x_n^{(1)}$ that a normal r.v. takes in n trials. We wish to estimate the mean μ of the pdf. An estimate of μ is furnished by

$$\hat{\mu}^{(1)} = \frac{1}{n} \sum_{i=1}^{n} x_i^{(1)}.$$

Note that if we collected a second set of samples $x_1^{(2)}, \ldots, x_n^{(2)}$, then the estimate of μ based on this second set, that is,

$$\hat{\mu}^{(2)} = \frac{1}{n} \sum_{i=1}^{n} x_i^{(2)}$$

would probably be different from $\hat{\mu}_i^{(1)}$.

In Examples 5.8-1 and 5.8-2 we can expect different sets of n sample values to yield different estimates. This suggests that we should view the estimate as a particular value of a random variable (r.v.) called an *estimator* and the n sample values $x_1, \ldots, x_n$ for a particular measurement as the values assumed by a set of n random variables $X_1, \ldots, X_n$ where X_i is the outcome of the ith measurement. We can think of each measurement as an observation on a generic r.v. X with some underlying pdf $f_X(x)$. Then the X_i, $i = 1, \ldots, n$ represent n independent and identically distributed (i.i.d.) observations on X, and each has pdf $f_{X_i}(x_i) = f_X(x_i), i = 1, \ldots, n$. Thus the estimate $\hat{\theta}$ is a particular value of the *sample mean estimator*

$$\hat{\Theta} \triangleq \frac{1}{n} \sum_{i=1}^{n} X_i. \tag{5.8-3}$$

The estimator in Equation 5.8-3 is often used to estimate $E[X]$.[†] Other estimators are used to estimate $\text{Var}[X]$, the covariance matrix $\mathbf{K}$, and so on. Some estimators are better than others. To evaluate estimators we need some definitions.[‡]

Definition 5.8-1   An estimator $\hat{\Theta}$ is a function of the observation vector $\mathbf{X} = (X_1, \ldots, X_n)^T$ that estimates θ but is not a function of θ.   ■

Definition 5.8-2   An estimator $\hat{\Theta}$ for θ is said to be unbiased if and only if $E[\hat{\Theta}] = \theta$. The bias in estimating θ with $\hat{\Theta}$ is[§]

$$|E[\hat{\Theta}] - \theta|. \quad ■$$

Definition 5.8-3   An estimator $\hat{\Theta}$ is said to be a *linear estimator* of θ if it is a linear function of the observation vector $\mathbf{X} \triangleq (X_1, \ldots, X_n)^T$, that is,

$$\hat{\Theta} = \mathbf{b}^T \mathbf{X}. \tag{5.8-4}$$

The vector $\mathbf{b}$ is an $n \times 1$ vector of coefficients that do not depend on $\mathbf{X}$.   ■

Definition 5.8-4   Let $\hat{\Theta}_n$ be an estimator computed from n samples $X_1, \ldots, X_n$ for every $n \geq 1$. Then $\hat{\Theta}_n$ is said to be *consistent* if

$$\lim_{n \to \infty} P[|\hat{\Theta}_n - \theta| > \varepsilon] = 0 \quad \text{for every} \quad \varepsilon > 0. \tag{5.8-5}$$

The condition in Equation 5.8-5 is often referred to as *convergence in probability*.   ■

Definition 5.8-5   An estimator $\hat{\Theta}$ is called *minimum-variance unbiased* if

$$E[(\hat{\Theta} - \theta)^2] \leq E[(\hat{\Theta}' - \theta)^2] \tag{5.8-6}$$

where $\hat{\Theta}'$ is any other estimator and $E[\hat{\Theta}'] = E[\hat{\Theta}] = \theta$.   ■

Definition 5.8-6   An estimator $\hat{\Theta}$ is called a *minimum mean-square error* (MMSE) estimator if

$$E[(\hat{\Theta} - \theta)^2] \leq E[(\hat{\Theta}' - \theta)^2], \tag{5.8-7}$$

where $\hat{\Theta}'$ is any other estimator.   ■

[†]The validity of estimating parameters as well as other objects, for instance probabilities, from repeated observations is based, fundamentally, on the law of large numbers and the Chebyshev inequality.

[‡]The definitions are given for scalar estimators. They are easily extended to vector estimators.

[§]The bias is often defined without the magnitude sign.

There are several other properties of estimators that are deemed desirable such as efficiency, completeness, and invariance. These properties are discussed in books on statistics[†] and will not be discussed further here.

Estimation of $E[X]$

Let X be a r.v. with pdf $f_X(x)$ and finite variance σ^2. We wish to estimate $\mu = E[X]$ from n i.i.d. observations on X, that is, we repeat the experiment n times with X_i denoting the ith outcome. Sometimes it is said that the X_i are *drawn independently* from $f_X(x)$. Then $f_{X_1}(x) = f_X(x), i = 1, \ldots, n$.

The sample mean estimator is

$$\hat{\Theta} \triangleq \frac{1}{n} \sum_{i=1}^{n} X_i \tag{5.8-8}$$

We now show that $\hat{\Theta}$ is an unbiased, consistent estimator of μ,

Unbiasedness of $\hat{\Theta}$ of Equation 5.8-8.   Since $E[X_i] = \mu, i = 1, \ldots, n$ we have

$$E[\hat{\Theta}] = E\left[\frac{1}{n} \sum_{i=1}^{n} X_i\right]$$

$$= \frac{1}{n} \sum_{i=1}^{n} E[X_i]$$

$$= \frac{1}{n} \cdot n\mu = \mu.$$

Consistency of $\hat{\Theta}$.   By the Chebyshev inequality, Equation 4.4-1, we obtain

$$P[|\hat{\Theta} - \mu| > \varepsilon] \leq \frac{\text{Var}[\hat{\Theta}]}{\varepsilon^2} \tag{5.8-9}$$

The variance of $\hat{\Theta}$ is obtained as

$$\text{Var}[\hat{\Theta}] = E\left[\left(\frac{1}{n} \sum_{i=1}^{n} X_i\right) - \mu\right]^2$$

$$= E\left[\frac{1}{n^2} \sum_{i=1}^{n} X_i^2 + \frac{1}{n^2} \sum_{\substack{i \neq j}}^{n}\sum^{n} X_i X_j + \mu^2 - 2\mu\left(\frac{1}{n} \sum_{i=1}^{n} X_i\right)\right]$$

$$= \frac{1}{n^2}(n\sigma^2 + n\mu^2) + \frac{1}{n^2}n(n-1)\mu^2 + \mu^2 - 2\mu^2$$

$$= \frac{\sigma^2}{n}. \tag{5.8-10}$$

[†]See, for example, Reference [4-9] In Chapter 4.

Now using Equation 5.8-10 in Equation 5.8-9 we obtain

$$P[|\hat{\Theta} - \mu| > \varepsilon] \leq \frac{\sigma^2}{n\varepsilon^2}, \qquad (5.8\text{-}11)$$

which goes to zero for $n \to \infty$ and every $\varepsilon > 0.$[†]

Estimation of Var[X]. Let X be a random variable with pdf $f_X(x)$ with mean μ and variance σ^2. We wish to estimate σ^2 from n i.i.d. observations $X_1, \ldots, X_n$ on X. Consider the estimator

$$\hat{\Theta} \triangleq \frac{1}{n-1} \sum_{i=1}^{n} (X_1 - \hat{\mu})^2, \qquad (5.8\text{-}12)$$

where

$$\hat{\mu} \triangleq \frac{1}{n} \sum_{i=1}^{n} X_i. \qquad (5.8\text{-}13)$$

The following shows that $\hat{\Theta}$ in Equation 5.8-12 is an unbiased, consistent estimator of $\text{Var}(X)$.

Unbiasedness of $\hat{\Theta}$ of Equation 5.8-12. Consider

$$E\left[\sum_{i=1}^{n}\left(X_i - \frac{1}{n}\sum_{j=1}^{n} X_j\right)^2\right]$$

$$= E\left[\sum_{i=1}^{n}\left\{X_i^2 - \frac{2}{n}X_i^2 - \frac{2}{n}\sum_{\substack{j=1 \\ j\neq i}}^{n} X_i X_j + \frac{1}{n^2}\sum_{j=1}^{n} X_j^2 + \frac{1}{n^2}\sum_{j\neq 1}^{n}\sum^{n} X_j X_i\right\}\right]$$

$$= (n-1)\sigma^2. \qquad (5.8\text{-}14)$$

In obtaining Equation 5.8-14, we used the fact that $E[X_i^2] = \sigma^2 + \mu^2$ $i = 1, \ldots, n$. Clearly if

$$E\left[\sum_{i=1}^{n} (X_i - \hat{\mu})^2\right] = (n-1)\sigma^2$$

then

$$E\left[\frac{1}{n-1}\sum_{i=1}^{n} (X_i - \hat{\mu})^2\right] = \sigma^2. \qquad (5.8\text{-}15)$$

But the quantity inside the square brackets is $\hat{\Theta}$ of Equation 5.8-12. Hence this $\hat{\Theta}$ is unbiased for σ^2.

[†]This result, or variations thereof, is essentially a statement of the weak *law of large numbers*.

Consistency of $\hat{\Theta}$ of Equation 5.8-12.   To indicate the dependence of $\hat{\Theta}$ on the sample size n we add the subscript n. The variance of $\hat{\Theta}_n$ is given by

$$\text{Var}[\hat{\Theta}_n] = E[\hat{\Theta}_n - \sigma^2]^2$$

$$= E\left[\frac{1}{(n-1)^2}\left\{\sum_{i=1}^{n}(X_i - \hat{\mu})^4 + \sum_{i\neq j}^{n}\sum^{n}(X_i - \hat{\mu})^2(X_j - \hat{\mu})^2\right\}\right.$$

$$\left. + \sigma^4 - \frac{2\sigma^2}{n-1}\sum_{i=1}^{n}(X_i - \hat{\mu})^2\right].$$

A straightforward calculation shows that for $n \gg 1$

$$\text{Var}[\hat{\Theta}_n] \simeq \frac{1}{n}m_4, \tag{5.8-16}$$

where $m_4 \triangleq E[(X_1 - \mu)^4]$ (see Equation 4.3-2). Assuming that m_4 (the fourth-order central moment) exists, we once again use the Chebyshev inequality to write that

$$P(|\hat{\Theta}_n - \sigma^2| > \varepsilon) \leq \frac{\text{Var}[\hat{\Theta}_n]}{\varepsilon^2}$$

$$\simeq \frac{m_4}{n\varepsilon^2}, \tag{5.8-17}$$

which goes to zero as $n \to \infty$ for every $\varepsilon > 0$. Hence $\hat{\Theta}_n$ is consistent.

5.9 ESTIMATION OF VECTOR MEANS AND COVARIANCE MATRICES

Let $\mathbf{X} \triangleq (X_1, \ldots, X_p)^T$ be a p-component random vector with pdf $f_{\mathbf{X}}(\mathbf{x})$. Let $\mathbf{X}_1, \ldots, \mathbf{X}_n$ be n observations on $\mathbf{X}$, that is, the $\mathbf{X}_i, i = 1, \ldots, n$ are drawn from $f_{\mathbf{X}}(\mathbf{x})$. Then $\mathbf{X}_i, i = 1, \ldots, n$ are i.i.d. random vectors with pdf $f_{\mathbf{X}}(\mathbf{x}_i)$. We show below how to estimate[†]

$$\text{(i) } \boldsymbol{\mu}_{\mathbf{X}} \triangleq E[\mathbf{X}] = (\mu_1, \ldots, \mu_p)^T,$$

where

$$\mu_j \triangleq E[X_j] \quad j = 1, \ldots, p$$

and

$$\text{(ii) } \mathbf{K}_{\mathbf{XX}} \triangleq E[(\mathbf{X} - \boldsymbol{\mu})(\mathbf{X} - \boldsymbol{\mu})^T].$$

Estimation of μ

Consider the p-vector estimator $\hat{\boldsymbol{\Theta}}$ given by

$$\hat{\boldsymbol{\Theta}} \triangleq \frac{1}{n}\sum_{i=j}^{n}\mathbf{X}_i. \tag{5.9-1}$$

[†]In anticipation of the notation used in subsequent chapters, we use bold subscripts to associate the parameter vectors or matrices with the appropriate random vectors.

Table 5.9-1 Observed Data

	$\mathbf{X}_1 \dots \mathbf{X}_i \dots \mathbf{X}_n$		
$\mathbf{Y}_1$	X_{11}	X_{i1}	X_{n1}
$\vdots$	$\cdot$	$\vdots$	$\cdot$
$\mathbf{Y}_j$	$\cdot$	X_{ij}	$\cdot$	p rows
$\vdots$	$\cdot$	$\vdots$	$\cdot$
$\mathbf{Y}_p$	X_{1p}		X_{np}

$\underbrace{\qquad\qquad\qquad}_{n \text{ columns}}$

The components of $\mathbf{Y}_j$ are all that is necessary for estimating the jth component, μ_j, of the vector $\boldsymbol{\mu}$.

We shall show that $\hat{\boldsymbol{\Theta}}$ is unbiased and consistent for $\boldsymbol{\mu}$. We arrange the observations as in Table 5.9-1.

In Table 5.9-1 X_{ij} is the jth component of the random vector $\mathbf{X}_i$. The components of the vector $\mathbf{Y}_j, j = 1, \dots, p$ are n i.i.d. observations on the jth component of the random vector $\mathbf{X}$. From the scalar case we already know that

$$\hat{\Theta}_j \triangleq \frac{1}{n} \sum_{i=1}^{n} X_{ij} \triangleq \hat{\mu}_j \quad j = 1, \dots, p \tag{5.9-2}$$

is unbiased and consistent for $\mu_j \triangleq E[X_{ij}]$ $i = 1, \dots, n$. It follows therefore that the vector estimator $\hat{\boldsymbol{\Theta}} \triangleq (\hat{\Theta}_1, \dots, \hat{\Theta}_p)^T$ is unbiased and consistent for $\boldsymbol{\mu}$. The vector $\mathbf{Y}_j$ contains all the information for estimating μ_j. Thus $E[\mathbf{Y}_j] = \mu_j \mathbf{i}$, where $\mathbf{i} \triangleq (1, 1, \dots, 1, 1)^T$.

When $\mathbf{X}$ is normal, $\hat{\boldsymbol{\Theta}}$ is normal. Even when $\mathbf{X}$ is not normal, $\hat{\boldsymbol{\Theta}}$ tends to the normal for large n by the central limit theorem (Theorem 4.7-1).

Estimation of the Covariance K

If the mean $\boldsymbol{\mu}$ is known, then the estimator

$$\hat{\boldsymbol{\Theta}} \triangleq \frac{1}{n} \sum_{i=1}^{n} (\mathbf{X}_i - \boldsymbol{\mu})(\mathbf{X}_i - \boldsymbol{\mu})^T \tag{5.9-3}$$

is unbiased for $\mathbf{K}$. However, since the mean is generally estimated from the *sample mean* $\hat{\boldsymbol{\mu}}$, it turns out that the estimator

$$\hat{\boldsymbol{\Theta}} \triangleq \frac{1}{n-1} \sum_{i=1}^{n} (\mathbf{X}_i - \hat{\boldsymbol{\mu}})(\mathbf{X}_i - \hat{\boldsymbol{\mu}})^T \tag{5.9-4}$$

is unbiased for **K**. To prove this result requires some effort. First observe that the diagonal elements of $\hat{\Theta}$ are of the form

$$S_{jj} \triangleq \frac{1}{n-1} \sum_{i=1}^{n} (X_{ij} - \hat{\mu}_j)^2, \tag{5.9-5}$$

which we already know from the univariate case are unbiased for $\sigma_j^2 \triangleq E[(X_j - \mu_j)^2]$. Next consider the sequence $(l \neq m)$

$$X_{1l} + X_{1m}, X_{2l} + X_{2m}, \ldots, X_{nl} + X_{nm}, \tag{5.9-6}$$

which are n i.i.d. observations $Z_{lm}^{(i)}$, on a univariate r.v. $Z_{lm} \triangleq X_l + X_m$ with mean $\mu_l + \mu_m$ and variance

$$\text{Var}[Z_{lm}] = E[(X_l - \mu_l) + (X_m - \mu_m)]^2$$
$$= \sigma_l^2 + \sigma_m^2 + 2K_{lm}, \tag{5.9-7}$$

where $K_{lm} \triangleq E[(X_l - \mu_l)(X_m - \mu_m)]$ is the lmth element of $\mathbf{K_{XX}}$. Finally, consider

$$\hat{\Theta}_{lm} \triangleq \frac{1}{n-1} \sum_{l=1}^{n} [Z_{lm}^{(i)} - (\hat{\mu}_l + \hat{\mu}_m)]^2, \tag{5.9-8}$$

which, by Equation 5.8-15, is unbiased for $\sigma_l^2 + \sigma_m^2 + 2K_{lm}$. If we expand Equation 5.9-8 and use the fact that $Z_{lm} \triangleq X_l + X_m$, we obtain

$$\hat{\Theta}_{lm} \triangleq \frac{1}{n-1} \sum_{i=1}^{n} [(X_{il} - \hat{\mu}_l) + (X_{im} - \hat{\mu}_m)]^2$$

$$= \frac{1}{n-1} \sum_{i=1}^{n} (X_{il} - \hat{\mu}_l)^2 + \frac{1}{n-1} \sum_{i=1}^{n} (X_{im} - \hat{\mu}_m)^2$$

$$+ \frac{2}{n-1} \sum_{i=1}^{n} (X_{il} - \hat{\mu}_l)(X_{im} - \hat{\mu}_m). \tag{5.9-9}$$

In Equation 5.9-9, the first term is unbiased for σ_l^2, the second is unbiased for σ_m^2, and the sum of all three is unbiased by Equation 5.9-8 for $\sigma_l^2 + \sigma_m^2 + 2K_{lm}$. We therefore conclude that

$$S_{lm} \triangleq \frac{1}{n-1} \sum_{i=1}^{n} (X_{il} - \hat{\mu}_l)(X_{im} - \hat{\mu}_m) \tag{5.9-10}$$

is unbiased for $K_{lm}(= K_{ml})$. Hence every term of $\hat{\Theta}$ in Equation 5.9-4 is unbiased for every corresponding term in $\mathbf{K_{XX}}$. In this sense $\hat{\Theta} \triangleq \hat{\mathbf{K}}_{\mathbf{XX}}$ is unbiased for $\mathbf{K_{XX}}$.

By resorting again to the univariate case and assuming that all moments up to the fourth order exist, we can show consistency for every term in the estimator for $\mathbf{K_{XX}}$, that is, Equation 5.9-4. Hence without specifying the distribution, Equations 5.9-1 and 5.9-4 are unbiased and consistent estimators for $\boldsymbol{\mu}_{\mathbf{X}}$ and $\mathbf{K}_{\mathbf{XX}}$ respectively.

When $\mathbf{X}$ is normal, $\hat{\mathbf{K}}$ obeys a structurally complex probability law called the Wishart distribution (see Fukunaga, Reference [5-6, p. 126]). More generally, when the *form* of the pdf of $\mathbf{X}$ is known, one can use another method of estimating such parameters as $\mu'_{\mathbf{X}}, \sigma^2_{\mathbf{X}}, \boldsymbol{\mu}_{\mathbf{X}}$ and $\mathbf{K}_{\mathbf{XX}}$ called the method of *maximum likelihood*. Maximum likelihood estimators have several desirable properties as estimators and by some measures can be regarded as "best" estimators. However, the reader should be cautioned that *best* implies a criterion of performance and what may be best by one criterion may be far from best by another. For instance, the sample mean estimator $\hat{\Theta}$ given in Equation 5.8-8 may be best in the sense of being unbiased and consistent but is not best according to a criterion called minimum mean-square error (MMSE) (see Definition 5.8-6). We illustrate with an example.

Example 5.9-1
(Kendall and Stuart, [5-8], p. 21.) Consider the sample mean estimator from Equation 5.8-3, that is,

$$\hat{\mu} = \frac{1}{n} \sum_{i=1}^{n} X_i.$$

What constant a in $\hat{\Theta} \triangleq a\hat{\mu}$ will generate the MMSE estimator of μ? Recall the X_i $i = 1, \ldots, n$ are i.i.d. r.v.'s with $E[X_i] = \mu$ and $\text{Var}[X_i] = \sigma^2$.

Solution   We are seeking the value of a such that

$$E[a\hat{\mu} - \mu]^2 \qquad (5.9\text{-}11)$$

is a minimum. Clearly $\hat{\mu}$ is unbiased for μ, and it seems hard to believe that there may exist an $\hat{\Theta}$ with $a \neq 1$ that—though yielding a biased estimator—gives a lower MSE than $\hat{\Theta} = \hat{\mu}$. For *any* estimator $\hat{\Theta}$, the mean square error in estimating μ is

$$E[(\hat{\Theta} - \mu)^2] = E[\{(\hat{\Theta} - E[\hat{\Theta}]) + (E[\hat{\Theta}] - \mu)\}^2]$$
$$= \text{Var}[\hat{\Theta}] + (E[\hat{\Theta}] - \mu)^2. \qquad (5.9\text{-}12)$$

If $\hat{\Theta}$ is unbiased then the last term, which is the square of the bias (Definition 5.8-2), is zero. For the case at hand, $\hat{\Theta} = a\hat{\mu}$; thus

$$E[(\hat{\Theta} - \mu)^2] = a^2 \, \text{Var}[\hat{\mu}] + (a\mu - \mu)^2$$
$$= \frac{a^2\sigma^2}{n} + (a - 1)^2\mu^2. \qquad (5.9\text{-}13)$$

To find the MMSE estimator, we differentiate Equation 5.9-13 with respect to a and set to zero. This yields the optimum value of $a = a_0$, that is,

$$a_0 = \frac{\mu^2}{(\sigma^2/n) + \mu^2} = \frac{n}{(\sigma^2/\mu^2) + n}, \qquad (5.9\text{-}14)$$

which is not unity. As $n \to \infty, a_0 \to 1$ and the estimator becomes unbiased but for any finite $n, a_0 \neq 1$. Thus the MMSE estimator is biased.

The question as to why not use the MMSE approach all the time is a valid one. An examination of Equation 5.9-14 shows that we need to know σ^2 and μ^2 or at least their ratio. But these quantities are not generally known and must be estimated, for example, by using Equations 5.8-12 and 5.8-3.

5.10 MAXIMUM LIKELIHOOD ESTIMATORS

In the previous section we furnished estimators for the mean and covariance of random vectors. While these estimators enjoyed desirable properties, they seemed quite arbitrary in that they did not follow from any general principle. In this section, we discuss a somewhat general approach for finding estimators. This approach is called the *maximum likelihood* (ML) principle and the estimators so derived are known as *maximum likelihood estimators* (MLEs). The main drawback to the MLE approach is that the underlying *form* of the pdf of the observed data must be known. The idea behind the MLE approach is illustrated in the following example.

Example 5.10-1

Consider a Bernoulli r.v that has probability mass function (PMF) $P_X(k) = p^k(1-p)^{1-k}$, where $P[X = 1] = p$, and $P[X = 0] = 1 - p$. We would like to estimate the value of p with an estimator, say, $\hat{p}$, that is a function of the observations on X. Suppose we make n observations on X and we call these observations $X_1, X_2, \cdots, X_n$. Then $Y = \sum_{i=1}^{n} X_i$ is the number of times that a *one* was observed in n tries. For example, the experiment might consist of tossing a coin n times and counting the number of times it came up heads, that is, $\{X = 1\}$ when the probability of a head is p. Suppose this number is k_1. The a *priori* probability of observing k_1 heads is given by $P[Y = k_1; p] = \binom{n}{k_1} p^{k_1}(1-p)^{n-k_1}$.

We explicitly show the dependence of the result on p because p is assumed unknown. We now ask what *value of p was most likely to have yielded this result*? Since the term on the right is a continuous function of p, we can obtain this result by a differentiation. Setting the derivative to zero yields

$$\frac{dP[Y = k_1; p]}{dp} = \binom{n}{k_1} p^{k_1-1}(1-p)^{n-k_1-1}[k_1(1-p) - p(n-k_1)] = 0.$$

Thus there are three roots: $p = 0$, $p = 1$, and $p = k_1/n$. The first two roots yield a minimum while $p = k_1/n$ yields a maximum. Thus our estimate for the most likely value of p in this case is k_1/n. Had we performed the experiment a second time and observed k_2 heads, our estimate for p would have been k_2/n. These estimates are realizations of the MLE estimator for p:

$$\hat{p} = \frac{\sum_{i=1}^{n} X_i}{n}. \tag{5.10-1}$$

In the previous example we used the fact that the distribution of $\sum_{i=1}^{n} X_i$ is binomial. Could we have obtained the same result without this knowledge? After all, for some distributions it might be quite a bit of work to compute the distribution of the sum of r.v.'s. The answer is yes and the result is based on generation of the *likelihood function*.

Definition 5.10-1   The likelihood function[†] $L(\theta)$ of the random variables $X_1, X_2, \cdots, X_n$ is the joint pdf $f_{X_1 X_2 \ldots X_n}(x_1, x_2, \cdots, x_n; \theta)$ considered as a function of the unknown parameter θ. In particular if $X_1, X_2, \cdots, X_n$ are independent observations on a random variable X with pdf $f_X(x; \theta)$, then the likelihood function becomes

$$L(\theta) = \prod_{i=1}^{n} f_X(x_i; \theta) \tag{5.10-2}$$

since the $\{X_i\}$ are i.i.d random variables with pdf $f_X(x; \theta)$. If, for a given outcome $\mathbf{x} = (x_1, x_2, \cdots, x_n), \theta^*(x_1, x_2, \cdots, x_n)$ is the value of θ that maximizes $L(\theta)$, then $\theta^*(x_1, x_2, \cdots, x_n)$ is the ML estimate of θ (a number) and $\hat{\theta} = \theta^*(X_1, X_2, \cdots, X_n)$ is the ML estimator (a random variable) for θ. It is not uncommon to define the likelihood function as the random variable $\hat{L}(\theta) \triangleq \prod_{i=1}^{n} f_X(X_i; \theta)$. Then, maximizing with respect to θ, yields the ML estimator $\hat{\theta}(X_1, \cdots, X_n)$ directly.   ■

Example 5.10-2
We consider finding the ML estimation of p in Example 5.10-1 using the likelihood function. If we make n i.i.d observations $X_1, X_2, \cdots, X_n$ on a Bernoulli random variable X, the likelihood function becomes $L(\theta) = \prod_{i=1}^{n} p^{x_i}(1 - p)^{1-x_i} = p^{\sum_{i=1}^{n} x_i}(1 - p)^{n - \sum_{i=1}^{n} x_i}$. By setting $dL(\theta)/d\theta = 0$, we obtain three roots: $p = 0, p = 1$, and $p = \sum_{i=1}^{n} x_i/n$. The first two roots yield a minimum, while the last root yields a maximum. Thus $p^*(\mathbf{x}) = \sum_{i=1}^{n} x_i/n$ and the MLE of p is $\hat{p} = p^*(X_1, X_2, \cdots, X_n) = \sum_{i=1}^{n} X_i/n$.

In many cases the differentiation is more conveniently done on the logarithm of the likelihood function. The *log-likelihood* function is $\log L(\theta)$ (usually the natural log is used) and has its maximum at the same value of θ as that of $L(\theta)$. Another point is that the MLE cannot always be found by differentiation in which case we have to use other methods. Finally, *multiple-parameter ML estimation* can be done by solving simultaneous equations. We illustrate all three points in the next three examples, respectively.

Example 5.10-3
Assume $X: N(\mu, \sigma^2)$ where σ is known. Compute the MLE of the mean μ.

Solution   The likelihood function for n realizations of X is

$$L(\mu) = \left(\frac{1}{\sqrt{2\pi\sigma^2}}\right)^n \exp\left(-\frac{1}{2\sigma^2} \sum_{i=1}^{n} (x_i - \mu)^2\right). \tag{5.10-3}$$

[†]Strictly speaking we should write $L(\theta; x_1, x_2, \cdots, x_n)$ or, as some books have, $L(\theta; X_1, X_2, \cdots, X_n)$. However, we dispense with this awkward notation.

Since the log function is monotonic, the maximum of $L(\mu)$ is also that of $\log L(\mu)$. Hence

$$\log L(\mu) = -\frac{n}{2}\log(2\pi\sigma^2) - \frac{1}{2\sigma^2}\sum_{i=1}^{n}(x_i - \mu)^2$$

and set

$$\frac{\partial \log L(\mu)}{\partial \mu} = 0.$$

This yields

$$\sum_{i=1}^{n}(x_i - \mu) = 0.$$

Thus the value of μ, say μ^*, that maximizes $L(\mu)$ is

$$\mu^* = \frac{1}{n}\sum_{i=1}^{n}x_i,$$

which implies that the MLE of μ should be

$$\hat{\mu} = \frac{1}{n}\sum_{i=1}^{n}X_i. \tag{5.10-4}$$

Thus we see that in the normal case, the MLE of μ can be computed by differentiation and that it turns out to be the sample mean.

Example 5.10-4

Assume X is uniform in $(0, \theta)$, that is,

$$f_\mathbf{X}(x) = \begin{cases} \dfrac{1}{\theta}, & 0 < x \le \theta \\ 0, & x > \theta, \end{cases}$$

and we wish to compute the MLE for θ. Let a particular realization of the n observations $X_1, \ldots, X_n$ be $\mathbf{x} = (x_1, \ldots, x_n)^T$ and let $x_m \triangleq \max(x_1, \ldots, x_n)$. The likelihood function is

$$L(\theta) = \begin{cases} \dfrac{1}{\theta^n}, & x_m \le \theta \\ 0, & \text{otherwise.} \end{cases}$$

Clearly to maximize L we must make the estimate θ^* as small as possible. But θ^* cannot be smaller than x_m. Hence θ^* is x_m and the MLE is

$$\hat{\theta} = \max(X_1, \ldots, X_n). \tag{5.10-5}$$

The probability distribution function (PDF) of $\hat{\theta}$ for $n = 2$ is

$$F_{\hat{\theta}}(\alpha) = F_{X_1}(\alpha)F_{X_2}(\alpha) = F_X^2(\alpha). \tag{5.10-6}$$

We leave the computation of the PDF and pdf of $\hat{\theta}$ for arbitrary n as an exercise for the reader.

Example 5.10-5

Consider the normal pdf

$$f_{\mathbf{X}}(x; \mu, \sigma^2) = \frac{1}{\sqrt{2\pi}\sigma} \exp\left(-\frac{1}{2\sigma^2}(x-\mu)^2\right) \quad -\infty < x < \infty.$$

The log-likelihood function, for n observations, is

$$\tilde{L}(\mu, \sigma) \triangleq \log L = -\frac{n}{2}\log 2\pi - n\log\sigma$$

$$-\frac{1}{2\sigma^2}\sum_{i=1}^{n}(x_i - \mu)^2. \tag{5.10-7}$$

Now set

$$\frac{\partial\tilde{L}}{\partial\mu} = 0 \quad \frac{\partial\tilde{L}}{\partial\sigma} = 0$$

and obtain the simultaneous equations

$$\sum_{i=1}^{n}(x_i - \mu) = 0 \tag{5.10-8}$$

$$-\frac{n}{\sigma} + \frac{1}{\sigma^3}\sum_{i=1}^{n}(x_i - \mu)^2 = 0. \tag{5.10-9}$$

From Equation 5.10-8 we infer that

$$\hat{\mu} = \frac{1}{n}\sum_{i=1}^{n}X_i. \tag{5.10-10}$$

From Equation 5.10-9 we infer that, using the result from Equation 5.10-10

$$\hat{\sigma}^2 = \frac{1}{n}\sum_{i=1}^{n}(X_i - \hat{\mu})^2. \tag{5.10-11}$$

Maximum likelihood estimators have a number of desirable properties including squared-error consistency and *invariance*. Invariance is that property that says that if $\hat{\theta}$ is the MLE for θ, then $h(\hat{\theta})$ is the MLE for $h(\theta)$. However, as seen in Example 5.10-5, (Equation 5.10-11) ML estimators cannot be counted on to be unbiased. We complete this section with an example that illustrates the invariance property.

Example 5.10-6

Consider n observations on a normal r.v. It is known that the mean is zero. The MLE of the variance is $\hat{\sigma}^2 = \frac{1}{n}\sum_{i=1}^{n}X_i^2$. The standard deviation σ is the square root of the variance. Hence the MLE of the standard deviation is the square root of the MLE for the variance, that is, $\hat{\sigma} = \left(\frac{1}{n}\sum_{i=1}^{n}X_i^2\right)^{1/2}$.

5.11 LINEAR ESTIMATION OF VECTOR PARAMETERS

A great many measurement problems in the real world are described by the following model:

$$y(t) = \int_T h(t,\tau)\theta(\tau)d\tau + n(t), \tag{5.11-1}$$

where $y(t)$ is the *observation* or *measurement*, T is the integration set, $\theta(\tau)$ is the unknown *parameter* function, $h(t,\tau)$ is a function that is characteristic of the system and links the parameter function to the measurement but is itself independent of $\theta(\tau)$, and $n(t)$ is the inevitable error in the measurement due to noise. For computational purposes Equation 5.11-1 must be reduced to its discrete form

$$\mathbf{Y} = \mathbf{H}\boldsymbol{\theta} + \mathbf{N}, \tag{5.11-2}$$

where $\mathbf{Y}$ is an $n \times 1$ vector of observations, $\mathbf{H}$ is an $n \times k$ matrix $(n > k)$, $\boldsymbol{\theta}$ is a $k \times 1$ parameter vector, and $\mathbf{N}$ is an $n \times 1$ random vector whose components $N_i, i = 1, \ldots, n$ are the errors or noise associated with the ith observation Y_i. We shall assume without loss of generality that $E[\mathbf{N}] = \mathbf{0}$.[†]

Equation 5.11-2 is known as the *linear model*. We now ask the following question: How do we extract a "good" estimate of $\boldsymbol{\theta}$ from the observed values of $\mathbf{Y}$ if we restrict our estimator $\hat{\boldsymbol{\Theta}}$ be a linear function of $\mathbf{Y}$? By a linear function we mean

$$\hat{\boldsymbol{\Theta}} = \mathbf{B}\mathbf{Y}, \tag{5.11-3}$$

where $\mathbf{B}$, which *does not* depend on $\mathbf{Y}$, is to be determined. The problem posed here is of great practical significance. It is one of the most fundamental problems in parameter estimation theory and covered in great detail in numerous books, for example, Kendall and Stuart [5-8] and Lewis and Odell [5-9]. It also is an immediate application of the probability theory of random vectors and is useful for understanding various topics in the subsequent chapters.

Before computing the matrix $\mathbf{B}$ in Equation 5.11-3, we must first furnish some results from matrix calculus.

Derivative of a scalar function of a vector. Let $q(\mathbf{x})$ be a scalar function of the vector $\mathbf{x} = (x_1, \ldots, x_n)^T$. Then

$$\frac{dq(\mathbf{x})}{d\mathbf{x}} \triangleq \left(\frac{\partial q}{\partial x_1}, \ldots, \frac{\partial q}{\partial x_n} \right)^T. \tag{5.11-4}$$

Thus the derivative of $q(\mathbf{x})$ with respect to $\mathbf{x}$ is a *column vector* whose ith component is the partial derivative of $q(\mathbf{x})$ with respect to x_i.

[†]The symbol $\mathbf{0}$ here stands for the zero vector, that is, the vector whose components are all zero.

Derivative of quadratic forms. Let $\mathbf{A}$ be a real-symmetric $n \times n$ matrix and let $\mathbf{x}$ be an arbitrary n-vector. Then the derivative of the quadratic form

$$q(\mathbf{x}) \stackrel{\Delta}{=} \mathbf{x}^T \mathbf{A} \mathbf{x}$$

with respect to $\mathbf{x}$ is

$$\frac{dq(\mathbf{x})}{d\mathbf{x}} = 2\mathbf{A}\mathbf{x}. \tag{5.11-5}$$

The proof of Equation 5.11-5 is obtained by writing

$$q(\mathbf{x}) = \sum_{i=1}^{n} \sum_{j=1}^{n} x_i a_{ij} x_j$$

$$= \sum_{i=1}^{n} x_i^2 a_{ii} + \sum_{i \neq j}^{n} \sum^{n} a_{ij} x_i x_j.$$

Hence

$$\frac{\partial q(\mathbf{x})}{\partial x_k} = 2 x_k a_{kk} + 2 \sum_{i \neq k} a_{ki} x_i$$

$$= 2 \sum_{i=1}^{n} a_{ki} x_i$$

or

$$\frac{dq(\mathbf{x})}{d\mathbf{x}} = 2\mathbf{A}\mathbf{x}. \tag{5.11-6}$$

Derivative of scalar products. Let $\mathbf{a}$ and $\mathbf{x}$ be two n-vectors. Then with $y = \mathbf{a}^T \mathbf{x}$, we obtain

$$\frac{dy}{d\mathbf{x}} = \mathbf{a}. \tag{5.11-7}$$

Let $\mathbf{x}$, $\mathbf{y}$, and $\mathbf{A}$ be two n-vectors and an $n \times n$ matrix, respectively. Then with $q \stackrel{\Delta}{=} \mathbf{y}^T \mathbf{A} \mathbf{x}$,

$$\frac{\partial q}{\partial \mathbf{x}} = \mathbf{A}^T \mathbf{y}. \tag{5.11-8}$$

We return now to Equation 5.11-2:

$$\mathbf{Y} = \mathbf{H}\boldsymbol{\theta} + \mathbf{N}$$

and assume that (recall $E[\mathbf{N}] = \mathbf{0}$)

$$\mathbf{K} \stackrel{\Delta}{=} E[\mathbf{N}\mathbf{N}^T] = \sigma^2 \mathbf{I} \tag{5.11-9}$$

where $\mathbf{I}$ is the identity matrix. Equation 5.11-9 is equivalent to stating that the measurement errors N_i, that is, $i = 1, \ldots, n$ are uncorrelated, and their variances are the same and equal to σ^2. This situation is sometimes called *white* noise.

A reasonable choice for estimating $\boldsymbol{\theta}$ is to find a $\hat{\boldsymbol{\Theta}}$ that *minimizes* the sum squares S defined by

$$S \triangleq (\mathbf{Y} - \mathbf{H}\hat{\boldsymbol{\Theta}})^T (\mathbf{Y} - \mathbf{H}\hat{\boldsymbol{\Theta}}) \triangleq \|\mathbf{Y} - \mathbf{H}\hat{\boldsymbol{\Theta}}\|^2. \tag{5.11-10}$$

Note that by finding $\hat{\boldsymbol{\Theta}}$ that best fits the measurement $\mathbf{Y}$ in the sense of minimizing $\|\mathbf{Y} - \mathbf{H}\hat{\boldsymbol{\Theta}}\|^2$, we are realizing what is commonly called a *least-squares* fit to the data. For this reason finding $\hat{\boldsymbol{\Theta}}$ that minimizes S in Equation 5.11-10 is called the least-squares (LS) method. To find the minimum of S with respect to $\hat{\boldsymbol{\Theta}}$, write

$$S = \mathbf{Y}^T \mathbf{Y} + \hat{\boldsymbol{\Theta}}^T \mathbf{H}^T \mathbf{H} \hat{\boldsymbol{\Theta}} - \hat{\boldsymbol{\Theta}}^T \mathbf{H}^T \mathbf{Y} - \mathbf{Y}^T \mathbf{H} \hat{\boldsymbol{\Theta}}$$

and compute (use Equation 5.11-4 on the LHS and Equations 5.11-5 and 5.11-8 on the RHS)

$$\frac{\partial S}{\partial \hat{\boldsymbol{\Theta}}} = 0 = 2[\mathbf{H}^T \mathbf{H}]\hat{\boldsymbol{\Theta}} - 2\mathbf{H}^T \mathbf{Y},$$

whence (assuming $\mathbf{H}^T \mathbf{H}$ has an inverse)

$$\hat{\boldsymbol{\Theta}}_{LS} = (\mathbf{H}^T \mathbf{H})^{-1} \mathbf{H}^T \mathbf{Y}. \tag{5.11-11}$$

Comparing our result with Equation 5.11-3 we see that $\mathbf{B}$ in Equation 5.11-3 is given by $\mathbf{B}_0 \triangleq (\mathbf{H}^T \mathbf{H})^{-1} \mathbf{H}^T$ in the LS method. Equation 5.11-11 is the LS estimator of $\boldsymbol{\theta}$ based on the measurement $\mathbf{Y}$.

The astute reader will have noticed that we never involved the fact that $\mathbf{K} = \sigma^2 \mathbf{I}$. Indeed, in arriving at Equation 5.11-11 we essentially treated $\mathbf{Y}$ as deterministic and merely obtained $\hat{\boldsymbol{\Theta}}_{LS}$ as the *generalized inverse* (see Lewis and Odell [5-9, p. 6]) of the system of equations $\mathbf{Y} = \mathbf{H}\boldsymbol{\theta}$. As it stands, the estimator $\hat{\boldsymbol{\Theta}}_{LS}$ given in Equation 5.11-11 has no claim to being optimum. However, when the covariance of the noise $\mathbf{N}$ is as in Equation 5.11-9, then $\hat{\boldsymbol{\Theta}}_{LS}$ does indeed have optimal properties in an important sense. We leave it to the reader to show that $\hat{\boldsymbol{\Theta}}_{LS}$ is unbiased and is a minimum variance estimator.

Example 5.11-1

We are given the following data:

$$6.2 = 3\theta + n_1$$
$$7.8 = 4\theta + n_2$$
$$2.2 = \theta + n_3.$$

Find the LS estimate of θ.

Solution   The data can be put in the form

$$\mathbf{y} = \mathbf{H}\boldsymbol{\theta} + \mathbf{n},$$

where $\mathbf{y} = (6.2, 7.8, 2.2)^T$ is a realization of $\mathbf{Y}$, $\mathbf{H} = \text{diag}(3, 4, 1)$ and $\mathbf{n} = (n_1, n_2, n_3)^T$ is a realization of $\mathbf{N}$. Hence $\mathbf{H}^T\mathbf{H} = \sum H_i^2 = 26$ and $\mathbf{H}^T\mathbf{y} = \sum_{i=1}^{3} H_i y_i = 52$. Thus

$$
\hat{\theta}_{LS} = (\mathbf{H}^T\mathbf{H})^{-1}\mathbf{H}^T\mathbf{y} = \frac{\displaystyle\sum_{i=1}^{3} H_i y_i}{\displaystyle\sum_{i=1}^{3} H_i^2} = \frac{52}{26} = 2.
$$

Example 5.11-2 _____

(Reference [5-8], p. 77.) Let $\boldsymbol{\theta} = (\theta_1, \theta_2)^T$ be a two-component parameter vector to be estimated, and let $\mathbf{H}$ be a $n \times 2$ matrix of coefficients partitioned into column vectors as $\mathbf{H} = (\mathbf{H}_1\mathbf{H}_2)$ where $\mathbf{H}_i, i = 1, 2$ is an n-vector. Then with the n-vector $\mathbf{Y}$ representing the observation data, the linear model assumes the form

$$
\mathbf{Y} = (\mathbf{H}_1\mathbf{H}_2)\boldsymbol{\theta} + \mathbf{N}
$$

and the LS estimator of $\boldsymbol{\theta}$ is

$$
\hat{\boldsymbol{\Theta}}_{LS} = \begin{bmatrix} \mathbf{H}_1^T\mathbf{H}_1 & \mathbf{H}_1^T\mathbf{H}_2 \\ \mathbf{H}_2^T\mathbf{H}_1 & \mathbf{H}_2^T\mathbf{H}_2 \end{bmatrix}^{-1} \begin{bmatrix} \mathbf{H}_1^T\mathbf{Y} \\ \mathbf{H}_2^T\mathbf{Y} \end{bmatrix}.
$$

5.12 SUMMARY

In this chapter we studied the calculus of multiple random variables. We found it convenient to organize multiple random variables into random vectors and treat these as single entities. Because in practice it is often difficult to describe the joint probability law of n random variables, we argued that in the case of random vectors we often settle for a less complete but more available characterization than that furnished by the pdf. We focused on the characterizations furnished by the lower order moments, especially the mean and covariance. In particular, because of the great importance of covariances matrices in signal processing, communication theory, pattern recognition, multiple regression analysis, and other areas of engineering and science, we made use of numerous results from matrix theory and linear algebra to reveal the properties of these matrices. To illustrate the usefulness of the mathematical techniques introduced in this chapter (the techniques generally requiring somewhat greater mathematical exposure than was required for Chapters 1 to 4) we solved a well-known problem in the field of applied probability called statistical pattern recognition (Example 5.5-3).

We discussed the multidimensional Gaussian (Normal) law and characteristic functions of random vectors. We demonstrated that under linear transformations Gaussian random vectors map into Gaussian random vectors. We showed how to derive a transformation that can convert correlated random variables into uncorrelated ones. The characteristic function (c.f.) of random vectors in general was defined and shown to be useful in computing moments and solving problems involving the sums of independent random variables; these assertions were illustrated with examples. Finally, using vector and matrix techniques we derived the c.f. for the Gaussian random vector and showed that it too had a Gaussian shape.

The last part of the chapter dealt with estimators, their properties, and realizations. Emphasis was on the maximum likelihood (ML) method for finding estimators with desirable properties. A number of worked-out examples showed that maximum likelihood estimators could often, but not always, be found by differentiation. A topic not covered here but in Chapter 9 deals with finding ML estimators when the observations from which the estimators are formed are incomplete in some sense. Likewise, when the functional relations between the parameters of the observations and the parameters of the original data are complicated, a direct approach to finding the ML estimators, as presented here, may not be feasible. In both these cases, an iterative technique known as the *expectation-maximization algorithm* (E-M algorithm) is useful. However, without the introductory material presented here, the E-M algorithm might be more difficult to understand by readers not familiar with estimation methods.

PROBLEMS

5.1. Let $f_{\mathbf{X}}(\mathbf{x})$ be given as

$$f_{\mathbf{X}}(\mathbf{x}) = Ke^{-\mathbf{x}^T \mathbf{\Lambda}}u(\mathbf{x}),$$

where $\mathbf{\Lambda} = (\lambda_1, \ldots, \lambda_n)^T$ with $\lambda_i > 0$ all i, $\mathbf{x} = (x_1, \ldots, x_n)^T$, $u(\mathbf{x}) = 1$ if $x_i \geq 0$ $i = 1, \ldots, n$ and zero otherwise, and K is a constant to be determined. What value of K will enable $f_{\mathbf{X}}(\mathbf{x})$ to be a pdf?

5.2. Let $B_i, i = 1, \ldots, n$ be n disjoint and exhaustive events. Show that the probability distribution function (PDF) of $\mathbf{X}$ can be written as

$$F_{\mathbf{X}}(\mathbf{x}) = \sum_{i=1}^{n} F_{\mathbf{X}|B_i}(\mathbf{x}|B_i)P[B_i].$$

5.3. For $-\infty < x_i < \infty$, $i = 1, 2, \ldots, n$, let

$$f_{\mathbf{X}}(\mathbf{x}) = \frac{1}{(2\pi)^{n/2}\sigma_1 \ldots \sigma_n} \exp\left\{-\frac{1}{2}\left(\sum_{i=1}^{n}\left(\frac{x_i}{\sigma_i}\right)^2\right)\right\}.$$

Show that all the marginal pdf's are Gaussian.

5.4. Show that any matrix $\mathbf{M}$ generated by an outer product of two vectors, that is, $\mathbf{M} = \mathbf{X}\mathbf{X}^T$ has rank at most unity. Explain why $\mathbf{R} \triangleq E[\mathbf{X}\mathbf{X}^T]$ can be of full rank.

5.5. Show that the two r.v.'s X_1 and X_2 with joint pdf

$$f_{X_1 X_2}(x_1, x_2) = \begin{cases} \frac{1}{16} & |x_1| < 4, \quad 2 < x_2 < 4 \\ 0 & \text{otherwise} \end{cases}$$

are independent and orthogonal.

5.6. Let $\mathbf{X}_i, i = 1, \ldots, n$ be n mutually orthogonal random vectors. Show that

$$E\left[\left\|\sum_{i=1}^{n} \mathbf{X}_i\right\|^2\right] = \sum_{i=1}^{n} E\left[\|\mathbf{X}_i\|^2\right].$$

(*Hint*: Use the definition $\|\mathbf{X}\|^2 \triangleq \mathbf{X}^T\mathbf{X}$.)

5.7. Let $\mathbf{X}_i, i = 1, \ldots, n$ be n mutually uncorrelated random vectors with means $\boldsymbol{\mu}_i \triangleq E[\mathbf{X}_i]$. Show that

$$E\left[\left\|\sum_{i=1}^{n} (\mathbf{X}_i - \boldsymbol{\mu}_i)\right\|^2\right] = \sum_{i=1}^{n} E\left[\|\mathbf{X}_i - \boldsymbol{\mu}_i\|^2\right].$$

5.8. Let $\mathbf{X}_i, i = 1, \ldots, n$ be n mutually uncorrelated random vectors with $E[\mathbf{X}_i] = \boldsymbol{\mu}_i$, $i = 1, \ldots, n$. Show that

$$E\left[\sum_{i=1}^{n} (\mathbf{X}_i - \boldsymbol{\mu}_i) \sum_{j=1}^{n} (\mathbf{X}_j - \boldsymbol{\mu}_j)^T\right] = \sum_{i=1}^{n} \mathbf{K}_i,$$

where $\mathbf{K}_i \triangleq E[(\mathbf{X}_i - \boldsymbol{\mu}_i)(\mathbf{X}_i - \boldsymbol{\mu}_i)^T]$.

5.9. Explain why none of the following matrices can be covariance matrices associated with real random vectors.

$$\begin{bmatrix} 2 & -4 & 0 \\ -4 & 3 & 1 \\ 0 & 1 & 2 \end{bmatrix} \quad \begin{bmatrix} 4 & 0 & 0 \\ 0 & 6 & 0 \\ 0 & 0 & -2 \end{bmatrix} \quad \begin{bmatrix} 6 & 1+j & 2 \\ 1-j & 5 & -1 \\ 2 & -1 & 6 \end{bmatrix} \quad \begin{bmatrix} 4 & 6 & 2 \\ 6 & 9 & 3 \\ 9 & 12 & 16 \end{bmatrix}$$

$$\text{(a)} \qquad\qquad \text{(b)} \qquad\qquad\quad \text{(c)} \qquad\qquad\quad \text{(d)}$$

5.10. (a) Let a vector $\mathbf{X}$ have $E[\mathbf{X}] = \mathbf{0}$ with covariance $\mathbf{K_{XX}}$ given by

$$\mathbf{K_{XX}} = \begin{bmatrix} 3 & \sqrt{2} \\ \sqrt{2} & 4 \end{bmatrix}.$$

Find a linear transformation $\mathbf{C}$ such that $\mathbf{Y} = \mathbf{CX}$ will have

$$\mathbf{K_{YY}} = \begin{bmatrix} 1 & 0 \\ 0 & 1 \end{bmatrix}.$$

Is $\mathbf{C}$ a unitary transformation?

(b) Consider the two real symmetric matrices $\mathbf{A}$ and $\mathbf{A}'$ given by

$$\mathbf{A} \triangleq \begin{bmatrix} a & b \\ b & c \end{bmatrix} \qquad \mathbf{A}' \triangleq \begin{bmatrix} a' & b' \\ b' & c' \end{bmatrix}.$$

Show that when $a = c$ and $a' = c'$, the product $\mathbf{AA}'$ is real symmetric. More generally, show that if $\mathbf{A}$ and $\mathbf{A}'$ are any real symmetric matrices then $\mathbf{AA}'$ will be symmetric if $\mathbf{AA}' = \mathbf{A}'\mathbf{A}$.

5.11. (K. Fukunaga [5-6, p. 33].) Let $\mathbf{K}_1$ and $\mathbf{K}_2$ be positive definite covariance matrices and form

$$\mathbf{K} = a_1\mathbf{K}_1 + a_2\mathbf{K}_2 \qquad \text{where } a_1, a_2 > 0.$$

Let $\mathbf{A}$ be a transformation that achieves

$$\mathbf{A}^T\mathbf{K}\mathbf{A} = \mathbf{I} \qquad \mathbf{A}^T\mathbf{K}_1\mathbf{A} = \mathbf{\Lambda}^{(1)} = \operatorname{diag}(\lambda_1^{(1)}, \ldots, \lambda_n^{(1)}).$$

(a) Show that $\mathbf{A}$ satisfies

$$\mathbf{K}^{-1}\mathbf{K}_1\mathbf{A} = \mathbf{A}\mathbf{\Lambda}^{(1)}.$$

(b) Show that $\mathbf{A}^T\mathbf{K}_2\mathbf{A} \triangleq \mathbf{\Lambda}^{(2)}$ is also diagonal, that is, $\mathbf{\Lambda}^{(2)} \triangleq \operatorname{diag}(\lambda_1^{(2)}, \ldots, \lambda_n^{(2)})$.

(c) Show that $\mathbf{A}^T\mathbf{K}_1\mathbf{A}$ and $\mathbf{A}^T\mathbf{K}_2\mathbf{A}$ share the same eigenvectors.

(d) Show that the eigenvalues of $\mathbf{\Lambda}^{(2)}$ are related to the eigenvalues of $\mathbf{\Lambda}^{(1)}$ as

$$\lambda_i^{(2)} = \frac{1}{a_2}[1 - a_1\lambda_i^{(1)}]$$

and therefore are in inverse order from those of $\mathbf{\Lambda}^{(1)}$.

5.12. (J. A. McLaughlin [5-7].) Consider the m vectors $\mathbf{X}_i = (X_{i1}, \ldots, X_{in})^T$, $i = 1, \ldots, m$ where $n > m$. Consider the $n \times n$ matrix $\mathbf{S} = \frac{1}{m}\sum_{i=1}^{m}\mathbf{X}_i\mathbf{X}_i^T$.

(a) Show that with $\mathbf{W} \triangleq (\mathbf{X}_1 \ldots \mathbf{X}_m)$, $\mathbf{S}$ can be written as

$$\mathbf{S} = \frac{1}{m}\mathbf{W}\mathbf{W}^T.$$

(b) What is the maximum rank of $\mathbf{S}$?

(c) Let $\mathbf{S}' \triangleq \frac{1}{m}\mathbf{W}^T\mathbf{W}$. What is the size of $\mathbf{S}'$? Show that the first m nonzero eigenvalues of $\mathbf{S}$ can be computed from

$$\mathbf{S}'\mathbf{\Phi} = \mathbf{\Phi}\mathbf{\Lambda},$$

where $\mathbf{\Phi}$ is the eigenvector matrix of $\mathbf{S}'$ and $\mathbf{\Lambda}$ is the matrix of eigenvalues. What are the relations between the eigenvectors and eigenvalues of $\mathbf{S}$ and $\mathbf{S}'$?

(d) What is the advantage of computing the eigenvectors from $\mathbf{S}'$ rather than $\mathbf{S}$?

5.13. (a) Let $\mathbf{K}$ be an $n \times n$ covariance matrix and let $\Delta\mathbf{K}$ be a real symmetric perturbation matrix. Let λ_i, $i = 1, \ldots, n$ be the eigenvalues of $\mathbf{K}$ and ϕ_i the associated eigenvectors. Show that the first-order approximation to the eigenvalues λ_i' of $\mathbf{K} + \Delta\mathbf{K}$ yields

$$\lambda_i' = \phi_i^T(\mathbf{K} + \Delta\mathbf{K})\phi_i, \qquad i = 1, \ldots, n.$$

(b) Show that the first-order approximation to the eigenvectors is given by

$$\Delta\phi_i = \sum_{j=1}^{n} b_{ij}\phi_j,$$

where $b_{ij} = \phi_j^T \Delta\mathbf{K}\phi_i/(\lambda_i - \lambda_j)$ $i \neq j$ and $b_{ii} = 0$.

5.14. (a) Let $\lambda_1 \geq \lambda_2 \geq \ldots \geq \lambda_n$ be the eigenvalues of a real symmetric matrix $\mathbf{M}$. For $i \geq 2$, let $\phi_1, \phi_2, \ldots, \phi_{i-1}$ be mutually orthogonal unit eigenvectors belonging to $\lambda_1, \ldots, \lambda_{i-1}$. Prove that the *maximum value* of $\mathbf{u}^T\mathbf{M}\mathbf{u}$ subject to $\|\mathbf{u}\| = 1$ and $\mathbf{u}^T\phi_1 = \ldots = \mathbf{u}^T\phi_{i-1} = 0$ is λ_i, that is, $\lambda_i = \max(\mathbf{u}^T\mathbf{M}\mathbf{u})$.

(b) Prove Theorem 5.5-3 using Theorem 5.5-2.

5.15. Let $\mathbf{X} = (X_1, X_2, X_3)^T$ be a random vector with $\mu \triangleq E[\mathbf{X}]$ given by

$$\mu = (5, -5, 6)^T$$

and covariance given by

$$\mathbf{K} = \begin{bmatrix} 5 & 2 & -1 \\ 2 & 5 & 0 \\ -1 & 0 & 4 \end{bmatrix}.$$

Calculate the mean and variance of

$$Y = \mathbf{A}^T\mathbf{X} + B,$$

where $\mathbf{A} = (2, -1, 2)^T$ and $B = 5$.

5.16. Two jointly normal r.v.'s X_1 and X_2 have joint pdf given by

$$f_{X_1 X_2}(x_1, x_2) = \frac{2}{\pi\sqrt{7}} \exp[-\tfrac{8}{7}(x_1^2 + \tfrac{3}{2}x_1 x_2 + x_2^2)].$$

Find a nontrivial transformation $\mathbf{A}$ in

$$\begin{pmatrix} Y_1 \\ Y_2 \end{pmatrix} = \mathbf{A} \begin{pmatrix} X_1 \\ X_2 \end{pmatrix}$$

such that Y_1 and Y_2 are independent. Compute the joint pdf of Y_1, Y_2.

5.17. Show that if $\mathbf{X} = (X_1, \ldots, X_n)^T$ has mean $\mu = (\mu_1, \ldots, \mu_n)^T$ and covariance

$$\mathbf{K} = \{K_{ij}\}_{n\times n},$$

then the scalar r.v. Y given by

$$Y \triangleq p_1 X_1 + \ldots + p_n X_n$$

has mean

$$E[Y] = \sum_{i=1}^{n} p_i \mu_i$$

and variance

$$\sigma_Y^2 = \sum_{i=1}^{n}\sum_{j=1}^{n} p_i p_j K_{ij}.$$

5.18. Compute the joint characteristic function of $\mathbf{X} = (X_1, \ldots, X_n)^T$, where the $X_i, i = 1, \ldots, n$ are mutually independent and identically distributed Cauchy r.v.'s, that is,

$$f_{X_i}(x) = \frac{\alpha}{\pi(x^2 + \alpha^2)}.$$

Use this result to compute the pdf of $Y = \sum_{i=1}^n X_i$.

5.19. Compute the joint characteristic function of $\mathbf{X} = (X_1, \ldots, X_n)^T$, where the X_i, $i = 1, \ldots, n$ are mutually independent and identically distributed binomial r.v.'s. Use this result to compute the PMF of $Y = \sum_{i=1}^n X_i$.

5.20. Let $\mathbf{X} = (X_1, \ldots, X_4)$ be a Gaussian random vector with $E[\mathbf{X}] = 0$. Show that

$$E[X_1 X_2 X_3 X_4] = K_{12}K_{34} + K_{13}K_{24} + K_{14}K_{23},$$

where the K_{ij} are elements of the covariance matrix $\mathbf{K} = \{K_{ij}\}_{4\times 4}$ of $\mathbf{X}$.

5.21. Let the joint pdf of X_1, X_2, X_3 be given by $f_{\mathbf{X}}(x_1, x_2, x_3) = 2/3 \cdot (x_1 + x_2 + x_3)$ over the region $S = \{(x_1, x_2, x_3) : 0 < x_i \leq 1, i = 1, 2, 3\}$ and zero elsewhere. Compute the covariance matrix and show that the random variables X_1, X_2, X_3, although not independent, are essentially uncorrelated.

5.22. Let X_1, X_2 be jointly Normal, zero-mean random, variables with covariance matrix

$$\mathbf{K} = \begin{bmatrix} 2 & -1.5 \\ -1.5 & 2 \end{bmatrix}.$$

Find a whitening transformation for $\mathbf{X} = (X_1 X_2)^T$. Write a MATLAB program to show a scatter diagram, that is, x_2 versus x_1 where the latter are realizations of X_2, X_1, respectively. Do this for the whitened variables as well. Choose between a hundred and a thousand realizations.

5.23. (Linear Transformations) Let $Y_k = \sum_{j=1}^n a_{kj}X_j, k = 1, \ldots, n$, where the a_{kj} are real constants, the matrix $\mathbf{A} = [a_{ij}]_{N \times N}$ is nonsingular, and the $\{X_j\}$ are random variables. Let $\mathbf{B} = \mathbf{A}^{-1}$. Show that the pdf of $\mathbf{Y}$, $f_{\mathbf{Y}}(y_1, \ldots, y_n)$ is given by

$$f_{\mathbf{Y}}(y_1, \ldots, y_n) = |\det \mathbf{B}| f_{\mathbf{X}}(x_1^*, \ldots, x_n^*), \quad \text{where } x_i^* = \sum_{k=1}^n b_{ik}y_k \text{ for } i = 1, \ldots, n.$$

5.24. (Auxiliary Variables) Let $Y_1 = \sum_{i=1}^n X_i$ and $Y_2 = \sum_{i=2}^n X_i$. Compute the joint pdf, $f_{Y_1 Y_2}(y_1, y_2)$, by introducing the auxiliary variables $Y_k = \sum_{i=k}^n X_i$, $k = 3, \ldots, n$ and integrating over the range of each auxiliary r.v. Show the $f_{\mathbf{Y}}(y_1, \ldots, y_n) = f_{\mathbf{X}}(y_1 - y_2, \ldots, y_{n-1} - y_n, y_n)$. (This problem and the previous are adapted from Example 4.9, p. 190, in *Probability and Stochastic Processes for Engineers*, C.W. Helstrom, Macmillan, 1984).

5.25. Show that the mean estimator enjoys the property of squared-error consistency.

5.26. Compute the MLE for the parameter λ in the exponential pdf. Show that the likelihood function is indeed a maximum at the MLE value.

5.27. Compute the MLE for the parameter p (the probability of a success) in the binomial pdf.

5.28. Compute the MLE for the parameters $0 < a < b$ in the uniform pdf. What is the pdf of the MLE for a?

REFERENCES

5-1. W. B. Davenport, Jr., *Probability and Random Processes*. New York: McGraw-Hill, 1970.

5-2. J. N. Franklin, *Matrix Theory*. Upper Saddle River, N.J.: Prentice Hall, 1968.

5-3. H. Stark and R. O'Toole, "Statistical Pattern Recognition Using Optical Fourier Transform Features," Chapter 11 in *Applications of Optical Fourier Transforms*, H. Stark, ed. New York: Academic Press, 1982.

5-4. R. O. Duda and P. E. Hart, *Pattern Classification and Scene Analysis*. New York: John Wiley, 1973.

5-5. K. S. Miller, *Multidimensional Gaussian Distributions*. New York: John Wiley, 1964.

5-6. K. Fukunaga, *Introduction to Statistical Pattern Recognition*. 2nd edition. New York: Academic, 1990.

5-7. J. A. McLaughlin and J. Raviv, "Nth Order Autocorrelations in Pattern Recognition," *Information and Control*, 12, pp. 121–142, Chapter 2, 1968.

5-8. M. G. Kendall and A. Stuart, *The Advanced Theory of Statistics*, Vol. 2, 3rd Ed. London, England: Charles Griffin and Co., 1951.

5-9. T. O. Lewis and P. L. Odell, *Estimation in Linear Models*, Upper Saddle River, N.J.: Prentice Hall, 1971.

ADDITIONAL READING

S. M. Kay, *Fundamentals of Statistical Signal Processing*, Upper Saddle River, N.J.: Prentice Hall, 1993.

G. Straud, *Linear Algebra and its Applications*, 3rd edition. New York: Saunders, 1988.

6

Random Sequences

Random sequences are used as models of sampled data arising in signal and image processing, digital control, and communications. They also arise as nonsampled data such as economic variables, the content of a register in a digital computer, something as simple as coin flipping (Bernoulli trials), or the number of packets on a link in a computer network. In each case, the random sequence models the unpredictable behavior of these sources from the user's perspective. In this chapter we will study the random sequence and some of its important properties. As we will see, a random (stochastic) sequence can be thought of as an infinite dimensional vector of random variables. As such it stands between finite dimensional random vectors (cf. Chapter 5) and continuous-time random functions, called random processes, to be studied in the next chapter.

Another way to generalize the random vector is by doubling the number of index parameters to two, thereby creating random matrices, which have been found useful as mathematical models in image processing. When these random matrices grow in size, in the infinite limit we have a two-dimensional random sequence, used in many theoretical studies in image and geophysical signal processing. While we will not study image processing here, many of the basic concepts of random sequences carry over to the two-dimensional case. Three- and four-dimensional random sequences have been found useful models of unpredictable aspects in video and other spatiotemporal signals.

6.1 BASIC CONCEPTS

In the course of developing this material we will have need to review and extend some of the basic material presented in Chapter 1 on the axioms of probability. This is because we must now routinely deal with an infinite number of random variables at one time, i.e., a random sequence. We start out this study by offering a definition of the random sequence followed by a few simple examples.

Definition 6.1-1   Let $(\Omega, \mathscr{F}, P)$ be a probability space. Let $\zeta \in \Omega$. Let $X[n, \zeta]$ be a mapping of the sample space Ω into a space of complex-valued sequences on some index set Z. If, for each fixed integer $n \in Z$, $X[n, \zeta]$ is a random variable, then $X[n, \zeta]$ is a *random (stochastic) sequence*. The index set Z is usually *all* the integers, $-\infty < n < +\infty$, but can be just a subset of the integers.   ■

See Figure 6.1-1 for an illustration for sample space $\Omega = \{1, \ldots, 10\}$. We see that $X[n, \zeta]$ for a *fixed* outcome ζ is an ordinary sequence of numbers, that is, a deterministic (nonrandom) function of the discrete parameter n. We often refer to these ordinary sequences as *realizations* of the random sequence, or as *sample sequences* and denote them by $X_\zeta[n]$ or merely by $x[n]$ when there is no confusion. Thus 10 sample sequences are plotted in Figure 6.1-1, one for each outcome $\zeta \in \Omega$. On the other hand, for n *fixed* and ζ variable, $X[n, \zeta]$ is a random variable.[†] Thus the *collection* of all these realizations, $-\infty < n < +\infty$, along with the probability space, *is* the random sequence. We shall often, but not always, denote the random sequence by just $X[n]$. We retain the notation $X[n, \zeta]$ when its use helps to clarify a point on the outcomes ζ of the underlying sample space Ω.

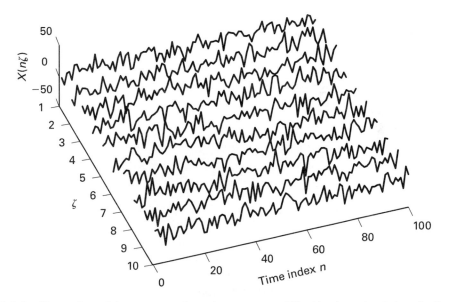

Figure 6.1-1   Illustration of the concept of *random sequence* $X(n, \zeta)$ where the ζ domain (i.e., the sample space Ω) consists of just 10 values. (Samples connected for plot.)

[†]Elementary probability books talk about a sequence of random variables, denoted $X_n(\zeta)$. Usually they are independent. Our random sequence is the same thing, but with the independence assumption lifted.

We give the following simple examples:

Example 6.1-1 _____

(Separable random sequence.) Let $X[n, \zeta] \triangleq X(\zeta)f[n]$ where $X(\zeta)$ is a random variable and $f[n]$ is a given deterministic (ordinary) sequence. Such a random sequence is the separable product of a random variable (function) and an ordinary sequence. We will also write $X[n] = Xf[n]$, suppressing the ζ variable, as is the custom for random variables.

Example 6.1-2 _____

(Sinusoid with random amplitude and phase.) Let $X[n, \zeta] \triangleq A(\zeta)\sin(\pi n/10 + \Theta(\zeta))$, where A and Θ are random variables, alternately written $X[n] = A\sin(\pi n/10 + \Theta)$.

These two simple random sequences are made from deterministic components, but they are also "deterministic" in another way. They have the unusual property that their future values are exactly determined from their present and past values. In Example 6.1-1, once we observe $X[n]$ at any fixed value of n, say $n = 0$, then, since the ordinary sequence $f[n]$ is assumed to be known and nonrandom, all of the random sequence $X[n]$ becomes known. We see that the random sequence $X[n]$ is *conditionally known* given its value at $n = 0$. The situation in Example 6.1-2 is just slightly more complicated but the same approach suffices to show that given two (nondegenerate) observations, say at $n = 0$ and $n = 5$, one can determine the values taken on by the random variables A and Θ; then the sequence $X[n]$ becomes conditionally known or perfectly predictable given these observations at $n = 0$ and $n = 5$. These deterministic random sequences would not be good models for noise on a communications channel because real noise is not so easily foiled.

    In the next example we see how a more general but still "deterministic" random sequence can be made out of a random vector.

Example 6.1-3 _____

(Random sequence with finite support.) Let $X[n, \zeta]$ be given by

$$X[n, \zeta] \triangleq \begin{cases} X_n(\zeta) & 1 \le n \le N \\ 0 & \text{else.} \end{cases}$$

Since $X[n] = 0$ except for $n \in [1, N]$, we say $X[n]$ has *finite support*. Because of this finite support property, we can model this random sequence by a random vector $\mathbf{X} = (X_1, X_2, \ldots, X_N)^T$ and then use the rich calculus of matrix algebra, e.g., covariance matrices and linear transformations. Many random sequences can be approximated this way, although note that we would have to consider the limiting behavior of $\mathbf{X}$ to model a general random process.

Example 6.1-4 _____

(Tree diagram for random sequence.) Let the random sequence $X[n]$ be defined over $n \ge 0$, and take on only M discrete values, $0, 1, 2, \ldots, M - 1$. Further assume the starting value is pinned at $X[0] = 0$. Then we can illustrate the evolution of the sample functions of this

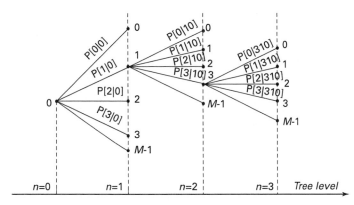

Figure 6.1-2   Tree diagram for discrete amplitude random sequence.

random sequence with a tree diagram, with branching factor M at each node $n = 0, 1, 2, \ldots$ as illustrated in Figure 6.1-2.

At each level n, of the tree, the node values give possible sample function values $x[n]$, with branch index $i = 0, \ldots, M - 1$. The sample functions are identified by the sequence of node values of a path through the tree starting from the root node $n = 0$. If we identify the path string $i_1 i_2 i_3 \ldots$ with the base-M number $0.i_1 i_2 i_3 \ldots$, we can call this point the outcome $\zeta \in [0, 1] = \Omega$, the sample space.[†] Finally we can label the branches with the conditional probability $P[X[n] = m_i | \{X[k] \text{ on same path for } k \leq n - 1\}]$, which in Figure 6.1-2 is denoted as $P[i_n | i_{n-1} i_{n-2} \ldots i_1 0]$. Then the probability of any node value at tree level n is just given by the product of all the probability branch labels back to the root node along this path. Note that all sample sequences that agree up to time n, will correspond to a neighborhood in the sample space $\Omega = [0, 1]$ of radius $\frac{1}{2} M^{-n}$.

This example also has shown how to construct a consistent underlying sample space in the common case where we are given just the probability distribution information about the set of random variables that make up the random sequence. Note that when the random variables are all independent of one another, i.e., jointly independent, and this probability distribution doesn't change with time, then the branch labels in the tree are all the same, and in effect, the tree collapses to one stage. This is the situation called a sequence of *independent and identically distributed* (IID) random variables in probability theory. Generalizing this slightly we have the following definition

Definition 6.1-2   An *independent random sequence* is one whose random variables at any time $n_1, n_2, \ldots, n_N$ are jointly independent for all positive integers N.   ∎

Independent random sequences play a key role in our theory because they are relatively easy to analyze, they form the basis of more complicated and accurate models, and it is

---

[†]For example let $M = 8$ and consider the base 8 number $0.1200 \ldots 0 \ldots$. This implies that $X[1] = 1, X[2] = 2$, and all subsequent values are 0.

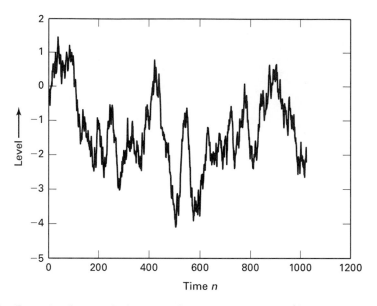

Figure 6.1-3   Example of a sample function of a random sequence. (Samples connected in plot.)

easy to get approximate sample sequences using random number generators on computers. Also when the discrete data arises sampling through continuous time data, statistical independence often is a good approximation if the samples are far apart.

Figure 6.1-3 shows a segment from a real noise sequence, and Figure 6.1-4 shows a close-up portion revealing its discrete-time nature and detailed "randomness." This segment could have been taken from anywhere in the noise sequence and the statistical properties would have been the same. This remarkable property hints at some form of "stationarity" which will shortly be defined (Definition 6.1-5). Note that successive random variables, making up this segment, do not appear to be independent. Rather they are evidently correlated, necessitating in general an Nth-order probability distribution to statistically describe just this segment of this noise sequence. Continuing in this way, we would need an infinite order probability distribution function to characterize the whole random sequence!

In order to deal with infinite length random sequences, we may have to be able to compute the probabilities of *infinite intersections*[†] of events, for example, the event $\{X[n] < 5$ for all positive $n\}$, which can be written as either $\bigcap_{n=1}^{+\infty}\{X[n] < 5\}$ or, by De Morgan's laws, in terms of the infinite union $\left(\bigcup_{n=1}^{\infty}\{X[n] \geq 5\}\right)^c$. This requires that we can *define* and work with the probabilities of infinite collections of events, which presents a problem with Axiom 3 of probability measure: That is, for $AB = \phi$ the null set,

$$P[A \cup B] = P[A] + P[B] \quad (\text{Axiom 3}). \tag{6.1-1}$$

[†]Please review Section 1.4 on the definition of infinite intersections and unions. The concept is simple but often misunderstood.

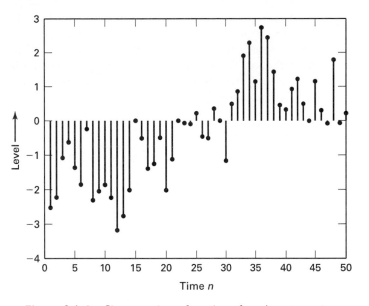

Figure 6.1-4   Close-up view of portion of random sequence.

By iteration we could build this result up to the result

$$P\left[\bigcup_{n=1}^{N} A_n\right] = \sum_{n=1}^{N} P[A_n],$$

for any finite positive N, assuming $A_i A_j = \phi$ for all $i \neq j$. This is called *finite additivity*. It will permit us to evaluate $\lim_{N\to\infty} P[\bigcup_{n=1}^{N} A_n]$, but what we need above is $P[\bigcup_{n=1}^{\infty} A_n]$ where $A_n \triangleq \{X[n] \geq 5\}$. For general functions these two quantities might not be the same, i.e., $\lim_{N\to\infty} f(x_N) \neq f(\lim_{N\to\infty} x_N)$. For this interchange of limiting operations to be valid, we need some kind of continuity built into probability measure P. This can be achieved by augmenting or replacing Axiom 3 by the stronger infinitely (countably) additive Axiom 4 given us.

Axiom 4   (Countable Additivity)

$$P\left[\bigcup_{n=1}^{\infty} A_n\right] = \sum_{n=1}^{+\infty} P[A_n], \tag{6.1-2}$$

for an infinite collection of events satisfying $A_i A_j = \phi$ for $i \neq j$.   ∎

Fortunately, in the branch of mathematics called *measure theory* [6-1] (see also Appendix D) it is shown that it is always possible to construct probability measures satisfying the stronger Axiom 4. Moreover if one has defined a probability measure P satisfying

Axiom 3, that is, it is finitely additive, then the Russian mathematician Kolmogorov [6-2], often referred to as the father of modern probability, has shown that it is always possible to extend the measure P to satisfy the countably additive Axiom 4. We pause now for an example, after which we will show that Axiom 4 is equivalent to the desired continuity of the probability measure P. Henceforth, we will assume that our probability measures satisfy Axiom 4, and say they are *countably additive*.

Infinite-length Bernoulli Trials

Let $\Omega = \{H, T\}$ with $P[H] = p$, with $0 < p < 1$, and $P[T] = q \triangleq 1 - p$. Define the random variable W by $W(H) \triangleq 1$ and $W(T) \triangleq 0$, indicative of successes and failures in coin flipping.

Let Ω_n be the sample space on the nth flip (the nth copy of Ω) and define a new event space as the infinite cross-product[†] $\Omega_\infty \triangleq \times_{n=1}^\infty \Omega_n$. This would be the sample space associated with an infinite sequence of flips, each with sample space Ω_n. We then define the random sequence $W[n, \zeta] \triangleq W(\zeta_n)$, thus generating the *Bernoulli random sequence* $W[n]$, $n \geq 1$.

Consider the probability measure for the infinite dimensional sample space Ω_∞. Letting A_n denote an event[‡] at trial n, that is, $A_n \in \mathscr{F}_n$ where $\mathscr{F}_n$ is the field of events in the probability space $(\Omega_n, \mathscr{F}_n, P)$ of trial n, we need to have $\bigcap_{n=1}^\infty A_n$ as an event in $\mathscr{F}_\infty$, the σ-field of events in Ω_∞. To complete this field of events we will have to augment it with all the countable intersections and unions of such events. For example, we may want to calculate the probability of the event

$$\{W[1] = 1, W[2] = 0\} \cup \{W[1] = 0, W[2] = 1\},$$

which can be interpreted as the union of two events of the form $\bigcap_{n=1}^\infty A_n$; i.e., $\{W[1] = 1, W[2] = 0\} = \bigcap_{n=1}^\infty A_n$ with $A_1 = \{W[1] = 1\}$, $A_2 = \{W[2] = 0\}$, and $A_n = \Omega_n$ for $n \geq 3$. Hence $\mathscr{F}_\infty$ must include all such events for completeness. To construct a probability measure on Ω_∞ we start with sets of the form $\boldsymbol{A}_\infty = \cap_{n=1}^\infty A_n$ and define *in the case of independent trials*,

$$\mathbf{P}_\infty[\boldsymbol{A}_\infty] \triangleq \prod_{n=1}^\infty P[A_n].$$

We then extend this probability measure to all of $\mathscr{F}_\infty$ by using Axiom 4 and the fact that every member of $\mathscr{F}_\infty$ is expressible as the countable union and intersection of events of the form $\cap_{n=1}^\infty A_n$. We have in principle thus constructed the probability space $(\boldsymbol{\Omega}_\infty, \mathscr{F}_\infty, \boldsymbol{P}_\infty)$ corresponding to the infinite-length Bernoulli trials, with associated Bernoulli random sequence

$$W[n, \boldsymbol{\zeta}] = W(\zeta_n), \qquad n \geq 1.$$

[†]Here the infinite cross product $\times_{n=1}^\infty \Omega_n$ simply means that the points in Ω_∞ consist of all the infinite-length sequences of events, each one in Ω_n for some n. Thus if $\zeta \in \Omega_\infty$, then $\zeta = (\zeta_1, \zeta_2, \zeta_3, \ldots)$, where ζ_n is in Ω_n for each $n \geq 1$. (The finite-length case of Bernoulli trials was treated in Section 1.9.)

[‡]Most likely just a singleton event, i.e., just one outcome, in this binary case.

We have just seen how to construct the sample space $\mathbf{\Omega}_\infty$ for the (infinite-length) Bernoulli random sequence, where the outcomes $\boldsymbol{\zeta}$ are just infinite-length sequences of H and T. This $W[n]$ is thus our first nontrivial example of a random sequence. However, it may seem a bit artificial to regard each random variable $W[n, \boldsymbol{\zeta}]$ as a function of the infinite dimensional outcome vectors that make up the elements in the sample space $\mathbf{\Omega}_\infty$. It seems as though we have unnecessarily complicated the situation, after all $W[n, \boldsymbol{\zeta}]$ is just $W(\zeta_n)$. To see that this notational complication is unavoidable, let us turn to the commonly occurring model for correlated noise,

$$X[n] = \sum_{m=1}^{n} \alpha^{n-m} W[m], \text{ for } n \geq 1, \tag{6.1-3}$$

where $W[n]$ is the Bernoulli random sequence just created. Writing the filtered output $X[n]$ for each outcome $\boldsymbol{\zeta}$,

$$X[n, \boldsymbol{\zeta}] = \sum_{m=1}^{n} \alpha^{n-m} W(\zeta_m),$$

we see that each $X[n, \boldsymbol{\zeta}]$ is a function of an ever-increasing (with n) number of components of $\boldsymbol{\zeta}$, i.e., the value of $X[n]$ depends on outcomes $\zeta_1, \zeta_2, \ldots, \zeta_n$. If we just dealt with each fixed value of n as a separate problem, i.e., a separate sample space and probability measure, there would be the unanswered question of consistency. This is where, in practice, we would call on Kolmogorov's consistency theorem to show that our results are consistent with *one* sample space $\mathbf{\Omega}_\infty$ which has (infinite-length) outcomes $\boldsymbol{\zeta}$.[†]

Example 6.1-5 _____

(Correlated noise.) Considering the random sequence in Equation 6.1-3, with $|\alpha| < 1$. We take the Bernoulli random sequence $W[n]$ as input, i.e., $W[n] = 1$ with probability p, and $W[n] = 0$ with probability $q \triangleq 1 - p$. We want to find the mean of $X[n]$ at each positive n. Since the expectation operator is linear, we can write

$$E\{X[n]\} = E\left\{\sum_{m=1}^{n} \alpha^{n-m} W[m]\right\} = \sum_{m=1}^{n} \alpha^{n-m} E\{W[m]\}$$

$$= \sum_{m=1}^{n} \alpha^{n-m} p = p \sum_{m=1}^{n} \alpha^{n-m}$$

$$= p \sum_{m'=0}^{n-1} \alpha^{m'} = p \frac{(1 - \alpha^n)}{(1 - \alpha)}.$$

The random sequence $X[n]$ thus created is not a sequence of independent random variables, as we can see by calculating the correlation

---

[†] The use of bold notation for $\mathbf{\Omega}_\infty, \boldsymbol{\zeta}, \mathbf{P}_\infty$ is rather extravagant but was introduced to avoid confusion. Clearly, $\mathbf{\Omega}_\infty$ is not the same as $\lim_{n \to \infty} \Omega_n$. Each outcome in Ω_n is either a $\{H\}$ or a $\{T\}$ no matter how large n gets. On the other hand the outcomes in $\boldsymbol{\zeta} \in \mathbf{\Omega}_\infty$ are infinitely long strings of Hs and Ts. In the future we shall dispense with the bold notation even if Ω is generated by an infinite cross product and its elements (outcomes) are infinitely long strings.

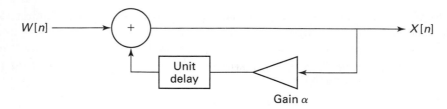

Figure 6.1-5   A feedback filter that generates correlated noise $X[n]$ from an uncorrelated sequence $W[n]$.

$$
\begin{aligned}
E\{X[2]X[1]\} &= E\{(\alpha W[1] + W[2])\,W[1]\} \\
&= \alpha E\{W^2[1]\} + E\{W[2]\}E\{W[1]\} \\
&= \alpha p + p^2 \\
&\neq (\alpha + 1)p^2 = E\{X[2]\}E\{X[1]\}.
\end{aligned}
$$

The random variables $X[2]$ and $X[1]$ must be dependent, since they are not even uncorrelated.

However, since the $W[n]$ are uncorrelated we can easily calculate the variance $\mathrm{var}\{X[n]\}$ as

$$
\begin{aligned}
\mathrm{var}\{X[n]\} &= \sum_{m=1}^{n} \mathrm{var}\{\alpha^{n-m}W[m]\} \\
&= \sum_{m=1}^{n} \alpha^{2(n-m)}\,\mathrm{var}\{W[m]\} \\
&= \frac{(1 - \alpha^{2n})}{(1 - \alpha^2)}\,pq.
\end{aligned}
$$

The dynamics of this random sequence can be modeled using a difference equation. Since $X[n-1] = \sum_{m=1}^{n-1} \alpha^{n-1-m}W[m]$, it follows that $X[n] = \alpha X[n-1] + W[n]$, a result that clearly exhibits the dependence of $X[n]$ on its immediate neighbor $X[n-1]$.[†] We leave the execution of the algebra to the reader. Thus correlated noise $X[n]$ can be generated from the independent sequence $W[n]$ by *filtering* with the configuration shown in Figure 6.1-5. From Equation 6.1-3 we see that for large n $X[n]$ is the sum of a large number of independent random variables. Hence by the central limit theorem it will tend to a Gaussian distribution, $n \to \infty$, with mean $p\frac{1-\alpha^n}{1-\alpha}$ and variance $pq\frac{1-\alpha^{2n}}{1-\alpha^2}$.

Zero-mean, correlated, precisely Gaussian noise can be generated using the same model. Thus with $W[1], W[2], \ldots, W[n], \ldots$ denoting zero-mean, independent, identically distributed, Gaussian random variables with $N(0, \sigma_W^2)$, the random sequence $X[n] = \sum_{m=1}^{n} \alpha^{n-m}W[m]$ will be zero mean, Gaussian with variance

$$
\mathrm{Var}\{X[n]\} = \frac{1 - \alpha^{2n}}{1 - \alpha^2}\sigma_W^2,
$$

[†]Such explicit dependence in the equation like this is sometimes called *direct dependence*.

where $\sigma_W^2 = \text{var}\{W[n]\}$. Here too, the sequence produced by the filter is correlated since $E\{X[2]X[1]\} = \alpha E\{W^2[1]\} = \alpha \sigma_W^2 \neq E\{X[2]\}E\{X[1]\} = 0$.

The next example gives a MATLAB method to construct realizations of the Bernoulli random sequence and then passes the resulting sample sequences through a first-order filter to generate sample sequences of a (more realistic) correlated random sequence.

Example 6.1-6

(Sample sequence construction.) We use MATLAB to construct a sample function of $W[n]$. The MATLAB program

```
u = rand(40,1);
w = 0.5 >= u;
stem (w),
```

uses the built-in function "rand" to generate a 40-element vector of uniform random variables. The second line sets the vector elements $w[n]$ to 1 if $u[n] > 0.5$, and to 0 if $u[n] < 0.5$. So $w[n]$ is a sample sequence of the Bernoulli random sequence with $p = 0.5$. The corresponding MATLAB plot is shown in Figure 6.1-6.

To model the sample sequences of $X[n]$, which we denote $x[n]$, we can filter the sequence $w[n]$ with the filter,

$$x[n] = \alpha x[n-1] + w[n],$$

which has impulse response $h[n] = \alpha^n u[n]$ to realize the linear operation of Equation 6.1-3. The corresponding MATLAB M-file fragment is

```
b = 1.0;
a = [1.0 -alpha];
```

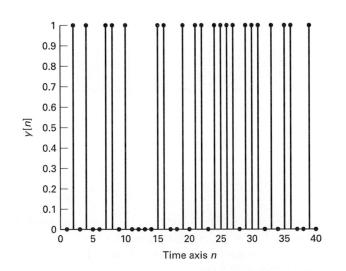

Figure 6.1-6 A sample sequence $w[n]$ for the Bernoulli random sequence $W[n]$.

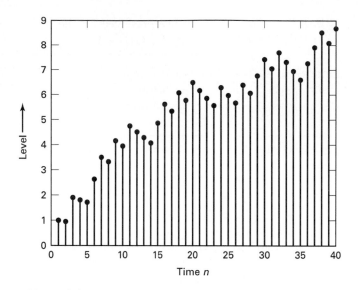

Figure 6.1-7   First 40 points illustrating startup transient.

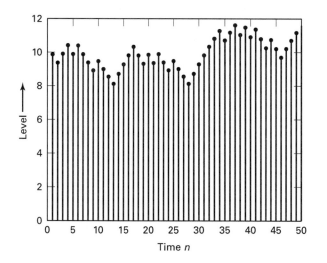

Figure 6.1-8   A segment of 50 points starting at $n = 350$.

```
x = filter(b,a,w);
stem (x)
```

The result for $\alpha = 0.95$ and a 400-element vector was computed. Figure 6.1-7 shows the startup transient for the first 40 values. Figure 6.1-8 shows a sample of the approximate steady-state behavior starting at $n = 350$ and plotted for 50 points. Note the sample average value that has built up in $x[n]$ over time.

Note that the random sequence $X[n]$ has real noise-like characteristics. The filter has correlated the random variables making up $X[n]$ so that sample sequences $x[n]$ look more "continuous." This simple example is called an *autoregressive (AR)* model and is widely used in signal processing to model both noises and signals. Note that the deterministic defect of the initial examples has now been removed. The reason is that the Bernoulli input sequence provides a new independent value for every sample, ensuring that the next sample cannot be perfectly predicted from the past.

Continuity of Probability Measure

When dealing with an infinite number of events, we have seen that continuity of the probability measure can be quite useful. Fortunately, the desired continuity is a direct consequence of the extended Axiom 4 on countable additivity (cf. Eq. 6.1-2).

Theorem 6.1-1   Consider an *increasing sequence* of events B_n, that is, $B_n \subset B_{n+1}$ for all $n \geq 1$ as shown in Figure 6.1-9. Define $B_\infty \triangleq \bigcup_{n=1}^{\infty} B_n$, then $\lim_{n \to \infty} P[B_n] = P[B_\infty]$.

Proof   Define the sequence of events A_n as follows:

$$A_1 \triangleq B_1$$
$$A_n \triangleq B_n B_{n-1}^c, \qquad n > 1.$$

The A_n are disjoint and $\bigcup_{n=1}^{N} A_n = \bigcup_{n=1}^{N} B_n$ for all N. Also $B_N = \bigcup_{n=1}^{N} B_n$ because the B_n are increasing. So

$$P[B_N] = P\left[\bigcup_{n=1}^{N} B_n\right] = P\left[\bigcup_{n=1}^{N} A_n\right] = \sum_{n=1}^{N} P[A_n],$$

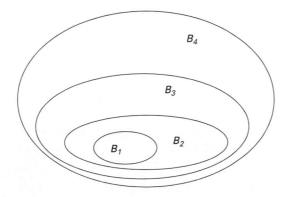

Figure 6.1-9   Illustrating an increasing sequence of events.

and

$$\lim_{n \to \infty} P[B_N] = \lim_{n \to \infty} \sum_{n=1}^{N} P[A_n]$$

$$= \sum_{n=1}^{+\infty} P[A_n] \quad \text{by definition of the limit of a sum,}$$

$$= P\left[\bigcup_{n=1}^{\infty} A_n\right] \quad \text{by Axiom 4,}$$

$$= P[B_\infty] \quad \text{by definition of the } A_n.$$

This last step results from $\bigcup_{n=1}^{\infty} A_n = \bigcup_{n=1}^{\infty} B_n \triangleq B_\infty$. ■

Corollary 6.1-1   Let B_n be a *decreasing sequence* of events, that is, $B_n \supset B_{n+1}$ for all $n \geq 1$. Then

$$\lim_{n \to \infty} P[B_n] = P[B_\infty],$$

where

$$B_\infty \triangleq \cap_{n=1}^{\infty} B_n.$$

Proof   Similar to proof of Theorem 6.1-1 and left to the student. ■

Example 6.1-7

Let $B_n \triangleq \{X[k] < 2 \text{ for } 0 < k < n\}$, for $n = 0, 1, 2, \ldots$. In words, B_n is the event that $X[k]$ is less than 2 for the indicated range of k. Clearly B_{n+1} is a subset of B_n, i.e., $B_{n+1} \subset B_n$ for all $n = 0, 1, 2, \ldots$. Also if we set $B_\infty \triangleq \{X[k] < 2 \text{ for all } k \geq 0\}$, then $B_\infty = \cap_{n=1}^{\infty} B_n$. So we can write, by the above corollary,

$$P[B_\infty] = \lim_{n \to \infty} P[B_n]$$

$$= \lim_{n \to \infty} P[X[0] < 2, \ldots, X[n] < 2].$$

Thus the corollary provides a way of calculating events involving an infinite number of random variables by just taking the limit of the probability involving a finite number of random variables. This type of limiting calculation is often performed in engineering analyses, and typically without explicit justification (i.e., without worrying about the consistency problem mentioned earlier). In this section we have seen that the correctness of the approach rests on a fundamental axiom of probability theory, Axiom 4 (countable additivity).

We next use the continuity of the probability measure P to prove an elementary fact about probability distribution functions (PDFs).

Example 6.1-8

The PDF is continuous from the *right*; that is, for $F_X(x) = P[X(\zeta) \leq x]$ [cf. Property (iii) of F_X in Section 2.3] we have

$$\lim_{n \to \infty} F_X\left(x + \frac{1}{n}\right) = F_X(x).$$

To show this, we define

$$B_n \triangleq \left\{\varsigma : X(\varsigma) \le x + \frac{1}{n}\right\}$$

and note that B_n is a decreasing sequence of events, where $B_\infty \triangleq \bigcap_{n=1}^{\infty} B_n = \{\varsigma : X(\varsigma) \le x\}$ and

$$F_X\left(x + \frac{1}{n}\right) = P[B_n].$$

By application of Corollary 6.1-1, we get

$$\lim_{n \to \infty} F_X\left(x + \frac{1}{n}\right) = \lim_{n \to \infty} P[B_n] = P[B_\infty]$$

$$= F_X(x).$$

Statistical Specification of a Random Sequence

A random sequence $X[n]$ is said to be *statistically specified* by knowing its Nth-order probability distribution functions (Nth-order PDFs) for all integers $N \ge 1$, and for all times, $n, n+1, \ldots, n+N-1$, that is, if we know

$$F_X(x_n, x_{n+1}, x_{n+2}, \ldots, x_{n+N-1}; n, n+1, \ldots, n+N-1)$$

$$\triangleq P\{X[n] \le x_n, X[n+1] \le x_{n+1}, \ldots, X[n+N-1] \le x_{n+N-1}\}, \qquad (6.1\text{-}4)$$

where the variables after the semicolon, $n, n+1, \ldots, n+N-1$, indicate the location of the N random variables in this joint PDF. Note that this is an infinite set of PDFs for *each* order N, because we must know the joint PDF at *all* times n, $-\infty < n < +\infty$. Incurring some penalty in notational clarity, we often write the joint PDFs more simply as

$$F_X(x_n, x_{n+1}, \ldots, x_{n+N-1}), \quad \text{for all } n, \text{ and for all } N \ge 1. \qquad (6.1\text{-}5)$$

We also define Nth-order PDFs for nonconsecutive time parameters,

$$F_X(x_{n_1}, x_{n_2}, \ldots, x_{n_N}; n_1, n_2, \ldots, n_N).$$

It may seem that this statistical specification is some distance from a complete description of the entire random sequence since no one distribution function in this infinite set of finite-order PDFs describes the *entire* random sequence. Nevertheless, if we specify all these finite-order joint distributions at all finite times, using continuity of the probability measure that we have just shown, we can calculate the probabilities of events involving infinite numbers of random variables via limiting operations involving the finite order PDFs. Of course, we do have to make sure that our set of Nth-order PDFs is consistent within itself! Sometimes it is trivial, for example the case where all the random variables that make up the random sequence are independent of one another, e.g., a Bernoulli random sequence.

Example 6.1-9 ———

(Consistency.) For consistency, the low-order PDFs must agree with the higher-order PDFs. For example, considering just $N = 2$ and 3, we must have

$$F_X \left(x_n, x_{n+2}; n, n+2 \right) = F_X \left(x_n, \infty, x_{n+2}; n, n+1, n+2 \right),$$

for all n, and for all values of x_n and x_{n+2}. Likewise, the $N = 1$ PDFs must be consistent with those of $N = 2$. Further the consistency must extend to all higher orders N.

———

Consistency can be *guaranteed* by construction, as in the case of the filtered Bernoulli random sequence of Example 6.1-6 above. If we were faced with a suspect set of Nth-order PDFs of unknown origin, it would be a daunting task, indeed, to show that they were consistent. Hence, the important role of constructive models in stochastic sequences and processes.

In summary, we have seen two ways to specify a random sequence: the statistical characterization (Equation 6.1-4) and the direct specification in terms of the random functions $X[n, \zeta]$. We use the word *statistical* to indicate that the former information can be obtained, at least conceptually, by estimating the Nth-order PDFs for $N = 1, 2, 3, \ldots$ and so forth, i.e., by using *statistics*.

The Nth-order probability density functions (pdf's) are given for differentiable F_X as

$$f_X(x_n, x_{n+1}, \ldots, x_{n+N-1}; n, n+1, \ldots, n+N-1)$$
$$= \frac{\partial^N F_X(x_n, x_{n+1}, \ldots, x_{n+N-1}; n, n+1, \ldots, n+N-1)}{\partial x_n \partial x_{n+1} \ldots \partial x_{n+N-1}}, \qquad (6.1\text{-}6)$$

for every integer (time) n and positive integer (order) N. Sometimes we will omit the subscript X when only one random sequence is under consideration. Also, we may drop the explicit time notation and write

$$f_X(x_n, x_{n+1}, \ldots, x_{n+N-1}) \quad \text{for} \quad f_X(x_n, x_{n+1}, \ldots, x_{n+N-1}; n, n+1, \ldots, n+N-1).$$

We will sometimes want to deal with complex random variables and sequences. By this we mean an ordered pair of real random variables, that is, $X = (X_R, X_I)$ often written as $X = X_R + jX_I$ with PDF

$$F_X(x_R, x_I) \triangleq P[X_R \leq x_R, X_I \leq x_I].$$

The corresponding pdf is then

$$f_X(x_R, x_I) = \frac{\partial^2 F_X(x_R, x_I)}{\partial x_R \partial x_I}.$$

To simplify notation we will write $f_X(x)$ for $f_X(x_R, x_I)$ in what follows, with the understanding that the respective integrals (sums for discrete valued complex case) are really double integrals on the (x_R, x_I) plane if the random variable is complex.[†]

The *moments* of a random sequence play an important role in most applications. In part this is because for a large class of random sequences (so-called *ergodic* sequences, see Section 8.4 in Chapter 8) they can be easy to estimate from just one sample sequence. The first moment or *mean function* of a random sequence is

$$\mu_X[n] \triangleq E\{X[n]\} = \int_{-\infty}^{+\infty} x f_X(x; n) dx$$

$$= \int_{-\infty}^{+\infty} x_n f_X(x_n) dx_n$$

for a continuous-valued random sequence $X[n]$. The mean function for a discrete-valued random sequence, taking on values from the set $\{x_k, -\infty < k < +\infty\}$ at time n, is evaluated as

$$\mu_X[n] = E\{X[n]\} = \sum_{k=-\infty}^{+\infty} x_k P[X[n] = x_k]. \tag{6.1-7}$$

In the case of a mixed random sequence, as in the case of mixed random variables, it is convenient to write

$$\mu_X[n] = \int_{-\infty}^{+\infty} x f_X(x; n) dx + \sum_{k=-\infty}^{+\infty} x_k P[X[n] = x_k]. \tag{6.1-8}$$

Actually using the concept of the Stieltjes integral [6-3] both terms can be rewritten in the one form

$$\mu_X[n] = \int_{-\infty}^{+\infty} x \, dF_X(x; n),$$

in terms of the PDF $F_X(x; n)$.

The expected value of the product of the random sequence evaluated at two times $X[k]X^*[l]$ is called the *autocorrelation function* and is a two-parameter function of *both* times k and l, where $-\infty < k, l < +\infty$,

$$R_{XX}[k, l] \triangleq E\{X[k]X^*[l]\}$$

$$= \int_{-\infty}^{+\infty} \int_{-\infty}^{+\infty} x_k x_l^* f_X(x_k, x_l; k, l) dx_k dx_l, \tag{6.1-9}$$

when the autocorrelation function exists (the usual case, but of course, in some cases the integral might not converge!). Most of the time we will deal with *second-order* random

---

[†]Complex random sequences are used as equivalent baseband models of certain bandpass signals and noises. The resulting complex valued simulation can be then simulated at a much lower sample rate.

sequences, defined by their property of having finite *average power* $E\{|X[n]|^2\} < \infty$. The correlation function will always exist in this case. Later we shall see that the conjugate on the second factor in the autocorrelation function definition results in some notational simplicities for complex-valued random sequences. We will also define the *centered* random sequence $X_c[n \triangleq X[n] - \mu_X[n]$, which is zero-mean, and consider its autocorrelation function, called the *autocovariance function* of the original sequence $X[n]$. It is defined as

$$K_{XX}[k,l] \triangleq E\{(X[k] - \mu_X[k])(X[l] - \mu_X[l])^*\}. \tag{6.1-10}$$

Directly from these definitions, we note the following symmetry conditions must hold:

$$R_{XX}[k,l] = R_{XX}^*[l,k], \tag{6.1-11}$$

$$K_{XX}[k,l] = K_{XX}^*[l,k], \tag{6.1-12}$$

called *Hermitian symmetry*. Also note that

$$K_{XX}[k,l] = R_{XX}[k,l] - \mu_X[k]\mu_X^*[l]. \tag{6.1-13}$$

The *variance function* is defined $\sigma_X^2[n] \triangleq K_{XX}[n,n]$ and denotes the average power in $X_c[n]$. The *power* of $X[n]$ itself has been given above and equals $R_{XX}[n,n]$.

Example 6.1-10

The mean function of $X[n]$ as given in Example 6.1-1 is

$$\mu_X[n] = E\{X[n]\} = E\{Xf[n]\} = \mu_X f[n],$$

where μ_X is the mean of the random variable X. The autocorrelation function is

$$R_{XX}[k,l] = E\{X[k]X^*[l]\} = E\{Xf[k]X^*f^*[l]\}$$
$$= E\{|X|^2\}f[k]f^*[l],$$

and so the autocovariance function is given as

$$K_{XX}[k,l] = E\{|X|^2 f[k]f^*[l]\} - |\mu_X|^2 f[k]f^*[l]$$
$$= E\{|X|^2 - |\mu_X|^2\}f[k]f^*[l],$$
$$= E\{|X - \mu_X|^2\}f[k]f^*[l],$$
$$= \sigma_X^2 f[k]f^*[l],$$

where $\sigma_X^2 = \text{var}(X)$. We thus see that the variance $\sigma_X^2[n]$ is just $\sigma_X^2|f[n]|^2$.

We look at a sequence which fits our notions of randomness better in the next example.

Example 6.1-11

(Waiting times.) Consider the random sequence consisting of independent random variables $\tau[n]$ for $n \geq 1$, each identically distributed with the exponential pdf of Equation 2.4-16,

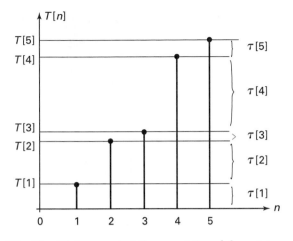

Figure 6.1-10   The $T[n]$ are arrival times and the $\tau[n]$ are interarrival times.

that is,

$$f_\tau(t;n) = f_\tau(t) = \lambda \exp(-\lambda t)u(t), \qquad n = 1, 2, \ldots$$

Write the running sum of the $\tau[k]$ up to time n, defined as

$$T[n] \triangleq \sum_{k=1}^{n} \tau[k], \qquad (6.1\text{-}14)$$

and consider $T[n]$ as a second random sequence for $n = 1, 2, \ldots.$ It turns out that the arrival of random events in time is often modeled in this way. We say that $T[n]$ is the *time to the nth arrival* or *waiting time* and we call the $\tau[n]$ the *interarrival times.*[†] See Figure 6.1-10.

Later, in Chapter 7, we shall see that the important Poisson random process can be constructed in this way. Here we want to determine the pdf of $T[n]$ at each n based on the definition in Equation 6.1-14. Using the fact that the $\tau[k]$ are independent, we can apply Equation 4.7-3 and conclude that the pdf of $T[n]$ will be the $(n-1)$-fold convolution product of exponential pdf's. Using convolution to determine the pdf of $T[2]$ we get

$$f_T(t;2) = f_\tau(t) * f_\tau(t) = \lambda^2 t \exp(-\lambda t)u(t).$$

Convolving this result with the exponential pdf a second time we get

$$f_T(t;3) = \frac{1}{2}\lambda^3 t^2 \exp(-\lambda t)u(t).$$

It turns out that the general form is the Erlang pdf,

$$f_T(t;n) = \frac{(\lambda t)^{n-1}}{(n-1)!}\lambda \exp(-\lambda t)u(t). \qquad (6.1\text{-}15)$$

[†]Please regard τ as a "capital tau" to continue our distinction between the random variable and the value it takes on, i.e., $X = x$.

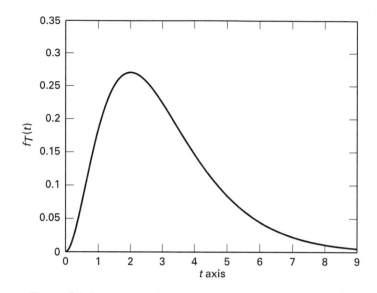

Figure 6.1-11   A plot of the Erlang pdf for $\lambda = 1$ and $n = 3$.

The Erlang or Gamma pdf [6-7] is widely used in waiting-time problems in telecommunications networks and is plotted via MATLAB in Figure 6.1-11 for $n = 3$ and $\lambda = 1.0$, which is the waiting time for $n = 3$ arrivals.

We can establish this density's correctness by the *Principle of Mathematical Induction.* (see Section A.4 in Appendix A.) It is composed of two steps: (1) First show the formula is correct at $n = 1$; (2) then show that **if** the formula is true at $n - 1$, it **must** also be true at n. Combining these two steps, we have effectively proved the result for all positive integers n.

We see that $f_T(t; 1)$ in Equation 6.1-15 is correct, so we proceed by assuming Equation 6.1-15 is true at $n - 1$. By convolving with the exponential, we can show that it is true at n as follows:

$$f_T(t; n) = f_T(t; n - 1) * \lambda \exp(-\lambda t)\, u(t)$$

$$= \int_0^t \exp(-\lambda\tau)\frac{(\lambda\tau)^{n-2}}{(n-2)!}\lambda^2 \exp(-\lambda(t-\tau))d\tau\, u(t)$$

$$= \lambda^n \exp(-\lambda t)\int_0^t \frac{\tau^{n-2}}{(n-2)!}d\tau\, u(t)$$

$$= \lambda^n \exp(-\lambda t)\frac{t^{n-1}}{(n-1)!}\, u(t).$$

Using the independence of the $\tau[n]$, we can also compute the mean as

$$\mu_T[n] = n\mu_\tau = n(1/\lambda) = n/\lambda$$

and variance of the sum $T[n]$ by repeated use of property (A) of Equation 4.3-20.

$$\text{Var}\left[T[n]\right] = n\text{Var}[\tau] = n/\lambda^2.$$

We next introduce the most widely used random model in electrical engineering, communications, and control: the Gaussian random sequence. Its wide popularity stems from two important facts: (1) the Central Limit theorem (Theorem 4.7-2) assures that many processes occurring in practice are approximately Gaussian; and (2) the mathematics is especially tractable in problems involving detection, estimation, filtering, and control theory.

Definition 6.1-3 A random sequence $X[n]$ is called a *Gaussian random sequence* if its Nth-order PDFs (pdf's) are jointly Gaussian, for all $N \geq 1$.

We note that the mean and covariance function will specify a Gaussian random sequence in the same way that the mean vector and covariance matrix determine a Gaussian random vector (see Section 5.5). This is because each Nth-order distribution function is just the PDF of a Gaussian random vector whose mean vector and covariance matrix are expressible in terms of the mean and covariance functions of the Gaussian random sequence.

Example 6.1-12 _____

Let $W[n]$ be a Gaussian random sequence with mean $\mu_W[n] = 0$ for all n and autocorrelation function $R_W[k, l] = \sigma^2 \delta[k - l]$, $\sigma > 0$, where δ is the *discrete-time impulse*

$$\delta[n] \triangleq \begin{cases} 1 & n = 0 \\ 0 & n \neq 0. \end{cases}$$

If we form a covariance matrix, then, for a vector of any N distinct samples, it will be diagonal. So, by Gausianity, each Nth-order pdf will factor into a product of N first-order pdf's. Hence the elements of this random sequence are jointly independent, or what we call an *independent (Gaussian) random sequence* cf. Def. 6.1-2. Next we create the random sequence $X[n]$ by taking the sum of the current and previous $W[n]$ values,

$$X[n] \triangleq W[n] + W[n - 1] \quad \text{for } -\infty < n < +\infty.$$

Here $X[n]$ is also Gaussian in all its Nth-order distributions (since a linear transformation of a Gaussian random vector produces a Gaussian vector by Theorem 5.6-1); hence $X[n]$ is also a Gaussian random sequence. We can easily evaluate the mean of $X[n]$ as

$$\mu_X[n] = E\{X[n]\} = E\{W[n]\} + E\{W[n - 1]\}$$
$$= 0,$$

and its correlation function as

$$R_{XX}[k, l] = E\{X[k]X^*[l]\}$$
$$= E\{(W[k] + W[k - 1])\,(W[l] + W^*[l - 1])\}$$
$$= E\{W[k]W^*[l]\} + E\{W[k]W^*[l - 1]\}$$
$$\quad + E\{W[k - 1]W^*[l]\} + E\{W[k - 1]W^*[1 - 1]\}$$

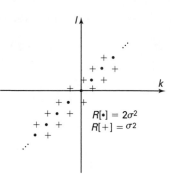

$R[\bullet] = 2\sigma^2$
$R[+] = \sigma^2$

Figure 6.1-12   Diagram of the tri-diagonal correlation function of Example 6.1-12.

$$= R_{WW}[k,l] + R_{WW}[k, l-1] + R_{WW}[k-1, l] + R_{WW}[k-1, l-1]$$
$$= \sigma^2 \big(\delta[k-l] + \delta[k-l+1] + \delta[k-l-1] + \delta[k-l] \big).$$

We can plot this autocorrelation in the (k, l) plane as shown in Figure 6.1-12 and see the time extent of the dependence of the random sequence $X[n]$.

From this figure, we see that the autocorrelation has value $2\sigma^2$ on the diagonal line $l = k$ and has value σ^2 on the diagonal lines $l = k \pm 1$. It should be clear from Figure 6.1-12 that $X[n]$ is **not** an independent random sequence. However, the banded support of this covariance function signifies that dependence is limited to shifts $(k - l) = \pm 1$ in time. Beyond this lag we have uncorrelated, and hence in this Gaussian case, independent random variables.

Example 6.1-13
(The random walk sequence.) Continuing with infinite-length Bernoulli trials, we now define a random sequence $X[n]$ as the running sum of the number of successes (heads) minus the number of failures (tails) in n trials times a step size s,

$$X[n] = \sum_{k=1}^{n} W[k] \qquad \text{with} \quad X[0] = 0,$$

where we redefine $W[H] = +s$ and $W[T] = -s$.

The resulting sequence then models a *random walk* on the integers starting at position $X[0] = 0$. At each succeeding time unit a step of size s is taken either to the right or to the left. After n steps we will be at a position rs for some integer r. This is illustrated in Figure 6.1-13.

If there are k successes and necessarily $(n - k)$ failures, then we have the following relation:

$$rs = ks - (n - k)s$$
$$= (2k - n)s,$$

which implies that $k = (n + r)/2$, for those values of r that make the right-hand side an integer. Then with $P[\text{success}] = P[\text{failure}] = \frac{1}{2}$, we have

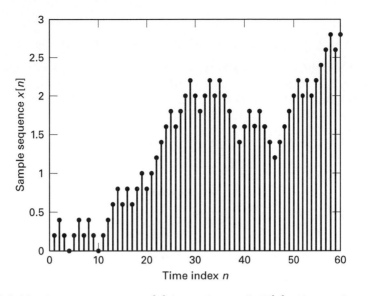

Figure 6.1-13   A sample sequence $x[n]$ for random walk $X[n]$ with step size $s = 0.2$.

$$P\{X[n] = rs\} = P\left[(n+r)/2 \text{ successes}\right]$$

$$= \begin{cases} \binom{n}{(n+r)/2}, & (n+r)/2 \text{ an integer} \\ 0, & \text{else.} \end{cases}$$

Using the fact that $X[n] = W[1] + W[2] + \ldots + W[n]$ and that the W's are jointly independent, we can compute the mean and variance of the random walk as follows:

$$E\{X[n]\} = \sum_{k=1}^{n} E\{W[k]\} = \sum_{k=1}^{n} 0 = 0,$$

and

$$E\{X^2[n]\} = \sum_{k-1}^{n} E\{W^2[k]\}$$

$$= \sum_{k=1}^{n} 0.5[(+s)^2 + (-s)^2]$$

$$= ns^2.$$

If we normalize $X[n]$ by dividing $\sqrt{n}$ and define

$$\tilde{X}[n] \triangleq \frac{1}{\sqrt{n}} X[n],$$

then by the Central Limit Theorem 4.7-2 we have that the PDF of $\tilde{X}[n]$ converges to the Gaussian (Normal) distribution $N(0, s^2)$. Thus for n large enough, we can approximate the probabilities

$$P[a < \tilde{X}[n] \le b] = P[a\sqrt{n} < X[n] \le b\sqrt{n}] \simeq \mathrm{erf}(b/s) - \mathrm{erf}(a/s).$$

Note, however, that when this probability is small, very large values of n might be required to keep the percentage error small because small errors in the PDF may be comparable to the required probability value. In practice this means that the Normal approximation will not be dependable on the tails of the distribution but only in the central part, hence the name Central Limit Theorem.

Note also that while $X[n]$ can never be considered approximately Gaussian for any n, (for example, if n is even, $X[n]$ can only be an even multiple of s) still we can approximately calculate the probability

$$P[(r-2)s < X[n] \le rs] = P\left[\frac{(r-2)s}{\sqrt{n}} < \tilde{X}[n] \le \frac{rs}{\sqrt{n}}\right]$$

$$= \frac{1}{\sqrt{2\pi}} \int_{(r-2)/\sqrt{n}}^{r/\sqrt{n}} \exp(-0.5v^2)dv$$

$$\approx 1/\sqrt{\pi(n/2)} \exp(-r^2/2n),$$

where r is small with respect to $\sqrt{n}$. See Section 1.11 for a similar result. In obtaining the last line, we assumed that the integrand was approximately constant over the interval $[(r-2)/\sqrt{n}, r/\sqrt{n}]$.

The waiting-time sequence in Example 6.1-11 and the random walk in Example 6.1-13 both have the property with that they build up over time from independent components or increments. More generally we can define an independent-increments property.

Definition 6.1-4 A random sequence is said to have *independent increments* if for all integer parameters $n_1 < n_2 < \ldots < n_N$, the increments $X[n_1], X[n_2] - X[n_1], X[n_3] - X[n_2], \ldots, X[x_N] - X[n_{N-1}]$ are jointly independent for all integers $N > 1$. ∎

If a random sequence has independent increments, one can build up its Nth-order probabilities (PMFs and pdf's) as products of the probabilities of its increments. (See Problem 6.10.)

In contrast to the evolving nature of independent increments, many random sequences have constant statistical properties that are invariant with respect to the index parameter n, normally time or distance. When this is valid, the random model is simplified in two ways: First, it is time-invariant, and second, the usually small number of model parameters can be estimated from available data.

Definition 6.1-5 If for all orders N and for all shift parameters k, the joint PDFs of $(X[n], X[n+1], \ldots, X[n+N-1])$ and $(X[n+k], X[n+k+1], \ldots, X[n+k+N-1])$ are the same functions, then the random sequence is said to be *stationary*, i.e., for all $N \ge 1$,

$$F_X(x_n, x_{n+1}, \ldots, x_{n+N-1}; n, n+1, \ldots, n+N-1)$$

$$= F_X(x_n, x_{n+1}, \ldots, x_{n+N-1}; n+k, n+1+k, \ldots, n+N-1+k) \quad (6.1\text{-}16)$$

for all $-\infty < k < +\infty$ and for all x_n through x_{n+N-1}. This definition also holds for pdf's when they exist and PMFs in the discrete amplitude case. ■

If we look back at Example 6.1-11 we see that $X[n]$ and $W[n]$ are both stationary random sequences. The same was true of the interarrival times $\tau[n]$ in Example 6.1-11, but the random arrival or waiting time sequence $T[n]$ was clearly nonstationary, since its mean and variance increase with time n.

Note that stationarity does not mean that the sample functions all look "similar," or even that they all look "noisy."[†] Also, unlike the concept of stationarity in mathematics and physics, we don't directly characterize the realizations of the random sequence as stationary, just the deterministic functions that characterize their behavior, i.e., PDF, PMF, and pdf.

It is often desirable to partially characterize a random sequence based on knowledge of only its first two moments, that is, its mean function and covariance function. This has already been encountered for random vectors in Chapter 5. We will encounter this for random sequences when we present a discussion of linear estimation in the DSP applications of Chapter 9. In anticipation we define a weakened kind of stationarity that involves only the mean and covariance (or correlation) functions. Specifically, if these two functions are consistent with stationarity, then we say that the random sequence is *wide-sense stationary* (WSS).

Definition 6.1-6   A random sequence $X[n]$ defined for $-\infty < n < +\infty$, is called *wide-sense stationary* (WSS) if

(1) The mean function of $X[n]$ is constant for all integers n, $-\infty < n < +\infty$,

$$\mu_X[n] = \mu_X[0] \quad \text{and}$$

(2) For all times $k, l, -\infty < k, l < +\infty$, and integers n, $-\infty < n < +\infty$, the covariance (correlation) function is independent of the shift n,

$$K_{XX}[k, l] = K_{XX}[k + n, l + n]. \quad ■ \tag{6.1-17}$$

We will call such a covariance (correlation) function *shift-invariant*. If we think of $[k, l]$ as a constellation or set of two samples on the time line, then we are translating this constellation up and down the time line, and saying that the covariance function does not change. When the mean function is constant, then shift invariance of the covariance and correlation functions is equivalent. Otherwise it is not. For a constant mean function, we can check property (2) for either the covariance or correlation function.

It should be fairly clear that all stationary sequences are wide-sense stationary but the reverse is not true. For example, the third moment could be shift-variant in a manner

---

[†]For example, suppose we do the Bernouilli experiment of flipping a fair coin once and generate a random sequence as follows: if the outcome is *heads* then $X[n] = 1$ for all n. If the outcome is tails then $X[n] = W[n]$ i.e., stationary white noise again for all n. Thus the sample functions look quite dissimilar, but the random sequence is easily seen to be stationary. In Chapter 8, we discuss the property of *ergodicity*, which, loosely speaking, enables expectations (ensemble averages) to be computed from time averages. In this case the sample functions would tend to have the same features, i.e., a viewer would subjectively feel that they come from the same source.

not consistent with stationarity even though the first moment is constant and the second moment is shift-invariant. Then the random sequence would be wide-sense stationary but not stationary. To further distinguish them, sometimes we refer to stationarity as *strict-sense stationary* to avoid confusion with the weaker concept of wide-sense stationarity. Below we prove that all stationary random sequences are also wide-sense stationary.

Theorem 6.1-2   All stationary random sequences are wide-sense stationary.

Proof   We first show that the mean is constant for a stationary random sequence. Let n be arbitrary

$$\mu_X[n] = E\{X[n]\} = \int_{-\infty}^{+\infty} x f_X(x; n) dx = \int_{-\infty}^{+\infty} x\, f_X(x; 0) dx = \mu_X[0],$$

since $f_X(x; n)$ does not depend on n. Next we show that the covariance function is shift-invariant by first showing that the correlation is shift-invariant:

$$
\begin{aligned}
R_{XX}[k, l] &= E\{X[k]X^*[l]\} \\
&= \int_{-\infty}^{+\infty} \int_{-\infty}^{+\infty} x_k x_l^* f_X(x_k, x_l) dx_k dx_l \\
&= \int_{-\infty}^{+\infty} \int_{-\infty}^{+\infty} x_{n+k} x_{n+l}^* f_X(x_{n+k}, x_{n+l}) dx_{n+k} dx_{n+l},^{\dagger} \\
&= R_{XX}[n+k, n+l],
\end{aligned}
$$

since $f_X(x_k, x_l)$ doesn't depend on the shift n, and the x_i's are dummy variables. Finally, we use Equation 6.1-13 and the result on the mean functions to conclude that the covariance function is also shift-invariant.

Since the covariance function is shift-invariant for any wide-sense stationary random sequence, we can define a one-parameter covariance function to simplify the notation for WSS sequences

$$
\begin{aligned}
K_{XX}[m] &\triangleq E\{X_c[k+m]X_c^*[k]\} = K_{XX}[k+m, k] \\
&= K_{XX}[m, 0].
\end{aligned}
\tag{6.1-18}
$$

We also do the same for correlation functions. Writing the one-parameter correlation function in terms of the corresponding two-parameter correlation function, we have

$$R_{XX}[m] = R_{XX}[k+m, k] = R_{XX}[m, 0]. \quad \blacksquare$$

Example 6.1-14 ───

(WSS covariance function.) The covariance function of Example 6.1-12 is shift-invariant and so we can take advantage of the simplified notation. We can thus write $K_{XX}[m] = \sigma^2(2\delta[m] + \delta[m-1] + \delta[m+1])$.

───────────

†Note that these last two lines use our simplified notation. They are not trivially equal because $f_X(x_k, x_l)$ and $f_X(x_{k+n}, x_{l+n})$ are really the joint densities at two different pairs of time.

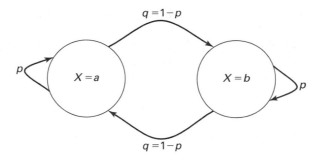

Figure 6.1-14   State-transition diagram of two-state (binary) random sequence with memory.

Example 6.1-15

(Two-state random sequence with memory.) We construct the two-level (binary) random sequence $X[n]$ on $n \geq 0$ as follows. Recursively, and for each $n(> 0)$ in succession, we set $X[n] = X[n-1]$ with probability p, for some given $0 < p < 1$. Otherwise, and with probability $q \triangleq 1 - p$, we set $X[n]$ to the "other" value (level). Let the two levels a, b satisfy $b > a$ and start off the sequence with $X[0] = a$. When $p = 0.5$, this is a special case of the Bernoulli random sequence. When $p \neq 0.5$, this is not an independent random sequence, since $P_X(x_n|x_{n-1}; n, n-1) \neq P_X(x_n; n)$. We say the random sequence has *memory*. To see this, consider the case where $p \approx 1.0$, then set x_n to the level other than x_{n-1}, and note that the conditional transition probability $P_X(x_n|x_{n-1}; n, n-1) \approx 0$, while the unconditional probability $P_X(x_n; n)$ is not so constrained. Intuitively, at least, it makes sense to call $X[n-1]$ the *state* at time $n-1$. In fact, the rules for generating this random sequence can be summarized in the *state-transition diagram* shown in Figure 6.1-14, where the directed branches are labeled by the relevant probabilities for the next state, given the present state, as can easily be verified by inspection. We can refer to p as the *no-transition* probability. This is a first example of a Markov random sequence which will be studied in Section 6.5.

The following MATLAB m-file can generate sample functions for these random sequences on $n \geq 1$:

```
function[w]=randmemseq(p,N,w0,a,b)
w=a*ones(1,N);
w(1)=w0;
for i=2:N
    rnum=rand;
    if rnum <p;
     w(i)=w(i-1);
     else
    if w(i-1)==a;
       w(i)=b;
    else
       w(i)=a;
    end
```

```
end
stem(1:N,w)
title('random sequence with memory')
xlabel('discrete time')
ylabel('level')
end
```

Sample waveforms are given in Figures 6.1-15 to 6.1-17 corresponding to level values $b = 1$, $a = 0$, and several values of p. We note that when p is near 1, there are few transitions. For p near 0.5, there will be many transitions displaying little memory. When $p = 0$, there is a transition every time.

Example 6.1-16 _____

(Correlation function of random sequence with memory.) Assume that the random sequence with memory of the last example has been running for a very long time. Later on we will show that in this case, a steady state develops wherein the probabilities of the two levels are constant with time and independent of the starting state (level). Here we assume that the steady state holds for all finite time. Clearly from the symmetry shown in the state diagram, it must be that $P_X(a) = P_X(b) = 0.5$. Now assume that the lower level $a = 0$ and the upper level is b as before, and consider the correlation at two distinct times n and $n + k$. We can write

$$R_{XX}[n, n+k] = b^2 P_X(b, b; n, n+k)$$
$$= b^2 P\{X[n] = b\} P\{X[n+k] = b | X[n] = b\}$$
$$= b^2/2\, P\{X[n+k] = b | X[n] = b\}$$

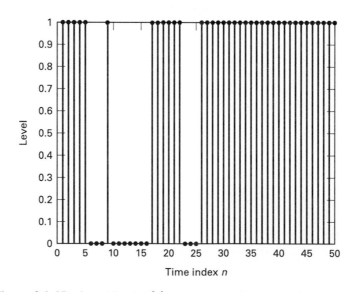

Figure 6.1-15   Initial level $X[1] = 1$, no-transition probability $p = 0.8$.

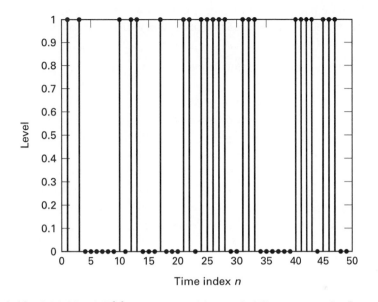

Figure 6.1-16   Initial level $X[1] = 1$, no-transition probability $p = 0.5$, the Bernoulli case.

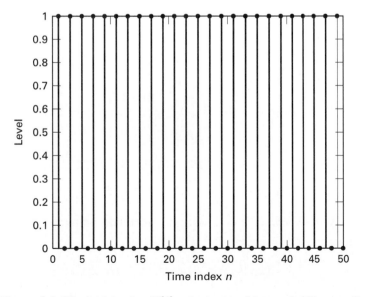

Figure 6.1-17   Initial value $X[1] = 1$, no-transition probability $p = 0$.

where the first equality holds since all the terms involving a are zero since $a = 0$. Now the only way that X can equal b at both times n and $n+k$ is for an even number of transitions to occur between these two times, and the probability of this is given by

$$P\{\text{even number of transitions}\} = \sum_{l=0,2,4,\dots}^{k} \binom{k}{l} (1-p)^l p^{k-l}$$

$$\stackrel{\triangle}{=} A_e,$$

which follows from the fact that this is just Bernoulli trials with "success" = "transition" and "failure" = "no transition." Thus interchanging the usual role of p and q in Bernoulli trials, we just add up the probability of an even number of successes (transitions). It turns out that A_e can be evaluated in closed form by the following "trick." Define

$$A_o \stackrel{\triangle}{=} \sum_{l=1,3,5,\dots}^{k} \binom{k}{l} (1-p)^l p^{k-l} (-1)^l.$$

Clearly we have $A_e - A_o = 1$ since l is always odd valued in the sum A_o. Similarly we note that

$$A_e = \sum_{l=0,2,4,\dots}^{k} \binom{k}{l} (1-p)^l p^{k-l} (-1)^l$$

$$= \sum_{l=0,2,4,\dots}^{k} \binom{k}{l} (p-1)^l p^{k-l},$$

where the first equality holds because l is always even in A_e. We now can see that

$$A_e + A_o = \sum_{l=0}^{k} \binom{k}{l} (p-1)^l p^{k-l}$$

$$= (2p-1)^k,$$

by the Binomial Theorem. It follows at once that $A_e = 1/2\left[(2p-1)^k + 1\right]$, so that

$$R_{XX}[n, n+k] = b^2/4\left[(2p-1)^k + 1\right],$$

which shows that $X[n]$ is WSS. We can write this correlation function more cleanly for the case $p > 1/2$. On defining $\alpha \stackrel{\triangle}{=} \ln(2p-1)|$, we have

$$R_{XX}[k] = b^2/4\left[\exp(-\alpha |k|) + 1\right].$$

Also since the mean value of $X[n]$ is easily seen to be $b/2$, we get the autocovariance function

$$K_{XX}[k] = b^2/4 \exp(-\alpha |k|).$$

A MATLAB m file for displaying the covariance functions of these sequences, for three values of p, is shown below:

```
function[mc1,mc2,mc3]=markov(b,p1,p2,p3,N)
mc1=0*ones(1,N);
mc2=0*ones(1,N);
mc3=0*ones(1,N);
for i=1:N
    mc1(i)=0.25*(b^2)*(((2*p1-1)^(i-1)));
    mc2(i)=0.25*(b^2)*(((2*p2-1)^(i-1)));
    mc3(i)=0.25*(b^2)*(((2*p3-1)^(i-1)));
end
x=linspace(0,N-1,N);
plot(x,mc1,mc2,x,mc3)
title('covariance of Markov Sequences')
xlabel('Lag interval')
ylabel('covariance value')
```

The normalized covariances for $p = 0.8$, 0.5, and 0.2 and $b = 2$ are shown in Figure 6.1-18.

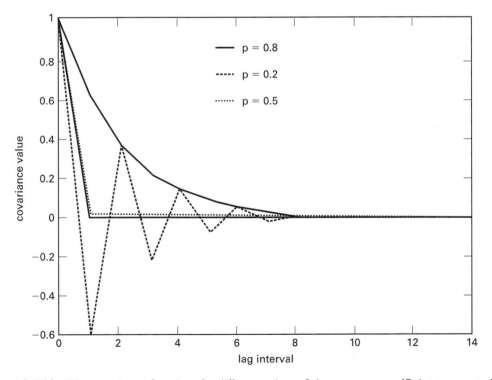

Figure 6.1-18   The covariance functions for different values of the parameter p. (Points connected by straight lines.)

This type of random sequence, which exhibits a one-step memory, is called a *Markov* random sequence (there are variations on the spelling of Markov) in honor of the mathematician A.A. Markov (1856–1922). In Section 6.5 we shall discuss this class of random sequences in greater detail. In the meanwhile we note that the system discussed in Example 6.1-5 i.e., $X[n] = \alpha X[n-1] + W[n]$ also exhibited a one-step memory and hence, could also be regarded as a Markov sequence, when $W[n]$ is an independent random sequence.

In Section 6.2, we provide a review or summary of the theory of linear systems for sequences, that is, discrete-time linear system theory. Readers with adequate background may skip this section. In Section 6.3, we will apply this theory to study the effect of linear systems on random sequences, an area rich in applications in communications, signal processing, and control systems.

6.2 BASIC PRINCIPLES OF DISCRETE-TIME LINEAR SYSTEMS

In this section we present some fundamental material on discrete-time linear system theory. This will then be extended in the next section to the case of random sequence inputs and outputs. This material is very similar to the continuous-time linear system theory including the topics of differential equations, Fourier transforms, and Laplace transforms. The corresponding quantities in the discrete-time theory are difference equations, Fourier transforms (for discrete-time signals), and Z-transforms.

With reference to Figure 6.2-1 we see that a linear system can be thought of as having an infinite-length sequence $x[n]$ as input with a corresponding infinite-length sequence $y[n]$ as output. Representing this linear operation in equation form we have

$$y[n] = L\{x[n]\}, \tag{6.2-1}$$

where the linear operator L is defined to satisfy the following definition adapted to the case of discrete-time signals. This notation might appear to indicate that $x[n]$ at time n is the only input value that affects the output $y[n]$ at time n. In fact, all input values can potentially affect the output at any time n. This is why we call L an operator[†] and not merely a function. The examples below will make this point clear. Mathematicians use the operator notation $y = L\{x\}$ which avoids this difficulty but makes the functional dependence of x and y on the (time) parameter n less clear than in our engineering notation.

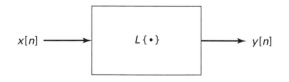

Figure 6.2-1   System diagram for generic linear system $L\{\cdot\}$ with input $x[n]$ and output $y[n]$ and time index parameter n.

[†]Operators map functions (sequences) into functions (sequences).

Definition 6.2-1   We say a system with operator L is *linear* if for all permissible input sequences $x_1[n]$ and $x_2[n]$, and for all permissible pairs of scalar gains a_1 and a_2 we have

$$L\left\{a_1 x_1[n] + a_2 x_2[n]\right\} = a_1 L\{x_1[n]\} + a_2 L\{x_2[n]\}. \quad \blacksquare$$

In words, the response of a linear system to a weighted sum of inputs is the weighted sum of the individual outputs. Examples of linear systems would include *moving averages* such as

$$y[n] = 0.33(x[n+1] + x[n] + x[n-1]), \qquad -\infty < n < +\infty,$$

and *autoregressions* such as,

$$y[n] = ay[n-1] + by[n-2] + cx[n], \qquad 0 \le n < +\infty,$$

when the initial conditions are *zero*. Both these equations are special cases of the more general *linear constant-coefficient difference equation* (LCCDE),

$$y[n] = \sum_{k=1}^{M} a_k y[n-k] + \sum_{k=0}^{N} b_k x[n-k]. \tag{6.2-2}$$

Example 6.2-1
(Solution of difference equations.) Consider the following second-order LCCDE,

$$y[n] = 1.7y[n-1] - 0.72y[n-2] + u[n], \tag{6.2-3}$$

with $y[-1] = y[-2] = 0$ and $u[n]$ the unit-step function. To solve this equation for $n \ge 0$ we first find the general solution to the *homogeneous equation*

$$y_h[n] = 1.7y_h[n-1] - 0.72y_h[n-2].$$

We try $y_h[n] = Ar^n$, where A and r are to be determined[†] and obtain

$$A(r^n - 1.7r^{n-1} + 0.72r^{n-2}) = 0$$

or

$$Ar^{n-2}(r^2 - 1.7r + 0.72) = 0.$$

We thus see that any value of r satisfying the *characteristic equation*

$$r^2 - 1.7r + 0.72 = 0$$

will give a general solution to the homogeneous equation. In this case there are two roots at $r_1 = 0.8$ and $r_2 = 0.9$. By linear superposition the general *homogeneous solution* must be of the form[‡]

$$y_h[n] = A_1 r_1^n + A_2 r_2^n,$$

where the constants A_1 and A_2 may be determined from the initial conditions.

[†]A thorough treatment of the solution of linear difference equations may be found in Reference [6-4].

[‡]Since the two roots are less than one in magnitude, the solution will be *stable* when run forward in time index n (cf. [6-4].)

To obtain the *particular solution*, we first observe that the input sequence $u[n]$ equals 1 for $n \geq 0$. Thus we try as a particular solution a constant, that is, following standard practice,

$$y_p[n] = B \qquad \text{for } n \geq 0$$

and obtain

$$B - 1.7B + 0.72B = 1$$

or

$$B = 1/(1 - 1.7 + 0.72) = 1/(0.02) = 50.$$

More generally this method can be modified for any input function of the form $C\rho^n$ over adjoining time intervals $[n_1, n_2 - 1]$. One just assumes the corresponding form for the solution and determines the constant C as shown. In this approach, we would solve the difference equation for each time interval separately, piecing the solution together at the boundaries by carrying across final conditions to become the initial conditions for the next interval. We illustrate our approach here for the time interval starting at $n = 0$. The total solution is

$$y[n] = y_h[n] + y_p[n]$$
$$= A_1(0.8)^n + A_2(0.9)^n + 50 \qquad \text{for } n \geq 0.$$

To determine A_1 and A_2 we first evaluate Equation 6.2-3 at $n = 0$ and $n = 1$ using $y[-1] = y[-2] = 0$ to carry across the initial conditions to obtain $y[0] = 1$ and $y[1] = 2.7$, from which we obtain the linear equations

$$A_1 + A_2 + 50 = 1 \quad (\text{at } n = 0)$$

and

$$A_1(0.8) + A_2(0.9) + 50 = 2.7 \quad (\text{at } n = 1).$$

This can be put in matrix form

$$\begin{bmatrix} 1.0 & 1.0 \\ 0.8 & 0.9 \end{bmatrix} \begin{bmatrix} A_1 \\ A_2 \end{bmatrix} = \begin{bmatrix} -49.0 \\ -47.3 \end{bmatrix}$$

and solved to yield

$$\begin{bmatrix} A_1 \\ A_2 \end{bmatrix} = \begin{bmatrix} 32 \\ -81 \end{bmatrix}.$$

Thus the complete solution, valid for $n \geq 0$, is

$$y[n] = 32(0.8)^n - 81(0.9)^n + 50.$$

We could then write the solution for all time, if the system was at rest for $n < 0$, as

$$y[n] = (32(0.8)^n - 81(0.9)^n + 50)\, u[n].$$

Note that the LCCDE in the previous example is a linear system because the initial conditions, i.e., $y[-1], y[-2]$ were zero, often called the *initial rest condition*. Without initial rest, an LCCDE is not a linear system. More generally, linear systems are described by *superposition* with a possibly *time-variant impulse response*

$$h[n, k] \triangleq L\{\delta[n - k]\}.$$

In words we call $h[n, k]$ the response at time n to an impulse applied at time k. We derive the result by simply writing the input as $x[n] = \sum x[k]\delta[n - k]$, and then using linearity to conclude

$$y[n] = L\left\{\sum_{k=-\infty}^{+\infty} x[k]\delta[n - k]\right\}$$

$$= \sum_{k=-\infty}^{+\infty} x[k]L\{\delta[n - k]\}$$

$$= \sum_{k=-\infty}^{+\infty} x[k]\, h[n, k],$$

which is called the *superposition summation* representation for linear systems.

Many linear systems are made of constant components and have an effect on input signals that is invariant to when the signal arrives at the system. A linear system is called *linear time-invariant* (LTI) or equivalently *linear shift-invariant* (LSI) if the response to a delayed (shifted) input is just the delayed (shifted) response. More precisely we have the following.

Definition 6.2-2   A linear system L is called *shift-invariant* if for all integer shifts k, $-\infty < k < +\infty$, we have

$$y[n + k] = L\{x[n + k]\} \quad \text{for all } n. \qquad \blacksquare \qquad (6.2\text{-}4)$$

An important property of LSI systems is that they are described by *convolution*,[†] that is, L is a convolution operator,

$$y[n] = h[n] * x[n] = x[n] * h[n],$$

where

$$h[n] * x[n] \triangleq \sum_{k=-\infty}^{+\infty} h[k]x[n - k], \qquad (6.2\text{-}5)$$

[†]We earlier encountered convolution for the sum of two independent random variables. (Ref. Equations 3.3-14 and 3.3-26b.) In the case of probability mass functions (pmf's) it was discrete parameter convolution. The index parameter was not time there.

and the sequence

$$h[n] \triangleq L\{\delta[n]\},$$

is called the *impulse response*. With relation to the time-varying impulse response $h[n, k]$, we can see that $h[n] = h[n, 0]$ when a linear system is shift-invariant.

In words we can say that—just as for continuous-time systems—if we know the impulse response of an LSI system, then we can compute the response to any other input by carrying out the convolution operation. In the discrete-time case this convolution operation is a summation rather than an integration, but the operation is otherwise the same.

While in principle we could determine the output to any input, given knowledge of the impulse response, in practice the calculation of the convolution operation may be tedious and time consuming. To facilitate such calculations and also to gain added insight, we turn to a frequency-domain characterization of LSI systems. We begin by defining the *Fourier transform* (*FT*) for sequences as follows.

Definition 6.2-3   The *Fourier transform* for a discrete-time signal or sequence is defined by the infinite sum (if it exists)

$$X(\omega) = FT\{x[n]\} \triangleq \sum_{n=-\infty}^{+\infty} x[n]e^{-j\omega n}, \quad \text{for } -\pi \leq \omega \leq +\pi,$$

and the function $X(\omega)$ is periodic with period 2π outside this range. The *inverse Fourier transform* is given as

$$x[n] = IFT\{X(\omega)\} = \frac{1}{2\pi} \int_{-\pi}^{+\pi} X(\omega)e^{j\omega n}d\omega. \quad \blacksquare$$

One can see that the Fourier transform and its inverse for sequences are really just the familiar Fourier series with the sequence x playing the role of the Fourier coefficients and the Fourier transform X playing the role of the periodic function. Thus the existence and uniqueness theorems of Fourier series are immediately applicable here to the Fourier transform for discrete-time signals. Note that the frequency variable ω is sometimes called *normalized frequency* because, if the sequence $x[n]$ arose from sampling, the period of such sampling has been lost. It is as though the sample period were $T = 1$, as would be consistent with the $[-\pi, +\pi]$ frequency range of the Fourier transform $X(\omega)$.[†]

For an LSI system the Fourier transform is particularly significant owing to the fact that complex exponentials are the *eigenfunctions* of discrete-time linear systems, that is,

$$L\{e^{j\omega n}\} = H(\omega)e^{j\omega n}, \tag{6.2-6}$$

as long as the impulse response h is absolutely summable. For LSI systems this absolute summability can easily be seen to be equivalent to bounded-input bounded-output (BIBO) stability [6-4].

[†]If the sequence arose from sampling with sample period T, the (true) radian frequency $\omega_r = \omega/T$.

Just as in continuous-time system theory, multiplication of Fourier transforms corresponds to convolution in the time (or space) domain.

Theorem 6.2-1   The convolution,

$$y[n] = x[n] * h[n], \qquad -\infty < n < +\infty,$$

is equivalent in the transform domain to

$$Y(\omega) = X(\omega)H(\omega), \qquad -\pi \leq \omega \leq +\pi.$$

Proof

$$Y(\omega) = \sum_{-\infty}^{+\infty} y[n]e^{j\omega n} = \sum_{-\infty}^{+\infty} (x[n] * h[n]) \, e^{-j\omega n},$$

$$= \sum_{n}\sum_{k} x[k]h[n-k]e^{-j\omega n} = \sum_{n}\sum_{k} x[k]h[n-k]e^{-j\omega(n-k+k)},$$

$$= \sum_{n}\sum_{k} [x[k]e^{-j\omega k}h[n-k]e^{-j\omega(n-k)}]$$

$$= \sum_{k} x[k]e^{-j\omega k} \left(\sum_{n} h[n-k]e^{-j\omega(n-k)} \right),$$

$$= \sum_{k} x[k]e^{-j\omega k} H(\omega)$$

$$= X(\omega)H(\omega).$$

Thus discrete-time linear shift-invariant systems are easily understood in the frequency domain similar to the situation for continuous-time LSI systems. Analogous to the Laplace transform for continuous-time signals, there is the Z-transform for discrete-time signals. It is defined as follows.   ■

Definition 6.2-4   The Z-transform of a discrete-time signal or sequence is defined as the infinite summation (if it exists)

$$\mathsf{X}(z) \triangleq \sum_{-\infty}^{+\infty} x[n]z^{-n}, \tag{6.2-7}$$

where z is a complex variable in the *region of absolute convergence* of this infinite sum.[†]   ■

Note that $\mathsf{X}(z)$ is a function of a *complex variable*, while $X(\omega)$ is a function of a *real variable*. The two are related by $\mathsf{X}(z)|_{z=e^{j\omega}} = X(\omega)$. We thus see that, if the Z-transform

---

[†]Note the sans serif font to distinguish between the Z-transform and the Fourier transform.

exists, the Fourier transform is just the restriction of the Z-transform to the unit circle in the complex z-plane. Similarly to the proof of Theorem 6.2-1, it is easy to show that the convolution-multiplication property Equation 6.2-1 is also true for Z-transforms. Analogous to continuous-time theory, the Z-transform $\mathsf{H}(z)$ of the impulse response $h[n]$ of an LSI system is called the *system function*. For more information on discrete-time signals and systems, the reader is referred to Reference [6-4].

6.3 RANDOM SEQUENCES AND LINEAR SYSTEMS

In this section we look at the topic of linear systems with random sequence inputs. In particular we will look at how the mean and covariance functions are transformed by both linear and LSI systems. We will do this first for the general case of a nonstationary random sequence and then specialize to the more common case of a stationary sequence. The topics of this section are perhaps the most widely used concepts from the theory of random sequences. Applications arise in communications when analyzing signals and noise in linear filters, in digital signal processing for the analysis of quantization noise in digital filters, and in control theory to find the effect of disturbance inputs on an otherwise deterministic control system.

The first issue is the meaning of inputing a random sequence to a linear system. The problem is that a random sequence is not just one sequence but a whole family of sequences indexed by the parameter ζ, a point in the sample space. As such for each fixed ζ, the random sequence is just an ordinary sequence that may be a permissible input for the linear system. Thus when we talk about a linear system with a random sequence input, it is natural to say that for each point in the sample space Ω, we input the corresponding realization, that is, the sample sequence $x[n]$. We would therefore regard the corresponding output $y[n]$ as a sample sequence[†] corresponding to the same point ζ in the sample space, thus collectively defining the output random sequence $Y[n]$.

Definition 6.3-1   When we write $Y[n] = L\{X[n]\}$ for a random sequence $X[n]$ and a linear system L, we mean that for each $\zeta \in \Omega$ we have

$$Y[n, \zeta] = L\left\{X[n, \zeta]\right\}.$$

Equivalently, for each sample function $x[n]$ taken on by the input random sequence $X[n]$, we set $y[n]$ as the corresponding sample sequence of the output random sequence $Y[n]$, i.e., $y[n] = L\{x[n]\}$. ■

This is the simplest way to treat systems with random inputs. A difficulty arises when the input sample sequences do not "behave well," in which case it may not be possible to define the system operation for every one of them.[‡] In Chapter 8 we will generalize this definition and discuss a so-called mean-square description of the system operation, which avoids such problems, although of necessity it will be more abstract.

[†]Recall that $x[n]$, $y[n]$ denote $X[n, \zeta]$, $Y[n, \zeta]$, respectively, for *fixed* ζ.

[‡]For example suppose the system is an accumulator i.e., $y[n] = \sum_{k=-\infty}^{n} x[k]$. Then an infinite length random sequence $X[n]$ might not be well-behaved with respect to this system since for any n either the mean or the variance of $Y[n]$ might be infinite.

In most cases it is very hard to find the probability distribution of the output from the probabilistic description of the input to a linear system. The reason is that since the impulse response is often very long (or infinitely long), high-order distributions of the input sequence would be required to determine the output PDF. In other words, if $Y[n]$ depends on the most recent k input values $X[n], \ldots, X[n-k+1]$, then the kth-order pdf of X has to be known in order to compute even the first-order pdf of Y. The situation with moment functions is different. The moments of the output random sequence can be calculated from equal- or lower-order moments of the input, when the system is linear. Partly for this reason, it is of considerable interest to determine the output moment functions in terms of the input moment functions. In the practical and important case of the Gaussian random sequence, we have seen that the entire probabilistic description depends only on the mean and covariance functions. In fact because the linear system is in effect performing a linear transformation on the infinite-dimensional vector that constitutes the input sequence, we can see that the output sequence will also obey the Gaussian law in its nth-order distributions if the input sequence is Gaussian. Thus the determination of the first- and second-order moment functions of the output is particularly important when the input sequence is Gaussian.

Theorem 6.3-1   For a linear system L and a random sequence $X[n]$, the mean of the output random sequence $Y[n]$ is

$$E\{Y[n]\} = L\{E\{X[n]\}\} \tag{6.3-1}$$

as long as both sides are well defined.

Proof (formal).   Since L is a linear operator, we can write

$$y[n] = \sum_{k=-\infty}^{+\infty} h[n,k]x[k]$$

for each sample sequence input–output pair, or

$$Y[n,\zeta] = \sum_{k=-\infty}^{+\infty} h[n,k]X[k,\zeta],$$

where we explicitly indicate the outcome ζ. If we operate on both sides with the expectation operator E we get

$$E\{Y[n]\} = E\left\{ \sum_{k=-\infty}^{+\infty} h[n,k]X[k] \right\}.$$

Now, assuming it is valid to bring the operator E inside the infinite sum, we get

$$E\{Y[n]\} = \sum_{k=-\infty}^{+\infty} h[n,k]E\{X[k]\}$$

$$= L\{E\{X[n]\}\},$$

which can be written as

$$\mu_Y[n] = \sum_{k=-\infty}^{+\infty} h[n,k]\mu_X[k],$$

i.e., the mean function of the output is the response of the linear system to the mean function of the input. ■

Some comments are necessary with regard to this interchange of the expectation and linear operator. It cannot always be done! For example, if the input has a nonzero mean function and the linear system is a *running sum*, that is,

$$y[n] = \sum_{k=0}^{+\infty} x[n-k]$$

the running sum of the mean may not converge. Then such an interchange is not valid. We will come back to this point when we study stochastic convergence in Section 6.7. We will see then that a sufficient condition for an LSI system to satisfy Equation 6.3-1 is that its impulse response $h[n]$ be absolutely summable.

There are special cases of Equation 6.3-1 depending on whether the input sequence is WSS and whether the system is LSI. If the system is LSI and the input is at least WSS, then the mean of the output is given as

$$E\{Y[n]\} = \sum_{k=-\infty}^{+\infty} h[n-k]\mu_X.$$

Now because μ_X is a constant, we can take it out of the sum and obtain

$$E\{Y[n]\} = \left[\sum_{k=-\infty}^{+\infty} h[k]\right]\mu_X \qquad (6.3\text{-}2)$$

$$= \mathsf{H}(z)|_{z=1}\,\mu_X, \qquad (6.3\text{-}3)$$

at least whenever $\sum_{k=-\infty}^{+\infty}|h[k]|$ exists, that is, for any BIBO stable system (cf. Section 6.2).

Thus we observe that in this case the mean of the output random sequence is a constant equal to the product of the *dc gain* or *constant gain* of the LSI system times the mean of the input sequence.

Example 6.3-1

Let the system be a *lowpass* filter with system function

$$\mathsf{H}(z) = 1/(1+az^{-1}),$$

where we require $|a| < 1$ for stability of this assumed causal filter (that is, the region of convergence is $|z| > |a|$, which includes the unit circle). Then if a WSS sequence is the input to this filter, the mean of the output will be

$$E\{Y[n]\} = \mathsf{H}(z)|_{z=1}E\{X[n]\}$$

$$= (1+a)^{-1}\mu_X.$$

We now turn to the problem of calculating the output covariance and correlation of the general linear system whose operator is L:

$$Y[n] = L\{X[n]\}.$$

We will find it convenient to introduce a cross-correlation function between the input and output,

$$R_{XY}[m,n] \triangleq E\{X[m]Y^*[n]\} \tag{6.3-4}$$

$$= E\{X[m]\,(L\{X[n]\})^*\}. \tag{6.3-5}$$

Now, in order to factor out the operator, we introduce the operator L_n^*, with impulse response $h^*[n,k]$, which operates on time index n, but treats time index m as a constant. We can then write $R_{XY}[m,n] = E\{X[m]L_n^*\{X[n]\}\} = L_n^*\{E\{X[m]X[n]\}\}$. Similarly we denote with L_m the linear operator with time index m, that treats n as a constant. The operator L_n^* is related to the *adjoint operator* studied in linear algebra.

Theorem 6.3-2   Let $X[n]$ and $Y[n]$ be two random sequences that are the input and output, respectively, of the linear operator L_n. Let the input correlation function be $R_{XX}[m,n]$. Then the cross- and output-correlation functions are respectively given by

$$R_{XY}[m,n] = L_n^* \{R_{XX}[m,n]\}$$

and

$$R_{YY}[m,n] = L_m \{R_{XY}[m,n]\}.$$

Proof   Write

$$X[m]Y^*[n] = X[m]L_n^*\{X^*[n]\}$$

$$= L_n^*\{X[m]X^*[n]\}.$$

Then

$$R_{XY}[m,n] = E\{X[m]Y^*[n]\} = E\{L_n^*\{X[m]X^*[n]\}\},$$

$$= L_n^*\{E\{X[m]X^*[n]\}\},$$

$$= L_n^*\{R_{XX}[m,n]\},$$

thus establishing the first part of the theorem. To show the second part, we proceed analogously by multiplying $Y[m]$ by $Y^*[n]$ to get

$$E\{Y[m]Y^*[n]\} = E\{L_m\{X[m]Y^*[n]\}\},$$

$$= L_m\{E\{X[m]Y^*[n]\}\},$$

$$= L_m\{R_{XY}[m,n]\},$$

as was to be shown.   ∎

If we combine both parts of Theorem 6.3-2 we get an operator expression for the output correlation in terms of the input correlation function:

$$R_{YY}[m,n] = L_m\{L_n^*\{R_{XX}[m,n]\}\}, \tag{6.3-6}$$

which can be put into the form of a superposition summation for a system with time-variant impulse response $h[n,k]$ as

$$R_{YY}[m,n] = \sum_{k=-\infty}^{+\infty} h[m,k] \left(\sum_{l=-\infty}^{+\infty} h^*[n,l]R_{XX}[k,l] \right). \tag{6.3-7}$$

Here the superposition summation representation for $R_{XY}[m,n]$ is

$$R_{XY}[m,n] = L_n^*\{R_{XX}[m,n]\}$$

$$= \sum_{l=-\infty}^{+\infty} h^*[n,l]R_{XX}[m,l],$$

and that for $R_{YX}[m,n]$ is

$$R_{YX}[m,n] = \sum_{k=-\infty}^{+\infty} h[m,k]R_{XX}[k,n].$$

To find the corresponding results for covariance functions, we note that the centered output sequence is the output due to the centered input sequence, due to the linearity of the system and Equation 6.3-1. Then applying Theorem 6.3-2 to these zero-mean sequences, we have immediately that, for covariance functions,

$$K_{XY}[m,n] = L_n^*\{K_{XX}[m,n]\} \tag{6.3-8}$$

$$K_{YY}[m,n] = L_m\{K_{XY}[m,n]\} \tag{6.3-9}$$

and

$$K_{YY}[m,n] = L_m\{L_n^*\{K_{XX}[m,n]\}\}, \tag{6.3-10}$$

which becomes the following superposition summation

$$K_{YY}[m,n] = \sum_{k=-\infty}^{+\infty} h[m,k] \left(\sum_{l=-\infty}^{+\infty} h^*[n,l]K_{XX}[k,l] \right). \tag{6.3-11}$$

Example 6.3-2 _____

(Edge detector.) Let $Y[n] \triangleq X[n] - X[n-1] = L\{X[n]\}$, an operator that represents a first-order (backward) difference. See Figure 6.3-1. This linear operator could be applied to locate an impulse noise spike in some random data. The output mean is $E[Y[n]] = L\{E[X[n]]\} = \mu_X[n] - \mu_X[n-1]$. The cross-correlation function is

$$R_{XY}[m,n]] = L_n\{R_{XX}[m,n]\}$$

$$= R_{XX}[m,n] - R_{XX}[m,n-1].$$

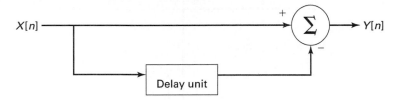

Figure 6.3-1   An edge detector that gives nearly zero output when $X[n] \approx X[n-1]$ and a large output when $|X[n] - X[n-1]|$ is large.

The output autocorrelation function is

$$R_{YY}[m,n] = L_m\{R_{XY}[m,n]\}$$
$$= R_{XY}[m,n] - R_{XY}[m-1,n]$$
$$= R_{XX}[m,n] - R_{XX}[m-1,n] - R_{XX}[m,n-1] + R_{XX}[m-1,n-1].$$

If the input random sequence were WSS with autocorrelation function,

$$R_{XX}[m,n] = a^{|m-n|}, \qquad 0 < a < 1,$$

then the above example would specialize to

$$\mu_Y[n] = 0,$$
$$R_{XY}[m,n] = a^{|m-n|} - a^{|m-n+1|}$$

and

$$R_{YY}[m,n] = 2a^{|m-n|} - a^{|m-1-n|} - a^{|m-n+1|},$$

which depends on only $m-n$. Hence the output random sequence is WSS and we can write (with $k = m - n$)

$$R_{YY}[k] = 2a^{|k|} - a^{|k-1|} - a^{|k+1|}.$$

For the input autocorrelation with $a = 0.7$ as shown in Figure 6.3-2, the output autocorrelation function is shown in Figure 6.3-3. Note that the edge detector has a strong tendency to *decorrelate* the input sequence.

Example 6.3-3 _____
(Covariance functions of a recursive system.) With $|\alpha| < 1$, let

$$Y[n] = \alpha Y[n-1] + (1-\alpha)W[n] \qquad (6.3\text{-}12)$$

for $n \geq 0$ subject to $Y[-1] = 0$. Since the initial condition (i.c.) is zero, the system is equivalently LSI for $n \geq 0$, so we can represent L by convolution, where

$$h[n] = (1-\alpha)\alpha^n u[n].$$

Here $h[n]$ is the impulse response of the corresponding deterministic first-order difference equation, i.e., $h[n]$ is the solution to

$$h[n] = \alpha h[n-1] + (1-\alpha)\delta[n]$$

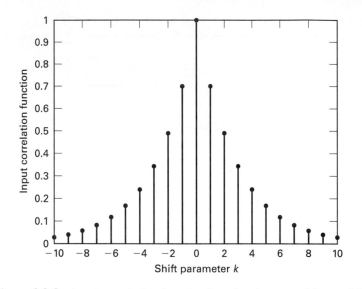

Figure 6.3-2   Input correlation function for edge detector with $a = 0.7$.

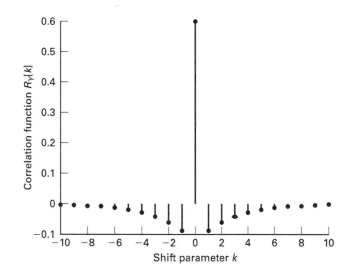

Figure 6.3-3   Correlation function $R_{YY}[k]$ for backward difference example, plot has $a = 0.7$.

where $\delta[n]$ is the discrete-time impulse sequence. This solution can be obtained easily by recursion or by using the Z-transform.[†] Then specializing Equation 6.3-1 we obtain

---

[†]Taking the Z-transform of both sides of the above equation, and noting that the Z-transform of the impulse sequence is 1, we obtain $\mathsf{H}(z) = (1 - \alpha)/(1 - \alpha z^{-1})$. Upon inverse Z-transform, one gets the $h[n]$ given above. (For help with the inverse Z-transform, see Appendix A).

$$\mu_Y[n] = \sum_{k=0}^{\infty} (1-\alpha)\alpha^k \mu_W[n-k], \text{ where } \mu_W[n] = 0 \text{ for } n < 0.$$

Applying Equations 6.3-8 and 6.3-9 to this case enables us to write, for α real,

$$K_{WY}[m,n] = \sum_{k=0}^{\infty} (1-\alpha)\alpha^k K_{WW}[m,n-k]$$

and

$$K_{YY}[m,n] = \sum_{l=0}^{\infty} (1-\alpha)\alpha^l K_{WY}[m-l,n],$$

which can be combined to yield

$$K_{YY}[m,n] = \sum_{k=0}^{\infty}\sum_{l=0}^{\infty} (1-\alpha)^2 \alpha^k \alpha^l K_{WW}[m-l,n-k].$$

Now if the input sequence $W[n]$ has covariance function

$$K_{WW}[m,n] = \sigma_W^2 \delta[m-n] \text{ for } m,n \geq 0,$$

then the output covariance is calculated as

$$K_{YY}[m,n] = \sum_{k=0}^{n} (1-\alpha)^2 \alpha^k \alpha^{(m-n)+k} \sigma_W^2 \quad \text{for} \quad m \geq n \geq 0,$$

$$= \alpha^{(m-n)}(1-\alpha)^2 \sum_{k=0}^{n} \alpha^{2k} \sigma_W^2$$

$$= \alpha^{(m-n)}\left[(1-\alpha)^2/(1-\alpha^2)\right]\sigma_W^2(1-\alpha^{2n+2}) \quad \text{for} \quad m \geq n \geq 0$$

$$= [(1-\alpha)/(1+\alpha)]\alpha^{|m-n|}\sigma_W^2(1-\alpha^{2\min(m,n)+2}) \quad \text{for all } m,n \geq 0$$

where the last step follows from the required symmetry in (m,n). Note that the term $\alpha^{2\min(m,n)+2}$ is a transient that dies away as $m,n \to \infty$, since $|\alpha| < 1$, so that asymptotically we have the *steady-state* answer

$$K_{YY}[m,n] = \left(\frac{1-\alpha}{1+\alpha}\right)\sigma_W^2 \, \alpha^{|m-n|}, \quad m,n \to \infty$$

a shift-invariant covariance function. If the mean function $\mu_Y[n]$ is found to be asymptotic to a constant, then the random sequence $Y[n]$ is said to be *asymptotically WSS*. We discuss WSS random sequences in greater detail in the next section.

As an alternative to this method of solution, one can take the expectation of Equation 6.3-12 to directly obtain a recursive equation for the output mean sequence which can be solved by the methods of Section 6.2:

$$\mu_Y[n] = \alpha\mu_Y[n-1] + (1-\alpha)\mu_W[n], \qquad n \geq 0,$$

with an appropriate initial condition. For example, if $\mu_Y[-1] = 0$ and $\mu_W[n] = \mu_W$, a given constant, then the solution is

$$\mu_Y[n] = (1 - \alpha^{n+1})\mu_W u[n].$$

We can also use this method to calculate the cross-correlation function between input and output. First we conjugate Equation 6.3-12, then multiply by $W[m]$, and finally take the expectation to yield, for α real,

$$R_{WY}[m, n] = \alpha R_{WY}[m, n-1] + (1 - \alpha)R_{WW}[m, n], \qquad (6.3\text{-}13)$$

which can be solved directly for R_{WY} in terms of R_{WW}. The partial difference equation for the output correlation R_{YY} is obtained by re-expressing Equation 6.3-12 as a function of m, multiplying by $Y^*[n]$, and then taking the expectation to yield

$$R_{YY}[m, n] = \alpha R_{YY}[m-1, n] + (1 - \alpha)R_{WY}[m, n]. \qquad (6.3\text{-}14)$$

These two difference equations can be solved by the methods of Section 6.2 since they can each be seen to be one-dimensional difference equations with constant coefficients in one index, with the other index simply playing the role of an additional parameter. Thus, for example, one must solve Equation 6.3-13 as a function of n for each value of m in succession.

6.4 WSS RANDOM SEQUENCES

In this section we will assume that the random sequences of interest are all WSS, i.e.,

$$(1) \qquad E\{X[n]\} = \mu_X, \text{ a constant,}$$

$$(2) \quad R_{XX}[k + m, k] = E\{X[k + m]X^*[k]\}$$

$$= R_{XX}[m],$$

and of *second order*, i.e., $E\{|X[n]|^2\} < \infty$.

Some important properties of the autocorrelation function of stationary random sequences are presented below. They also hold for covariance functions, since they are just the autocorrelation function of the *centered* random sequence $X_c[n] \triangleq X[n] - \mu_X$.

1. For arbitrary m, $|R_{XX}[m]| \leq R_{XX}[0] \geq 0$, which follows directly from $E\{|X[m] - X[0]|^2\} \geq 0$.
2. $|R_{XY}[m]| \leq \sqrt{R_{XX}[0]R_{YY}[0]}$, which is derived using the Schwartz inequality. (cf. Equation 4.3-17)
3. $R_{XX}[m] = R_{XX}^*[-m]$ since $R_{XX}[m] = E\{X[m + l]X^*[l]\} = E\{X[l]X^*[l - m]\} = E^*\{X[l - m]X^*[l]\} = R_{XX}^*[-m]$.
4. For all $N > 0$ and all complex $a_1, a_2, \ldots, a_N$ we must have

$$\sum_{n=1}^{N} \sum_{k=1}^{N} a_n a_k^* R_{XX}[n - k] \geq 0.$$

Property 4 is the *positive semidefinite* property of autocorrelation functions. It is a necessary and sufficient property for a function to be a valid autocorrelation function of a random sequence. In general it is very difficult to directly apply property 4 to test a function to see if it qualifies as a valid autocorrelation function. However, we soon will introduce an equivalent frequency domain function called *power spectral density*, which furnishes an easy test of validity.

Many of the input–output relations derived in the previous section take a surprisingly simple form in the case of WSS random sequences and LSI systems described via convolution. For example, starting with

$$Y[n] = \sum_{k=-\infty}^{+\infty} h[n-k]X[k],$$

we obtain

$$R_{XY}[m,n] = E\{X[m]Y^*[n]\}$$

$$= \sum_{k=-\infty}^{+\infty} h^*[n-k]E\{X[m]X^*[k]\}$$

$$= \sum_{k=-\infty}^{+\infty} h^*[n-k]R_{XX}[m-k]$$

$$= \sum_{k=-\infty}^{+\infty} h^*[-l]R_{XX}[(m-n)-l], \quad \text{with } l \triangleq k-n,$$

if the input random sequence $X[n]$ is WSS. So, the output cross-correlation function $R_{XY}[m,n]$ is shift-invariant, and we can make use of the one-parameter cross-correlation function $R_{XY}[m] \triangleq R_{XY}[m,0]$ to write

$$R_{XY}[m] = \sum_{k=-\infty}^{+\infty} h^*[-l]R_{XX}[m-l]$$

$$= h^*[-m] * R_{XX}[m],$$

in terms of the one-parameter autocorrelation function $R_{XX}[m]$. Likewise, recalling that

$$R_{YY}[n+m,n] \triangleq E\{Y[n+m]Y^*[n]\}$$

$$= \sum_{k=-\infty}^{+\infty} h[k]E\{X[n+m-k]Y^*[n]\}$$

$$= \sum_{k=-\infty}^{+\infty} h[k]R_{XY}[m-k]$$

$$= h[m] * R_{XY}[m],$$

we see that the autocorrelation function of the output is shift-invariant, and so making use of the one-parameter autocorrelation function $R_{YY}[m] \triangleq R_{YY}[m, 0]$, we have

$$R_{YY}[m] = h[m] * R_{XY}[m].$$

Combining both equations, we get

$$
\begin{aligned}
R_{YY}[m] &= h[m] * h^*[-m] * R_{XX}[m] \\
&= (h[m] * h^*[-m]) * R_{XX}[m] \\
&= g[m] * R_{XX}[m], \qquad \text{with } g[m] \triangleq h[m] * h^*[-m]
\end{aligned}
\tag{6.4-1}
$$

where $g[m]$ is sometimes called the *autocorrelation impulse response* (AIR). Note that if the input random sequence is WSS and independent, then its autocorrelation function would be a positive constant times $\delta[m]$, so that taking this constant to be unity, we would have the output autocorrelation function equal to $g[m]$ itself. Therefore $g[m]$ must possess all the properties of autocorrelation functions, i.e., $g[l] = g^*[-l], g[0] \geq g[l]$ for all l, and positive semidefiniteness. The AIR g depends only on the impulse response h of the LSI system; however in the absence of other information, we cannot uniquely determine h from g. In astronomy, crystallography, and other fields the problem of estimating h from the AIR is an important problem known by various names including *phase recovery* and *deconvolution*.

Example 6.4-1
We cannot in general calculate the impulse response from the AIR. To show this, first take the Fourier transform of $g[m]$ to obtain $G(\omega) = H(\omega)H^*(\omega) = |H(\omega)|^2$. Then note that $|H(\omega)| = \sqrt{G(\omega)}$. Thus the phase of $H(\omega)$ is lost in the AIR, but the magnitude of $H(\omega)$ is preserved. Often there is some information available that can narrow down or possibly pinpoint the phase, for example, the support of $h[n]$ in an image application, or causality for a time-based signal. For the interested reader, the literature contains many articles on this subject; see for example [6-10].

Example 6.4-2
(Correlation function analysis of the edge detector using impulse response.) In the edge detector of Example 6.3-2, the linear transformation was given as

$$Y[n] = L\{X[n]\} \triangleq X[n] - X[n - 1],$$

an LSI operation with impulse response $h[n] = \delta[n] - \delta[n - 1]$, and input autocorrelation function $R_{XX}[m] = a^{|m|}$, with $|a| < 1$. We can easily calculate the AIR as

$$
\begin{aligned}
g[m] &= h[m] * h[-m] \\
&= (\delta[m] - \delta[m - 1]) * (\delta[-m] - \delta[-m - 1]) \\
&= (\delta[m] - \delta[m - 1]) * (\delta[m] - \delta[m + 1]) \\
&= \delta[m] - \delta[m - 1] - \delta[m + 1] + \delta[m] \\
&= 2\delta[m] - \delta[m - 1] - \delta[m + 1].
\end{aligned}
$$

We then calculate the output autocorrelation function in this WSS case as

$$R_{YY}[m] = g[m] * R_{XX}[m]$$
$$= (2\delta[m] - \delta[m-1] - \delta[m+1]) * a^{|m|}$$
$$= 2a^{|m|} - a^{|m-1|} - a^{|m+1|}, \quad \text{for } -\infty < m < +\infty,$$

which agrees with the answer in Example 6.3-2, where the result was plotted for $a = 0.7$.

Power Spectral Density

We define *power spectral density* (psd) as the *FT* (cf. Definition 6.2-3) of the one-parameter discrete-time autocorrelation function of a WSS random sequence $X[n]$:

$$S_{XX}(\omega) \triangleq \sum_{m=-\infty}^{+\infty} R_{XX}[m] \exp(-j\omega m), \quad \text{for } -\pi \le \omega \le +\pi. \tag{6.4-2a}$$

Now by taking the *FT* of Equation 6.4-1, we get the following important psd input/output relation for an LSI system excited by a WSS random sequence:

$$S_{YY}(\omega) = |H(\omega)|^2 S_{XX}(\omega) = G(\omega) S_{XX}(\omega), \tag{6.4-2b}$$

where the various frequency-domain quantities are discrete-time Fourier transforms. Equation 6.4-2b is a central result in the theory of WSS random sequences in that it enables the computation of the output psd directly from knowledge of the input psd and the transfer function magnitude. By using the *IFT*, we can calculate the autocorrelation function as

$$R_{XX}[m] = IFT\{S_{XX}(\omega)\} = \frac{1}{2\pi} \int_{-\pi}^{+\pi} S_{XX}(\omega) e^{j\omega m} d\omega,$$

so that knowledge of the psd implies knowledge of the autocorrelation function.

As to the name power spectral density, note that $R_{XX}[0] = E\{|X[n]|^2\}$ is the ensemble average power in $X[n]$ and so by the above relation, we see that

$$E\{|X[n]|^2\} = R_{XX}[0] = \frac{1}{2\pi} \int_{-\pi}^{+\pi} S_{XX}(\omega) d\omega,$$

so that the integral average of the psd over its frequency range $[-\pi, +\pi]$ is indeed average power. To pursue this further, we consider a WSS random sequence $X[n]$ input to an LSI system consisting of a narrow band filter $H(\omega)$, with very small bandwidth 2ϵ, centered at frequency ω_o, where $|\omega_o| < \pi$, and with unity passband gain. Writing $S_{XX}(\omega)$ for the input psd, we have for the output ensemble average power, approximately

$$R_{YY}[0] = \frac{1}{2\pi} \int_{\omega_o-\epsilon}^{\omega_o+\epsilon} S_{XX}(\omega) \, d\omega \simeq S_{XX}(\omega_0) \frac{\epsilon}{\pi},$$

thus showing that $S_{XX}(\omega)$ can be interpreted as a density function in frequency for ensemble average power.

Some important properties of the psd are given below:

1. The function $S_{XX}(\omega)$ is real-valued since $R_{XX}[m]$ is conjugate-symmetric.
2. If $X[n]$ is a real-valued random sequence, then $S_{XX}(\omega)$ is an even function of ω.
3. The function $S_{XX}(\omega) \geq 0$ for every ω, whether $X[n]$ is real- or complex-valued.
4. If $R_{XX}[m] = 0$ for all $|n| > N$ for some finite integer $N > 0$ (i.e., it has *finite support*), then $S_{XX}(\omega)$ is an *analytic function* in ω. This means that $S_{XX}(\omega)$ can be represented in a Taylor series given its value and that of all its derivatives at a single point ω_o.

Since $S_{XX}(\omega)$ is the Fourier transform of a (autocorrelation) sequence, it is periodic with period 2π. This is why the inverse Fourier transform, which recovers the autocorrelation function, only integrates over $[-\pi, +\pi]$, the *primary period*. We also define the Fourier transform of the cross-correlation function of two jointly WSS random sequences:

$$S_{XY}(\omega) \triangleq \sum_{m=-\infty}^{+\infty} R_{XY}[m] \exp(-j\omega m), \quad \text{for } -\pi \leq \omega \leq +\pi,$$

called the *cross-power spectral density* between random sequences X and Y. In general, this cross-power spectral density can be complex, negative, and lacking in symmetry. Its main use is as an intermediate step in calculation of psd's.

Interpretation of the PSD

From its name, we expect that the psd should be related to some kind of average of the magnitude-square of the Fourier transform of the random signal. Now since a WSS random signal $X[n]$ has constant average power $R_{XX}[0]$ for all time, we cannot define its FT; however, we can define the following transform quantity,

$$X_N(\omega) \triangleq FT\{w_N[n]X[n]\}$$

with aid of the rectangular window function

$$w_N[n] \triangleq \begin{cases} 1 & |n| \leq N \\ 0 & \text{else.} \end{cases}$$

Then, taking the expectation of the magnitude square $|X_N(\omega)|^2$, and dividing by $2N + 1$, we get

$$\frac{1}{2N+1} E\{|X_N(\omega)|^2\} = \frac{1}{2N+1} E\left\{ \sum_{k=-N}^{+N} \sum_{l=-N}^{+N} X[k]X^*[l] \exp(-j\omega k) \exp(+j\omega l) \right\}$$

$$= \frac{1}{2N+1} \sum_{k=-N}^{+N} \sum_{l=-N}^{+N} E\{X[k]X^*[l]\} \exp(-j\omega k) \exp(+j\omega l)$$

$$= \frac{1}{2N+1} \sum_{k=-N}^{+N} \sum_{l=-N}^{+N} R_{XX}[k-l] \exp[-j\omega(k-l)]$$

$$= \sum_{m=-2N}^{+2N} R_{XX}[m] \left(1 - \frac{|m|}{2N+1}\right) \exp(-j\omega m),$$

where the last line comes from the fact that $R_{XX}[k-l]$ is constant along diagonals $k-l = m$ of the $(2N+1) \times (2N+1)$ point square in the (k,l) plane.

Now as $N \to \infty$, the triangular function $(1 - \frac{|m|}{2N+1})$ has less and less effect if $|R_{XX}[m]| \to 0$ as $|m| \to \infty$, as it must for the Fourier transform, that is, $S_{XX}(\omega)$ to exist. In fact, if we assume that $|R_{XX}[m]|$ decays fast enough to satisfy $\sum_{m=-\infty}^{+\infty} |m||R_{XX}[m]| < \infty$, then we have

$$S_{XX}(\omega) = \lim_{N \to \infty} \frac{1}{2N+1} E\{|X_N(\omega)|^2\}. \tag{6.4-2c}$$

In words we have that the ensemble average of the power at frequency ω in the windowed random sequence is given by the psd $S_{XX}(\omega)$. Note that we have said nothing about the *variance* of the random variable $\frac{1}{2N+1}|X_N(\omega)|^2$, but just that its *mean value* converges to the psd. We will see later (cf. Chapter 9) that the variance does not get small as N gets large, so that $\frac{1}{2N+1}|X_N(\omega)|^2$ *cannot* be considered a good estimate of the psd without first doing some averaging.

Example 6.4-3
Here is a MATLAB m-file to compute the psd's of the random sequences with memory in Example 6.1-16 for $p = 0.8$, 0.5, and 0.2.

```
function[psd1,psd2,psd3]=psdmarkov2(N,p1,p2,p3)}
mc1=0*ones(1,N);
mc2=0*ones(1,N);
mc3=0*ones(1,N);
for i=1:N
 mc1(i)=0.25*(((-1)*(2*p1-1))^(i-1));% The (-1)^(i-1) factor shifts the
spectrum to yield
 mc2(i)=0.25*(((-1)*(2*p2-1))^(i-1));%an even function of frequency.
Otherwise
 mc3(i)=0.25*(((-1)*(2*p3-1))^(i-1));%the highest frequency components appear

end
x=linspace(-pi,pi,N);%at pi and the lowest at 2*pi.
psd1=abs(fft(mc1));
psd2=abs(fft(mc2));
psd3=abs(fft(mc3));
```

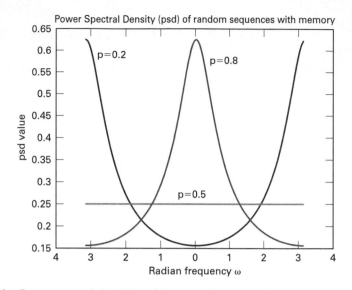

Figure 6.4-1   Power spectral densities of three stationary random sequences with memory.

```
plot(x,psd1,x,psd2,x,psd3)
title('Power Spectral Density (psd) of random sequences with memory')
xlabel('radian frequency')
ylabel('psd value')
end
```

See the three plots in Figure 6.4-1.

Example 6.4-4 ———

A stationary random sequence $X[n]$ has power spectral density $S_{XX}(\omega) = N_0 w(3\omega/4\pi)$, where the rectangular window function w is given as

$$w(x) \triangleq \begin{cases} 1 & |x| \leq 1/2 \\ 0 & \text{else.} \end{cases}$$

It is desired to produce an output random sequence $Y[n]$ with the psd $S_{YY}(\omega) = N_0 w(\omega/\pi)$. An LSI system (not necessarily causal) with impulse response $h[n]$ is proposed. Which of the following impulse responses should be used. (Note that $\text{sinc}(x) \triangleq \sin(\pi x)/\pi x$.)

    (a) $2\text{sinc}(n/2)$,
    (b) $\frac{1}{2}\text{sinc}((n-10)/2)$,
    (c) $1.5e^{-|n|}u[n]$,
    (d) $u[n+2] - u[n-2]$,
    (e) $(1 - |n|)w(n/2)$.

Solution   Clearly what is needed is an $H(\omega)$ with transfer-function magnitude $|H(\omega)| = w(\omega/\pi)$. Choices (c) through (e) are ruled out immediately because their Fourier transforms

Table 6.4-1 Input/Output Relations for WSS Sequences and Linear Systems

Random Sequence:
$$Y[n] = h[n] * X[n]$$

Output Mean:
$$\mu_Y = H(0)\mu_X$$

Crosscorrelations:
$$R_{XY}[m] = R_{XX}[m] * h^*[-m]$$
$$R_{YX}[m] = h[m] * R_{XX}[m]$$
$$R_{YY}[m] = R_{YX}[m] * h^*[-m]$$

Cross–Power Spectral Densities:
$$S_{XY}(\omega) = S_{XX}(\omega)H^*(\omega)$$
$$S_{YX}(\omega) = H(\omega)S_{XX}(\omega)$$
$$S_{YY}(\omega) = S_{YX}(\omega)H^*(\omega)$$

Autocorrelation:
$$R_{YY}[m] = h[m] * h^*[-m] * R_{XX}[m]$$
$$= g[m] * R_{XX}[m]$$

Power Spectral Density:
$$S_{YY}(\omega) = |H(\omega)|^2 S_{XX}(\omega)$$
$$= G(\omega) S_{XX}(\omega)$$

Output Power and Variance:

$$E\{|Y[n]|^2\} = R_{YY}[0] = \frac{1}{2\pi}\int_{-\pi}^{+\pi} |H(\omega)|^2 S_{XX}(\omega)d\omega$$

$$\sigma_Y^2 = R_{YY}[0] - |\mu_Y|^2$$

do not have constant magnitude inside any frequency band. Since the IFT of $w(\omega/\pi)$ is $\frac{1}{2}\operatorname{sinc}(n/2)$, we choose (b) since its 10-sample delay does not affect the magnitude $|H(\omega)|$.

A useful summary of input/output relations for random sequences is presented in Table 6.4-1.

Synthesis of Random Sequences and Discrete-Time Simulation

Here we consider the problem of finding the appropriate transfer function $H(\omega)$ to generate a random sequence with a specified psd or correlation function. Consider Equation 6.2-2, repeated here for convenience:

$$y[n] = \sum_{k=1}^{M} a_k y[n-k] + \sum_{k=0}^{N} b_k x[n-k] \tag{6.4-3}$$

where the coefficients are real valued. The transfer function $H(\omega)$ is given by

$$H(\omega) = \frac{Y(\omega)}{X(\omega)} = \frac{B(\omega)}{A(\omega)},$$

where $B(\omega) \triangleq \sum_{k=0}^{N} b_k e^{-j\omega k}$ and $A(\omega) \triangleq 1 - \sum_{k=1}^{M} a_k e^{-j\omega k}$. When driven by a white-noise sequence, $W[n]$, with power $E\{|W[n]|^2\} = \sigma_W^2$, the output psd, $S_{YY}(\omega)$, is given by

$$S_{YY}(\omega) = \frac{|B(\omega)|^2}{|A(\omega)|^2}\sigma_W^2 = \frac{B(\omega)B^*(\omega)}{A(\omega)A^*(\omega)}\sigma_W^2. \tag{6.4-4}$$

Now, recalling that $B(z) \triangleq B(\omega)$ at $z = e^{j\omega}$ and similarly for $A(z)$, $H(z)$, that $e^{-j\omega} = z^{-1}$, and that $B^*(e^{j\omega}) = B(e^{-j\omega})$ we obtain for a LCCDE with real coefficients[†]

$$S_{YY}(z) = \frac{B(z)B(z^{-1})}{A(z)A(z^{-1})}\sigma_W^2 = H(z)H(z^{-1})\sigma_W^2, \qquad (6.4\text{-}5)$$

where up to this point we have confined z to the *unit circle*. For the purpose of further analysis, it is of interest to extend Equation 6.4-5 to the whole z-*plane*.

This last step is called *analytic continuation* and simply amounts to finding a rational function of z which agrees with the given psd information on the unit circle $z = e^{j\omega}$.

Given any rational $S_{XX}(z)$, that is, one with a finite number of poles and zeros in the finite z-plane, then one can find such a *spectral factorization* as Equation 6.4-5 by defining $H(z)$ to have all the poles and zeros that are inside the unit circle, $\{|z| < 1\}$, and then $H(z^{-1})$ will necessarily have all the poles and zeros outside the unit circle, $\{|z| > 1\}$.

Example 6.4-5

Consider the psd

$$S_{XX}(\omega) = \frac{\sigma_W^2}{1 - 2\rho\cos\omega + \rho^2} \qquad \text{with } |\rho| < 1.$$

We want to first extend $S_{XX}(\omega)$ to all of the z-plane. Now $\cos\omega = \frac{1}{2}(e^{+j\omega} + e^{-j\omega})$ which can be extended as $\frac{1}{2}(z + z^{-1})$ and satisfies the symmetry condition $S_{XX}(z) = S_{XX}(z^{-1})$ of a real-valued random sequence. Then

$$S_{XX}(z) = \frac{\sigma_W^2}{1 - \rho(z + z^{-1}) + \rho^2} \qquad \text{for } |\rho| < |z| < \frac{1}{|\rho|}$$

$$= \frac{\sigma_W^2}{(1 - \rho z)(1 - \rho z^{-1})}$$

$$= \sigma_W^2 H(z)H(z^{-1})$$

with $H(z) = \dfrac{1}{1 - \rho z^{-1}}$ for region of convergence $|\rho| < |z|$.

Since $|\rho| < 1$, the region of convergence (ROC) includes the unit circle and so H is both stable and causal. Indeed the system with $h[n] = \rho^n u[n]$ will yield $S_{XX}(\omega)$ from an independent sequence.

If a zero occurs on the unit circle, then it must be of even order, since otherwise one can easily show that $S_{XX}(e^{j\omega})$ must go through zero and hence be negative in its vicinity. Thus we can assign half the zeros to $H(z)$ and the other half to $H(z^{-1})$. Since $H(z)$ contains only poles inside the unit circle, it will be BIBO stable [6-4]. Except in the case of a zero on the unit circle, its inverse will also be stable. The other factor $H(z^{-1})$ has all its poles outside the unit circle, so it is stable in the anticausal sense. Denoting the largest pole

---

[†]Only when, as here, the impulse response coefficients are real-valued. This is true here since the numerator and denominator coefficients are real numbers.

magnitude inside the unit circle by p_{max}, we thus have that $S_{XX}(z)$ is analytic, that is, free of singularities in the annular region of convergence $\{p_{max} < |z| < 1/p_{max}\}$.

Following the above procedures, we can obtain the system function $H(z)$ that, when driven by a white noise $W[n]$, will generate random sequence $X[n]$ with special psd $S_{XX}(\omega)$. This can be the basis for a discrete-time simulation on a computer. The white random sequence $W[n]$ is easily obtained by using the computer's random number generator. Then one specifies appropriate initial conditions and proceeds to recursively calculate $X[n]$ using the LCCDE of the system function $H(z)$.

To achieve a Gaussian distribution for X, one could transform the output of the random number generator to achieve a Gaussian distribution for W, which would carry across to X. An approximate method that is often used is to average 6 to 10 calls to the random number generator to obtain an approximate Gaussian distribution for W via the Central Limit theorem. When simulating a non-Gaussian random variable, the distribution for X and W is not the same. Thus the preceding method will not work. One possibility is to use the LCCDE to generate samples of $W[n]$ from some real data and then use the resulting distribution for $W[n]$ in the simulation.

Example 6.4-6

In order to simulate a zero-mean random sequence with average power $R_{XX}[0] = \sigma^2$ and nearest neighbor correlation $R_{XX}[1] = \rho\sigma^2$, we want to find the parameters of a first-order stochastic difference equation to achieve these values. Thus consider

$$X[n] = aX[n-1] + bW[n], \tag{6.4-6}$$

where $W[n]$ is a zero-mean white noise source with unit power. Computing the impulse response, we get

$$h[n] = ba^n u[n]$$

and the corresponding system function

$$H(z) = \frac{b}{1 - az^{-1}}.$$

Since the mean is zero, we calculate the covariance of the output $X[n]$ of Equation 6.4-6:

$$K_{XX}[m] = h[m] * h[-m] * K_{WW}[m]$$

$$= h[m] * h[-m]$$

$$= b^2 \left(a^m u[m]\right) * \left(a^{-m} u[-m]\right)$$

$$= b^2 \sum_{k=-\infty}^{+\infty} a^k u[k] a^{m+k} u[m+k]$$

$$= b^2 a^m \sum_{k=\max(0,-m)}^{+\infty} a^{2k}$$

$$= \frac{b^2}{1 - a^2} a^{|m|}, \qquad -\infty < m < +\infty.$$

From the specifications at $m = 0$ and $m = 1$, we need

$$K_{XX}[0] = \sigma^2 = b^2/(1 - a^2),$$
$$K_{XX}[1] = \rho\sigma^2 = ab^2/(1 - a^2).$$

Thus

$$a = \rho \text{ and } b^2 = \sigma^2(1 - \rho^2).$$

To compute the resulting psd we use Equation 6.4-2 to get

$$S_{XX}(\omega) = \frac{b^2}{|1 - ae^{-j\omega}|^2}$$
$$= \frac{\sigma^2(1 - \rho^2)}{1 - 2\rho\cos\omega + \rho^2}.$$

Example 6.4-7

(Decimation and interpolation.) Let $X[n]$ be a WSS random sequence. We consider what happens to its stationarity and psd when we subject it to decimation or interpolation as occur in many signal processing systems.

Decimation

Set $Y[n] \triangleq X[2n]$, called *decimation* by the factor 2, thus throwing away every odd indexed sample of $X[n]$ (Figure 6.4-2). We easily calculate the mean function as $\mu_Y[n] \triangleq E\{Y[n]\} = E\{X[2n]\} = \mu_X[2n] = \mu_X$, a constant. For the correlation,

$$R_{YY}[n + m, n] = E\{X[2n + 2m]X^*[2n]\}$$
$$= R_{XX}[2n + 2m, 2n]$$
$$= R_{XX}[2m],$$

thus showing that the WSS property of the original random sequence $X[n]$ is preserved in the decimated random sequence. The psd of $Y[n]$ can be computed as

$$S_{YY}(\omega) \triangleq \sum_{m=-\infty}^{+\infty} R_{YY}[m]\exp[-j\omega m]$$
$$= \sum_{m=-\infty}^{+\infty} R_{XX}[2m]\exp[-j\omega m]$$
$$= \sum_{m \text{ even}} R_{XX}[m]\exp\left[-j\frac{\omega}{2}m\right] = \sum_{m \text{ even}} R_{XX}[m]\exp\left[-j\frac{\omega}{2}m\right](-1)^m.$$

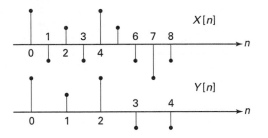

Figure 6.4-2   In decimation every other value of $X[n]$ is discarded.

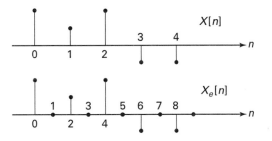

Figure 6.4-3   In interpolation, the expansion step inserts zeros between adjacent values of the $X[n]$ sequence, to get the expanded sequence $X_e[n]$.

Now, define $A_e \triangleq S_{YY}(\omega)$, and $A_o \triangleq \sum_{m \text{ odd}} R_{XX}[m] \exp[-j\frac{\omega}{2}m]$. Then, clearly $A_e + A_o = S_{XX}(\frac{\omega}{2})$ and $A_e - A_o = S_{XX}(\frac{\omega - 2\pi}{2})$, so that

$$S_{YY}(\omega) = \frac{1}{2}\left[S_{XX}\left(\frac{\omega}{2}\right) + S_{XX}\left(\frac{\omega - 2\pi}{2}\right) \right],$$

which displays an *aliasing* [6-4] of higher-frequency terms.

Interpolation

For interpolation by the factor 2, we do the opposite of decimation. First we perform an *expansion* by setting

$$X_e[n] \triangleq \begin{cases} X[\frac{n}{2}], & n = \text{even} \\ 0, & n = \text{odd} \end{cases}.$$

The resulting expanded random sequence is clearly nonstationary, because of the zero insertions. See Figure 6.4-3. Formally the psd of $X_e[n]$ doesn't exist since the psd is defined only for WSS sequences (Figure 6.4-4). We encounter such problems with a broad class of random sequences and processes[†] classified as being *cyclostationary* (Section 7.6 in Chapter 7) to which $X_e[n]$ belongs to. It is easy to convert such sequences to WSS by

---

[†] *Random processes* are continuous-time random waveforms discussed in Chapter 7.

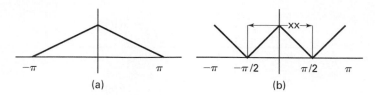

Figure 6.4-4   (a) The original psd of $X[n]$: (b) the psd of $X_e[n]$ (not drawn to scale). Note the "leakage" of power density from the secondary periods into the primary period. An ideal lowpass filter with support $[-\pi/2, \pi/2]$ will eliminate the contribution from the secondary periods.

randomizing their start times and then averaging over the start time (Example 7.6-1). However, we leave this approach to the next chapter and instead compute the power spectral density using Equation 6.4-2c, which is permissible for cyclostationary waveforms. Thus we write

$$E\{|Y_N(\omega)|^2\} = E\left\{ \left| \sum_{\mu=-N}^{N} X_e[n]e^{-j\omega n} \right|^2 \right\}$$

and take the limit of $E\{|X_e^{(N)}(\omega)|^2\}/(2N+1)$ as $N \to \infty$. This quantity can be interpreted as the psd, $S_{X_e X_e}(\omega)$, of the random sequence $X_e[n]$. If the algebra is carried out and we assume that $R_{XX}[m]$ is absolutely summable, we find that $S_{X_e X_e}(\omega) = \frac{1}{2}S_{XX}(2\omega)$. For further discussion of the expansion step see Problems 6.47 and 6.48.

Next we put $X_e[n]$, sometimes called an *upsampled version* of $X[n]$, through an ideal lowpass filter with bandwidth $[-\frac{\pi}{2}, +\frac{\pi}{2}]$ and gain of 2, to produce the "ideal" interpolated output $Y[n]$ as

$$Y[n] \triangleq h[n] * X_e[n].$$

The impulse response of such a filter is

$$h[n] = \frac{\sin(\pi n/2)}{(\pi n/2)},$$

thus

$$Y[n] = \sum_{k=-\infty}^{+\infty} X_e[k] \frac{\sin(n-k)\pi/2}{(n-k)\pi/2}$$

$$= \sum_{k=-\infty}^{+\infty} X[k] \frac{\sin(n-2k)\pi/2}{(n-2k)\pi/2}.$$

First we calculate the mean function of $Y[n]$,

$$\mu_Y[n] \triangleq E\{Y[n]\}$$

$$= E\left\{ \sum_{k=-\infty}^{+\infty} X[k] \frac{\sin(n-2k)\pi/2}{(n-2k)\pi/2} \right\}$$

$$= \sum_{k=-\infty}^{+\infty} \mu_X[k] \frac{\sin(n-2k)\pi/2}{(n-2k)\pi/2}$$

$$= \mu_X \sum_{k=-\infty}^{+\infty} \frac{\sin(n-2k)\pi/2}{(n-2k)\pi/2},$$

the last step being allowed since μ_X is a constant. Now sampling theory can be used to show that the infinite sum is 1, so that $\mu_Y[n] = \mu_X$. To see this we write the sampling theorem representation for a bandlimited function $g(t)$ and sampling period $T = 2$ as [6-4]

$$g(t) = \sum_{k=-\infty}^{+\infty} g(k2) \frac{\sin(t-2k)\pi/2}{(t-2k)\pi/2}. \tag{6.4-7}$$

Then we simply choose $t = n$ for the bandlimited function $g(t) = 1$ with bandwidth 0 to see that

$$1 = \sum_{k=-\infty}^{+\infty} \frac{\sin(n-2k)\pi/2}{(n-2k)\pi/2}.$$

To find the correlation function, we proceed to calculate

$$R_{YY}[n+m,n] = E\{Y[n+m]Y^*[n]\}$$

$$= \sum_{k_1,k_2=-\infty}^{+\infty} E\{X[k_1]X^*[k_2]\}h[n+m-2k_1][h[n-2k_2]$$

$$= \sum_{l_1=\text{even}} R_{XX}[l_1] \sum_{l_2=\text{even}} h[n+m-l_1-l_2]h[n+l_1-l_2]$$

$$+ \sum_{l_1=\text{odd}} R_{XX}[l_1] \sum_{l_2=\text{odd}} h[n+m-l_1-l_2]h[n+l_1-l_2]$$

with $l_1 \overset{\Delta}{=} k_1 - k_2$ and $l_2 \overset{\Delta}{=} k_1 + k_2$. Considering the above sums

$$\sum_{l_2=\text{even or odd}} h[n+m-l_1-l_2]h[n+1_1-l_2]$$

and using Equation 6.4-7 with $g(t) \overset{\Delta}{=} \frac{\sin \pi t/2}{\pi t/2}$ and setting $t = m$, we find each sum (even and odd) equal to $h[m - 2l_1]$. Thus

$$R_{YY}[m+n,n] = R_{YY}[m] = \sum_{l_1} R_{XX}[l_1]h[m-2l_1].$$

We thus see that $Y[n]$ is WSS, that $R_{YY}[m]$ interpolates $R_{XX}[m]$, i.e.

$$R_{YY}[2m] = \sum_{l_1} R_{XX}[l_1]h[2m-2l_1]$$

$$= R_{XX}[2m],$$

and calculating the psd

$$S_{YY}(\omega) = \sum_m R_{YY}[m]e^{-j\omega m}$$

$$= \sum_m \sum_{l_1} R_{XX}[l_1]h[m-2l_1]e^{-j\omega m}$$

$$= \sum_{l_1} R_{XX}[l_1] \sum_m h[m-2l_1]e^{-j\omega m}$$

$$= \sum_{l_1} R_{XX}[l_1]H(\omega)e^{-j2\omega l_1}$$

$$= H(\omega)S_{XX}(2\omega)$$

$$= \begin{cases} 2S_{XX}(2\omega) & |\omega| \le \pi/2 \\ 0 & \dfrac{\pi}{2} < |\omega| \le \pi \end{cases}.$$

6.5 MARKOV RANDOM SEQUENCES

We have already encountered some examples of Markov random sequences. Such sequences were loosely said to have a *memory* and to possess a *state*. Here we make these concepts more precise. We start with a definition.

Definition 6.5-1　(Markov random sequence.)

(a) A *continuous-valued Markov random sequence* $X[n]$, defined for $n \ge 0$, satisfies the conditional pdf expression

$$f_X(x_{n+k}|x_n, x_{n-1}, \ldots, x_0) = f_X(x_{n+k}|x_n)$$

for all $x_0, \ldots, x_n, x_{n+k}$, for all $n > 0$, and for all integers $k \ge 1$.

(b) A *discrete-valued Markov random sequence* $X[n]$, defined for $n \ge 0$, satisfies the conditional PMF expression

$$P_X(x_{n+k}|x_n, \ldots, x_0) = P_X(x_{n+k}|x_n)$$

for all $x_0, \ldots, x_n, x_{n+k}$, for all $n > 0$, and for all $k \ge 1$.　■

It is sufficient for the above properties to hold for just $k = 1$, which is the so-called *one-step* case, as the general property can be built up from it. The *discrete-valued* Markov random sequence is also called a Markov chain and will be covered in the next section. Here we consider the continuous-valued case.

To check the meaning and usefulness of the Markov concept, consider the general Nth-order pdf $f_X(x_N, x_{N-1}, \ldots, x_0)$ of random sequence X, and repeatedly use conditioning to obtain the *chain rule* of probability

$$f_X(x_0, x_1, \ldots, x_N) = f_X(x_0)f_X(x_1|x_0)f_X(x_2|x_1, x_0)\ldots f_X(x_N|x_{N-1}, \ldots, x_0). \qquad (6.5\text{-}1)$$

Now substitute the basic one-step $(k = 1)$ version of the Markov definition to obtain

$$f_X(x_0, x_1, \ldots, x_N) = f_X(x_0)f_X(x_1|x_0)f_X(x_2|x_1)\ldots f_X(x_N|x_{N-1})$$

$$= f_X(x_0)\prod_{k=1}^{N} f_X(x_k|x_{k-1}).$$

Next we present two examples of continuous-valued Markov random sequences which are Gaussian distributed.

Example 6.5-1
(Gauss Markov random sequence.) Let $X[n]$ be a random sequence defined for $n \geq 1$, with initial pdf

$$f_X(x; 0) = N(0, \sigma_0^2)$$

for a given $\sigma_0 > 0$ and transition pdf

$$f_X(x_n|x_{n-1}; n, n-1) \sim N(\rho x_{n-1}, \sigma_W^2)$$

with $|\rho| < 1$ and $\sigma_W > 0$. We want to determine the unconditional density of $X[n]$ at an arbitrary time $n \geq 1$ and proceed as follows.

In general, one would have to advance recursively from the initial density by performing the integrals (cf. Equation 2.6-50)

$$f_X(x; n) = \int_{-\infty}^{+\infty} f_X(x|\xi; n, n-1)f_X(\xi; n-1)d\xi \tag{6.5-2}$$

for $n = 1, 2, 3$, and so forth. However, in this example we know that the unconditional first-order density will be Gaussian because each of the pdf's in Equation 6.5-2 is Gaussian, and the Gaussian density "reproduces itself" in this context, i.e., the product of two exponential functions is still exponential. Hence the pdf $f_X(x; n)$ is determined by its first two moments. We first calculate the mean function

$$\mu_X[n] = E\{X[n]\}$$
$$= E[E\{X[n]|X[n-1]\}]$$
$$= E[\rho X[n-1]]$$
$$= \rho\mu_X[n-1],$$

where the outer expectation is over the values of $X[n-1]$. We thus obtain the recursive equation

$$\mu_X[n] = \rho\mu_X[n-1], \qquad n \geq 1,$$

with prescribed initial condition $\mu_X[0] = 0$. Hence $\mu_X[n] = 0$ for all n.

We also need the variance function $\sigma_X^2[n]$, which in this case is just $E[X^2[n]]$ since the mean is zero. Calculating, we obtain

$$E\{X^2[n]\} = E[E\{X^2[n]|X[n-1]\}]$$
$$= E[\sigma_W^2 + \rho^2 X^2[n-1]]$$
$$= \sigma_W^2 + \rho^2 E\{X^2[n-1]\}$$

or

$$\sigma_X^2[n] = \rho^2 \sigma_X^2[n-1] + \sigma_W^2, \qquad n \geq 1.$$

This is a first-order difference equation, which can be solved for $\sigma_X^2[n]$ given the condition $\sigma_X^2[0] = \sigma_0^2$ supplied by the initial pdf. The solution then is

$$\sigma_X^2[n] = [1 + \rho^2 + \rho^4 + \ldots + \rho^{2(n-1)}]\sigma_W^2 + \rho^{2(n-1)}\sigma_0^2$$

$$\rightarrow \frac{1}{1-\rho^2}\sigma_W^2 \quad \text{as } n \rightarrow \infty.$$

Example 6.5-2

Consider the difference equation

$$X[n] = \rho X[n-1] + W[n],$$

where $W[n]$ is an independent random sequence (cf. Def. 6.1-2). Let $n > 0$; then

$$f_X(x_n, x_{n-1}, \ldots, x_0) = f_X(x_n|x_{n-1})f_X(x_{n-1}|x_{n-2}) \ldots f_X(x_1|x_0)f_W(x_0)$$

$$= \left(\prod_{k=1}^{n} f_W(x_k - \rho x_{k-1}) \right) f_W(x_0),$$

where $x[n] = x_n$ and $w[n] = w_n$ are the sample function values taken on by the random sequences $X[n]$ and $W[n]$, respectively. Clearly $X[n]$ is a Markov random sequence. If $W[n]$ is an independent and Gaussian random sequence, then this is just the case of Example 6.5-1 above. Otherwise, the Markov sequence $X[n]$ will be non-Gaussian.

The Markov property can be generalized to cover higher-order dependence and higher-order difference equations, thus extending the *direct dependence* concept to more than one-sample distance.

Definition 6.5-2   (Markov-p random sequence.) Let the positive integer p be called the *order* of the Markov-p random sequence. A continuous-valued Markov-p random sequence $X[n]$, defined for $n \geq 0$, satisfies the conditional pdf equations

$$f_X(x_{n+k}|x_n, x_{n-1}, \ldots, x_0) = f_X(x_{n+k}|x_n, x_{n-1}, \ldots, x_{n-p+1})$$

for all $k \geq 1$ and for all $n \geq p$.   ■

Returning to look at Equation 6.5-1, we can see that as the Markov order p increases, the modeling error in approximating a general random sequence by a Markov random sequence should get better.

$$f_X(x_0, x_1, \ldots, x_N) = f_X(x_0)f_X(x_1|x_0)f_X(x_2|x_1, x_0) \ldots f_X(x_N|x_{N-1}, \ldots, x_0)$$

$$\approx f_X(x_0)f_X(x_1|x_0)f_X(x_2|x_1, x_0) \ldots f_X(x_p|x_{p-1}, \ldots, x_0)$$

$$\times \prod_{k=p+1}^{N} f_X(x_k|x_{k-1}, \ldots, x_{k-p+1}).$$

This approximation would be expected to hold for the usual case where the strongest dependence is on the nearby values, say $X[n-1]$ and $X[n-2]$, with the conditional dependence on far away values being generally negligible. When the Markov-p model is used in signal processing, one of the most important issues is determining an appropriate model order p so that statistics like the joint pdf's (Equation 6.5-1) of the *original data* are adequately approximated by those of the Markov-p model. In Chapter 9 on applications in statistical signal processing, we will see that Markov-p random sequences are quite useful in modern spectral estimation. The celebrated Kalman filter for the recursive linear estimation of distorted signals in noise is based on the Markov models.

ARMA Models

A class of linear constant coefficient difference equation models are called ARMA for *autoregressive moving average*. Here the input is an independent random sequence $W[n]$ with mean $\mu_W = 0$ and variance $\sigma_W^2 = 1$. The LCCDE model then takes the form

$$X[n] = \sum_{k=1}^{M} c_k X[n-k] + \sum_{k=0}^{L} d_k W[n-k].$$

If the model is BIBO stable and $-\infty < n < +\infty$, then a WSS output sequence results with psd

$$S_{XX}(\omega) = \frac{\left| \sum_{k=0}^{L} d_k \exp -j\omega k \right|^2}{\left| 1 - \sum_{k=1}^{M} c_k \exp -j\omega k \right|^2}.$$

The ARMA sequence is not Markov, but when $L = 0$, the sequence is Markov-M, and the resulting model is called *autoregressive* (AR). On the other hand when $M = 0$, i.e., there are no feedback coefficients c_k; then the equation becomes just

$$X[n] = \sum_{k=0}^{L} d_k W[n-k],$$

and the model is called *moving average* (MA). The MA model is often used to estimate the time-average value over a data window, as shown in the next example.

Example 6.5-3 _____

(Running time average.) Consider a sequence of independent random variables $W[n]$ on $n \geq 1$. Denote their running time average as

$$\widehat{\mu}_W[n] = \frac{1}{n} \sum_{k=1}^{n} W[k].$$

Since we can write $\widehat{\mu}_W[n]$ equivalently as satisfying the time-varying AR equation,

$$\widehat{\mu}_W[n] = \frac{n-1}{n} \widehat{\mu}_W[n-1] + \frac{1}{n} W[n],$$

it follows from the joint independence of the input $W[n]$, that $\hat{\mu}_W[n]$ is a nonstationary Markov random sequence.[†]

Markov Chains

A Markov random sequence can take on either continuous or discrete values and then be represented either by probability density functions (pdf's) or probability mass functions (PMFs) accordingly. In the discrete-valued case, we call the random sequence a Markov chain. Applications occur in buffer occupancy, computer networks, and discrete-time approximate models for the continuous-time Markov chains (cf. Chapter 7).

Definition 6.5-3   A discrete-time Markov chain is a random sequence $X[n]$ whose Nth-order conditional PMFs satisfy

$$P_X(x[n]|x[n-1], \ldots, x[n-N]) = P_X(x[n]|x[n-1]) \qquad (6.5\text{-}3)$$

for all n, for all values of $x[k]$, and for all integers $N > 1$. ∎

The value of $X[n]$ at time n is called *"the state."* This is because this current value, i.e., the value at time n, determines future conditional PMFs, independent of the past values taken on by $X[n]$.

A practical case of great importance is when the range of values taken on by $X[n]$ is finite, say M. The discrete range of $X[n]$, i.e., the values that X takes on, is sometimes referred to as a set of *labels*. The usual choices for the label set are either the integers $\{1, M\}$, or $\{0, M-1\}$. Such a Markov chain is said to have a finite *state space*, or is simply a *finite-state Markov chain*. In this case, and when the random sequence is stationary, we can represent the statistical transition information in a matrix $\mathbf{P}$ with entries

$$p_{ij} = P_{X[n]|X[n-1]}(j|i), \qquad (6.5\text{-}4)$$

for $1 \leq i, j \leq M$. The matrix $\mathbf{P}$ is referred to as the *state-transition matrix*. Its defining property is that it is a matrix with nonnegative entries, whose rows sum to 1. Usually, and again without loss of generality, we can consider that the Markov chain starts at time index $n = 0$. Then we must specify the set of initial probabilities of the states at $n = 0$, i.e., $P_X(i; 0), 1 \leq i \leq M$, which can be stored in the initial probability vector $\mathbf{p}[0]$, a *row vector* with elements $(\mathbf{p}[0])_i = P_X(i; 0), 1 \leq i \leq M$.

The following example re-introduces the useful concept of *state-transition diagram*, already seen in Example 6.1-15.

Example 6.5-4 ───

Let $M = 2$; then we can summarize transition probability information about a two-state Markov chain in Figure 6.5-1. The only addition needed is the set of initial probabilities, $P_X(1; 0)$ and $P_X(2; 0)$.

─────────

[†]Note that the variance of $\hat{\mu}_W[n]$ decreases with n.

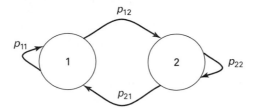

Figure 6.5-1   The state-transition diagram of a two-state Markov chain.

Possible questions might be: Given that we are in state 1 at time 4, what is the probability we end up in state 2 at time 6? Or given a certain probability distribution over the two states at time 3, what is the probability distribution over the two states at time 7? Note that there are several ways or paths to go from one state at one time to another state several time units later. The answers to these questions thus will involve a summation over these mutually exclusive outcomes.

Here we have $M = 2$, and the two-element probability *row vector* $\mathbf{p}[n] = (p_0[n], p_1[n])$. Using the state-transition matrix we then have,

$$\mathbf{p}[1] = \mathbf{p}[0]\mathbf{P}$$
$$\mathbf{p}[2] = \mathbf{p}[1]\mathbf{P}$$
$$= \mathbf{p}[0]\mathbf{P}^2$$

or, in general,

$$\mathbf{p}[n] = \mathbf{p}[0]\mathbf{P}^n.$$

In a statistical steady state, if one exists, we would have

$$\mathbf{p}[\infty] = \mathbf{p}[\infty]\mathbf{P}, \text{ where } \mathbf{p}[\infty] = \lim_{n \to \infty} \mathbf{p}[n].$$

Writing $\mathbf{p} \triangleq \mathbf{p}[\infty]$, we have $\mathbf{p}(\mathbf{I} - \mathbf{P}) = \mathbf{0}$, which furnishes $M - 1$ independent linear equations. Then with help of the additional equation $\mathbf{p1} = 1$, where $\mathbf{1}$ is a size M column vector of all ones, we can solve for the M values in $\mathbf{p}$. The existence of a steady state, or equivalently asymptotic stationarity, will depend on the eigenvalues of the state-transition matrix $\mathbf{P}$.

Example 6.5-5
(Asymmetric two-state Markov chain.) Here we consider an example of a two-state, asymmetric Markov chain (AMC), with state labels zero and one, and state transition matrix,

$$\mathbf{P} = \begin{bmatrix} p_{00} & p_{01} \\ p_{10} & p_{11} \end{bmatrix} = \begin{bmatrix} 0.9 & 0.1 \\ 0.2 & 0.8 \end{bmatrix}.$$

See Figure 6.5-2.

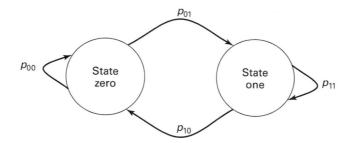

Figure 6.5-2 State-transition diagram of general (asymmetric) two-state Markov random sequence, with state labels *zero* and *one*.

Note that in this model there is no requirement that $p_{00} = p_{11}$ and the steady-state probabilities, if they exist, are given by the solution of

$$\mathbf{p}[n+1] = \mathbf{p}[n]\mathbf{P}, \tag{6.5-5}$$

if we let $n \to \infty$. Denoting these probabilities by $p_0[\infty]$ and $p_1[\infty]$, and using $p_0[\infty] + p_1[\infty] = 1$, we obtain

$$p_0[\infty] = \frac{1 - p_{11}}{2 - p_{00} - p_{11}}, \quad \text{and}$$

$$p_1[\infty] = \frac{1 - p_{00}}{2 - p_{00} - p_{11}},$$

which, using the $\mathbf{P}$ matrix from Example 6.5-4, yields $p_0[\infty] = \frac{2}{3}$ and $p_1[\infty] = \frac{1}{3}$.

The *steady-state* autocorrelation function of the AMC of this example can be computed from the Markov state probabilities. For example, assuming asymptotic stationarity,

$$R_{XX}[m] \approx P\{X[k] = 1, X[m+k] = 1\} \quad \text{for sufficiently large } k$$
$$= P\{X[k] = 1\}P\{X[m+k] = 1 | X[k] = 1\}$$
$$= p_1[\infty]\, P\{X[m] = 1 | X[0] = 1\}, \tag{6.5-6}$$

where the last factor is an m-step transition *from* state one *to* state one. It can be computed recursively from Equation 6.5-5, with the initial condition $\mathbf{p}[0] = [0, 1]$. The needed computation can also be illustrated in a trellis diagram as seen in the following example.

Example 6.5-6
(Trellis diagram for Markov chain.) Consider once again Example 6.1-15, where we introduced the state-transition diagram for what we now know as a Markov chain. Another useful diagram that shows allowable paths to reach a certain state, and the probability of those paths, is the *trellis diagram*, named for its resemblance to the common wooden garden trellis that supports some plants. See Figure 6.5-3 for the two-state case having labels zero

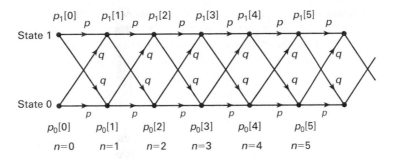

Figure 6.5-3 A trellis diagram of a two-state symmetric Markov chain with state labels 0 and 1. Here $p_i[n]$ is the probability of being in state i at time n.

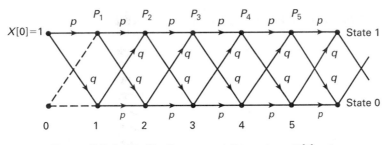

Figure 6.5-4 Trellis diagram conditioned on $X[0] = 1$.

and one, which also assumes symmetry, i.e., $p_{ij} = p_{ji}$. We see that this trellis is a collapsing of the more general tree diagram of Example 6.1-4. The collapse of the tree to the trellis is permitted because of the Markov condition on the conditional probabilities, that serve as the branch labels.

Each node represents the state at a given time instant. The node value (label) is its probability at time n. The links (directed branches) denote possible transitions and are labeled with their respective transition probabilities. Paths through the trellis then represent allowable multiple time-step transitions, with probability given as the product of the transition probabilities along the path.

If we know that the chain is in state one at time $n = 0$, then the *modified* trellis diagram simplifies to that of Figure 6.5-4 where we have labeled the state *one* nodes with $P_n \triangleq P\{X[n] = 1 | X[0] = 1\}$, and we can use this trellis to calculate the probabilities $P\{X[n] = 1 | X[0] = 1\}$ needed in Equation 6.5-6. The first few P_n values are easily calculated as $P_1 = p$, $P_2 = p^2 + q^2$, $P_3 = p^3 + 3pq^2$, etc. For the case $p_0[\infty] = p_1[\infty] = \frac{1}{2}$, and $p = 0.8$, the *asymptotically stationary autocorrelation* (ASA) function $R_{XX}[m]$ then becomes $R_{XX}[0] = 0.5$, $R_{XX}[\pm 1] = 0.4$, $R_{XX}[\pm 2] = 0.34$, $R_{XX}[\pm 3] = 0.304$, and so forth.[†]

[†]The ASA is computed as $R_{XX}[m] = E\{X[k+m]X[k]\}$, where $k \to \infty$. For levels of 0 and 1, $R_{XX}[m] = P\{X[m+k] = 1 | X[k] = 1\} \times 0.5$. Then clearly $R_{XX}[0] = 1 \times 0.5 = 0.5$, $R_{XX}[1] = 0.8 \times 0.5 = 0.4$, $R_{XX}[2] = [(0.8)^2 + (0.2)^2] \times 0.5 = 0.34$, etc.

The trellis diagram shows that, except in trivial cases, there are many allowable paths to reach a certain node, i.e., a given state at a given time. This raises the question of which path is most probable (most likely) to make the required multistep traversal. In the previous example, and with $p > q$, it is just a matter of finding the path with the most p's. In general, however, finding the most likely path is a time consuming problem and, if left to trial-and-error techniques, would quickly exhaust the capabilities of most computers. Much research has been done on this problem because of its many engineering applications, one being speech recognition by computer. In Chapter 9, we discuss the efficient *Viterbi algorithm* for finding the most likely path.

Example 6.5-7

Consider the Markov chain as a model for a communications buffer with $M+1$ states, with labels 0 to M indicating buffer fullness. In other words, the state label is the number of bytes currently stored in the M byte capacity buffer. Assume that transitions can occur only between neighboring states, i.e., the fullness can change at most by one byte in each time unit. The state-transition diagram then appears as shown in Figure 6.5-5.

If we let M go to infinity in Example 6.5-7, we have what is called the general *birth-death* chain, which was first used to model the size of a population over time. In each time unit, there can be at most one birth and at most one death.

Solving the equations. Consider a two-state Markov chain with transition probability matrix

$$\mathbf{P} = \begin{bmatrix} p_{00} & p_{01} \\ p_{10} & p_{11} \end{bmatrix}.$$

We can write the equation relating $\mathbf{p}[n]$ and $\mathbf{p}[n+1]$ then as follows:

$$[p_0[n+1], \ p_1[n+1]] = [p_0[n], \ p_1[n]] \begin{bmatrix} p_{00} & p_{01} \\ p_{10} & p_{11} \end{bmatrix}. \tag{6.5-7}$$

This vector equation is equivalent to two scalar difference equations which we have to solve together, i.e., two *simultaneous* difference equations. We try a solution of the form

$$p_0[n] = C_0 z^n, \qquad p_1[n] = C_1 z^n.$$

Inserting this attempted solution into Equation 6.5-7 and canceling the common term z^n, we obtain

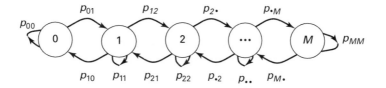

Figure 6.5-5 Markov chain model for $M+1$ state communications buffer.

$$C_0 z = C_0 p_{00} + C_1 p_{10},$$

$$C_1 z = C_0 p_{01} + C_1 p_{11},$$

which implies the following necessary conditions:

$$C_1 = C_0 \left(\frac{z - p_{00}}{p_{10}} \right) = C_0 \left(\frac{p_{01}}{z - p_{11}} \right).$$

This gives a constraint relation between the constants C_0 and C_1 as well as a necessary condition on z, the latter being called the *characteristic equation*

$$(z - p_{00})(z - p_{11}) - p_{10} p_{01} = 0.$$

It turns out that the characteristic equation (CE) can be written, using the determinant function, as

$$\det(z\mathbf{I} - \mathbf{P}) = 0.$$

Solving our two-state equation, we obtain just two solutions z_1 and z_2, one of which must equal 1. (Can you see this? Note that $1 - p_{00} = p_{01}$.) The solutions we have obtained thus far can be written as

$$p_0[n] = C_0 z_i^n, \, p_1[n] = C_0 \left(\frac{z_i - p_{00}}{p_{10}} \right) z_i^n, \, i = 1, 2.$$

Since the vector difference equation is linear, we can add the two solutions corresponding to the different values of z_i, to get the general solution, written in row vector form

$$\mathbf{p}[n] = A_1 \left[1, \frac{z_1 - p_{00}}{p_{10}} \right] z_1^n + A_2 \left[1, \frac{z_2 - p_{00}}{p_{10}} \right] z_2^n,$$

where we have introduced two new constants A_1 and A_2 for each of the two linearly independent solutions. These two constants must be evaluated from the initial probability vector $\mathbf{p}[0]$ and the necessary conditions on the probability row vector at time index n, i.e., $\sum_{i=0}^{1} p_i[n] = 1$ for all $n \geq 0$.

Example 6.5-8 _____
Let

$$\mathbf{P} = \begin{bmatrix} 0.9 & 0.1 \\ 0.2 & 0.8 \end{bmatrix}, \text{ with } \mathbf{p}[0] = \begin{bmatrix} 1/2, & 1/2 \end{bmatrix},$$

and solve for the complete solution, including the startup transient and the steady-state values for $\mathbf{p}[n]$.

The first step is to find the eigenvalues of $\mathbf{P}$, which are the roots of the characteristic equation (CE)

$$\det(s\mathbf{I} - \mathbf{P}) = \det \begin{pmatrix} z - 0.9 & -0.1 \\ -0.2 & z - 0.8 \end{pmatrix} = z^2 - 1.7z + 0.7 = 0.$$

This gives roots $z_1 = 0.7$ and $z_2 = 1.0$. Thus we can write

$$\mathbf{p}[n] = C_1 [1, -1]\, 0.7^n + C_2 [1, 0.5]\, 1^n.$$

From steady-state requirement that the components of $\mathbf{p}$ sum to 1.0, we get $C_2 = \frac{2}{3}$. So we can further write

$$\mathbf{p}[n] = C_1 [1, -1]\, 0.7^n + \left[\tfrac{2}{3}, \tfrac{1}{3}\right].$$

Finally we invoke the specified initial conditions $\mathbf{p}[0] = [1/2, 1/2]$ to obtain $C_1 = -\frac{1}{6}$ and

$$\mathbf{p}[n] = [-\tfrac{1}{6}, \tfrac{1}{6}]\, 0.7^n + [\tfrac{2}{3}, \tfrac{1}{3}], \text{ or in scalar form,}$$

$$\begin{aligned} p_0[n] &= -\tfrac{1}{6}\, 0.7^n + \tfrac{2}{3} \\ p_1[n] &= \tfrac{1}{6}\, 0.7^n + \tfrac{1}{3} \end{aligned} \quad \text{for } n \geq 0.$$

Here we see that the steady-state probabilities exist and are $p_0[\infty] = \frac{2}{3}$ and $p_1[\infty] = \frac{1}{3}$. The next example shows that such steady-state probabilities do not always exist.

Example 6.5-9 _____

Consider the two-state Markov chain with transition probability matrix $\mathbf{P} = \begin{bmatrix} 0 & 1 \\ 1 & 0 \end{bmatrix}$. The characteristic equation becomes

$$\det(s\mathbf{I} - \mathbf{P}) = \det \begin{pmatrix} z & -1 \\ -1 & z \end{pmatrix} = z^2 - 1 = 0,$$

with two roots $z_{1,2} = \pm 1$. Thus there is no steady state in this case, even though one of the eigenvalues of $\mathbf{P}$ is one. Indeed, direct from the state-transition diagram, we can see that the random sequence will forever cycle back and forth between states zero and one with each successive time tick. The phase can be irrevocably set by the initial probability vector $\mathbf{p}[0] = [1, 0]$.

While we cannot always assume a steady state exists, note that this example is degenerate in that the transition probabilities into and out of states are either zero or one. Another problem for existence of the steady state is a so-called *trapping state*. This is a state with transitions into, but not out of, itself. In most cases of interest in communications and signal processing, a steady state will exist, independent of where the chain starts.

6.6 VECTOR RANDOM SEQUENCES AND STATE EQUATIONS

The scalar random sequence concepts we have seen thus far can be extended to vector random sequences. They will be used in Chapter 9 to derive linear estimators for signals in noise (Kalman filter). They are also used in models of sensor arrays, e.g., seismic, acoustic, and radar. This section will introduce difference equations for random vectors and the concept of vector Markov random sequence. Interestingly, a high-order Markov-p scalar random sequence can be represented as a first-order vector Markov sequence.

Definition 6.6-1   A *vector random sequence* is a mapping from a probability sample space Ω, corresponding to probability space $(\Omega, \mathscr{F}, P)$, into the space of vector-valued sequences over complex numbers.   ■

Thus for each $\zeta \in \Omega$ and fixed time n, we generate a vector $\mathbf{X}(n, \zeta)$. The vector random sequence is usually written $\mathbf{X}[n]$, suppressing the outcome ζ.

For example the first-order PDF for a random vector sequence $\mathbf{X}[n]$, would be given as

$$F_{\mathbf{X}}(\mathbf{x}; n) \overset{\Delta}{=} P\{\mathbf{X}[n] \le \mathbf{x}\},$$

where $\{\mathbf{X}[n] \le \mathbf{x}\}$ means every element satisfies the inequality, i.e., $\{X_1[n] \le x_1, X_2[n] \le x_2, \dots, X_N[n] \le x_N\}$. Second- and higher-order probabilities would be specified accordingly. The vector random sequence is said to be *statistically specified* by the set of all its first- and higher-order PDFs (or pdf's or PMFs).

The following example treats the correlation analysis for a vector random sequence input to a vector LCCDE

$$\mathbf{y}[n] = \mathbf{A}\mathbf{y}[n-1] + \mathbf{B}\mathbf{x}[n],$$

with N-dimensional coefficient matrices $\mathbf{A}$ and $\mathbf{B}$. In this vector case, BIBO stability is assured when the eigenvalues of $\mathbf{A}$ are less than one in magnitude.

Example 6.6-1

In the vector case, the scalar first-order LCCDE model, excited by column vector random sequence $\mathbf{X}[n]$, becomes

$$\mathbf{Y}[n] = \mathbf{A}\mathbf{Y}[n-1] + \mathbf{B}\mathbf{X}[n], \tag{6.6-1}$$

which is a first-order vector difference equation in the sample vector sequences. The vector impulse response is the column vector sequence

$$\mathbf{h}[n] = \mathbf{A}^n \mathbf{B} u[n],$$

and the zero initial-condition response to the sequence $\mathbf{X}[n]$ is

$$\mathbf{Y}[n] = \sum_{k=0}^{n} \mathbf{A}^{n-k} \mathbf{B}\mathbf{X}[k]$$

$$\overset{\Delta}{=} \mathbf{h}[n] * \mathbf{X}[n].$$

The matrix system function is

$$\mathbf{H}(z) = (\mathbf{I} - \mathbf{A}z^{-1})^{-1}\mathbf{B},$$

as can easily be verified. The WSS cross-correlation matrices $\mathbf{R}_{\mathbf{YX}}[m] \overset{\Delta}{=} E\{\mathbf{Y}[n+m]\mathbf{X}^{\dagger}[n]\}$ and $\mathbf{R}_{\mathbf{XY}}[m] \overset{\Delta}{=} E\{\mathbf{X}[n+m]\mathbf{Y}^{\dagger}[n]\}$, between an input WSS random vector sequence and its WSS output random sequence, become

$$\mathbf{R}_{\mathbf{YX}}[m] = \mathbf{h}[m] * \mathbf{R}_{\mathbf{XX}}[m],$$

$$\mathbf{R}_{\mathbf{XY}}[m] = \mathbf{R}_{\mathbf{XX}}[m] * \mathbf{h}^{\dagger}[-m].$$

Parenthetically, we note that for a causal $\mathbf{h}$, such as would arise from recursive solution of the above vector LCCDE, we have the output $\mathbf{Y}[n]$ uncorrelated with the future values of the input $\mathbf{X}[n]$, when the input $\mathbf{X} = \mathbf{W}$ is assumed a white noise vector sequence.

The output correlation matrix is

$$\mathbf{R_{YY}}[m] = \mathbf{h}[m] * \mathbf{R_{XX}}[m] * \mathbf{h}^\dagger[-m]$$

and the output psd matrix becomes upon Fourier transformation

$$\mathbf{S_{YY}}(\omega) = \mathbf{H}(\omega)\mathbf{S_{XX}}(\omega)\mathbf{H}^\dagger(\omega).$$

The total solution of Equation 6.6-1 for any $n \geq n_0$ can be written as

$$\mathbf{Y}[n] = \mathbf{A}^{n-n_0}\mathbf{Y}[n_0] + \sum_{k=n_0}^{n} \mathbf{h}[n-k]\mathbf{X}[k], \qquad n \geq n_0$$

in terms of the initial condition $\mathbf{Y}[n_0]$ that must be specified at n_0. In the limit as $n_0 \to -\infty$, and for a stable system matrix $\mathbf{A}$, this then becomes the convolution summation

$$\mathbf{Y}[n] = \mathbf{h}[n] * \mathbf{X}[n], \qquad -\infty < n < +\infty.$$

Definition 6.6-2   A vector random sequence $\mathbf{Y}[n]$ is *vector Markov* if for all $K > 0$ and for all $n_K > n_{K-1} > \ldots > n_1$, we have

$$P\{\mathbf{Y}[n_K] \leq \mathbf{y}_K | \mathbf{y}[n_{K-1}], \ldots, \mathbf{y}[n_1]\} = P\{\mathbf{Y}[n_K] \leq \mathbf{y}_K | \mathbf{y}[n_{K-1}]\}$$

for all real values of the vector $\mathbf{y}_K$, and all conditioning vectors $\mathbf{y}[n_{K-1}], \ldots, \mathbf{y}[n_1]$. (cf. Definition 6.5-2 of Markov-p property.)   ■

We can now state the following theorem for vector random sequences:

Theorem 6.6-1   In the state equation

$$\mathbf{X}[n] = \mathbf{A}\mathbf{X}[n-1] + \mathbf{B}\mathbf{W}[n], \text{for } n > 0, \text{ with } \mathbf{X}[0] = \mathbf{0},$$

let the input $\mathbf{W}[n]$ be a white Gaussian random sequence. Then the output $\mathbf{X}[n]$ for $n > 0$ is a vector Markov random sequence.

The proof is left to the reader as an exercise.   ■

Example 6.6-2 _____

(Relation between scalar Markov-p and vector Markov.) Let $X[n]$ be a Markov-p random sequence satisfying the pth order difference equation

$$X[n] = a_1 X[n-1] + \ldots + a_p X[n-p] + bW[n].$$

Defining the p-dimensional vector random sequence $\mathbf{X}[n] = [X[n], \ldots, X[n-p+1]]^T$, and coefficient matrix

$$\mathbf{A} = \begin{bmatrix} a_1 & a_2 & \cdots & \cdots & a_p \\ 1 & 0 & 0 & \cdots & 0 \\ 0 & 1 & \cdot & \cdot & \cdot \\ \cdot & \cdot & \cdot & \cdot & 0 \\ 0 & \cdots & 0 & 1 & 0 \end{bmatrix},$$

we have

$$\mathbf{X}[n] = \mathbf{A}\mathbf{X}[n-1] + \mathbf{b}W[n].$$

Thus $\mathbf{X}[n]$ is a vector Markov random sequence with $\mathbf{b} = [b, 0, \ldots, 0]^T$. Such a vector transformation of a scalar equation is called a *state-variable representation* [6-8].

6.7 CONVERGENCE OF RANDOM SEQUENCES

Some nonstationary random sequences may converge to a limit as the sequence index goes to infinity. This asymptotic behavior is evidenced in probability theory by convergence of the fraction of successes in an infinite Bernoulli sequence, where the relevant theorems are called the laws of large numbers. Also, when we study the convergence of random processes in Chapter 8 we will sometimes make a sequence of finer and finer approximations to the output of a random system at a given time, say, t_0 , that is, $Y_n(t_0)$. The index n then defines a random sequence, which should converge in some sense to the true output. In this section we will look at several types of convergence for random sequences, that is, sequences of random variables.

We start by reviewing the concept of convergence for deterministic sequences. Let x_n be a sequence of complex (or real) numbers; then convergence is defined as follows.

Definition 6.7-1   A sequence of complex (or real) numbers x_n converges to the complex (or real) number x if given any $\varepsilon > 0$, there exists an integer n_0 such that whenever $n > n_0$, we have

$$|x_n - x| < \varepsilon. \quad \blacksquare$$

Note that in this definition the value n_0 may depend on the value ε; that is, when ε is made smaller then most likely n_0 will need to be made larger. Sometimes this dependence is formalized by writing $n_0(\varepsilon)$ in place of n_0 in this definition. This is often written as

$$\lim_{n \to \infty} x_n = x \qquad \text{or as} \qquad x_n \to x \text{ as } n \to \infty.$$

A practical problem with this definition is that one must have the limit x to test for convergence. For simple cases one can often guess what the limit is and then use the definition to verify that this limit indeed exists. Fortunately, for more complex situations there is an alternative in the *Cauchy criterion* for convergence, which we state as a theorem without proof.

Theorem 6.7-1   (Cauchy criterion [6-5].) A sequence of complex (or real) numbers x_n converges to a limit *if and only if* (iff)

$$|x_n - x_m| \to 0 \text{ as both } n \text{ and } m \to \infty.$$

The reason that this works for complex (or real) numbers is that the set of all complex (or real) numbers is *complete*, meaning that it contains all its limit points. For example, the set $\{0 < x < 1\} = (0, 1)$ is not complete, but the set $\{0 \le x \le 1\} = [0, 1]$ is complete because sequences x_n in these sets and tending to 0 or 1 have a *limit point* in the set $[0, 1]$ but have no limit point in the set $(0, 1)$. In fact, the set of all complex (or real) numbers is complete as well as n-dimensional linear vector spaces over both the real and complex number fields. Thus the Cauchy criterion for convergence applies in these cases also. For more on numerical convergence see [6-5].

Convergence for functions is defined using the concept of convergence of sequences of numbers. We say the sequence of functions $f_n(x)$ converges to the function $f(x)$ if the corresponding sequence of numbers converges for each x. It is stated more formally in the following definition.

Definition 6.7-2   The sequence of functions $f_n(x)$ converges (pointwise) to the function $f(x)$ if for each x_0 the sequence of complex numbers $f_n(x_0)$ converges to $f(x_0)$. ∎

The Cauchy criterion for convergence applies to pointwise convergence of functions if the set of functions under consideration is complete. The set of continuous functions is not complete because a sequence of continuous functions may converge to a discontinuous function (cf. item (d) in Example 6.7-1). However, the set of bounded functions is complete [6-5].

The following are some examples of convergent sequences of numbers and functions. We leave the demonstration of these results as exercises for the reader.

Example 6.7-1 _____

(Some convergent sequences.)

(a) $x_n = (1 - 1/n)a + (1/n)b \to a$ as $n \to \infty$.
(b) $x_n = \sin(\omega + e^{-n}) \to \sin \omega$ as $n \to \infty$.
(c) $f_n(x) = \sin[(\omega + 1/n)x] \to \sin(\omega x)$, as $n \to \infty$ for any (fixed) x.
(d) $f_n(x) = \left\{ \begin{array}{l} e^{-n^2 x}, \text{ for } x > 0 \\ \quad 1, \quad \text{ for } x \le 0 \end{array} \right\} \to u(-x)$, as $n \to \infty$ for any (fixed) x.

The reader should note that in the convergence of the functions in (c) and (d), the variable x is held constant as the limit is being taken. The limit is then repeated for each such x value to find the limiting function.

_____

Since a random variable is a function, a sequence of random variables (also called a random sequence) is a sequence of functions. Thus we can define the first and strongest type of convergence for random variables.

Definition 6.7-3   (Sure convergence.) The random sequence $X[n]$ *converges surely* to the random variable X if the sequence of functions $X[n, \zeta]$ converges to the function $X(\zeta)$ as $n \to \infty$ for all $\zeta \in \Omega$. ∎

As a reminder, the functions $X(\zeta)$ are not arbitrary. They are random variables and thus satisfy the condition that the set $\{\zeta : X(\zeta) \leq x\} \subset \mathscr{F}$ for all x, that is, that this set be an event for all values of x. This is in fact necessary for the calculation of probability since the probability measure P is only defined for events. Such functions X are more generally called *measurable functions* and in a course on real analysis it is shown that the space of measurable functions is complete [6-1]. If we have a Cauchy sequence of measurable functions (random variables), then one can show that the limit function exists and is also measurable (a random variable). Thus the Cauchy convergence criterion also applies for random variables.

Most of the time we are not interested in precisely defining random variables for sets in Ω of probability zero because it is thought that these events will never occur. In this case, we can weaken the concept of sure convergence to the still very strong concept of almost-sure convergence.

Definition 6.7-4 (Almost-sure convergence.) The random sequence $X[n]$ converges *almost surely* to the random variable X if the sequence of functions $X[n, \zeta]$ converges for all $\zeta \in \Omega$ except possibly on a set of probability zero. ■

This is the strongest type of convergence normally used in probability theory. It is also called *probability-1* convergence. It is sometimes written

$$P\left\{\lim_{n \to \infty} X[n, \zeta] = X(\zeta)\right\} = 1,$$

meaning simply that there is a set A such that $P[A] = 1$ and $X[n]$ converges to X for all $\zeta \in A$. In particular $A \triangleq \{\zeta : \lim_{n \to \infty} X[n, \zeta] = X(\zeta)\}$. Here the set A^c is the probability-zero set mentioned in this definition. As shorthand notation we also use

$$X[n] \to X \text{ a.s.} \quad \text{and} \quad X[n] \to X \qquad \text{pr.1,}$$

where the abbreviation "a.s." stands for *almost surely*, and "pr.1" stands for *probability one*.

An example of probability-1 convergence is the Strong Law of Large Numbers to be proved in the next section. Three examples of random sequences are next evaluated for possible convergence.

Example 6.7-2 _____
(Convergence of random sequences.) For each of the following three random sequences, we assume that the probability space $(\Omega, \mathscr{F}, P)$ has sample space $\Omega = [0, 1]$. $\mathscr{F}$ is the family of Borel subsets of Ω and the probability measure P is Lebesgue measure, which on a real interval $(a, b]$, is just its length l, that is,

$$l(a, b] \triangleq b - a \quad \text{for } b \geq a.$$

(a) $X[n, \zeta] = n\zeta$.
(b) $X[n, \zeta] = \sin(n\zeta)$.
(c) $X[n, \zeta] = \exp[-n^2(\zeta - \frac{1}{n})]$.

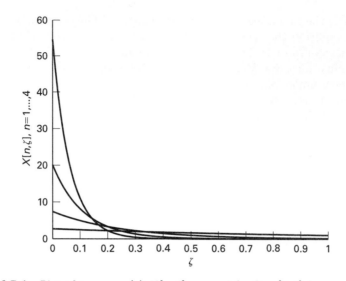

Figure 6.7-1   Plot of sequence (c) $X[n, \zeta]$ versus ζ for $\Omega = [0, 1]$ for $n = 1, \dots, 4$.

The sequence in (a) clearly diverges to $+\infty$ for any $\zeta \neq 0$. Thus this random sequence does not converge. The sequence in (b) does not diverge, but it oscillates between -1 and $+1$ except for the one point $\zeta = 0$. Thus this random sequence does not converge either.

Considering the random sequence in (c), the graph in Figure 6.7-1 shows that this sequence converges as follows:

$$\lim_{n \to \infty} X[n, \zeta] = \begin{cases} \infty \text{ for } \zeta = 0 \\ 0 \text{ for } \zeta > 0 \end{cases}.$$

Thus we can say that the random sequence converges to the (degenerate) random variable $X = 0$ with probability-1. We simply take $A = (0, 1]$ and note that $P[A] = 1$ and that $X[n, \zeta] \simeq 0$ for every ζ in A for sufficiently large n. We write $X[n] \to 0$ a.s. However $X[n]$ clearly does not converge surely to zero.

Thus far we have been discussing pointwise convergence of sequences of functions and random sequences. This is similar to considering a space of *bounded* functions $\mathscr{B}$ with the norm

$$\|f\|_\infty \overset{\Delta}{=} \sup_x |f(x)|.^\dagger$$

When we write $f_n \to f$ in the function space $\mathscr{B}$, we mean that $\|f_n - f\|_\infty = \sup_x |f_n(x) - f(x)| \to 0$, giving us pointwise convergence. The space of continuous bounded functions is denoted L_∞ and is known to be complete ([6-1], p. 115).

---

†The supremum or sup operator is similar to the max operator. The supremum of a set of numbers is the smallest number greater than or equal to each number in the set, for example, $\sup\{0 < x < 1\} = 1$. Note the difficulty with max in this example since 1 is not included in the open interval $(0, 1)$, thus the max does not exist here!

Another type of function space of great practical interest uses the *energy norm* (cf. Eq. 4.4-6):

$$||f||_2 \triangleq \left(\int_{-\infty}^{+\infty} |f(x)|^2 dx \right)^{1/2}.$$

The space of integrable (measurable) functions with *finite energy norm* is denoted L^2. When we say a sequence of functions converges in L^2, that is, $||f_n - f||_2 \to 0$, we mean that

$$\left(\int_{-\infty}^{+\infty} |f_n(x) - f(x)|^2 dx \right)^{1/2} \to 0.$$

This space of integrable functions is also complete [6-1]. A corresponding concept for random sequences is given by mean-square convergence.

Definition 6.7-5   (Mean-square convergence.) A random sequence $X[n]$ converges in the *mean-square sense* to the random variable X if $E\{|X[n] - X|^2\} \to 0$ as $n \to \infty$.   ■

This type of convergence depends only on the second-order properties of the random variables and is thus often easier to calculate than a.s. convergence. A second benefit of the mean-square type of convergence is that it is closely related to the physical concept of power. If $X[n]$ converges to X in the mean-square sense, then we can expect that the variance of the error $\varepsilon[n] \triangleq X[n] - X$ will be small for large n. If we look back at Example 6.7-2c, we can see that this random sequence does not converge in the mean-square sense, so that the error variance or power as defined here would not ever be expected to be small. To see this, consider possible mean-square convergence to zero (since $X[n] \to 0$ a.s.),

$$E\{|X[n] - 0|^2\} = E\{X[n]^2\}$$

$$= \int_0^1 \exp(-2n^2 \zeta) \exp 2n d\zeta$$

$$= \exp(2n) \int_0^1 \exp(-2n^2 \zeta) d\zeta$$

$$= \exp(2n) \left[\frac{1 - \exp(-2n^2)}{2n^2} \right] \to \infty \quad \text{as } n \to \infty,$$

hence $X[n]$ does not converge in the mean-square sense to 0.

Still another type of convergence that we will consider is called *convergence in probability*. It is weaker than probability-1 convergence and also weaker than mean-square convergence. This is the type of convergence displayed in the Weak Laws of Large Numbers to be discussed in the next section. It is defined as follows:

Definition 6.7-6   (Convergence in probability.) Given the random sequence $X[n]$ and the limiting random variable X, we say that $X[n]$ converges in probability to X if for every $\varepsilon > 0$,

$$\lim_{n \to \infty} P\left[|X[n] - X| > \varepsilon\right] = 0. \quad ■$$

We sometimes write $X[n] \to X(p)$, where (p) denotes the type of convergence. Also convergence in probability is sometimes called p-convergence.

One can use Chebyshev's inequality (Theorem 4.4-1), $P[|Y| > \varepsilon] \le E[|Y|^2]/\varepsilon^2$ for $\varepsilon > 0$, to show that mean-square convergence implies convergence in probability. For example, let $Y \triangleq X[n] - X$; then the preceding inequality becomes

$$P[|X[n] - X| > \varepsilon] \le E\left[|X[n] - X|^2\right]/\varepsilon^2.$$

Now mean-square convergence implies that the right-hand side goes to zero as $n \to \infty$, for any fixed $\varepsilon > 0$, which implies that the left-hand side must also go to zero which is the definition of convergence in probability. Thus we have proved the following result.

Theorem 6.7-2   Convergence of a random sequence in the mean-square sense implies convergence in probability.   ■

The relation between convergence with probability-1 and convergence in probability is more subtle. The main difference between them can be seen by noting that the former talks about the probability of the limit while the latter talks about the limit of the probability. Further insight can be gained by noting that a.s. convergence is concerned with convergence of the entire sample sequences while p-convergence is concerned only with the convergence of the random variable at an individual n. That is to say, a.s. convergence is concerned with the joint events at an infinite number of times, while p-convergence is concerned with the simple event at time n, albeit large. One can prove the following theorem.

Theorem 6.7-3   Convergence with probability-1 implies convergence in probability.

Proof   (Adapted from Gnedenko [6-6].) Let $X[n] \to X$ a.s. and define the set A,

$$A \triangleq \bigcap_{k=1}^{\infty} \bigcup_{n=1}^{\infty} \bigcap_{m=1}^{\infty} \{\zeta : |X[n+m, \zeta] - X(\zeta)| < 1/k\}.$$

Then it must be that $P[A] = 1$. To see this we note that A is the set of ζ such that starting at some n and for all later n we have $|X[n, \zeta] - X(\zeta)| < 1/k$ and furthermore this must hold for all $k > 0$. Thus A is precisely the set of ζ on which $X[n, \zeta]$ is convergent. So $P[A]$ must be one. Eventually for n large enough and $1/k$ small enough we get $|X[n, \zeta] - X(\zeta)| < \varepsilon$, and the error stays this small for all larger n. Thus

$$P\left[\bigcup_{n=1}^{\infty} \bigcap_{m=1}^{\infty} \{|X[n+m] - X| < \varepsilon\}\right] = 1 \qquad \text{for all } \varepsilon > 0,$$

which implies by the continuity of probability,

$$\lim_{n \to \infty} P\left[\bigcap_{m=1}^{\infty} \{|X[n+m] - X| < \varepsilon\}\right] = 1 \qquad \text{for all } \varepsilon > 0,$$

which in turn implies the greatly weakened result

$$\lim_{n\to\infty} P[|X[n+m] - X| < \varepsilon] = 1 \qquad \text{for all } \varepsilon > 0, \qquad (6.7\text{-}1)$$

which is equivalent to the definition of p-convergence. ■

Because of the gross weakening of the a.s. condition, that is, the enlargement of the set A in the foregoing proof, it can be seen that p-convergence does not imply a.s. convergence. We note in particular that Equation 6.7-1 may well be true even though no single sample sequence stays close to X for all $n + m > n$. This is in fact the key difference between these two types of convergence.

Example 6.7-3

Define a random pulse sequence $X[n]$ on $n \geq 0$ as follows: Set $X[0] = 1$. Then for the *next two points* set exactly one of the $X[n]$'s to 1, the other to 0, equally likely. For the *next three points* set exactly one of the $X[n]$'s to 1 equally likely and set the others to 0. Continue this procedure for the *next four points*, setting exactly one of the $X[n]$'s to 1 equally likely, and so forth. A sample function would look like Figure 6.7-2.

Obviously this random sequence is slowly converging to zero in some sense as $n \to \infty$. In fact a simple calculation would show p-convergence and also m.s. convergence due to the growing distance between pulses as $n \to \infty$. In fact at $n \simeq \frac{1}{2}l^2$, the probability of a one (pulse) is only $1/l$. However, we do not have a.s. convergence, since *every* sample sequence has ones appearing arbitrarily far out on the n axis. Thus no sample sequences converge to zero.

One final type of convergence that we consider is not a convergence for random variables at all! Rather it is a type of convergence for distribution functions.

Definition 6.7-7 A random sequence $X[n]$ with probability distribution function $F_n(x)$ converges in distribution to the random variable X with probability distribution function $F(x)$ if

$$\lim_{n\to\infty} F_n(x) = F(x)$$

at all x for which F is continuous. ■

Note that in this definition we are not really saying anything about the random variables themselves, just their probability distribution functions (PDFs). Convergence in distribution

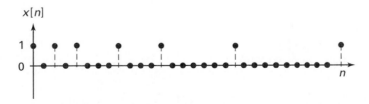

Figure 6.7-2   A sequence that is converging in probability but not with probability-1.

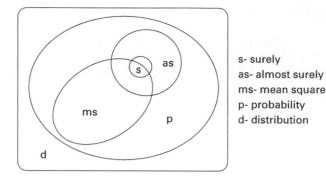

Figure 6.7-3   Venn diagram illustrating relationships of various possible convergence modes for random sequences.

just means that as n gets large the PDFs are converging or becoming alike. For example, the sequence $X[n]$ and the variable X can be jointly independent even though $X[n]$ converges to X in distribution. This is radically different from the four earlier types of convergence, where as n gets large the random variables $X[n]$ and X are becoming very dependent because some type of "error" between them is going to zero. Convergence in distribution is the type of convergence that occurs in the Central Limit Theorem (Section 4.7). The relationships between these five types of convergence are shown diagrammatically in Figure 6.7-3, where we have used the fact that p-convergence implies convergence in distribution, which is shown below. Note that even sure convergence may not imply mean square convergence. This because the integral of the square of the limiting random variable, with respect to the probability measure, may diverge.

To see that p-convergence implies convergence in distribution, assume that the limiting random variable X is continuous so that it has a pdf. First we consider the conditional distribution function

$$F_{X[n]|X}(y|x) = P\{X[n] \leq y | X = x\}.$$

From the definition of p-convergence, it should be clear that

$$F_{X[n]|X}(y|x) \rightarrow \begin{cases} 1, y > x \\ 0, y < x \end{cases}, \qquad \text{as } n \rightarrow \infty,$$

so that

$$F_{X[n]|X}(y|x) \rightarrow u(y - x), \qquad \text{except possibly at the one point } y = x,$$

and hence

$$F_{X[n]}(y) = P\{X[n] \leq y\} = \int_{-\infty}^{+\infty} F_{X[n]|X}(y|x) f_X(x) dx$$

$$\rightarrow \int_{-\infty}^{+\infty} u(y - x) f_X(x) dx$$

$$= \int_{-\infty}^{y} f_X(x)dx$$

$$= F_X(y),$$

as was to be shown. In the case where the limiting random variable X is not continuous, we must exercise more care but the result is still true at all points x for which $F_X(x)$ is continuous. (See Problem 6.43.)

6.8 LAWS OF LARGE NUMBERS

The Laws of Large Numbers have to do with the convergence of a sequence of estimates of the mean of a random variable. As such they concern the convergence of a random sequence to a constant. The Weak Laws obtain convergence in probability, while the Strong Laws yield convergence with probability-1. A version of the Weak Law has already been demonstrated in Example 4.4-3. We restate it here for convenience.

Theorem 6.8-1   (Weak Law of Large Numbers.) Let $X[n]$ be an independent random sequence with mean μ_X and variance σ_X^2 defined for $n \geq 1$. Define another random sequence as

$$\hat{\mu}_X[n] \triangleq (1/n) \sum_{k=1}^{n} X[k] \qquad \text{for } n \geq 1.$$

Then $\hat{\mu}_X[n] \to \mu_x \quad (p) \quad$ as $n \to \infty.$ ■

Remember, an independent random sequence is one whose terms are all jointly independent. Another version of the Weak Law allows the random sequence to be of nonuniform variance.

Theorem 6.8-2   (Weak Law—nonuniform variance.) Let $X[n]$ be an independent random sequence with constant mean μ_X and variance $\sigma_X^2[n]$ defined for $n \geq 1$. Then if

$$\sum_{n=1}^{\infty} \sigma_X^2[n]/n^2 < \infty,$$

$$\hat{\mu}_X[n] \to \mu_X \quad (p) \quad \text{as } n \to \infty. \quad ■$$

Both of these theorems are also true for convergence with probability-1, in which case they become Strong Laws. The theorems concerning convergence with probability-1 are best derived using the concept of a Martingale sequence. By introducing this concept we can also get another useful result called the Martingale convergence theorem, which is helpful in estimation and decision/detection theory.

Definition 6.8-1   †A random sequence $X[n]$ defined for $n \geq 0$ is called a *Martingale* if the conditional expectation

$$E\{X[n]|X[n-1], X[n-2], \ldots, X[0]\} = X[n-1] \qquad \text{for all } n \geq 1. \quad ■$$

---

†The material dealing with Martingale sequences can be omitted on a first reading.

Viewing the conditional expectation as an estimate of the present value of the sequence based on the past, then for a Martingale this estimate is just the most recent past value. If we interpret $X[n]$ as an amount of capital in a betting game, then the Martingale condition can be regarded as necessary for fairness of the game, which in fact is how it was first introduced [6-1].

Example 6.8-1 _____

(Binomial counting sequence.) Let $W[n]$ be a Bernoulli random sequence taking values ± 1 with equal probability and defined for $n \geq 0$. Let $X[n]$ be the corresponding Binomial counting sequence

$$X[n] \triangleq \sum_{k=0}^{n} W[k], \qquad n \geq 0.$$

Then $X[n]$ is a Martingale, which can be shown as follows:

$$
\begin{aligned}
E\{X[n]|X[n-1],\ldots,X[0]\} &= E\left\{\sum_{k=0}^{n} W[k]|X[n-1],\ldots,X[0]\right\} \\
&= \sum_{k=0}^{n} E\{W[k]|X[n-1],\ldots,X[0]\} \\
&= \sum_{k=0}^{n} E\{W[k]|W[n-1],\ldots,W[0]\} \\
&= \sum_{k=0}^{n-1} W[k] + E\{W[n]\} \\
&= X[n-1].
\end{aligned}
$$

The first equality follows from the definition of $X[n]$. The third equality follows from the fact that knowledge of the first $(n-1)$ Xs is equivalent to knowledge of the first $(n-1)$ Ws. The next-to-last equality follows from $E[W|W] = W$. The last equality follows from the fact that $E\{W[n]\} = 0$.

Example 6.8-2 _____

(Independent-increments sequences.) Let $X[n]$ be an independent-increments random sequence (see Definition 6.1-4) defined for $n \geq 0$. Then $X_c[n] \triangleq X[n] - \mu_X[n]$ is a Martingale. To show this we write $X_c[n] = (X_c[n] - X_c[n-1]) + X_c[n-1]$ and note that by independent increments and the fact that the mean of X_c is zero, we have

$$
\begin{aligned}
E\{X_c[n]|X_c[n-1],\ldots,X_c[0]\} &= E\{X_c[n] - X_c[n-1]|X_c[n-1],\ldots,X_c[0]\} \\
&\quad + E\{X_c[n-1]|X_c[n-1],\ldots,X_c[0]\} \\
&= E\{X_c[n] - X_c[n-1]\} + X_c[n-1] \\
&= X_c[n-1].
\end{aligned}
$$

The next theorem shows the connection between the Strong Laws, which have to do with the convergence of sample sequences, and Martingales. It provides a kind of Chebyshev inequality for the maximum term in an n-point Martingale sequence.

Theorem 6.8-3   Let $X[n]$ be a Martingale sequence defined on $n \geq 0$. Then for every $\varepsilon > 0$ and for any positive n,

$$P\left[\max_{0 \leq k \leq n} |X[k]| \geq \varepsilon\right] \leq E\{X^2[n]\}/\varepsilon^2.$$

Proof   For $0 \leq j \leq n$, define the mutually exclusive events,

$$A_j \triangleq \{|X[k]| \geq \varepsilon \text{ for the } \textit{first time} \text{ at } j\}.$$

Then the event $\{\max_{0 \leq k \leq n} |X[k]| \geq \varepsilon\}$ is just a union of these events. Also define the random variables,

$$I_j \triangleq \begin{cases} 1 & \text{if } A_j \text{ occurs} \\ 0 & \text{otherwise} \end{cases},$$

called the *indicators* of the events A_j. Then

$$E\{X^2[n]\} \geq \sum_{j=0}^{n} E\{X^2[n]I_j\} \tag{6.8-1}$$

since $\sum_{j=0}^{n} I_j \leq 1$. Also $X^2[n] = (X[j] + (X[n] - X[j]))^2$, so expanding and inserting into Equation 6.8-1 we get

$$E\{X^2[n]\} \geq \sum_{j=0}^{n} E\{X^2[j]I_j\} + 2\sum_{j=0}^{n} E\{X[j]\,(X[n] - X[j])\,I_j\}$$

$$+ \sum_{j=0}^{n} E\{(X[n] - X[j])^2\,I_j\}$$

$$\geq \sum_{j=0}^{n} E\{X^2[j]I_j\} + 2\sum_{j=0}^{n} E\{X[j]\,(X[n] - X[j])\,I_j\}. \tag{6.8-2}$$

Letting $Z_j \triangleq X[j]I_j$, we can write the second term in Equation 6.8-2 as $E\{Z_j\,(X[n] - X[j])\}$ and noting that Z_j depends only on $X[0], \ldots, X[j]$, we then have

$$E\{Z_j\,(X[n] - X[j])\} = E\{E[Z_j\,(X[n] - X[j])\,|X[0], \ldots, X[j]]\}$$

$$= E\{Z_j E\,[X[n] - X[j]|X[0], \ldots, X[j]]\}$$

$$= E\{Z_j\,(X[j] - X[j])\}$$

$$= 0.$$

Thus Equation 6.8-2 becomes

$$E\{X^2[n]\} \geq \sum_{j=0}^{n} E\{X^2[j]I_j\}$$

$$\geq \varepsilon^2 E\left\{\sum_{j=0}^{n} I_j\right\}$$

$$= \varepsilon^2 P\left\{\bigcup_{j=0}^{n} A_j\right\}$$

$$= \varepsilon^2 P\left\{\max_{0 \leq k \leq n} |X[k]| \geq \varepsilon\right\}. \quad \blacksquare$$

Theorem 6.8-4   (Martingale Convergence theorem.) Let $X[n]$ be a Martingale sequence on $n \geq 0$, satisfying

$$E\{X^2[n]\} \leq C < \infty \quad \text{for all } n \text{ for some } C.$$

Then

$$X[n] \to X \quad \text{(a.s.)} \quad \text{as } n \to \infty,$$

where X is the limiting random variable.

Proof   Let $m \geq 0$ and define $Y[n] \triangleq X[n + m] - X[m]$ for $n \geq 0$. Then $Y[n]$ is a Martingale, so by Theorem 6.8-3

$$P\left[\max_{0 \leq k \leq n} |X[m + k] - X[m]| \geq \varepsilon\right] \leq \frac{1}{\varepsilon^2} E\left\{Y^2[n]\right\},$$

where

$$E\{Y^2[n]\} = E\{(X[n + m] - X[m])^2\}$$

$$= E\{X^2[n + m]\} - 2E\{X[n + m]X[m]\} + E\{X^2[m]\}.$$

Rewriting the middle term, we have

$$E\{X[m]X[n + m]\} = E\{X[m]E[X[n + m]|X[m], \ldots, X[0]]\}$$

$$= E\{X[m]X[m]\}$$

$$= E\{X^2[m]\} \quad \text{since } X \text{ is a Martingale,}$$

so

$$E\{Y^2[n]\} = E\{X^2[n + m]\} - E\{X^2[m]\} \geq 0 \quad \text{for all } m, n \geq 0. \tag{6.8-3}$$

Therefore $E\{X^2[n]\}$ must be monotonic nondecreasing. Since it is bounded from above by $C < \infty$, it must converge to a limit. Since it has a limit, then by Equation 6.8-3, the $E\{Y^2[n]\} \to 0$ as m and $n \to \infty$. Thus

$$\lim_{m \to \infty} P\left[\max_{k \geq 0} |X[m+k] - X[m]| > \varepsilon\right] = 0,$$

which implies $P[\lim_{n \to \infty} \max_{k \geq 0} |X[m+k] - X[m]| > \varepsilon] = 0$ by the continuity of the probability measure P (cf. Corollary to Theorem 6.1-1). Finally by the Cauchy convergence criteria, there exists a random variable X such that

$$X[n] \to X \quad \text{(a.s.).} \quad \blacksquare$$

Theorem 6.8-5 (Strong Law of Large Numbers.) Let $X[n]$ be a WSS independent random sequence with mean μ_X and variance σ_X^2 defined for $n \geq 1$. Then as $n \to \infty$

$$\hat{\mu}_X[n] = \frac{1}{n} \sum_{k=1}^{n} X[k] \to \mu_X \quad \text{(a.s.).}$$

Proof Let $Y[n] \triangleq \sum_{k=1}^{n} \frac{1}{k} X_c[n]$; then $Y[n]$ is a Martingale on $n \geq 1$. Since

$$E\{Y^2[n]\} = \sum_{k=1}^{n} \frac{1}{k^2} \sigma_X^2 \leq \sigma_X^2 \sum_{k=1}^{\infty} \frac{1}{k^2} = C < \infty,$$

we can apply Theorem 6.8-4 to show that $Y[n] \to Y$ (a.s.) for some random variable Y. Next noting that $X_c[k] = k\left(Y[k] - Y[k-1]\right)$, we can write.

$$\frac{1}{n} \sum_{k=1}^{n} X_c[k] = \frac{1}{n} \left[\sum_{k=1}^{n} kY[k] - \sum_{k=1}^{n} kY[k-1]\right]$$

$$= -\frac{1}{n} \sum_{k=1}^{n} Y[k] + \frac{n+1}{n} Y[n]$$

$$\to -Y + Y = 0 \quad \text{(a.s.)}$$

so that

$$\hat{\mu}_X[n] \to \mu_X \quad \text{(a.s.).} \quad \blacksquare$$

6.9 SUMMARY

In this chapter we introduced the concept of random sequence and studied some of its properties. We defined the random sequence as a family of sample sequences. We introduced several important random sequences. Then we reviewed linear discrete-time theory and considered the practical problem of finding out how the mean and covariance function are transformed by a linear system. We then considered the special but important case of stationary and wide-sense stationary random sequences and introduced the concept of power spectral density for them. We looked at convergence of random sequences and learned to appreciate the variety of modes of convergence that are possible. We then applied some of these results to the laws of large numbers and used Martingale properties to prove the important strong law of large numbers.

In the next chapter we will discover that many of these results extend to the case of continuous time as we continue our study with random processes.

PROBLEMS

6.1.   Prove the chain rule for the probability of the intersection of N events, $\{A_n\}_{n=1}^N$. For example, for $N = 3$ we have,

$$P[A_1 A_2 A_3] = P[A_1]P[A_2|A_1]P[A_3|A_1 A_2].$$

Interpret this result for joint PDFs and joint pdf's.

6.2.   Consider an N-dimensional random vector $\mathbf{X}$. Show that pairwise independence of its random variable components does not imply that the components are jointly independent.

6.3.   Let $\mathbf{X} = (X_1, X_2, \ldots, X_5)^T$ be a random vector whose components satisfy the equations

$$X_i = X_{i-1} + B_i,$$

where the B_i, $1 \leq i \leq 5$ are jointly independent and Bernoulli distributed, taking on values 0 and 1, with mean value $1/2$. The first value is $X_1 = B_1$. Put the B_i together to make a random vector $\mathbf{B}$.

    (a)   Write $\mathbf{X} = \mathbf{AB}$ for some constant matrix $\mathbf{A}$ and determine $\mathbf{A}$.
    (b)   Find the mean vector $\boldsymbol{\mu}_{\mathbf{X}}$.
    (c)   Find the covariance matrix $\mathbf{K_B}$.
    (d)   Find the covariance matrix $\mathbf{K_X}$.

[For parts (b) through (d), express your answers in terms of the matrix $\mathbf{A}$].

6.4.   Often one is given a problem statement starting as follows: "Let X be a real-valued random variable with pdf $f_X(x)\ldots$." Since a r.v. is a mapping from a sample space Ω with field of events $\mathscr{F}$ and a probability measure P, evidently the existence of an underlying probability space $(\Omega, \mathscr{F}, P)$ is assumed by such a problem statement. Show that a suitable underlying probability space $(\Omega, \mathscr{F}, P)$ can always be created, thus legitimizing problem statements such as the one above.

6.5.   Let T be a continuous random variable denoting the time at which the first photon is emitted from a light source; T is measured from the instant the source is energized. Assume that the probability density function for T is $f_T(t) = \lambda e^{-\lambda/t} u(t)$.

    (a)   What is the probability that at least one photon is emitted prior to time t_2 if it is known that none was emitted prior to time t_1, where $t_1 < t_2$?

(b) What is the probability that at least one photon is emitted prior to time t_2 if three independent sources of this type are energized simultaneously?

6.6. Let X be a conditionally Normal random variable, with conditional density function $N(\mu, \sigma^2)$, given the values of $M = \mu$ and $\Sigma^2 = \sigma^2$.

(a) Assume σ^2 is a known constant but that M is a random variable having the probability distribution function (PDF)

$$F_M(m) = [1 - e^{-\lambda m}]u(m).$$

(note that variable m is *continuous* here!)

where λ is a known positive value. Determine the characteristic function for X (*Hint*: First define a conditional characteristic function.)

(b) Now assume both Σ^2 and M are independent random variables. Let their distributions be arbitrary, but assume both have a finite mean and variance. Determine the mean and variance for X in terms of μ_M, μ_{Σ^2}, and σ_M^2.

6.7. Let X and Y be independent identically distributed (IID) random variables with the exponential probability density functions

$$f_X(w) = f_Y(w) = \lambda e^{-\lambda w}u(w).$$

(a) Determine the probability density function for the ratio

$$0 \le R \triangleq \frac{X}{X+Y} \le 1, \quad \text{i.e., } f_R(r), \qquad 0 < r \le 1.$$

(b) Let A be the event $X < 1/Y$. Determine the conditional pdf of X given that A occurs and that $Y = y$; that is, determine

$$f_X(x|A, Y = y).$$

(c) Using the definitions of (b), what is the minimum mean-square error estimate of X given that the event A occurs and that $Y = y$?

6.8. Let $\mathbf{X} = (X_1, X_2, \ldots, X_{10})^T$ be a random vector whose components satisfy the equations,

$$X_i = \frac{2}{5}(X_{i-1} + X_{i+1}) + W_i \quad \text{for } 2 \le i \le 9,$$

where the W_i are independent and Laplacian* distributed with mean zero and variance σ^2 for $i = 1$ to 10, and $X_1 = \frac{1}{2}X_2 + \frac{5}{4}W_1$ and $X_{10} = \frac{1}{2}X_9 + \frac{5}{4}W_{10}$.

(a) Find the mean vector $\mu_{\mathbf{X}}$
(b) Find the covariance matrix $\mathbf{K_{XX}}$.
(c) Write an expression for the multidimensional pdf of the random vector $\mathbf{X}$.

[*Hint:*

Matrix identity: if $\mathbf{A} \triangleq \begin{bmatrix} 1 & \rho & \rho^2 & \cdots & \rho^N \\ \rho & 1 & \rho & \rho^2 & \cdots \\ \rho^2 & \rho & 1 & \cdots & \cdots \\ \cdots & \rho^2 & \cdots & \cdots & \rho \\ \rho^N & \cdots & \cdots & \rho & 1 \end{bmatrix}$,

then $\mathbf{A}^{-1}$ is given as

$$\beta^2 \mathbf{A}^{-1} = \begin{bmatrix} 1-\rho\alpha & -\alpha & 0 & \cdots & 0 \\ -\alpha & 1 & -\alpha & 0 & \cdots \\ 0 & -\alpha & 1 & \cdots & 0 \\ \cdots & 0 & \cdots & \cdots & -\alpha \\ 0 & \cdots & 0 & -\alpha & 1-\rho\alpha \end{bmatrix}$$

with $\alpha \triangleq \frac{\rho}{1+\rho^2}$ and $\beta^2 \triangleq \frac{1-\rho^2}{1+\rho^2}$. The Laplacian pdf is given as

$$f_W(w) = \frac{1}{\sqrt{2}\sigma} \exp\left(-\sqrt{2}\frac{|w|}{\sigma}\right), \qquad -\infty \le w \le +\infty.$$

6.9. Prove Corollary 6.1-1.

6.10. Let $\{X_i\}$ be a sequence of independent and identically distributed (IID) Normal random variables with zero-mean and unit variance. Let

$$S_k \triangleq X_1 + X_2 + \ldots + X_k \quad \text{for } k \ge 1.$$

Determine the joint probability density function for S_n and S_m where $1 \le m < n$.

6.11. In Example 6.1-8 we saw that probability distribution functions (PDFs) are continuous from the right. Are they continuous from the left also? Either prove or give a counterexample.

6.12. Let the probability space $(\Omega, \mathscr{F}, P)$ be given as follows:

$$\Omega = \{a, b, c\}, \text{i.e., the outcome } \zeta = a \text{ or } b \text{ or } c,$$

$$\mathscr{F} = \text{all subsets of } \Omega,$$

$$P[\{\zeta\}] = 1/3 \text{ for each outcome } \zeta.$$

Let the random sequence $X[n]$ be defined as follows:

$$X[n, a] = 3\delta[n]$$
$$X[n, b] = u[n-1]$$
$$X[n, c] = \cos \pi n/2.$$

(a) Find the mean function $\mu_X[n]$.
(b) Find the correlation function $R_X[m, n]$.
(c) Are $X[1]$ and $X[0]$ independent? Why?

6.13. Consider a random sequence $X[n]$ as the input to a linear filter with impulse response

$$h[n] = \begin{cases} 1/2, & n = 0 \\ 1/2, & n = 1 \\ 0, & \text{else.} \end{cases}$$

We denote the output random sequence $Y[n]$, i.e., for each outcome ζ,

$$Y[n, \zeta] = \sum_{k=-\infty}^{k=+\infty} h[k]X[n - k, \zeta].$$

Assume the filter runs for all time, $-\infty < n < +\infty$. We are given the mean function of the input $\mu_X[n]$ and correlation function of the input $R_{XX}[n_1, n_2]$. Express your answers in terms of these assumed known functions.

(a) Find the mean function of the output $\mu_Y[n]$.
(b) Find the autocorrelation function of the output $R_{YY}[n_1, n_2]$.
(c) Write the autocovariance function of the output $K_{YY}[n_1, n_2]$ in terms of your answers to parts (a) and (b).
(d) Now assume that the input $X[n]$ is a Gaussian random sequence, and write the corresponding joint pdf of the output $f_Y(y_1, y_2; n_1, n_2)$ at two arbitrary times $n_1 \neq n_2$ in terms of $\mu_Y[n]$ and $K_{YY}[n_1, n_2]$.

6.14. The random arrival time sequence $T[n]$, defined for $n \geq 1$, was found to have the Erlang type pdf, for some $\lambda > 0$,

$$f_T(t; n) = \frac{(\lambda t)^{n-1}}{(n - 1)!} \lambda \exp(-\lambda t) \, u(t).$$

Find the joint pdf $f_T(t_2, t_1; 10, 5)$. Recall that $T[n]$ is the sum up to time n of the IID interarrival time sequence with exponential pdf's.

6.15. This problem considers a random sequence model for a charge coupled device (CCD) array with very "leaky" cells. We start by defining the width-3 pulse function:

$$h[n] = \begin{cases} 1/4 & n = -1 \\ 1/2 & n = 0 \\ 1/4 & n = +1 \\ 0 & \text{else,} \end{cases}$$

and as illustrated in Figure P6.15, which we will use to account for 25% of the charge in a cell that leaks out to its right and left neighboring cells. We assume that the one-dimensional CCD array is infinitely long and represents the array contents by the random sequence X:

$$X[n, \zeta] = \sum_{i=-\infty}^{i=+\infty} A(\zeta_i)h[n - i],$$

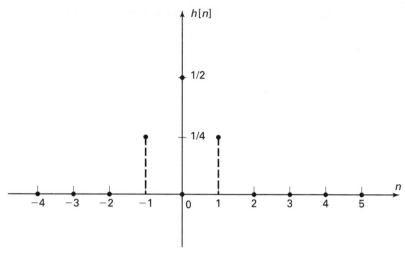

Figure P6.15

where ζ_i is the ith component of ζ, the infinite dimensional outcome of the experiment. The random variables $A(\zeta_i)$ are jointly independent and Gaussian distributed with mean λ and variance λ.

    (a) Find the mean function $\mu_X[n]$.
    (b) Find the first-order pdf $f_X(x;n)$.
    (c) Find the joint pdf $f_X(x_1, x_2; n, n+1)$.

6.16. We are given a random sequence $X[n]$ for $n \geq 0$ with conditional pdf's

$$f_X(x_n|x_{n-1}) = \alpha \exp[-\alpha(x_n - x_{n-1})]\, u(x_n - x_{n-1}) \quad \text{for } n \geq 1,$$

with $u(x)$ the unit-step function and initial pdf $f_X(x_0) = \delta(x_0)$. Take $\alpha > 0$.

    (a) Find the first-order pdf $f_X(x_n)$ for $n = 2$.
    (b) Find the first-order pdf $f_X(x_n)$ for arbitrary $n > 1$ using mathematical induction.

6.17. Let $x[n]$ be a deterministic input to the linear shift invariant (LSI) discrete-time system H shown in Figure P6.17.

Figure P6.17   LSI system with impulse response $h[n]$.

(a) Use linearity and shift-invariance properties to show that

$$y[n] = x[n] * h[n] \triangleq \sum_{k=-\infty}^{+\infty} x[k]h[n-k] = h[n] * x[n].$$

(b) Define the Fourier transform of a sequence $a[n]$ as

$$A(\omega) \triangleq \sum_{n=-\infty}^{\infty} a[n]e^{-j\omega n} \qquad -\pi \le \omega \le +\pi$$

and show that the inverse Fourier transform is

$$a[n] = \frac{1}{2\pi} \int_{-\pi}^{+\pi} A(\omega)e^{+j\omega n} d\omega, \qquad -\infty < n < +\infty.$$

(c) Using the results in (a) and (b), show that

$$Y(\omega) = H(\omega)X(\omega), \qquad -\pi \le \omega \le +\pi$$

for an LSI discrete-time system.

6.18. Consider the difference equation

$$y[n] + \alpha y[n-1] = x[n], \qquad -\infty < n < +\infty$$

where $-1 < \alpha < +1$.

(a) Let the input be $x[n] = \beta^n u[n]$ for $-1 < \beta < +1$. Find the solution for $y[n]$ assuming causality applies, i.e., $y[n] = 0$ for $n < 0$.
(b) Let the input be $x[n] = \beta^{-n} u[-n]$ for $-1 < \beta < +1$. Find the solution for $y[n]$ assuming anticausality applies,[†] i.e., $y[n] = 0$ for $n > 0$.

6.19. Let $W[n]$ be an independent random sequence with mean 0 and variance σ_W^2 defined for $-\infty < n < +\infty$. For appropriately chosen ρ, let the stationary random sequence $X[n]$ satisfy the causal LCCDE

$$X[n] = \rho X[n-1] + W[n], \qquad -\infty < n < +\infty.$$

(a) Show that $X[n-1]$ and $W[n]$ are independent at time n.
(b) Derive the characteristic function equation

$$\Phi_X(\omega) = \Phi_X(\rho\omega)\Phi_W(\omega).$$

(c) Find the continuous solution to this functional equation for the unknown function Φ_X when $W[n]$ is assumed to be Gaussian. [*Note:* $\Phi_X(0) = 1$.]
(d) What is σ_X^2?

[†]This part requires more detailed knowledge of the z-transform. (cf. Appendix A.)

6.20. Consider the LSI system shown in Figure P6.20, whose *deterministic input* $x[n]$ is contaminated by noise (a random sequence) $W[n]$. We wish to determine the properties of the output random sequence $Y[n]$. The noise $W[n]$ has mean $\mu_W[n] = 2$ and autocorrelation $E\{W[m]W[n]\} = \sigma_W^2 \delta[m - n] + 4$. The impulse response is $h[n] = \rho^n u[n]$ with $|\rho| < 1$. The deterministic input $x[n]$ is given as $x[n] = 3$ for all n.

    (a) Find the output mean $\mu_Y[n]$.
    (b) Find the output power $E\{Y^2[n]\}$.
    (c) Find the output covariance $K_{YY}[m, n]$.

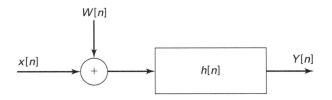

Figure P6.20   LSI system with deterministic-plus-noise input.

6.21. Show that the random sequence $X[n]$ generated in Example 6.1-15 is not an independent random sequence.

6.22. Let $W[n]$ be an independent random sequence with mean $\mu_W = 0$ and variance σ_W^2. Define a new random sequence $X[n]$ as follows:

$$X[0] = 0$$

$$X[n] = \rho X[n - 1] + W[n] \quad \text{for} \quad n \geq 1.$$

    (a) Find the mean value of $X[n]$ for $n \geq 0$.
    (b) Find the covariance of $X[n]$, denoted $K_{XX}[m, n]$.
    (c) For what values of ρ does $K_{XX}[m, n]$ tend to $G[m-n]$ (for some finite-valued function G) as m and n become large? This situation is called *asymptotic stationarity*.

6.23. Let the random variables A and B be independent and indentically distributed (IID) with mean 0, variance σ^2, and third-order moment $E[A^3] = E[B^3] = \xi_3 \neq 0$. Consider the random sequence $X[n] = A \cos \omega_0 n + B \sin \omega_0 n$, $-\infty < n < +\infty$, where ω_0 is a fixed radian frequency.

    (a) Show that $X[n]$ is wide-sense stationary (WSS).
    (b) Prove that $X[n]$ is not stationary by presenting a counterexample.

6.24. Consider a WSS random sequence $X[n]$ with mean $\mu_X[n] = \mu$, a constant, and correlation function $R_{XX}[m] = p^2 \delta[m]$ with $p^2 > 0$. In such a case μ must be zero, as you will show in this problem. Note that the covariance function here is $K_{XX}[m] = p^2 \delta[m] - \mu^2$.

   (a) Take $m = 0$ and conclude that $p^2 \geq \mu^2$.

   (b) Take a vector $\mathbf{X}$ of length N out of the random sequence $X[n]$. Show that the corresponding covariance matrix $\mathbf{K_{XX}}$ will be positive semidefinite only if $\mu^2 \leq \sigma^2/(N-1)$, where $\sigma^2 \overset{\Delta}{=} p^2 - \mu^2$. (*Hint:* Take coefficient vector $\mathbf{a} = \mathbf{1}$, i.e., all $1'$s.)

   (c) Let $N \to \infty$ and conclude that μ must be zero for the stationary white noise sequence $X[n]$.

6.25. For the linear filter of Problem 6.13, assume the input random sequence is WSS, and write the psd of the output sequence $S_{YY}(\omega)$ in terms of the psd of the input $S_{XX}(\omega)$.

   (a) Show that the psd is a real-valued function, even if $X[n]$ is a complex-valued random sequence.

   (b) Show that if $X[n]$ is real-valued, then $S_{XX}(\omega) = S_{XX}(-\omega)$.

   (c) Show that $S_{XX}(\omega) \geq 0$ for every ω whether $X[n]$ is complex-valued or not.

6.26. Let the WSS random sequence X have correlation function

$$R_{XX}[m] = 10e^{-\lambda_1|m|} + 5e^{-\lambda_2|m|}$$

with $\lambda_1 > 0$ and $\lambda_1 > 0$. Find the corresponding psd $S_{XX}(\omega)$ for $|\omega| \leq \pi$.

6.27. The psd of a certain random sequence is given as $S_{XX}(\omega) = 1/[(1+\alpha^2) - 2\alpha \cos \omega]^2$ for $-\pi \leq \omega \leq +\pi$, where $|\alpha| < 1$. Find the random sequence's correlation function $R_X[m]$.

6.28. Let the input to system $H(\omega)$ be $W[n]$, a white noise random sequence with $\mu_W[n] = 0$ and $K_{WW}[m] = \delta[m]$. Let $X[n]$ denote the corresponding output random sequence. Find $K_{XW}[m]$ and $S_{XW}(\omega)$.

6.29. Consider the system shown in Figure P6.29. Let $X[n]$ and $V[n]$ be WSS and mutually uncorrelated with zero-mean and psd's $S_{XX}(\omega)$ and $S_{VV}(\omega)$, respectively.

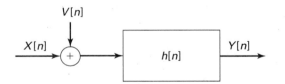

Figure P6.29 LSI system with random signal-plus-noise input.

   (a) Find the psd of the output $S_{YY}(\omega)$.

   (b) Find the cross-power spectral density between the input X and the output Y, i.e., find $S_{XY}(\omega)$.

6.30. Consider the discrete-time system with input random sequence $X[n]$ and output $Y[n]$ given as

$$Y[n] = \frac{1}{5} \sum_{k=-2}^{+2} X[n-k].$$

Assume that the input sequence $X[n]$ is WSS with psd $S_{XX}(\omega) = 2$.

    (a) Find the psd of the output random sequence $S_{YY}(\omega)$.

    (b) Find the output correlation function $R_{YY}[m]$.

6.31. Let $X[n]$ be a Markov chain on $n \geq 0$ taking values 1 and 2 with *one-step transition probabilities*,

$$\mathbf{P}_{ij} \triangleq P\{X[n] = j | X[n-1] = i\}, \qquad 1 \leq i, j \leq 2,$$

given in matrix form as

$$\mathbf{P} = \begin{bmatrix} 0.9 & 0.1 \\ 0.2 & 0.8 \end{bmatrix} = (\mathbf{P}_{i,j}).$$

We describe the state probabilities at time n by the vector

$$\mathbf{p}[n] \triangleq [P\{X[n] = 1\}, P\{X[n] = 2\}].$$

    (a) Show that $\mathbf{p}[n] = \mathbf{p}[0]\mathbf{P}^n$.

    (b) Draw a two-state transition diagram and label the branches with the one-step transition probabilities $\mathbf{P}_{ij}$. Don't forget the $\mathbf{P}_{ii}$ or *self-transitions*. (See Figure 6.5-1 for state transition diagram of a Markov chain.)

    (c) Given that $X[0] = 1$, find the probability that the first transition to state 2 occurs at time n.

6.32. We defined a Markov random sequence $X[n]$ in this chapter as being specified by its first-order pdf $f_X(x; n)$ and its one-step conditional pdf

$$f_X(x_n | x_{n-1}; n, n-1) = f_X(x_n | x_{n-1}) \quad \text{for short.}$$

    (a) Find the two-step pdf for a Markov random sequence $f_X(x_n | x_{n-2})$ in terms of the above functions. Here, take $n \geq 2$ for a random sequence starting at $n = 0$.

    (b) Find the N-step pdf $f_X(x_n | x_{n-N})$ for arbitrary positive integer N, where here we only need consider $n \geq N$.

6.33. Consider a Markov random sequence $X[n]$ running on $1 \leq n \leq 100$. It is statistically described by its first-order pdf $f_X(x; 1)$ and its one-step transition (conditional) pdf $f_X(x_n | x_{n-1}; n, n-1)$. By the Markov definition, we have (suppressing the time variables) that

$$f_X(x_n | x_{n-1}) = f_X(x_n | x_{n-1}, x_{n-2}, \ldots, x_1) \quad \text{for } 2 \leq n \leq 100.$$

Show that a Markov random sequence is also Markov *in the reverse order*, i.e.,

$$f_X(x_n | x_{n+1}) = f_X(x_n | x_{n+1}, x_{n+2}, \ldots, x_{100}) \quad \text{for } 1 \leq n \leq 99,$$

and so one can alternatively statistically describe a Markov random sequence by the one-step backward pdf $f_X(x_{n-1} | x_n; n-1, n)$ and first-order pdf $f_X(x; 100)$.

6.34. Given a Markov chain $X[n]$ on $n \geq 1$, with transition probabilities given as $P(x[n]|x[n-1])$, find an expression for the two-step transition probabilities $P(x[n]|x[n-2])$. Also show that

$$P(x[n+1]|x[n-1], x[n-2], \ldots, x[1]) = P(x[n+1]|x[n-1]), \text{ for } n \geq 1.$$

6.35. Write a MATLAB function called `triplemarkov` that will compute and plot the autocorrelation functions for the asymmetric, two-state Markov model in Example 6.1-16 for any three sets of parameters $\{p_{00}, p_{11}\}$. Denote the maximum lag interval as N. Run your routine for $\{0.2, 0.8\}$, $\{0.2, 0.5\}$, and $\{0.2, 0.2\}$. Repeat for $\{0.8, 0.2\}$, $\{0.8, 0.5\}$, and $\{0.8, 0.8\}$. Describe what you observe.

6.36. Consider the probability space $(\Omega, \mathscr{F}, P)$ with $\Omega = [0, 1]$, $\mathscr{F}$ defined to be the Borel sets of Ω, and $P[(0, \zeta] = \zeta$ for $0 < \zeta \leq 1$.

    (a) Show that $P[\{0\}] = 0$ by using the axioms of probability.

    (b) Determine in what *senses* the following random sequences converge:

        (i) $X[n, \zeta] = e^{-n\zeta}, n \geq 0$

        (ii) $X[n, \zeta] = \sin\left(\zeta + \frac{1}{n}\right), n \geq 1$

        (iii) $X[n, \zeta] = \cos^n(\zeta), n \geq 0$.

    (d) If the preceding sequences converge, what are the limits?

6.37. The members of the sequence of jointly independent random variables $X[n]$ have pdf's of the form

$$f_X(x; n) = \left(1 - \frac{1}{n}\right) \frac{1}{\sqrt{2\pi}\sigma} \exp\left[-\frac{1}{2\sigma^2}\left(x - \frac{n-1}{n}\sigma\right)^2\right]$$

$$+ \frac{1}{n}\sigma \exp(-\sigma x)u(x).$$

Determine whether or not the random sequence $X[n]$ converges in

    (i) the mean-square sense,

    (ii) probability,

    (iii) distribution.

6.38. The members of the random sequence $X[n]$ have joint pdf's of the form

$$f_X(\alpha, \beta; m, n) = \frac{mn}{2\pi\sqrt{1 - \rho^2}} \exp\left(-\frac{1}{2(1 - \rho^2)}[m^2\alpha^2 - 2\rho mn\alpha\beta + n^2\beta^2]\right)$$

for $m \geq 1$ and $n \geq 1$ where $-1 < \rho < +1$.

    (a) Show that $X[n]$ converges in the mean-square sense as $n \to \infty$ for all $-1 < \rho < +1$.

    (b) Specify the probability distribution function of the mean-square limit $X \triangleq \lim_{n\to\infty} X[n]$.

6.39. State conditions under which the mean-square limit of a sequence of Gaussian random variables is also Gaussian.

6.40. Let $X[n]$ be a real-valued random sequence on $n \geq 0$, made up from stationary and *independent increments*, i.e., $X[n] - X[n-1] = W[n]$, "the increment" with $W[n]$ being a stationary and independent random sequence. The random sequence always starts with $X[0] = 0$. We also know that at time $n = 1$, $E\{X[1]\} = \eta$ and $var\{X[1]\} = \sigma^2$.

(a) Find $\mu_X[n]$ and $\sigma_X^2[n]$, the mean and variance functions of the random sequence X at time n for any time $n > 1$.

(b) Prove that $X[n]/n$ converges in probability to η as the time n approaches infinity.

6.41. This problem demonstrates that p-convergence implies convergence in distribution even when the limiting pdf does not exist.

(a) For any real number x and any positive ε, show that

$$P[X \leq x - \varepsilon] \leq P[X[n] \leq x] + P[|X[n] - X| \geq \varepsilon].$$

(b) Similarly show that

$$P[X > x + \varepsilon] \leq P[X[n] > x] + P[|X[n] - X| \geq \varepsilon].$$

For part (c), assume the random sequence $X[n]$ converges to the random variable X in probability.

(c) Let $n \to \infty$ and conclude that

$$\lim_{n \to \infty} F_X(x; n) = F_X(x)$$

at points of continuity of F_X.

6.42. Let $X[n]$ be a second-order random sequence. Let $h[n]$ be the impulse response of an LSI system. We wish to define the output of the system $Y[n]$ as a mean-square limit.

(a) Show that we can define the mean-square limit

$$Y[n] \triangleq \sum_{k=-\infty}^{+\infty} h[k]X[n-k], \qquad -\infty < n < +\infty, \text{ (m.s.)}$$

if

$$\sum_k \sum_l h[k]h^*[l]R_{XX}[n-k, n-l] < \infty \text{ for all } n.$$

(*Hint:* Set $Y_N[n] \triangleq \sum_{k=-N}^{+N} h[k]X[n-k]$ and show that m.s. limit of $Y_N[n]$ exists by using the Cauchy convergence criteria.)

(b) Find a simpler condition for the case when $X[n]$ is a wide-sense stationary random sequence.

(c) Find the necessary condition on $h[n]$ when $X[n]$ is (stationary) white noise.

6.43. If $X[n]$ is a Martingale sequence on $n \geq 0$, show that

$$E\{X[n+m]|X[m], \ldots, X[0]\} = X[m] \qquad \text{for all } n \geq 0.$$

6.44. Let $Y[n]$ be a random sequence and X a random variable and consider the conditional expectation

$$E\{X|Y[0], \ldots, Y[n]\} \triangleq G[n].$$

Show that the random sequence $G[n]$ is a Martingale.

6.45. We can enlarge the concept of Martingale sequence somewhat as follows. Let $G[n] \triangleq g(X[0], \ldots, X[n])$ for each $n \geq 0$ for measurable functions g. We say G is a *Martingale with respect to (wrt)* X if

$$E\{G[n]|X[0], \ldots, X[n-1]\} = G[n-1].$$

(a) Show that Theorem 6.8-3 holds for G a Martingale wrt X. Specifically, substitute G for X in the statement of the theorem. Then make necessary changes to the proof.

(b) Show that the Martingale convergence Theorem 6.8-4 holds for G a Martingale with respect to (wrt) X.

6.46. Consider the hypothesis-testing problem involving $(n+1)$ observations $X[0], \ldots, X[n]$ of the random sequence X. Define the likelihood ratio

$$L_X[n] \triangleq \frac{f_X(X[0], \ldots, X[n]|H_1)}{f_X(X[0], \ldots, X[n]|H_0)}, \qquad n \geq 0,$$

corresponding to two hypotheses H_1 and H_0. Show that $L_X[n]$ is a Martingale wrt X under hypothesis H_0.

6.47. In the discussion of interpolation in Example 6.4-7, work-out the algebra needed to arrive at the psd of the up-sampled random sequence $X_e[n]$.

6.48. The up-sampled sequence $X_e[n]$ in the interpolation process is clearly not WSS, even if $X[n]$ is WSS. Create an up-sampled random sequence that is WSS by randomizing the start-time of the sequence $X[n]$. That is, define a binary random variable Θ with $P[\Theta = 0] = P[\Theta = 1] = 0.5$. Define the start-time randomized sequence by $X_r[n] \triangleq X[n + \Theta]$. Then the resulting up-sampled sequence is $X_{er}[n] = X\left[\frac{n+\Theta}{2}\right]$. Show that $R_{X_r X_r}[k] = R_{XX}[k]$ and $R_{X_{er} X_{er}}[m, m+k] = R_{X_{er} X_{er}}[k] = 0.5 R_{XX}[k/2]$ for k even, and zero for k odd.

REFERENCES

6-1.   R. B. Ash, *Real Analysis and Probability.* New York: Academic Press, 1972, pp. 1–53 and 115.

6-2.   A. Kolmogorov, *Foundations of the Theory of Probability*. New York: Chelsea, 1950.

6-3.   T. M. Apostol, *Mathematical Analysis*. Reading, MA: Addison-Wesley, 1957, pp. 192–202.

6-4.   A. V. Oppenheim and R. W. Schafer, *Discrete-Time Signal Processing*. Upper Saddle River, NJ: Prentice Hall, 1989, Chapters 2–3.

6-5.   W. Rudin, *Principles of Mathematical Analysis*. New York: McGraw-Hill, 1964, pp. 45–48.

6-6.   B. V. Gnedenko, *The Theory of Probability* (translated by B. D. Seckler). New York: Chelsea, 1963, p. 237.

6-7.   Y. Viniotis, *Probability and Random Processes*. Boston: WCB/Mcgraw-Hill, 1998, p. 103.

6-8.   S. S. Soliman and M. D. Srinath, *Continuous and Discrete Signals and Systems*. N.J.: Prentice Hall, 1998.

6-9.   E. Wong and B. Hajek, *Stochastic Processes in Engineering Systems*. New York: Springer-Verlag, 1985, p. 3.

6-10.   H. Stark and Y. Yang, *Vector Space Projections*. New York: Wiley, 1998.

6-11.   P. P. Vaidyanathan, *Multirate Systems of Fitter Banks*, New Jersey: Prentice-Hall, 1993.

7

Random Processes

In the last chapter, we learned how to generalize the concept of random variable to that of random sequence. We did this by associating a sample sequence with each outcome $\zeta \in \Omega$, thereby generating a family of sequences collectively called a random sequence. In this chapter we generalize further by considering random functions of a *continuous* parameter. We consider this continuous parameter time, but it could equally well be position, or angle, or some other continuous parameter. The collection of all these continuous time functions is called a random process. Random processes will be perhaps the most useful objects we study because they can be used to model physical processes directly without any intervening need to sample the data. Even when of necessity one is dealing with sampled data, the concept of random process will give us the ability to reference the properties of the sample sequence to those of the limiting continuous process so as to be able to judge the adequacy of the sampling rate.

Random processes find a wide variety of applications. Perhaps the most common use is as a model for noise in physical systems, this modeling of the noise being the necessary first step in deciding on the best way to mitigate its negative effects. A second class of applications concerns the modeling of random phenomena that are not noise but are nevertheless unknown to the system designer. An example would be a multimedia signal (audio, image, or video) on a communications link. The signal is not noise, but it is unknown from the viewpoint of a distant receiver and can take on many (an enormous number of) values. Thus we model such signals as random processes, when some statistical description of the source is available. Situations such as this arise in other contexts also, such as control systems, pattern recognition, etc. Indeed from an information theory viewpoint, any waveform that communicates information must have at least some degree of randomness in it.

We start with a definition of random process and study some of the new difficulties to be encountered with continuous time. Then we look at the moment functions for random processes and generalize the correlation and covariance functions from Chapter 6 to this continuous parameter case. We also look at some basic random processes of practical importance. We then begin our study of linear systems and random processes. Indeed, this topic is central to our study of random processes and is most widely used in applications. Then we present some classifications of random processes based on general statistical properties. Finally, we introduce stationary and wide-sense stationary random processes and their analysis for linear systems.

7.1 BASIC DEFINITIONS

It is most important to fully understand the basic concept of the random process and its associated moment functions. The situation is analogous to the discrete-time case treated in Chapter 6. The main new difficulty is that the time axis has now become uncountable. We start with the basic definition.

Definition 7.1-1 Let $(\Omega, \mathscr{F}, P)$ be a probability space. Then define a mapping X from the sample space Ω to a space of continuous time functions. The elements in this space will be called *sample functions*. This mapping is called a *random process* if at each fixed time the mapping is a random variable, that is[†], $X(t, \zeta) \in \mathscr{F}$ for each fixed t on the real line $-\infty < t < +\infty$. ∎

Thus we have a multidimensional function $X(t, \zeta)$, which for each fixed ζ is an ordinary time function and for each fixed t is a random variable. This is shown diagrammatically in Figure 7.1-1 for the special case $\Omega = [0, \infty]$. We see a family of random variables indexed by t when we look along the time axis, and we see a family of time functions indexed by ζ when we look along the outcome "axis."

We have the following easy examples of random processes:

Example 7.1-1 _____
$X(t, \zeta) = X(\zeta)f(t)$ where X is a random variable and f is a deterministic function of the parameter t. We also write $X(t) = Xf(t)$.

Example 7.1-2 _____
$X(t, \zeta) = A(\zeta)\sin(\omega_0 t + \Theta(\zeta))$ where A and Θ are random variables. We also write $X(t) = A\sin(\omega_0 t + \Theta)$, suppressing the outcome ζ.

More typical examples of random processes can be constructed from random sequences.

Example 7.1-3 _____
$X(t) = \sum_n X[n]p_n(t - T[n])$ where $X[n]$ and $T[n]$ are random sequences and the functions $p_n(t)$ are deterministic waveforms that can take on various shapes. For example, the $p_n(t)$

[†]$X \in \mathscr{F}$ is shorthand for $\{\zeta : X(\zeta) \leq x\} \subset \mathscr{F}$ for all x. This condition permits us to measure the probability of events of this kind and hence define PDFs.

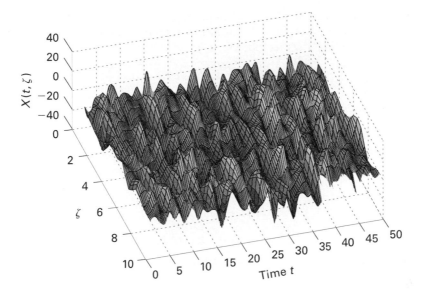

Figure 7.1-1 A random process for a continuous sample space $\Omega = [0, 10]$.

might be ideal unit-step functions that could provide a model for a so-called *jump process*. In this interpretation the T would be the times of the *arrivals* and the $X[n]$ would be the *amplitudes* of the jumps. Then $X(t)$ would indicate the total amplitude up to time t. If all the $X[n]$'s were 1, we would have a *counting process* in that $X(t)$ would count the arrivals prior to time t.

If we sample the random process at n times t_1 through t_n, we get an n-dimensional random vector. If we know the probability distribution of this vector for all times t_1 through t_n and for all positive n, then clearly we know a lot about the random process. If we know all this information, we say that we have *statistically specified* (statistically determined) the random process in a fashion that is analogous to the corresponding case for random sequences.

Definition 7.1-2 A random process $X(t)$ is statistically specified by its complete set of nth order PDFs (pdf's or PMFs) for all positive integers n, i.e., $F_X(x_1, x_2, \ldots, x_n; t_1, t_2, \ldots, t_n)$ for all $x_1, x_2, \ldots, x_n$ and for all $t_1 < t_2 < \ldots < t_n$. ■

The term "statistical" comes from the fact that this is the limit of the information that could be obtained from accumulating relative frequencies of events determined by the random process $X(t)$ at all finite collections of time instants. Clearly, this is all we could hope to determine by measurements on an observed process that we may wish to model. However, the question arises: Is this enough information to completely determine the random process? Unfortunately the general answer is no. We need to impose a continuity requirement on the sample functions $x(t)$. To see this the following simple example suffices.

Example 7.1-4
(From Karlin [7-1], p. 32.) Let U be a uniform random variable on $[0, 1]$ and define the random processes $X(t)$ and $Y(t)$ as follows:

$$X(t) \triangleq \begin{cases} 1 & \text{for } t = U \\ 0, & \text{else,} \end{cases}$$

and

$$Y(t) \triangleq 0 \quad \text{for all } t.$$

Then $Y(t)$ and $X(t)$ have the same finite-order distributions, yet obviously the probability of the following two events is not the same:

$$\{X(t) \leq 0.5 \text{ for all } t\}$$

and

$$\{Y(t) \leq 0.5 \text{ for all } t\}.$$

To show that $Y(t)$ and $X(t)$ have the same nth-order pdf's, find the conditional nth-order pdf of X given $U = u$, then integrate out the conditioning on U. We leave this as an exercise to the reader.

The problem in Example 7.1-4 is really that the complementary event $\{X(t) > 0.5$ for some $t \in [0, 1]\}$ involves an *uncountable* number of random variables. Yet the statistical determination and the extended additivity Axiom 4 (see Section 6.1) only allow us to evaluate probabilities corresponding to *countable* numbers of random variables. In what follows, we will generally assume that we always have a process "continuous enough" that the family of finite-order distribution functions suffices to determine the process for all time.[†] Such processes are called *separable*. The random process $X(t)$ of the above example is obviously not separable.

As in the case of random sequences, the moment functions play an important role in practical applications. The *mean function*, denoted $\mu_X(t)$, is given as

$$\mu_X(t) \triangleq E[X(t)], \qquad -\infty < t < +\infty. \tag{7.1-1}$$

Similarly the *correlation function* is defined as the expected value of the conjugate product,

$$R_{XX}(t_1, t_2) \triangleq E[X(t_1)X^*(t_2)], \qquad -\infty < t_1, t_2 < +\infty. \tag{7.1-2}$$

The *covariance function* is defined as the expected value of the conjugate product of the *centered process* $X_c(t) \triangleq X(t) - \mu_X(t)$ at times t_1 and t_2

$$K_{XX}(t_1, t_2) \triangleq E[X_c(t_1)X_c^*(t_2)]$$

$$\triangleq E[(X(t_1) - \mu_X(t_1))(X(t_2) - \mu_X(t_2))^*]. \tag{7.1-3}$$

[†]An exception is white noise to be introduced in Section 7.3.

Clearly these three functions are not unrelated and in fact we have,

$$K_{XX}(t_1, t_2) = R_{XX}(t_1, t_2) - \mu_X(t_1)\mu_X^*(t_2). \tag{7.1-4}$$

We also define the *variance function* as $\sigma_X^2(t) \triangleq K_{XX}(t, t) = E\left[|X_c(t)|^2\right]$, and the *power function* $R_{XX}(t, t) = E\left[|X(t)|^2\right]$.

Example 7.1-5 _____

Consider the random process

$$X(t) = A \sin(\omega_0 t + \Theta),$$

where A and Θ are independent, real-valued random variables and Θ is uniformly distributed over $[-\pi, +\pi]$.

For this sinusoidal random process, we will find the mean function $\mu_X(t)$ and correlation function $R_{XX}(t_1, t_2)$. First

$$\mu_X(t) = E[A \sin(\omega_0 t + \Theta)]$$

$$= E[A]E[\sin(\omega_0 t + \Theta)]$$

$$= \mu_A \cdot \frac{1}{2\pi} \int_{-\pi}^{+\pi} \sin(\omega_0 t + \theta) d\theta$$

$$= \mu_A \cdot 0 = 0.$$

Then for the correlation,

$$R_{XX}(t_1, t_2) = E[X(t_1)X^*(t_2)]$$

$$= E[A^2 \sin(\omega_0 t_1 + \Theta) \sin(\omega_0 t_2 + \Theta)]$$

$$= E[A^2]E[\sin(\omega_0 t_1 + \Theta) \sin(\omega_0 t_2 + \Theta)].$$

Now, the second factor can be rewritten as

$$\tfrac{1}{2}\{E[\cos(\omega_0(t_1 - t_2))] - E[\cos(\omega_0(t_1 + t_2) + 2\Theta)]\} \tag{7.1-5}$$

by applying the trigonometric identity

$$\sin(B)\sin(C) = \tfrac{1}{2}\{\cos(B - C) - \cos(B + C)\},$$

and bringing the expectation operator inside. Then, since Θ is uniformly distributed over $[-\pi, +\pi]$, the integral arising from the second expectation in Equation 7.1-5 is zero, and we finally obtain

$$R_{XX}(t_1, t_2) = \tfrac{1}{2}E[A^2] \cos \omega_0(t_1 - t_2).$$

We note that $\mu_X = 0$ (a constant) and $R_{XX}(t_1, t_2)$ depends only on $t_1 - t_2$. Such processes will be classified as *wide-sense stationary* in Section 7.4.

As in the discrete-time case, the correlation and covariance functions are Hermitian symmetric, that is,

$$R_{XX}(t_1, t_2) = R_{XX}^*(t_2, t_1),$$

$$K_{XX}(t_1, t_2) = K_{XX}^*(t_2, t_1),$$

which directly follow from the linearity of the expectation operator E.

If we sample the random process at N times $t_1, t_2, \ldots, t_N$, we form a *random vector*. We have already seen that the correlation or covariance matrix of a random vector must be positive semidefinite (cf. Chapter 5). This, then, imposes certain requirements on the respective correlation and covariance function of the random process. Specifically, every correlation (covariance) matrix that can be formed from a correlation (covariance) function must be positive semidefinite. We next define *positive semidefinite* functions.

Definition 7.1-3   The two-dimensional function $g(t, s)$ is *positive semidefinite* if for all $N > 0$, and all $t_1 < t_2 < \ldots < t_N$, and for all complex constants $a_1, a_2, \ldots, a_N$, we have

$$\sum_{i=1}^{N} \sum_{j=1}^{N} a_i a_j^* g(t_i, t_j) \geq 0. \quad \blacksquare$$

Using this definition, we can thus say that all correlation and covariance functions must be positive semidefinite. Later we will see that this necessary condition is also sufficient. Although positive semidefiniteness is an important constraint, it is difficult to apply this condition in a test of the legitimacy of a proposed correlation function.

Another fundamental property of correlation and covariance functions is *diagonal dominance*,

$$|R_{XX}(t, s)| \leq \sqrt{R_{XX}(t, t) R_{XX}(s, s)} \quad \text{for all } t, s,$$

which follows from the Cauchy-Schwarz inequality (cf. Equation 4.3-17). Diagonal dominance is implied by positive semidefiniteness but is a much weaker condition.

7.2 SOME IMPORTANT RANDOM PROCESSES

In this section we introduce several important random processes. We start with the asynchronous binary signaling process (ABS) and the random telegraph signal (RTS). We continue with the Poisson counting process; the phase-shift keying (PSK) random process, an example of digital modulation; the Wiener process, which is obtained as a continuous limit of a random walk sequence, and lastly the broad class of Markov processes.

Asynchronous Binary Signaling

A sample function of the asynchronous binary signaling (ABS) process (important in the age of digital modulation and computers) is shown in Figure 7.2-1. Each pulse has width T

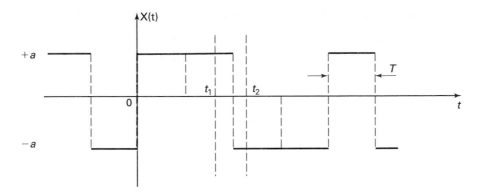

Figure 7.2-1   Sample function realization of the asynchronous binary signaling (ABS) process.

with the random variable X_n indicating the height of the nth pulse, taking on values $\pm a$ with equal probability.

The sequence is *asynchronous* because the start time of the nth pulse, or equivalently, the displacement D of the 0th pulse is a Uniform random variable $U[-\frac{T}{2}, \frac{T}{2})$. For $|t_2 - t_1| < T$, the sampling instant t_2 could be on the same pulse containing the sampling instant t_1 or on a different pulse.

The ABS process can thus be described mathematically by

$$X(t) = \sum_n X_n \, w \left[\frac{t - D - nT}{T} \right],$$

where the pulse (rectangular window) function $w(t)$ is defined as

$$w(t) \triangleq \begin{cases} 1 & \text{for } |t| \leq \frac{1}{2} \\ 0 & \text{else.} \end{cases}$$

The correlation function for this real-valued process is given as

$$R_{XX}(t_1, t_2) = E[X(t_1)X(t_2)]$$

$$= E \left[\sum_n \sum_l X_n X_l \, w \left(\frac{t_1 - D - nT}{T} \right) w \left(\frac{t_2 - D - lT}{T} \right) \right].$$

In the ABS process it is assumed the levels of different pulses are independent random variables and that these, in turn, are independent of the random displacement D. Since $E[X_n X_l] = E[X_n]E[X_l]$ for $n \neq l$ and $E[X_n^2] = a^2$, we obtain

$$R_{XX}(t_1, t_2) = a^2 \sum_n E \left[w \left(\frac{t_1 - D - nT}{T} \right) w \left(\frac{t_2 - D - nT}{T} \right) \right]$$

$$+ \sum_{n \neq l} \sum_l E[X_n]E[X_l] \, E \left[w \left(\frac{t_1 - D - nT}{T} \right) w \left(\frac{t_2 - D - lT}{T} \right) \right].$$

Now, this second term on the right is zero because $E[X_n] = E[X_l] = 0$. Also

$$\sum_n E\left[w\left(\frac{t_1 - D - nT}{T}\right) w\left(\frac{t_2 - D - nT}{T}\right)\right]$$

$$= \sum_n \frac{1}{T} \int_{t_2 - \frac{T}{2} - nT}^{t_2 + \frac{T}{2} - nT} w\left(\frac{\alpha}{T}\right) w\left(\frac{\alpha - (t_2 - t_1)}{T}\right) d\alpha \text{ with } \alpha = t_2 - d - nT,$$

$$= \left(1 - \frac{(t_2 - t_1)}{T}\right) w\left(\frac{t_2 - t_1}{2T}\right) \text{ for } t_2 > t_1.$$

More generally, and for $\tau \overset{\Delta}{=} t_2 - t_1 \lessgtr 0$, we can write that

$$R_{XX}(\tau) = a^2 \left(1 - \frac{|\tau|}{T}\right) w\left(\frac{\tau}{2T}\right). \tag{7.2-1}$$

Equation 7.2-1 is directly extended to the case of equiprobable transitions between two arbitrary levels, say a and b. The required modification is

$$R_{XX}(\tau) = \frac{1}{4}(a - b)^2 \left(1 - \frac{\tau}{T}\right) w\left(\frac{\tau}{2T}\right) + \left(\frac{a + b}{2}\right)^2.$$

We leave the derivation of this result as an exercise for the reader. In Figure 7.2-2 we show the ABS correlation function $R_{XX}(\tau)$ for $a = 1, b = 0$ and $T = 1$.

Poisson Counting Process

Let the process $N(t)$ represent the total number of counts (arrivals) up to time t. Then we can write

$$N(t) \overset{\Delta}{=} \sum_{n=1}^{\infty} u(t - T[n]),$$

where $u(t)$ is the unit-step function and $T[n]$, *the time to the nth arrival*, is the random sequence of times considered in Example 6.1-11. There we showed that the $T[n]$ obeyed the

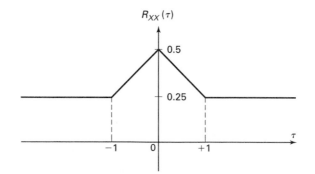

Figure 7.2-2   Autocorrelation function of ABS random process.

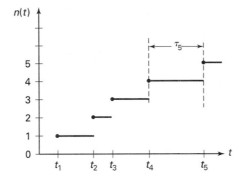

Figure 7.2-3 A sample function of the Poisson process running on $[0, \infty)$.

nonstationary first-order Erlang density,

$$f_T(t; n) = \frac{(\lambda t)^{n-1}}{(n-1)!} \lambda e^{-\lambda t} u(t), \qquad n \geq 0, \tag{7.2-2}$$

which was obtained as an n-fold convolution of exponential pdf's. A typical sample function is shown in Figure 7.2-3 where $T[n] = t_n$ and $\tau[n] = \tau_n$. Note that the time between the arrivals,

$$\tau[n] \overset{\Delta}{=} T[n] - T[n-1],$$

the *interarrival times*, are jointly independent and identically distributed, having the exponential pdf,

$$f_\tau(t) = \lambda e^{-\lambda t} u(t),$$

as in Example 6.1-11. Thus $T[n]$ denotes the total time until the nth arrival if we begin counting at the reference time $t = 0$.

Now by the construction involving the unit-step function, the value $N(t)$ is the number of arrivals up to and including time t, so

$$P\{N(t) = n\} = P\{T[n] \leq t, T[n+1] > t\},$$

because the only way that $N(t)$ can equal n is if the random variable $T[n]$ is less than or equal to t and the random variable $\tau[n+1]$ is greater than t. If we bring in the independent interarrival times, we can re-express this probability as

$$P\{T[n] \leq t, \tau[n+1] > t - T[n]\},$$

which can be easily calculated using the statistical independence of the arrival time $T[n]$ and the interarrival time $\tau[n+1]$ as follows:

$$\int_0^t f_T(\alpha; n) \left[\int_{t-\alpha}^\infty f_\tau(\beta) d\beta \right] d\alpha = \int_0^t \frac{\lambda^n \alpha^{n-1} e^{-\lambda \alpha}}{(n-1)!} \left(\int_{t-\alpha}^\infty \lambda e^{-\lambda \beta} d\beta \right) d\alpha \cdot u(t)$$

$$= \left(\int_0^t \alpha^{n-1} d\alpha \right) \lambda^n e^{-\lambda t} / (n-1)! \quad u(t),$$

or, with $P_N(n;t) \triangleq P\{N(t) = n\}$,

$$P_N(n;t) = \frac{(\lambda t)^n}{n!} e^{-\lambda t} u(t) \quad \text{for } t \geq 0, \qquad n \geq 0. \tag{7.2-3}$$

We have thus arrived at the probability mass function (PMF) of the *Poisson counting process* and we note that it is the PMF of a Poisson random variable (cf. Equation 1.10-5) with mean λt, that is,

$$E[N(t)] = \lambda t. \tag{7.2-4}$$

We call λ the *mean-arrival rate* (sometimes called *intensity*). It is intuitively satisfying that the average value of the process at time t is the mean arrival rate λ multiplied by the length of the time interval $(0, t]$. We leave it as an exercise for the reader to consider why this is so.

Since the random sequence $T[n]$ has independent increments (cf. Definition 6.1-4) and the unit-step function used in the definition of the Poisson process is causal, it seems reasonable that the Poisson process $N(t)$ would also have independent increments. However, this result is not clear because one of the jointly independent interarrival times $\tau[n]$ may be partially in two disjoint intervals, hence causing a dependency in neighboring increments. Nevertheless, using the *memoryless property* of the exponential pdf (see Problem 7.8), one can show that the independent increments property does hold for the Poisson process.

Using independent increments we can evaluate the PMF of the increment in the Poisson counting process over an interval $(t_a, t_b]$ as

$$P[N(t_b) - N(t_a) = n] = \frac{[\lambda(t_b - t_a)]^n}{n!} e^{-\lambda(t_b - t_a)} u(n), \tag{7.2-5}$$

where we have used the fact that the interarrival sequence is stationary, that is, that λ is a constant. We formalize this somewhat in the following definition.

Definition 7.2-1   A random process has *independent increments* when the set of n random variables,

$$X(t_1), X(t_2) - X(t_1), \ldots, X(t_n) - X(t_{n-1}),$$

are jointly independent for all $t_1 < t_2 < \ldots < t_n$ and for all $n \geq 1$. ∎

This just says that the increments are statistically independent when the corresponding intervals do not overlap. Just as in the random sequence case, the independent-increment property makes it easy to get the higher-order distributions. For example, in the case at hand, the Poisson counting process, we can write for $t_2 > t_1$,

$$P_N(n_1, n_2; t_1, t_2) = P[N(t_1) = n_1]P[N(t_2) - N(t_1) = n_2 - n_1]$$

$$= \frac{(\lambda t_1)^{n_1}}{n_1!} e^{-\lambda t_1} \frac{[\lambda(t_2 - t_1)]^{n_2 - n_1}}{(n_2 - n_1)!} e^{-\lambda(t_2 - t_1)} u(n_1) u(n_2 - n_1),$$

which simplifies to

$$\frac{\lambda^{n_2} t_1^{n_1} (t_2 - t_1)^{n_2 - n_1}}{n_1!(n_2 - n_1)!} e^{-\lambda t_2} u(n_1) u(n_2 - n_1), \qquad 0 \leq t_1 < t_2.$$

See also Problem 1.39. Using the independent-increments property we can formulate the following alternative definition of a Poisson counting process.

Definition 7.2-2   A Poisson counting process is the independent-increments process whose increments are Poisson distributed as in Equation 7.2-5. ∎

Concerning the moment function of the Poisson process, the first-order moment has been shown to be λt. This is the mean function of the process. Letting $t_2 \geq t_1$, we can calculate the correlation function using the independent increments property as

$$E[N(t_2)N(t_1)] = E[(N(t_1) + [N(t_2) - N(t_1)])N(t_1)]$$
$$= E[N^2(t_1)] + E[N(t_2) - N(t_1)]E[N(t_1)]$$
$$= \lambda t_1 + \lambda^2 t_1^2 + \lambda(t_2 - t_1)\lambda t_1$$
$$= \lambda t_1 + \lambda^2 t_1 t_2.$$

If $t_2 < t_1$, we merely interchange t_1 and t_2 in the preceding formula. Thus the general result for all t_1 and t_2 is

$$R_{NN}(t_1, t_2) = E[N(t_1)N(t_2)]$$
$$= \lambda \min(t_1, t_2) + \lambda^2 t_1 t_2. \tag{7.2-6}$$

If we evaluate the covariance using Equations 7.2-4 and 7.2-6 we obtain

$$K_{NN}(t_1, t_2) = \lambda \min(t_1, t_2). \tag{7.2-7}$$

We thus see that the variance of the process is equal to λt and is the same as its mean, a property inherited from the Poisson random variable. Also we see that the covariance depends only on the earlier of the two times involved. The reason for this is seen by writing $N(t)$ as the value at an earlier time plus an increment, and then noting that the independence of this increment and $N(t)$ at the earlier time implies that the covariance between them must be zero. Thus the covariance of this independent-increments process is just the variance of the process at the earlier of the two times.

Example 7.2-1 _____

(Radioactivity monitor.) In radioactivity monitoring, the particle-counting process can often be adequately modeled as Poisson. Let the counter start to monitor at some arbitrary time t and then count for T_0 seconds. If the count is above a threshold, say N_0, an alarm will be sounded. Assuming the arrival rate to be λ, we want to know the probability that the alarm will not sound when radioactive material is present.

Since the process is Poisson, we know it has independent increments that satisfy the Poisson distribution. Thus the count ΔN in the interval $(t, t + T]$, that is, $\Delta N \triangleq N(t + T) - N(t)$, is Poisson distributed with mean λT_0 independent of t. The probability of N_0 or fewer counts is thus

$$P[\Delta N \leq N_0] = \sum_{k=0}^{N_0} \frac{(\lambda T_0)^k}{k!} e^{-\lambda T_0},$$

If N_0 is small we can calculate the sum directly. If $\lambda T_0 \gg 1$, we can use the Gaussian approximation (Equation 1.11-9) to the Poisson distribution.

Example 7.2-2

(Sum of two independent Poisson processes.) Let $N_1(t)$ be a Poisson counting process with rate λ_1. Let $N_2(t)$ be a second Poisson counting process with rate λ_2, where N_2 is independent of N_1. The sum of the two processes, $N(t) \triangleq N_1(t) + N_2(t)$, could model the total number of failures of two separate machines, whose failure rate are λ_1 and λ_2, respectively. It is a remarkable fact that $N(t)$ is also a Poisson counting process with rate $\lambda = \lambda_1 + \lambda_2$.

To see this we use Definition 7.2-2 of the Poisson counting process and verify these conditions for $N(t)$. First, it is clear with a little reflection that the sum of two independent-increments processes will also be an independent-increments process *if* the processes are jointly independent. Second, for any increment $N(t_b) - N(t_a)$ with $t_b > t_a$, we can write

$$N(t_b) - N(t_a) = N_1(t_b) - N_1(t_a) + N_2(t_b) - N_2(t_a).$$

Thus the increment in N is the sum of two corresponding increments in N_1 and N_2. The desired result then follows from the fact that the sum of two independent Poisson random variables is also Poisson distributed with the sum of the two parameters (cf. Example 3.3-6). Thus the parameter of the increment in $N(t)$ is

$$\lambda_1(t_b - t_a) + \lambda_2(t_b - t_a) = (\lambda_1 + \lambda_2)(t_b - t_a)$$

as desired.

The Poisson counting process $N(t)$ can be generalized in several ways. We can let the arrival rate, sometimes called *intensity*, be a function of time. The arrival rate $\lambda(t)$ must satisfy $\lambda(t) \geq 0$. The average value of the resulting *nonuniform Poisson counting process* then becomes

$$\mu_X(t) = \int_0^t \lambda(\tau)d\tau, \qquad t \geq 0. \tag{7.2-8}$$

Alternative Derivation of Poisson Process

It may be interesting to rederive the Poisson counting process from the elementary properties of random points in time listed in Chapter 1, Section 1.10. They are repeated here in a notation consistent with that used in this chapter. For Δt small:

(1) $P_N(1; t, t + \Delta t) = \lambda(t)\Delta t + o(\Delta t)$.
(2) $P_N(k; t, t + \Delta t) = o(\Delta t), \qquad k > 1$.
(3) Events in nonoverlapping time intervals are statistically independent.

Here the notation $o(\Delta t)$, read "little oh," denotes any quantity that goes to zero at a faster than linear rate, so that

$$\lim_{\Delta t \to 0} \frac{o(\Delta t)}{\Delta t} = 0,$$

and $P_N(k; t, t + \Delta t) = P[N(t + \Delta t) - N(t) = k]$.

We note that property (3) is just the independent increments property for the counting process $N(t)$ which counts the number of events occurring in $(0, t]$.

We can compute the probability $P_N(k; t, t+\tau)$ of k events in $(t, t+\tau)$ as follows. Consider $P_N(k; t, t+\tau+\Delta t)$; if Δt is very small, then in view of properties (1) and (2) there are only the following two possibilities for getting k events in $(t, t+\tau+\Delta t)$:

$$E_1 = \{k \text{ in } (t, t+\tau) \text{ and zero in } (t+\tau, t+\tau+\Delta t)\} \quad \text{or}$$

$$E_2 = \{k - 1 \text{ in } (t, t+\tau) \text{ and one in } (t+\tau, t+\tau+\Delta t)\}.$$

Since events E_1 and E_2 are disjoint events, their probabilities add and we can write

$$P_N(k; t, t+\tau+\Delta t) = P_N(k; t, t+\tau)P_N(0; t+\tau, t+\tau+\Delta t)$$

$$+ P_N(k - 1; t, t+\tau)P_N(1; t+\tau, t+\tau+\Delta t)$$

$$= P_N(k; t, t+\tau)[1 - \lambda(t+\tau)\Delta t]$$

$$+ P_N(k - 1; t, t+\tau)\lambda(t+\tau)\Delta t.$$

If we rearrange terms, divide by Δt and take limits, we obtain the linear differential equations (LDEs),

$$\frac{dP_N(k; t, t+\tau)}{d\tau} = \lambda(t+\tau)[P_N(k - 1; t, t+\tau) - P_N(k; t, t+\tau)].$$

Thus we obtain a set of recursive first-order differential equations from which we can solve for $P_N(k; t, t+\tau), k = 0, 1, \ldots$. We set $P_N(-1; t, t+\tau) = 0$, since this is the probability of the impossible event. Also, to shorten our notation, we temporarily write $P_N(k) \triangleq P_N(k; t, t+\tau)$; thus the dependences on t and τ are submerged but of course are still there.

When $k = 0$,

$$\frac{dP_N(0)}{d\tau} = -\lambda(t+\tau)P_N(0).$$

This is a simple first-order, homogeneous differential equation for which the solution is

$$P_N(0) = C \exp\left[-\int_t^{t+\tau} \lambda(\xi)d\xi\right].$$

Since $P_N(0; t, t) = 1, C = 1$ and

$$P_N(0) = \exp\left[-\int_t^{t+\tau} \lambda(\xi)d\xi\right].$$

Let us define μ by

$$\mu \triangleq \int_t^{t+\tau} \lambda(\xi)d\xi.$$

Then

$$P_N(0) = e^{-\mu}.$$

When $k = 1$, the differential equation is now

$$\frac{dP_N(1)}{d\tau} + \lambda(t+\tau)P_N(1) = \lambda(t+\tau)P_N(0)$$

$$= \lambda(t+\tau)e^{-\mu}. \tag{7.2-9}$$

This elementary first-order, inhomogeneous equation has a solution that is the sum of the homogeneous and particular solutions. For the homogeneous solution, P_h, we already know from the $k = 0$ case that

$$P_h = C_2 e^{-\mu}.$$

For the particular solution P_p we use the method of *variation of parameters* to assume that

$$P_p = v(t+\tau)e^{-\mu},$$

where $v(t+\tau)$ is to be determined. By substituting this equation into Equation 7.2-9 we readily find that

$$P_p = \mu e^{-\mu}.$$

The complete solution is $P_N(1) = P_h + P_p$. Since $P_N(1;t,t) = 0$, we obtain $C_2 = 0$ and thus

$$P_N(1) = \mu e^{-\mu}.$$

General case.   The LDE in the general case is

$$\frac{dP_N(k)}{d\tau} + \lambda(t+\tau)P_N(k) = \lambda(t+\tau)P_N(k-1)$$

and, proceeding by induction, we find that

$$P_N(k) = \frac{\mu^k}{k!}e^{-\mu} \qquad k = 0, 1, \ldots$$

which is the key result. Recalling the definition of μ, we can write

$$P_N(k; t, t+\tau) = \frac{1}{k!}\left[\int_t^{t+\tau}\lambda(\xi)d\xi\right]^k \exp\left[-\int_t^{t+\tau}\lambda(\xi)d\xi\right]. \tag{7.2-10}$$

We thus obtain the nonuniform Poisson counting process.

Another way to generalize the Poisson process is to use a different pdf for the independent interarrival times. With a nonexponential density the more general process is called a *renewal process* [7-3]. The word "renewal" can be related to the interpretation of the arrival times as the failure times of certain equipment, thus the value of the counting process $N(t)$ models the number of renewals that have had to be made up to the present time.

Random Telegraph Signal

When all the information in a random waveform is contained in the *zero crossings*, a so-called "hard clipper" is often used to generate a simpler yet equivalent two-level waveform

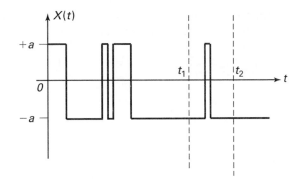

Figure 7.2-4 Sample function of the random telegraph signal (RTS).

that is free of unwanted random amplitude variation. A special case is when the number of zero crossings in a time interval follows the Poisson law, and the resulting random process is called the *random telegraph signal* (RTS). A sample function of the RTS is shown in Figure 7.2-4

We construct the RTS on $t \geq 0$ as follows: Let $X(0) = \pm a$ with equal probability. Then take the Poisson arrival time sequence $T[n]$ of Chapter 6 and use it to switch the level of the RTS, i.e., at $T[1]$ switch the sign of $X(t)$, and then at $T[2]$, and so forth. Clearly from the symmetry and the fact that the interarrival times $\tau[n]$ are stationary and form an independent random sequence, we must have that $\mu_X(t) = 0$ and that the PMF $P_X(a) = P_X(-a) = 1/2$. Let $t_2 > t_1 > 0$, and consider $P_X(x_1, x_2) \triangleq P[X(t_1) = x_1, X(t_2) = x_2]$ along with $P_X(x_2 \mid x_1) \triangleq P[X(t_2) = x_2 \mid X(t_1) = x_1]$. Then we can write the correlation function as

$$
\begin{aligned}
R_{XX}(t_1, t_2) &= E[X(t_1)X(t_2)] \\
&= a^2 P_X(a, a) + (-a)^2 P_X(-a, -a) + a(-a)P_X(a, -a) - a(a)P_X(-a, a) \\
&= \frac{1}{2}a^2(P_X(a|a) + P_X(-a|-a) - P_X(-a|a) - P_X(a|-a)),
\end{aligned}
$$

since $P_X(a) = P_X(-a) = 1/2$. But $P_X(-a|-a) = P_X(a|a)$ is just the probability of an even number of zero crossings in the time interval $(t_1, t_2]$, while $P_X(-a|a) = P_X(a|-a)$ is the probability of an odd number of crossings of 0. Hence, writing the arrival rate as λ, and substituting $\tau \triangleq t_2 - t_1$, we get

$$
\begin{aligned}
R_{XX}(t_1, t_2) &= a^2 \sum_{\text{even } k \geq 0} e^{-\lambda\tau}\frac{(\lambda\tau)^k}{k!} - \sum_{\text{odd } k \geq 0} e^{-\lambda\tau}\frac{(\lambda\tau)^k}{k!} \\
&= a^2 e^{-\lambda\tau} \sum_{\text{all } k \geq 0} (-1)^k \frac{(\lambda\tau)^k}{k!},
\end{aligned}
$$

where we have combined the two sums by making use of the function $(-1)^k$, since $(-1)^k = 1$ for k even and $(-1)^k = -1$ for k odd. Thus we now have

$$R_{XX}(t_1, t_2) = a^2 e^{-\lambda\tau} \sum_{\text{all } k \geq 0} \frac{(-\lambda\tau)^k}{k!}$$

$$= a^2 e^{-2\lambda\tau}$$

for the case when $\tau > 0$. Since the correlation function of a real-valued process must be symmetric, we have $R_{XX}(t_1, t_2) = R_{XX}(t_2, t_1)$, so that when $\tau \leq 0$, we can substitute $-\tau$ into the above equation to get $R_{XX}(t_1, t_2) = a^2 e^{+2\lambda\tau}$. Thus overall we have, valid for all interval lengths τ,

$$R_{XX}(t_1, t_2) = a^2 e^{-2\lambda|\tau|}.$$

A plot of this correlation function is shown in Figure 7.2-5.

Digital Modulation Using Phase-Shift Keying

Digital computers generate many binary sequences (data) to be communicated to other distant computers. Binary modulation methods frequency-shift this data to a region of the electromagnetic spectrum which is well suited to the transmission media, for example, a telephone line. A basic method for modulating binary data is phase-shift keying (PSK). In this method binary data, modeled by the random sequence $B[n]$, are mapped bit-by-bit into a phase-angle sequence $\Theta[n]$, which is used to modulate the *carrier signal* $\cos(2\pi f_c t)$.

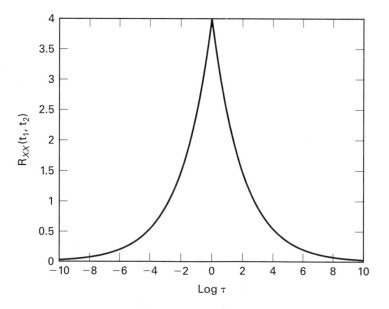

Figure 7.2-5   The symmetric exponential correlation function of an RTS process ($a = 2.0, \lambda = 0.5$).

Specifically let $B[n]$ be a Bernoulli random sequence taking on the values 0 and 1 with equal probability. Then define the random phase sequence $\Theta[n]$ as follows:

$$\Theta[n] \triangleq \begin{cases} +\pi/2 & \text{if } B[n] = 1, \\ -\pi/2 & \text{if } B[n] = 0. \end{cases}$$

Using $\Theta_a(t)$ to denote the analog angle process, we define

$$\Theta_a(t) \triangleq \Theta[k] \quad \text{for } kT \leq t < (k+1)T,$$

and construct the modulated signal as

$$X(t) = \cos(2\pi f_c t + \Theta_a(t)). \tag{7.2-11}$$

Here T is a constant time for the transmission of one bit. Normally, T is chosen to be a multiple of $1/f_c$ so that there are an integral number of carrier cycles per bit time T. The reciprocal of T is called the baud rate. The overall modulator is shown in Figure 7.2-6. The process $X(t)$ is the PSK process.

Our goal here is to evaluate the mean function and correlation function of the random PSK process. To help in the calculation we define two basis functions,

$$s_I(t) \triangleq \begin{cases} \cos(2\pi f_c t) & 0 \leq t \leq T \\ 0 & \text{else,} \end{cases}$$

and

$$s_Q(t) \triangleq \begin{cases} \sin(2\pi f_c t) & 0 \leq t \leq T \\ 0 & \text{else,} \end{cases}$$

which together with Equation 7.2-11 imply

$$\cos[2\pi f_c t + \Theta(t)] = \cos(\Theta_a(t)) \cos 2\pi f_c t - \sin(\Theta_a(t)) \sin 2\pi f_c t$$

$$= \sum_{k=-\infty}^{+\infty} \cos(\Theta[k]) s_I(t - kT) - \sum_{k=-\infty}^{+\infty} \sin(\Theta[k]) s_Q(t - kT), \tag{7.2-12}$$

by use of the sum of angles formula for cosines.

The mean of $X(t)$ can then be obtained in terms of the means of the random sequences $\cos(\Theta[n])$ and $\sin(\Theta[n])$. Because of the definition of $\Theta[n]$, in this particular case $\cos(\Theta[n]) = 0$ and $\sin(\Theta[n]) = \pm 1$ with equal probability so that mean of $X(t)$ is zero, that is, $\mu_X(t) = 0$.

Using Equation 7.2-12 we can calculate the correlation function

$$R_{XX}(t_1, t_2) = \sum_{k,l} E\{\sin \Theta[k] \sin \Theta[l]\} s_Q(t_1 - kT) s_Q(t_2 - lT),$$

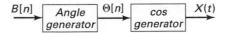

Figure 7.2-6 System for PSK modulation of Bernoulli random sequence $B[n]$.

which involves the correlation function of the random sequence $\sin(\Theta[n])$,

$$R_{\sin\Theta,\sin\Theta}[k,l] = \delta[k-l].$$

Thus the overall correlation function then becomes

$$R_{XX}(t_1, t_2) = \sum_{k=-\infty}^{+\infty} s_Q(t_1 - kT)s_Q(t_2 - kT). \tag{7.2-13}$$

Since the support of s_Q is only of width T, there is no overlap in (t_1, t_2) between product terms in Equation 7.2-13. So for any fixed (t_1, t_2), only one of the product terms in the sum can be nonzero. Also if t_1 and t_2 are not in the same period, then this term is zero also. More elegantly, using the notation,

$$(t) \overset{\Delta}{=} t \bmod T \quad \text{and} \quad \lfloor t/T \rfloor \overset{\Delta}{=} \text{integer part } (t/T),$$

we can write that

$$R_{XX}(t_1, t_2) = \begin{cases} s_Q((t_1))s_Q((t_2)) & \text{for } \lfloor t_1/T \rfloor = \lfloor t_2/T \rfloor \\ 0 & \text{else.} \end{cases}$$

In particular for $0 \le t_1 \le T$ and $0 \le t_2 \le T$, we have

$$R_{XX}(t_1, t_2) = s_Q(t_1)s_Q(t_2).$$

Wiener Process or Brownian Motion

In Chapter 6 we considered a random sequence $X[n]$ called the random walk. See Example 6.1-13. Here we construct an analogous *random process* that is piecewise constant for intervals of length T as follows:

$$X_T(t) \overset{\Delta}{=} \sum_{k=1}^{\infty} W[k]u(t - kT),$$

where

$$W[k] \overset{\Delta}{=} \begin{cases} +s & \text{with } p = 0.5 \\ -s & \text{with } p = 0.5. \end{cases}$$

Then $X_T(nT) = X[n]$ the random-walk sequence, since

$$X_T(nT) = \sum_{k=1}^{n} W[k] = X[n].$$

Hence we can evaluate the PMFs and moments of this random process by employing the known results for the corresponding random-walk sequence. Now the Wiener[†] process, sometimes also called *Wiener–Levy* or Brownian motion, is the process whose distribution is obtained as a limiting form of the distribution of the above piecewise constant process as

---

[†] After Norbert Wiener, American mathematician (1894–1964), a pioneer in communication and estimation theories.

the interval T shrinks to zero. We let s, the jump size, and the interval T shrink to zero in a precise way to obtain a *continuous random process* in the limit, that is, a process whose sample functions are continuous functions of time. In letting s and T tend to zero we must be careful to make sure that the limit of the variance stays finite and nonzero. The resulting Wiener process will inherit the independent increments property.

The original motivation for the Wiener process was to develop a model for the chaotic random motion of gas molecules. Modeling the basic discrete collisions with a random walk, one then finds the asymptotic process when an infinite (very large) number of molecules interact on an infinitesimal (very small) time scale.

As in Example 6.1-3, we let n be the number of trials, k be the number of successes, and $n - k$ be the number of failures. Also $r \triangleq k - (n - k) = 2k - n$ denotes the excess number of successes over failures. Then $2k = n + r$ or $k = (n + r)/2$ and must be an integer e.g., you cannot have 2.5 "successes". Thus $n + r$ must be even and the probability that $X_T(nT) = rs$ is the probability that there are $0.5(n + r)$ successes $(+s)$ and $0.5(n - r)$ failures $(-s)$ out of a total of n trials. Thus by the binomial PMF,

$$P[X_T(nT) = rs] = \binom{n}{\frac{n+r}{2}} 2^{-n} \quad \text{for } n + r \text{ even.}$$

If $n + r$ is odd, then $X_T(nT)$ cannot equal rs.

The mean and variance can be most easily calculated by noting that the random variable $X[n]$ is the sum of n independent Bernoulli random variables defined in Section 6.1. Thus

$$E[X_T(nT)] = 0$$

and

$$E[X_T^2(nT)] = ns^2.$$

On expressing the variance in terms of $t = nT$, we have

$$\text{Var}[X_T(t)] = E[X_T^2(nT)] = t\frac{s^2}{T}.$$

Thus we need s^2 proportional to T to get an interesting limiting distribution.[†] We set $s^2 = \alpha T$, where $\alpha > 0$. Now as T goes to zero we keep the variance constant at αt. Also, by an elementary application of the Central Limit theorem (cf. Section 4.7) we get a limiting Gaussian distribution. We take the limiting random process (convergence in the distribution sense) to be an independent-increments process since all the above random-walk processes had independent increments for all T, no matter how small. Hence we arrive at the following specification for the limiting process, which is termed the *Wiener process*:

$$\mu_X(t) = 0, \qquad \text{Var}[X(t)] = \alpha t$$

[†]The physical implication of having s^2 proportional to T is that if we take $v \triangleq s/T$ as the speed of the particle, then the particle speed goes to infinity as the displacement s goes to zero such as to keep the product of the two a constant.

and

$$f_X(x;t) = \frac{1}{\sqrt{2\pi\alpha t}} \exp\left(-\frac{x^2}{2\alpha t}\right), \qquad t > 0. \tag{7.2-14}$$

The pdf of the increment $\Delta \triangleq X(t) - X(\tau)$ for all $t > \tau$ is given as

$$f_\Delta(\delta;t-\tau) = \frac{1}{\sqrt{2\pi\alpha(t-\tau)}} \exp\left(-\frac{\delta^2}{2\alpha(t-\tau)}\right), \tag{7.2-15}$$

since

$$E[X(t) - X(\tau)] = E[\Delta] = 0, \tag{7.2-16}$$

and

$$E\left[(X(t) - X(\tau))^2\right] = \alpha(t-\tau) \quad \text{for } t > \tau. \tag{7.2-17}$$

Example 7.2-3 _____
We can use MATLAB to visually investigate the sample functions typical of the Wiener process. Since it is a computer simulation, we also can evaluate the effect of the limiting sequence occurring as $s = \sqrt{\alpha T}$ approaches 0 for fixed $\alpha > 0$.

We start with a 1000-element vector that is a realization of the Bernoulli random vector W with $p = 0.5$ generated as

```
u = rand(1000,1)
w = 0.5 >= u
```

The following line then converts the range of w to $\pm s$ for a prespecified value s:

```
w = s*(2*w - 1.0)
```

and then we generate a segment of a sample function of $X_T(nT) = X[n]$ as elements of the random vector

```
x = cumsum(w)
```

For the numerical experiment let $\alpha = 1.0$ and set $T = 0.01$ $(s = 0.1)$. Using a computer variable x with dimension 1000 for $T = 0.01$, we get the results shown in Figure 7.2-7. Note particularly in this near limiting case, the effects of increasing variance with time. Also note that *trends* or *long-term waves* appear to develop as time progresses.

_____

From the first-order pdf of X and the density of the increment Δ, it is possible to calculate a complete set of consistent nth-order pdf's as we have seen before. It thus follows that all nth-order pdf's of a Wiener process are Gaussian.

Definition 7.2-3   If for all positive integers n, the nth-order pdf's of a random process are all jointly Gaussian, then the process is called a *Gaussian random process.*   ■

The Wiener process is thus an example of a Gaussian random process. The covariance function of the Wiener process (which is also its correlation function because $\mu_X(t) = 0$) is given as

$$K_{XX}(t_1, t_2) = \alpha \min(t_1, t_2), \quad \alpha > 0. \tag{7.2-18}$$

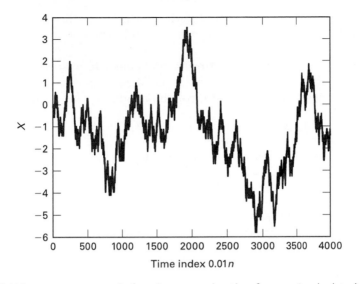

Figure 7.2-7   A Wiener process sample function approximation for $\alpha = 1$ calculated with $T = 0.01$.

To show this we take $t_1 \geq t_2$, and noting that the (forward) increment $X(t_1) - X(t_2)$ is independent of $X(t_2)$ and that they both have zero-mean,

$$E[(X(t_1) - X(t_2))X(t_2)] = E[X(t_1) - X(t_2)]E[X(t_2)]$$

$$= 0$$

or

$$E[X(t_1)X(t_2)] = E[X^2(t_2)]$$

$$= \alpha t_2.$$

If $t_2 > t_1$, we get $E[X(t_2)X(t_1)] = \alpha t_1$, thus establishing Equation 7.2-18.

Note that the Wiener process has the same variance function as the Poisson process, even though the two processes are dramatically different. While the Poisson process consists solely of jumps separated by constant values, the Wiener process has no jumps and can in fact be proven to be a.s. continuous, that is, the sample functions are continuous with probability-1. Later, we will show that the Wiener process is continuous in a weaker mean-square sense (to be specified more precisely in Chapter 8).

Markov Random Processes

We have discussed five random processes thus far. Of these, the Wiener and Poisson are fundamental in that many other rather general random processes have been shown to be obtainable by nonlinear transformations on these two basic processes. In both cases, the difficulty of specifying a consistent set of nth-order distributions from processes with dependence was overcome by use of the independent-increments property. In fact, this is quite a general approach in that we can start out with some arbitrary first-order distribution and then specify a distribution for the increment, thereby obtaining a consistent set of nth-order distributions that exhibit dependence.

Another way of going from the first-order probability to a consistent set of nth-order probabilities, which has proved quite useful, is the Markov process approach. Here we start with a first-order density (or PMF) and a conditional density (or conditional PMF)

$$f_X(x;t) \quad \text{and} \quad f_X(x_2|x_1;t_2,t_1), \qquad t_2 > t_1,$$

and then build up the nth-order pdf $f(x_1,\ldots,x_n;t_1,\ldots,t_n)$ (or PMF) as the product,

$$f(x_1;t)f(x_2|x_1;t_2,t_1)\ldots f(x_n|x_{n-1};t_n,t_{n-1}). \tag{7.2-19}$$

We ask the reader to show that this is a valid nth order pdf (that is, that this function is nonnegative and integrates to one) whenever the conditional and first-order pdf's are well defined.

Conversely, if we start with an arbitrary nth-order pdf and repeatedly use the definition of conditional probability we obtain,

$$f(x_1,\ldots,x_n;t_1,\ldots,t_n) = f(x_1;t_1)f(x_2|x_1;t_2,t_1)f(x_3|x_2,x_1;t_3,t_2,t_1)$$
$$\ldots\ f(x_n|x_{n-1},\ldots,x_1;t_n,\ldots,t_1), \tag{7.2-20}$$

which can be made equivalent to Equation 7.2-19 by constraining the conditional densities to depend only on the most recent conditioning value. This motivates the following definition of a Markov random process.

Definition 7.2-4   (Markov random process.)

(a) A *continuous-valued (first-order) Markov process* $X(t)$ satisfies the conditional pdf expression

$$f_X(x_n|x_{n-1},x_{n-2},\ldots,x_1;t_n,\ldots,t_1) = f_X(x_n|x_{n-1};t_n,t_{n-1}),$$

for all $x_1, x_2, \ldots, x_n$, for all $t_1 < t_2 < \ldots < t_n$ and for all integers $n > 0$.

(b) A *discrete-valued (first-order) Markov random process* satisfies the conditional PMF expression

$$P_X(x_n|x_{n-1},\ldots,x_1;t_n,\ldots,t_1) = P_X(x_n|x_{n-1};t_n,t_{n-1})$$

for all $x_1,\ldots,x_n$ and for all $t_1 < \ldots < t_n$, and for all integers $n > 0$. ■

The value of the process $X(t)$ at a given time t thus determines the conditional probabilities for future values of the process. The values of the process are called the *states of the process*, and the conditional probabilities are thought of as *transition probabilities* between the states. If only a finite or countable set of values x_i is allowed, the discrete-valued Markov process is called a *Markov chain*. An example of a Markov chain would be the Poisson counting process studied earlier. The Wiener process is an example of a continuous-valued Markov process. Both these processes are Markov because of their independent-increments property. In fact, any independent-increment process is also Markov. To see this note that, for the discrete-valued case, by example,

$$P_X(x_n|x_{n-1}, \ldots, x_1; t_n, \ldots, t_1)$$

$$= P[X(t_n) = x_n|X(t_{n-1}) = x_{n-1}, \ldots, X(t_1) = x_1]$$

$$= P[X(t_n) - X(t_{n-1}) = x_n - x_{n-1}|X(t_{n-1}) = x_{n-1}, \ldots, X(t_1) = x_1]$$

$$= P[X(t_n) - X(t_{n-1}) = x_n - x_{n-1}] \quad \text{by the independent increments property}$$

$$= P[X(t_n) - X(t_{n-1}) = x_n - x_{n-1}|X(t_{n-1}) = x_{n-1}] \quad \text{again by independent increments}$$

$$= P[X(t_n) = x_n|X(t_{n-1}) = x_{n-1}]$$

$$= P_X(x_n|x_{n-1}; t_n, t_1).$$

Note, however, that the inverse argument is not true. A Markov random process does not necessarily have independent increments. (See Problem 7.17.)

Markov random processes find application in many areas including signal processing and control systems. Markov chains are used in communications, computer networks, and reliability theory.

Example 7.2-4

(Multiprocessor reliability.) Given a computer with two independent processors, we can model it as a three-state system: 0—both processors down; 1—exactly one processor up; and 2—both processors up. We would like to know the probabilities of these three states. A common probabilistic model is that the processors will fail randomly with time-to-failure, the *failure time*, exponentially distributed with some parameter $\lambda > 0$. Once a processor fails, the time to service it, the *service time*, will be assumed to be also exponentially distributed with parameter $\mu > 0$. Furthermore, we assume that the processor's failures and servicing are independent; thus we make the failure and service times in our probabilistic model jointly independent.

If we define $X(t)$ as the state of the system at time t, then X is a continuous-time Markov chain. We can show this by first showing that the times between state transitions of X are exponentially distributed and then invoking the memoryless property of the exponential distribution (see Problem 7.8). Analyzing the *transition times* (either failure times or service times), we proceed as follows. The transition time for going from state $X = 0$ to $X = 1$ is the minimum of two exponentially distributed services times, which are assumed to be independent. By Problem 3.19, this time will be also exponentially distributed with parameter 2μ. The expected time for this transition will thus be $1/(2\mu) = \frac{1}{2}(1/\mu)$ that is, one-half the average time to service a single processor. This is quite reasonable since both processors are down in state $X = 0$ and hence both are being serviced independently and simultaneously. The rate parameter for the transition, 0 to 1, is thus 2μ. The transition, 1 to 2, awaits one exponential service time at rate μ. Thus its rate is also μ. Similarly, the state transition, 1 to 0, awaits only one failure at rate λ, while the transition,

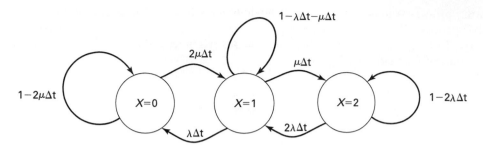

Figure 7.2-8   Short-time state-transition diagram with indicated transition probabilities.

2 to 1, awaits the minimum of two exponentially distributed failure times. Thus its rate is 2λ. Simultaneous transitions from 0 to 2 or 2 to 0 are of probability 0 and hence are ignored.

This Markov chain model is summarized in the short-time *state-transition diagram* of Figure 7.2-8. In this diagram the directed branches represent short-time i.e., as $\Delta t \to 0$, transition probabilities between the states. The transition times are assumed to be exponentially distributed with the parameter given by the branch label. These transition times might be more properly called intertransition times and are analogous to the interarrival times of the Poisson counting process, which are also exponentially distributed.

Consider the probability of being in *state two* at $t + \Delta t$, having been in *state one* at time t. This requires that the service time T_s to lie in the interval $(t, t + \Delta t]$ conditional on $T_s \geq t$. Let $P_i(t) \triangleq P[X(t) = i]$ for $0 \leq i \leq 2$. Then

$$P_2(t + \Delta t) = P_1(t)P[t < T_s \leq t + \Delta t | T_s \geq t],$$

where

$$P[t < T_s \leq t + \Delta t | T_s \geq t] = \frac{F_{T_s}(t + \Delta t) - F_{T_s}(t)}{1 - F_{T_s}(t)} = \mu \Delta t + o(\Delta t).$$

Using this type of argument for connecting the probability of transitions from states at time t to states at time $t + \Delta t$ and ignoring transitions from state two to state zero and vice versa, enables as to write the state probability at time $t + \Delta t$ in terms of the state probability at t in vector matrix form:

$$\begin{bmatrix} P_0(t + \Delta t) \\ P_1(t + \Delta t) \\ P_2(t + \Delta t) \end{bmatrix} = \begin{bmatrix} 1 - 2\mu\,\Delta t & \lambda\,\Delta t & 0 \\ 2\mu\,\Delta t & 1 - (\lambda + \mu)\,\Delta t & 2\lambda\,\Delta t \\ 0 & \mu\,\Delta t & 1 - 2\lambda\,\Delta t \end{bmatrix} \begin{bmatrix} P_0(t) \\ P_1(t) \\ P_2(t) \end{bmatrix} + \mathbf{o}(\Delta t),$$

where $\mathbf{o}(\Delta t)$ denotes a quantity of lower order than Δt.

Rearranging, we have

$$\begin{bmatrix} P_0(t + \Delta t) - P_0(t) \\ P_1(t + \Delta t) - P_1(t) \\ P_2(t + \Delta t) - P_2(t) \end{bmatrix} = \begin{bmatrix} -2\mu & \lambda & 0 \\ 2\mu & -(\lambda + \mu) & 2\lambda \\ 0 & \mu\,\Delta t & -2\lambda \end{bmatrix} \begin{bmatrix} P_0(t) \\ P_1(t) \\ P_2(t) \end{bmatrix} \Delta t + \mathbf{o}(\Delta t).$$

Dividing both sides by Δt and using an obvious matrix notation we obtain

$$\frac{d\mathbf{P}(t)}{dt} = \mathbf{A}\mathbf{P}(t). \tag{7.2-21}$$

The matrix $\mathbf{A}$ is called the *generator* of the Markov chain X. This first-order vector differential equation can be solved for an initial probability vector, $\mathbf{P}(0) \triangleq \mathbf{P}_0$, using methods of linear-system theory [7-2]. The solution is expressed in terms of the matrix exponential

$$e^{\mathbf{A}t} \triangleq \mathbf{I} + \mathbf{A}t + \frac{1}{2!}(\mathbf{A}t)^2 + \frac{1}{3!}(\mathbf{A}t)^3 + \ldots,$$

which converges for all finite t. The solution $\mathbf{P}(t)$ is then given as

$$\mathbf{P}(t) = e^{\mathbf{A}t}\mathbf{P}_0, \qquad t \geq 0.$$

For details on this method as well as how to obtain an explicit solution see Reference [7-11].

For the present we content ourselves with the steady-state solution obtained by setting the time derivative in Equation 7.2-21 to zero, thus yielding $\mathbf{A}\mathbf{P}=0$. From the first and last rows we get

$$-2\mu P_0 + \lambda P_1 = 0$$

and

$$+\mu P_1 - 2\lambda P_2 = 0.$$

From this we obtain $P_1 = (2\mu/\lambda)P_0$ and $P_2 = (\mu/2\lambda)P_1 = (\mu/\lambda)^2 P_0$. Then invoking $P_0 + P_1 + P_2 = 1$, we obtain $P_0 = \lambda^2/(\lambda^2 + 2\mu\lambda + \mu^2)$ and finally

$$\mathbf{P} = \frac{1}{\lambda^2 + 2\mu\lambda + \mu^2}[\lambda^2, 2\mu\lambda, \mu^2]^T.$$

Thus the steady-state probability of both processors being down is $P_0 = [\lambda/(\lambda + \mu)]^2$. Incidentally, if we had used only one processor modeled by a two-state Markov chain we would have obtained $P_0 = \lambda/(\lambda + \mu)$.

Clearly we can generalize this example to any positive integer number of states n with independent exponential interarrival times between theses states. Such a process is called a *queueing process*. Other examples are the number of toll booths busy on a superhighway and congestion states in a computer or telephone network. For more on queueing systems see Reference [7-3]. An important point to notice in the last example is that the exponential transition times were crucial in showing the Markov property. In fact, any other distribution but exponential would not be memoryless, and the resulting state-transition process would not be a Markov chain.

Birth–Death Markov Chains

A Markov chain in which transitions are permissible only between adjacent states is called a birth–death chain. We first deal with the case where the number of states is infininite and then treat the finite-state case.

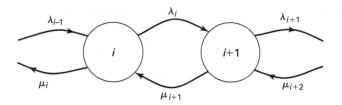

Figure 7.2-9 Markov state diagram for the birth–death process showing transition rate parameters.

1. Infinite-length queues. The state-transition diagram for the infinite-length queue is shown in Figure 7.2-9.[†] In going from state i to state $i+1$, we say that a *birth* has occurred. Likewise, in going from state i to state $i-1$ we say a *death* has occurred. At any time t, $P_j(t)$ is the probability of being in state j, that is of having a "population" of size j; in other words the excess of the number of births over deaths. In this model, births are generated by a Poisson process. The times between births τ_B, and the time between deaths τ_D, depend on the states but obey the exponential distribution with parameters λ_i and μ_i, respectively. The model is used widely in queuing theory where a birth is an arrival to the queue and a death is a departure of one from the queue because of the completion of service. An example is people waiting in line to purchase a ticket at a single-server ticket booth. If the theater is very large and there are no restrictions on the length of the queue (e.g., the queue may block the sidewalk and create a hazard) overflow and saturation can be disregarded. Then the dynamics of the queue are described by the basic equation $W_n = \max\{0, W_{n-1} + \tau_s - \tau_i\}$, where W_n is the waiting time in the queue for the nth arrival, τ_s is the service time for the $(n-1)$st arrival, and τ_i is the interarrival time between the nth and $(n-1)$st arrivals. This is an example of unrestricted queue length. On the other hand data packets stored in a finite-size memory present a different problem. When the buffer is filled (saturation), a new arrival must be turned away (in this case we say the datum packet is "lost").

Following the procedure in Example 7.2-4, we can write that

$$\mathbf{P}(t + \Delta t) = \mathbf{B}\mathbf{P}(t),$$

where

$$\mathbf{B} = \begin{bmatrix} 1 - \lambda_0 \Delta t & \mu_1 \Delta t & 0 & \cdots & \\ \lambda_0 \Delta t & 1 - (\lambda_1 + \mu_1)\Delta t & \mu_2 \Delta t & 0 & \cdots \\ 0 & \lambda_1 \Delta t & 1 - (\lambda_2 + \mu_2)\Delta t & \mu_2 \Delta t & 0 \\ \vdots & \vdots & \vdots & \vdots & \vdots \end{bmatrix}.$$

Rearranging and dividing by Δt and letting $\Delta t \to 0$, we get

$$d\mathbf{P}(t)/dt = \mathbf{A}\mathbf{P}(t)$$

[†]In keeping with standard practice, we draw the diagram *showing only the transition rate parameters* i.e., the μ_i's and λ_i's over the links between states. This type of diagram does not show explicitly, for example, that in the Poisson case the short time probability of staying in state i is $1 - (\lambda_i + \mu_i)\Delta t$. While this type of diagram is less clear it is less crowded than, say the nonstandard short-time transition probability diagram in Figure 7.2-8.

where $\mathbf{P}(t) = [P_0(t), P_1(t), \ldots, P_j(t), \ldots]^T$, and $\mathbf{A}$, the *generator matrix* for the Markov chain is given by

$$\mathbf{A} = \begin{bmatrix} -\lambda_0 & \mu_1 & 0 & \cdots & \\ \lambda_0 & -(\lambda_1 + \mu_1) & \mu_2 & 0 & \cdots \\ 0 & \lambda_1 & -(\lambda_2 + \mu_2) & \mu_2 & 0 \\ \vdots & \vdots & \vdots & \vdots & \vdots \end{bmatrix}.$$

In the steady state $\mathbf{P}'(t) = \mathbf{0}$. Thus, we obtain from $\mathbf{AP} = \mathbf{0}$,

$$P_1 = \rho_1 P_0,$$

$$P_2 = \rho_2 P_1 = \rho_1 \rho_2 P_0,$$

$$\vdots$$

$$P_j = \rho_j P_{j-1} = \rho_j \cdots \rho_2 \rho_1 P_0,$$

where $\rho_j \triangleq \lambda_{j-1}/\mu_j$, for $j \geq 1$.

Assuming that the series converges, we require that $\sum_{i=0}^{\infty} P_i = 1$. With the notation $r_j \triangleq \rho_j \cdots \rho_2 \rho_1$, and $r_0 = 1$, this means $P_0 \sum_{i=0}^{\infty} r_i = 1$ or $P_0 = 1/\sum_{i=0}^{\infty} r_i$. Hence the steady-state probabilities for the birth-death Markov chain are given by

$$P_j = r_j \Big/ \sum_{i=0}^{\infty} r_i, \quad j \geq 0.$$

Failure of the denominator to converge implies that there is no steady state and therefore the steady-state probabilities are zero. This model is often called the *M/M/1 queue*.

2. $M/M/1$ Queue with constant birth and death parameters and finite storage L.

Here we assume that $\lambda_i = \lambda$ and $\mu_i = \mu$, for all i and that the queue length cannot exceed L. This stochastic model can apply to the analysis of a finite buffer as shown in Figure 7.2-10. The dynamical equations are

$$dP_0(t)/dt = -\lambda P_0(t) + \mu P_1(t)$$

$$dP_1(t)/dt = +\lambda P_0(t) - (\lambda + \mu)P_1(t) + \mu P_2(t)$$

$$\vdots \quad \vdots \quad \vdots$$

$$dP_L(t)/dt = +\lambda P_{L-1}(t) - \mu P_L(t).$$

Note that the first and last equations contain only two terms, since a death cannot occur in an empty queue and a birth cannot occur when the queue has its maximum size L. From these equations, we easily obtain that the steady-state solution is $P_i = \rho^i P_0$, for $0 \leq i \leq L$, where $\rho \triangleq \lambda/\mu$. From the condition that the buffer must be in some state, we obtain that

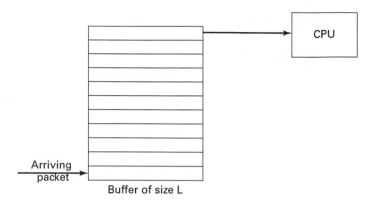

Figure 7.2-10   Illustration of packet arriving at buffer of finite size L.

$\sum_{i=0}^{L} \rho^i P_0 = 1$, or that $P_0 = (1 - \rho)/(1 - \rho^{L+1})$ *Saturation* occurs when the buffer is full. The steady-state probability of this event is $P_L = \rho^L (1 - \rho)/(1 - \rho^{L+1})$. Thus for a birth rate which is half the death rate, and a buffer of size of 10, the probability of saturation is, approximately, 5×10^{-4}.

Example 7.2-5

In computer and communication networks, *packet switching* refers to the transmission of blocks of data called packets from node to node. At each node the packets are processed with a view toward determining the next link in the source-to-destination route. The arrival time of the packets, the amount of time they have to wait in a buffer, and the service time in the CPU (the central processing unit) are random variables.

Assume a first-come, first served, infinite-capacity buffer, with exponential service time with parameter μ, and Poisson-distributed arrivals with a Poisson rate parameter of λ arrivals per unit time. We know from earlier in this section that the interarrival times of the Poisson process are IID exponential random variables with parameter λ. The state diagram for this case is identical to that of Figure 7.2-9 except that $\mu_1 = \mu_2 = \ldots = \mu$ and $\lambda_0 = \lambda_1 = \ldots = \lambda$. Then specializing the results of the previous discussion to this example, we find that $P_i = \rho^i P_0$, for $0 \leq i$, where $\rho \triangleq \lambda/\mu$ and $P_0 = (1 - \rho)$. Thus, in the steady state $P_i = \rho^i (1 - \rho)$, and the average number of packets in the queue $E[N]$, is computed from

$$E[N] = \sum_{i=0}^{\infty} i P_i = \frac{\rho}{1 - \rho}.$$

We leave the details of this elementary calculation as an exercise.

Example 7.2-6

(Finite capacity buffer.) We revisit Example 7.2-5 except that now the arriving data packets are stored in a buffer of size L. Consider the following set-up: The data stored in the buffer is processed by a CPU on a first-come, first-service basis.

Assume that, say at time t, the buffer is filled to capacity, that there is a packet being processed in the CPU, and an arriving packet on the way to the buffer. If the interarrival time τ_i, between this packet and the previous one is less than τ_s, the service time for the packet in the CPU, the arriving packet will be lost. The probability of this event is

$$P[\text{`packet loss'}] = P[\text{`saturation'} \cap \{\tau_s > \tau_i\}]$$
$$= \rho^L(1-\rho)/(1-\rho^{L+1}) \times P[\tau_s - \tau_i > 0],$$

since the event's "saturation" and $\{\tau_s > \tau_i\}$ are independent. Since τ_s and τ_i are independent, the probability $P[\tau_s - \tau_i > 0]$ can easily be computed by convolution. The result is $P[\tau_s - \tau_i > 0] = \lambda/(\lambda + \mu)$. The probability of loosing the incoming packet is then

$$P[\text{`packet loss'}] = \rho^L(1-\rho)/(1-\rho^{L+1}) \times \rho/(1+\rho),$$

which, for $\rho = 0.5$, yields $P[\text{`packet loss'}] = 1.6 \times 10^{-4}$ for the buffer of size 10, with arrival rate equal to half the service rate.

Chapman–Kolmogorov Equations

In the examples of a Markov random sequence in Chapter 6, we specified the transition density as a one-step transition, that is, from $n-1$ to n. More generally, we can specify the transition density from time n to time $n+k$ where $k \geq 0$, as in the general definition of a Markov random sequence. However, in this more general case we must make sure that this multistep transition density is consistent, that is, that there exists a one-step density that would sequentially yield the same results. This problem is even more important in the random process case, where due to continuous time one is always effectively considering multistep transition densities, i.e., between any two times $t_2 \neq t_1$, there is a time in between.

For example, given a continuous time transition density $f_X(x_2|x_1; t_2, t_1)$, how do we know that an unconditional pdf $f_X(x; t)$ can be found to satisfy the equation,

$$f_X(x_2; t_2) = \int_{-\infty}^{+\infty} f_X(x_2|x_1; t_2, t_1) f_X(x_1; t_1) dx_1$$

for all $t_2 > t_1$, and all x_1 and x_2?

The Chapman–Kolmogorov equations supply both necessary and sufficient conditions for these general transition densities. There is also a version of the Chapman-Kolmogorov equations for the discrete-valued case involving PMFs of multistep transitions.

Consider three times $t_3 > t_2 > t_1$ and the Markov process random variables at these three times $X(t_3)$, $X(t_2)$, and $X(t_1)$. We wish to compute the conditional density of $X(t_3)$ given $X(t_1)$. First, we write the joint pdf

$$f_X(x_3, x_1; t_3, t_1) = \int_{-\infty}^{+\infty} f_X(x_3|x_2, x_1; t_3, t_2, t_1) f_X(x_2, x_1; t_2, t_1) dx_2.$$

If we now divide both sides of this equation by $f(x_1; t_1)$, we obtain

$$f_X(x_3|x_1) = \int_{-\infty}^{+\infty} f_X(x_3|x_2, x_1) f_X(x_2|x_1) dx_2,$$

where we have suppressed the times t_i for notational simplicity. Then using the Markov property the above becomes

$$f_X(x_3|x_1) = \int_{-\infty}^{+\infty} f_X(x_3|x_2) f_X(x_2|x_1) dx_2, \tag{7.2-22}$$

which is known as the *Chapman–Kolmogorov equation* for the transition density $f_X(x_3|x_1)$ of a Markov process. This equation must hold for all $t_3 > t_2 > t_1$ and for all values of x_3 and x_1. It can be proven that the Chapman–Kolmogorov condition expressed in Equation 7.2-22 is also sufficient for the existence of the transition density in question [7-5].

Random Process Generated from Random Sequences

We can obtain a Markov random process as the limit of an infinite number of simulations of Markov random sequences. For example, consider the random sequence generated by the equation

$$X[n] = \rho X[n-1] + W[n], \quad -\infty < n < +\infty,$$

as given in Example 6.4-6 of Chapter 6, where $|\rho| < 1.0$ to insure stability. There we found that the correlation function of $X[n]$ was

$$R_{XX}[m] = \sigma_W^2 \, \rho^{|m|},$$

where σ_W^2 is the variance of the independent random sequence $W[n]$. Replacing $X[n]$ with $X(nT)$, and setting $X(t) = X[nT]$ for $nT \le t < (n+1)T$, we get

$$R_{XX}(t+\tau, t) = \sigma_W^2 \, \rho^{\left|\frac{\tau}{T}\right|} = \sigma_W^2 \, \exp(-\alpha|\tau|),$$

where $\alpha \triangleq \frac{1}{T} \ln \frac{1}{\rho}$ or alternatively $\rho = \exp(-\alpha T)$. Thus if we generate a set of simulations with $T_k \triangleq T_0/k$ for $k = 1, 2, 3, \ldots$, and then for each simulation set $\rho_k \triangleq \sqrt[k]{\exp(-\alpha T_0)}$, we will get a set of denser and denser approximations to a limiting random process $X(t)$, that is WSS with correlation function

$$R_{XX}(t+\tau, t) = \sigma_W^2 \, \exp(-\alpha|\tau|).$$

7.3 CONTINUOUS-TIME LINEAR SYSTEMS WITH RANDOM INPUTS

In this section we look at transformations of stochastic processes. We concentrate on the case of linear transformations with memory, since the memoryless case can be handled by the transformation of random variables method of Chapter 3. The definition of a linear continuous-time system is recalled first.

Definition 7.3-1   Let $x_1(t)$ and $x_2(t)$ be two deterministic time functions and let a_1 and a_2 be two scalar constants. Let the linear system be described by the operator equation $y = L\{x\}$. Then the system is linear if

$$L\{a_1 x_1(t) + a_2 x_2(t)\} = a_1 L\{x_1(t)\} + a_2 L\{x_2(t)\} \qquad (7.3\text{-}1)$$

for all admissible functions x_1 and x_2 and all scalars a_1 and a_2.   ■

This amounts to saying that the response to a weighted sum of inputs must be the weighted sum of the responses to each one individually. Also, in this definition we note that the inputs must be in the allowable input space for the system (operator) L. When we think of generalizing L to allow a random process input, the most natural choice is to input the sample functions of X and find the corresponding sample functions of the output, which thereby define a new random process Y. Just as the original random process X is a mapping from the sample space to a function space, the linear system in turn maps this function space to a new function space. The cascade or composition of the two maps thus defines an output random process. This is depicted graphically in Figure 7.3-1. Our goal in this section will be to find out how the first- and second-order moments, that is, the mean and correlation (and covariance) are transformed by a linear system.

Theorem 7.3-1   Let the random process $X(t)$ be the input to a linear system L with output process $Y(t)$. Then the mean function of the output is given as

$$E[Y(t)] = L\{E[X(t)]\}$$
$$= L\{\mu_X(t)\}. \qquad (7.3\text{-}2)$$

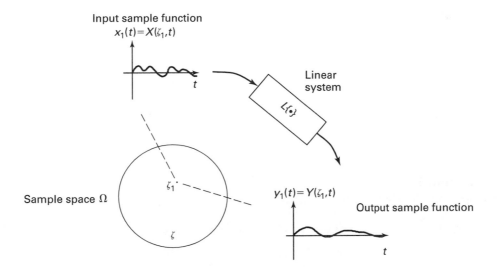

Figure 7.3-1   Interpretation of applying a random process to a linear system.

Proof (formal).   By definition we have for each sample function

$$Y(t, \zeta) = L\{X(t, \zeta)\}$$

so

$$E[Y(t)] = E[L\{X(t)\}].$$

If we can interchange the two operators, we get the result that the mean function of the output is just the result of L operating on the mean function of the input. This can be heuristically (formally) justified as follows, if we assume the operator L can be represented by the superposition integral:

$$Y(t) = \int_{-\infty}^{+\infty} h(t, \tau) X(\tau) d\tau.$$

Taking the expectation, we obtain

$$E[Y(t)] = E\left[\int_{-\infty}^{+\infty} h(t, \tau) X(\tau) d\tau\right]$$

$$= \int_{-\infty}^{+\infty} h(t, \tau) E[X(\tau)] d\tau$$

$$= L\{\mu_X(t)\}. \quad \blacksquare$$

We present a rigorous proof of this theorem after we study the mean-square stochastic integral in Chapter 8. We next look at how the correlation function is transformed by a linear system. There are now two stochastic processes to consider, the input and the output, and the cross-correlation function $E[X(t_1) Y^*(t_2)]$ comes into play. We thus define the *cross-correlation function*

$$R_{XY}(t_1, t_2) \triangleq E[X(t_1) Y^*(t_2)].$$

From the autocorrelation function of the input $R_{XX}(t_1, t_2)$, we first calculate the cross-correlation function $R_{XY}(t_1, t_2)$ and then the autocorrelation function of the output $R_{YY}(t_1, t_2)$. If the mean is zero for the input process, then by Theorem 7.3-1 the mean of the output process is also zero. Thus the following results can be seen also to hold for covariance functions by changing the input to the centered process $X_c(t) \triangleq X(t) - \mu_X(t)$, which produces the centered output $Y_c(t) \triangleq Y(t) - \mu_Y(t)$.

Theorem 7.3-2   Let $X(t)$ and $Y(t)$ be the input and output random processes of the linear operator L. Then the following hold:

$$R_{XY}(t_1, t_2) = L_2^*\{R_{XX}(t_1, t_2)\}, \tag{7.3-3}$$

$$R_{YY}(t_1, t_2) = L_1\{R_{XY}(t_1, t_2)\}, \tag{7.3-4}$$

where L_i means the time variable of the operator L is t_i.

Proof (formal).   Write

$$X(t_1)Y^*(t_2) = X(t_1)L_2^*\{X^*(t_2)\}$$
$$= L_2^*\{X(t_1)X^*(t_2)\},$$

where we have used the *adjoint operator* L^* whose impulse response is $h^*(t,\tau)$. Then

$$E[X(t_1)Y^*(t_2)] = E\left[L_2^*\{X(t_1)X^*(t_2)\}\right]$$
$$= L_2^*\{E[X(t_1)X^*(t_2)]\} \quad \text{by interchanging } L_2^* \text{ and } E,$$
$$= L_2^*\{R_{XX}(t_1,t_2)\},$$

which is Equation 7.3-3.

Similarly, to prove Equation 7.3-4, we multiply by $Y^*(t_2)$ and get

$$Y(t_1)Y^*(t_2) = L_1\{X(t_1)Y^*(t_2)\}$$

so that

$$E[Y(t_1)Y^*(t_2)] = E\left[L_1\{X(t_1)Y^*(t_2)\}\right]$$
$$= L_1\{E[X(t_1)Y^*(t_2)]\} \quad \text{by interchanging } L_1 \text{ and } E,$$
$$= L_1\{R_{XY}(t_1,t_2)\},$$

which is Equation 7.3-4.

If we combine Equation 7.3-3 and Equation 7.3-4 we get

$$R_{YY}(t_1,t_2) = L_1 L_2^*\{R_{XX}(t_1,t_2)\}. \quad \blacksquare \tag{7.3-5}$$

Example 7.3-1 $\rule{8cm}{0.4pt}$

(Edge or "change" detector.) Let $X(t)$ be a real-valued random process, modeling a certain sensor signal, and define $Y(t) \triangleq L\{X(t)\} \triangleq X(t) - X(t-1)$ so

$$E[Y(t)] = L\{\mu_X(t)\} = \mu_X(t) - \mu_X(t-1).$$

Also

$$R_{XY}(t_1,t_2) = L_2\{R_{XX}(t_1,t_2)\}$$
$$= R_{XX}(t_1,t_2) - R_{XX}(t_1,t_2-1)$$

and

$$R_{YY}(t_1,t_2) = L_1\{R_{XY}(t_1,t_2)\}$$
$$= R_{XY}(t_1,t_2) - R_{XY}(t_1-1,t_2)$$
$$= R_{XX}(t_1,t_2) - R_{XX}(t_1-1,t_2)$$
$$-R_{XX}(t_1,t_2-1) + R_{XX}(t_1-1,t_2-1).$$

To be specific, if we take $\mu_X(t) = 0$ and

$$R_{XX}(t_1, t_2) \triangleq \sigma_X^2 \exp(-\alpha|t_1 - t_2|),$$

then

$$E[Y(t)] = 0 \qquad \text{since } \mu_X = 0,$$

and

$$R_{YY}(t_1, t_2) = \sigma_X^2 \left(2 \exp(-\alpha|t_1 - t_2|) - \exp(-\alpha|t_1 - t_2 - 1|) \right.$$
$$\left. - \exp(-\alpha|t_1 - t_2 + 1|) \right).$$

We note that both R_{XX} and R_{XY} are functions only of the difference of the two observation times t_1 and t_2. The input correlation function R_{XX} is plotted in Figure 7.3-2, for $\alpha = 2$ and $\sigma_X^2 = 2$. Note the negative correlation values in output correlation function R_{YY}, shown in Figure 7.3-3, introduced by the difference operation of the edge detector. The variance of $Y(t)$ is constant and is given as

$$\sigma_Y^2(t) = \sigma_Y^2 = 2\sigma_X^2[1 - \exp(-\alpha)].$$

We see that as α tends to zero, the variance of Y goes to zero. This is because as α tends to zero, $X(t)$ and $X(t-1)$ become very positively correlated, and hence there is very little power in their difference.

Example 7.3-2 _____

Let $X(t)$ be a real-valued random process with constant mean function $\mu_X(t) = \mu$ and covariance function

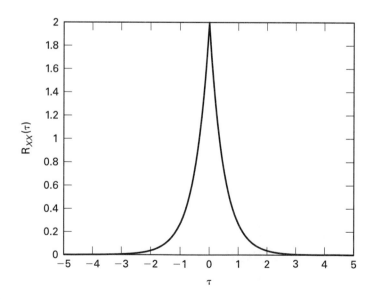

Figure 7.3-2 Plot of input correlation function R_{XX} versus $\tau = t_1 - t_2$.

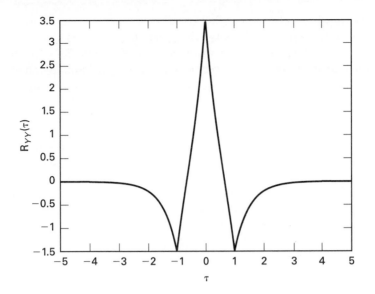

Figure 7.3-3   Plot of output correlation function R_{YY} versus $\tau = t_1 - t_2$.

$$K_{XX}(t, s) = \sigma^2 \cos\omega_0(t - s).$$

We wish to determine the mean and covariance function of the derivative process $X'(t)$. Here the linear operator is $d(\cdot)/dt$. First we determine the mean,

$$\mu_{X'}(t) = E[X'(t)] = \frac{d}{dt}E[X(t)]$$

$$= \frac{d}{dt}\mu_X(t)$$

$$= \frac{d}{dt}\mu = 0.$$

Now, for this real-valued random process, the covariance function of $X'(t)$ is

$$K_{X'X'}(t_1, t_2) = E[X'(t_1)X'(t_2)],$$

since $\mu'_X(t) = 0$. Thus by Equation 7.3-5, with $X'(t) = Y(t)$,

$$K_{X'X'}(t_1, t_2) = \frac{\partial}{\partial t_1}\left(\frac{\partial}{\partial t_2}K_{XX}(t_1, t_2)\right)$$

$$= \frac{\partial}{\partial t_1}\left(\frac{\partial}{\partial t_2}\sigma^2\cos\omega_0(t_1 - t_2)\right)$$

$$= \frac{\partial}{\partial t_1}(\omega_0\sigma^2\sin\omega_0(t_1 - t_2))$$

$$= (\omega_0\sigma)^2\cos\omega_0(t_1 - t_2).$$

We note that the result is just the original covariance function scaled up by the factor ω_0^2. This similarity-in-form happened because the given $K_{XX}(t,s)$ is the covariance function of a sine wave with random amplitude and phase (cf. Example 7.1-5). Since the phase is random, the sine and its derivative the cosine are indistinguishable by shape.

White Noise

Let the random process under consideration be the Wiener process of Section 7.2. Here we consider the derivative of this process. For any $\alpha > 0$, the covariance function of the Wiener process is $K_{XX}(t_1, t_2) = \alpha \min(t_1, t_2)$ and its mean function $\mu_X = 0$. Let $W(t) = dX(t)/dt$. Then proceeding as in the above example, we can calculate $\mu_W(t) = E[dX(t)/dt] = d\mu_X(t)/dt = 0$. For the covariance,

$$
\begin{aligned}
K_{WW}(t_1, t_2) &= \frac{\partial}{\partial t_1}\left(\frac{\partial}{\partial t_2} K_{XX}(t_1, t_2)\right) \\
&= \frac{\partial}{\partial t_1}\left(\frac{\partial}{\partial t_2}\alpha \min(t_1, t_2)\right) \\
&= \frac{\partial}{\partial t_1}\left(\frac{\partial}{\partial t_2}\begin{cases}\alpha t_2, & t_2 < t_1 \\ \alpha t_1, & t_2 \geq t_1\end{cases}\right) \\
&= \frac{\partial}{\partial t_1}\begin{cases}\alpha, & t_2 < t_1 \\ 0, & t_2 > t_1\end{cases} \\
&= \frac{\partial}{\partial t_1}\begin{cases}0, & t_1 < t_2 \\ \alpha, & t_1 > t_2\end{cases} \\
&= \frac{\partial}{\partial t_1}\alpha u(t_1 - t_2) \\
&= \alpha \delta(t_1 - t_2).
\end{aligned}
$$

Thus the covariance function of white noise is the impulse function. Since white noise always has zero-mean, the correlation function too is an impulse. Often one writes

$$
R_{WW}(t_1, t_2) = \sigma^2 \delta(t_1 - t_2) = K_{WW}(t_1, t_2), \tag{7.3-6}
$$

replacing α by σ^2, but one should note that the power in this process $E[|W(t)|^2] = \sigma^2 \delta(0)$ is infinite, not σ^2. In fact, σ^2 is a power density for the white noise process.

Note that the sample functions are highly discontinuous and the white noise process is not *separable*.[†]

[†]The idea of separability (cf. Section 7.1) is to make a countable set of points on the t-axis (e.g. time-axis) determine the properties of the process. In effect it says that knowing the PDF, or pdf over a countable set of points implies knowing the PDF or pdf everywhere. See [7-12].

7.4 SOME USEFUL CLASSIFICATIONS OF RANDOM PROCESSES

Here we look at several classes of random processes and pairs of processes. These classifications also apply to the random sequences studied earlier.

Definition 7.4-1   Let X and Y be random processes. They are

(a) *Uncorrelated* if $R_{XY}(t_1, t_2) = \mu_X(t_1)\mu_Y^*(t_2)$, for all t_1 and t_2;
(b) *Orthogonal* if $R_{XY}(t_1, t_2) = 0$ for all t_1 and t_2;
(c) *Independent* if for all positive integers n, the nth order PDF of X and Y factors, that is,

$$F_{XY}(x_1, y_1, x_2, y_2, \ldots, x_n, y_n; t_1, \ldots, t_n)$$
$$= F_X(x_1, \ldots, x_n; t_1, \ldots, t_n)F_Y(y_1, \ldots, y_n; t_1, \ldots, t_n),$$

for all x_i, y_i and for all $t_1, \ldots, t_n$.   ■

Note that two random processes are orthogonal if they are uncorrelated and at least one of their mean functions is zero. Actually, the orthogonality concept is useful only when the random processes under consideration are zero-mean, in which case it becomes equivalent to the uncorrelated condition. The orthogonality concept was introduced for random vectors in Chapter 5. This concept will prove useful for estimating random processes and sequences in Chapter 9.

A random process may be uncorrelated, orthogonal, or independent of *itself* at earlier and/or later times. For example, we may have $R_{XX}(t_1, t_2) = 0$ for all $t_1 \neq t_2$, in which case we call X an *orthogonal random process*. Similarly $X(t)$ may be independent of $\{X(t_1), \ldots, X(t_n)\}$ for all $t \notin \{t_1, \ldots, t_n\}$ and for all $t_1, \ldots, t_n$ and for all $n \geq 1$. Then we say $X(t)$ is an *independent random process*. Clearly, the sample functions of such processes will be quite rough, since arbitrarily small changes in t yield complete independence.

Stationarity

We say a random process is stationary when its statistics do not change with the continuous parameter, often time. The formal definition is:

Definition 7.4-2   A random process $X(t)$ is *stationary* if it has the same nth-order probability distribution function (PDF) as $X(t + T)$, i.e., the two n-dimensional functions

$$F_X(x_1, \ldots, x_n; t_1, \ldots, t_n) = F_X(x_1, \ldots, x_n; t_1 + T, \ldots, t_n + T)$$

are identically equal for all T, for all positive integers n, and for all $t_1, \ldots, t_n$.   ■

When the PDF is differentiable, we can equivalently write this in terms of the probability density function (pdf) as

$$f_X(x_1, \ldots, x_n; t_1, \ldots, t_n) = f_X(x_1, \ldots, x_n; t_1 + T, \ldots, t_n + T),$$

and this is the form of the stationarity condition that is most often used. This definition implies that the mean of a stationary process is a constant. To prove this note that $f(x;t) = f(x;t+T)$ for all T implies $f(x;t) = f(x;0)$ by taking $T = -t$, which in turn implies that $E[X(t)] = \mu_X(t) = \mu_X(0)$, a constant.

Since the second-order density is also shift invariant, that is,

$$f(x_1, x_2; t_1, t_2) = f(x_1, x_2; t_1 + T, t_2 + T),$$

we have, on choosing $T = -t_2$, that

$$f(x_1, x_2; t_1, t_2) = f(x_1, x_2; t_1 - t_2, 0),$$

which implies $E[X(t_1)X^*(t_2)] = R_{XX}(t_1 - t_2, 0)$. In the stationary case, therefore, the notation for correlation function can be simplified to a function of just the shift $\tau \triangleq t_1 - t_2$ between the two sampling instants or parameters. Thus we can define the one-parameter correlation function

$$R_{XX}(\tau) \triangleq R_{XX}(\tau, 0)$$
$$= E[X(t + \tau)X^*(t)], \tag{7.4-1}$$

which is functionally independent of the parameter t. Examples of this sort of correlation function were seen in Section 7.3.

A weaker form of stationarity exists which does not directly constrain the nth-order PDFs, but rather just the first- and second-order moments. This property, which is easier to check, is called wide-sense stationarity and will be quite useful in what follows.

Definition 7.4-3   A random process X is *wide-sense stationary* (WSS) if $E[X(t)] = \mu_X$, a constant, and $E[X(t + \tau)X^*(t)] = R_{XX}(\tau)$ for all $-\infty < \tau + \infty$, independent of the time parameter t. ∎

Example 7.4-1 _____

(A random complex exponential.) Let $X(t) \triangleq A \exp(j2\pi ft)$ with f a known real constant and A a *real-valued* random variable with mean $E[A] = 0$ and finite average power $E[A^2]$. Calculating the mean and correlation of $X(t)$, we obtain

$$E[X(t)] = E[A \exp(j2\pi ft)] = E[A] \exp(j2\pi ft) = 0,$$

and

$$E[X(t + \tau)X^*(t)] = E[A \exp(j2\pi f(t + \tau))A \exp(-j2\pi ft)]$$
$$= E[A^2] \exp(j2\pi f\tau) = R_{XX}(\tau).$$

Note that $E[A] = 0$ is a necessary condition here. *Question:* Would this work with a cosine function in place of the complex exponential?

The process in Example 7.4-1, while shown to be wide-sense stationary, is clearly not stationary. Consider, for example, that $X(0)$ must be pure real while $X(\frac{1}{4f})$ must always be pure imaginary. We thus conclude that the WSS property is considerably weaker than stationarity.

We can generalize this example to have M complex sinusoids and obtain a rudimentary frequency domain representation for zero-mean random processes. Consider

$$X(t) = \sum_{k=1}^{M} A_k \exp(j2\pi f_k t)$$

where the generally complex random variables A_k are uncorrelated with mean zero and variances σ_k^2. Then the resulting random process is WSS with mean zero and autocorrelation (or autocovariance) equal to

$$R_{XX}(\tau) = \sum_{k=1}^{M} \sigma_k^2 \exp(j2\pi f_k \tau). \tag{7.4-2}$$

For such random processes $X(t)$, the set of random coefficients $\{A_k\}$ constitutes a frequency domain representation. From our experience with Fourier analysis of deterministic functions, we can expect that as M became large and as the f_k became dense, that is, the spacing between the f_k became small and they covered the frequency range of interest, then most random processes would have such an approximate representation. Such is the case (cf. Section 8.6).

7.5 WIDE-SENSE STATIONARY PROCESSES AND LSI SYSTEMS

In this section we treat random processes that are jointly stationary and of *second order*, that is,

$$E[|X(t)|^2] < \infty.$$

Some important properties of the auto- and cross-correlation functions of stationary second-order processes are summarized as follows. They, of course, also hold for the respective covariance functions.

1. $|R_{XX}(\tau)| \leq R_{XX}(0)$, which directly follows from $E[|X(t+\tau) - X(t)|^2] \geq 0$.
2. $|R_{XY}(\tau)| \leq \sqrt{R_{XX}(0)R_{YY}(0)}$, which is derived using the Schwarz inequality. (cf. Section 4.3. Also called diagonal dominance.)
3. $R_{XX}(\tau) = R_{XX}^*(-\tau)$, since $E[X(t+\tau)X^*(t)] = E[X(t)X^*(t-\tau)] = E^*[X(t-\tau) X^*(t)]$ for WSS random processes, which is called the *conjugate symmetry property*. In the special case of a real-valued process, this property becomes that of even symmetry, that is,
3a. $R_{XX}(\tau) = R_{XX}(-\tau)$. Another important property of the autocorrelation function of a complex valued, stationary random process is that it must be *positive semidefinite*, i.e.

4. for all $N > 0$, all $t_1 < t_2 < \ldots < t_N$ and all complex $a_1, a_2, \ldots, a_N$,

$$\sum_{k=1}^{N} \sum_{l=1}^{N} a_k a_l^* R_{XX}(t_k - t_l) \geq 0.$$

This was shown in Section 7.1 to be a necessary condition for a given function $g(t, s) = g(t - s)$ to be an autocorrelation function. We will show that this property is also a sufficient condition, so that positive semidefiniteness actually *characterizes* autocorrelation functions. In general, however, it is very difficult to check property 4 directly.

To start off, we can specialize the results of Theorems 7.3-1 and 7.3-2, which were derived for the general case, to LSI systems. Rewriting Equation 7.3-2 we have

$$E[Y(t)] = L\{\mu_X(t)\}$$

$$= \int_{-\infty}^{\infty} \mu_X(\tau) h(t - \tau) \, d\tau$$

$$\overset{\Delta}{=} \mu_X(t) * h(t).$$

Using Theorem 7.3-2 and Equations 7.3-3 and 7.3-4, we get also

$$R_{XY}(t_1, t_2) = \int_{-\infty}^{+\infty} h^*(\tau_2) R_{XX}(t_1, t_2 - \tau_2) d\tau_2,$$

and

$$R_{YY}(t_1, t_2) = \int_{-\infty}^{+\infty} h(\tau_1) R_{XY}(t_1 - \tau_1, t_2) d\tau_1,$$

which can be written in convolution operator notation as

$$R_{XY}(t_1, t_2) = h^*(t_2) * R_{XX}(t_1, t_2),$$

where the convolution is along the t_2-axis, and

$$R_{YY}(t_1, t_2) = h(t_1) * R_{XY}(t_1, t_2),$$

where the convolution is along the t_1-axis. Combining these two equations, we get $R_{YY}(t_1, t_2) = h(t_1) * R_{XX}(t_1, t_2) * h^*(t_2)$.

Wide-Sense Stationary Case

If we input the stationary random process $X(t)$ to an LSI system with impulse response $h(t)$, then the output random process can be expressed as the convolution integral,

$$Y(t) = \int_{-\infty}^{+\infty} h(\tau) X(t - \tau) d\tau, \tag{7.5-1}$$

when this integral exists. Computing the mean of the output process $Y(t)$, we get

$$E[Y(t)] = \int_{-\infty}^{+\infty} h(\tau)E[X(t-\tau)]d\tau \quad \text{by Theorem 7.3-1,}$$

$$= \int_{-\infty}^{+\infty} h(\tau)\mu_X d\tau = \mu_X \int_{-\infty}^{+\infty} h(\tau)d\tau,$$

$$= \mu_X H(0), \tag{7.5-2}$$

where $H(\omega)$ is the system's frequency response.

We thus see that the mean of the output is constant and equals the mean of the input times the system function evaluated at $\omega = 0$, the so-called "dc gain" of the system. If we compute the cross-correlation function between the input process and the output process we find that

$$R_{YX}(\tau) = E[Y(t+\tau)X^*(t)]$$

$$= E[Y(t)X^*(t-\tau)] \quad \text{by substituting } t-\tau \text{ for } t,$$

$$= \int_{-\infty}^{+\infty} h(\alpha)E[X(t-\alpha)X^*(t-\tau)]d\alpha,$$

and bringing the operator E inside the integral by Theorem 7.3-2,

$$= \int_{-\infty}^{+\infty} h(\alpha)R_{XX}(\tau-\alpha)d\alpha$$

which can be rewritten as

$$R_{YX}(\tau) = h(\tau) * R_{XX}(\tau). \tag{7.5-3}$$

Thus the cross-correlation R_{YX} equals h convolved with the autocorrelation R_{XX}. This fact can be used to identify unknown systems (see Problem 7.28).

The output autocorrelation function $R_{YY}(\tau)$ can now be obtained from $R_{YX}(\tau)$ as follows:

$$R_{YY}(\tau) = E[Y(t+\tau)Y^*(t)]$$

$$= E[Y(t)Y^*(t-\tau)] \quad \text{by substituting } t \text{ for } t-\tau,$$

$$= \int_{-\infty}^{+\infty} h^*(\alpha)E[Y(t)X^*(t-\tau-\alpha)]d\alpha$$

$$= \int_{-\infty}^{+\infty} h^*(\alpha)E[Y(t)X^*(t-(\tau+\alpha))]d\alpha$$

$$= \int_{-\infty}^{+\infty} h^*(\alpha)R_{YX}(\tau+\alpha)d\alpha$$

$$= \int_{-\infty}^{+\infty} h^*(-\alpha)R_{YX}(\tau-\alpha)d\alpha$$

$$= h^*(-\tau) * R_{YX}(\tau).$$

Combining both equations, we get

$$R_{YY}(\tau) = h(\tau) * h^*(-\tau) * R_{XX}(\tau). \tag{7.5-4}$$

We observe that when $R_{XX}(\tau) = \delta(\tau)$, then the output correlation function is $R_{YY}(\tau) = h(\tau) * h^*(-\tau)$, which is sometimes called the *autocorrelation impulse response* denoted as $g(\tau) = h(\tau) * h^*(-\tau)$. Note that $g(\tau)$ must be positive semidefinite, and indeed $FT\{g(\tau)\} = |H(\omega)|^2 \geq 0$.

Similarly, we also find (proof left as an exercise for the reader)

$$R_{XY}(\tau) = \int_{-\infty}^{+\infty} h^*(-\alpha) R_{XX}(\tau - \alpha) d\alpha$$

$$= h^*(-\tau) * R_{XX}(\tau), \tag{7.5-5}$$

and

$$R_{YY}(\tau) = \int_{-\infty}^{+\infty} h(\alpha) R_{XY}(\tau - \alpha) d\alpha.$$

$$= h(\tau) * R_{XY}(\tau)$$

$$= h(\tau) * h^*(-\tau) * R_{XX}(\tau)$$

$$= g(\tau) * R_{XX}(\tau).$$

Example 7.5-1

(Derivative of wide-sense stationary process.) Let the second-order random process $X(t)$ be stationary with one parameter correlation function $R_X(\tau)$ and constant mean function $\mu_X(t) = \mu_X$. Consider the system consisting of a derivative operator, i.e.,

$$Y(t) = \frac{dX(t)}{dt}.$$

Using the above equations we find $\mu_Y(t) = \frac{d\mu_X(t)}{dt} = 0$ and cross-correlation function

$$R_{XY}(\tau) = \delta_1^*(-\tau) * R_{XX}(\tau)$$

$$= -\frac{dR_{XX}(\tau)}{d\tau},$$

since the impulse response of the derivative operator is $h(t) = \delta_1(t)$, the (formal) derivative of the Dirac delta function or impulse $\delta(t)$.

$$R_{YY}(\tau) = \delta_1(\tau) * R_{XY}(\tau)$$

$$= \frac{dR_{XY}(\tau)}{d\tau}$$

$$= -\frac{d^2 R_{XX}(\tau)}{d\tau^2}.$$

Notice the autocorrelation impulse response here is $g(\tau) = -\delta_2(\tau)$, or minus the second (formal) derivative of $\delta(\tau)$.

Power Spectral Density

For wide-sense stationary, and hence for stationary processes, we can define a useful density for average power versus frequency, called the *power spectral density* (psd).

Definition 7.5-1   Let $R_{XX}(\tau)$ be an autocorrelation function. Then we define the *power spectral density* $S_{XX}(\omega)$ to be its Fourier transform (if it exists), that is,

$$S_{XX}(\omega) \triangleq \int_{-\infty}^{+\infty} R_{XX}(\tau)e^{-j\omega\tau}\,d\tau. \quad \blacksquare \qquad (7.5\text{-}6)$$

Under quite general conditions one can define the inverse Fourier transform, which equals $R_{XX}(\tau)$ at all points of continuity,

$$R_{XX}(\tau) = \frac{1}{2\pi}\int_{-\infty}^{+\infty} S_{XX}(\omega)e^{+j\omega\tau}\,d\omega. \qquad (7.5\text{-}7)$$

In operator notation we have,

$$S_{XX} = FT\{R_{XX}\}$$

and

$$R_{XX} = IFT\{S_{XX}\}$$

where FT and IFT stand for the respective Fourier operators.

The name power spectral density (psd) will be justified later. All that we have done thus far is define it as the Fourier transform of $R_{XX}(\tau)$. We can also define the Fourier transform of the cross-correlation function $R_{XY}(\tau)$ to obtain a frequency function called the *cross-power spectral density*,

$$S_{XY}(\omega) \triangleq \int_{-\infty}^{+\infty} R_{XY}(\tau)e^{-j\omega\tau}\,d\tau. \qquad (7.5\text{-}8)$$

We will see later that the power spectral density or psd, $S_{XX}(\omega)$, is real and everywhere nonnegative and in fact, as the name implies, has the interpretation of a density function for average power versus frequency. By contrast, the cross-power spectral density has no such interpretation and is generally complex valued.

We next list some properties of the psd $S_{XX}(\omega)$:

1. $S_{XX}(\omega)$ is real-valued since $R_{XX}(\tau)$ is conjugate symmetric.
2. If $X(t)$ is a real-valued WSS process, then $S_{XX}(\omega)$ is an even function since $R_{XX}(\tau)$ is real and even. Otherwise $S_{XX}(\omega)$ may not be an even function of ω.
3. $S_{XX}(\omega) \geq 0$ (to be shown in Section in Theorem 7.5-1).

Additional properties of the psd are shown in Table 7.5-1. One could continue with such a table, but it will suit our purposes to stop at this point. One comment is in order: We note the simplicity of these operations in the frequency domain. This suggests that for LSI systems and stationary or WSS random processes, we should solve for output correlation functions by first transforming the input correlation function into the frequency domain, carry out the indicated operations, and then transform back to the correlation domain.

Table 7.5-1 Correlation Function Properties of Corresponding Power Spectral Densities

Random Process	Correlation Function	Power Spectral Density
$X(t)$	$R_{XX}(\tau)$	$S_{XX}(\omega)$
$aX(t)$	$|a|^2 R_{XX}(\tau)$	$|a|^2 S_{XX}(\omega)$
$X_1(t) + X_2(t)$ with X_1 and X_2 orthogonal	$R_{X_1 X_1}(\tau) + R_{X_2 X_2}(\tau)$	$S_{X_1 X_1}(\omega) + S_{X_2 X_2}(\omega)$
$X'(t)$	$-d^2 R_{XX}(\tau)/d\tau^2$	$\omega^2 S_{XX}(\omega)$
$X^{(n)}(t)$	$(-1)^n d^{2n} R_{XX}(\tau)/d\tau^{2n}$	$\omega^{2n} S_{XX}(\omega)$
$X(t)\exp(j\omega_0 t)$	$\exp(j\omega_0 \tau) R_{XX}(\tau)$	$S_{XX}(\omega - \omega_0)$
$X(t)\cos(\omega_0 t + \Theta)$ with independent Θ uniform on $[-\pi, +\pi]$	$\frac{1}{2} R_{XX}(\tau)\cos(\omega_0 \tau)$	$\frac{1}{4}[S_{XX}(\omega + \omega_0) + S_{XX}(\omega - \omega_0)]$
$X(t) + b \quad (\overline{X} = 0)$	$R_{XX}(\tau) + |b|^2$	$S_{XX}(\omega) + 2\pi|b|^2 \delta(\omega)$

This is completely analogous to the situation in deterministic linear system theory for shift-invariant systems.

Another comment would be that if the interpretation of $S_{XX}(\omega)$ as a density of average power is correct, then the constant or mean component has all its average power concentrated at $\omega = 0$ by the last entry in the table. Also by the next-to-last two entries in the table, modulation by the frequency ω_0 shifts the distribution of average power up in frequency by ω_0. Both of these results should be quite intuitive.

Example 7.5-2

(*Power spectral density of white noise.*) The correlation function of a white noise process $W(t)$ with parameter σ^2 is given by $R_{WW}(\tau) = \sigma^2 \delta(\tau)$. Hence the power spectral density (psd), its Fourier transform, is just

$$S_{WW}(\omega) = \sigma^2, \quad -\infty < \omega < +\infty.$$

The psd is thus flat, and hence the name, *white noise*, by analogy to white light, which contains equal power at every wavelength. Just like white light, white noise is an idealization that cannot really occur, since as we have seen earlier $R_{WW}(0) = \infty$, necessitating infinite power.

An Interpretation of the psd

Given a WSS process $X(t)$, consider the finite support segment,

$$X_T(t) \triangleq X(t) I_{[-T, +T]}(t),$$

where $I_{[-T, +T]}$ is an indicator function equal to 1 if $-T \le t \le +T$ and equal to 0 otherwise, and $T > 0$. We can compute the Fourier transform of X_T by the integral

$$FT\{X_T(t)\} = \int_{-T}^{+T} X(t) e^{-j\omega t}\, dt.$$

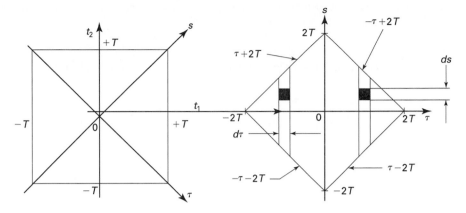

Figure 7.5-1 a) Square region in (t_1, t_2) plane; b) Integration in diamond-shaped region created by the transformation $s = t_1 + t_2, \tau = t_1 - t_2$.

The magnitude squared of this random variable is

$$|FT\{X_T(t)\}|^2 = \int_{-T}^{+T} \int_{-T}^{+T} X(t_1)X^*(t_2)e^{-j\omega(t_1-t_2)}\,dt_1\,dt_2.$$

Dividing by $2T$ and taking the expectation, we get

$$\frac{1}{2T}E\left[|FT\{X_T(t)\}|^2\right] = \frac{1}{2T}\int_{-T}^{+T}\int_{-T}^{+T}R_{XX}(t_1-t_2)e^{-j\omega(t_1-t_2)}\,dt_1\,dt_2, \quad (7.5\text{-}9a)$$

To evaluate the double integral on the right introduce the new coordinate system $s = t_1 + t_2, \tau = t_1 - t_2$. The relationship between the (s, τ) and (t_1, t_2) coordinate systems is shown in Figure 7.5-1a. The Jacobian (scale-change) of this transformation is $1/2$ and the region of integration is the diamond-shaped surface $\wp$ shown in Figure 7.5-1b, which is Figure 7.5-1a rotated counterclockwise 45 degrees and whose sides have length $T\sqrt{2}$. The double integral in Equation 7.5-9a then becomes

$$\frac{1}{4T}\iint_{\wp} R_{XX}(\tau)e^{-j\omega\tau}\,d\tau\,ds$$

$$= \frac{1}{4T}\left\{\int_{-2T}^{0}R_{XX}(\tau)e^{-j\omega\tau}\left[\int_{-(2T+\tau)}^{2T+\tau}ds\right]d\tau\right\}$$

$$+\frac{1}{4T}\left\{\int_{0}^{2T}R_{XX}(\tau)e^{-j\omega\tau}\left[\int_{-(2T-\tau)}^{2T-\tau}ds\right]d\tau\right\} = \int_{-2T}^{+2T}\left[1 - \frac{|\tau|}{2T}\right]R_{XX}(\tau)e^{-j\omega\tau}d\tau.$$

In the limit as $T \to +\infty$, this integral tends to Equation 7.5-6 for an integrable R_{XX}, thus

$$S_{XX}(\omega) = \lim_{T\to\infty}\frac{1}{2T}E\left[|FT\{X_T(t)\}|^2\right] \quad (7.5\text{-}9b)$$

so that $S_{XX}(\omega)$ is real and nonnegative and is related to average power at frequency ω.

We next look at two examples of the computation of power spectral densities corresponding to correlation functions we have seen earlier.

Example 7.5-3 _____

Find the power spectral density for the following exponential autocorrelation function with parameter $\alpha > 0$:

$$R_{XX}(\tau) = \exp(-\alpha |\tau|), \quad -\infty < \tau < +\infty.$$

This is the autocorrelation function of the random telegraph signal (RTS) discussed above. Its psd is computed as

$$S_{XX}(\omega) = \int_{-\infty}^{+\infty} R_{XX}(\tau)e^{-j\omega\tau}d\tau = \int_{-\infty}^{+\infty} e^{-\alpha|\tau|}e^{-j\omega\tau}\,d\tau$$

$$= \int_{-\infty}^{0} e^{(\alpha-j\omega)\tau}\,d\tau + \int_{0}^{\infty} e^{-(\alpha+j\omega)\tau}\,d\tau$$

$$= 2\alpha/[\alpha^2 + \omega^2], \quad -\infty < \omega < +\infty.$$

This function is plotted in Figure 7.5-2 for $\alpha = 3$. We see that the peak value is at the origin and equal to $2/\alpha$. The "bandwidth" of the process is seen to be α on a 3 dB basis (if S_{XX} is indeed a power density, to be shown). We note that while there is a cusp at the origin of the correlation function R_{XX}, there is no cusp in its spectral density S_{XX}. In fact S_{XX} is continuous and differentiable everywhere. (It is true that S_{XX} will always be continuous if R_{XX} is absolutely integrable.)

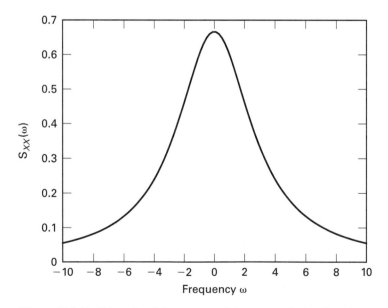

Figure 7.5-2   Plot of psd for exponential autocorrelation function.

Figure 7.5-2 was created using MATLAB with the short m-file:

```
clear alpha=3;
b = [1.0 0.0 alpha^2];
w = linspace(-10,+10);
den = polyval(b,w);
num = 2*alpha;
S = num./den;
plot (w,S)
```

We note that the psd decays rather slowly, and thus the RTS process requires a significant amount of bandwidth. The reason the tails of the psd are so long is due to the jumps in the RTS sample functions.

Example 7.5-4 _____

(psd of triangular autocorrelation.) Consider an autocorrelation function that is triangular in shape such that the correlation goes to zero at shift $T > 0$,

$$R_{XX}(\tau) = \max\left[1 - \frac{|\tau|}{T}, 0\right].$$

One way this could arise is the asynchronous binary signaling (ABS) process introduced above. This function is plotted as Figure 7.5-3. If we realize that this triangle can be written as the convolution of two rectangular pulses, each of width T and height $1/\sqrt{T}$, then we see that the psd of the triangular correlation function is just the square of the Fourier transform of the rectangular pulse, that is, the *sinc* function. The transform of the rectangular pulse is

$$\sqrt{T}\frac{\sin(\omega T/2)}{(\omega T/2)},$$

and the power spectral density S_{XX} of the triangular correlation function is thus

$$S_{XX}(\omega) = T\left(\frac{\sin(\omega T/2)}{\omega T/2}\right)^2. \tag{7.5-10}$$

As a check we note that $S_{XX}(0)$ is just the area under the correlation function, that in the triangular case is easily seen to be T. Thus checking,

$$S_{XX}(0) = \int_{-\infty}^{+\infty} R_{XX}(\tau)\,d\tau = 2 \cdot \frac{1}{2} \cdot 1 \cdot T.$$

_____

Another way the triangular correlation function can arise is the running integral average operating on white noise. Consider

$$X(t) \triangleq \frac{1}{\sqrt{T}}\int_{t-T}^{t} W(\tau)\,d\tau,$$

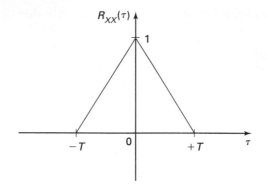

Figure 7.5-3   A triangular autocorrelation function.

with $W(t)$ a white noise with zero-mean and correlation function $R_{WW}(\tau) = \delta(\tau)$. Then $\mu_X(t) = 0$ and $E[X(t_1)X(t_2)]$ can be computed as

$$R_{XX}(t_1, t_2) = \frac{1}{T} \int_{t_1-T}^{t_1} \int_{t_2-T}^{t_2} R_{WW}(s_1 - s_2) ds_1 \, ds_2$$

$$= \frac{1}{T} \int_{t_1-T}^{t_1} \left[\int_{t_2-T}^{t_2} \delta(s_2 - s_1) \, ds_2 \right] ds_1.$$

Now defining the inner integral as

$$g_{t_2}(s_1) \triangleq \int_{t_2-T}^{t_2} \delta(s_2 - s_1) ds_2 = \begin{cases} 1, & t_2 - T \le s_1 \le t_2 \\ 0 & \text{else,} \end{cases}$$

which as a function of s_1 looks as shown in Figure 7.5-4, so

$$R_{XX}(t_1, t_2) = \frac{1}{T} \int_{t_1-T}^{t_1} g_{t_2}(s_1) ds_1$$

$$= \max \left[1 - \frac{|t_1 - t_2|}{T}, 0 \right].$$

More on White Noise

The correlation function of white noise is an impulse (Equation 7.3-6), so its psd is a constant

$$S_{WW}(\omega) = \sigma^2, \qquad -\infty < \omega < +\infty.$$

The name white noise thus arises out of the fact that the power spectral density is constant at all frequencies just as in white light, which contains all wavelengths in equal amounts.[†]

[†]A mathematical idealization! Physics tells us that, for realistic models, the power density must tend toward zero as $\omega \to \infty$.

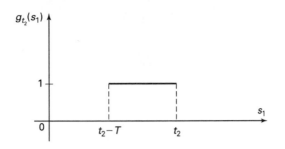

Figure 7.5-4   Plot of equation versus s_1 for $t_2 > T$.

Here we look at the white noise process as a limit approached by a sequence of second-order processes. To this end consider an independent increment process (cf. Definition 7.2-1) with zero mean such as the Wiener process $(R_{XX}(t_1, t_2) = \sigma^2 \min(t_1, t_2))$ or a centered Poisson process, that is, $N_c(t) = N(t) - \lambda t$, with correlation $R_{N_c N_c}(t_1, t_2) = \lambda \min(t_1, t_2)$. Actually we need only uncorrelated increments here; thus $X(t)$ needs only to have uncorrelated increments. For such processes we have by Equation 7.2-17,

$$E\left[(X(t + \Delta) - X(t))^2\right] = \alpha \Delta,$$

where α is the variance parameter.

Thus upon letting $X_\Delta(t)$ denote the first-order difference divided by Δ,

$$X_\Delta(t) \stackrel{\Delta}{=} [X(t + \Delta) - X(t)]/\Delta,$$

we have

$$E[X_\Delta^2(t)] = \alpha/\Delta$$

and

$$E[X_\Delta(t_1)X_\Delta(t_2)] = 0 \quad \text{for} \quad (t_1, t_1 + \Delta] \cap (t_2, t_2 + \Delta] = \phi.$$

If we consider t_1 and t_2 closer than Δ, we can do the following calculation, which shows that the resulting correlation function is triangular, just as in Example 7.5-4. Since $X(t_1 + \Delta) - X(t_1)$ is $N(0, \Delta)$, taking $t_1 < t_2$ and shifting t_1 to 0, and t_2 to $t_2 - t_1$, the expectation becomes

$$\frac{1}{\Delta^2} E\left[X(\Delta)\left(X(t_2 - t_1 + \Delta) - X(t_2 - t_1)\right)\right]$$

$$= \frac{1}{\Delta^2} [X(\Delta)(X(\Delta) - X(t_2 - t_1))] \quad \text{since } (\Delta, t_2 - t_1 + \Delta] \cap (0, \Delta] = \phi,$$

$$= \frac{1}{\Delta^2}[\alpha\Delta - \alpha(t_2 - t_1)] = \frac{\alpha}{\Delta}[1 - (t_2 - t_1)/\Delta].$$

Thus the process generated by the first-order difference is wide-sense stationary (the mean is zero) and has correlation function $R_{\Delta\Delta}(\tau)$ given as

$$R_{\Delta\Delta}(\tau) = \frac{\alpha}{\Delta} \max\left[1 - \frac{|\tau|}{\Delta}, 0\right].$$

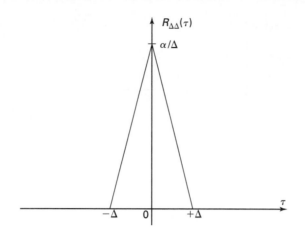

Figure 7.5-5   Correlation function of $X_\Delta(t)$.

We note from Figure 7.5-5 that as Δ goes to zero this correlation function tends to a delta function.

Since we just computed the Fourier transform of a triangular function in Example 7.5-4, we can write the psd by inspection as

$$S_{\Delta\Delta}(\omega) = \alpha \left(\frac{\sin(\omega\Delta/2)}{\omega\Delta/2} \right)^2.$$

This psd is approximately flat out to $|\omega| = \pi/(3\Delta)$. As $\Delta \to 0$, $S_{\Delta\Delta}(\omega)$ approaches the constant α everywhere. Thus as $\Delta \to 0$, $X_\Delta(t)$ "converges" to white noise, the *formal* derivative of an uncorrelated increments process,

$$R_{\dot{X}X}(t_1, t_2) = \frac{\partial^2}{\partial t_1 \partial t_2} [\sigma^2 \min(t_1, t_2)]$$

$$= \frac{\partial}{\partial t_1} [\sigma^2 u(t_1 - t_2)]$$

$$= \sigma^2 \delta(t_1 - t_2).$$

If one has a system that is continuous in its response to stimuli, then we say that the *system is continuous*; that is, the system operator is a continuous operator. This would mean, for example, that the output would change only slightly if the input changed slightly. A stable differential or difference equation is an example of such a continuous operator. We will see that for linear shift-invariant systems that are described by system functions, the response to the random process $X_\Delta(t)$ will change only slightly when Δ changes, if Δ is small and if the systems are lowpass in the sense that the system function tends to zero as $|\omega| \to +\infty$. Thus the white noise can be seen as a convenient artifice for more easily constructing this limiting output. (See Problem 7.37.)

If we take Fourier transforms of both sides of Equation 7.5-3 we obtain the *cross-power spectral density*,

$$S_{YX}(\omega) = H(\omega)S_{XX}(\omega). \tag{7.5-11}$$

Since S_{YX} is a frequency-domain representation of the cross-correlation function R_{YX}, Equation 7.5-11 tells us that $Y(t)$ and $X(t)$ will have high cross correlation at those frequencies ω where the product of $H(\omega)$ and $S_{XX}(\omega)$ is large. Similarly, from Equation 7.5-5, we can obtain

$$S_{XY}(\omega) = H^*(\omega)S_{XX}(\omega). \tag{7.5-12}$$

From the fundamental Equation 7.5-4, repeated here for convenience,

$$R_{YY}(\tau) = h(\tau) * R_{XX}(\tau) * h^*(-\tau), \tag{7.5-13}$$

we get, upon Fourier transformation, in the spectral density domain,

$$S_{YY}(\omega) = |H(\omega)|^2 S_{XX}(\omega) = G(\omega)S_{XX}(\omega) \tag{7.5-14}$$

These two equations are among the most important in the theory of stationary random processes. In particular, Equation 7.5-14 shows how the average power in the output process is composed solely as the average input power at that frequency multiplied by $|H(\omega)|^2$, the power gain of the LSI system. We can call $G(\omega) = |H(\omega)|^2$ the *psd transfer function*.

Example 7.5-5
The transfer function of an LSI system is given by

$$H(\omega) = \text{sgn}(\omega)\left(\frac{\omega}{2\pi}\right)^2 \exp\left[-j\left(\omega \cdot \frac{8}{\pi}\right)\right] W(\omega),$$

where $\text{sgn}(\cdot)$ is the algebraic sign function, and where

$$W(\omega) \triangleq \begin{cases} 1 & \text{for } |\omega| \leq 40\pi \\ 0 & \text{else.} \end{cases}$$

Let the WSS input random process have autocorrelation function,

$$R_{XX}(\tau) = \frac{5}{2}\delta(\tau) + 2.$$

Compute the average measurable power in the band 0.0 to 1.0 Hertz (single-sided). In radians, this is the double-sided range -2π to 2π. Now $S_{XX}(\omega) = \frac{5}{2} + 4\pi\delta(\omega)$. Now the psd transfer function $G(\omega) = |H(\omega)|^2 = \left(\frac{\omega}{2\pi}\right)^4 W(\omega)$. The output psd then is

$$S_{YY}(\omega) = \frac{5}{2}\left(\frac{\omega}{2\pi}\right)^4 W(\omega),$$

and the total average output power would be calculated as

$$R_{YY}(0) = \frac{1}{2\pi}\int_{-40\pi}^{+40\pi} \frac{5}{2}\left(\frac{\omega}{2\pi}\right)^4 d\omega,$$

while the power in the band $[-2\pi, +2\pi]$ is

$$P = \frac{1}{2\pi} \int_{-2\pi}^{+2\pi} \frac{5}{2} \left(\frac{\omega}{2\pi} \right)^4 d\omega$$

$$= 1 \text{ watt.}$$

The following comment on Equations 7.5-3 through 7.5-14 may help you keep track of the conjugates and minus signs. Notice that the conjugate and negative argument on the impulse response, which becomes simply a conjugate in the frequency domain, arises in connection with the second factor in the correlation. The $h(\tau)$ without the conjugate or negative time argument comes from the linear operation implied by the first subscript, that is, the first factor in the correlation.

With reference to Equation 7.5-11 we see that the cross-spectral density function can be complex and hence has no positivity or conjugate symmetry properties, since those that S_{XX} has will be lost upon multiplication with an arbitrary, generally complex H. On the other hand, as shown in Equation 7.5-14, the psd of the output will share the real and nonnegative aspects of the psd of the input, since multiplication with $|H|^2$ will not change these properties. Table 7.5-2 sets forth all the above relations for easy reference.

We are now in a position to show that the psd $S(\omega)$ has a precise interpretation as a density for average power versus frequency. We will show directly that $S(\omega) \geq 0$ for all ω and that the average power in the frequency band (ω_1, ω_2) is given by the integral of $S(\omega)$ over that frequency band.

Theorem 7.5-1   Let $X(t)$ be a stationary, second-order random process with correlation function $R_{XX}(\tau)$ and power spectral density $S_{XX}(\omega)$. Then $S_{XX}(\omega) \geq 0$ and

Table 7.5-2   Input/Output Relations for Linear Systems with WSS Inputs

WSS Random Process:	Output Mean:
$Y(t) = h(t) * X(t)$	$\mu_Y = H(0)\mu_X$
Crosscorrelations:	Cross–Power Spectral Densities:
$R_{XY}(\tau) = R_{XX}(\tau) * h^*(-\tau)$	$S_{XY}(\omega) = S_{XX}(\omega)H^*(\omega)$
$R_{YX}(\tau) = h(\tau) * R_{XX}(\tau)$	$S_{YX}(\omega) = H(\omega)S_{XX}(\omega)$
$R_{YY}(\tau) = R_{YX}(\tau) * h^*(-\tau)$	$S_{YY}(\omega) = S_{YX}(\omega)H^*(\omega)$
Autocorrelation:	Power Spectral Density:
$R_{YY}(\tau) = h(\tau) * R_{XX}(\tau) * h^*(-\tau)$	$S_{YY}(\omega) = |H(\omega)|^2 S_{XX}(\omega)$
$\quad = g(\tau) * R_{XX}(\tau)$	$\quad = G(\omega)S_{XX}(\omega)$

Output Power and Variance:

$$E\{|Y(t)|^2\} = R_{YY}(0) = \frac{1}{2\pi} \int_{-\infty}^{+\infty} |H(\omega)|^2 S_{XX}(\omega) d\omega$$

$$\sigma_Y^2 = R_{YY}(0) - |\mu_Y|^2$$

for all $\omega_2 \geq \omega_1$,

$$\frac{1}{2\pi} \int_{\omega_1}^{\omega_2} S_{XX}(\omega)d\omega$$

is the average power in the frequency band (ω_1, ω_2).

Proof   Let $\omega_2 > \omega_1$ both be real numbers. Define a filter transfer function as follows:

$$H(\omega) \triangleq \begin{cases} 1, & \omega \in (\omega_1, \omega_2) \\ 0, & \text{else,} \end{cases}$$

and note that it passes signals only in the band (ω_1, ω_2). If $X(t)$ is input to this filter, the psd of the output $Y(t)$ is (by Equation 7.5-14)

$$S_{YY}(\omega) = \begin{cases} S_{XX}(\omega), & \omega \in (\omega_1, \omega_2) \\ 0, & \text{else.} \end{cases}$$

Now the output power in $Y(t)$ has average value $E[|Y(t)|^2] = R_{YY}(0)$,

$$R_{YY}(0) = \frac{1}{2\pi} \int_{-\infty}^{+\infty} S_{YY}(\omega)d\omega = \frac{1}{2\pi} \int_{\omega_1}^{\omega_2} S_{XX}(\omega)d\omega \geq 0,$$

and this holds for all $\omega_2 > \omega_1$. So by choosing $\omega_2 \simeq \omega_1$ we can conclude that $S_{XX}(\omega) \geq 0$ for all ω and that the function S_{XX} thus has the interpretation of a power density in the sense that if we integrate this function across a frequency band, we get the average power in that band.   ■

We saw earlier that the conditions that a function must meet to be a valid correlation or covariance function are rather strong. In fact, we have seen that the function must be positive semidefinite, although we have not in fact shown that this condition is sufficient. It turns out that one more advantage of working in the frequency domain is the ease with which we can specify when a given frequency function qualifies as a power spectral density. The function simply must be real and nonnegative, that is, $S(\omega) \geq 0$. We can see this for a given function $F(\omega) \geq 0$ by taking a filter with transfer function $H(\omega) = \sqrt{F(\omega)}$ and letting the input be white noise with $S_{WW} = 1$. Then by Equation 7.5-14 the output psd is $S_{XX}(\omega) = F(\omega)$, thus showing that F is a valid psd. If the random process is real-valued, as it frequently is, then we also need $F(\omega)$ to be an even function to satisfy psd property 2 listed just after Definition 7.5-1. All this can be formalized as follows.

Theorem 7.5-2   Let $F(\omega)$ be an integrable function that is real and nonnegative; that is, $F(\omega) \geq 0$ for all ω. Then there exists a stationary random process with power spectral density $S(\omega) = F(\omega)$. In particular, if the random process is real-valued then $F(\omega)$ is even.   ■

We now see that the test for a valid spectral density function is much easier than the condition of positive semidefiniteness for the correlation function. In fact, it is relatively easy to show that the positive semidefinite condition on a function is equivalent to the nonnegativity of its transform and hence that positive semidefiniteness is the sufficient condition for a function to be a valid correlation or covariance function. First, by Theorem 7.5-2 we know that the positive semidefinite condition is implied by the nonnegativity of $S(\omega)$. To show equivalence, what remains is to show that the positive semidefinite condition on a function $f(\tau)$ implies that its Fourier transform $F(\omega)$ is nonnegative. We proceed as follows: Since $f(\tau)$ is positive semidefinite we have,

$$\sum_{n=1}^{N} \sum_{m=1}^{N} a_n a_m^* f(\tau_n - \tau_m) \geq 0.$$

Also since

$$f(\tau) = \frac{1}{2\pi} \int_{-\infty}^{+\infty} F(\omega) e^{+j\omega\tau} d\omega,$$

we have

$$\frac{1}{2\pi} \sum_n \sum_m \left(a_n a_m^* \int_{-\infty}^{+\infty} F(\omega) e^{+j\omega(\tau_n - \tau_m)} \right) d\omega \geq 0,$$

which can be rewritten as

$$\frac{1}{2\pi} \int_{-\infty}^{+\infty} F(\omega) \left[\sum_n \sum_m a_n a_m^* e^{+j\omega(\tau_n - \tau_m)} \right] d\omega = \frac{1}{2\pi} \int_{-\infty}^{+\infty} F(\omega) \left| \sum_{n=1}^{N} a_n e^{+j\omega\tau_n} \right|^2 d\omega \geq 0,$$

where we recognize the term inside the magnitude square sign as a so-called transversal or tapped delay-line filter. Thus by choosing N large enough, with the τ_n equally spaced, we can select the a_n's to arbitrarily approximate any ideal filter transfer function $H(\omega)$. Then by choosing H to be very narrow bandpass filters centered at each value of ω, we can eventually conclude that $F(\omega) \geq 0$ for all ω, $-\infty < \omega < +\infty$. We have thereby established the following theorem.

Theorem 7.5-3   A necessary and sufficient condition for $f(\tau)$ to be a correlation function is that it be positive semidefinite.   ■

Incidentally, there is an analogy here for probability density functions, which can be regarded as the Fourier transforms of their characteristic functions. As we know, nonnegativity is the sufficient condition for a function to be a valid probability density function (assuming that it is normalized to integrate to one); thus the probability density is analogous to the power spectral density; and in fact one can consider a spectral distribution function [7-10] analogous to the probability distribution function. Thus the characteristic function and the correlation function are also analogous and so both must be positive semidefinite to be valid for their respective roles. Also for the characteristic function the normalization of the probability density to integrate to one imposes the condition $\Phi(0) = 1$, which is easily met by scaling an arbitrary positive semidefinite function that is not identically zero.

Stationary Processes and Differential Equations

We shall now examine stochastic differential equations with a stationary or at least WSS input, and also with the linear constant-coefficient differential equation (LCCDE) holding for all time. We assume that the equation is stable in the bounded-input, bounded-output (BIBO) sense, so that the resulting output process is also stationary (or WSS if that is the condition on the input process).

Thus consider the following general LCCDE:

$$a_N Y^{(N)}(t) + a_{N-1} Y^{(N-1)}(t) + \ldots + a_0 Y(t)$$
$$= b_M X^{(M)}(t) + b_{M-1} X^{(M-1)}(t) + \ldots + b_0 X(t), \qquad -\infty < t < +\infty.$$

This represents the relationship between output $Y(t)$ and input $X(t)$ in a linear system with frequency response

$$H(\omega) = B(\omega)/A(\omega), \quad \text{with } a_0 \neq 0,$$

where

$$B(\omega) \triangleq \sum_{m=0}^{M} b_m (j\omega)^m$$

and

$$A(\omega) \triangleq \sum_{n=0}^{N} a_n (j\omega)^n,$$

which is a *rational function* with *numerator polynomial* $B(\omega)$ and *denominator polynomial* $A(\omega)$. Because the system is stable, we can apply the results of the previous section to obtain

$$\mu_Y = \mu_X H(0)$$
$$S_{YX}(\omega) = H(\omega) S_{XX}(\omega),$$

and

$$S_{YY}(\omega) = |H(\omega)|^2 S_{XX}(\omega),$$

where

$$H(0) = b_0/a_0 \quad \text{and} \quad |H(\omega)|^2 = |B(\omega)|^2/|A(\omega)|^2.$$

So

$$\mu_Y = (b_0/a_0)\,\mu_X \quad \text{and} \quad S_{YY}(\omega) = \left(|B(\omega)|^2/|A(\omega)|^2\right) S_{XX}(\omega).$$

This frequency-domain analysis method is generally preferable to the time-domain approach but is restricted to the case where both the input and output processes are at least WSS. After we obtain the various spectral densities, then we can use the IFT to obtain the correlation and covariance functions if they are desired. The calculation of the required IFTs is often easier if viewed as an inverse two-sided *Laplace transform*. The Laplace transform of Equation 7.5-3 is

$$S_{YX}(s) = H(s)S_{XX}(s) \tag{7.5-15}$$

while the Laplace transform of Equation 7.5-13 is written

$$S_{YY}(s) = H(s)H(-s)S_{XX}(s) \tag{7.5-16}$$

in light of $h^*(-\tau) \leftrightarrow H(-s)$. Recalling the definition of the two-sided Laplace transform [7-2], for any $f(\tau)$

$$F(s) \triangleq \int_{-\infty}^{+\infty} f(\tau)e^{-s\tau}d\tau,$$

we note that such a function of the complex variable s may be obtained from the Fourier transform $F(\omega)$, a function of the real variable ω, by a two-step procedure. First set

$$F(s)_{s=j\omega} \triangleq F(\omega)$$

and then replace $j\omega$ by s. An analogous extention method was used earlier for the discrete-time case in Chapter 6 where the Fourier transform was extended to the entire complex plane by the Z-transform.

Example 7.5-6

Consider the first-order differential equation

$$Y'(t) + \alpha Y(t) = X(t), \qquad \alpha > 0,$$

with stationary input $X(t)$ with mean $\mu_X = 0$ and impulse covariance function $K_{XX}(\tau) = \delta(\tau)$. The system function is easily seen to be

$$H(\omega) = \frac{1}{\alpha + j\omega},$$

and the psd of the input process is

$$S_{XX}(\omega) = 1,$$

so we have the following cross- and output-power spectral densities:

$$S_{YX}(\omega) = H(\omega)S_{XX}(\omega) = \frac{1}{\alpha + j\omega},$$

$$S_{YY}(\omega) = |H(\omega)|^2 S_{XX}(\omega) = \frac{1}{|\alpha + j\omega|^2} = \frac{1}{\alpha^2 + \omega^2}.$$

We now convert to Laplace transforms, with $s = j\omega$,

$$S_{YY}(j\omega) = \frac{1}{\left(\alpha^2 - (j\omega)^2\right)}$$

$$= \frac{1}{(a + j\omega)(a - j\omega)}$$

so that

$$S_{YY}(s) = \frac{1}{(s+a)(-s+a)}.$$

Using the residue method (or partial fraction expansions), one can then directly obtain the following output correlation function by inverse Laplace transform:

$$R_{YY}(\tau) = \frac{1}{2\alpha} \exp(-\alpha|\tau|), \qquad -\infty < \tau < +\infty,$$

which is also the output covariance function since $\mu_Y = 0$. By the above equation for $S_{YX}(\omega)$ we also obtain the cross-correlation function $R_{YX}(\tau) = \exp(-\alpha\tau)u(\tau)$.

In Example 7.5-6 it is interesting that $R_{YX}(\tau)$ is 0 for $\tau < 0$. This means that the output Y is orthogonal to all future values of the input X, which is a white noise. This occurs because of two reasons: The system is causal and the input is a white noise process. The system causality requires that the output not depend *directly* on (that is, not be a function of) future inputs but only depend directly on present and past inputs. The whiteness of the input X guarantees that the past and present inputs will be uncorrelated with future inputs. Combining both conditions we see that there will be no cross-correlation between the present output and the future inputs. If we assume additionally that the input is Gaussian, then the input process is an independent process and the output becomes independent of all future inputs. Then we can say that the causality of the system prevents the *direct dependence* of the present output on future inputs, and the independent process input prevents any *indirect dependence*. These concepts are important to the theory of Markov processes as used in estimation theory (cf. Chapter 9).

Example 7.5-7

Consider the following second-order LCCDE:

$$\frac{d^2Y(t)}{dt^2} + 3\frac{dY(t)}{dt} + 2Y(t) = 5X(t),$$

again with white noise input as in the previous example. Here the system function is

$$H(\omega) = \frac{5}{(j\omega)^2 + 3j\omega + 2} = \frac{5}{(2 - \omega^2) + j3\omega}.$$

Thus analogously to Example 7.5-6 the output psd becomes

$$S_{YY}(\omega) = \frac{25}{(2 - \omega^2)^2 + (3\omega)^2} = \frac{25}{\omega^4 + 5\omega^2 + 4}.$$

Applying the residue method to evaluate the IFT, we define the function of a complex variable $S_{YY}(s)|_{s=j\omega} \triangleq S_{YY}(\omega)$ and rewrite the right-hand side in terms of the complex variable $j\omega$ to obtain

$$S_{YY}(j\omega) = \frac{25}{(j\omega)^4 - 5(j\omega)^2 + 4}.$$

Substituting $s = j\omega$, we get

$$S_{YY}(s) = \frac{25}{s^4 - 5s^2 + 4},$$

which factors as

$$\frac{5}{(s+2)(s+1)} \cdot \frac{5}{(-s+2)(-s+1)} = \mathsf{H}(s)\mathsf{H}(-s),$$

where $\mathsf{H}(s)$ is the Laplace transform system function. Then the inverse Laplace transform yields the output correlation function

$$R_{YY}(\tau) = 25 \left[\frac{1}{6} \exp(-|\tau|) - \frac{1}{12} \exp(-2|\tau|) \right], \qquad -\infty < \tau < +\infty.$$

We leave the details of the calculation to the interested reader.

7.6 PERIODIC AND CYCLOSTATIONARY PROCESSES

Besides stationarity and its wide-sense version, two other classes of random processes are often encountered. They are periodic and cyclostationary processes and are here defined.

Definition 7.6-1   A random process $X(t)$ is *wide-sense periodic* if there is a $T > 0$ such that

$$\mu_X(t) = \mu_X(t+T) \qquad \text{for all } t$$

and

$$K_{XX}(t_1, t_2) = K_{XX}(t_1+T, t_2) = K_{XX}(t_1, t_2+T) \qquad \text{for all } t_1, t_2.$$

The smallest such T is called the *period*. Note that $K_{XX}(t_1, t_2)$ is then periodic with period T along both axes.   ■

An example of a wide-sense periodic random process is the random complex exponential of Example 7.4-1. In fact, the random Fourier series representation of the process:

$$X(t) = \sum_{k=1}^{\infty} A_k \exp\left(j\frac{2\pi kt}{T} \right) \tag{7.6-1}$$

with random variable coefficients A_k, would also be wide-sense periodic. A wide-sense periodic process can also be WSS, in which case we call it *wide-sense periodic stationary*. We will consider these processes further in Chapter 8, where we also refer to them as *mean-square periodic*. The covariance function of a wide-sense periodic process is generically sketched in Figure 7.6-1. We see that $K_{XX}(t_1, t_2)$ is doubly periodic with a two-dimensional period of (T, T). In Chapter 8 we will see that the sample functions of a wide-sense periodic random process are periodic with probability-1, that is,

$$X(t) = X(t+T) \quad \text{for all } t,$$

except for a set of outcomes, i.e., an event, of probability zero.

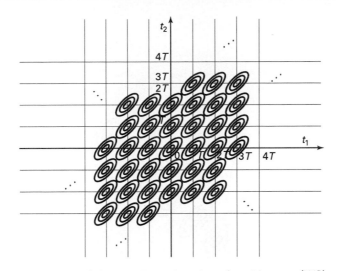

Figure 7.6-1 Possible contours of the covariance function of a wide-sense (WS) periodic random process.

A final classification that we mention here is *cyclostationarity*. It is only partially related to periodicity and is often confused with it. The reader should carefully note the difference in the following definition. Roughly speaking, cyclostationary processes have *statistics* that are periodic, while periodic processes have *sample functions* that are periodic.

Definition 7.6-2   A random process $X(t)$ is wide-sense cyclostationary if there exists a positive value T such that

$$\mu_X(t) = \mu_X(t + T) \quad \text{for all } t$$

and

$$K_{XX}(t_1, t_2) = K_{XX}(t_1 + T, t_2 + T) \quad \text{for all } t_1 \text{ and } t_2. \quad \blacksquare$$

An example of cyclostationarity is the random PSK process of Equation 7.2-11. Its mean function is zero and hence trivially periodic. Its covariance function (Equation 7.2-13) is invariant to a shift by T in *both* its arguments. Note that Equation 7.2-13 is *not* doubly–periodic since $R_{XX}(0, T) = 0 \neq R_{XX}(0,0)$. Also note that the sample functions of $X(t)$ are *not periodic* in any sense.

The constant-value contours of the covariance function of a typical cyclostationary random process are shown in Figure 7.6-2. Note the difference between this configuration and that of a periodic random process, as shown in Figure 7.6-1. Effectively, cyclostationarity means that the statistics are periodic, but the process itself is not periodic.

By averaging along 45° lines (i.e., $t_1 = t_2$), we can get the wide-sense stationary (WSS) versions of both types of processes. The contours of constant density of the periodic process then become the straight lines of the WSS periodic process shown in Figure 7.6-3. The WSS version of a cyclostationary process just becomes an ordinary WSS process, because of the lack of any periodic structure along 135° (anti-diagonal) lines (i.e., $t_1 = -t_2$).

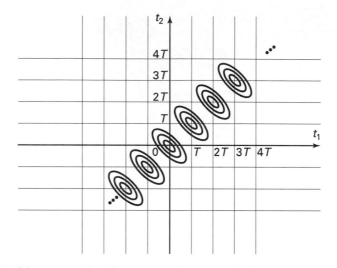

Figure 7.6-2   Possible contour plot of covariance function of WS cyclostationary random process.

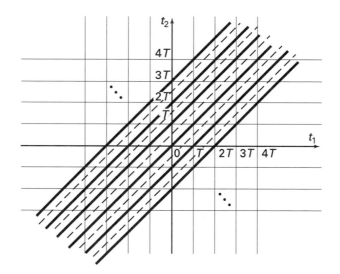

Figure 7.6-3   Possible contour plot of covariance function of WSS periodic random process. (Solid lines are maxima; dashed lines are minima.)

In addition to modulators, scanning sensors tend to produce cyclostationary processes. For example, the line-by-line scanning in television transforms the random image field into a one-dimensional random process that has been modeled as cyclostationary. In communications, cyclostationarity often arises due to waveform repetition at the baud or symbol rate.

A place where cyclostationarity arises in signal processing is when a stationary random sequence is analyzed by a filter bank and subsampled. The subsequent filter bank synthesis involves upsampling and reconstruction filters. If the *subsampling* period is N, then the resulting synthesized random sequence will be cyclostationary with period N. When perfect reconstruction filters are used, then true stationarity be achieved for the synthesized output.

While cyclostationary processes are not stationary or WSS except in trivial cases, it is sometimes appropriate to convert a cyclostationary process into a stationary process as in the following example.

Example 7.6-1

(Wide-sense stationary PSK.) We have seen that the PSK process of Section 7.2 is cyclostationary and hence not wide-sense stationary. This is easily seen with reference to Equation 7.2-13. This cyclostationarity arises from the fact that the analog angle process $\Theta_a(t)$ is stepwise constant and changes only at $t = nT$ for integer n. In many real situations the modulation process starts at an arbitrary time t, which in fact can be modeled as random from the viewpoint of the system designer. Thus in this practical case, the modulated signal process (Equation 7.2-11) is converted to

$$\tilde{X}(t) = \cos\left(2\pi f_c t + \Theta_a(t) + 2\pi f_c T_0\right), \tag{7.6-2}$$

by the addition of a random variable T_0, which is uniformly distributed on $[0, T]$ and independent of the angle process $\Theta_a(t)$. It is then easy to see that the mean and covariance functions need only to be modified by an ensemble average over T_0, which by the uniformity of T_0, is just an integral over $[0, T]$. We thus obtain

$$R_{\tilde{X}}(t_1 + \tau, t_1) = \frac{1}{T} \int_0^T R_X(t_1 + \tau + t, t_1 + t)dt$$

$$= \frac{1}{T} \int_{-\infty}^{+\infty} s_Q(t_1 + t + \tau)s_Q(t_1 + t)dt$$

$$= \frac{1}{T} s_Q(\tau) * s_Q(-\tau), \tag{7.6-3}$$

which is just a function of the shift τ. Thus $\tilde{X}(t)$ is a WSS random process.

Example 7.6-2

(Power spectral density of PSK.) A wide-sense stationary version of the random PSK signal was defined in Example 7.6-1 through an averaging process, where the average was taken over the message time or baud interval T. The resulting WSS random process $\tilde{X}(t)$ had correlation function (Equation 7.6-3) given as

$$R_{\tilde{X}\tilde{X}}(\tau) = \frac{1}{T} s_Q(\tau) * s_Q(-\tau),$$

where $s_Q(\tau)$ was given as

$$s_Q(\tau) = \begin{cases} \sin(2\pi f_c \tau), & 0 \le \tau \le T, \\ 0, & \text{else.} \end{cases}$$

Then the psd of this wide-sense stationary version of PSK can be calculated as

$$S_{\tilde{X}\tilde{X}}(\omega) = FT\{R_{\tilde{X}\tilde{X}}(\tau)\}$$

$$= \frac{1}{T}|FT\{s_Q(\tau)\}|^2$$

$$\approx (T/4)\left\{ \left(\frac{\sin(\omega - 2\pi f_c)\frac{T}{2}}{(\omega - 2\pi f_c)\frac{T}{2}} \right)^2 + \left(\frac{\sin(\omega + 2\pi f_c)\frac{T}{2}}{(\omega + 2\pi f_c)\frac{T}{2}} \right)^2 \right\},$$

for $f_c T >> 1$, \hfill (7.6-4)

which can be plotted using MATLAB. The file `psd_PSK.m` included on this book's Web site.[†]

Some plots were made using `psd_PSK.m`, for two different sets of values for f_c and T. First we look at the psd plot in Figure 7.6-4 for $f_c = 2.5$ and $T = 0.5$, which gives considerable overlap of the positive and negative frequency lobes of $S_{\tilde{X}\tilde{X}}(\omega)$. The lack of power concentration at the carrier frequency f_c is not surprising, since there is little over

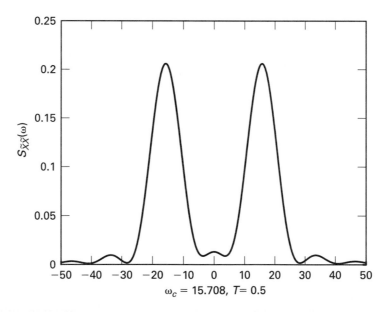

Figure 7.6-4   Power spectral density of PSK plotted for $f_c = 2.5$ and $T = 0.5$.

[†]The reason for the approximate equals sign is that we have neglected the cross-term in Eq. 7.6-4 between the two sinc terms at $\pm f_c$, as is appropriate for $f_c T >> 1$.

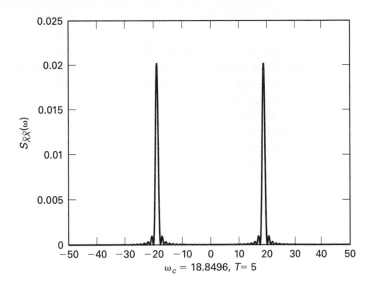

Figure 7.6-5   Power spectral density of PSK plotted for $f_c = 3$ and $T = 5$.

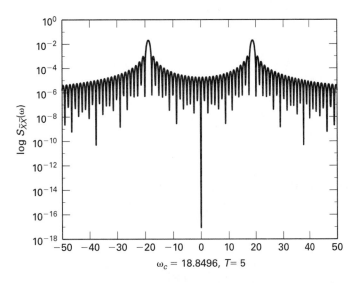

Figure 7.6-6   Log of power spectral density of PSK plotted for $f_c = 3$ and $T = 5$.

one period of $s_Q(t)$ in the baud interval T. The next pair of plots show a quite different case with power strongly concentrated at ω_c. This plot was computed with the values $f_c = 3.0$ and $T = 5.0$, thus giving 15 periods of the sine wave in the baud interval T. Figure 7.6-5 is a linear plot, while Figure 7.6-6 shows $S_{\tilde{X}\tilde{X}}(\omega)$ on a logarithmic scale.

7.7 VECTOR PROCESSES AND STATE EQUATIONS

In this section we will generalize some of the results of Section 7.5 to the important class of vector random processes. This will lead into a brief discussion of state equations and vector Markov processes. Vector random processes occur in two-channel systems that are used in communications to model the in-phase and quadrature components of bandpass signals. Vector processes are also used extensively in control systems to model plants with several inputs and outputs. Also, vector models are created artificially from high-order scalar models in order to employ the useful concept of *state* in control theory.

Let $X_1(t)$ and $X_2(t)$ be two jointly stationary random processes that are input to the systems H_1 and H_2, respectively. Call the outputs Y_1 and Y_2, as shown in Figure 7.7-1.

From earlier discussions we know how to calculate $R_{X_1 Y_1}$, $R_{X_2 Y_2}$, $R_{Y_1 Y_1}$, $R_{Y_2 Y_2}$. We now look at how to calculate the correlations *across* the systems, that is, $R_{X_1 Y_2}$, $R_{X_2 Y_1}$ and $R_{Y_1 Y_2}$. Given $R_{X_1 X_2}$, we first calculate,

$$R_{X_1 Y_2}(\tau) = E[X_1(t+\tau)Y_2^*(t)]$$

$$= \int_{-\infty}^{+\infty} E[X_1(t+\tau)X_2^*(t-\beta)]h_2^*(\beta)d\beta$$

$$= \int_{-\infty}^{+\infty} R_{X_1 X_2}(\tau+\beta)h_2^*(\beta)d\beta$$

$$= \int_{-\infty}^{+\infty} R_{X_1 X_2}(\tau-\beta')h_2^*(-\beta')d\beta', \qquad (\beta' = -\beta),$$

so

$$R_{X_1 Y_2}(\tau) = R_{X_1 X_2}(\tau) * h_2^*(-\tau),$$

and by symmetry

$$R_{X_2 Y_1}(\tau) = R_{X_2 X_1}(\tau) * h_1^*(-\tau).$$

The cross-correlation at the outputs is

$$R_{Y_1 Y_2}(\tau) = h_1(\tau) * R_{X_1 X_2}(\tau) * h_2^*(-\tau).$$

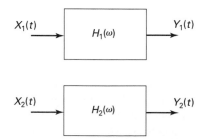

Figure 7.7-1   A generic (uncoupled) two-channel LSI system.

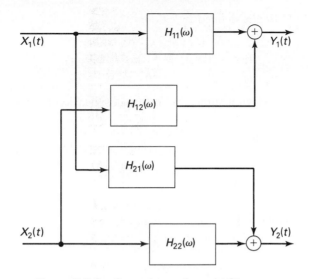

Figure 7.7-2   General two-channel LSI system.

Expressing these results in the spectral domain, we have

$$S_{X_1 Y_2}(\omega) = S_{X_1 X_2}(\omega) H_2^*(\omega)$$

and

$$S_{Y_1 Y_2}(\omega) = H_1(\omega) H_2^*(\omega) S_{X_1 X_2}(\omega).$$

In passing, we note the following important fact: If the *supports*[†] of the two system functions H_1 and H_2 do not overlap, then Y_1 and Y_2 are orthogonal random processes independent of any correlation in the input processes. We can generalize the above to a two-channel system with internal coupling as seen in Figure 7.7-2. Here two additional system functions have been added to cross-couple the inputs and outputs. They are denoted by H_{12} and H_{21}.

This case is best treated with vector notation; thus we define

$$\mathbf{X}(t) \triangleq [X_1(t), X_2(t)]^T, \qquad \mathbf{Y}(t) \triangleq [Y_1(t), Y_2(t)]^T,$$

and

$$\mathbf{h}(t) \triangleq \begin{bmatrix} h_{11}(t) & h_{12}(t) \\ h_{21}(t) & h_{22}(t) \end{bmatrix},$$

where $h_{ij}(t)$ is the impulse response of the subsystem with frequency response $H_{ij}(\omega)$. We then have

$$\mathbf{Y}(t) = \mathbf{h}(t) * \mathbf{X}(t), \tag{7.7-1}$$

[†]We recall that the support of a function g is defined as

$$\text{supp}(g) \triangleq \{x | g(x) \neq 0\}.$$

where the *vector convolution* is defined by

$$(\mathbf{h}(t) * \mathbf{X}(t))_i \triangleq \sum_{j=1}^{N} h_{ij}(t) * X_j(t).$$

If we define the following relevant input and output correlation matrices

$$\mathbf{R_{XX}}(\tau) \triangleq \begin{bmatrix} R_{X_1 X_1}(\tau) & R_{X_1 X_2}(\tau) \\ R_{X_2 X_1}(\tau) & R_{X_2 X_2}(\tau) \end{bmatrix} \tag{7.7-2}$$

$$\mathbf{R_{YY}}(\tau) \triangleq \begin{bmatrix} R_{Y_1 Y_1}(\tau) & R_{Y_1 Y_2}(\tau) \\ R_{Y_2 Y_1}(\tau) & R_{Y_2 Y_2}(\tau) \end{bmatrix}, \tag{7.7-3}$$

one can show that (Problem 7.44)

$$\mathbf{R_{YY}}(\tau) = \mathbf{h}(\tau) * \mathbf{R_{XX}}(\tau) * \mathbf{h}^\dagger(-\tau), \tag{7.7-4}$$

where the † indicates the Hermitian (or conjugate) transpose.

Taking the matrix Fourier transformation, we obtain

$$\mathbf{S_{YY}}(\omega) = \mathbf{H}(\omega)\mathbf{S_{XX}}(\omega)\mathbf{H}^\dagger(\omega) \tag{7.7-5}$$

with

$$\mathbf{H}(\omega) = FT\{\mathbf{h}(t)\},$$

and

$$\mathbf{S}(\omega) = FT\{\mathbf{R}(\tau)\},$$

where this notation indicates an *element-by-element Fourier transform*. This multichannel generalization clearly extends to the M input and N output case by just enlarging the matrix dimensions accordingly.

State Equations

As shown in Problem 7.43, it is possible to rewrite an Nth order LCCDE in the form of a first-order vector differential equation where the dimension of the output vector is equal to N,

$$\dot{\mathbf{Y}}(t) = \mathbf{A}\mathbf{Y}(t) + \mathbf{B}\mathbf{X}(t), \qquad -\infty < t < +\infty. \tag{7.7-6}$$

This is just a multichannel system as seen in Equation 7.7-1 and can be interpreted as a set of N coupled first-order LCCDEs. We can take the vector Fourier transform and calculate the system function

$$\mathbf{H}(\omega) = (j\omega\mathbf{I} - \mathbf{A})^{-1}\mathbf{B} \tag{7.7-7}$$

to specify this LSI operation in the frequency domain. Here $\mathbf{I}$ is the identity matrix. Alternately, we can express the operation in terms of a matrix convolution

$$\mathbf{Y}(t) = \mathbf{h}(t) * \mathbf{X}(t),$$

where we assume the multichannel system is stable; that is, all the impulse responses h_{ij} are BIBO stable. The solution proceeds much the same as in the scalar case for the first-order equation; in fact, it can be shown that

$$\mathbf{h}(t) = \exp(\mathbf{A}t)\,\mathbf{B}u(t). \tag{7.7-8}$$

The matrix exponential function $\exp(\mathbf{A}t)$ was encountered earlier in this chapter in the solution of the probability vector for a continuous-time Markov chain. This function is widely used in linear system theory, where its properties have been studied extensively [7-2].

If we compute the cross-correlation matrices in the WSS case, we obtain

$$\mathbf{R_{YX}}(\tau) = \exp(\mathbf{A}\tau)\,\mathbf{B}u(\tau) * \mathbf{R_{XX}}(\tau)$$

and

$$\mathbf{R_{XY}}(\tau) = \mathbf{R_{XX}}(\tau) * \mathbf{B}^\dagger \exp(-\mathbf{A}^\dagger\tau)u(-\tau),$$

with output correlation matrix, as before,

$$\mathbf{R_{YY}}(\tau) = \mathbf{h}(\tau) * \mathbf{R_{XX}}(\tau) * \mathbf{h}^\dagger(-\tau). \tag{7.7-9}$$

Upon vector Fourier transformation this becomes

$$\mathbf{S_{YY}}(\omega) = (j\omega\mathbf{I} - \mathbf{A})^{-1}\mathbf{B}\mathbf{S_{XX}}(\omega)\mathbf{B}^\dagger(-j\omega\mathbf{I} - \mathbf{A}^\dagger)^{-1}. \tag{7.7-10}$$

If $\mathbf{R_{XX}}(\tau) = \mathbf{Q}\delta(\tau)$, then since the system $\mathbf{H}$ is assumed causal, that is, $\mathbf{h}(t) = \mathbf{0}$ for $t < 0$, we have that the cross-correlation matrix $\mathbf{R_{YX}}(\tau) = \mathbf{0}$ for $\tau < 0$; that is, $E[\mathbf{Y}(t+\tau)\mathbf{X}^\dagger(t)] = \mathbf{0}$ for $\tau < 0$. In words we say that $\mathbf{Y}(t + \tau)$ is orthogonal to $\mathbf{X}(t)$ for $\tau < 0$. Thus, the past of $\mathbf{Y}(t)$ is orthogonal to the present and future of $\mathbf{X}(t)$. If we additionally assume that the input process $\mathbf{X}(t)$ is a Gaussian process, then the uncorrelatedness condition becomes an independence condition. Under the Gaussian assumption then, the output $\mathbf{Y}(t)$ is independent of the present and future of $\mathbf{X}(t)$. A similar result was noted earlier in the scalar-valued case. We can use this result to show that the solution to a first-order vector LCCDE is a vector Markov random process with the following definition.

Definition 7.7-1 A random process $\mathbf{Y}(t)$ is *vector Markov* if for all $n > 0$ and for all $t_n > t_{n-1} > \ldots > t_1$, and for all values $\mathbf{y}(t_{n-1}), \ldots, \mathbf{y}(t_1)$, we have

$$P[\mathbf{Y}(t_n) \le \mathbf{y}_n | \mathbf{y}(t_{n-1}), \ldots, \mathbf{y}(t_1)] = P[\mathbf{Y}(t_n) \le \mathbf{y}_n | \mathbf{y}(t_{n-1})]$$

for all values of the real vector $\mathbf{y}_n$. Here $\mathbf{A} \le \mathbf{a}$ means

$$(A_n \le a_n, A_{n-1} \le a_{n-1}, \ldots, A_1 \le a_1). \quad \blacksquare$$

Before discussing vector differential equations we briefly recall a result for *deterministic* vector LCCDEs. The first-order vector equation,

$$\dot{\mathbf{y}}(t) = \mathbf{A}\mathbf{y}(t) + \mathbf{B}\mathbf{x}(t), \qquad t \ge t_0,$$

subject to the initial condition $\mathbf{y}(t_0)$, can be shown to have solution, employing the matrix exponential

$$\mathbf{y}(t) = \exp[\mathbf{A}(t - t_0)]\mathbf{y}(t_0) + \int_{t_0}^{t} \mathbf{h}(t - v)\mathbf{x}(v)dv, \qquad t \geq t_0,$$

thus generalizing the scalar case. This deterministic solution can be found in any graduate text on linear systems theory, for example in Reference [7-2]. The first term is called the *zero-input solution* and the second term is called the *zero-state* (or *driven*) *solution* analogously to the solution for scalar LCCDEs.

We can extend this theory to the stochastic case by considering the differential Equation 7.7-6 over the semi-infinite domain $t_0 \leq t < \infty$ and replacing the above deterministic solution with the following stochastic solution, expressed with the help of an integral:

$$\mathbf{Y}(t) = \exp[\mathbf{A}(t - t_0)]\mathbf{Y}(t_0) + \int_{t_0}^{t} \mathbf{h}(t - v)\mathbf{X}(v)dv. \qquad (7.7\text{-}11)$$

If the LCCDE is BIBO stable, i.e., the real parts of the eigenvalues of $\mathbf{A}$ are all negative, in the limit as $t_0 \to -\infty$, we get the solution for all time, that is $t_0 = -\infty$,

$$\mathbf{Y}(t) = \int_{-\infty}^{t} \mathbf{h}(t - v)\mathbf{X}(v)dv = \mathbf{h}(t) * \mathbf{X}(t), \qquad (7.7\text{-}12)$$

which is the same as already derived for the stationary infinite time-interval case. In effect, we use the stability of the system to conclude that the resulting zero-input part of the solution must be zero at any finite time.

The following theorem shows a method to generate a vector Gauss–Markov random process using the above approach. The input is now a white Gaussian vector process $\mathbf{W}(t)$ and the output vector Markov process is denoted $\mathbf{X}(t)$.

Theorem 7.7-1   Let the input to the state equation

$$\dot{\mathbf{X}}(t) = \mathbf{A}\mathbf{X}(t) + \mathbf{B}\mathbf{W}(t)$$

be the white Gaussian process $\mathbf{W}(t)$. Then the output $\mathbf{X}(t)$ is a vector Gauss-Markov random process.

Proof   We write the solution at t_n in terms of the solution at an earlier time t_{n-1} as

$$\mathbf{X}(t_n) = \exp[\mathbf{A}(t_n - t_{n-1})]\mathbf{X}(t_{n-1}) + \int_{t_{n-1}}^{t_n} \mathbf{h}(t_n - v)\mathbf{W}(v)dv.$$

Then we write the integral term as $\mathbf{I}(t_n)$ and note that it is independent of $\mathbf{X}(t_{n-1})$. Thus we can deduce that

$$P[\mathbf{X}(t_n) \leq \mathbf{x}_n | \mathbf{x}(t_{n-1}), \ldots, \mathbf{x}(t_1)]$$

$$= P[\mathbf{I}(t_n) \leq \mathbf{x}_n - e^{\mathbf{A}(t_n - t_{n-1})}\mathbf{x}(t_{n-1}) | \mathbf{x}(t_{n-1}), \ldots, \mathbf{x}(t_1)]$$

$$= P[\mathbf{I}(t_n) \leq \mathbf{x}_n - e^{\mathbf{A}(t_n - t_{n-1})}\mathbf{x}(t_{n-1}) | \mathbf{x}(t_{n-1})]$$

and hence that $\mathbf{X}(t)$ is a vector Markov process.   ■

If in Theorem 7.7-1 we did not have the Gaussian condition on the input $\mathbf{W}(t)$ but just the white noise condition, then we could not conclude that the output was Markov. This is because we would not have the independence condition required in the proof but only the weaker uncorrelatedness condition. On the other hand, if we relax the Gaussian condition but require that the input $\mathbf{W}(t)$ be an independent random process, then the process $\mathbf{X}(\mathbf{t})$ would still be Markov. We use $\mathbf{X}$ for the process in this theorem rather than $\mathbf{Y}$ to highlight the fact that LCCDEs are often used to model input processes too.

7.8 SUMMARY

In this chapter we introduced the concept of the random process, an ensemble of functions of a continuous parameter. The parameter is most often time but can be position or another continuous variable. Most topics in this chapter generalize to two- and three-dimensional parameters. Many modern applications, in fact, require a two-dimensional parameter, for example, the intensity function $i(t_1, t_2)$ of an image. Such random functions are called *random fields* and can be analyzed using extensions of the methods of this chapter. See Reference [7-6] or Chapter 7 in Reference [7-5].

We introduced a number of important processes: asynchronous binary signaling; the Poisson counting process; the random telegraph signal; phase-shift keying, which is basic to digital communications; the Wiener process, our first example of a Gaussian random process and a basic building block process in nonlinear filter theory; and the Markov process, which is widely used for its efficiency and tractability and is the signal model in the Kalman–Bucy filter of Chapter 9.

We considered the effect of linear systems on the second-order properties of random processes. We specialized our results to the useful subcategory of stationary and WSS processes and introduced the power spectral density and the corresponding analysis for LSI systems. We also briefly considered the classes of wide-sense periodic and cyclostationary processes and introduced random vector processes and systems and extended the Markov model to them.

PROBLEMS

7.1. Let $X[n]$ be a real valued stationary random sequence with mean $E\{X[n]\} = \mu_X$ and autocorrelation function $E\{X[n+m]X[n]\} = R_{XX}[m]$. If $X[n]$ is the input to a D/A converter, the continuous-time output can be idealized as the *analog* random process $X_a(t)$ with

$$X_a(t) \triangleq X[n] \quad \text{for } n \leq t < n+1, \quad \text{for all } n,$$

as shown in Figure P7.1.

    (a) Find the mean $E[X_a(t)] = \mu_a(t)$ as a function of μ_X.
    (b) Find the correlation $E[X_a(t_1)X_a(t_2)] = R_{X_a X_a}(t_1, t_2)$ in terms of $R_{XX}[m]$.

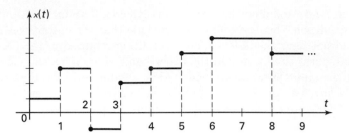

Figure P7.1   Typical output of sample-hold D/A converter.

7.2.   Consider a WSS random sequence $X[n]$ with mean function μ_X, a constant, and correlation function $R_{XX}[m]$. Form a random process as

$$X(t) \triangleq \sum_{n=-\infty}^{+\infty} X[n]\frac{\sin \pi(t-nT)/T}{\pi(t-nT)/T} \quad -\infty < t < +\infty.$$

In what follows, we assume the infinite sums converge and so, do not worry about stochastic convergence issues.

    (a) Find $\mu_X(t)$ in terms of μ_X. Simplify your answer as much as possible.

    (b) Find $R_{XX}(t_1,t_2)$ in terms of $R_{XX}[m]$. Is $X(t)$ WSS?

Hint: The sampling theorem from *Linear Systems Theory* states that any bandlimited deterministic function $g(t)$ can be recovered exactly from its evenly spaced samples, i.e.,

$$g(t) = \sum_{n=-\infty}^{+\infty} g(nT)\frac{\sin \pi(t-nT)/T}{\pi(t-nT)/T},$$

when the radian bandwidth of the function $g(t)$ is π/T or less.

7.3.   Let $B[n]$ be a Bernoulli random sequence equally likely taking on values ± 1. Then define the random process,

$$X(t) \triangleq \sqrt{p}\sin\left(2\pi f_0 t + B[n]\frac{\pi}{2}\right) \quad \text{for } nT \le t < (n+1)T, \quad \text{for all } n,$$

where $\sqrt{p}$ and f_0 are real numbers.

    (a) Determine the mean function $\mu_X(t)$.

    (b) Determine the covariance function $K_{XX}(t_1,t_2)$.

7.4.   The output $Y(t)$ of a tapped delay line filter shown in Figure P7.4, with input $X(t)$ and N taps, is given by

$$Y(t) = \sum_{n=0}^{N-1} A_n X(t-nT).$$

The input $X(t)$ is a stationary Gaussian random process with zero-mean and autocorrelation function $R_{XX}(\tau)$ having the property that $R_{XX}(nT) = 0$ for every integer

$n \neq 0$. The tap gains $A_n, n = 0, 1, \ldots, N - 1$, are zero-mean, uncorrelated Gaussian random variables with common variance σ_A^2. Every tap gain is independent of the input process $X(t)$.

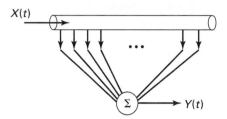

Figure P7.4   Tapped delay line filter.

(a) Find the autocorrelation function of $Y(t)$.

(b) For a given value of t, find the characteristic function of $Y(t)$. Justify your steps.

(c) For fixed t, what is the asymptotic PDF of $\frac{1}{\sqrt{N}} Y(t)$, asymptotic as $N \to \infty$? Explain.

(d) Suppose now that the number of taps N is a Poisson random variable with mean $\lambda(> 0)$. Find the answers to parts (a) and (b) now.

(*Note*: You may need to use the following: $e^{-x} \approx \frac{1}{1+x}$ for $|x| << 1$, and $e^x = \sum_{n=0}^{\infty} \frac{x^n}{n!}$.)

7.5. Let $N(t)$ be a Poisson random process defined on $0 \leq t < \infty$ with $N(0) = 0$ and mean arrival rate $\lambda = 0$.

(a) Find the joint probability $P[N(t_1) = n_1, N(t_2) = n_2]$ for $t_2 > t_1$.

(b) Find an expression for the Kth order joint PMF,

$$P_N(n_1, \ldots, n_K; t_1, \ldots, t_K),$$

with $0 \leq t_1 < t_2 < \ldots < t_K < \infty$. Be careful to consider the relative values of $n_1, \ldots, n_K$.

7.6. The *nonuniform Poisson counting process* $N(t)$ is defined for $t \geq 0$ as follows:

(a) $N(0) = 0$.

(b) $N(t)$ has independent increments.

(c) For all $t_2 \geq t_1$,

$$P[N(t_2) - N(t_1) = n] = \frac{\left[\int_{t_1}^{t_2} \lambda(v)dv \right]^n}{n!} \exp\left(-\int_{t_1}^{t_2} \lambda(v)dv \right), \quad \text{for } n \geq 0.$$

The function $\lambda(t)$ is called the *intensity function* and is everywhere nonnegative, that is, $\lambda(t) \geq 0$ for all t.

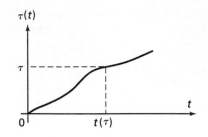

Figure P7.6   Plot of τ versus t.

    (a) Find the mean function $\mu_N(t)$ of the nonuniform Poisson process.
    (b) Find the correlation function $R_{NN}(t_1, t_2)$ of $N(t)$.
Define a *warping* of the time axis as follows:

$$\tau(t) \triangleq \int_0^t \lambda(v)\,dv.$$

Now $\tau(t)$ is monotonic increasing if $\lambda(v) > 0$ for all v, so we can then define
the inverse mapping $t(\tau)$ as shown in Figure P7.6.

    (c) Assume $\lambda(t) > 0$ for all t and define the counting process,

$$N_u(\tau) \triangleq N(t(\tau)).$$

Show that $N_u(\tau)$ is a uniform Poisson counting process with rate $\lambda = 1$;
that is, show for $\tau \geq 0$

    (1) $N_u(0) = 0$.
    (2) $N_u(\tau)$ has independent increments.
    (3) For all $\tau_2 \geq \tau_1$,

$$P[N_u(\tau_2) - N_u(\tau_1) = n] = \frac{(\tau_2 - \tau_1)^n}{n!} e^{-(\tau_2 - \tau_1)} \qquad n \geq 0.$$

7.7.   A nonuniform Poisson process $N(t)$ has intensity function (mean arrival rate)

$$\lambda(t) = 1 + 2t,$$

for $t \geq 0$. Initially $N(0) = 0$.

    (a) Find the mean function $\mu_N(t)$.
    (b) Find the correlation function $R_{NN}(t_1, t_2)$.
    (c) Find an expression for the probability that $N(t) \geq t$, that is, find $P[N(t) \geq t]$
        for any $t > 0$.
    (d) Give an approximate answer for (c) in terms of the error function $\mathrm{erf}(x)$.

7.8.   This problem concerns the construction of the Poisson counting process as given in
Section 7.2.

(a) Show the density for the nth arrival time $T[n]$ is

$$f_T(t; n) = \frac{\lambda^n t^{n-1}}{(n-1)!} e^{-\lambda t} u(t), \qquad n > 0.$$

In the derivation of the property that the increments of a Poisson process are Poisson distributed, that is,

$$P[X(t_a) - X(t_b) = n] = \frac{[\lambda(t_a - t_b)]^n}{n!} e^{-\lambda(t_a - t_b)} u[n], \qquad t_a > t_b,$$

we implicitly use the fact that the first interarrival time in $(t_b, t_a]$ is exponentially distributed. Actually, this fact is not clear as the interarrival time in question is only partially in the interval $(t_b, t_a]$. A pictorial diagram is shown in Figure P7.8.

Define $\tau'[i] \triangleq T[i] - t_b$ as the *partial interarrival time*. We note $\tau'[i] = \tau[i] - T$ where the random variable $T \triangleq t_b - T[i-1]$ and $\tau[i]$ denotes the (full) interarrival time.

(b) Fix the random variable $T = t$ and find the PDF

$$F_{\tau'[i]}(\tau'|T = t) = P\{\tau[i] \leq \tau' + t | \tau[i] \geq t\}.$$

(c) Modify the result of part (b) to account for the fact that T is a random variable, and find the unconditional PDF of τ'. (*Hint*: This part does not involve a lot of calculations.)

Because of the preceding properties, the exponential distribution is called *memoryless*. It is the only continuous distribution with this property.

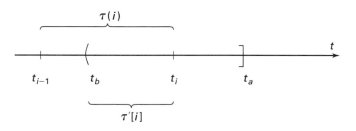

Figure P7.8 Illustrative example of relation of arrival times to arbitrary observation interval.

7.9. Let $N(t)$ be a counting process on $[0, \infty)$ whose average rate $\lambda(t)$ depends on another positive random process $S(t)$, specifically $\lambda(t) = S(t)$. We assume that $N(t)$ given $\{S(t)$ on $[0, \infty)\}$ is a *nonuniform* Poisson process. We know $\mu_S(t) = \mu_0 > 0$ and also know $K_{SS}(t_1, t_2)$.

(a) Find $\mu_N(t)$ for $t \geq 0$ in terms of μ_0.
(b) Find $\sigma_N^2(t)$ for $t \geq 0$ in terms of $K_{SS}(t_1, t_2)$.

7.10. Let the random process $K(t)$ (not a covariance!) depend on a uniform Poisson process $N(t)$, with mean arrival rate $\lambda > 0$, as follows: Starting at $t = 0$, both $N(t) = 0$ and $K(t) = 0$. When an arrival occurs in $N(t)$, an independent Bernoulli trial takes place with probability of success p, where $0 < p < 1$. On success, $K(t)$ is incremented by 1, otherwise $K(t)$ is left unchanged. This arrangement is shown in Figure P7.10. Find the first-order PMF of the discrete-valued random process $K(t)$ at time t, i.e., $P_K(k;t)$, for $t \geq 0$.

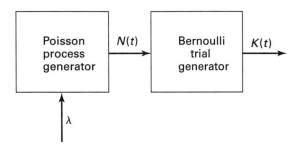

Figure P7.10   Poisson modulated Bernoulli trial process.

7.11. Let the scan-line of an image be described by the spatial random process $S(x)$, which models the ideal gray level at the point x. Let us transmit each point independently with an optical channel by modulating the intensity of a photon source:

$$\lambda(t, x) = S(x) + \lambda_0, \qquad 0 \leq t \leq T.$$

In this way we create a family of random processes, indexed by the continuous parameter x,

$$\{N(t, x)\}.$$

For each x, $N(t, x)$ given $S(x)$ is a uniform Poisson process. At the end of the observation interval, we store $N(x) \triangleq N(T, x)$ and inquire about the statistics of this spatial process.

To summarize, $N(x)$ is an integer-valued *spatial* random process that depends on the value of another random process $S(x)$, called the signal process. The spatial random process $S(x)$ is stationary with zero-mean and covariance function

$$K_{SS}(x) = \sigma_S^2 \exp(-\alpha|x|),$$

where $\alpha > 0$. The conditional distribution of $N(x)$, given $S(x)$, is Poisson with mean $\lambda(x) \triangleq (S(x) + \lambda_0)T$ where λ_0 is a positive constant; that is,

$$P[N(x) = n|S(x)] = \frac{\lambda^n(x)}{n!} e^{-\lambda(x)} u[n].$$

The random variables $N(x)$ are conditionally independent from point to point.

(a) Find the (unconditional) mean and variance

$$\mu_N(x) = E[N(x)] \quad \text{and} \quad E\left[(N(x) - \mu_N(x))^2\right].$$

(*Hint:* First find the conditional mean and conditional mean square.)

(b) Find $R_{NN}(x_1, x_2) \triangleq E[N(x_1)N(x_2)]$.

7.12. Let $X(t)$ be a random telegraph signal (RTS) defined on $t \geq 0$. Fix $X(0) = +1$. The RTS uses a Poisson random arrival time sequence $T[n]$ to switch the value of $X(t)$ between ± 1. Take the average arrival rate as $\lambda (> 0)$. Thus we have

$$X(t) \triangleq \begin{cases} 1, & 0 \leq t < T[1] \\ -1, & T[1] \leq t < T[2] \\ +1, & T[2] \leq t < T[3] \\ \cdots, & \cdots \end{cases}.$$

(a) Argue that $X(t)$ is a Markov process and draw and label the state transition diagram.
(b) Find the steady-state probability that $X(t) = +1$, i.e., $P_X(1, \infty)$, in terms of the rate parameter λ.
(c) Write the differential equations for the state probabilities $P_X(1, t)$ and $P_X(-1, t)$.

7.13. A uniform Poisson process $N(t)$ with rate $\lambda (> 0)$ is an infinite state Markov chain with the state-transition diagram in Figure P7.13a

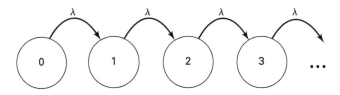

Figure P7.13a   Poisson process represented as Markov chain.

Here the state labels are the values of the process (chain) $N(t)$ between the transitions. Also the independent interarrival times $\tau[n]$ are exponentially distributed with parameter λ.

We make the following modifications to the above scenario. Replace the independent interarrival times $\tau[n]$ by an arbitrary nonnegative, stationary, and independent random sequence, still denoted $\tau[n]$, resulting in the generalization called a *renewal process* in the literature. See Figure P7.13b.

(a) Show that the PMF $P_N(n; t) = P[N(t) = n]$ of a renewal process is given, in terms of the PDF of the independent arrival times $F_T(t; n)$, as

$$P_N(n; t) = F_T(t; n) - F_T(t; n + 1), \quad \text{when } n \geq 1,$$

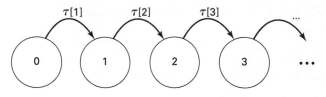

Figure P7.13b   More general (renewal) process chain.

where the arrival time $T[n] = \sum_{k=1}^{n} \tau[k]$ and $F_T(t; n)$ is the corresponding PDF of the arrival time $T[n]$.

(b) Let $\tau[n]$ be $U[0, 1]$, i.e., uniformly distributed over $[0, 1]$, and find $P_N(n; t)$ for $n = 0, 1,$ and 2, for this specific renewal process.

(c) Find the characteristic function of the renewal process of part (b).

(d) Find an approximate expression for the PDF $F_T(t; n)$ of the renewal process in part (b), that is good for large n, and not too far from the $T[n]$ mean value. (*Hint*: For small x we have the trigonometric series approximation $\sin x \approx x - x^3/3!$)

7.14. Let $W(t)$ be a *standard* Wiener process, defined over $[0, \infty)$ (that is, distributed as $N(0, t)$ at time t). Find the joint density $f_W(a_1, a_2; t_1, t_2)$ for $0 < t_1 < t_2$.

7.15. Let $W_1(t)$ and $W_2(t)$ be two Wiener processes, independent of one another, both defined on $t \geq 0$, with variance parameters α_1 and α_2, respectively. Let the process $X(t)$ be defined as their algebraic difference, i.e., $X(t) \triangleq W_1(t) - W_2(t)$.

   (a) What is $R_{XX}(t_1, t_2)$ for $t_1, t_2 \geq 0$?

   (b) What is the pdf $f_X(x; t)$ for $t \geq 0$?

7.16. Let the random process $X(t)$ have mean function $\mu_X(t) = 4$ and covariance function $K_{XX}(t_1, t_2) = 5[\min(t_1, t_2)]^2$. Let the derivative process be denoted at $Y(t) = X'(t)$ for $t \geq 0$.

   (a) Find the mean function of $Y(t)$.

   (b) Find the correlation function $R_{YY}(t_1, t_2)$.

   (c) Is the derivative process $Y(t)$ wide-sense stationary (WSS)?

   (d) Extra: Show that the above $X(t)$ process actually exists by constructing it from standard Wiener process(s).

7.17. Let $W(t)$ be a standard Wiener process, i.e., $\alpha = 1$, and define

$$X(t) \triangleq W^2(t) \quad \text{for} \quad t \geq 0.$$

   (a) Find the probability density $f_X(x; t)$.

   (b) Find the conditional probability density $f_X(x_2 | x_1; t_2, t_1), \qquad t_2 > t_1.$

   (c) Is $X(t)$ Markov? Why?

   (d) Does $X(t)$ have independent increments? Justify.

7.18. Let $X(t)$ be a Markov random process on $[0, \infty)$ with initial density $f_X(x; 0) = \delta(x - 1)$ and conditional pdf

$$f_X(x_2 | x_1; t_2, t_1) = \frac{1}{\sqrt{2\pi(t_2 - t_1)}} \exp\left(-\frac{1}{2}\frac{(x_2 - x_1)^2}{t_2 - t_1}\right), \qquad \text{for all } t_2 > t_1.$$

(a) Find $f_X(x; t)$ for all t.

(b) Repeat part (a) for $f_X(x; 0) \sim N(0, 1)$.

7.19. Consider the three-state Markov chain $N(t)$ with the state-transition diagram shown in Figure P7.19. Here the state labels are the actual outputs, i.e. $N(t) = 3$, while the chain is in state 3. The state transitions are governed by jointly independent, exponentially distributed interarrival times, with average rates as indicated on the branches.

(a) *Given* that we start in state 2 at time $t = 0$, what is the probability (conditional probability) that we *remain* in this state until time t, for some arbitrary $t > 0$? (*Hint*: There are two ways to leave state 2. So you will leave at the lesser of the two independent exponential random variables with rates μ_2 and λ_2.)

(b) Write the differential equations for the probability of being in state i at time $t \geq 0$, denoting them as $p_i(t)$, $i = 1, 2, 3$. [*Hint*: First write $p_i(t + \delta t)$ in terms of the $p_i(t)$, $i = 1, 2, 3$, only keeping terms up to order $O(\delta t)$.]

(c) Find the steady-state solution for $p_i(t)$ for $i = 1, 2, 3$, i.e., $p_i(\infty)$.

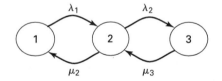

Figure P7.19   A three-state continuous-time Markov chain.

7.20. Let a certain wireless communication binary channel be in a *good* state or *bad* state, described by the continuous-time Markov chain with transition rates as shown in Figure P7.20. Here we are given that the exponentially distributed state transitions have rates $\lambda_1 = 1$ and $\lambda_2 = 9$. The value of ϵ for each state is given in part (b) below.

(a) Find the steady-state probability that the channel is in *good* state. Label $P\{X(t) = \text{good}\} = P_G$, and $P\{X(t) = \text{bad}\} = P_B$. (*Hint*: Assume the steady state exists and then write P_G at time t in terms of the two possibilities at time $t - \delta$, keeping only terms to first order in δ, taken as very small.)

(b) Assume that in the good state, there are no errors on the binary channel, but in the bad state the probability of error is $\epsilon = 0.01$ Find the average error probability on the channel. (Assume that the channel does not change state during the transmission of each single bit.)

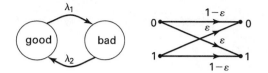

Figure P7.20   Model of two-state wireless communication channel.

7.21. This problem concerns the Chapman–Kolmogorov equation (cf. Equation 7.2-22) for a continuous-amplitude Markov random process $X(t)$,

$$f_X(x(t_3)|x(t_1)) = \int_{-\infty}^{+\infty} f_X(x(t_3)|x(t_2)) f_X(x(t_2)|x(t_1)) \, dx(t_2),$$

for the conditional pdf at three increasing observation times $t_3 > t_2 > t_1 > 0$. You will show that the pdf of the Wiener process with covariance function $K_{XX}(t,s) = \alpha \min(t,s)$, $\alpha > 0$ solves the above equation.

(a) Write the first-order pdf $f_X(x(t))$ of this Wiener process for $t > 0$.
(b) Write the first-order conditional pdf $f_X(x(t)|x(s))$, $t > s > 0$.
(c) Referring back to the Chapman-Kolmogorov equation, set $t_3 - t_2 = t_2 - t_1 = \delta$ and use x_3, x_2, and x_1 to denote the values taken on. Then verify that your conditional pdf from part (b), satisfies the resulting equation

$$f_X(x_3|x_1) = \int_{-\infty}^{+\infty} f_X(x_3|x_2) f_X(x_2|x_1) \, dx_2.$$

7.22. Is the random process $X'(t)$ of Example 7.3-2 stationary? Why?

7.23. Let A and B be independent and identically distributed (IID) random variables with mean 0, variance σ^2, and third moment $\mu \triangleq E[A^3] = E[B^3] \neq 0$. Consider the random process

$$X(t) = A\cos(2\pi ft) + B\sin(2\pi ft), \qquad -\infty < t < +\infty,$$

where f is a given frequency.

(a) Show that the random process $X(t)$ is WSS.
(b) Show that $X(t)$ is not strictly stationary.

7.24. Earlier we proved Theorem 7.3-2, thus deriving Equation 7.3-5. State and prove a corresponding theorem for covariance functions. Do not assume $\mu_X(t) = 0$.

7.25. Let $X(t)$ be a stationary random process with mean μ_X and covariance $K_{XX}(\tau) = \delta(\tau)$. Let the sample functions of $Y(t)$ drive the differential equation

$$\dot{Y} + aY(t) = X(t), \qquad a > 0, -\infty < t < +\infty.$$

    (a) Find $\mu_Y(t) = E[Y(t)]$.
    (b) Find $R_{YY}(\tau)$.
    (c) Find $\sigma_Y^2(t)$.

7.26. Is the random process $X(t)$ generated by D/A conversion in Problem 7.1 wide-sense stationary?

7.27. The psd of a random process is given as $S_{XX}(\omega) = 1/(\omega^2 + 9)$ for $-\infty < \omega < +\infty$. Find its autocorrelation function $R_{XX}(\tau)$.

7.28. Consider the LSI system shown in Figure P7.28, whose input is the zero-mean random process $W(t)$ and whose output is the random process $X(t)$. The frequency response of the system is $H(\omega)$. Given $K_{WW}(\tau) = \delta(\tau)$, find $H(\omega)$ in terms of the cross-covariance $K_{XW}(\tau)$ or its Fourier transform.

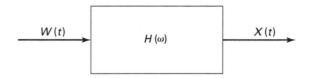

Figure P7.28   LSI system with white noise input.

7.29. Let the random process $Y(t)$ be given as

$$Y(t) = X(t) + 0.3\,\frac{dX(t)}{dt},$$

where $X(t)$ is a random process with mean function $\mu_X(t) = 5t$, and covariance function

$$K_{XX}(t_1, t_2) = \frac{\sigma^2}{1 + \alpha(t_1 - t_2)^2}, \quad \alpha > 0.$$

    (a) Find the mean function $\mu_Y(t)$.
    (b) Find the covariance function $K_{YY}(t_1, t_2)$.
    (c) Is the random process $Y(t)$ wide-sense stationary (WSS) ? Why?

7.30. Consider the first-order stochastic differential equation

$$\frac{dX(t)}{dt} + X(t) = W(t)$$

driven by the zero-mean white noise $W(t)$ with correlation function $R_{WW}(t, s) = \delta(t - s)$.

    (a) If this differential equation is valid for all time, $-\infty < t < +\infty$, find the psd of the resulting wide-sense stationary process $X(t)$.
    (b) Using residue theory (or any other method), find the inverse Fourier transform of $S_{XX}(\omega)$, the autocorrelation function $R_{XX}(\tau)$, $-\infty < \tau < +\infty$.

(c) If the above differential equation is run only for $t > 0$, is it possible to choose an initial condition random variable $X(0)$ such that $X(t)$ is wide-sense stationary for all $t \geq 0$? If such an random variable exists, find its mean and variance. *Justify your answer.* You may assume that the random variable $X(0)$ is orthogonal to $W(t)$ on $t \geq 0$; that is, $X(0) \perp W(t)$. [*Hint:* Express $X(t)$ for $t > 0$ in terms of the initial condition and a stochastic integral involving $W(t)$.]

7.31. Show that $h(\tau) * h^*(-\tau)$ is a positive semidefinite function by working directly with the definition and exclusively in the time-domain. Assume that the function $h(t)$ is square integrable, i.e., $\int_{-\infty}^{+\infty} |h(t)|^2 dt < \infty$.

7.32. Let the random process $X(t)$ with mean value 128 and covariance function

$$K_{XX}(\tau) = 1000 \exp(-10|\tau|)$$

be filtered by the lowpass filter

$$H(\omega) = \frac{1}{1 + j\omega}$$

to produce the output process $Y(t)$.

    (a) Find the mean function $\mu_Y(t)$.
    (b) Find the covariance $K_{YY}(\tau)$.

7.33. Consider the continuous-time system with input random process $X(t)$ and output process $Y(t)$:

$$Y(t) = \frac{1}{4} \int_{-2}^{+2} X(t - s) \, ds.$$

Assume that the input $X(t)$ is WSS with psd $S_{XX}(\omega) = 2$ for $-\infty < \omega < +\infty$.

    (a) Find the psd of the output $S_{YY}(\omega)$.
    (b) Find $R_{YY}(\tau)$, the correlation function of the output.

7.34. A WSS and zero-mean random process $Y(t)$ has sample functions consisting of successive rectangular pulses of random amplitude and duration as shown in Figure P7.34. The pdf for the pulse width is:

$$f_W(w) = \begin{cases} \lambda e^{-\lambda w}, & w \geq 0, \\ 0, & w < 0, \end{cases}$$

with $\lambda > 0$. The amplitude of each pulse is a random variable X (independent of W) with mean 0 and variance σ_X^2. Successive amplitudes and pulse widths are independent.

    (a) Find the autocorrelation function $R_{YY}(\tau) = E[Y(t + \tau)Y(t)]$.
    (b) Find the corresponding psd $S_{YY}(\omega)$.

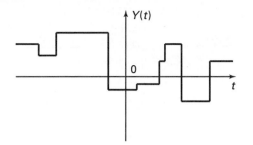

Figure P7.34   Random amplitude pulse train.

[*Hint*: First find the *conditional* autocorrelation function $E[Y(t + \tau)Y(t)|W = w]$, where t is assumed to be at the start of a pulse (ω log per WSS hypothesis for $Y(t)$).]

7.35. Let $X(t)$ be a WSS random process with mean μ_X and covariance function $K_{XX}(\tau) = 5\cos(10\tau)e^{-2|\tau|}$. The process $X(t)$ is now input to the linear system with system function

$$H(s) = \frac{s + 4}{s^2 + 11s + 30},$$

yielding the output process $Y(t)$.

   (a) First, find the input psd $S_{XX}(\omega)$. Sketch your answer.
   (b) Write an expression for the average power of $X(t)$ in the frequency range $\omega_1 \leq |\omega| \leq \omega_2$. You may leave the expression in integral form.
   (c) Find the output psd $S_{YY}(\omega)$. Sketch your answer.

7.36. In this problem we consider using white noise as an approximation to a smoother process (cf. More on White Noise in Section 7.5), which is the input to a lowpass filter. The output process from the filter is then investigated to determine the error resulting from the white noise approximation. Let the stationary random process $X(t)$ have zero-mean and autocovariance function

$$K_{XX}(\tau) = \frac{1}{2\tau_0} \exp(-|\tau|/\tau_0),$$

which can be written as $h(\tau)*h(-\tau)$ with $h(\tau) = \frac{1}{\tau_0}e^{-\tau/\tau_0}u(\tau)$.

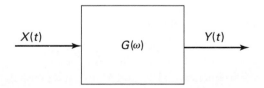

Figure P7.36a   Approximation to white noise input to filter.

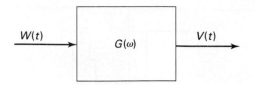

Figure P7.36b   White noise input to filter.

(a) Let $X(t)$ be input to the lowpass filter shown in Figure P7.36a, with output $Y(t)$. Find the output psd $S_Y(\omega)$, for

$$G(\omega) \triangleq \begin{cases} 1, & |\omega| \le \omega_0 \\ 0, & \text{else.} \end{cases}$$

(b) Alternatively we may, at least formally, excite the system directly with a standard white noise $W(t)$, with mean zero and $K_{WW}(\tau) = \delta(\tau)$. Call the output $V(t)$ as shown in Figure P7.36b. Find the output psd $S_{VV}(\omega)$.

(c) Show that for $|\omega_0 \tau_0| << 1$, $S_{YY} \simeq S_{VV}$ and find an upper bound on the *power error*

$$|R_{VV}(0) - R_{YY}(0)|.$$

7.37. Consider the LSI system shown in Figure P7.37. Let $X(t)$ and $N(t)$ be WSS and mutually uncorrelated with power spectral densities $S_{XX}(\omega)$ and $S_{NN}(\omega)$ and zero-means.

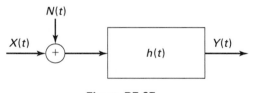

Figure P7.37

(a) Find the psd of the output $Y(t)$.

(b) Find the cross-power spectral density of X and Y, i.e., find $S_{XY}(\omega)$ and $S_{YX}(\omega)$.

(c) Define the error $\varepsilon(t) \triangleq Y(t) - X(t)$ and evaluate the psd of $\varepsilon(t)$.

(d) Assume that $h(t) = a\delta(t)$ and choose the value of a which minimizes $E[\varepsilon^2(t)] = R_{\varepsilon\varepsilon}(0)$.

7.38. Let $X(t)$ be a random process defined by

$$X(t) \triangleq N \cos(2\pi f_0 t + \Theta),$$

where f_0 is a known frequency and N and Θ are independent random variables. The characteristic function for N is

$$\Phi_N(\omega) = E[e^{+j\omega N}] = \exp\{\lambda[e^{j\omega} - 1]\},$$

where λ is a given positive constant (that is, N is a Poisson random variable). The random variable Θ is uniformly distributed on $[-\pi, +\pi]$.

    (a) Determine the mean function $\mu_X(t)$.
    (b) Determine the covariance function $K_{XX}(t, s)$.
    (c) Is $X(t)$ WSS? Justify your answer.
    (d) Is $X(t)$ stationary? Justify your answer.

7.39. Let $X(t)$ be an independent-increment random process defined on $t \geq 0$ with initial value $X(0) = X_0$, a random variable. Assume the following characteristic functions exist: $E[e^{j\omega X_0}] \triangleq \Phi_{X_0}(\omega)$ and

$$E[e^{j\omega(X(t)-X(s))}] \triangleq \Phi_{X(t)-X_0(s)}(\omega) \quad \text{for} \quad t \geq s.$$

    (a) On defining $E[e^{j\omega X(t)}] \triangleq \Phi_{X(t)}(\omega)$, show that

$$\Phi_{X(t)}(\omega) = \Phi_{X_0}(\omega)\Phi_{X(t)-X_0}(\omega).$$

    (b) Show that for all $t_2 \geq t_1$, the joint characteristic function of $X(t_2)$ and $X(t_1)$ is given by

$$\Phi_{X(t_2),X(t_1)}(\omega_2, \omega_1) = \Phi_{X_0}(\omega_1 + \omega_2)\Phi_{X(t_1)-X_0}(\omega_1 + \omega_2)\Phi_{X(t_2)-X(t_1)}(\omega_2).$$

    (c) Apply part (a) to Problem 7.18(b) by using Gaussian characteristic functions.

7.40. Express the answers to the following questions in terms of probability density functions (pdf's).

    (a) State the definition of an *independent-increments* random process.
    (b) State the definition of a Markov random process.
    (c) Prove that any random process that has independent increments also has the Markov property.

7.41. Let $X(t)$ defined over $t \geq 0$, have independent increments with mean function $\mu_X(t) = \mu_0$ and covariance function

$$K_{XX}(t_1, t_2) = \sigma_X^2(\min(t_1, t_2)),$$

where $\sigma_X^2(t)$ is an increasing function, i.e., $d\sigma_X^2(t)/dt > 0$ for all $t \geq 0$, called the *variance function*. Note that $\text{var}[X(t)] = \sigma_X^2(t)$. Fix $T > 0$ and find the mean and covariance functions of $Y(t) \triangleq X(t) - X(T)$ for all $t \geq T$. (*Note*: For the covariance function take t_1 and $t_2 \geq T$.)

7.42. Following Example 7.2-3, use MATLAB to compute a 1000-element sample function of the Wiener process $X(t)$ for $\alpha = 2$ and $T = 0.01$.

    (a) Use the MATLAB routine `hist.m` to compute the histogram of $X(10)$ and compare it with the ideal Gaussian pdf.

(b) Estimate the mean of $X(10)$ using `mean.m` and the standard deviation using `std.m` and compare them to theoretical values.
[*Hint*: Use `Wiener.m`[†] in a `for` loop to calculate 100 realizations of `x(1000)`. Then use `hist`. Question: Why can't you just use the last 100 elements of the vector `x` to approximately obtain the requested statistics?]

7.43. Let the WSS random process $X(t)$ be the input to the third-order differential equation,

$$\frac{d^3 Y}{dt^3} + a_2 \frac{d^2 Y}{dt^2} + a_1 \frac{dY}{dt} + a_0 Y(t) = X(t),$$

with WSS output random process $Y(t)$.

(a) Put this equation into the form of a first-order vector differential equation,

$$\frac{d\mathbf{Y}}{dt} = \mathbf{A}\mathbf{Y}(t) + \mathbf{B}\mathbf{X}(t),$$

by defining $\mathbf{Y}(t) \triangleq \begin{bmatrix} Y(t) \\ Y'(t) \\ Y''(t) \end{bmatrix}$ and $\mathbf{X}(t) \triangleq [X(t)]$ and evaluating the matrices **A** and **B**.

(b) Find a first-order matrix-differential equation for $\mathbf{R_{XY}}(\tau)$ with input $\mathbf{R_{XX}}(\tau)$.

(c) Find a first-order matrix-differential equation for $\mathbf{R_{YY}}(\tau)$ with input $\mathbf{R_{XY}}(\tau)$.

(d) Using matrix Fourier transforms, show that the output psd matrix $\mathbf{S_{YY}}$ is given as

$$\mathbf{S_{YY}}(\omega) = (j\omega\mathbf{I} - \mathbf{A})^{-1}\mathbf{B}\mathbf{S_{XX}}(\omega)\mathbf{B}^{\dagger}(-j\omega\mathbf{I} - \mathbf{A}^{\dagger})^{-1}.$$

7.44. Let $\mathbf{X}(t)$ be a WSS vector random process, which is input to the LSI system with impulse response matrix $\mathbf{h}(t)$.

(a) Show that the correlation matrix of the output $\mathbf{Y}(t)$ is given by Equation 7.7-4.

(b) Derive the corresponding equation for matrix covariance functions.

7.45. In geophysical signal processing one often has to simulate a multichannel random process. The following problem brings out an important constraint on the power spectral density matrix of such a vector random process. Let the N-dimensional vector random process $\mathbf{X}(t)$ be wide-sense stationary (WSS) with correlation matrix

$$\mathbf{R_{XX}}(\tau) \triangleq E[\mathbf{X}(t + \tau)\mathbf{X}^{\dagger}(t)]$$

and power spectral density matrix

$$\mathbf{S_{XX}}(\omega) \triangleq FT\{\mathbf{R_{XX}}(\tau)\}.$$

[†]`Wiener.m` is provided on this book's Web site.

Here $FT\{\cdot\}$ denotes the matrix Fourier transform, that is, the (i,j)th component of $\mathbf{S_{XX}}$ is the Fourier transform of the (i,j)th component of $\mathbf{R_{XX}}$, which is $E[X_i(t+\tau)X_j^*(t)]$ where $X_i(t)$ is the ith component of $\mathbf{X}(t)$.

   (a) For constants $a_1,\dots,a_N$ define the WSS scalar process

$$Y(t) \triangleq \sum_{i=1}^{N} a_i X_i(t).$$

       Find the power spectral density of $Y(t)$ in terms of the components of the matrix $\mathbf{S_{XX}}(\omega)$.

   (b) Show that the power spectral density matrix $\mathbf{S_{XX}}(\omega)$ must be a positive semidefinite matrix for each fixed ω; that is, we must have $\mathbf{a}^T \mathbf{S_{XX}}(\omega)\mathbf{a}^* \geq 0$ for all complex column vectors $\mathbf{a}$.

7.46. Consider the linear system shown in Figure P7.46 excited by the two *orthogonal*, zero-mean, jointly wide-sense stationary random processes $X(t)$, "the signal," and $U(t)$, "the noise." Then the input to the system G is

$$Y(t) = h(t) * X(t) + U(t),$$

which models a distorted-signal-in-noise estimation problem. If we pass this $Y(t)$, "the received signal" through the filter G, we get an estimate $\hat{X}(t)$. Finally $\varepsilon(t)$ can be thought of as the "estimation error"

$$\varepsilon(t) = \hat{X}(t) - X(t).$$

In this problem we will calculate some relevant power spectral densities and cross-power spectral density.

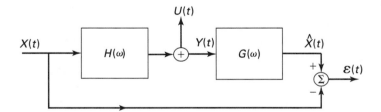

Figure P7.46   System for evaluating estimation error.

   (a) Find $S_{YY}(\omega)$.
   (b) Find $S_{\hat{X}X}(\omega) = S_{\hat{X}X}^*(\omega)$, in terms of H, G, S_{XX} and S_{UU}.
   (c) Find $S_{\varepsilon\varepsilon}(\omega)$.
   (d) Use your answer to part (c) to show that to minimize $S_{\varepsilon\varepsilon}(\omega)$ at those frequencies where

$$S_{XX}(\omega) >> S_{UU}(\omega),$$

we should have $G \approx H^{-1}$ and where

$$S_{XX}(\omega) << S_{UU}(\omega)$$

we should have $G \approx 0$.

7.47. Let $X(t)$, the input to the system in Figure P7.47 be a stationary Gaussian random process. The power spectral density of $Z(t)$ is measured experimentally and found to be

$$S_{ZZ}(\omega) = \pi\delta(\omega) + \frac{2\beta}{(\omega^2 + \beta^2)(\omega^2 + 1)}.$$

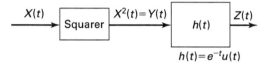

Figure P7.47   Squarer nonlinearity followed by linear filter.

(a) Find the correlation function of $Y(t)$ in terms of β.
(b) Find the correlation function of $X(t)$.

REFERENCES

7-1.   S. Karlin and H. M. Taylor, *A First Course in Stochastic Processes*. New York: Academic Press, 1975.

7-2.   T. Kailath, *Linear Systems*. Upper Saddle River, NJ: Prentice Hall, 1980.

7-3.   L. Kleinrock, *Queueing Systems*, Vol. 1: *Theory*. New York: John Wiley, 1975.

7-4.   H. L. Van Trees, *Detection, Estimation and Modulation Theory: Part I*. New York: John Wiley, 1968, p. 142.

7-5.   E. Wong and B. Hajek, *Stochastic Processes in Engineering Systems*. New York: Springer-Verlag, 1985, pp. 62–63.

7-6.   P. Whittle, "On Stationary Process in the Plane," *Biometrika*, Vol. 41 (1954), pp. 434–449.

7-7.   A. V. Oppenheim, R. W. Schafer, and J. R. Buck, *Discrete-Time Signal Processing*, 2nd edition. Upper Saddle River, NJ: Prentice Hall, 1999, Chapter 3.

7-8.   N. S. Jayant & P. Noll, *Digital Coding of Waveforms*. Prentice Hall, 1984.

7-9.   C. W. Therrien, *Discrete Random Signals and Statistical Signal Processing*. Upper Saddle River, NJ: Prentice Hall, 1992.

7-10. J. L. Doob, *Stochastic Processes*. New York: John Wiley, 1953.

7-11. E. W. Kamen, *Introduction to Signals and Systems*, 2nd edition. New York: Macmillan, 1990, p. 172.

7-12. P. Billingsley, *Probability and Measure*, New York: John Wiley, 1979, pp. 467–477.

8

Advanced Topics in Random Processes

In this chapter, we reconsider some of the topics in Chapters 6 and 7 from a more advanced or sophisticated viewpoint. In particular we introduce the mean-square derivative and integral to provide an extension of the sample function stochastic integral and derivative of Chapter 7. This will increase the scope of our linear system analysis of random processes to a much broader class, called second-order processes, that are most often encountered in more advanced studies, as well as in routine practice.

8.1 MEAN-SQUARE (m.s.) CALCULUS

From our work with limits in Chapter 6, we expect that the mean-square derivative and integral will be weaker concepts than the sample-function derivative and integral that we looked at in Chapter 7. The reason this added abstractness is necessary is that many useful random processes do not have sample-function derivatives. Furthermore, this defect cannot be determined from just examining the mean and correlation or covariance functions. In the first section we begin by looking at the various concepts of continuity for random processes.

Stochastic Continuity and Derivatives [8-1]

We will consider random processes that may be real- or complex-valued. The concept of continuity for random processes relies on the concept of limit for random processes just the same as in the case of ordinary functions. However, in the case of random processes, as for random sequences, there are four concepts for limit, which implies that there are four types of stochastic continuity. The strongest continuity would correspond to sure convergence of the sample function limits,

$$sample - function\ continuity, \quad \lim_{\varepsilon \to 0} X(t + \varepsilon, \zeta) = X(t, \zeta) \quad \text{for all} \quad \zeta \in \Omega.$$

The next strongest situation would be to disregard those sample functions in a set of probability zero that are discontinuous at time t. This would yield continuity almost surely (a.s.),

$$a.s.\ continuity, \quad P\left[\left(\lim_{s \to t} X(s)\right) \neq X(t)\right] = 0.$$

Corresponding to the concept of limit in probability, we could study the concept of continuity in probability.

$$p - continuity, \quad \lim_{s \to t} P\left[|X(s) - X(t)| > \varepsilon\right] = 0 \quad \text{for each } \varepsilon > 0.$$

The most useful and tractable concept of continuity turns out to be a mean-square-based definition. This is the concept that we will use almost exclusively.

Definition 8.1-1   A random process $X(t)$ is continuous in the mean-square sense at the point t if

$$\text{as } \varepsilon \to 0 \text{ we have } E[|X(t + \varepsilon) - X(t)|^2] \to 0.$$

If the above holds for all t, we say $X(t)$ is *mean-square* (m.s.) *continuous.*   ■

One advantage of this definition is that it is readily expressible in terms of correlation functions. By expanding out the expectation of the square of the difference, it is seen that we just require a certain continuity in the correlation function.

Theorem 8.1-1   The random process $X(t)$ is m.s. continuous at t if $R_{XX}(t_1, t_2)$ is continuous at the point $t_1 = t_2 = t$.

Proof   Expand the expectation in Definition 8.1-1 to get an expression involving R_{XX},

$$E[|X(t + \varepsilon) - X(t)|^2] = R_{XX}(t + \varepsilon, t + \varepsilon) - R_{XX}(t, t + \varepsilon)$$
$$-R_{XX}(t + \varepsilon, t) + R_{XX}(t, t).$$

Clearly the right-hand side goes to zero as $\varepsilon \to 0$ if the two-dimensional function R_{XX} is continuous at $t_1 = t_2 = t$.   ■

Example 8.1-1 _____
(Standard Wiener process.) We investigate the m.s. continuity of the Wiener process of Chapter 7. By Equation 7.2-18 we have

$$R_{XX}(t_1, t_2) = \min(t_1, t_2), \quad t_1, t_2 \geq 0.$$

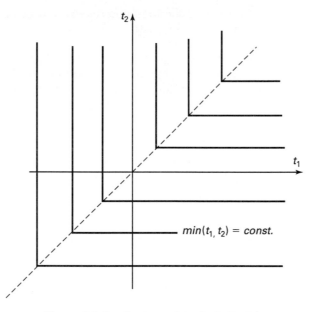

Figure 8.1-1   Contour plot of $\min(t_1, t_2)$.

The problem thus reduces to whether the function $\min(t_1, t_2)$ is continuous at the point (t, t). (See Figure 8.1-1.) The value of the function $\min(t_1, t_2)$ at (t, t) is t so we consider

$$|\min(t_1, t_2) - t|$$

for $t_1 = t + \varepsilon_1$ and $t_2 = t + \varepsilon_2$,

$$|\min(t + \varepsilon_1, t + \varepsilon_2) - t|.$$

But

$$|\min(t + \varepsilon_1, t + \varepsilon_2) - t| \le \max(\varepsilon_1, \varepsilon_2),$$

so this magnitude can be made arbitrarily small by choice of $\varepsilon_1 > 0$ and $\varepsilon_2 > 0$. Thus the Wiener process is m.s. continuous.

Lest the reader feel overly confident at this point, note that the Poisson counting process has the same correlation function when centered at its continuous mean function; thus the Poisson process is also m.s. continuous even though the sample functions of this jump process are clearly not continuous! Evidently m.s. continuity does not mean that the sample functions are continuous.

We next look at a special case of Theorem 8.1-1 for wide-sense stationary random processes.

Corollary 8.1-1   A wide-sense stationary random process $X(t)$ is m.s. continuous for all t if $R_{XX}(\tau)$ is continuous at $\tau = 0$.   ∎

Proof   By Theorem 8.1-1, we need continuity on $t_1 = t_2$, but this is the same as $\tau = 0$. Hence $R_{XX}(\tau)$ must be continuous at $\tau = 0$.   ∎

We note that in the stationary case we get m.s. continuity of the random process for all time after verifying only the continuity of a one-dimensional function $R_{XX}(\tau)$ at the origin, $\tau = 0$. Continuity is a necessary condition for the existence of the derivative of an ordinary function. However, considering the case where the difference

$$x(t + \varepsilon) - x(t) = O(\sqrt{\varepsilon}), \tag{8.1-1}$$

we see that it is not a *sufficient* condition for a derivative to exist, even in the ordinary calculus. Similarly, in the mean-square calculus, we find that m.s. continuity is not a sufficient condition for the existence of the m.s. derivative, which is defined as follows.

Definition 8.1-2   The random process $X(t)$ has a *mean-square derivative at t* if the mean-square limit of $[X(t + \varepsilon) - X(t)]/\varepsilon$ exists as $\varepsilon \to 0$.   ∎

If it exists, we denote this m.s. derivative by X', $X^{(1)}$, dX/dt or $\dot{X}$. Generally, we do not know X' when we are trying to determine whether it exists, so we turn to the Cauchy convergence criterion (ref. Section 6.7). In this case the test becomes

$$E\left\{ |[X(t + \varepsilon_1) - X(t)]/\varepsilon_1 - [X(t + \varepsilon_2) - X(t)]/\varepsilon_2|^2 \right\}$$

$$\to 0 \text{ as } \varepsilon_1 \text{ and } \varepsilon_2 \to 0. \tag{8.1-2}$$

As was the case for continuity, we can express this condition in terms of the correlation function, making it easier to apply. This generally useful condition is stated in the following theorem.

Theorem 8.1-2   A random process $X(t)$ with autocorrelation function $R_{XX}(t_1, t_2)$ has a m.s. derivative at time t if $\partial^2 R_{XX}(t_1, t_2)/\partial t_1 \partial t_2$ exists at $t_1 = t_2 = t$.

Proof   Expand the square inside the expectation in Equation 8.1-2 to get three terms, the first and last of which look like

$$E\left[|(X(t + \varepsilon) - X(t))/\varepsilon|^2 \right]$$

$$= [R_{XX}(t + \varepsilon, t + \varepsilon) - R_{XX}(t, t + \varepsilon) - R_{XX}(t + \varepsilon, t) + R_{XX}(t, t)]/\varepsilon^2$$

which converges to

$$\partial^2 R_{XX}(t_1, t_2)/\partial t_1 \partial t_2,$$

if the second mixed partial derivative exists at the point $(t_1, t_2) = (t, t)$. The middle or cross-term is

$$-2E\left\{ [X(t + \varepsilon_1) - X(t)]/\varepsilon_1 \cdot [X(t + \varepsilon_2) - X(t)]^*/\varepsilon_2 \right\}$$

$$= -2\{ R_{XX}(t + \varepsilon_1, t + \varepsilon_2) - R_{XX}(t, t + \varepsilon_2) - R_{XX}(t + \varepsilon_1, t) + R_{XX}(t, t) \}/\varepsilon_1 \varepsilon_2$$

$$= -2\{[R_{XX}(t+\varepsilon_1, t+\varepsilon_2) - R_{XX}(t+\varepsilon_1, t)]/\varepsilon_2 - [R_{XX}(t, t+\varepsilon_2) - R_{XX}(t, t)]/\varepsilon_2\}/\varepsilon_1$$

$$\rightarrow -2\frac{\partial}{\partial t_1}\left(\frac{\partial R_{XX}(t_1, t_2)}{\partial t_2}\right)\Bigg|_{(t_1, t_2)=(t,t)}$$

$$= -2\partial^2 R_{XX}(t_1, t_2)/\partial t_1 \partial t_2\big|_{(t_1, t_2)=(t,t)}$$

if this second mixed partial derivative exists at the point $(t_1, t_2) = (t, t)$. Combining all three of these terms, we get convergence to

$$2\partial^2 R_{XX}/\partial t_1 \partial t_2 - 2\partial^2 R_{XX}/\partial t_1 \partial t_2 = 0. \quad \blacksquare$$

In the preceding theorem the reader should note that we are talking about a two-dimensional function $R_{XX}(t_1, t_2)$ and its second mixed partial derivative evaluated on the diagonal points $(t_1, t_2) = (t, t)$. This is clearly not the same as the second derivative of the one-dimensional function $R_{XX}(t, t)$, which is the restriction of $R_{XX}(t_1, t_2)$ to the diagonal line $t_1 = t_2$. In some cases the derivative of $R_{XX}(t, t)$ will exist while the partial derivative of $R_{XX}(t_1, t_2)$ will not.

Example 8.1-2

(Derivative of Wiener process.) Let $W(t)$ be a Wiener process with correlation function $R_{WW}(t_1, t_2) = \sigma^2 \min(t_1, t_2)$.[†] We enquire about the behavior of $E[|(W(t+\varepsilon) - W(t))/\varepsilon|^2]$ when ε is near zero. Assuming that ε is positive, we have by calculation that $E[|(W(t+\varepsilon) - W(t))|^2] = \sigma^2 \varepsilon$ which shows that the Wiener process is mean-square continuous, as we already found in Example 8.1-1. But now when we divide by ε^2 inside the expectation, as required by $E[|(W(t+\varepsilon) - W(t))/\varepsilon|^2]$ we end up with σ^2/ε which goes to infinity as ε approaches zero. So the mean-square derivative of the Wiener process does not exist, at least in an ordinary sense. Looking at the above equation, we see the problem is that in some sense at least, the sample functions $w(t)$ of the Wiener process have increments typically on the order of $w(t+\varepsilon) - w(t) = \sigma\sqrt{\varepsilon}$.

Example 8.1-3

Let $X(t)$ be a random process with correlation function $R_{XX}(t_1, t_2) = \sigma^2 \exp(-\alpha|t_1 - t_2|)$. To test for the existence of a m.s. derivative X', we attempt to compute the second mixed partial derivative of R_{XX}. We first compute

$$\partial R_{XX}/\partial t_2 = \begin{cases} \dfrac{\partial}{\partial t_2}\left[\sigma^2 \exp(-\alpha(t_2 - t_1))\right], & t_1 < t_2 \\[2mm] \dfrac{\partial}{\partial t_2}\left[\sigma^2 \exp(-\alpha(t_1 - t_2))\right], & t_1 \geq t_2, \end{cases} \tag{8.1-3}$$

$$= \begin{cases} -\alpha\,\sigma^2 \exp(-\alpha(t_2 - t_1)), & t_1 < t_2 \\ +\alpha\,\sigma^2 \exp(-\alpha(t_1 - t_2)), & t_1 \geq t_2. \end{cases} \tag{8.1-4}$$

[†]While it is conventional to use σ^2 as the parameter of the Wiener process, please note that σ^2 is not the variance! Earlier we used α for this parameter.

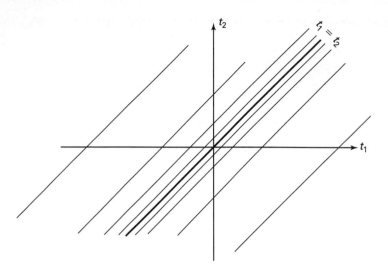

Figure 8.1-2   Contour plot of $R_{XX}(t_1, t_2)$ of Example 8.1-3.

Graphing the function $R_{XX}(t_1, t_2)$ as shown in Figure 8.1-2, we see that there is a cusp on the diagonal line $t_1 = t_2$. Thus there is no partial derivative there for any t. So the second mixed partial cannot exist there either, and we conclude that no m.s. derivative exists for an $X(t)$ with this correlation function. Evidently such random processes are not smooth enough.

Example 8.1-4 _____

We look at another random process $X(t)$ with mean function $\mu_X = 5$ and correlation function,

$$R_{XX}(t_1, t_2) = \sigma^2 \exp(-\alpha(t_1 - t_2)^2) + 25,$$

which is smooth on the diagonal line. The first partial with respect to t_2 is

$$\frac{\partial R_{XX}}{\partial t_2} = 2\alpha(t_1 - t_2)\sigma^2 \exp(-\alpha(t_1 - t_2)^2).$$

Then the second mixed partial becomes

$$\frac{\partial^2 R_{XX}}{\partial t_1 \partial t_2} = 2\alpha\sigma^2 \left[1 - 2\alpha(t_1 - t_2)^2\right] \exp(-\alpha(t_1 - t_2)^2),$$

which evaluated at $t_1 = t_2 = t$ becomes

$$\left. \frac{\partial^2 R_{XX}}{\partial t_1 \partial t_2} \right|_{(t_1, t_2) = (t, t)} = 2\alpha\sigma^2,$$

so that in this case the m.s. derivative $X'(t)$ exists for all t.

Given the existence of the m.s. derivative, the next question we might be interested in is: What is its probability law? Or more simply: (a) What are its mean and correlation function?; and (b) How is X' correlated with X? To answer (a) and (b), we start by considering the expectation

$$E[X'(t)] = E[dX(t)/dt].$$

Now assuming that the m.s. derivative is a linear operator, we would expect to be able to interchange the derivative and the expectation under certain conditions to obtain,

$$E[X'(t)] = dE[X(t)]/dt = d\mu_X(t)/dt. \tag{8.1-5}$$

We can show that this equation is indeed true for the m.s. derivative by making use of the inequality

$$|E[Z]|^2 \leq E[|Z|^2],$$

for a complex random variable Z, a consequence of the nonnegativity of $\text{Var}[Z]$. First we set $A_n \triangleq n[X(t + 1/n) - X(t)]$ and note that the right-hand side of Equation 8.1-5 is just $\lim_{n \to \infty} E[A_n]$. Thus Equation 8.1-5 will be true if $\lim_{n \to \infty} E[X'(t) - A_n] = 0$. Then making use of the above inequality with $Z \triangleq X'(t) - A_n$, we get

$$|E[X'(t) - A_n]|^2 \leq E[|X'(t) - A_n|^2],$$

where the right-hand side goes to zero by the definition of m.s. derivative. Thus the left-hand side must also go to zero, and hence we are free to interchange the order of m.s. differentiation and mathematical expectation, that is, Equation 8.1-5 has been shown to be correct.

To calculate the correlation function of the mean-square derivative process we first define,

$$R_{X'X'}(t_1, t_2) = E[X'(t_1)X'^*(t_2)], \tag{8.1-6}$$

and formally compute, interchanging the order of expectation and differentiation,

$$R_{X'X'}(t_1, t_2) = \lim_{m,n \to \infty} E\left[n\left(X\left(t_1 + \frac{1}{n}\right) - X(t_1)\right) \cdot m\left(X\left(t_2 + \frac{1}{m}\right) - X(t_2)\right)^*\right]$$

$$= \lim_{n \to \infty} n\left[\frac{\partial R_{XX}\left(t_1 + \frac{1}{n}, t_2\right)}{\partial t_2} - \frac{\partial R_{XX}(t_1, t_2)}{\partial t_2}\right]$$

$$= \partial^2 R_{XX}(t_1, t_2)/\partial t_1 \partial t_2.$$

This is really just the first step in the proof of Theorem 8.1-2 generalized to allow $t_1 \neq t_2$. To justify this interchange of the m.s. derivative and expectation operators, we make use of the Schwarz inequality (Section 4.4 in Chapter 4), which has been derived for real-valued random variables in Chapter 4. (It also holds for the complex case as will be shown in the next section.)

We first define

$$R_{XX'}(t_1, t_2) \triangleq E[X(t_1)X'^*(t_2)]. \tag{8.1-7}$$

We use the Schwarz inequality $|E[AB_n^*]| \leq \sqrt{E[|A|^2]E[|B_n|^2]}$ with $A \triangleq X(t_1)$ and

$$B_n \triangleq X'(t_2) - [X(t_2 + \varepsilon_n) - X(t_2)]/\varepsilon_n,$$

to obtain

$$
\begin{aligned}
R_{XX'}(t_1, t_2) &= \lim_{n \to \infty} E\left[X(t_1)\left(X(t_2 + \varepsilon_n) - X(t_2)\right)^* / \varepsilon_n\right] \\
&= \lim_{n \to \infty} [R_{XX}(t_1, t_2 + \varepsilon_n) - R_{XX}(t_1, t_2)]/\varepsilon_n \\
&= \frac{\partial R_{XX}(t_1, t_2)}{\partial t_2}.
\end{aligned}
$$

Then $E[X'(t_1)X'^*(t_2)]$ is obtained similarly as

$$
\lim_{n \to \infty} E\left[\left(\frac{X(t_1 + \varepsilon_n) - X(t_1)}{\varepsilon_n}\right) X'^*(t_2)\right] = \lim_{n \to \infty} [R_{XX'}(t_1 + \varepsilon_n, t_2) - R_{XX'}(t_1, t_2)]/\varepsilon_n
$$

$$
= \partial R_{XX'}(t_1, t_2)/\partial t_1.
$$

Thus we have finally obtained the following theorem.

Theorem 8.1-3   If a random process $X(t)$ with mean function $\mu_X(t)$ and correlation function $R_{XX}(t_1, t_2)$ has an m.s. derivative $X'(t)$, then the mean and correlation functions of $X'(t)$ are given by

$$\mu_{X'}(t) = d\mu_X(t)/dt$$

and

$$R_{X'X'}(t_1, t_2) = \partial^2 R_{XX}(t_1, t_2)/\partial t_1 \partial t_2.$$

Example 8.1-5 —————————————————————————————————————

We now continue to study the m.s. derivative of the process $X(t)$ of Example 8.1-4 by calculating its mean function and its correlation function. We obtain

$$\mu_{X'}(t) = d\mu_X(t)/dt = 0,$$

and

$$
\begin{aligned}
R_{X'X'}(t_1, t_2) &= \partial^2 R_{XX}(t_1, t_2)/\partial t_1 \partial t_2 \\
&= 2\alpha\sigma^2[1 - 2\alpha(t_1 - t_2)^2] \exp(-\alpha(t_1 - t_2)^2).
\end{aligned}
$$

We note that in the course of calculating $R_{X'X'}$, we are effectively verifying existence of the m.s. derivative by noting whether this deterministic partial derivative exists at $t_1 = t_2$.

Generalized m.s. derivatives.   In Example 8.1-3 we found that the correlation function had a cusp at $t_1 = t_2$ which precluded the existence of a m.s. derivative. However, this second mixed partial derivative does exist in the sense of singularity functions. In earlier experience with linear systems we have worked with singularity functions and have found them operationally very elegant and simple to use. They are properly treated mathematically through the rather abstract theory of *generalized functions*. If we proceed to take the partial derivative, in this generalized sense, with respect to t_1 of $\partial R_{XX}/\partial t_2$ in Equation 8.1-4, the step discontinuity on the diagonal gives rise to an impulse in t_1. We obtain

$$\frac{\partial^2 R_{XX}(t_1, t_2)}{\partial t_1 \partial t_2} = \left\{ \begin{array}{ll} -\alpha^2 \sigma^2 \exp(-\alpha(t_2 - t_1)), & t_1 < t_2 \\ -\alpha^2 \sigma^2 \exp(-\alpha(t_1 - t_2)), & t_1 \geq t_2 \end{array} \right\} + 2\alpha\sigma^2 \delta(t_1 - t_2)$$

$$= 2\alpha\sigma^2 \delta(t_1 - t_2) - \alpha^2 \sigma^2 \exp(-\sigma|t_1 - t_2|). \tag{8.1-8}$$

We can call the random process with this autocorrelation function a *generalized random process* and say it is the *generalized m.s. derivative* of the conventional process $X(t)$. When we say this we mean that the defining mean-square limit (Equation 8.1-2) is zero in the sense of generalized functions. A more detailed justification of generalized random processes may be found in [8-2]. In this text we will be content to use it formally as a notation for a limiting behavior of conventional random processes.

The surprising thing about the autocorrelation function in Equation 8.1-8 is the term $2\alpha\sigma^2 \delta(t_1 - t_2)$. In fact, if we single out this term and consider the autocorrelation function

$$R_{XX}(t_1, t_2) = \sigma^2 \delta(t_1 - t_2), \tag{8.1-9}$$

we can show that it corresponds to the generalized m.s. derivative of the Wiener process defined in Chapter 7. By definition $\mu_X(t) = 0$ and $R_{XX}(t_1, t_2) = \alpha \min(t_1, t_2)$ for the Wiener process. We proceed by calculating,

$$\partial R_{XX}/\partial t_2 = \frac{\partial}{\partial t_2} \left\{ \begin{array}{ll} \alpha t_2, & t_2 < t_1 \\ \alpha t_1, & t_1 \geq t_2 \end{array} \right.$$

$$= \left\{ \begin{array}{ll} \alpha, & t_2 < t_1 \\ 0, & t_1 \geq t_2. \end{array} \right.$$

Then

$$\frac{\partial^2 R_{XX}}{\partial t_1 \partial t_2} = \frac{\partial}{\partial t_1} \left\{ \begin{array}{ll} 0, & t_2 < t_1 \\ \alpha, & t_1 \geq t_2 \end{array} \right.$$

$$= \frac{\partial}{\partial t_1} [\alpha u(t_1 - t_2)]$$

$$= \alpha \delta(t_1 - t_2),$$

which is the same as Equation 8.1-9 if we set $\sigma^2 = \alpha$.[†]

[†]Please note again here that σ^2 is merely a positive parameter here. It is *not* the variance of the white noise. We use this notation only because it has become standard to do so.

The generalized m.s. derivative of the Wiener process is called *white Gaussian noise*. It is not a random process in the conventional sense since, for example, its mean-square value $R_{X'X'}(t,t)$ at time t is infinite. Nevertheless, it is the formal limit of approximating conventional processes whose correlation function is a narrow pulse of area α. These approximating processes will often yield almost the same system outputs (up to mean-square equivalence); thus the white noise can be used to simplify the analysis in these cases in essentially the same way that impulse functions are used in deterministic system analysis. In terms of power spectral density, the white noise is the idealization of having a flat power spectral density function over all frequencies, $S_{X'X'}(\omega) = \alpha$, $-\infty < \omega < +\infty$.

Many random processes are stationary or approximately so. In the stationary case we have seen that we can write the correlation function as a 1-D function $R_{XX}(\tau)$. Then we can express the conditions and results of Theorems 8.1-2 and 8.1-3 in terms of this function. Unfortunately, the resulting formulas are not as intuitive and tend to be somewhat confusing. For the special case of a stationary or wide-sense stationary random process we get the following:

Theorem 8.1-4 The m.s. derivative of a WSS random process $X(t)$ exists at time t iff (if and only if) the autocorrelation $R_{XX}(\tau)$ has derivatives up to order two at $\tau = 0$.

Proof By the previous result we need $\partial^2 R_{XX}(t_1,t_2)/\partial t_1 \partial t_2 \big|_{(t_1,t_2)=(t,t)}$. Now

$$R_{XX}(\tau) = R_{XX}(t+\tau,t), \text{ functionally independent of } t,$$

so

$$\frac{\partial R_{XX}(t_1,t_2)}{\partial t_1}\bigg|_{(t_1,t_2)=(t+\tau,t)} = dR_{XX}(\tau)/d\tau$$

since $t_2 = t$ is held constant, and

$$\frac{\partial R_{XX}(t_1,t_2)}{\partial t_2}\bigg|_{(t_1,t_2)=(t,t-\tau)} = \partial R_{XX}(t,t-\tau)/\partial(-\tau)$$

$$= -dR_{XX}(\tau)/d\tau,$$

since $t_1 = t$ is held constant here; thus

$$\frac{\partial^2 R_{XX}(t_1,t_2)}{\partial t_1 \partial t_2}\bigg|_{(t_1,t_2)=(t+\tau,t)} = -d^2 R_{XX}(\tau)/d\tau^2. \quad \blacksquare$$

Calculating the second-order properties of X', we have

$$E[X'(t)] = \mu_{X'}(t) = 0,$$

$$E[X'(t+\tau)X^*(t)] = R_{X'X}(\tau) = +dR_{XX}(\tau)/d\tau, \tag{8.1-10}$$

$$E[X(t+\tau)X'^*(t)] = R_{XX'}(\tau) = -dR_{XX}(\tau)/d\tau, \tag{8.1-11}$$

and

$$E[X'(t+\tau)X'^*(t)] = R_{X'X'}(\tau) = -d^2 R_{XX}(\tau)/d\tau^2,$$

which follow from the formulas used in the above proof.

One can also derive Equations 8.1-11 directly, for example,

$$E[X(t+\tau)X'^*(t)] = \lim_{\varepsilon \to 0} \left(\frac{E[X(t+\tau)X^*(t+\varepsilon)] - E[X(t+\tau)X^*(t)]}{\varepsilon} \right)$$

$$= \lim_{\varepsilon \to 0} \left(\frac{R_{XX}(\tau - \varepsilon) - R_{XX}(\tau)}{\varepsilon} \right)$$

$$= -dR_{XX}(\tau)/d\tau = R_{XX'}(\tau)$$

and similarly for $R_{X'X}(\tau)$.

Example 8.1-6

Let $X(t)$ have zero-mean and correlation function

$$R_{XX}(\tau) = \sigma^2 \exp(-\alpha^2 \tau^2).$$

Here the m.s. derivative exists because $R(\tau)$ is infinitely differentiable at $\tau = 0$. Computing the first and second derivatives, we get

$$dR_{XX}/d\tau = -2\alpha^2 \tau \sigma^2 \exp(-\alpha^2 \tau^2).$$

Then

$$R_{X'X'}(\tau) = -d^2 R_{XX}/d\tau^2 = 2\alpha^2 \sigma^2 (1 - 2\alpha^2 \tau^2) \exp(-\alpha^2 \tau^2).$$

Further Results on m.s. Convergence [8-1]

We now consolidate and present further results on mean-square convergence that are helpful in this chapter. We will have use for two inequalities for the moments of complex random variables: the Schwarz inequality and another inequality called the triangle inequality.

Complex Schwarz inequality.   Let X and Y be second-order complex random variables, that is,

$$E[|X|^2] < \infty \quad \text{and} \quad E[|Y|^2] < \infty;$$

then we have the inequality

$$|E[XY^*]| \le \sqrt{E[|X|^2]E[|Y|^2]}.$$

Proof   Consider $W \triangleq aX + Y$. Then minimize $E[|W|^2]$ as a function of a, where a is a complex variable. Then, since the minimum must be nonnegative, the preceding inequality is obtained as follows. First we write the function,

$$f(a) = E[|W|^2]$$

$$= E[(aX + Y)(a^* X^* + Y^*)].$$

Now, we want to minimize this function with respect to (wrt) the complex variable a. If a were a real variable, we could just take the derivative and set it equal to zero and then solve for the minimizing a. Here, things are not so simple. The most straightforward approach would be to express $a = a_r + ja_i$ and then consider partial derivatives wrt the real part a_r and imaginary part a_i separately. A more elegant approach is to take partials wrt a and a^*, as though they were independent variables, which they are not! Using this latter method we get

$$0 = \frac{\partial f(a)}{\partial a} = E[X(a^*X^* + Y^*)]$$

$$0 = \frac{\partial f(a)}{\partial a^*} = E[(aX + Y)X^*],$$

either of which yields the desired result that the minimizing value of a is

$$a_{\min} = -\frac{E[X^*Y]}{E[|X|^2]}.$$

Finally, evaluating $f(a_{\min})$, and noting that $f(a)$ is always nonnegative, we get

$$E[|Y|^2] - \frac{|E[X^*Y]|^2}{E[|X|^2]} \geq 0,$$

which is equivalent to the above stated Cauchy–Schwarz inequality.   ■

An interesting aside here is that we can regard $-a_{\min}X$ as a best linear estimate[†] of Y, which we can denote as $\widehat{Y}$. We have the result

$$\widehat{Y} = \frac{E[X^*Y]}{E[|X|^2]}X.$$

Another inequality that will be useful is the triangle inequality, which shows that the root mean-square (rms) value $\sqrt{E[|X|^2]}$ can be considered as a "length" for random variables!

Triangle inequality.   Let X and Y be second-order complex random variables; then

$$\sqrt{E[|X + Y|^2]} \leq \sqrt{E[|X|^2]} + \sqrt{E[|Y|^2]}.$$

Proof

$$E[|X + Y|^2] = E[(X + Y)(X^* + Y^*)]$$
$$= E[|X|^2] + E[XY^*] + E[X^*Y] + E[|Y|^2]$$

[†]Warning: A better "linear" estimate would be $aX + b$, and this would give a lower error when the mean of X or Y is not zero.

$$\leq E[|X|^2] + 2|E[XY^*]| + E[|Y|^2]$$

$$\leq E[|X|^2] + 2\sqrt{E[|X|^2]E[|Y|^2]} + E[|Y|^2]$$

$$\text{(by the Schwarz inequality)}$$

$$= \left(\sqrt{E[|X|^2]} + \sqrt{E[|Y|^2]}\right)^2. \quad \blacksquare$$

Note that the quantity $\sqrt{E[|X|^2]}$ thus obeys the equation for a distance or *norm*,

$$||X + Y|| \leq ||X|| + ||Y||, \qquad (8.1\text{-}12)$$

and hence the name "triangle inequality," as pictured in Figure 8.1-3. We "see" that the "length" of the vector sum $X + Y$ can never exceed the sum of the "length" of X plus the "length" of Y.

The norm $||X||$ also satisfies the equation

$$||aX|| = |a| \cdot ||X||$$

for any complex scalar a. In fact, one can then define the linear space of second-order complex random variables with *norm*

$$||X|| \triangleq \sqrt{E[|X|^2]}.$$

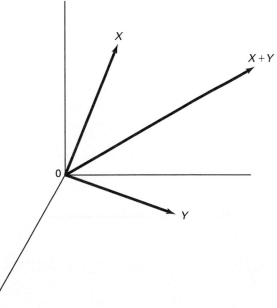

Figure 8.1-3   Conceptualization of triangle inequality in linear space of random variables.

This space is then a Hilbert Space with *inner product*

$$\langle X, Y \rangle \triangleq E[XY^*].$$

Writing the Cauchy–Schwarz inequality in this geometric notation, we have

$$|\langle X, Y \rangle| \le ||X|| \, ||Y||,$$

and the best linear estimate becomes

$$\widehat{Y} = \frac{\langle Y, X \rangle}{||X||^2} X,$$

$$= \frac{\langle Y, X \rangle}{||X|| \, ||Y||} \frac{X}{||X||} ||Y||,$$

three terms with the following interpretations: the first term is the orthogonal projection onto Y; the second term is the normalized vector X; and the third term is a scaling up to the norm of Y. This orthogonal projection of Y onto X is illustrated in Figure 8.1-4.

If $\langle Y, X \rangle = 0$, then we write $X \perp Y$, and say "X and Y are orthogonal random variables." Note that in this case $\widehat{Y}$ would be zero. If $\langle Y, X \rangle$ would take its largest value $||X|| \, ||Y||$, by the Cauchy–Schwarz inequality then we would have the estimate $\widehat{Y} = \frac{X}{||X||} ||Y||$, which is just a scaling of the random variable X, by the ratio of the norms (standard deviations). Continuing this geometric viewpoint, we can get Pythagorean's theorem if $X \perp Y$, because then by the above proof, the triangle inequality Equation 8.1-12 takes on its maximum value and $||X + Y||^2 = ||X||^2 + ||Y||^2$.

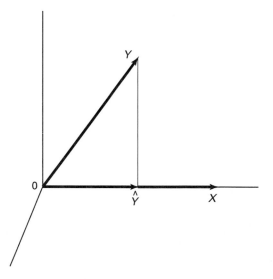

Figure 8.1-4   Illustration of orthogonal projection of random variable Y onto random variable X.

Later in this chapter we will be working with m.s. derivatives and integrals and also m.s. differential equations. As such, we will have need of the following general results concerning the moments of mean-square limits.

Theorem 8.1-5   Let $X[n] \to X$ and $Y[n] \to Y$ in the m.s. sense with $E[|X|^2] < \infty$ and $E[|Y|^2] < \infty$. Then we have the properties

(1) $\lim_{n \to \infty} E\left[X[n]\right] = E[X]$,

(2) $\lim_{n \to \infty} E\left[X[n]Y^*\right] = E[XY^*]$,

(3) $\lim_{n \to \infty} E\left[|X[n]|^2\right] = E[|X|^2]$,

(4) $\lim_{n \to \infty} E\left[X[n]Y^*[n]\right] = E[XY^*]$,

(5) if $X = X_1$ (m.s.), then $P[X \neq X_1] = 0$, i.e., $X = X_1$ (a.s.).

Proof

(1) Let $A_n \triangleq X - X[n]$; then we need $\lim_{n\to\infty} E[A_n] = 0$. But $|E[A_n]|^2 \leq ||A_n||^2$ where $||A_n|| \triangleq \sqrt{E[|A_n|^2]}$ and $||A_n|| \to 0$ by the definition of m.s. convergence; thus $\lim_{n\to\infty} E[A_n] = 0$.

(2) We note $XY^* - X[n]Y^* = A_n Y^*$. Then by the Schwarz inequality,

$$|E[A_n Y^*]|^2 \leq ||A_n||^2 \cdot ||Y||^2 \to 0 \quad \text{as} \quad n \to \infty,$$

since Y is second order, i.e., $||Y||^2$ is finite.

(3) This is the same as $||X[n]|| \to ||X||$ in terms of the norm notation. Now by the triangle inequality Equation 8.1-12,

$$||X[n]|| = ||X[n] - X + X|| \leq ||X[n] - X|| + ||X||.$$

This tells us that $\lim_{n\to\infty} ||X[n]|| \leq ||X||$. But also $||X|| = ||X - X[n] + X[n]|| \leq ||X - X[n]|| + ||X[n]||$, which implies $\lim_{n\to\infty} ||X[n]|| \geq ||X||$. Therefore, $\lim_{n\to\infty} ||X[n]|| = ||X||$.

(4) Consider $XY^* - X[n]Y^*[n]$, and add and subtract $X[n]Y^*$ to obtain

$$(X - X[n])\, Y^* + X[n]\, (Y - Y[n])^*\,.$$

Next use the triangle inequality followed by the Schwarz inequality to obtain,

$$|E\left[XY^* - X[n]Y^*[n]\right]| \leq ||X - X[n]|| \cdot ||Y|| + ||X[n]|| \cdot ||Y - Y[n]||$$

$$\to 0 \quad \text{by definition of m.s. convergence and by property (3)}.$$

(5) Let $\varepsilon > 0$; then by the Chebyshev inequality we have,

$$P\left[|X - X_1| > \varepsilon\right] \leq E[|X - X_1|^2]/\varepsilon^2 = 0.$$

Since ε is arbitrary, letting the event $B_n \triangleq \{\zeta | \ |X - X_1| > 1/n\}$, we have $B_n \nearrow$ $B_\infty \triangleq \{X \neq X_1\}$, and so

$$P[X \neq X_1] = 0,$$

by the continuity of P-measure (ref. Theorem 6.1-1). ∎

Another useful property for our work later in this chapter will be the fact that the m.s. limit is a linear operator.

Theorem 8.1-6   Let $X[n] \to X$ (m.s.) and $Y[n] \to Y$ (m.s.) with both X and Y second-order random variables, i.e., their second moments are finite. Then

$$\lim_{n \to \infty} \{aX[n] + bY[n]\} = aX + bY. \qquad \text{(m.s.)}$$

Proof   We have to show that

$$||aX[n] + bY[n] - (aX + bY)|| \to 0.$$

By the triangle inequality Equation 8.1-12,

$$\begin{aligned}
||a\,(X[n] - X) + b\,(Y[n] - Y)\,|| &\leq ||a\,(X[n] - X)\,|| + ||b\,(Y[n] - Y)\,|| \\
&= |a| \cdot ||X[n] - X|| + |b| \cdot ||Y[n] - Y|| \\
&\to 0 \text{ as } n \to \infty,
\end{aligned}$$

by assumed m.s. convergence of $X[n]$ and $Y[n]$. ∎

Corollary 8.1-2   The m.s. derivative is a linear operator, that is,

$$\frac{d}{dt}[aX(t) + bY(t)] = aX'(t) + bY'(t). \qquad \text{(m.s.)} \quad ∎$$

Proof   We leave this as an exercise to the reader. ∎

Example 8.1-7 _____

Let the random sequence $X[n]$ be given by the second-order random variables X_1 and X_2 as $X[n] \triangleq (1 - 1/n)X_1 + (1/n)X_2$. By m.s. convergence theory we know that the m.s. limit of $X[n]$ as $n \to \infty$ is $X = X_1$. Then by Theorem 8.1-5 (a) we have that the mean value of X is $E[X] = \lim E[X[n]] = E[X_1]$, and by part (c) that the average power of X is $E[|X|^2] = \lim E[|X[n]|^2] = E[|X_1|^2]$.

8.2 MEAN-SQUARE STOCHASTIC INTEGRALS

In this section we continue our study of the calculus of random processes by considering the stochastic integral. Stochastic integrals are important in applications for representing linear operators such as convolution, which arise when random processes are passed through linear systems. Our discussion will be followed in the next section by a look at stochastic differential equations, which of course will involve the stochastic integral in their solution. Another application of the stochastic integral is in forming averages for estimating the moments and probability functions of stationary processes. This topic is related to ergodic theory and will be introduced in Section 8.4.

We will be interested in the mean-square stochastic integral. It is defined as the mean-square limit of its defining sum as the partition of the integration interval gets finer and finer. We first look at the integration of a random process $X(t)$ over a finite interval (T_1, T_2). The operation of the integral is then just a simple averager. First we create a partition of (T_1, T_2) consisting of n points using $(t_1, t_2, \ldots, t_n)$. Then the approximate integral is the sum

$$I_n \triangleq \sum_{i=1}^{n} X(t_i)\Delta t_i. \tag{8.2-1}$$

On defining the m.s. limit random variable as I, we have the following definition of the mean-square integral.

Definition 8.2-1   The *mean-square integral* of $X(t)$ over the interval (T_1, T_2) is denoted I. It exists when

$$\lim_{n \to \infty} E\left[\left| I - \sum_{i=1}^{n} X(t_i)\Delta t_i \right|^2 \right] = 0. \tag{8.2-2}$$

We give I the following symbol:

$$I = \int_{T_1}^{T_2} X(t)dt \triangleq \lim_{n \to \infty} I_n \quad \text{(m.s.)} \quad \blacksquare$$

Because the m.s. limit is a linear operator by Theorem 8.1-6, it follows that the integral just defined is linear, that is, it obeys

$$\int_{T_1}^{T_2} [aX_1(t) + bX_2(t)] \, dt = a \int_{T_1}^{T_2} X_1(t)dt + b \int_{T_1}^{T_2} X_2(t)dt,$$

whenever the integrals on the right exist.

To study the existence of the mean-square integral, as before we make use of the Cauchy criterion for the existence of a limit. Thus the integral I exists iff $\lim_{m,n \to \infty} E[|I_n - I_m|^2] = 0$. If we expand this expression into three terms, we get $E[|I_n|^2] - 2Re\{E[I_n I_m^*]\} + E[|I_m|^2]$. Without loss of generality we concentrate on the cross-term and evaluate

$$E[I_n I_m^*] = \sum_{i,,j} R_{XX}(t_i, t_j)\Delta t_i \Delta t_j,$$

where the sums range over $1 \le i \le n$ and $1 \le j \le m$. As $m, n \to +\infty$, this converges to the deterministic integral

$$\int_{T_1}^{T_2} \int_{T_1}^{T_2} R_{XX}(t_1, t_2)dt_1 dt_2, \tag{8.2-3}$$

if this deterministic integral exists. Clearly, the other two terms $E[|I_n|^2]$ and $E[|I_m|^2]$ converge to the same double integral. Thus we see that the m.s. integral exists whenever the double integral of Equation 8.2-3 exists in the sense of the ordinary calculus.

The mean and mean-square (power) of I are directly computed (via Theorem 8.1-5) as

$$E[I] = E\left[\int_{T_1}^{T_2} X(t)dt\right] = \int_{T_1}^{T_2} E[X(t)]dt,$$

and

$$E[|I|^2] = E\left[\int_{T_1}^{T_2} \int_{T_1}^{T_2} X(t_1)X^*(t_2)dt_1 dt_2\right]$$

$$= \int_{T_1}^{T_2} \int_{T_1}^{T_2} R_{XX}(t_1, t_2)dt_1 dt_2. \tag{8.2-4}$$

The variance of I is given as

$$\sigma_I^2 = \int_{T_1}^{T_2} \int_{T_1}^{T_2} K_{XX}(t_1, t_2)dt_1 dt_2.$$

Example 8.2-1

(Integral of white noise over $(0, t]$.) Let the random process $X(t)$ be zero mean with covariance function $K_{XX}(\tau) = \sigma^2\delta(\tau)$ and define the running m.s. integral as

$$Y(t) \triangleq \int_0^t X(s)ds, \qquad t \geq 0.$$

For fixed t, $Y(t)$ is a random variable; therefore $Y(t)$ is a stochastic process. Its mean is given as $E[Y(t)] = \int_0^t \mu_X(s)ds$ and equals 0 since $\mu_X = 0$. The covariance is calculated for $t_1 \geq 0$, $t_2 \geq 0$ as

$$K_{YY}(t_1, t_2) = \int_0^{t_1} \int_0^{t_2} K_{XX}(s_1, s_2)ds_1 ds_2,$$

$$= \sigma^2 \int_0^{t_1} \int_0^{t_2} \delta(s_1 - s_2)ds_1 ds_2,$$

$$= \sigma^2 \int_0^{t_2} u(t_1 - s_2)ds_2,$$

$$= \sigma^2 \int_0^{\min(t_1, t_2)} ds_2,$$

$$= \sigma^2 \min(t_1, t_2),$$

which we recognize as the covariance of the Wiener process (Section 7.2). Thus the m.s. integral of white Gaussian noise is the Wiener process. Note that $Y(t)$ must be Gaussian if $X(t)$ is Gaussian, since $Y(t)$ is the m.s. limit of a weighted sum of samples of $X(t)$ (see Problem 8.13).

We can generalize this integral by including a weighting function $h(t)$ multiplying the random process $X(t)$. It would thus be the mean-square limit of the following generalization of Equation 8.2-1:

$$I_n \triangleq \sum_{i=1}^{n} h(t_i)X(t_i)\Delta t_i.$$

The amounts to the following definition for the weighted integral.

Definition 8.2-2   The weighted mean-square integral of $X(t)$ over the interval (T_1, T_2) is defined by

$$\lim_{n\to\infty} E\left[\left|I - \sum_{i=1}^{n} h(t_i)X(t_i)\Delta t_i\right|^2\right] = 0$$

when the limit exists. We give it the following symbol:

$$I \triangleq \int_{T_1}^{T_2} h(t)X(t)dt. \quad \blacksquare$$

Example 8.2-2 ————————————————————————————————
(Application to linear systems.) A linear system L has response $h(t, s)$ at time t to an impulse applied at time s. Then for a deterministic function $x(t)$ as input, we have the output $y(t)$ given as

$$y(t) = L\{x(t)\} = \int_{-\infty}^{+\infty} h(t, s)x(s)ds, \tag{8.2-5}$$

whenever the foregoing integral exists. If $x(t)$ is bounded and integrable, one condition for the existence of Equation 8.2-5 would be

$$\int_{-\infty}^{+\infty} |h(t, s)|ds < +\infty \qquad \text{for all } -\infty < t < +\infty. \tag{8.2-6}$$

We can generalize this integral to a m.s. stochastic integral if the following double integral exists:

$$\int_{-\infty}^{+\infty} \int_{-\infty}^{+\infty} h(t_1, s_1)h^*(t_2, s_2)R_{XX}(s_1, s_2)ds_1ds_2,$$

in the ordinary calculus. A condition for this, in the case where R_{XX} is bounded and integrable, is Equation 8.2-6. Given the existence of such a condition, $Y(t)$ exists as an m.s. limit and defines a random process,

$$Y(t) \triangleq \int_{-\infty}^{+\infty} h(t, s)X(s)ds,$$

whose mean and covariance functions are

$$\mu_Y(t) = \int_{-\infty}^{+\infty} h(t,s)\mu_X(s)ds,$$

and

$$R_{YY}(t_1, t_2) = \int_{-\infty}^{+\infty} \int_{-\infty}^{+\infty} h(t_1, s_1)h^*(t_2, s_2)R_{XX}(s_1, s_2)ds_1 ds_2.$$

8.3 MEAN-SQUARE STOCHASTIC DIFFERENTIAL EQUATIONS

Having introduced stochastic derivatives and integrals, we now turn to the subject of stochastic differential equations. The simplest stochastic differential equation is

$$dY(t)/dt = X(t),$$

where the derivative on the left is an m.s. derivative. The solution turns out to be

$$Y(t) = \int_{t_0}^{t} X(s)ds + Y(t_0), \qquad t \geq t_0,$$

where the integral is an m.s. integral.

Using the general linear, constant-coefficient differential equation (LCCDE) as a model, we form the general stochastic LCCDE,

$$a_n Y^{(n)}(t) + a_{n-1}Y^{(n-1)}(t) + \ldots + a_0 Y(t) = X(t), \qquad (8.3\text{-}1)$$

for $t \geq 0$ with prescribed initial conditions,

$$Y(0), Y^{(1)}(0), \ldots, Y^{(n-1)}(0),$$

where the equality is in the m.s. sense.

To appreciate the meaning of Equation 8.3-1 more fully, we point out that it is a mean-square equality at each t separately. Thus Equation 8.3-1 is an equality at each t with probability-1 and hence at any countable collection of t's by Property 5 of Theorem 8.1-5. However, this does not say that the sample functions of $Y(t)$, which in fact may not even be differentiable, satisfy the differential equation driven by the sample functions of $X(t)$. The sample function interpretation of Equation 8.3-1 would require that it hold with probability-1 for *all* $t \geq 0$, which is an *uncountable* collection of times.

We can only think of the m.s. differential equation as an idealization approached in the limit, much like the impulse and the ideal lowpass filter. Also, the m.s. differential equation must be treated with care because it is not quite what it seems. Just as we would not put an impulse into a linear circuit in reality, similarly if we simulate a stochastic differential equation we would not use a random process lacking sample function derivatives as the input. Instead, we would use a smoother approximation to the process and if this smoothing is slight, we would expect that the idealized solution obtained from our m.s. differential equation would have similar properties. This, of course, needs further justification, but for the present we will assume that it can be made precise.

One may raise the question: Why work with such extreme processes, that is, processes without sample-function derivatives? The answer is that the analysis of these m.s. differential equations can proceed very well using the basic methods of linear system analysis. If we had instead included the extra "smoothing" to guarantee sample-function derivatives, then the analysis would be more complicated. For comparison, imagine trying to find the exact response of a linear system to a very narrow pulse of area one versus finding the ideal impulse response.

We proceed by finding the mean function $\mu_Y(t)$ and correlation function $R_{YY}(t_1, t_2)$ of the m.s. Equation 8.3-1. Since we have equality with probability-1 for each t, we can compute the expectations of both sides of Equation 8.3-1 and use

$$d^i E[Y(t)]/dt^i = E[Y^{(i)}(t)],$$

to obtain

$$a_n \mu_Y^{(n)}(t) + a_{n-1}\mu_Y^{(n-1)}(t) + \ldots + a_0 \mu_Y(t) = \mu_X(t) \tag{8.3-2}$$

with prescribed initial conditions at $t = 0$,

$$\mu_Y^{(i)}(0) = E[Y^{(i)}(0)] \qquad \text{for } i = 0, 1, \ldots, n-1.$$

Thus the mean function of $Y(t)$ is the solution to this linear differential equation, whose input is the mean function of $X(t)$. So knowledge of $\mu_X(t)$ is sufficient to determine $\mu_Y(t)$, that is, we do not have to know any of the higher moment functions of $X(t)$. Note that this would not be true if we were considering nonlinear differential equations. However, essentially no change would be necessary to accommodate time-varying coefficients, but of course the resulting equations would be much harder to solve. Thus we will stay with the constant-coefficient case. Parenthetically, we note that if $\mu_X(t) = 0$ for all $t > 0$ and the initial conditions are zero, then clearly $\mu_Y(t) = 0$ for all $t > 0$.

Next we determine the cross-correlation function $R_{XY}(t_1, t_2)$, which is the correlation between the input to Equation 8.3-1 at time t_1 and the output at time t_2. This quantity can be useful in system identification studies. We will assume for simplicity that the coefficients a_i are real. Then we substitute t_2 for t in Equation 8.3-1, conjugate, and multiply both sides by $X(t_1)$ to obtain

$$X(t_1)\left[\sum_{i=0}^{n} a_i Y^{(i)*}(t_2)\right] = X(t_1)X^*(t_2), \qquad t_1 \geq 0, \quad t_2 \geq 0,$$

which holds with probability-1 by Theorem 8.1-5, item (5). Taking expectations we obtain

$$\sum_{i=0}^{n} a_i E[X(t_1)Y^{(i)*}(t_2)] = E[X(t_1)X^*(t_2)],$$

or

$$\sum_{i=0}^{n} a_i R_{XY^{(i)}}(t_1, t_2) = R_{XX}(t_1, t_2),$$

which, using $R_{XY^{(i)}} = \partial^{(i)} R_{XY}/\partial t_2^i$, is the same as

$$\sum_{i=0}^{n} a_i \partial^{(i)} R_{XY}(t_1, t_2)/\partial t_2^i = R_{XX}(t_1, t_2), \qquad t_2 \geq 0, \tag{8.3-3}$$

for each $t_1 \geq 0$, subject to the initial conditions

$$\partial^{(i)} R_{XY}(t_1, 0)/\partial t_2^i \quad \text{for} \quad i = 0, 1, \ldots, n-1.$$

To obtain a differential equation for $R_{YY}(t_1, t_2)$, we multiply both sides of Equation 8.3-1 by $Y^*(t_2)$ and similarly obtain, for each $t_2 \geq 0$,

$$\sum_{i=0}^{n} a_i \partial^{(i)} R_{YY}(t_1, t_2)/\partial t_1^i = R_{XY}(t_1, t_2) \qquad \text{for } t_1 \geq 0, \tag{8.3-4}$$

with initial conditions $\partial^{(i)} R_{YY}(0, t_2)/\partial t_1^i$, for $i = 0, 1, \ldots, n-1$.

One can obtain equations identical to Equations 8.3-3 and 8.3-4 for the covariance functions K_{XY} and K_{YY} by noting that equation 8.3-2 can be used to center Equation 8.3-1 at the means of X and Y. This follows from the linearity of Equation 8.3-1 and converts it to an m.s. differential equation in Y_c and X_c. This then yields Equations 8.3-3 and 8.3-4 for the covariance functions.

We now turn to the solution of these partial differential equations. We will solve Equation 8.3-3 first, followed by Equation 8.3-4. We also note that Equation 8.3-3 is not really a partial differential equation since t_1 just plays the role of a constant parameter. Thus we must first solve Equation 8.3-3, an LCCDE in time variable t_2, for each t_1, thereby obtaining the cross-correlation R_{XY}. Then we use this function as input to Equation 8.3-4, which in turn is solved in time variable t_1, for each value of the parameter t_2. What remains is the problem of obtaining the appropriate initial conditions for these two deterministic LCCDEs from the given stochastic LCCDE Equation 8.3-1. This is illustrated by the following example, which also shows the formal advantages of working with the idealization called white noise.

Example 8.3-1

Let $X(t)$ be a stationary random process with mean μ_X and covariance function $K_{XX}(\tau) = \sigma^2 \delta(\tau)$.[†] Let $Y(t)$ be the solution to the differential equation

$$dY(t)/dt + \alpha Y(t) = X(t), \qquad t \geq 0, \tag{8.3-5}$$

[†]Note that the parameter σ^2 used here is *not* the variance of this white noise process. In fact $K_{XX}(0) = \infty$ for white noise.

subject to the initial condition $Y(0) = 0$. We assume $\alpha > 0$ for stability. Then the mean μ_Y is the solution to the first-order differential equation

$$\mu_Y'(t) + \alpha\mu_Y(t) = \mu_X, \qquad t \geq 0,$$

subject to $\mu_Y(0) = 0$. This initial condition comes from the fact that the initial random variable $Y(0)$ equals the constant 0 and therefore $E[Y(0)] = 0$. The solution is then easily obtained as

$$\mu_Y(t) = (\mu_X/\alpha)(1 - \exp(-\alpha t)) \qquad \text{for } t \geq 0.$$

Next we use the covariance version of Equation 8.3-3 specialized to the first-order differential Equation 8.3-5 to obtain the cross-covariance differential equation

$$\partial K_{XY}(t_1, t_2)/\partial t_2 + \alpha K_{XY}(t_1, t_2) = \sigma^2\delta(t_1 - t_2),$$

to be solved for $t_2 \geq 0$ subject to the initial condition, $K_{XY}(t_1, 0) = 0$, which follows from $Y(0) = 0$. For $0 \leq t_2 < t_1$, the solution is just 0 since the input is zero for $t_2 < t_1$. For the interval $t_2 \geq t_1$, we get the delayed impulse response $\sigma^2 \exp(-\alpha(t_2 - t_1))$ since the input is a delayed impulse occurring at $t_2 = t_1$. Thus the overall solution is

$$K_{XY}(t_1, t_2) = \begin{cases} 0, & 0 \leq t_2 < t_1, \\ \sigma^2 \exp(-\alpha(t_2 - t_1)), & t_2 \geq t_1. \end{cases}$$

This cross-covariance function is plotted in Figure 8.3-1 for $\alpha = 0.7$ and $\sigma^2 = 3$.

Next we obtain the differential equation for the output covariance K_{YY} by specializing the covariance version of Equation 8.3-4 to the first-order m.s. differential Equation 8.3-5:

$$\frac{\partial K_{YY}(t_1, t_2)}{\partial t_1} + \alpha K_{YY}(t_1, t_2) = K_{XY}(t_1, t_2),$$

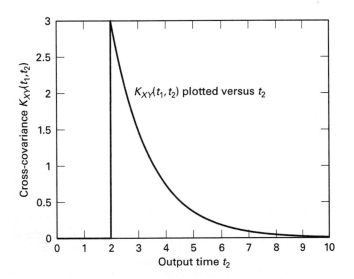

Figure 8.3-1   Cross-covariance plotted versus t_2 for the fixed value $t_1 = 2$.

subject to the initial condition at $t_1 = 0$ (for each t_2):

$$K_{YY}(0, t_2) = 0$$

for the interval $0 < t_1 \leq t_2$,

$$K_{XY}(t_1, t_2) = \sigma^2 \exp(-\alpha(t_2 - t_1)),$$

so

$$\partial K_{YY}(t_1, t_2)/\partial t_1 + \alpha K_{YY}(t_1, t_2) = \sigma^2 \exp(-\alpha(t_2 - t_1)),$$

which has solution

$$K_{YY}(t_1, t_2) = (\sigma^2/2\alpha)e^{-\alpha t_2}(e^{\alpha t_1} - e^{-\alpha t_1}).$$

For $t_1 \geq t_2$, $K_{XY}(t_1, t_2) = 0$, we then have to solve

$$\partial K_{YY}(t_1, t_2)/\partial t_1 + \alpha K_{YY}(t_1, t_2) = 0,$$

subject to $K_{YY}(t_2, t_2) = (\sigma^2/2\alpha)[1 - \exp(-2\alpha t_2)]$. We obtain,

$$K_{YY}(t_1, t_2) = \frac{\sigma^2}{2\alpha}(1 - \exp(-2\alpha t_2)) \exp(-\alpha(t_1 - t_2)), \qquad t_1 \geq t_2.$$

The overall function is plotted versus t_1 in Figure 8.3-2 for the same α and σ^2 as in Figure 8.3-1.

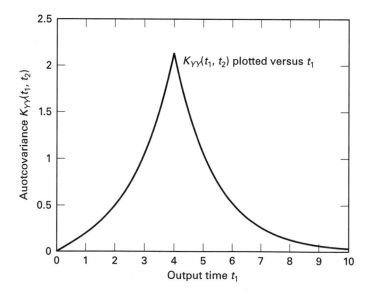

Figure 8.3-2   Plot of output covariance versus t_1 with fixed $t_2 = 4$.

We note that as t_1 and $t_2 \to +\infty$, the variance of $Y(t)$ tends to a constant, that is, $K_{YY}(t,t) \to \sigma^2/2\alpha$. In fact with $t_1 = t + \tau$ and $t_2 = t$, the covariance of $Y(t)$ becomes

$$K_{YY}(t+\tau, t) = \begin{cases} \dfrac{\sigma^2}{2\alpha}(1 - \exp(-2\alpha t))\exp(-\alpha\tau), & \tau \geq 0, \\[2ex] \dfrac{\sigma^2}{2\alpha}(\exp \alpha\tau - e^{-2\alpha t}\exp(-\alpha\tau)), & \tau < 0. \end{cases}$$

Now if we let $t \to +\infty$ for any fixed value of τ, we obtain

$$K_{YY}(\tau) = (\sigma^2/2\alpha)\exp(-\alpha|\tau|).$$

This is an example of what is called *asymptotic wide-sense stationarity*. It happens here because the input random process is WSS and the LCCDE is stable. In fact the only thing creating the nonstationarity is the zero initial condition, the effect of which decays away with time due to the stability assumption.

The reader may wonder at this point whether the random process $Y(t)$ with correlation R_{YY} and cross-correlation R_{XY} given by Equations 8.3-3 and 8.3-4 actually satisfies Equation 8.3-1 in the m.s. sense. This necessary question is taken up in Problem 8.18 at the end of this chapter.

8.4 ERGODICITY [8-3]

Until now we have generally assumed that a statistical description of the random process is available. Of course, this is seldom true in practice; thus we must develop some way of learning the needed statistical quantities from the observed sample functions of the random processes of interest. Fortunately, for many stationary random processes, we can substitute time averages for the unknown ensemble averages. We can use the stochastic integral defined in Section 8.2 to form a time average of the random process $X(t)$ over the interval $[-T, T]$,

$$\frac{1}{2T}\int_{-T}^{+T} X(t)dt.$$

In many cases this average will tend to the ensemble mean as T goes to infinity. When this happens for a random process, we say the random process is ergodic in the mean. Other types of ergodicity can be defined, such as ergodic in power or ergodic in correlation function or ergodic in the probability distribution function. Each type of ergodicity means that the corresponding time average converges to the named ensemble average.

This gives us a way of learning the probability and moment functions, which can then be used to statistically characterize the random processes of interest. To study convergence of the respective time averages and hence determine whether the required ergodicity holds, we need to decide on a type of convergence for the random variables in question. For most of our work we will adopt the relatively simple concept of mean-square convergence. For example, we might say that the process $X(t)$ is mean-square ergodic in both the mean function and the covariance function.

The property of mean-square ergodicity occurs when the random process decorrelates sufficiently rapidly with time shift so that the time average in question looks like the average of many almost uncorrelated random variables, which in turn—by appropriate forms of the weak Law of Large Numbers—will converge to the appropriate ensemble average. That is, we can write upon setting $\Delta T = T/N$,

$$\frac{1}{2T}\int_{-T}^{+T} X(t)dt = \frac{1}{2N\Delta T}\int_{-N\Delta T}^{+N\Delta T} X(t)dt$$

$$= \frac{1}{2N}\sum_{n=-N}^{+(N-1)}\left(\frac{1}{\Delta T}\int_{n\Delta T}^{(n+1)\Delta T} X(t)dt\right),$$

where the terms in the sum are approximately uncorrelated if ΔT is large enough. If the random process stays highly correlated over arbitrarily long time intervals, then we would not expect such behavior, and indeed ergodicity would not hold. Two simple examples are

(1) $X(t) = A$ where A is a random variable;
(2) $X(t) = A\cos 2\pi ft + B\sin 2\pi ft$, where A and B are random variables with $E[A] = E[B] = 0$, $E[A^2] = E[B^2] > 0$, and $E[AB] = 0$.

Example 1 is clearly not ergodic because any time average of $X(t)$ is just A, a random variable Thus there is no convergence to the ensemble mean $E[A]$. Example 2 can be shown to be WSS and ergodic in the mean but is not ergodic in power.

Definition 8.4-1   A wide-sense stationary (WSS) random process is *ergodic in the mean* if the time average of $X(t)$ converges to the ensemble average $E[X(t)] = \mu_X$, that is,

$$\hat{M}(T) \triangleq \frac{1}{2T}\int_{-T}^{+T} X(t)dt \to \mu_X \quad \text{(m.s.)} \quad \text{as } T \to \infty. \quad \blacksquare$$

In the above equation we observe that the time average $\hat{M}$ is a random variable. Hence we can compute its mean and variance using the theory of m.s. integrals obtaining

$$E[\hat{M}] = \frac{1}{2T}\int_{-T}^{+T} E[X(t)]dt = \mu_X, \tag{8.4-1}$$

and

$$\sigma_{\hat{M}}^2 = \frac{1}{(2T)^2}\int_{-T}^{+T}\int_{-T}^{+T} K_{XX}(t_1 - t_2)dt_1 dt_2. \tag{8.4-2}$$

The mean of $\hat{M}$ is thus the ensemble mean of $X(t)$, so if the variance is small the estimate will be good. Estimates that have the correct mean value are said to be *unbiased* (cf. Definition 5.8-2). Noting the mean value from Equation 8.4-1 we see that

$$\sigma_{\hat{M}}^2 = E[|\hat{M} - \mu_X|^2];$$

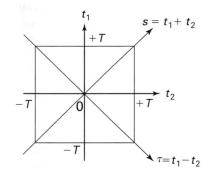

Figure 8.4-1 Square region in t_1, t_2 plane for integral in Equation 8.4-3.

thus convergence of the integral in Equation 8.4-2 to zero is the same as the convergence of $\hat{M}$ to μ_X in the m.s. sense. Since we will mostly deal with m.s. ergodicity, we will omit its mention with the understanding that unless otherwise stated time averages are computed in the m.s. sense.

To evaluate the integral in Equation 8.4-2 we look at Figure 8.4-1, which shows the area over which the integration is performed. Realizing that the random process $X(t)$ is WSS and the covariance function is thus only a function of the difference between t_1 and t_2, we can make the following change of variables that simplifies the integral:

$$
\begin{aligned}
s &\triangleq t_1 + t_2 \\
\tau &\triangleq t_1 - t_2
\end{aligned}
\quad \text{with Jacobian } |\mathbf{J}| = \text{abs}
\begin{vmatrix}
\dfrac{\partial s}{\partial t_1} & \dfrac{\partial s}{\partial t_2} \\[2mm]
\dfrac{\partial \tau}{\partial t_1} & \dfrac{\partial \tau}{\partial t_2}
\end{vmatrix}
= \text{abs}
\begin{vmatrix}
1 & 1 \\
1 & -1
\end{vmatrix}
= |-2| = 2,
$$

so $dt_1 dt_2 = |\mathbf{J}|^{-1} ds d\tau = \frac{1}{2} ds d\tau$, and Equation 8.4-2 becomes

$$
\frac{1}{(2T)^2} \int_{-2T}^{+2T} \left[\int_{-(2T-|\tau|)}^{+(2T-|\tau|)} \frac{1}{2} K_{XX}(\tau) ds \right] d\tau = \frac{1}{(2T)^2} \int_{-2T}^{+2T} K_{XX}(\tau)(2T - |\tau|) d\tau. \quad (8.4\text{-}3)
$$

Thus we arrive at the equivalent condition for the ergodicity in the mean of a WSS random process:

$$
\frac{1}{2T} \int_{-2T}^{+2T} \left(1 - \frac{|\tau|}{2T} \right) K_{XX}(\tau) d\tau \to 0 \qquad T \to \infty.
$$

Note: A sufficient condition for ergodicity in the mean is the stronger condition

$$
\int_{-\infty}^{+\infty} |K_{XX}(\tau)| d\tau < \infty,
$$

that is, that the covariance function be absolutely integrable.

Theorem 8.4-1 A WSS random process $X(t)$ is ergodic in the mean iff its covariance function $K_{XX}(\tau)$ satisfies

$$\lim_{T \to \infty} \left[\frac{1}{2T} \int_{-2T}^{+2T} \left(1 - \frac{|\tau|}{2T} \right) K_{XX}(\tau) d\tau \right] = 0.$$

We note that this is the same as saying that the estimate $\hat{M}$ converges to μ_X in the mean-square sense, hence the name m.s. ergodicity. Since m.s. convergence also implies convergence in probability, we have that $\hat{M}$ also converges to the ensemble mean μ_X in probability. This is analogous to the weak Law of Large Numbers for uncorrelated random sequences as studied in Section 7.6.

We can also define ergodicity in the higher moments, for example, mean square or power.

Definition 8.4-2 A WSS random process $X(t)$ is *ergodic in mean square* if

$$\lim_{T \to \infty} \left[\frac{1}{2T} \int_{-T}^{+T} X^2(t) dt \right] = R_{XX}(0) \qquad \text{(m.s.).} \quad \blacksquare$$

Similarly, we can define ergodicity in correlation and covariance.

Definition 8.4-3 A WSS random process $X(t)$ is *ergodic in correlation* for shift (lag) λ iff

$$\lim_{T \to \infty} \left[\frac{1}{2T} \int_{-T}^{+T} X(t + \lambda) X^*(t) dt \right] = R_{XX}(\lambda). \qquad \text{(m.s.).} \quad \blacksquare$$

If this condition is true for all λ, we say X is *ergodic in correlation*. The conditions for the preceding two types of ergodicities are covered by the following theorem on ergodicity in correlation, where we have defined the random process for each λ,

$$\Phi_\lambda(t) \triangleq X(t + \lambda) X^*(t).$$

Theorem 8.4-2 The WSS random process $X(t)$ is ergodic in correlation at the shift λ iff

$$\lim_{T \to \infty} \left[\frac{1}{2T} \int_{-T}^{+T} \left(1 - \frac{|\tau|}{2T} \right) K_{\Phi_\lambda \Phi_\lambda}(\tau) d\tau \right] = 0. \tag{8.4-4}$$

Proof The time-average estimate of the correlation function found in Definition 8.4-3 is just the time average of the random process $\Phi_\lambda(t)$ for each fixed λ. Thus we can apply Theorem 8.4-1 to the ergodicity in the mean problem for $\Phi_\lambda(t)$ as long as this process is WSS, that is, $X(t)$ has stationary second and fourth moments. The preceding condition is then seen to be the same as in Theorem 8.4-1 with the substitution of the covariance function $K_{\Phi_\lambda \Phi_\lambda}(\tau)$ for $K_{XX}(\tau)$. $\blacksquare$

Note that

$$K_{\Phi_\lambda \Phi_\lambda}(\tau) = R_{\Phi_\lambda \Phi_\lambda}(\tau) - |R_{XX}(\tau)|^2,$$

where

$$R_{\Phi_\lambda \Phi_\lambda}(\tau) = E\left[\Phi_\lambda(t+\tau)\Phi_\lambda^*(t)\right]$$
$$= E\left[X(t+\tau+\lambda)X^*(t+\tau)X^*(t+\lambda)X(t)\right],$$

which shows explicitly that $X(t)$ must have the fourth-moment stationarity here denoted for this theorem to apply. Some examples of WSS random processes that are ergodic in various senses follow.

Example 8.4-1 _____

Consider the WSS process of Example 7.1-5, which is a random amplitude cosine at frequency f_0 with random phase,

$$X(t) = A\cos(2\pi f_0 t + \Theta), \qquad -\infty < t < +\infty.$$

Here A is $N(0,1)$, Θ is uniformly distributed over $[-\pi, +\pi]$, and both A and Θ are independent. Then

$$E[X(t)] = E[A]E[\cos(2\pi f_0 t + \Theta)] = 0$$

and

$$E[X(t+\tau)X(t)] = \frac{1}{2\pi}\int_{-\pi}^{+\pi} \cos(2\pi f_0 t + 2\pi f_0\tau + \theta)\cos(2\pi f_0 t + \theta)d\theta$$

$$= \frac{1}{2}\cos(2\pi f_0\tau), \qquad -\infty < \tau < +\infty,$$

so that $X(t)$ is indeed wide-sense stationary. (In fact it can be shown that the process is also strict-sense stationary.) We first inquire whether $X(t)$ is ergodic in the mean; hence we compute

$$\sigma_{\hat{M}}^2 = \frac{1}{2t}\int_{-2T}^{+2T}\left[1 - \frac{|\tau|}{2T}\right]\frac{1}{2}\cos(2\pi f_0\tau)\,d\tau.$$

If we realize that the triangular term in the square brackets can be expressed as the convolution of two rectangular pulses, as shown in Figure 8.4-2, then we can write the Fourier transform of the triangular pulse as the square of the transform of one rectangular pulse of half the width:

$$\left(\sqrt{2T}\frac{\sin 2\pi fT}{2\pi fT}\right)^2 = 2T\left(\frac{\sin 2\pi fT}{2\pi fT}\right)^2,$$

and then use Parseval's theorem [8-4] to evaluate the above variance as

$$\frac{1}{2T}\int_{-2T}^{+2T}\left[I - \frac{|\tau|}{2T}\right]\frac{1}{2}\cos 2\pi f_0\tau d\tau = \frac{1}{2T}\int_{-\infty}^{+\infty} 2T\left(\frac{\sin 2\pi fT}{2\pi fT}\right)^2 \pi[\delta(f+f_0) + \delta(f-f_0)]df.$$

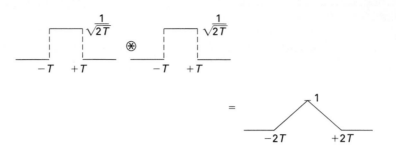

Figure 8.4-2   Illustration of the convolution of two rectangular pulses.

Thus we have

$$\sigma_{\hat{M}}^2 = \frac{1}{2}\left(\frac{\sin 2\pi f_0 T}{2\pi f_0 T}\right) \to 0 \quad \text{as } T \to \infty \text{ for } f_0 \neq 0.$$

Hence this random process is ergodic in the mean for any $f_0 \neq 0$.

To determine whether it is ergodic in power, we could use the condition of Equation 8.4-3. However, in this simple case we can obtain the result by examining the time average directly. Thus

$$\frac{1}{2T}\int_{-T}^{+T} X^2(t)dt = A^2 \frac{1}{2T}\int_{-T}^{+T}\cos^2(2\pi f_0 t + \Theta)dt.$$

Clearly, this time average will converge to $A^2/2$ not to $E[A^2]/2$ since for any Θ, the time average of the cosine squared will converge to $1/2$ for $f_0 \neq 0$. Thus this random cosine is *not* ergodic in power and hence *not ergodic in correlation* either. This is not unexpected since $K_{XX}(\tau)$ does not tend to zero as $|\tau|$ tends to infinity. ($K_{XX}(\tau)$ is in fact periodic!)

Another useful type of ergodicity is ergodicity in distribution function. Here we consider using time averages to estimate the distribution function of some random process X, which is at least stationary to first order; that is, the first-order PDF is shift-invariant. We can form such a time average by first forming the *indicator process* $I_x(t)$,

$$I_x(t) \triangleq \begin{cases} 1, & \text{if } X(t) \leq x, \\ 0, & \text{else.} \end{cases}$$

Thus the random process $I_x(t)$ for each fixed x is one if the event $\{X(t) \leq x\}$ occurs and zero if the event $\{X(t) \leq x\}$ does not occur. The function $I_x(t)$ thus "indicates" the event in the sense of Boolean logic. Since $I_x(t)$ is a function of the random process $X(t)$, it in turn is a random process. The time average value of $I_x(t)$ can be used to estimate the distribution function $F_X(x;t) = P[X(t) \leq x]$ as

$$\hat{F}_X(x) \triangleq \frac{1}{2T}\int_{-T}^{+T} I_x(t)dt. \tag{8.4-5}$$

First we consider the mean of this estimate. It is directly seen that

$$E[\hat{F}_X(x)] = E[I_x(t)] = 1 \cdot P[X(t) \leq x] = F_X(x;t).$$

Next we consider the correlation function of $I_x(t)$,

$$E[I_x(t_1)I_x(t_2)] = P[X(t_1) \leq x, X(t_2) \leq x]$$
$$= F_X(x, x; t_1, t_2)$$
$$= F_X(x, x; t_1 - t_2),$$

where the last line follows if X is stationary of order 2.[†] Thus we can say that $I_x(t)$ will be a WSS random process iff $X(t)$ is stationary of order 2. In this case we can apply Theorem 8.4-1 to $I_x(t)$ to get the following result.

Theorem 8.4-3 A random process $X(t)$, stationary up to order 2, is ergodic in distribution iff

$$\lim_{T \to \infty} \left[\frac{1}{2T} \int_{-T}^{+T} \left[1 - \frac{|\tau|}{2T} \right] K_{I_x I_x}(\tau) d\tau \right] = 0,$$

where

$$K_{I_x I_x}(\tau) = E[I_x(t + \tau)I_x(t)] - E^2[I_x(t)]$$
$$= F_X(x, x; \tau) - F_X^2(x; 0)$$
$$= F_X(x, x; \tau) - F_X^2(x).$$

Thus $K_{I_x}(\tau)$ must generally decay to zero as $|\tau| \to +\infty$ for the foregoing condition to be met; that is, we generally need

$$F_X(x, x; \tau) \to F_X^2(x) \quad \text{as } |\tau| \to +\infty. \tag{8.4-6}$$

This would be saying that $X(t + \tau)$ and $X(t)$ are asymptotically independent as $|\tau| \to +\infty$.

Example 8.4-2 _____
Let $X(t)$ be a random process with covariance function

$$K_{XX}(\tau) = \sigma_X^2 \exp(-\alpha|\tau|).$$

Then $X(t)$ is ergodic in the mean since $K_{XX}(\tau)$ is absolutely integrable. If we further assume that $X(t)$ is a Gaussian random process, then using the Gaussian fourth-order moment property (cf. Problem 5.20), we can show ergodicity in correlation and hence in power and covariance. Also, again invoking the Gaussian property, we can conclude ergodicity in distribution since the Gaussian PDF is a continuous function of its mean and covariance so that Equation 8.4-6 is satisfied.

_____

[†]We recall that "stationary of order 2" means that the second-order PDFs and pdf's are time-invariant, i.e., they only depend on the difference of the two times t_1 and t_2.

We note that the three theorems of this section give the same kind of condition for the three types of ergodicity considered. The difference between them lies in the three different covariance functions. The general form of the condition is

$$\lim_{T \to \infty} \frac{1}{2T} \int_{-2T}^{+2T} \left(1 - \frac{|\tau|}{2T}\right) K(\tau) d\tau = 0. \tag{8.4-7}$$

We now present an equivalent simpler condition for the case when $K(\tau)$ has a limit as $|\tau| \to +\infty$.

Theorem 8.4-4   If a covariance $K(\tau)$ has a limit as $|\tau| \to +\infty$, then Equation 8.4-7 is equivalent to $\lim_{|\tau| \to \infty} K(\tau) = 0$.

Proof   If $K(\tau)$ tends to a nonzero value, then clearly Equation 8.4-7 will not hold. So assume $\lim_{\tau \to \infty} K(\tau) = 0$; then for T large enough, $|K| < \varepsilon$ so that

$$\frac{1}{2T} \left| \int_{-2T}^{+2T} \left(1 - \frac{|\tau|}{2T}\right) K(\tau) d\tau \right| \leq \frac{1}{2T}[4T\varepsilon + M], \qquad \text{where } M < \infty,$$

$$\to 2\varepsilon; \text{ as } T \to \infty$$

which was to be shown. Here, M is a bound on $|K(\tau)|$ over $[-T, +T]$.   ■

8.5 KARHUNEN–LOÈVE EXPANSION [8-5]

Another application of the stochastic integral is to the Karhunen–Loève expansion. The idea is to decompose a general second-order random process into an orthonormal expansion whose coefficients are uncorrelated random variables. The expansion functions are just deterministic functions that have been orthonormalized to serve as a basis set for this decomposition. The Karhunen–Loève (K–L) expansion has proved to be a very useful theoretical tool in such diverse areas as detection theory, pattern recognition, and image coding. It is often used as an intermediate step in deriving general results in these areas. We will present an example to show how the K–L expansion is used in optimal detection of a known signal in white Gaussian noise (Example 8.5-4).

Theorem 8.5-1   Let $X(t)$ be a zero-mean, second-order random process defined over $[-T/2, +T/2]$ with continuous covariance function $K_{XX}(t_1, t_2) = R_{XX}(t_1, t_2)$, because of zero-mean. Then we can write

$$X(t) = \sum_{n=1}^{\infty} X_n \phi_n(t) \text{ (m.s.)} \qquad \text{for } |t| \leq T/2 \tag{8.5-1}$$

with

$$X_n \triangleq \int_{-T/2}^{+T/2} X(t) \phi_n^*(t) dt, \tag{8.5-2}$$

where the set of functions $\{\phi_n(t)\}$ is a complete orthonormal set of solutions to the integral equation

$$\int_{-T/2}^{T/2} K_{XX}(t_1, t_2)\phi_n(t_2)dt_2 = \lambda_n\phi_n(t_1), \qquad |t_1| \leq T/2, \tag{8.5-3}$$

and the coefficients X_n are statistically[†] orthogonal; that is,

$$E[X_n X_m^*] = \lambda_n\delta_{mn}, \tag{8.5-4}$$

with δ_{mn} the Kronecker delta function.

The functions $\phi_n(t)$ are orthonormal in the sense that

$$\int_{-T/2}^{T/2} \phi_n(t)\phi_m^*(t)dt = \delta_{mn}. \tag{8.5-5}$$

In fact it is easy to show that any two normalized solutions $\phi_n(t)$ and $\phi_m(t)$ to the integral Equation 8.5-3 must be orthonormal if $(\lambda_n \neq \lambda_m)$ and both λ's are not zero. See Problem 8.27. Note that we could just as well have used the correlation function $R_{XX}(t_1, t_2)$ here in the K–L theorem because the mean of the random process $X(t)$ is assumed zero.

The interesting thing about this expansion is that the coefficients are uncorrelated or statistically orthogonal. Otherwise any expansion such as the Fourier series expansion would suffice.[‡] The point of this theorem is that there exists at least one set of orthonormal functions with the special property that the coefficients in its expansion are uncorrelated random variables. We break up the proof of this important theorem into two steps or lemmas. We also need a result from the theory of integral equations known as Mercer's theorem, which states that

$$K_{XX}(t_1, t_2) = \sum_{n=1}^{\infty} \lambda_n\phi_n(t_1)\phi_n^*(t_2). \tag{8.5-6}$$

This result is derived in the appendix at the end of this chapter. Also, a constructive method is shown in Facts 1 to 3 of this appendix to find the $\{\phi_n(t)\}$, the set of orthonormal solutions to Equation 8.5-3.

Lemma 1   If $X(t) = \sum_{n=1}^{\infty} X_n\phi_n(t)$ (m.s.) and the X_n's are statistically orthogonal, then the $\phi_n(t)$ must satisfy integral Equation 8.5-3.

Proof   We compute $X(t)X_n^* = \sum_m X_m X_n^*\phi_m(t)$, thus

$$E[X(t)X_n^*] = \sum_m E[X_m X_n^*]\phi_m(t) = E[|X_n|^2]\phi_n(t),$$

[†]We say statistically orthogonal coefficients X_n to avoid confusion with the deterministic orthogonality of the basis functions $\phi_n(t)$.

[‡]Incidentally, it can be shown that the m.s. Fourier series coefficients of $X(t)$ become asymptotically uncorrelated as $T \to \infty$ and further that the K–L eigenfunctions approach complex exponentials as $T \to \infty$ [8-6]. We will establish this result at the end of this chapter.

but also

$$E[X(t)X_n^*] = E\left[X(t)\int_{-T/2}^{T/2} X^*(s)\phi_n(s)ds\right]$$

$$= \int_{-T/2}^{T/2} K_{XX}(t,s)\phi_n(s)ds,$$

hence

$$\int_{-T/2}^{T/2} K_{XX}(t,s)\phi_n(s)ds = \lambda_n\phi_n(t) \quad \text{with} \quad \lambda_n \triangleq E[|X_n|^2]. \quad \blacksquare$$

We next show a partial converse to this lemma.

Lemma 2   If the orthonormal set $\{\phi_n(t)\}$ satisfies integral Equation 8.5-3, then the random-variable coefficients X_n given in Equation 8.5-2 must be statistically orthogonal.

Proof   By Equation 8.5-2

$$X_m = \int_{-T/2}^{T/2} X(t)\phi_m^*(t)dt,$$

so

$$E[X(t)X_m^*] = \int_{-T/2}^{T/2} E[X(t)X^*(s)]\phi_m(s)ds = \lambda_m\phi_m(t),$$

because the ϕ_n's satisfy Equation 8.5-3. Thus

$$E[X_n X_m^*] = E\left[\int_{-T/2}^{T/2} X(t)\phi_n^*(t)dt\, X_m^*\right]$$

$$= \int_{-T/2}^{T/2} E[X(t)X_m^*]\phi_n^*(t)dt$$

$$= \int_{-T/2}^{T/2} \lambda_m\phi_m(t)\phi_n^*(t)dt = \lambda_m\delta_{mn}. \quad \blacksquare$$

By combining the results in the above two lemmas we can see that the K–L coefficients will be statistically orthogonal if and only if the orthonormal basis functions are chosen as solutions to Equation 8.5-3. What remains to show is that the K–L expansion does in fact produce a mean-square equality, which we show below with the help of Mercer's theorem.

Proof of Theorem 8.5-1   Define $\hat{X}(t) \triangleq \sum_{n=1}^{\infty} X_n\phi_n(t)$ and consider $E[|X(t) - \hat{X}(t)|^2] = E[X(t)(X(t) - \hat{X}(t))^*] - E[\hat{X}(t)(X(t) - \hat{X}(t))^*]$. The second term is zero since

$$E[\hat{X}(t)(X(t) - \hat{X}(t))^*] = E\left[\sum_m X_m X^*(t)\phi_m(t) - \sum_{m,n} X_m X_n^* \phi_m(t)\phi_n^*(t)\right]$$

$$= \sum_m E[X_m X^*(t)]\phi_m(t) - \sum_{m,n} E[X_m X_n^*]\phi_m(t)\phi_n^*(t)$$

$$= \sum_m E[X_m X^*(t)]\phi_m(t) - \sum_{m,n} E[X_m X_n^*]\phi_m(t)\phi_n^*(t)$$

$$= \sum_m \lambda_m |\phi_m(t)|^2 - \sum_m \lambda_m |\phi_m(t)|^2 = 0,$$

by the first step in the proof of Lemma 2, so evaluating the first term, we get

$$E[|X(t) - \hat{X}(t)|^2] = K_{XX}(t,t) - \sum_{n=1}^{\infty} \lambda_n \phi_n(t)\phi_n^*(t),$$

which is zero by Equation 8.5-6, Mercer's theorem.   ■

We now present some examples of the calculation of the K–L expansion.

Example 8.5-1
(K–L expansion of white noise.) Let $K_{WW}(t_1, t_2) = \sigma^2 \delta(t_1 - t_2)$. Then we must find the ϕ_n's to satisfy the K–L integral equation,

$$\sigma^2 \int_{-T/2}^{T/2} \delta(t_1 - t_2)\phi(t_2)dt_2 = \lambda\phi(t_1), \qquad -T/2 \le t_1 \le +T/2,$$

or

$$\sigma^2 \phi(t) = \lambda\phi(t).$$

Thus in this case the $\phi(t)$ functions are arbitrary and all the λ_n's equal σ^2. So the expansion functions $\phi_n(t)$ can be any complete orthonormal set with corresponding eigenvalues taken to be $\lambda_n = \sigma^2$.

Note that this example, though easy, violates the second-order constraint on the covariance function in the K–L Theorem 8.5-1. Nevertheless, the resulting expansion can be shown to be valid in the sense of generalized random processes (cf. definition of white Gaussian noise in Section 8.1).

Example 8.5-2
(Random process plus white noise.) Here we look at what happens if we add a random process to the white noise of the previous example and then want to know the K–L expansion for the noisy process:

$$Y(t) = X(t) + W(t),$$

where $W(t)$ is white noise and X and W are orthogonal (Definition 7.4-1), which we denote as $X \perp W$. Plugging into the K–L equation as before, we obtain

$$\int_{-T/2}^{T/2} [K_{XX}(t,s) + \sigma^2 \delta(t-s)]\phi^{(y)}(s)ds = \lambda^{(y)}\phi^{(y)}(t)$$

$$= \int_{-T/2}^{T/2} K_{XX}(t,s)\phi^{(y)}(s)ds + \sigma^2\phi^{(y)}(t),$$

so

$$\int_{-T/2}^{T/2} K_{XX}(t,s)\phi^{(y)}(s)ds = (\lambda^{(y)} - \sigma^2)\phi^{(y)}(t),$$

where we use the superscripts x and y to denote the respective eigenvalues and eigenfunctions of $X(t)$ and $Y(t)$. But since

$$\int_{-T/2}^{T/2} K_X(t,s)\phi^{(x)}(s)ds = \lambda^{(x)}\phi^{(x)}(t),$$

we immediately obtain

$$\phi^{(y)}(t) = \phi^{(x)}(t)$$

and

$$\lambda^{(y)} - \sigma^2 = \lambda^{(x)}.$$

We see that the eigenfunctions, that is, the K–L basis functions, are the same for both the X and Y processes. The K–L coefficients $Y_n = (X_n + W_n)$, then have variances $\lambda_n^{(y)} = \lambda_n^{(x)} + \sigma^2$.

Example 8.5-3
(K–L expansion for Wiener process.) For this example let the time interval be $(0,T)$ to match our definition of the Wiener process. Using Equation 7.2-18 in the K–L integral Equation 8.5-3, we obtain,

$$\sigma^2 \int_0^T \min(t,s)\phi(s)ds = \lambda\phi(t), \qquad 0 < t < T. \tag{8.5-7}$$

or

$$\sigma^2 \left[\int_0^t s\phi(s)ds + t \int_t^T \phi(s)ds \right] = \lambda\phi(t).$$

We temporarily agree to set $\sigma^2 = 1$ to simplify the equations. The standard method of solution of Equation 8.5-7 is to differentiate it as many times as necessary to convert it to a differential equation. We then evaluate the boundary conditions needed to solve the differential equation by plugging the general solution back into the integral equation. Here we take one derivative with respect to t and obtain

$$\int_t^T \phi(s)ds = \lambda\dot{\phi}(t). \tag{8.5-8}$$

Taking a second derivative, we obtain a differential equation,

$$-\phi(t) = \lambda \ddot{\phi}(t),$$

with general solution,

$$\phi(t) = A \sin \frac{t}{\sqrt{\lambda}} + B \cos \frac{t}{\sqrt{\lambda}}.$$

Next we use the boundary conditions at 0 and T to determine the coefficients A, B, and λ. From Equation 8.5-7 at $t = 0+$ we get $\phi(0+) = 0$, which implies $B = 0$. From Equation 8.5-8 at $t = T-$ we get $\dot{\phi}(T-) = 0$, which implies that

$$\cos(T/\sqrt{\lambda}) = 0, \quad \text{i.e. } \lambda = \lambda_n = \left(\frac{T}{(n - \frac{1}{2})\pi} \right)^2 \text{ for } n \geq 1.$$

Finally, A is chosen such that ϕ is normalized, that is,

$$\int_0^T \phi_n^2(t) dt = 1, \text{ which implies } A = \sqrt{2/T}.$$

Thus we get the following solution for the K–L basis functions,

$$\phi_n(t) = \sqrt{2/T} \sin \left[\left(n - \tfrac{1}{2} \right) \pi t/T \right], n \geq 1.$$

Now by Problem 8.27, the $\phi_n(t)$ must satisfy $\phi_n \perp \phi_m$ for $n \neq m$ since the eigenvalues $\lambda_n = [T/(n - \frac{1}{2})\pi]^2$ are distinct. This is the K–L expansion for the *standard* Wiener process, that is, one with $\sigma = 1$. If $\sigma \neq 1$, then λ_n is replaced by with $\sigma^2 \lambda_n$.

We now present an application of the Karhunen–Loève expansion to a simple problem from the area of communication theory known as Detection Theory.

Example 8.5-4 _____

(Application to Detection Theory.) Assume we observe a waveform $X(t)$ over the interval $[-T/2, +T/2]$ and wish to decide whether it contains a signal buried in noise or just noise alone. To be more precise we define two hypotheses, H_1 and H_0, and consider the decision theory problem:

$$X(t) = \begin{cases} m(t) + W(t): & H_1 \\ W(t): & H_0 \end{cases}$$

where $m(t)$ is a deterministic function, that is the *signal*, and $W(t)$ is the *noise* modeled by a zero-mean, white Gaussian process. Using the K–L expansion, we can simplify the preceding decision problem by replacing this waveform problem by a sequence of simpler scalar problems,

$$X_n = \begin{cases} m_n + W_n: & H_1 \\ W_n: & H_0 \end{cases}$$

where m_n and W_n are the respective K–L coefficients.

Effectively we take the K–L transform of the original received signal $X(t)$. The transform space is then just the space of sequences of K–L coefficients. Using the fact that the noise is zero-mean Gaussian and that the expansion coefficients are orthogonal, we conclude that the random variables W_n are jointly independent, that is, W_n is an independent random sequence. The problem can be simplified even further by observing that $K_{WW}(t_1, t_2) = \sigma_W^2 \delta(t_1 - t_2)$ permits the $\phi_n(t)$'s to be any complete set of orthonormal solutions to the K–L integral Equation 8.5-3. It is convenient to take $\phi_1(t) = cm(t)$ where c is the normalizing constant

$$c \triangleq \left[\int_{-T/2}^{T/2} m^2(t)dt \right]^{-1/2},$$

and then complete the orthonormal set in any valid way. We then notice that all the m_k will be zero except for $k = 1$; thus only X_1 is affected by the presence or absence of the signal. One can then show that this detection problem can finally be reduced to the scalar problem,

$$X_1 = \begin{cases} c^{-1} + W_1: & H_1 \\ W_1: & H_0. \end{cases}$$

To compute X_1 we note that it is just the stochastic integral

$$X_1 = c \int_{-T/2}^{T/2} X(t)m(t)dt$$

that is often referred to as a matching operation. In fact, it can be performed by sampling the output of a filter whose impulse response is $h(t) = cm(T - t)$, where T is chosen large enough to make this impulse response causal. The filter output at time T is then X_1. This filter is called a *matched filter* and is widely used in communications and pattern recognition.

Another application of the K–L expansion is in the derivation of important results in linear estimation theory. Analogously to the preceding example, the approach is to reduce the waveform estimation problem to the simpler one of estimating the individual K–L coefficients (cf. Problems 8.29 and 8.30).

8.6 REPRESENTATION OF BANDLIMITED AND PERIODIC PROCESSES

Here we consider expansions of random processes in terms of sets of random variables. An example that we have already seen would be the Karhunen–Loève expansion of Section 8.5. In general, the sets of random variables will contain an infinite number of elements; thus we are equivalently representing a random process by a random sequence. This representation is essential for digital processing of waveforms. Also, when the coefficients in the representation or expansion are uncorrelated or independent, then important additional simplifications result. We start out by considering WSS processes whose power spectral densities have finite support; that is, the respective correlation functions are bandlimited. We then develop an m.s. sampling theorem.

Bandlimited Processes

Definition 8.6-1   A random process $X(t)$ that is WSS is said to be *bandlimited* to $[\omega_1, \omega_2]$ if $S_{XX}(\omega) = 0$ for $|\omega| \notin [\omega_1, \omega_2]$. When $\omega_1 = 0$ we say the process is *lowpass*, and we set $\omega_2 = \omega_c$, called the *cutoff* frequency.   ∎

In the case of a lowpass random process $X(t)$ we can use the ordinary sampling theorem for deterministic signals [8-4] to write the following representation for the lowpass function correlation function $R_{XX}(\tau)$ in terms of the infinite set of samples $R_{XX}(nT)$ taken at spacing $T = \pi/\omega_c$:

$$R_{XX}(\tau) = \sum_{n=-\infty}^{+\infty} R_{XX}(nT) \frac{\sin \omega_c(\tau - nT)}{\omega_c(\tau - nT)}. \tag{8.6-1}$$

It turns out that one can define a mean-square sampling theorem for WSS random processes, which we next state and prove.

Theorem 8.6-1   If a second-order WSS random process $X(t)$ is lowpass with cutoff frequency ω_c, then upon setting $T \overset{\Delta}{=} \pi/\omega_c$,

$$X(t) = \sum_{n=-\infty}^{+\infty} X(nT) \frac{\sin \omega_c(t - nT)}{\omega_c(t - nT)}. \qquad \text{(m.s.)}$$

We point out that the foregoing equality is in the sense of a m.s. limit, that is, with

$$X_N(t) \overset{\Delta}{=} \sum_{n=-N}^{+N} X(nT) \frac{\sin \omega_c(t - nT)}{\omega_c(t - nT)},$$

then $\lim_{N \to \infty} E\left[|X(t) - X_N(t)|^2\right] = 0$ for each t.

Proof   First we observe that

$$E\left[|X(t) - X_N(t)|^2\right] = E\left[(X(t) - X_N(t)) X^*(t)\right] - E\left[(X(t) - X_N(t)) X_N^*(t)\right]. \tag{8.6-2}$$

Since $X_N^*(t)$ is just a weighted sum of the $X^*(mT)$, we begin by obtaining the preliminary result that $E\left[(X(t) - X_N(t)) X^*(mT)\right] \to 0$:

$$E\left[\left(X(t) - \sum_n X(nT) \frac{\sin \omega_c(t - nT)}{\omega_c(t - nT)}\right) X^*(mT)\right]$$

$$= R_{XX}(t - mT) - \sum_{n=-N}^{+N} R_{XX}(nT - mT) \frac{\sin \omega_c(t - nT)}{\omega_c(t - nT)}$$

$$\overrightarrow{N \to \infty} \; R_{XX}(t - mT) - R_{XX}(t - mT) = 0,$$

where the last equality follows by replacing τ with $t - mT$ in $R_{XX}(\tau)$ and writing the sampling expansion for this bandlimited function of t. Setting $\hat{X}(t) \triangleq \lim X_N(t)$ in the m.s. sense, we get

$$E[(X(t) - \hat{X}(t))X^*(mT)] = 0,^\dagger$$

that is, the error $X(t) - \hat{X}(t)$ is orthogonal to $X(mT)$ for all m. We write this symbolically as

$$(X(t) - \hat{X}(t)) \perp X(mT), \qquad \forall m.$$

But then we also have that $X(t) - \hat{X}(t) \perp X_N(t)$ because $X_N(t)$ is just a weighted sum of $X(mT)$. Then letting $N \to +\infty$, we get

$$(X(t) - \hat{X}(t)) \perp \hat{X}(t), \quad \forall t,$$

which just means $E[(X(t) - \hat{X}(t))\hat{X}^*(t)] = 0$; thus the second term in Equation 8.6-2 is asymptotically zero. Considering the first term in Equation 8.6-2 we get

$$E\left[(X(t) - X_N(t))\, X^*(t)\right] = R_{XX}(0) - \sum_{n=-N}^{+N} R_{XX}(nT - t)\frac{\sin \omega_c(t - nT)}{\omega_c(t - nT)},$$

which tends to zero as $n \to +\infty$ by virtue of the representation

$$R_{XX}(0) = \sum_{n=-\infty}^{+\infty} R_{XX}(nT - t)\frac{\sin \omega_c(t - nT)}{\omega_c(t - nT)}$$

obtained by right-shifting the bandlimited $R_{XX}(\tau)$ in Equation 8.6-1 by the shift t, thereby obtaining

$$R_{XX}(\tau - t) = \sum_{n=-\infty}^{+\infty} R_{XX}(nT - t)\frac{\sin \omega_c(t - nT)}{\omega_c(\tau - nT)}$$

and then setting the free parameter $\tau = t$. Thus $E\left[|X(t) - X_N(t)|^2\right] \to 0$ as $N \to +\infty$. ∎

In words the m.s. sampling theorem tells us that knowledge of the sampled sequence is sufficient for determining the random process at time t up to an event of probability zero since two random variables equal in the m.s. sense are equal with probability-1. (See Theorem 8.1-5.)

To consider digital processing of the resulting random sequence, we change the notation slightly by writing X_a for the random process and use X to denote the corresponding random sequence. Then the mean of the random sequence $X[n] \triangleq X_a(nT)$ is $\mu_X = E[X_a(t)]$, and

---

†We have used the fact that $X(t)$ is second order, i.e., $E[|X(t)|^2] < \infty$, so that Theorem 8.1-5 applies and allows us to interchange the m.s. limit with the expectation operator.

the correlation function is $R_{XX}[m] = R_{X_a X_a}(mT)$. This then gives the following power spectral density for the random sequence:

$$S_{XX}(\omega) = \frac{1}{T} S_{X_a X_a} \left(\frac{\omega}{T} \right), \qquad |\omega| \leq \pi,$$

if $X(t)$ is lowpass with cutoff $\omega_c = \pi/T$. After digital processing we can restore the continuous-time random process using the m.s. sampling expansion. Note that we assume perfect sampling and reconstruction. Here the WSS random process becomes a stationary random sequence and the reconstructed process is again wide-sense stationary. If sample-and-hold type suboptimal reconstruction is used (as in Problem 7.1), then the reconstructed random process will not even be WSS.

We see that the coefficients in this expansion, that is the samples of $X_a(t)$ at spacing T, most often will be correlated. However, there is one case where they will be uncorrelated. That case is when the process has a flat power spectral density and is lowpass.

Example 8.6-1

Let $X_a(t)$ be WSS and bandlimited to $(-\omega_c, +\omega_c)$ with flat psd

$$S_{X_a X_a}(\omega) = S_{X_a X_a}(0) I_{(-\omega_c, +\omega_c)}(\omega)$$

as seen in Figure 8.6-1. Then $R_{X_a X_a}(\tau)$ is given as

$$R_{X_a X_a}(\tau) = R_{X_a X_a}(0) \frac{\sin \omega_c \tau}{\omega_c \tau} \quad \text{with} \quad R_{X_a X_a}(0) = \frac{\omega_c}{\pi} S_{X_a X_a}(0).$$

Since $R_{X_a X_a}(mT) = 0$ for $m \neq 0$ we see that the samples are orthogonal, all with the same average power $R_{X_a X_a}(0)$. Thus the random sequence $X[n]$ is a white noise and its psd is flat:

$$S_{XX}(\omega) = \frac{1}{T} S_{X_a X_a}(0), \qquad |\omega| < \pi$$

with correlation function a discrete-time impulse

$$R_{XX}[m] = R_{X_a X_a}(0) \, \delta[m].$$

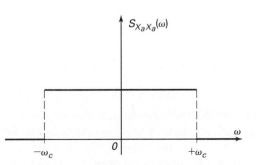

Figure 8.6-1   The psd of a bandlimited white random process.

Bandpass Random Processes

Next we consider the treatment of bandpass random processes. Such processes are used as models for random signals or noise that have been modulated by a carrier wave to enable long-distance transmission. Also, sometimes the frequency selective nature of the transmission medium converts a wideband signal into an approximately narrowband signal. Common examples are radio waves in the atmosphere and pressure waves in the ground or underwater.

First we show that we can construct a WSS bandpass random process $U(t)$ using two lowpass WSS processes. Thus consider a real-valued bandpass random process whose psd, for positive frequencies, is centered at ω_0:

$$U(t) = X(t)\cos(\omega_0 t) + Y(t)\sin(\omega_0 t), \qquad (8.6\text{-}3)$$

where the lowpass processes X and Y are real-valued, have zero-means, are wide-sense stationary, and satisfy

$$K_{XX}(\tau) = K_{YY}(\tau), \qquad (8.6\text{-}4)$$

$$K_{XY}(\tau) = -K_{YX}(\tau). \qquad (8.6\text{-}5)$$

In a representation such as Equations 8.6-3 thru 8.6-5, X is called the *in-phase component* of U, and Y is called it's *quadrature component*.

The symmetry conditions of Equations 8.6-4 and 8.6-5, while heavily constraining the component lowpass processes $\{X(t), Y(t)\}$, are sufficient to guarantee that Equation 8.6-3 is a WSS process (cf. Problem 8.37). We show below that this model is general enough to describe an arbitrary WSS bandpass noise process. In particular, we will see that the cross-covariance terms $K_{XY}(\tau)$ model the part of a general psd $S_{UU}(\omega)$ which is *odd* or nonsymmetrical about the center frequency ω_0, while the covariance terms $K_{XX}(\tau)$ model the part that is *even* with respect to the center frequency.

From Equations 8.6-4 and 8.6-5, it follows that

$$K_{UU}(\tau) = K_{XX}(\tau)\cos(\omega_0 \tau) + K_{YX}(\tau)\sin(\omega_0 \tau)$$

or

$$S_{UU}(\omega) = \frac{1}{2}\left[S_{XX}(\omega - \omega_0) + S_{XX}(\omega + \omega_0)\right] + \frac{1}{2j}\left[S_{YX}(\omega - \omega_0) - S_{YX}(\omega + \omega_0)\right].$$

Now $K_{XY}(\tau) = -K_{YX}(\tau)$. Also, $K_{YX}(\tau) = K_{XY}^*(-\tau) = K_{XY}(-\tau)$ since X and Y are real-valued, so

$$S_{YX}(\omega) = S_{XY}^*(\omega) = S_{XY}(-\omega) = -S_{XY}(+\omega).$$

From $S_{XY}^*(\omega) = -S_{XY}(\omega)$ we get that S_{XY} is pure imaginary. From $S_{XY}(-\omega) = -S_{XY}(\omega)$ we get that S_{XY} is an odd function of ω. The same holds for S_{YX} since $K_{XY}(\tau) = K_{YX}^*(-\tau) = K_{YX}(-\tau)$; thus $\frac{1}{2}S_{XX}(\omega - \omega_0)$ is the in-phase or even part of the psd at ω_0, and $(1/2j)S_{YX}(\omega - \omega_0)$ is the quadrature or odd part. These properties are illustrated in Figure 8.6-2.

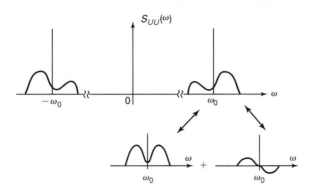

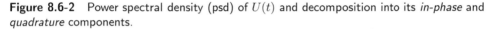

Figure 8.6-2   Power spectral density (psd) of $U(t)$ and decomposition into its *in-phase* and *quadrature* components.

Thus we can conclude that Equation 8.6-3 together with the symmetry conditions of Equations 8.6-4 and 8.6-5 on the component lowpass processes $X(t)$ and $Y(t)$ are sufficiently general to model an arbitrary WSS bandpass process $U(t)$.

To find the lowpass components, we could decompose the process $U(t)$ as shown in Figure 8.6-3; however, the random processes obtained after multiplication by cos and sin are not WSS so that the system of the figure may not be analyzed in the frequency domain. An alternative approach is through the *Hilbert transform operator*, defined as filtering with the system function

$$H(\omega) = -j\,\mathrm{sgn}(\omega).$$

Using this operator we can define an *analytic signal process* $Z(t)$,

$$Z(t) \overset{\Delta}{=} U(t) + j\check{U}(t), \tag{8.6-6}$$

where the superscript ˘ indicates a Hilbert transformation. Then it turns out that we can take

$$X(t) = Re[Z(t)e^{-j\omega_0 t}]$$

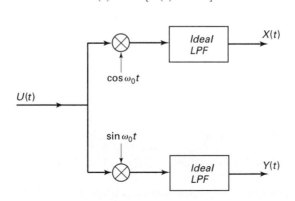

Figure 8.6-3   Decomposition of bandpass random process.

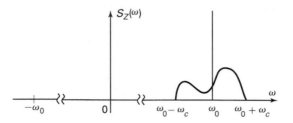

Figure 8.6-4   psd of analytic signal random process.

and

$$Y(t) = -Im[Z(t)e^{-j\omega_0 t}],$$

to achieve the desired representation (Equation 8.6-3). These X and Y are actually the same as in Figure 8.6-3. The psd of $Z(t)$ is

$$S_{ZZ}(\omega) = 4S_{UU}(\omega)\ u(\omega),\tag{8.6-7}$$

where $u(\omega)$ is the unit-step function. This psd is sketched in Figure 8.6-4, where we note that its support is restricted to positive ω.

One can use this theory for computer simulation of bandpass processes by first decomposing the process into its X and Y lowpass components, or equivalently the complex lowpass process $X(t) - jY(t)$, and then representing these lowpass processes by their sampled equivalents. Thus one can simulate bandpass processes and bandpass systems by discrete-time processing of coupled pairs of random sequences, that is, vector random sequences of dimension two, at a greatly reduced sample rate because X and Y are lowpass.

WSS Periodic Processes

A wide-sense stationary random process may have a correlation function that is periodic; that is, $R(\tau) = R(\tau + T)$ for all τ. This is a special case of the general periodic process introduced in Chapter 7.

Definition 8.6-2   A WSS random process $X(t)$ is *mean-square periodic* if for some T we have $R_{XX}(\tau) = R_{XX}(\tau + T)$ for all τ. We call the smallest such $T > 0$ the *period*. ∎

In Chapter 7, we called such processes wide-sense periodic. In Problem 8.38 the reader is asked to show that a m.s. periodic process is also periodic in the stronger sense:

$$E\left[|X(t) - X(t+T)|^2\right] = 0.$$

We now show this directly for the WSS periodic case. Evaluating we get

$$E[|X(t) - X(t+T)|^2] = 2(R_{XX}(0) - Re[R_{XX}(T)]).$$

Now since $R_{XX}(0) = R_{XX}(T)$, it follows that $R_{XX}(T)$ is real and $Re\{R_{XX}(T)\} = R_{XX}(T) = R_{XX}(0)$. Hence $X(t) = X(t+T)$ (m.s.) and hence also with probability-1.

Turning to the psd of a WSS periodic process, we know that $R_{XX}(\tau)$ has a Fourier series

$$R_{XX}(\tau) = \sum_{n=-\infty}^{+\infty} \alpha_n e^{+j\omega_0 n\tau}, \quad \left(\omega_0 = \frac{2\pi}{T}\right)$$

with coefficients

$$\alpha_n = \frac{1}{T} \int_{-T/2}^{+T/2} R_{XX}(\tau) e^{-j\omega_0 n\tau} \, d\tau.$$

Thus the psd of a WSS periodic process is a line spectra of the form

$$S_{XX}(\omega) = 2\pi \sum_{n=-\infty}^{+\infty} \alpha_n \delta(\omega - n\omega_0), \tag{8.6-8}$$

which can be summarized in the following theorem. We note that the α_n's are thus necessarily nonnegative.

Theorem 8.6-2   If a WSS random process $X(t)$ is m.s. periodic, then its psd is a line spectra with impulses at multiples of the fundamental frequency ω_0. The impulse areas are given by the Fourier coefficients of the periodic correlation function $R_{XX}(\tau)$.

Example 8.6-2 _____

Let the input to an LSI system with frequency response $H(\omega)$ be the periodic random process $X(t)$ as indicated in Figure 8.6-5. In general we have

$$S_{YY}(\omega) = |H(\omega)|^2 S_{XX}(\omega).$$

Using Equation 8.6-7, we get

$$S_{YY}(\omega) = 2\pi \sum_{n=-\infty}^{+\infty} \alpha_n |H(\omega_0 n)|^2 \delta(\omega - n\omega_0).$$

Hence the output $Y(t)$ is m.s. periodic with the same period T and has correlation function given by

$$R_{YY}(\tau) = \sum_{n=-\infty}^{+\infty} \alpha_n |H(\omega_0 n)|^2 e^{+j\omega_0 n\tau}.$$

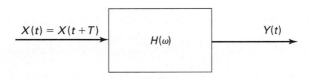

Figure 8.6-5   Periodic random process input to LSI system.

Deterministic functions that are periodic can be represented by Fourier series as we have done for the periodic correlation function. We now show that the WSS periodic process itself may be represented as a Fourier series in the mean-square sense.

Theorem 8.6-3   Let the WSS random process $X(t)$ be m.s. periodic with period T. Then we can write

$$X(t) = \sum_{n=-\infty}^{+\infty} A_n e^{+j\omega_0 nt}, \quad \text{(m.s.)}$$

where $\omega_0 = 2\pi/T$, with random Fourier coefficients

$$A_n \triangleq \frac{1}{T} \int_{-T/2}^{+T/2} X(\tau) e^{-j\omega_0 n\tau} d\tau, \qquad \text{(m.s.)}$$

with mean

$$E[A_n] = \mu_X \delta[n],$$

and correlation

$$E[A_n A_m^*] = \alpha_n \delta[m-n],$$

and mean-square value

$$\alpha_n \triangleq \frac{1}{T} \int_{-T/2}^{+T/2} R_{XX}(\tau) e^{-j\omega_0 n\tau} d\tau. \tag{8.6-9}$$

Thus the periodic random process can be expanded in a Fourier series whose coefficients are statistically orthogonal. Thus the Fourier series is the Karhunen–Loève expansion (see Section 8.5) for a WSS periodic process.

Proof   First we show that the A_n are statistically orthogonal. We readily see that

$$E[A_n] = \frac{1}{T} \int_{-T/2}^{+T/2} E[X(u)] e^{-j\omega_0 nu} du = \mu_X \delta[n].$$

Then

$$E[A_n^* X(t)] = \frac{1}{T} \int_{-T/2}^{+T/2} E[X^*(u)X(t)] e^{+j\omega_0 nu} du$$

$$= \frac{1}{T} \int_{-T/2}^{+T/2} R_{XX}(t-u) e^{+j\omega_0 nu} du$$

$$= \left[\frac{1}{T} \int_{-T/2+t}^{T/2+t} R_{XX}(\tau) e^{-j\omega_0 n\tau} d\tau \right] e^{+j\omega_0 nt}$$

$$= \alpha_n e^{+j\omega_0 nt},$$

since the integrand is periodic in τ with period T and also by Equation 8.6-9. Next we consider

$$E[A_k A_n^*] = \frac{1}{T} \int_{-T/2}^{+T/2} E[A_n^* X(u)] e^{-j\omega_0 k u} du.$$

$$= \alpha_n \left[\frac{1}{T} \int_{-T/2}^{+T/2} e^{+j\omega_0 n u} e^{-j\omega_0 k u} du \right]$$

$$= \alpha_n \delta[n - k].$$

It remains to show that

$$E\left[\left| X(t) - \sum_{n=-\infty}^{+\infty} A_n e^{+j\omega_0 n t} \right|^2 \right] = 0.$$

Expanding the left-hand side, we get

$$E[|X(t)|^2] - \sum_n E[A_n^* X(t)] e^{-j\omega_0 n t} - \sum_n E[A_n X^*(t)] e^{+j\omega_0 n t} + \sum_n \sum_k E[A_k A_n^*] e^{+j\omega_0 (k-n)t}$$

$$= R_{XX}(0) - \sum_n \alpha_n - \sum_n \alpha_n^* + \sum_n \alpha_n.$$

But the α_n are real since $\alpha_n = E[|A_n|^2]$, and also $R_{XX}(0) = \sum \alpha_n$ so that we finally have

$$\sum \alpha_n - \sum \alpha_n - \sum \alpha_n + \sum \alpha_n = 0. \quad \blacksquare$$

As shown in Problem 8.40, analogous results hold for WSS periodic random sequences; that is, $R[m] = R[m + T]$ or all m where $T = $ positive integer.

Fourier Series for WSS Processes

Of course, we can expand a general nonperiodic WSS random process in a Fourier series also. However, the corresponding expansion coefficients will not be statistically orthogonal. Yet in the limit as the length T of the expansion interval tends to infinity, the coefficients become asymptotically orthogonal (also asymptotically uncorrelated since zero-mean is assumed). Recalling the expansion and coefficient definition from Theorem 8.6-3, we can express the cross-correlation as

$$E[A_n A_m^*] = \frac{1}{T^2} \int_{-T/2}^{+T/2} \int_{-T/2}^{+T/2} R_{XX}(t_1 - t_2) e^{-j\omega_0 (n t_1 - m t_2)} \, dt_1 \, dt_2.$$

Next make the substitution $t = t_1 + t_2$ and $\tau = t_1 - t_2$ with Jacobian $|J| = 2$ as in the calculation of the psd in Section 7.5 to obtain

$$E[A_n A_m^*] = \frac{1}{T^2} \int_{-T}^{+T} R_{XX}(\tau) \exp\left(j \frac{\omega_0}{2}(n+m)\tau \right) \left(\frac{1}{2} \int_{-(T-|\tau|)}^{+(T-|\tau|)} \exp\left(-j \frac{\omega_0}{2}(n-m)t \right) dt \right) d\tau$$

$$= \frac{1}{T^2} \int_{-T}^{+T} R_{XX}(\tau) \exp\left(j\frac{\omega_0}{2}(n+m)\tau\right) \frac{\sin\left(\frac{\omega_0}{2}(n-m)(T-|\tau|)\right)}{\omega_0(n-m)} d\tau$$

$$= O\left(\frac{1}{T^2|n-m|}\right) \to 0 \quad \text{as} \quad T \to \infty.$$

If we let $n = m$, we get the coefficient variance (also mean-square value) as

$$E[|A_n|^2] = \frac{1}{T} \int_{-T}^{+T} \left(\frac{T-|\tau|}{T}\right) R_{XX}(\tau)e^{+j\omega_0 n\tau} d\tau \text{ converges to } \frac{1}{T}S_{XX}(n\omega_0) \quad \text{as} \quad T \to \infty.$$

Related to this Fourier series expansion are the discrete-time Fourier and Cosine transforms used in audio and video data compression [8-7]. These block transforms,[†] like the Fourier series for continuous expansion over a finite-length interval, asymptotically achieve the statistical orthogonality property of the exact Karhunen–Loève transform, when the random sequence is wide-sense stationary.

We still must show that the resulting expansion converges in the mean-square sense. Unlike the proof in the periodic case, we do not have the exact statistical orthogonality that was used in the proof of Theorem 8.6-3. Instead we proceed as follows:

$$E[|X - \hat{X}|^2] = E[(X - \hat{X})(X^* - \hat{X}^*)]$$

where

$$X(t_1) - \hat{X}(t_1) = X(t_1) - \frac{1}{T} \int_{-\frac{T}{2}}^{\frac{T}{2}} X(\tau_1) \left[\sum_{n_1=-N}^{+N} e^{+j\omega_0 n_1(t_1-\tau_1)}\right] d\tau_1$$

and

$$X^*(t_1) - \hat{X}^*(t_1) = X(t_1) - \frac{1}{T} \int_{-\frac{T}{2}}^{\frac{T}{2}} X^*(\tau_2) \left[\sum_{n_2=-N}^{+N} e^{-j\omega_0 n_2(t_2-\tau_2)}\right] d\tau_2.$$

Thus

$$E[(X(t_1) - \hat{X}(t_1))(X^*(t_2) - \hat{X}^*(t_2))]$$

$$= R_{XX}(t_1, t_2) - \frac{1}{T} \int_{-\frac{T}{2}}^{\frac{T}{2}} R_{XX}(\tau_1, t_2) \left[\sum_{n_1=-N}^{+N} e^{+j\omega_0 n_1(t_1-\tau_1)}\right] d\tau_1$$

$$- \frac{1}{T} \int_{-\frac{T}{2}}^{\frac{T}{2}} R_{XX}(t_1, \tau_2) \left[\sum_{n_2=-N}^{+N} e^{-j\omega_0 n_2(t_2-\tau_2)}\right] d\tau_2$$

$$+ \frac{1}{T^2} \int_{-\frac{T}{2}}^{\frac{T}{2}} \int_{-\frac{T}{2}}^{\frac{T}{2}} R_{XX}(\tau_1, \tau_2) \left[\sum_{n_1=-N}^{+N} e^{+j\omega_0 n_1(t_1-\tau_1)}\right] \left[\sum_{n_2=-N}^{+N} e^{-j\omega_0 n_2(t_2-\tau_2)}\right] d\tau_1\, d\tau_2.$$

[†]By *block transform*, we mean a transform of a finite block of data rather than the whole line.

Considering the third term,

$$\frac{1}{T} \int_{-\frac{T}{2}}^{\frac{T}{2}} R_{XX}(t_1, \tau_2) \left[\sum_{n_2=-N}^{+N} e^{-j\omega_0 n_2(t_2-\tau_2)} \right] d\tau_2,$$

the sum inside the square brackets tends to the impulse $T\delta(t_2 - \tau_2)$ as $N \to \infty$. Thus overall this term becomes asymptotically equal to $R_{XX}(t_1, t_2)$. The same is true for the second term. The last term becomes

$$\frac{1}{T^2} \int_{-\frac{T}{2}}^{\frac{T}{2}} \int_{-\frac{T}{2}}^{\frac{T}{2}} R_{XX}(\tau_1, \tau_2) \left[\sum_{n_1=-N}^{+N} e^{+j\omega_0 n_1(t_1-\tau_1)} \right] \left[\sum_{n_2=-N}^{+N} e^{-j\omega_0 n_2(t_2-\tau_2)} \right] d\tau_1 \, d\tau_2$$

$$\to \frac{1}{T^2} \int_{-\frac{T}{2}}^{\frac{T}{2}} \int_{-\frac{T}{2}}^{\frac{T}{2}} R_{XX}(\tau_1, \tau_2) T\delta(t_1 - \tau_1) T\delta(t_2 - \tau_2) \, d\tau_1 \, d\tau_2$$

$$= R_{XX}(t_1, t_2).$$

Since each of the four terms asymptotically approaches $R_{XX}(t_1, t_2)$, the mean-square value of $X - \hat{X}$ is zero, as was to be shown.

8.7 SUMMARY

We have studied an extension of many of the concepts of calculus to random processes by studying the mean-square calculus. This enabled us to define useful differential-equation and integral operators for a wide class of second-order random processes. In so doing, we derived further results on m.s. convergence, including the notion that random variables with finite mean-square value, that is, second-order random variables, can be viewed as vectors in a Hilbert space with the expectation of their conjugate product serving as an inner product. This viewpoint will be important for applications to linear estimation in Chapter 9.

We defined the m.s. stochastic integral and applied it to two problems: ergodicity, the problem of estimating the parameter functions of random processes, and the Karhunen–Loève expansion, an important theoretical tool that decomposes a possibly nonstationary process into a sum of products of orthogonal random variables and orthonormal basis functions.

8.8 APPENDIX: INTEGRAL EQUATIONS

In this appendix we look at some of the properties of the solution of integral equations that are needed to appreciate the Karhunen–Loève expansion presented in Section 8.5. In Facts 1 to 3 that follow, we develop a method to solve for a complete set of orthonormal solutions $\{\lambda_n, \phi_n(t)\}$. Consider the integral equation,

$$\int_a^b R(t, s)\phi(s)ds = \lambda\phi(t) \quad \text{on} = a \le t \le b, \tag{8.8-1}$$

where the *kernel* $R(t, s)$ is continuous, Hermitian, and positive semidefinite. Thus R fulfills the conditions to be a correlation function, although in the integral equation setting there may be no such interpretation. The solution to Equation 8.8-1 consists of the function $\phi(t)$, called an *eigenfunction*, and the scalar λ, called the corresponding *eigenvalue*. To avoid the trivial case, we rule out the solution $\phi(t) = 0$ as a valid eigenfunction.

A fundamental theorem concerns the existence of solutions to Equation 8.8-1. Its proof can be found in the book on functional analysis by F. Reisz and B. Sz-Nagy [8-8].

Existence Theorem

If the continuous and Hermitian kernel $R(t, s)$ in the integral Equation 8.8-1 is nonzero and positive semidefinite, then there exists at least one eigenfunction with nonzero eigenvalue.

While the proof of this theorem is omitted, it should seem reasonable based on our experience with computing the eigenvalues and eigenvectors of covariance matrices in Section 5.4. To see the connection, note that the integral in Equation 8.8-1 is the limit of a sum involving samples in s. If we require its solution only at samples in t, then we have equivalently the vector-matrix eigenvector problem of Section 5.4. So with some form of continuity in R and ϕ, the properties of the eigenvectors and eigenvalues should carry over to the present eigenfunctions and eigenvalues.

The existence theorem allows us to conclude several useful results concerning the solutions of Equation 8.8-1, culminating in Mercer's theorem, which is applied in Section 8.5 to prove the Karhunen–Loève representation. This method is adapted from Reference [8-2].

Fact 1.   All the eigenvalues must be real and nonnegative. Additionally, when the kernel $R(t, s)$ is positive definite over the interval (a, b), they must all be positive.

Proof   Since R is positive semidefinite,

$$\int_a^b \int_a^b \phi^*(t) R(t, s) \phi(s) dt ds \geq 0,$$

but

$$\int_a^b \phi^*(t) \left[\int_a^b R(t, s) \phi(s) ds \right] dt = \int_a^b \phi^*(t) \lambda \phi(t) dt$$

$$= \lambda \int_a^b |\phi(t)|^2 dt.$$

Thus λ is real and nonnegative since $\int_a^b |\phi(t)|^2 dt \neq 0$. Also if R is additionally positive definite, then λ cannot be zero.   ■

Fact 2.   Let $R(t, s)$ be Hermitian and positive semidefinite. Let ϕ_1, λ_1 be a corresponding eigenfunction and eigenvalue pair. Then

$$R_1(t, s) \triangleq R(t, s) - \lambda_1 \phi_1(t) \phi_1^*(s)$$

is also positive semidefinite.

Fact 3. Either $R_1(t, s) = 0$ for all t and s or else there is another eigenfunction and eigenvalue ϕ_2, λ_2 with[†] $\phi_2 \perp \phi_1$ such that

$$R_2(t, s) \overset{\Delta}{=} R_1(t, s) - \lambda_2 \phi_2(t) \phi_2^*(s)$$

is positive semidefinite.

Continuing with this procedure, we eventually obtain

$$R_N(t, s) \overset{\Delta}{=} R(t, s) - \sum_{n=1}^{N} \lambda_n \phi_n(t) \phi_n^*(s).$$

Now since

$$R_N(t, t) = R(t, t) - \sum_{n=1}^{N} \lambda_n |\phi_n(t)|^2 \geq 0,$$

we have that the increasing sum $\sum_{n=1}^{N} \lambda_n |\phi_n(t)|^2$ is bounded from above so that it must tend to a limit. Thus $R_\infty(t, s)$ exists for $s = t$. For $s \neq t$ consider the partial sum for $m > n$,

$$\Delta R_{m,n}(t, s) \overset{\Delta}{=} \sum_{k=n}^{m} \lambda_k \phi_k(t) \phi_k^*(s).$$

Then $|\Delta R_{m,n}(t, t)| \to 0$ as $m, n \to \infty$. Using the Schwarz inequality we conclude[‡]

$$|\Delta R_{m,n}(t, s)| \leq \sqrt{\Delta R_{m,n}(t, t) \cdot \Delta R_{m,n}(s, s)}.$$

Thus $|\Delta R_{m,n}(t, s)| \to 0$ as $m, n \to \infty$. Therefore we can write

$$R_\infty(t, s) \overset{\Delta}{=} R(t, s) - \sum_{n=1}^{\infty} \lambda_n \phi_n(t) \phi_n^*(s). \tag{8.8-2}$$

Mercer's theorem now can be seen as an affirmative answer to the question of whether or not $R_\infty(t, s) = 0$. We now turn to the proofs of Facts 2 and 3 prior to proving Mercer's theorem.

Proof of Fact 2 We must show that R_1 is positive semidefinite over (a, b). We do this by defining the random process

$$Y(t) \overset{\Delta}{=} X(t) - \phi_1(t) \int_a^b X(\tau) \phi_1^*(\tau) d\tau$$

[†]By $\phi_2 \perp \phi_1$ we mean $\int_a^b \phi_1(t) \phi_2^*(t) dt = 0$, so that upon normalization, assuming they are non-zero, ϕ_1 and ϕ_2 become orthonormal (Equation 8.5-5).

[‡]We can regard $\Delta R_{m,n}(t, s)$ as the correlation of two r.v.'s with variance $\Delta R_{m,n}(t, t)$ and $\Delta R_{m,n}(s, s)$. Alternatively we can use the Schwartz inequality for complex numbers: $|\Sigma a_i b_i|^2 \leq \left(\Sigma |a_i|^2 \right) \left(\Sigma |b_i|^2 \right)$.

and showing that R_1 is its correlation function. We have

$$E[Y(t)Y^*(s)] = E[X(t)X^*(s)] - \phi_1^*(s) \int_a^b E[X^*(\tau)X(t)]\phi_1(\tau)d\tau$$

$$-\phi_1(t) \int_a^b E[X(\tau)X^*(s)]\phi_1^*(\tau)d\tau$$

$$+\phi_1(t)\phi_1^*(s) \int_a^b \int_a^b E[X(\tau_1)X^*(\tau_2)]\phi_1^*(\tau_1)\phi_1(\tau_2)d\tau_1 d\tau_2$$

$$= R_X(t,s) - \phi_1^*(s) \int_a^b R_{XX}(t,\tau)\phi_1(\tau)d\tau$$

$$-\phi_1(t) \int_a^b R_{XX}(\tau,s)\phi_1^*(\tau)d\tau$$

$$+\phi_1(t)\phi_1^*(s) \int_a^b \int_a^b R_{XX}(\tau_1,\tau_2)\phi_1^*(\tau_1)\phi_2(\tau_2)d\tau_1 d\tau_2,$$

but $\lambda_1\phi_1(t) = \int_a^b R_{XX}(t,\tau)\phi_1(\tau)d\tau$ and using $R_{XX}^*(t,\tau) = R_{XX}(\tau,t)$ we get

$$\lambda_1\phi_1^*(s) = \int_a^b R_{XX}(\tau,s)\phi_1^*(\tau)d\tau,$$

so that the two cross-terms are each equal to $-\lambda_1\phi_1(t)\phi_1^*(s)$. Evaluating

$$\int_a^b \int_a^b R_{XX}(\tau_1,\tau_2)\phi_1^*(\tau_1)\phi_1(\tau_2)d\tau_1 d\tau_2 = \int_a^b \phi_1^*(\tau_1) \left[\int_a^b R_{XX}(\tau_1,\tau_2)\phi_1(\tau_2)d\tau_2 \right] d\tau_1$$

$$= \int_a^b \phi_1^*(\tau_1)\lambda_1\phi_1(\tau_1)d\tau_1$$

$$= \lambda_1 \int_a^b |\phi_1(t)|^2 dt$$

$$= \lambda_1,$$

and combining, we get

$$R_{YY}(t,s) = R_{XX}(t,s) - \lambda_1\phi_1(t)\phi_1^*(s),$$

which agrees with the definition of R_1 in Fact 2 with $R_{XX} = R$. ∎

Proof of Fact 3   Just repeat the proof of Fact 2, with R_1 in place of R to conclude that R_2 is positive semidefinite. To show that $\phi_2 \perp \phi_1$, we proceed as follows. We first note that

$$\lambda_2\phi_2(t) = \int_a^b R_1(t,s)\phi_2(s)ds$$

and then plug in the definition of R_1 to obtain

$$\lambda_2\phi_2(t) = \int_a^b R(t,s)\phi_2(s)ds - \lambda_1\phi_1(t)\int_a^b \phi_1^*(s)\phi_2(s)ds.$$

Then we multiply by $\phi_1^*(t)$ and integrate over (a,b) to obtain

$$\lambda_2\int_a^b \phi_1^*(t)\phi_2(t)dt = \int_a^b\int_a^b \phi_1^*(t)\phi_2(s)R(t,s)dtds$$

$$-\lambda_1\int_a^b |\phi_1(t)|^2 dt \cdot \int_a^b \phi_1^*(s)\phi_2(s)ds$$

$$= \int_a^b \phi_2(s)\left[\int_a^b R(t,s)\phi_1^*(t)dt\right]ds - \lambda_1\int_a^b \phi_1^*(s)\phi_2(s)ds$$

$$= \int_a^b \phi_2(s)\lambda_1^*\phi_1^*(s)ds - \lambda_1\int_a^b \phi_1^*(s)\phi_2(s)ds,$$

by the Hermitian symmetry of R, that is, $R(t,s) = R^*(s,t)$, we thus obtain

$$\lambda_2\int_a^b \phi_1^*(t)\phi_2(t)dt = (\lambda_1^* - \lambda_1)\int_a^b \phi_1^*(s)\phi_2(s)ds.$$

Now λ_1 is real so

$$\lambda_2\int_a^b \phi_1^*(t)\phi_2(t)dt = 0.$$

Thus either $\lambda_2 = 0$ or $\phi_1 \perp \phi_2$. We can reject the first possibility by the existence theorem and hence we are done. ■

Mercer's Theorem   Let the kernel $R(t,s)$ be continuous, Hermitian, and positive semidefinite. Let $\{\lambda_n, \phi_n(t)\}$ be the possibly infinite complete set of orthonormal solutions to the integral Equation 8.8-1. Then the following expansion holds for all t, $s \in [a,b]$,

$$R(t,s) = \sum_{n=1}^{\infty} \lambda_n\phi_n(t)\phi_n^*(s).$$

Proof   By Equation 8.8-2 we know that the question reduces to whether the positive semidefinite kernel R_∞ is equal to zero. If it is not zero, then by the existence theorem there is an eigenfunction and nonzero eigenvalue λ for R_∞. Since $\lambda > 0$, adding this new eigenfunction-eigenvalue pair to the right-hand side of Equation 8.8-2, we get a change in the value of R_∞, which contradicts the assumed convergence. Thus $R_\infty = 0$ and the theorem is proved. ■

PROBLEMS

8.1. Use Theorems 8.1-5 and 8.1-6 to show the following properties of the mean-square derivative.

(a) $\dfrac{d}{dt}(aX_1(t) + bX_2(t)) = a\dfrac{dX_1(t)}{dt} + b\dfrac{dX_2(t)}{dt}.$

(b) $E[X_1(t_1)\dot{X}_2^*(t_2)] = \dfrac{\partial}{\partial t_2}R_{X_1X_2}(t_1, t_2).$

8.2. Let $X(t)$ be a random process with constant mean $\mu_X(\neq 0)$ and covariance function

$$K_{XX}(t, s) = \sigma^2 \cos\omega_0(t - s).$$

(a) Show that the m.s. derivative $X'(t)$ exists here.
(b) Find the covariance function of the m.s. derivative $K_{X'X'}(t, s)$.
(c) Find the *correlation function* of the m.s. derivative $R_{X'X'}(t, s)$.

8.3. Let the random process $X(t)$ be wide-sense stationary (WSS) with correlation function

$$R_{XX}(\tau) = \sigma^2 e^{-(\tau/T)^2}.$$

Let $Y(t) = 3X(t) + 2X'(t)$, where the derivative is interpreted in the mean-square (m.s.) sense.

(a) State conditions for the m.s. existence of such a $Y(t)$ in terms of a *general* correlation function $R_{XX}(\tau)$.
(b) Find the correlation function $R_{YY}(\tau)$ for the given $R_{XX}(\tau)$ in terms of σ^2 and T.

8.4. Let $X(t)$ be a stationary random process with mean μ_X and covariance function

$$K_{XX}(\tau) = \frac{\sigma_X^2}{1 + \alpha^2\tau^2}, \qquad -\infty < \tau < +\infty.$$

(a) Show that a mean-square derivative exists for all t.
(b) Find $\mu_{\dot{X}}(t)$ and $K_{\dot{X}\dot{X}}(\tau)$ for all t and τ.

8.5. Carefully show from basic definitions that the mean-square integral is linear (in the integrand). That is, for a fixed interval $0 \leq t \leq T$, we wish to conclude that the following equality must hold in the mean-square sense:

$$\int_0^T (a_1X_1(t) + a_2X_2(t))\,dt = a_1\int_0^T X_1(t)\,dt + a_2\int_0^T X_2(t)\,dt.$$

Assume that each individual integral on the *right-hand side* exists in the mean-square sense and that the complex constants a_i are finite. (*Hint:* Use the triangle inequality for the norm $\|X(t)\| \triangleq \sqrt{E[|X|^2}$, i.e., $\|X + Y\| \leq \|X\| + \|Y\|$.)

8.6. Show from the basic definitions that the mean-square integral is linear in its upper and lower limits of integration. That is, for an arbitrary second-order random process $X(t)$, show the following mean-square equality:

$$\int_{t_1}^{t_3} X(t)\,dt = \int_{t_1}^{t_2} X(t)\,dt + \int_{t_2}^{t_3} X(t)\,dt,$$

for all $-\infty < t_1 < t_2 < t_3 < +\infty$. (*Hint*: Since second-order means $E[|X(t)|^2] < \infty$, we already know that each of the above m.s. integrals exists, because it must be that

$$\int_{t_i}^{t_j}\int_{t_i}^{t_j} R_{XX}(t,s)\,dt\,ds < \infty,$$

for any finite $t_i < t_j$. You may want to make use of the triangle inequality.)

8.7. To estimate the mean of a stationary random process, we often consider an integral average

$$I(T) \triangleq \frac{1}{T}\int_0^T X(t)\,dt, \qquad T > 0.$$

  (a) Find the mean of $I(T)$ as a function of T, denoted $\mu_I(T)$ for $T > 0$ as a function of the unknown mean μ_X.
  (b) Find the variance of $I(T)$, denoted $\sigma_I^2(T)$ for $T > 0$ as a function of the unknown covariance function $K_{XX}(\tau)$.

8.8. Let $X(t)$ be a WSS Gaussian random process. Show that the m.s. derivative of $Y(t) \triangleq X^2(t)$ is $\dot{Y}(t) = 2X(t)\dot{X}(t)$ and find the correlation function of $\dot{Y}$ in terms of R_{XX} and its derivatives. (*Hint*: Recall that for jointly Gaussian random variables $E[X_1 X_2 X_3 X_4] = R_{12}R_{34} + R_{13}R_{24} + R_{14}R_{23}$ with $R_{ij} = E[X_i X_j]$.)

8.9. Let $Y(t) = \cos(2\pi f_0 t)\,X(t)$, where $X(t)$ is a second-order random process possessing a mean-square derivative $X'(t)$ at each time t. It turns out that the product rule for derivatives is true for m.s. derivatives in this case. In particular we have the m.s. equality,

$$Y'(t) = -2\pi f_0\,\sin(2\pi f_0 t)\,X(t) + \cos(2\pi f_0 t)\,X'(t).$$

[*Hint*: Use the triangle inequality applied to the error for an approximate derivative calculated as a difference for times t and $t + \frac{1}{n}$. Then add and subtract $X(t + \frac{1}{n})$ inside. Letting $n \to \infty$ should yield the desired result.]

8.10. This problem concerns a property of the correlation function of a general independent increments random process.

  (a) If $U(t)$ is an independent increments random process, with zero-mean, show that $R_{UU}(t_1, t_2) = f(\min(t_1, t_2))$ where $f(t) \triangleq E[U^2(t)]$.
  (b) Using the definition of an independent process (see Section 7.4), show that the generalized m.s. derivative $U'(t)$ is an independent process.
  (c) What is the condition on the function $f(t)$ such that the random process $U'(t)$ would be wide-sense stationary?

8.11. Consider the running integral

$$Y(t) = \int_0^t a(t,\tau)X(\tau)\,d\tau, \quad \text{for} \quad t \geq 0.$$

    (a) State conditions for the m.s. existence of this integral in terms of the kernel function $a(t,\tau)$ and the correlation function of $X(t)$.

    (b) Find the mean function $\mu_Y(t)$ in terms of the kernel function $a(t,\tau)$ and the mean function $\mu_X(t)$.

    (c) Find the covariance function $K_{YY}(t,s)$ in terms of the kernel function $a(t,\tau)$ and the covariance function $K_{XX}(t,s)$.

8.12. This problem concerns the mean-square derivative. Let the random process $X(t)$ be second order, i.e., $E[|X(t)|^2] < \infty$, and with correlation function $R_{XX}(t_1,t_2)$. Let the random process $Y(t)$ be defined by the mean-square integral

$$Y(t) \triangleq \int_{-\infty}^t e^{-(t-s)}X(s)\,ds.$$

    (a) State the condition for the existence of the mean-square integral $Y(t)$ for any given t in terms of $R_{XX}(t_1,t_2)$.

    (b) Find the correlation function $R_{YY}(t_1,t_2)$ of $Y(t)$ in terms of $R_{XX}(t_1,t_2)$.

    (c) Consider the mean-square derivative $dY(t)/dt$ and find the condition on $R_{XX}(t_1,t_2)$ for its existence.

8.13.   (a) Let $X[n]$ be a sequence of Gaussian random variables. Let Y be the limit of this sequence where we assume Y exists in the mean-square sense. Use the fact that convergence in mean square implies convergence in distribution to conclude that Y is also Gaussian distributed.

    (b) Repeat the above argument for a sequence of Gaussian random vectors $\mathbf{X}[n] = [X_1[n],\ldots,X_K[n]]^T$.
(Note: Mean-square convergence for random vectors means

$$E[|\mathbf{X}[n] - \mathbf{X}|^2] \to 0$$

where $|\mathbf{X}|^2 \triangleq \sum_{i=1}^K X_i^2$, and Chebyshev's inequality for random vectors is

$$P[|\mathbf{X}| > \varepsilon] \leq \int \cdots \int \frac{|\mathbf{x}|^2}{\varepsilon^2} f(\mathbf{x})\,d\mathbf{x} = \frac{E[|\mathbf{X}|^2]}{\varepsilon^2}.)$$

    (c) Let $\mathbf{X}[n] \triangleq [X_n(t_1),\ldots,X_n(t_K)]^T$ and use the result of part (b) to conclude that an m.s. limit of Gaussian random processes is another Gaussian random process.

8.14. Let $X(t)$ be a Gaussian random process on some time interval. Use the result of the last problem to show that the m.s. derivative random process $X'(t)$ must also be a Gaussian random process.

8.15. Let the stationary random process $X(t)$ be bandlimited to $[-\omega_c, +\omega_c]$ where ω_c is a positive real number. Then define $Y(t)$ as the output of an ideal lowpass filter with passband $[-\omega_1, +\omega_1]$ and input random process $X(t)$.

    (a) Show that $X(t) = Y(t)$ in the m.s. sense if $\omega_1 > \omega_c$.
    (b) In the above case, also show that $X(t) = Y(t)$ with probability-1.

8.16. Consider the following mean-square differential equation,

$$\frac{dX(t)}{dt} + 3X(t) = U(t), \quad t \geq t_0,$$

driven by a WSS random process $U(t)$ with power spectral density

$$S_{UU}(\omega) = \frac{1}{\omega^2 + 4}.$$

The differential equation is subject to the initial condition $X(t_0) = X_0$, where the random variable X_0 has zero-mean, variance 5, and is *orthogonal* to the input random process $U(t)$.

    (a) As a preliminary step, express the *deterministic solution* to the above differential equation, now regarded as an ordinary differential equation with deterministic input $u(t)$ and initial condition x_0, not a random variable. Write your solution as the sum of a *zero-input* part and a *zero-state* part.
    (b) Now returning to the m.s. differential equation, write the solution random process $X(t)$ as a mean-square convolution integral of the input process $U(t)$ over the time interval $(t_0, t]$ plus a zero-input term due to the random initial condition X_0. Justify the mean-square existence of the terms in your solution.
    (c) Write the integral expression for the two-parameter output correlation function $R_{XX}(t, s)$ over the time intervals $t, s \geq t_0$. You do not have to evaluate the integral.

8.17. Consider the m.s. differential equation

$$\frac{dY(t)}{dt} + 2Y(t) = X(t)$$

for $t > 0$ subject to the initial condition $Y(0) = 0$. The input is

$$X(t) = 5\cos 2t + W(t),$$

where $W(t)$ is a Gaussian white noise with mean zero and covariance function $K_{WW}(\tau) = \sigma^2 \delta(\tau)$.

    (a) Find $\mu_Y(t)$ for $t > 0$.

(b) Find the covariance $K_{YY}(t_1, t_2)$ for t_1 and $t_2 > 0$.

(c) What is the maximum value of σ such that

$$P[|Y(t) - \mu_Y(t)| < 0.1] > 0.99 \qquad \text{for all } t > 0?$$

Use Table 2.4-1 for erf$(\cdot)$ in Chapter 2.

$$\left(\text{erf}(x) \triangleq \frac{1}{\sqrt{2\pi}} \int_0^x e^{-\frac{1}{2}u^2} du, \qquad x \geq 0 \right).$$

8.18. Show that the (m.s.) solution to

$$\dot{Y}(t) + \alpha Y(t) = X(t),$$

that is, the random process $Y(t)$ having $R_{YY}(t_1, t_2)$ and $R_{XY}(t_1, t_2)$ of Equations 8.3-3 and 8.3-4 actually satisfies the differential equation in the mean-square sense. More specifically, show that in the general complex-valued case,

$$E\left[\left| \dot{Y}(t) + \alpha Y(t) - X(t) \right|^2 \right] = 0.$$

8.19. Find the generalized m.s. differential equation that when driven by a standard white noise process $W(t)$, i.e., mean 0 and correlation function $\delta(\tau)$, will yield the solution random processes $X(t)$ with power spectral density (PSD)

$$S_{XX}(\omega) = \frac{1}{(1 + \omega^2)^2}.$$

(*Note*: We want the causal solution here, so differential equation poles should be all in the left half of the complex s-plane.)

8.20. Consider the following generalized mean-square differential equation driven by a white noise process $W(t)$ with correlation function $R_{WW}(\tau) = \delta(\tau)$:

$$\frac{dX(t)}{dt} + 3X(t) = \frac{dW(t)}{dt} + 2W(t), \qquad t \geq t_0,$$

subject to the initial condition $X(t_0) = X_0$, where X_0 is a zero-mean random variable orthogonal to the white noise process $W(t)$.

(a) Write the solution process $X(t)$ as an expression involving a generalized mean-square integral of the white noise $W(t)$ over the interval $(t_0, t]$.

(b) If $t_0 = -\infty$, then $X(t)$ becomes wide-sense stationary. Find the autocorrelation function $R_{XX}(\tau)$ of this WSS process of part (b) above.

(c) Is $X(t)$ a Markov random process? Justify your answer.

8.21. In the diagram below a WSS random signal $X(t)$ is corrupted by additive WSS noise $N(t)$ prior to entering the LTI system with impulse response $h(t)$.

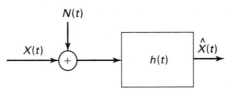

Figure P8.21   Linear estimator of $X(t)$ from signal plus noise observations.

The output of the system is intended as an estimate of the noise-free random signal $X(t)$ and is denoted $\widehat{X}(t)$. Here X and N are mutually orthogonal random processes. The frequency response $H(\omega)$ is specified as

$$H(\omega) = \frac{S_{XX}(\omega)}{S_{XX}(\omega) + S_{NN}(\omega)},$$

where S_{XX} and S_{NN} are the corresponding power spectral densities (psd's).

   (a) In terms of h, R_{XX}, and R_{NN}, write the integral expressing the condition for mean-square existence of the estimate $\widehat{X}(t)$.
   (b) Use PSDs and the given frequency response $H(\omega)$ to find a simpler frequency-domain version of this condition.
   (c) Show that the condition in (b) is always satisfied for a WSS second-order random process $X(t)$.

8.22. To detect a constant signal of amplitude A in white Gaussian noise of variance σ^2 and mean zero, we consider two hypotheses (that is, events):

$$\left.\begin{array}{ll} H_0: & R(t) = W(t) \\ H_1: & R(t) = A + W(t) \end{array}\right\} \quad \text{for } t \in [0, T].$$

It can be shown that the optimal detector, to decide on the correct hypothesis, first computes the integral

$$\Lambda \triangleq \int_0^T R(t)dt,$$

and then performs a threshold test.

   (a) Find the mean value of the integral Λ under each hypothesis, H_0 and H_1.
   (b) Find the variance of Λ under each hypothesis.
   (c) An optimal detector would compare Λ to the threshold $\Lambda_0 \triangleq AT/2$ in the case when each hypothesis is equally likely, that is, $P[H_0] = P[H_1] = 1/2$. Under these conditions, find

$$P[\Lambda \geq \Lambda_0 | H_0]$$

expressing your result in terms of the error function $\text{erf}(\cdot)$ defined in Chapter 2. [*Note:* By Problem 8.13(c), Λ is a Gaussian random variable.]

8.23. This problem concerns ergodicity for random processes.

  (a) State the general definition of "ergodic in the mean" for a wide-sense stationary process $X(t)$.

  (b) Let $X(t)$ be a wide-sense stationary Gaussian random process with zero-mean and correlation function

$$R_{XX}(\tau) = \sigma^2 e^{-\alpha|\tau|} \cos 2\pi f\tau,$$

  where σ^2, α, and f are all positive constants. Show that $X(t)$ is ergodic in the mean. (*Hint*: You may want to use a sufficient condition for ergodicity.)

8.24. Assume that a WSS random process $X(t)$ is ergodic in the mean and that the following limit exists: $\lim_{|\tau|\to\infty} K_{XX}(\tau)$. Show that the correlation function $R_{XX}(\tau)$ then has the asymptotic value $\lim_{|\tau|\to\infty} R_{XX}(\tau) = |\mu_X|^2$ by making use of results in this chapter.

8.25. Let the random sequence $X[n]$ be stationary over the range $0 \leq n < \infty$. Define the time average

$$\hat{M}[N] \triangleq \frac{1}{N} \sum_{n=1}^{N} X[n].$$

Analogously to the concept of "ergodic in the mean" for random processes, we have the following:

Definition. A stationary random sequence $X[n]$ is ergodic in the mean if $\hat{M}[N]$ converges to the ensemble mean μ_X in the mean-square sense as $N \to \infty$.

  (a) Find a suitable condition on the covariance function $K_{XX}[m]$ for $X[n]$ to be ergodic in the mean.

  (b) Show that this condition can be put in the form

$$\lim_{N\to\infty} \left\{ \frac{1}{N} \sum_{n=-N}^{+N} \left(1 - \frac{|n|}{N}\right) K_{XX}[n] \right\} = 0.$$

  (c) Is the stationary random sequence with covariance

$$K_{XX}[m] = 5(0.9)^{|m|} + 15(0.8)^{|m|}$$

  ergodic in the mean?

8.26. Let $X(t)$ be a random process with constant mean μ_X and covariance function $K_{XX}(t,s) = \sigma^2 \cos\omega_0(t-s)$. Does this process have an orthogonal Fourier series? Why? Over which intervals?

8.27. If the K–L integral Equation 8.5-3 has two solutions, $\phi_1(t)$ and $\phi_2(t)$ corresponding to the eigenvalues λ_1 and λ_2, then show that if $\lambda_1 \neq 0$ and $\lambda_2 \neq \lambda_1$ we must have

$$\int_{-T/2}^{T/2} \phi_1(t)\phi_2^*(t)dt = 0.$$

[*Hint*: Substitute for $\phi_1(t)$ in the above expression and use the Hermitian symmetry of $K(t,s)$, that is, $K(t,s) = K^*(s,t)$.]

8.28. This problem concerns the convergence of a commonly seen infinite-series estimate for signals observed in additive noise. We have the observation consisting of the *second-order* and zero-mean random process $X(t)$ over the interval $[0, T]$, where

$$X(t) = S(t) + N(t), \quad \text{with } S \perp N,$$

and we call S the "signal" and N the "noise." Consider the K–L integral equation for the observed process $X(t)$,

$$\int_0^T R_{XX}(t, s)\phi_n(s)\, ds = \lambda_n \phi_n(t), \qquad \text{for } 0 \leq t \leq T.$$

Denote the following three (infinite) sets of random coefficients, for $n \geq 1$:

$$X_n = \int_0^T X(t)\phi_n^*(t)\, dt,$$

$$S_n = \int_0^T S(t)\phi_n^*(t)\, dt,$$

$$N_n = \int_0^T N(t)\phi_n^*(t)\, dt.$$

Further, denote by $\sigma_{(\cdot)_n}^2$ the three sets of variance terms for these three sets of random coefficients.[†] We now want to make a linear estimate of the signal $S(t)$ from the coefficients X_n as follows:

$$\widehat{S}(t) = \sum_{n=1}^{\infty} \frac{\sigma_{S_n}^2}{\sigma_{S_n}^2 + \sigma_{N_n}^2} X_n \phi_n(t),$$

and based indirectly, through the X_n on the observed random process $X(t)$. The problem can now be stated as proving that this infinite sum converges in the mean-square sense. You should do this using Mercer's theorem:

$$R_{XX}(t, s) = \sum_{n=1}^{\infty} \lambda_n \phi_n(t)\phi_n^*(s),$$

and the fact that second order means that $R_{XX}(t, t) < \infty$. Show Cauchy convergence of the partial sums $n: 1, \ldots, N$.

8.29. In this problem we use the K–L expansion to get LMMSE estimates of a Gaussian signal in white Gaussian noise. We observe $X(t)$ on $[0, T]$ where

$$X(t) = S(t) + W(t), \quad \text{with } S \perp W.$$

Here $W(t)$ is zero-mean and has covariance $K_{WW}(\tau) = \sigma_W^2 \delta(\tau)$, and $S(t)$ is a zero-mean with covariance $K_{SS}(t_1, t_2)$.

[†]Note that S_n and N_n may not be the K–L coefficients of their respective processes.

(a) Show that any set of orthonormal functions $\{\phi_n(t)\}$ satisfies the K–L integral equation for the random process $W(t)$.

(b) Using part (a), show that the same set of orthonormal functions may be used for the K–L expansion of $X(t)$ and $S(t)$.

(c) Show that for $X_n = S_n + W_n$, the LMMSE estimate of S_n is given as

$$\hat{E}[S_n|X_n] = \frac{\sigma_{S_n}^2}{\sigma_{S_n}^2 + \sigma_W^2} X_n,$$

where X_n, S_n, and W_n are the K–L coefficients of the respective random processes.

(d) Using the above, argue that

$$\widehat{S}(t) = \sum_{n=1}^{\infty} \frac{\sigma_{S_n}^2}{\sigma_{S_n}^2 + \sigma_W^2} X_n \phi_n(t).$$

[*Hint*: Expand $S(t)$ in a K–L expansion, $S(t) = \sum_{n=1}^{\infty} S_n \phi_n(t)$.]

8.30. A certain continuous-time communications channel can be modeled as signal plus an independent additive noise

$$Y(t) = S(t) + N(t),$$

where the noise process $N(t)$ is Gaussian distributed with known K–L expansion $\{\lambda_k, \phi_k(t)\}$, where the λ_k are indexed in *decreasing* order, i.e., $\lambda_1 \geq \lambda_2 \geq \lambda_3 \geq \ldots \geq 0$. It is decided to structure a "digital" signal as $S(t) = M\phi_k(t)$, for some choice of k, say $k = k_0$, and the discrete-value message random variable M can take on any of eight values $0, \ldots, 7$ equally likely. Repeating this signaling strategy over successive intervals of length T, we can send a message random sequence $M[n]$ at the date rate of $3/T$ bits per second. We keep $k = k_0$ constant here.

(a) The receiver must process the received signal $Y(t)$ to determine the value of the message random variable M. Argue that a receiver which computes and bases its decision exclusively on, for all $k \geq 1$,

$$Y_k = \int_0^T Y(t)\phi_k^*(t)\,dt,$$

can safely ignore all $Y_k's$ except for Y_{k_0}. Assume that M is independent of the noise process $N(t)$.

(b) How should we determine which values of k_0 to use? Is there a best value?

8.31. Derive the Karhunen–Loève expansion for random sequences:
If $X[n]$ is a second-order, zero-mean random sequence with covariance $K_{XX}[n_1, n_2]$, then

$$X[n] = \sum_{k=1}^{N+1} X_k \phi_k[n] \quad \text{with } |n| \le N/2 \text{ and } N \text{ even}$$

$$\text{where } X_k \triangleq \sum_{n=-N/2}^{+N/2} X[n]\phi_k^*[n]$$

$$\text{and } E[X_k X_l^*] = \lambda_k \delta[k-l]$$

$$\text{and } \sum_{n=-N/2}^{+N/2} \phi_k[n]\phi_l^*[n] = \delta[k-l].$$

You may assume Mercer's theorem holds in the form

$$K_{XX}[n_1, n_2] = \sum_{k=1}^{N+1} \lambda_k \phi_k[n_1]\phi_k^*[n_2],$$

which is just the eigenvalue–eigenvector decomposition of the covariance matrix $\mathbf{K_{XX}}$ with entries $K_{XX}[i, j]$ for $i, j = -N/2, \ldots, +N/2$. (*Note*: It may be helpful to rewrite the above in matrix-vector form.)

8.32. Let the zero-mean random process $X(t)$, defined over the interval $[0, T]$, have covariance function:

$$K_{XX}(t, s) = 3 + 2\cos\left(\frac{2\pi t}{T}\right)\cos\left(\frac{2\pi s}{T}\right) + \cos\left(\frac{4\pi t}{T}\right)\cos\left(\frac{4\pi s}{T}\right).$$

   (a) Show that $K_{XX}(t, s)$ is positive semidefinite, and hence, a valid covariance function.
   (b) Find the Karhunen–Loève (K–L) expansion for the random process $X(t)$ valid over the interval $[0, T]$. (*Hint*: Use Mercer's theorem.)
   (c) Explain modifications to our *general* K–L expansion to include an $X(t)$ with nonzero mean $\mu_X(t)$. How does the K–L integral equation change?
   (d) Is the $X(t)$, with covariance given above, mean-square periodic? Explain on your answer.

8.33. Let the zero-mean, second-order random processes $X_1(t)$ and $X_2(t)$ be defined over the time interval $[0, T]$. Assume that they have the *same* Karhunen–Loève basis functions $\{\phi_n(t)\}$, but different random variable coefficients $X_{1,n}$ and $X_{2,n}$, where $X_{i,n} \triangleq \int_0^T X_i(t)\phi_n(t)dt$. Consider now a new quantity, called the *ensemble time-average correlation*

$$E\left[\int_0^T X_1(t)X_2(t)\, dt\right],$$

and find its equivalent expression in the KLT domain in terms of the KLT coefficient cross-correlations $E[X_{1,n}X_{2,n}^*]$. Don't worry about existence and convergence issues here.

8.34. Derive Equation 8.6-6 directly from the definition of the Hilbert-transform system function,

$$H(\omega) \overset{\Delta}{=} -j\,\text{sgn}(\omega),$$

where $j = \sqrt{-1}$ and $\text{sgn}(\omega) = 1$ for $\omega > 0$ and $= -1$ for $\omega < 0$.

8.35. If a stationary random process is periodic, then we can represent it by a Fourier series with orthogonal coefficients. This is not true in general when the random process, though stationary, is not periodic. Thus point out the fallacy in the following proposition, which purports to show that the Fourier series coefficients are always orthogonal: First take a segment of length T from a stationary random process $X(t)$. Repeat the corresponding segment of the correlation function periodically. This then corresponds to a periodic random process. If we expand this process in a Fourier series, its coefficients will be orthogonal. Furthermore, the periodic process and the original process will agree over the original time interval.

8.36. Prove Theorem 8.6-3 by showing that

$$\phi_n(t) = \frac{1}{\sqrt{T}}\exp(j2\pi f_0 nt)$$

are the Karhunen–Loève basis functions for a WSS periodic random process. (Refer to Section 8.5 and note $f_0 = 1/T$.)

8.37. Let the complex process $Z(t) \overset{\Delta}{=} X(t)+jY(t)$, where the real-valued random processes $X(t)$ and $Y(t)$ are jointly WSS. We now modulate $Z(t)$ to obtain a (possibly narrowband) process centered at frequency ω_0 as follows:

$$U(t) = \text{Re}\left\{Z(t)\,e^{-j\omega_0 t}\right\}.$$

Given the relevant correlation and cross-correlation functions, i.e., $R_{XX}(\tau), R_{XY}(\tau)$, $R_{YX}(\tau)$, and $R_{YY}(\tau)$, find general conditions *on them* such that the modulated process is also WSS. Also find the conditions needed on the mean functions μ_X and μ_Y. Show that the resulting process $U(t)$ is actually WSS.

8.38. Show that if a random process is wide-sense periodic with period T, then for any t, $E[|X(t+T)-X(t)|^2] = 0$, i.e., mean-square periodic. Then show, using Chebyshev's inequality, that also $X(t) = X(t+T)$ with probability-1.

8.39. Consider the bandpass random process, centered at ω_0,

$$U(t) = X(t)\cos(\omega_0 t) + Y(t)\sin(\omega_0 t),$$

where $X(t)$ and $Y(t)$ are jointly wide-sense stationary (WSS), with correlation functions satisfying the symmetry conditions of Equations 8.6-4 and 8.6-5. Show that the resulting bandpass random process $U(t)$ is also WSS and find its correlation function in terms of the auto- and cross-correlation functions of $X(t)$ and $Y(t)$. Take random processes X and Y to be zero-mean.[†]

[†]Some helpful trigonometric identities are $\cos(\alpha \pm \beta) = \cos\alpha\cos\beta \mp \sin\alpha\sin\beta$ and $\sin(\alpha \pm \beta) = \sin\alpha\cos\beta \pm \cos\alpha\sin\beta$.

8.40. Prove an analogous result to Theorem 8.6-3 for WSS periodic random sequences. (*Hint*: Perform the expansion for $X[n]$ over $0 \leq n \leq T - 1$ where T is an integer. Only use T Fourier coefficients $A_0, \ldots, A_{T-1}$.)

REFERENCES

8-1. W. B. Davenport, Jr., *Probability and Random Processes*. New York: McGraw-Hill, 1970.

8-2. E. Wong and B. Hajek, *Stochastic Processes in Engineering Systems*. New York: Springer-Verlag, 1985.

8-3. A. Papoulis, *Probability, Random Variables, and Stochastic Processes*. 3rd edition. New York: McGraw-Hill, 1991.

8-4. A. V. Oppenheim, A. S. Willsky, and I. T. Young, *Signals and Systems*. Upper Saddle River, NJ: Prentice Hall, 1983, p. 224.

8-5. M. Loève, *Probability Theory*. New York: Van Nostrand, 1963.

8-6. H. L. Van Trees, *Detection Estimation and Modulation Theory: Part I*. New York: John Wiley, 1968.

8-7. C. W. Therrien, *Discrete Random Signals & Statistical Signal Processing*. Prentice Hall, 1992.

8-8. F. Riesz and B. Sz.-Nagy, *Functional Analysis*. New York: Ungar, 1955.

9

Applications to Statistical Signal Processing

This chapter contains several applications of random sequences and processes in the area of digital signal processing. We start with the problem of linear estimation and then present the discrete-time *Wiener* and *Kalman filters*. The *E–M algorithm* is then presented, which can be used to estimate the model parameters used in these linear filters. A section on *hidden Markov models* follows. Estimating power spectral density functions, called *spectral estimation*, is presented next for both conventional and high-resolution (parameter-based) models. Finally we close with a section on *simulated annealing*, a powerful stochastic optimization method used for finding global optimal estimates for compound models, closes the chapter.

9.1 ESTIMATION OF RANDOM VARIABLES

The estimation of a random variable by observing other random variables is a central problem in signal processing, for example, predicting the position of an airplane at some future time from past and present observations on the position. In this section, we consider this problem as well as predicting one random vector from observing another. We introduce the basic ideas in this section; subsequent development and application of these ideas will be taken up in later sections.

To make clear what we mean by estimating one random variable from another, consider the following example: Let X_1 denote the barometric pressure (BP) and X_2 denote the rate of change of the BP at $t = 0$. Let Y denote the relative humidity one hour after measuring $\mathbf{X} = (X_1, X_2)^T$. Clearly, in this case, $\mathbf{X}$ and Y are dependent r.v.'s; then using $\mathbf{X}$ to estimate Y is a case of estimating one r.v. from another (actually $\mathbf{X}$ here is a random vector).[†]

In terms of the axiomatic theory we can describe the problem of estimating one r.v. with another in the following terms: Consider an underlying experiment $\mathscr{H}$ with probability space $\mathscr{P} \triangleq (\Omega, \mathscr{F}, P)$. Let X and Y be two r.v.'s defined on $\mathscr{P}$. For every $\zeta \in \Omega$, we generate the numbers $X(\zeta), Y(\zeta)$. Suppose we can observe only $X(\zeta)$; how do we proceed to estimate $Y(\zeta)$ in some optimum fashion?

At this point the reader may wonder why observing $X(\zeta)$ doesn't uniquely specify $Y(\zeta)$. After all, since X and Y are deterministic functions, why can't we reason that $X(\zeta)$ specifies ζ specifies $Y(\zeta)$? The answer is that observing $X(\zeta)$ does not, in general, uniquely specify the outcome ζ and therefore does not uniquely specify $Y(\zeta)$. For example, let $\Omega = \{-2, -1, 0, 1, 2\}$, $X(\zeta) \triangleq \zeta^2$, and $Y(\zeta) \triangleq \zeta$. Then the observation $X(\zeta) = 4$ is associated with the outcomes $\zeta = 2$ or $\zeta = -2$ (of course, these may not be equally probable) and $Y(2) = 2$ while $Y(-2) = -2$. Hence all we can say about $Y(\zeta)$ after observing $X(\zeta)$ is that $Y(\zeta)$ has value 2 or -2. If all outcomes $\zeta \in \Omega$ are equally likely, then the *a priori* probability $P[Y = 2] = \frac{1}{5}$ and $P[Y = 2 | X = 4] = \frac{1}{2}$.

Assume, at first, for simplicity that we are constrained to estimate Y[‡] by the linear function aX. Assume $E[X] = E[Y] = 0$. Note that a generalization of this problem has already been treated in Example 4.3-4. The mean-square error (MSE) in estimating Y by aX is given by

$$\varepsilon^2 \triangleq E[(Y - aX)^2]$$
$$= \sigma_Y^2 - 2a \, \mathrm{Cov}[X, Y] + a^2 \sigma_X^2 \tag{9.1-1}$$

Setting the first derivative with respect to a equal to zero to find the minimum, we obtain

$$a_o = \frac{\mathrm{Cov}[X, Y]}{\sigma_X^2}. \tag{9.1-2}$$

Equation 9.1-2 furnishes the value of a, which yields a minimum mean-square error (MMSE) if we restrict ourselves to linear estimates of the form $Y = aX$. We note in passing that the inner product[§]

$$(Y - a_o X, X) \triangleq E[(Y - a_o X)X]$$
$$= 0, \tag{9.1-3}$$

which suggests that the random error $\varepsilon \triangleq Y - a_o X$ is orthogonal to the datum. This interesting result is sometimes called the *orthogonality principle* and will shortly be generalized.

[†]The abbreviation r.v. can mean random variable or random vector without ambiguity.

[‡]All random variables in this section are initially assumed to be real. However, later in the discussion we shall generalize to include complex r.v.'s.

[§]For a discussion of inner products involving random variables see the last portion of Section 4.4.

Let us now remove the constraints of linear estimation and consider the more general problem of estimating Y with a (possibly nonlinear) function of X, that is, $g(X)$ so as to *minimize* the MSE. Thus we seek the function $g_o(X)$, which minimizes

$$\varepsilon^2 = E[(Y - g_o(X))^2].$$

The answer to this problem is surprisingly easy, although its implementation is, except for the Gaussian case, generally very difficult. The result is given in Theorem 9.1-1.

Theorem 9.1-1 The MMSE estimator of Y based on observing the r.v. X is the conditional mean, that is, $g_o(X) \triangleq E[Y|X].$[†]

Proof We write $g(X)$ as $g(X) = g_o(X) + \delta g$, that is, as a variation about the *assumed* optimal value. Then

$$\varepsilon = E[(Y - E[Y|X] - \delta g)^2]$$
$$= E[(Y - E[Y|X])^2] - 2E[(Y - E[Y|X])\delta g]$$
$$+ E[(\delta g)^2].$$

Now regarding the cross-term observe that

$$E[(Y - E[Y|X])\delta g] = E[Y \delta g] - E[E[Y|X]\delta g]$$
$$= E[Y \delta g] - E[Y \delta g]$$
$$= 0.$$

We leave it as an exercise to the reader to show how line 2 was obtained from line 1. With the result just obtained we write

$$\varepsilon^2 = E[(Y - E[Y|X])^2] + E[(\delta g)^2]$$
$$\geq E[(Y - E[Y|X])^2]$$
$$= \varepsilon_{\min}^2. \tag{9.1-4}$$

So to make ε a minimum, we set $\delta g = 0$, which implies that the MMSE estimator is $g_o(X) = E[Y|X].$ ∎

The preceding theorem readily generalizes to random vectors[‡] $\mathbf{X}$ and $\mathbf{Y}$, where we wish to minimize the individual component MSE. It follows readily from Theorem 9.1-1 (the

---

[†]The conditional mean estimator is a function of X and hence is itself a random variable. See the discussion on the conditional mean as a r.v. in Section 4.2.

[‡]In anticipation of the material in later sections, we let $\mathbf{X}, \mathbf{Y}$ be complex random vectors. A complex r.v. X is written $X = X_r + jX_i$; where X_r and X_i are real r.v.'s and represent the real and imaginary components of X, respectively, and $j = \sqrt{-1}$. The PDF of X is the joint PDF of X_r and X_i, that is, $F_{X_r X_i}(x_r, x_i)$.

actual demonstration is left as an exercise) that

$$\varepsilon^2_{min} = \sum_{i=1}^{N} E[|Y_i - g_o^{(i)}(\mathbf{X})|^2], \tag{9.1-5}$$

where

$$g_o^{(i)}(\mathbf{X}) \triangleq E[Y_i|\mathbf{X}].$$

Thus the MMSE estimate of a random vector $\mathbf{Y}$ after observing a random vector $\mathbf{X}$ is also the conditional mean that, in vector notation, becomes

$$\mathbf{g}_o(\mathbf{X}) \triangleq E[\mathbf{Y}|\mathbf{X}]. \tag{9.1-6}$$

We can now generalize the orthogonality property observed in Equation 9.1-3. We show that this is a property of the conditional mean.

Property 9.1-1   The MMSE *error vector* ε of the vector $\mathbf{Y}$ given the random vector $\mathbf{X}$, that is,

$$\varepsilon \triangleq \mathbf{Y} - E[\mathbf{Y}|\mathbf{X}]$$

is orthogonal to any measurable function $h(\mathbf{X})$ of the data, that is,

$$E[(\mathbf{Y} - E[\mathbf{Y}|\mathbf{X}])h^*(\mathbf{X})] = \mathbf{0} \text{ (the asterisk denotes conjugation).} \tag{9.1-7}$$

Proof   Use the same method as in Theorem 9.1-1, which showed that the error was orthogonal to δg. Then generalize the result to the vector case.   ■

Example 9.1-1

An r.v. Y is estimated by $\widehat{Y} = \sum_{i=1}^{n} a_i X_i$, where $\mu_Y = \mu_{X_i} = 0$ for $i = 1, \ldots, n$. Assume $E[X_i X_j] = 0\ i \neq j$, $E[X_i^2] = \sigma_i^2$. Compute the $a_i, i = 1, \ldots, n$ that minimize the MSE. Show that $Y - \sum_{i=1}^{n} a_i X_i$ is orthogonal to $\sum_{i=1}^{n} a_i X_i$. Assume real r.v.'s.

Solution   We wish to minimize

$$\varepsilon^2 = E\left[\left(Y - \sum_{i=1}^{n} a_i X_i\right)^2\right]$$

$$= \sigma_Y^2 - 2\sum_{i=1}^{n} a_i \operatorname{Cov}[X_i, Y] + \sum_{i=1}^{n} a_i^2 \sigma_i^2.$$

By differentiating ε with respect to each of the a_i, we obtain

$$\frac{\partial \varepsilon^2}{\partial a_i} = -2\operatorname{Cov}[X_i, Y] + 2a_i\sigma_i^2 = 0$$

or

$$a_{io} = \frac{\operatorname{Cov}[X_i, Y]}{\sigma_i^2}.$$

To show the required orthogonality:

$$E\left[\left(Y - \sum_{i=1}^{n} a_{io}X_i\right)\left(\sum_{i=1}^{n} a_{io}X_i\right)\right] = \sum_{i=1}^{n}\left(\frac{\text{Cov}^2[X_i, Y]}{\sigma_i^2} - \frac{\text{Cov}^2[X_i, Y]}{\sigma_i^4}\sigma_i^2\right)$$

$$= 0.$$

Theorem 9.1-2   Let $\mathbf{X} = (X_1, \ldots, X_N)^T$ and Y be jointly Gaussian distributed with *zero means*. The MMSE estimate is the conditional mean given as

$$E[Y|\mathbf{X}] = \sum_{i=1}^{N} a_i X_i,$$

where the a_i's are chosen such that

$$E\left[\left(Y - \sum_{i=1}^{N} a_i X_i\right)X_k^*\right] = 0 \quad \text{for} \quad k = 1, \ldots, N, \tag{9.1-8}$$

which is called the *orthogonality condition* and is written

$$\left(Y - \sum_{i=1}^{N} a_i X_i\right) \perp X_k \qquad k = 1, \ldots, N.$$

We see that this condition is a special case of Property 9.1-1.

Proof   The random variables

$$\left(Y - \sum_{i=1}^{N} a_i X_i\right), X_1, X_2, \ldots, X_N$$

are jointly Gaussian. Hence, since the first one is uncorrelated with all the rest, it is independent of them. Thus the error

$$Y - \sum_{i=1}^{N} a_i X_i$$

is independent of the random vector $\mathbf{X} = (X_i, \ldots, X_N)^T$, so

$$E\left[\left(Y - \sum_{i=1}^{N} a_i X_i\right)\bigg|\mathbf{X}\right] = E\left[Y - \sum_{i=1}^{N} a_i X_i\right]$$

$$= E[Y] - \sum_{i=1}^{N} a_i E[X_i]$$

$$= 0, \quad \text{since } E[Y] = E[X_i] = 0.$$

But

$$0 = E\left[\left(Y - \sum_{i=1}^{N} a_i X_i\right)\bigg|\mathbf{X}\right]$$

$$= E[Y|\mathbf{X}] - \sum_{i=1}^{N} a_i E[X_i|\mathbf{X}]$$

$$= E[Y|\mathbf{X}] - \sum_{i=1}^{N} a_i X_i.$$

Hence

$$E[Y|\mathbf{X}] = \sum_{i=1}^{N} a_i X_i. \quad \blacksquare \tag{9.1-9}$$

Theorem 9.1-2 points out a great simplification of the jointly Gaussian case, that is, that the conditional mean is linear and easily determined with linear algebraic methods as the solution to the *orthogonality equations*, which can be put into symbolic form as

$$(Y - \mathbf{a}^T\mathbf{X}) \perp \mathbf{X}$$

or, in matrix form, as

$$E[(Y - \mathbf{a}^T\mathbf{X})\mathbf{X}^\dagger] = \mathbf{0}^T \,(\dagger \text{ denotes transpose conjugate}).$$

Hence the optimum value of $\mathbf{a}$, denoted by $\mathbf{a}_o$, is given by

$$\mathbf{a}_o^T = \mathbf{k}_{Y\mathbf{X}}\mathbf{K}_{\mathbf{XX}}^{-1}, \tag{9.1-10}$$

where $\mathbf{k}_{Y\mathbf{X}} \triangleq E[Y\mathbf{X}^\dagger]$ and $\mathbf{K}_{\mathbf{XX}} \triangleq E[\mathbf{X}\mathbf{X}^\dagger]$.

If the means are not zero, the answer is slightly more complicated. If $\mathbf{X}$ and Y are jointly Gaussian with means $\mu_{\mathbf{X}}$ and μ_Y, respectively, we can define the zero-mean random variables

$$\mathbf{X}_c \triangleq \mathbf{X} - \mu_{\mathbf{X}}$$

$$Y_c \triangleq Y - \mu_Y.$$

Then Theorem 9.1-2 applies to them directly, and we can write

$$E[Y_c|\mathbf{X}_c] = \sum_{i=1}^{N} a_i X_{ci} = E[(Y - \mu_Y)|\mathbf{X}_c]. \tag{9.1-11}$$

But the conditional expectation is a linear operation so that

$$E[(Y - \mu_Y)|\mathbf{X}_c] = E[Y|\mathbf{X}_c] - \mu_Y. \tag{9.1-12}$$

Let us observe next that

$$E[Y|\mathbf{X}_c] = E[Y|\mathbf{X}], \tag{9.1-13}$$

a result that is intuitively agreeable since we do not expect the conditional expectation of Y to depend on whether the average value of $\mathbf{X}$ is included in the conditioning or not. A formal demonstration of Equation 9.1-13 is readily obtained by writing out the definition of conditional expectation using pdf's and considering the transformation $\mathbf{X}_c = \mathbf{X} - \mu_{\mathbf{X}}$; this is left as an exercise. Now using Equation 9.1-13 in Equation 9.1-12 and using the latter in Equation 9.1-11, we obtain our final result as

$$E[Y|\mathbf{X}] = \sum_{i=1}^{N} a_i X_{ci} + \mu_Y$$

$$= \sum_{i=1}^{N} a_i(X_i - \mu_{X_i}) + \mu_Y, \tag{9.1-14}$$

which is the general expression for the jointly Gaussian case when the means are nonzero. We see that the estimate is in the form of a linear transformation plus a bias. In passing, we note that the a_i's would be determined from Equation 9.1-10 using the correlation matrix and cross-correlation vector of the zero-mean random variables, which is the same as the covariance matrices of the original random variables $\mathbf{X}$ and Y.

More on the Conditional Mean

Above we introduced the concept of the minimum mean-square error (MMSE) estimator of one random vector from another. Theorem 9.1-1 showed that this optimum estimator is equal to the conditional mean. Equivalently the MMSE estimator of the random vector $\mathbf{X}$ based on observation of the random vector $\mathbf{Y}$ is $E[\mathbf{X}|\mathbf{Y}]$ which is identical to Equation 9.1-6 with the roles of $\mathbf{X}$ and $\mathbf{Y}$ interchanged. Here $\mathbf{X}$ represents the transmitted signal and $\mathbf{Y}$ represents the observed signal, which is a corrupted version of $\mathbf{X}$. In using the symbols this way, we follow the notation often used in the signal processing literature. The corruption of $\mathbf{X}$ can come through added external noise, or through other means such as random channel behavior. To extend the (scalar) conditional mean result to (possibly complex) random vectors, we define the total MMSE by

$$\varepsilon_{\min}^2 = E[\|\mathbf{X} - \mathbf{g}(\mathbf{Y})\|^2]$$

$$= \sum_{i=1}^{N} E[|X_i - g_i(\mathbf{Y})|^2], \tag{9.1-15}$$

and must now show that $g_i(\mathbf{Y}) = E[X_i|\mathbf{Y}]$ minimizes each term in this sum of square values. Thus we need a version of Theorem 9.1-1 for complex random variables. This can be established by first modifying the MSE in the proof of Theorem 9.1-1 to

$$\varepsilon^2 = E[|X - E[X|\mathbf{Y}] - \delta g(\mathbf{Y})|^2]$$

$$= E[|X - E[X|\mathbf{Y}]|^2] - 2\mathrm{Re}E[(X - E[X|\mathbf{Y}])\delta g^*] + E[|\delta g|^2],$$

and then proceeding to show that the cross-term is zero, using the smoothing property of conditional expectation cf. Equation 4.2-27, thus obtaining

$$\varepsilon^2 = E[|X - E[X|\mathbf{Y}]|^2] + E[|\delta g(\mathbf{Y})|^2]. \tag{9.1-16}$$

From this equation, we conclude that δg is zero just as before. Thus $g_i(\mathbf{Y}) = E[X_i|\mathbf{Y}]$ will minimize each term in Equation 9.1-15, thus establishing the conditional mean as the MSE optimal estimate for complex random vectors.

Assume that we observe a complex random sequence $\mathbf{Y}[n]$ over $n \geq 0$, and that we wish to estimate $X[n]$ based on the present and past of $Y[n]$. Based on the preceding vector results, we have immediately upon definition of the random vector,

$$\mathbf{Y}_n \triangleq [Y[n], Y[n-1], \dots, Y[0]]^T,$$

that the MMSE estimate of $X[n]$, denoted $\widehat{X}[n]$, is

$$\widehat{X}[n] = E[X[n]|\mathbf{Y}_n]$$

at each n. Such an estimate is called *causal* or sometimes a *filter* estimate since it does not involve the input sequence Y at future times $m > n$. For infinite-length observations Y, one can define the estimate of $X[n]$ based on the vector $\mathbf{Y}_N^{(n)} \triangleq [Y[n], Y[n-1], \dots, Y[n-N]]^T$ and then let N go to infinity to define the mean-square limit under appropriate convergence conditions,

$$\lim_{N \to \infty} E[X[n]|\mathbf{Y}_N^{(n)}], \tag{9.1-17}$$

thereby defining the expectation conditioned on infinite-length sequences. One can show that this limit exists with probability-1 by using Martingale theory (cf. Section 6.8.). We can show (see Problem 6.44) that for each fixed n, the random sequence with time parameter N,

$$G[N] \triangleq E[X[n]|\mathbf{Y}_N^{(n)}] \tag{9.1-18}$$

is a Martingale. We thus can use the Marginale Convergence theorem 6.8-4 to conclude the probability-1 existence of the limit (Equation 9.1-17) if the variance of the random sequence $G[N]$ is uniformly bounded in the sense of the theorem; that is, for all $N \geq 1$,

$$\sigma_G^2[N] \leq C < \infty \quad \text{for some finite } C. \tag{9.1-19}$$

In fact, this variance is bounded by the variance of $X[n]$. We leave the demonstration of this result to the reader (Problem 9.4).

Similar expressions can be obtained for random processes under appropriate continuity conditions. For instance, by conditioning on ever more dense samplings of a random process $Y(t)$ over ever larger intervals of the past, one can define the conditional mean of the random process $X(t)$ based on causal observation of the random process $Y(t)$,

$$E[X(t)|Y(\tau), \tau \leq t]. \tag{9.1-20}$$

We also learned that when the observations Y are Gaussian and zero-mean, then the conditional mean is linear in these conditioning random variables (compare to Theorem 9.1-2). For the case of estimating the random vector $\mathbf{X}$ from the random vector $\mathbf{Y}$, this becomes

$$E[\mathbf{X}|\mathbf{Y}] = \mathbf{AY}, \tag{9.1-21}$$

where the coefficients A_{ij} are determined by the orthogonality conditions

$$(\mathbf{X} - \mathbf{AY})_i \perp Y_k \qquad 1 \leq i, k \leq N, \tag{9.1-22}$$

which is just the orthogonality condition of Theorem 9.1-2 applied to the estimation of each component X_i of $\mathbf{X}$. The proof of this theorem goes through just the same with the proper definition of the complex Gaussian random vector. However, the definition is somewhat restrictive and the development would take us too far afield. The interested reader may consult [9-1] for the complex Gaussian theory which is often applied to model narrowband data, such as in Section 8.6.

Orthogonality and Linear Estimation

We have seen that the MMSE estimate is given by the conditional mean which is linear in the observations when the data is jointly Gaussian. Unfortunately, in the non-Gaussian case the conditional mean estimate can be nonlinear and it is often very difficult to obtain. In general, the derivation of an optimal nonlinear estimate will depend on higher-order moment functions that may not be available. For these reasons, in this section we concentrate on the best linear estimate for minimizing the MSE. We denote this estimate as LMMSE, standing for *linear* minimum mean-square error. Of course, for Gaussian data the LMMSE and the MMSE estimators are the same. Sometimes we will use the phrase *optimal linear* to describe the LMMSE estimate or estimator.

Consider the random sequence $Y[n]$ observed for $n \geq 0$. We wish to linearly estimate the random signal sequence $X[n]$. We assume both sequences are zero-mean to simplify the discussion, since the reader should be able to extend the argument to the nonzero-mean case with no difficulty. We denote the *LMMSE estimate* by

$$\widehat{E}[X[n]|Y[n], \ldots, Y[0]], \tag{9.1-23}$$

where the hat on $\widehat{E}$ distinguishes this linear estimate from the nonlinear conditional mean estimate $E[X[n]|Y[n], \ldots, Y[0]]$. For the moment, we will treat Equation 9.1-23 as just a notation for the LMMSE estimate, but at the end of the section we will introduce the $\widehat{E}$ operator. The following theorem establishes that the LMMSE estimate for a complex random sequence is determined by the orthogonality principle.

Theorem 9.1-3 The LMMSE estimate of the zero-mean random sequence $X[n]$ based on the zero-mean random sequence $Y[n]$, is given as

$$\widehat{E}[X[n] \mid Y[n], \ldots, Y[0]] = \sum_{i=0}^{n} a_i^{(n)} Y[i],$$

where the $a_i^{(n)}$'s satisfy the orthogonality condition,

$$\left[X[n] - \sum_{i=0}^{n} a_i^{(n)} Y[i]\right] \perp Y[k], \qquad 0 \le k \le n.$$

Furthermore, the LMMSE is given by

$$\varepsilon_{\min}^2 = E[|X[n]|^2] - \sum_{i=0}^{n} a_i^{(n)} E[Y[i]X^*[n]]. \tag{9.1-24}$$

Proof   Let the $a_i^{(n)}$'s be the coefficients determined by the orthogonality principle and let $b_i^{(n)}$ be some other set of coefficients. Then we can write the error using this other set of coefficients as

$$X[n] - \sum_{i=0}^{n} b_i^{(n)} Y[i] = \left[X[n] - \sum_{i=0}^{n} a_i^{(n)} Y[i]\right] + \sum_{i=0}^{n} \left(a_i^{(n)} - b_i^{(n)}\right) Y[i],$$

where we have both added and subtracted $\Sigma a_i^{(n)} Y[i]$. Because the first term on the right of the equal sign is orthogonal to $Y[i]$ for $i = 0, \dots, n$, we have

$$\left[X[n] - \sum_{i=0}^{n} a_i^{(n)} Y[i]\right] \perp \sum_{i=0}^{n} \left(a_i^{(n)} - b_i^{(n)}\right) Y[i],$$

which implies

$$E\left[\left|X[n] - \sum_{i=0}^{n} b_i^{(n)} Y[i]\right|^2\right] = E\left[\left|X[n] - \sum_{i=0}^{n} a_i^{(n)} Y[i]\right|^2\right]$$

$$+ E\left[\left|\sum_{i=0}^{n} (a_i^{(n)} - b_i^{(n)}) Y[i]\right|^2\right]$$

$$\ge E\left[\left|X[n] - \sum_{i=0}^{n} a_i^{(n)} Y[i]\right|^2\right]$$

with equality if and only if $a_i^{(n)} = b_i^{(n)}$ for $i = 0, \dots, n$.
   To evaluate the MSE, we compute

$$\varepsilon^2 = E\left[\left|X[n] - \sum a_i^{(n)} Y[i]\right|^2\right]$$

$$= E\left[\left(X[n] - \sum a_i^{(n)} Y[i]\right) X^*[n]\right]$$

$$- E\left[\left(X[n] - \sum a_i^{(n)} Y[i]\right)\left(\sum a_i^{(n)} Y[i]\right)^*\right].$$

We note also that

$$\left[X[n] - \sum_{i=0}^{n} a_i Y[i]\right] \perp \sum_{i=0}^{n} a_i^{(n)} Y[i],$$

and thus by this orthogonality we have

$$\varepsilon^2 = E[|X[n]|^2] - \sum_{i=0}^{n} a_i^{(n)} E[Y[i]X^*[n]]$$

$$= \sigma_X^2[n] - \sum_{i=0}^{n} a_i^{(n)} K_{YX}[i, n]. \quad \blacksquare$$

Proceeding to solve for $a_i^{(n)}$, we suppress the superscript (n) for notational simplicity and write out the orthogonality condition of the preceding theorem:

$$E[X[n]Y^*[k]] = \sum_{i=0}^{n} a_i E[Y[i]Y^*[k]], \qquad 0 \leq k \leq n. \tag{9.1-25}$$

We define the column vector,

$$\mathbf{a} \triangleq [a_0, a_1, \ldots, a_n]^T$$

and *row vector*,

$$\mathbf{k}_{XY} \triangleq E[X[n]Y^*[0], X[n]Y^*[1], \ldots, X[n]Y^*[n]]$$

$$= [K_{XY}[n, 0], \ldots, K_{XY}[n, n]]$$

and covariance matrix

$$\mathbf{K_{YY}} \triangleq E[\mathbf{Y}\mathbf{Y}^\dagger], \quad \text{with random column vector } \mathbf{Y} \triangleq (Y[0], \ldots, Y[n])^T$$

where

$$(\mathbf{K_{YY}})_{ij} = E[Y(i)Y^*[j]]$$

$$= K_{YY}[i, j].$$

Then Equation 9.1-25 becomes

$$\mathbf{k}_{XY} = \mathbf{a}^T \mathbf{K_{YY}},$$

with solution

$$\mathbf{a}^T = \mathbf{k}_{XY} \mathbf{K_{YY}}^{-1}. \tag{9.1-26}$$

The MSE of Equation 9.1-24 then becomes

$$\varepsilon_{\min}^2 = \sigma_X^2[n] - \mathbf{k}_{XY} \mathbf{K_{YY}}^{-1} \mathbf{k}_{XY}^\dagger. \tag{9.1-27}$$

One comment on the MSE expression is that the maximum possible error output from an LMMSE estimator is $\sigma_X^2[n]$, which happens when the $a_i^{(n)}$'s equal zero. The latter is optimal when there is no cross-covariance between the observations and the signal $X[n]$. Any nonzero cross-covariance causes the MSE to decrease from $\sigma_X^2[n]$ down to the value given in Equation 9.1-27, the amount of *decrease* given by

$$\mathbf{k}_{X\mathbf{Y}}\mathbf{K}_{\mathbf{Y}\mathbf{Y}}^{-1}\mathbf{k}_{X\mathbf{Y}}^{\dagger}.$$

We next look at an example of the above linear estimation procedure applied to the problem of estimating a signal contaminated by white noise.

Example 9.1-2
(Estimation of a signal in noise.) Assume we have a random signal sequence $X[n]$, which is immersed in white noise $V[n]$ of variance σ_V^2, where the signal and noise are uncorrelated and zero-mean. Let the observations be for $n \geq 0$,

$$Y[n] = X[n] + V[n], \qquad X \perp V.$$

We want to determine the causal LMMSE estimate $\widehat{X}[n]$,

$$\widehat{X}[n] = \widehat{E}[X[n]|Y[n], \dots, Y[0]]$$

$$= \sum_{i=0}^{n} a_i^{(n)} Y[i],$$

in terms of the covariance function of the signal X and the variance of the white noise V. From Theorem 9.1-3 the coefficients of the optimal linear estimator $a_i^{(n)}$ are determined by the orthogonality conditions (Equation 9.1-25) specialized to this example. The solution is thus Equation 9.1-26, which must be solved for each value of $n \geq 0$. It remains to determine the covariance matrix $\mathbf{K}_{\mathbf{YY}}$ and cross-covariance vector $\mathbf{k}_{X\mathbf{Y}}$. Looking at the ijth component of $\mathbf{K}_{\mathbf{YY}}$ we compute

$$(\mathbf{K}_{\mathbf{YY}})_{ij} = E[Y[i]Y^*[j]]$$

$$= E[X[i]X^*[j]] + E[V[i]V^*[j]]$$

$$= (\mathbf{K}_{\mathbf{XX}})_{ij} + \sigma_V^2 \delta_{ij}.$$

For the $\mathbf{k}_{X\mathbf{Y}}$ we obtain

$$(\mathbf{k}_{X\mathbf{Y}})_i = E[X[n]Y^*[i]] = E[X[n](X[i] + V[i])^*]$$

$$= E[X[n]X^*[i]] \triangleq (\mathbf{k}_{\mathbf{XX}})_i, \qquad 0 \leq i \leq n,$$

since the signal is orthogonal to the noise. Thus we obtain the estimator coefficient vector

$$\mathbf{a}^T = \mathbf{k}_{X\mathbf{X}}[\mathbf{K}_{\mathbf{XX}} + \sigma_V^2\mathbf{I}]^{-1}.$$

A special case that allows considerable simplification of this example is when the signal covariance is diagonal; that is, $X[n]$ is also a white noise,

$$\mathbf{K_{XX}} \triangleq \sigma_X^2 \mathbf{I} \quad \text{and} \quad \mathbf{k}_{XX} \triangleq \sigma_X^2[0, 0, \ldots, 0, 1].$$

Then the solution for the coefficient vector is

$$\mathbf{a}^T = \frac{\sigma_X^2}{\sigma_X^2 + \sigma_V^2}[0, 0, \ldots, 0, 1],$$

which means that $a_i^{(n)} = 0$ except for $i = n$, so that the LMMSE estimate is

$$\hat{X}[n] = [\sigma_X^2/(\sigma_X^2 + \sigma_V^2)]Y[n].$$

Actually this special case arose when we estimated the coefficients of a Karhunen–Loève expansion of a signal process from the corresponding expansion coefficients of the random signal process with additive white noise in Example 8.5-2.

In general, the signal random sequence is not white and therefore the $a_i^{(n)}$ are not zero and the estimate takes the *growing-memory* form:

$$\hat{X}[n] = \sum_{i=0}^{n} a_i^{(n)} Y[i].$$

It is called growing memory because all past data must be used in the current estimate. The MSE is given by the formula

$$E[|X[n] - \hat{X}[n]|^2] = \sigma_X^2[n] - \sum_{i=0}^{n} a_i^{(n)} K_{YX}[i, n].$$

We next turn to the application of these results to the prediction problem for random sequences.

Example 9.1-3 _____

(Normal equations for linear prediction.) Let $X(t)$ be a real-valued zero-mean random process. Assume that we wish to predict $X(t_{n+1})$ from a linear combination of n previous observations $X(t_n), X(t_{n-1}), X(t_{n-2}), \ldots, X(t_1)$, where $t_{n+1} > t_n > t_{n-1} > \cdots > t_1$. Denote the predicted value by $\hat{X}(t_{n+1})$. Then *linear prediction* implies that $\hat{X}(t_{n+1}) = \sum_{i=1}^{n} c_i X(t_i)$ and that the *prediction error* is $\varepsilon \triangleq X(t_{n+1}) - \hat{X}(t_{n+1})$. To find the LMMSE we adjust the coefficients $\{c_i\}$ so that the mean-square error $\varepsilon^2 \triangleq E\{\varepsilon^2\} = E\{[\hat{X}(t_{n+1}) - \sum_{i=1}^{n} c_i X(t_i)]^2\}$ is minimized. The minimum can be computed from

$$\partial \varepsilon^2/\partial c_j = 0, \quad \text{for} \quad j = 1, 2, \ldots, n.$$

Indeed, carrying out the differentiation furnishes a specialization of the orthogonality principle, namely,

$$E[\varepsilon X(t_j)] = 0 \quad \text{for} \quad j = 1, 2, \ldots, n.$$

So, as we already know, the data must be orthogonal to the error. The optimum coefficients satisfy

$$R_{XX}(t_{n+1}, t_j) - \sum_{i=1}^{n} c_i R_{XX}(t_i, t_j) = 0 \quad \text{for} \quad j = 1, 2, \ldots n.$$

In the WSS case, these equations take the form

$$R_{XX}(t_{n+1} - t_j) - \sum_{i=1}^{n} c_i R_{XX}(t_i - t_j) = 0.$$

Usually the sampling intervals are equally spaced so that $t_{n+1} - t_j = (n + 1 - j)\Delta t$. Letting $\Delta t = 1$, we obtain the Normal equations

$$R_{XX}(l) - \sum_{i=1}^{n} c_i' R_{XX}(l - i) = 0,$$

where the c_i' are in inverse order from the c_i, that is, $c_i' = c_{n+1-i}$. The solution to these equations can be obtained efficiently by the *Levinson–Durbin algorithm*, which exploits the special symmetry properties of the correlation or covariance matrix. The algorithm is discussed in advanced texts on signal processing, for example [9-2].

Example 9.1-4

(Linear prediction for a first-order WSS Markov process.) For the real-valued, first-order Markov process consider an increasing sequence of time $t_1 < t_2 < \ldots < t_n$ for some positive n. Then $f_X(x_{n+1}|x_n, x_{n-1}, x_{n-2}, \ldots, x_1; t_{n+1}, t_n, t_{n-1}, \ldots t_1) = f_X(x_{n+1}|x_n; t_{n+1}, t_n)$ and, hence, the predictability of $X(t_{n+1})$ depends only on observing $X(t_n)$. Thus, in this case, $\hat{X}(t_{n+1}) = cX(t_n)$, with prediction error $\varepsilon = \hat{X}(t_{n+1}) - cX(t_n)$. The Normal equations for this case are

$$R_{XX}(t_{n+1} - t_j) = cR_{XX}(t_n - t_j), j = 1, 2, \ldots, n.$$

With $\tau_i \overset{\Delta}{=} t_{i+1} - t_i \geq 0$, $i = 1, 2, \ldots, n$, we can rewrite the Normal equations in a more revealing form as

$$R_{XX}(\tau_n) = cR_{XX}(0), \quad j = n$$

$$R_{XX}(\tau_n + \tau_{n-1}) = cR_{XX}(\tau_{n-1}), \quad j = n - 1$$

$$R_{XX}(\tau_n + \tau_{n-1} + \tau_{n-2}) = cR_{XX}(\tau_{n-1} + \tau_{n-2}), \quad j = n - 2$$

$$\vdots$$

From these equations we can establish a form for the correlation function of a zero-mean, WSS, first-order Markov process. For example, if we divide the first equation by the second, we easily obtain

$$R_{XX}(\tau_n)R_{XX}(\tau_{n-1}) = R_{XX}(0)R_{XX}(\tau_n + \tau_{n-1})$$

which is satisfied by the form $R_{XX}(\tau) = b \exp[\pm a\tau]$. Now using the general results that $R_{XX}(0) \geq |R_{XX}(\tau)|$ and $R_{XX}(\tau) = R_{XX}(-\tau)$, we obtain that

$$R_{XX}(\tau) = b \exp(-a|\tau|), \quad -\infty < \tau < \infty, a > 0$$

implies a first-order process.

Example 9.1-5

(Optimum linear interpolation.) In Example 6.4-7 we constructed a sequence in which every other sample had value zero. The sequence so constructed retained all of the original samples, except that introducing alternate zeros between the original samples expanded it. In practice, interpolated values replace the zeros to eliminate expansion artifacts. There are a number of ways to construct these interpolated samples. One is to pass the sequence through an appropriate low pass filter whose impulse response is the interpolating function. Another is to create the interpolated sample $\hat{X}[n]$ by taking the sample mean of the adjacent samples on either side, that is, $\hat{X}_s[n] = \frac{1}{2}X[n-1] + \frac{1}{2}X[n+1]$.

A third is to apply optimum linear interpolation. We compare the last two methods here. In the case of optimum linear interpolation the orthogonality principle requires that

$$E\{(\hat{X}[n] - X[n])X[n-1]\} = 0$$

$$E\{(\hat{X}[n] - X[n])X[n+1]\} = 0,$$

where

$$\hat{X}[n] = aX[n-1] + bX[n+1]$$

and the coefficients a, b must be determined so that the orthogonality conditions are satisfied. We denote these (optimum) coefficients by a_0, b_0. From the orthogonality equations, in the real-valued case, we obtain that

$$\begin{bmatrix} R_{XX}[1] \\ R_{XX}[1] \end{bmatrix} = \begin{bmatrix} R_{XX}[0] & R_{XX}[2] \\ R_{XX}[2] & R_{XX}[0] \end{bmatrix} \begin{bmatrix} a_0 \\ b_0 \end{bmatrix}$$

which yields the solution

$$\begin{bmatrix} a_0 \\ b_0 \end{bmatrix} = \begin{bmatrix} R_{XX}[0] & R_{XX}[2] \\ R_{XX}[2] & R_{XX}[0] \end{bmatrix}^{-1} \begin{bmatrix} R_{XX}[1] \\ R_{XX}[1] \end{bmatrix}$$

or $a_0 = b_0 = R_{XX}[1]/(R_{XX}[0] + R_{XX}[2])$. For any estimator the estimation error is $\varepsilon[n] = \hat{X}[n] - X[n]$ and the mean-square error ε^2 can be written as $\varepsilon^2 = E\{\varepsilon^2[n]\}$, where ε^2 doesn't depend on n in the WSS case. After some simple algebraic manipulations we obtain, for this real-valued case,

$$\varepsilon_o^2 = R_{XX}[0] - \frac{2R_{XX}^2[1]}{R_{XX}[0] + R_{XX}[2]}$$

for the optimum interpolator, and

$$\varepsilon_s^2 = R_{XX}[0] - \left(2R_{XX}[1] - \frac{R_{XX}[0] + R_{XX}[2]}{2}\right)$$

for the sample mean interpolator. To show that $\varepsilon_0^2 \leq \varepsilon_s^2$ we must show that

$$\frac{2R_{XX}^2[1]}{R_{XX}[0] + R_{XX}[2]} \geq \left(2R_{XX}[1] - \frac{R_{XX}[0] + R_{XX}[2]}{2}\right).$$

But this result is equivalent to writing

$$\left[R_{XX}[1] - \left(\frac{R_{XX}[0] + R_{XX}[2]}{2}\right)\right]^2 \geq 0,$$

which is always true.

Example 9.1-6

In Example 9.1-5 we showed that optimum linear interpolation yields a smaller mean-square error (MSE) than the sample mean interpolator. How much smaller? Assume that the psd of the sequence $X[n]$ is $S_{XX}(\omega) = W(2\omega/\pi)$, where ω is a normalized frequency such that $|\omega| \leq \pi$ and $w(x)$ is the rectangular window function

$$W(x) = \begin{cases} 1 & |x| \leq 1 \\ 0 & \text{else.} \end{cases}$$

Then

$$R_{XX}[m] = \frac{\sin(\pi m/2)}{m\pi} \text{ and } R_{XX}[0] = 0.5, R_{XX}[1] = 1/\pi, \text{ and } R_{XX}[2] = 0.$$

Thus $a_0 = b_0 = R_{XX}[1]/(R_{XX}[0] + R_{XX}[2]) = 2/\pi$. Then $\varepsilon_0^2 \approx 0.0947$ while $\varepsilon_s^2 = 0.113$ and estimating $X[n]$ by μ_X (which is zero) would yield a MSE of 0.5. The fractional percent improvement in the MSE (optimum) compared to the MSE (sample-mean interpolator) is $100 \times \frac{\varepsilon_s^2 - \varepsilon_0^2}{\varepsilon_0^2} = 19\%$.

Example 9.1-7

If we set

$$Y[n] \triangleq X[n-1]$$

for all n, we can use the result of Theorem 9.1-3 to evaluate the LMMSE estimate

$$\hat{E}[X[n]|X[n-1], \ldots, X[0]],$$

which is called the LMMSE *one-step predictor* for $X[n]$:

$$\hat{X}[n] = \sum_{i=0}^{n-1} a_i^{(n)} X[i].$$

Specializing our results, we replace n by $n-1$ in Equation 9.1-25 and obtain

$$\mathbf{a}^{(n)T} = \mathbf{k}_{XX}\mathbf{K}_{XX}^{-1}$$

where in this case $\mathbf{k}_{XX} = (K_{XX}[n,0], K_{XX}[n,1], \ldots, K_{XX}[n, n-1])$ and $\mathbf{K}_{XX}$ is given by

$$\mathbf{K}_{XX} = \begin{bmatrix} K_{XX}[0,0] & \cdots & K_{XX}[0, n-1] \\ \vdots & & \vdots \\ K_{XX}[n-1,0] & \cdots & K_{XX}[n-1, n-1] \end{bmatrix}$$

Example 9.1-8

Let $\mu_X = 0$ and $K_{XX}[m] = \sigma_1^2 \, \rho_1^{|m|} + \sigma_2^2 \, \rho_2^{|m|}$. Given the observations $Y[n] = X[n] + W[n]$, where mean $\mu_W = 0$ and variance σ_W^2, find the following linear estimators.

(a) First we find the single observation conditional mean $\hat{E}\{X[n]|Y[n]\}$. Since the random sequences in question are WSS, we can set $n = 0$ in Theorem 9.1-3 and calculate $\hat{E}\{X[0]|Y[0]\}$ via Equation 9.1-26 with $n = 0$ to obtain

$$a = K_{XY}[0]K_{YY}^{-1}[0]$$

$$= \frac{E\{X[0](X[0] + Y[0])\}}{\sigma_1^2 + \sigma_2^2 + \sigma_W^2}$$

$$= \frac{E\{X^2[0]\}}{\sigma_1^2 + \sigma_2^2 + \sigma_W^2}$$

$$= \frac{\sigma_1^2 + \sigma_2^2}{\sigma_1^2 + \sigma_2^2 + \sigma_W^2}.$$

(b) Next, we find the two-point observation conditional mean $\hat{E}\{X[n]|Y[n], Y[n-1]\}$. Here we apply Theorem 9.1-3 with $n = 2$ to calculate $\hat{E}\{X[2]|Y[1], Y[0]\}$ for the WSS random sequences X and Y. Using Equation 9.1-26 we obtain

$$\begin{pmatrix} a_1 \\ a_2 \end{pmatrix}^{\mathrm{T}} = \begin{pmatrix} \sigma_1^2 + \sigma_2^2 \\ \sigma_1^2\rho_1 + \sigma_2^2\rho_2 \end{pmatrix}^{\mathrm{T}} \begin{pmatrix} \sigma_1^2 + \sigma_2^2 + \sigma_W^2 & \sigma_1^2\rho_1 + \sigma_2^2\rho_2 \\ \sigma_1^2\rho_1 + \sigma_2^2\rho_2 & \sigma_1^2 + \sigma_2^2 + \sigma_W^2 \end{pmatrix}^{-1}$$

In the wide-sense stationary case, the matrix $\mathbf{K_{XX}}$ is Toeplitz; that is,

$$(\mathbf{K_{XX}})_{ij} = g[i - j]$$

for some g since the covariance depends only on the difference of the two time parameters. Efficient algorithms exist for computing the inverse of Toeplitz matrices, which allow the recursive calculation of the coefficient vectors $\mathbf{a}^{(n)}$ for increasing n. Such an algorithm is the Levinson algorithm, which is described in [9-3, pp. 835–838]. Linear prediction, as the foregoing is called, is widely used in speech analysis, synthesis, and coding [9-3, pp. 828–834].

One difficulty in the preceding approach is that the resulting predictors and estimators, though linear, are nevertheless growing memory except in the simplest cases. We will overcome this problem in Section 9.2 by incorporating a Markov signal model. We first pause briefly to discuss the properties of the LMMSE operator we have just derived.

Some Properties of the Operator $\hat{E}$

We have introduced the symbol $\hat{E}$ in Equation 9.1-23 for the LMMSE linear estimator. Here we regard $\hat{E}$ as an operator and establish certain linearity properties of this operator that will be useful in the next section. We will use this operator later on to simplify the derivation of important results in linear estimation.

Theorem 9.1-4   The operator $\hat{E}$ has the following linearity properties:

(a) $\hat{E}[X_1 + X_2|\mathbf{Y}] = \hat{E}[X_1|\mathbf{Y}] + \hat{E}[X_2|\mathbf{Y}]$

and

(b) when $\mathbf{Y}_1$ and $\mathbf{Y}_2$ are (statistically) orthogonal then

$$\hat{E}[X|\mathbf{Y}_1, \mathbf{Y}_2] = \hat{E}[X|\mathbf{Y}_1] + \hat{E}[X|\mathbf{Y}_2].$$

Proof   To prove (a) we note that

$$\hat{E}[X|\mathbf{Y}] = \mathbf{c}^T\mathbf{Y},$$

where the vector $\mathbf{c}$ is given as

$$\mathbf{c}^T = \mathbf{k}_{X\mathbf{Y}}\mathbf{K}_{\mathbf{YY}}^{-1}.$$

Clearly $\mathbf{c} = \mathbf{c}_1 + \mathbf{c}_2$ since $\mathbf{k}_{X\mathbf{Y}} = \mathbf{k}_{X_1\mathbf{Y}} + \mathbf{k}_{X_2\mathbf{Y}}$; thus

$$\mathbf{c}^T\mathbf{Y} = \mathbf{c}_1^T\mathbf{Y} + \mathbf{c}_2^T\mathbf{Y}.$$

To show (b) we note that since $\mathbf{Y}_1$ and $\mathbf{Y}_2$ are statistically orthogonal; that is, $E[\mathbf{Y}_1\mathbf{Y}_2^{\dagger}] = \mathbf{0}$, which we write as $\mathbf{Y}_1 \perp \mathbf{Y}_2$, then the statistical orthogonalities of the individual estimates

$$(X - \mathbf{c}_1^T\mathbf{Y}_1) \perp \mathbf{Y}_1 \quad \text{and} \quad (X - \mathbf{c}_2^T\mathbf{Y}_2) \perp \mathbf{Y}_2$$

imply that

$$(X - \mathbf{c}_1^T\mathbf{Y}_1 - \mathbf{c}_2^T\mathbf{Y}_2) \perp \text{ both } \mathbf{Y}_1 \quad \text{and} \quad \mathbf{Y}_2,$$

which can be seen from Figure 9.1-1. Then (b) follows. Here $\mathbf{c}_1 = \mathbf{k}_{X\mathbf{Y}_1}\mathbf{K}_{\mathbf{Y}_1\mathbf{Y}_1}^{-1}$ and $\mathbf{c}_2 = \mathbf{k}_{X\mathbf{Y}_2}\mathbf{K}_{\mathbf{Y}_2\mathbf{Y}_2}^{-1}$.   ∎

With reference to Figure 9.1-1, we see that the operator $\hat{E}$ projects the signal X onto the linear subspace spanned by the observation vectors $\mathbf{Y}_1$. Thus $\hat{E}$ is sometimes referred to as an *orthogonal projection* operator. Geometrically, it is clear that such an orthogonal projection will minimize the error in an estimate that is constrained by linearity to lie in the linear subspace spanned by the observation vectors. Property (a) then says that the orthogonal projection of the sum of two vectors is the sum of their orthogonal projections, a result that is geometrically intuitive. Property (b) says that the orthogonal projection onto a linear subspace can be computed by summing the orthogonal projections onto each of its *orthogonal* basis vectors. This property will be quite useful in the next section on linear prediction.

All this reinforces the Hilbert or linear-space concept of random variables introduced in Section 8.2, where we defined the r.v. norm

$$\|X\|^2 \triangleq E[|X|^2]$$

and inner product

$$(X, Y) = E[XY^*].$$

This linear vector space of random variables must be distinguished from the random vectors $\mathbf{X}$ and $\mathbf{Y}$. To emphasize this difference we sometimes say "statistically orthogonal."

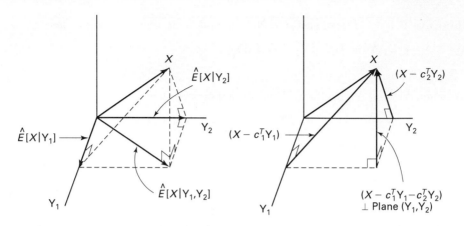

Figure 9.1-1 Illustration of orthogonal projection. The random variable X is shown as a "vector" in the Hilbert space of random variables (see Section 8.1).

9.2 INNOVATION SEQUENCES AND KALMAN FILTERING

In this section we look at the use of signal models to avoid the growing memory aspect of the prediction solution found in the last section. We do this by introducing a certain signal model, the vector difference equation driven by a white random input sequence $\mathbf{W}[n]$,

$$\mathbf{X}[n] = \mathbf{A}\mathbf{X}[n-1] + \mathbf{B}\mathbf{W}[n], \qquad n \geq 0, \tag{9.2-1}$$

with $\mathbf{X}[-1] = \mathbf{X}_{-1}$ given and

$$\mathbf{R}_{\mathbf{WW}}[m] = E[\mathbf{W}[m+n]\mathbf{W}^{\dagger}[n]] = \sigma_{\mathbf{w}}^2 \delta[m],$$

and $\mathbf{W}[n]$ is orthogonal to the past of $\mathbf{X}[n]$, including the initial condition $\mathbf{X}_{-1}$. In symbols this becomes

$$\mathbf{W}[n] \perp \mathbf{X}[m] \quad \text{for} \quad m < n \quad \text{and} \quad \mathbf{W}[n] \perp \mathbf{X}_{-1} \quad \text{for} \quad n \geq 0.$$

By Theorem 6.6-1 $\mathbf{X}[n]$ of Equation 9.2-1 is vector Markov. Using a technique of Example 6.6-2, any scalar LCCDE[†] driven by white noise can be put into the form Equation 9.2-1. The resulting matrix $\mathbf{A}$ will always be nonsingular if the dimension of the state vector is equal to the order of the scalar LCCDE. Thus we will assume $\mathbf{A}$ is a nonsingular matrix, (cf. Example 9.2-11 below).

Starting from the initial condition $\mathbf{X}_{-1}$ we can recursively compute the following forwards-in-time or causal solution,

$$\mathbf{X}[0] = \mathbf{A}\mathbf{X}_{-1} + \mathbf{B}\mathbf{W}[0],$$

$$\mathbf{X}[1] = \mathbf{A}\mathbf{X}[0] + \mathbf{B}\mathbf{W}[1]$$

[†]LCCDE here stands for Linear Constant Coefficient Difference Equation.

$$= \mathbf{A}^2\mathbf{X}_{-1} + \mathbf{A}\mathbf{B}\mathbf{W}[0] + \mathbf{B}\mathbf{W}[1],$$

$$\mathbf{X}[2] = \mathbf{A}^3\mathbf{X}_{-1} + \mathbf{A}^2\mathbf{B}\mathbf{W}[0] + \mathbf{A}\mathbf{B}\mathbf{W}[1] + \mathbf{B}\mathbf{W}[2],$$

and so forth.

We thus infer the general solution

$$\mathbf{X}[n] = \sum_{k=0}^{n} \mathbf{A}^{n-k}\mathbf{B}\mathbf{W}[k] + \mathbf{A}^{n+1}\mathbf{X}_{-1}, \qquad (9.2\text{-}2)$$

the first term of which is just a convolution of the vector input sequence with the matrix impulse response,

$$\mathbf{H}[n] \triangleq \mathbf{A}^n\mathbf{B}u[n].$$

It is important to note that Equation 9.2-2 is a causal and linear transformation from the space of input random sequences $\mathbf{W}[n]$ (including $\mathbf{X}_{-1}$) to the space of output random sequences $\mathbf{X}[n]$. If we use Equation 9.2-1 and assume that $\mathbf{B}^{-1}$ exists, we can also write the input sequence as a causal linear transformation on the output random sequence,

$$\mathbf{W}[n] = \mathbf{B}^{-1}[\mathbf{X}[n] - \mathbf{A}\mathbf{X}[n-1]], \qquad n \geq 0.$$

We say the white sequence $\mathbf{W}[n]$ is *causally equivalent* to $\mathbf{X}[n]$ and call it the *innovations sequence* associated with $\mathbf{X}$. The name comes from the fact that $\mathbf{W}$ contains the new information obtained when we observe $\mathbf{X}[n]$ given the past $[\mathbf{X}[n-1], \mathbf{X}[n-2], \ldots, \mathbf{X}[0]]$. This statement will subsequently become more clear. In general, we make the following definition.

Definition 9.2-1   The *innovations sequence* of a random sequence $\mathbf{X}[n]$ is defined to be a white random sequence, which is a causal and causally invertible[†] linear transformation of the sequence $\mathbf{X}[n]$.   ■

It follows immediately that we can write

$$\hat{E}[\mathbf{X}[n]|\mathbf{X}[n-1], \ldots, \mathbf{X}[0], \mathbf{X}_{-1}] = \hat{E}[\mathbf{X}[n]|\mathbf{W}[n-1], \ldots, \mathbf{W}[0], \mathbf{X}_{-1}],$$

because the LMMSE estimate can always undo a causally invertible linear transformation. That is, the required inverse can, if needed, be part of the general causal linear transformation $\hat{E}$. To see the benefit of the innovations concept consider evaluating

$$\hat{E}[\mathbf{X}[n]|\mathbf{W}[n-1], \ldots, \mathbf{W}[0], \mathbf{X}_{-1}].$$

Rewriting Equation 9.2-2 by isolating the $k = n$, term in the sum we see that

$$\mathbf{X}[n] = \sum_{k=0}^{n-1} \mathbf{A}^{n-k}\mathbf{B}\mathbf{W}[k] + \mathbf{B}\mathbf{W}[n] + \mathbf{A}^{n+1}\mathbf{X}_{-1}.$$

[†]The term "causally invertible" means that the linear transformation has an inverse which is causal.

Applying the $\hat{E}$ operator, and using linearity property (a) of Theorem 9.1-4, we obtain

$$\hat{E}[\mathbf{X}[n]|\mathbf{W}[n-1],\ldots,\mathbf{W}[0],\mathbf{X}_{-1}]$$

$$= \sum_{k=0}^{n-1} \mathbf{A}^{n-k}\mathbf{B}\hat{E}[\mathbf{W}[k]|\mathbf{W}[n-1],\ldots,\mathbf{W}[0],\mathbf{X}_{-1}]$$

$$+\mathbf{B}\hat{E}[\mathbf{W}[n]|\mathbf{W}[n-1],\ldots,\mathbf{W}[0],\mathbf{X}_{-1}]$$

$$+\mathbf{A}^{n+1}\hat{E}[\mathbf{X}_{-1}|\mathbf{W}[n-1],\ldots,\mathbf{W}[0],\mathbf{X}_{-1}].$$

Then repeatedly using linearity property (b) of the same theorem, since the innovations $[\mathbf{W}[n-1],\ldots,\mathbf{W}[0],\mathbf{X}_{-1}]$ are orthogonal, we have

$$\hat{E}[\mathbf{X}[n]|\mathbf{W}[n-1],\ldots,\mathbf{W}[0],\mathbf{X}_{-1}] = \sum_{k=0}^{n-1} \mathbf{A}^{n-k}\mathbf{B}\mathbf{W}[k] + \mathbf{A}^{n+1}\mathbf{X}_{-1}. \qquad (9.2\text{-}3)$$

From Equation 9.2-2 we also have

$$\mathbf{X}[n-1] = \sum_{k=0}^{n-1} \mathbf{A}^{n-1-k}\mathbf{B}\mathbf{W}[k] + \mathbf{A}^{n}\mathbf{X}_{-1},$$

so combining, we get

$$\hat{\mathbf{X}}[n] = \mathbf{A}\mathbf{X}[n-1]. \qquad (9.2\text{-}4)$$

This is the final form of the LMMSE predictor for the state equation model with a white random input sequence. (*Note*: We are assuming that the mean of the input sequence is zero. This is incorporated in the white noise definition.) The overall operation is shown in Figure 9.2-1.

Equation 9.2-4 can also be derived by applying the $\hat{E}$ operator directly to Equation 9.2-1 and using the fact that $\mathbf{X}[n-1]\perp\mathbf{W}[n]$. For example $\hat{E}\{\mathbf{X}[n]|\mathbf{X}[n-1]\} = \hat{E}\{\mathbf{A}\mathbf{X}[n-1]+\mathbf{B}\mathbf{W}[n]|\mathbf{X}[n-1]\} = \hat{E}\{\mathbf{A}\mathbf{X}[n-1]|\mathbf{X}[n-1]\} + \hat{E}\{\mathbf{B}\mathbf{W}[n]|\mathbf{X}[n-1]\}$. But as $\mathbf{W}[n] \perp \mathbf{X}[n-1]$, it follows that $\hat{E}\{\mathbf{X}[n]|\mathbf{X}[n-1]\} = \hat{E}\{\mathbf{A}\mathbf{X}[n-1]|\mathbf{X}[n-1]\} = \mathbf{A}\mathbf{X}[n-1]$. We can thus view Equation 9.2-1 as a one-step innovations representation in the same sense that Equation 9.2-2 is an $(n+1)$-step innovations representation. The innovations method is of quite general use in linear estimation theory, the basic underlying concept being a representation of the observed data as an orthogonal decomposition. We will make good use of the innovations method in deriving the Kalman filter below [9-3].

With reference to our state equation model (Equation 9.2-1) we note that if we added the condition that the driving noise $\mathbf{W}[n]$ be Gaussian, then $\mathbf{X}[n]$ would be both Gaussian and Markov. In this case the preceding LMMSE estimate would also be the MMSE estimate; that is, $\hat{E}$ would be E. This motivates the following weakening of the Markov property.

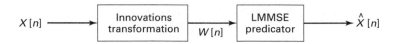

Figure 9.2-1   Innovations decomposition of LMMSE predictor.

Definition 9.2-2 A random sequence $\mathbf{X}[n]$ is called *wide-sense Markov* if for all n

$$\hat{E}[\mathbf{X}[n]|\mathbf{X}[n-1], \mathbf{X}[n-2], \ldots] = \hat{E}[\mathbf{X}[n]|\mathbf{X}[n-1]].$$

We note that this definition has built in the concept of limited memory in that the LMMSE prediction cannot be changed by the incorporation of earlier data. From Equation 9.2-4, it follows immediately that the solution to Equation 9.2-1 is a wide-sense Markov random sequence. One can prove that any wide-sense Markov random sequence would satisfy a first-order vector state equation with a white input that is uncorrelated with the past of $\mathbf{X}[n]$. ∎

Theorem 9.2-1 Let $\mathbf{X}[n]$ be a wide-sense Markov zero-mean sequence. Then there exists an innovations sequence $\mathbf{W}[n]$ such that Equation 9.2-1 is satisfied and the sequence $\mathbf{W}$ is orthogonal to the past of $\mathbf{X}$.

Proof Since $\mathbf{X}$ is wide-sense Markov and zero-mean,

$$\hat{E}[\mathbf{X}[n]|\mathbf{X}[n-1], \mathbf{X}[n-2], \ldots] = \hat{E}[\mathbf{X}[n]|\mathbf{X}[n-1]]$$
$$\triangleq \mathbf{A}\mathbf{X}[n-1],$$

where $\mathbf{A}$ is defined as the matrix of the indicated LMMSE one-step prediction. Next we define $\mathbf{W}$ as

$$\mathbf{W}[n] \triangleq \mathbf{X}[n] - \mathbf{A}\mathbf{X}[n-1].$$

It then follows that $E[\mathbf{W}[n]] = 0$ and

$$\mathbf{W}[n] \perp (\mathbf{X}[n-1], \mathbf{X}[n-2], \ldots)$$

because of the fact that $\mathbf{X}$ is wide-sense Markov and the prediction error must be orthogonal to the past data used in the prediction. Furthermore,

$$E[\mathbf{W}[n]\mathbf{W}^\dagger[m]] = 0 \quad \text{for} \quad m < n,$$

since $\mathbf{W}[m]$ is a linear function of the past and present of $\mathbf{X}[m]$, which is in the past of $\mathbf{X}[n]$ for $m < n$. Thus

$$E[\mathbf{W}[n]\ \mathbf{W}^\dagger[m]] = \sigma_{\mathbf{W}}^2[n]\delta[m-n],$$

where

$$\sigma_{\mathbf{W}}^2[n] \triangleq E[\mathbf{W}[n]\mathbf{W}^\dagger[n]]. \quad ∎$$

Example 9.2-1 _____

Consider the second-order *scalar* difference equation driven by white noise,

$$X[n] = \alpha X[n-1] + \beta X[n-2] + W[n],$$

where $X[-1] = X[-2] = 0$ and where $W[n]$ has zero-mean and variance $\text{Var}[W[n]] = \sigma_W^2$. To apply the vector wide-sense Markov prediction results, we construct the random vector

$$\mathbf{X}[n] \triangleq [X[n], X[n-1]]^T$$

and obtain the first-order vector equation

$$\mathbf{X}[n] = \mathbf{A}\mathbf{X}[n-1] + \mathbf{b}W[n]$$

on setting

$$\mathbf{A} \triangleq \begin{pmatrix} \alpha & \beta \\ 1 & 0 \end{pmatrix} \quad \text{and} \quad \mathbf{b} \triangleq \begin{pmatrix} 1 \\ 0 \end{pmatrix}.$$

Using Equation 9.2-4 we then have

$$\hat{E}[\mathbf{X}[n] | \mathbf{X}[n-1], \dots, \mathbf{X}[0]] = \mathbf{A}\mathbf{X}[n-1],$$

so that

$$\hat{E}\left[\begin{pmatrix} X[n] \\ X[n-1] \end{pmatrix} \bigg| X[n-1], \dots, X[0] \right] = \begin{pmatrix} \alpha & \beta \\ 1 & 0 \end{pmatrix} \begin{pmatrix} X[n-1] \\ X[n-2] \end{pmatrix}$$

or

$$\hat{X}[n] = \alpha X[n-1] + \beta X[n-2].$$

Sometimes such a scalar random sequence is called wide-sense *Markov of order 2*. More generally a pth order, scalar LCCDE driven by white noise generates a scalar wide-sense Markov random sequence of order p. (cf. Definition 6.5-2.)

Predicting Gaussian Random Sequences

Here we look at the special case where the random sequence is Gaussian so that the results of the last section having to do with wide-sense Markov become strict-sense Markov and the orthogonal random sequence $\mathbf{W}[n]$ becomes an independent random sequence. We start with a theorem, which is in part a restatement and in part a strengthening of some of the previous results.

Theorem 9.2-2   Let $\mathbf{X}[n]$ be a zero-mean, Gauss-Markov random sequence. Then $\mathbf{X}[n]$ satisfies the difference equation

$$\mathbf{X}[n] = \mathbf{A}_n \mathbf{X}[n-1] + \mathbf{B}_n \mathbf{W}[n]$$

for some $\mathbf{A}_n, \mathbf{B}_n$, and white Gaussian, zero-mean sequence $\mathbf{W}[n]$.

Proof   Since $\mathbf{X}[n]$ is Gaussian, the conditional mean $E[\mathbf{X}[n] | \mathbf{X}[n-1], \mathbf{X}[n-2], \dots]$ is a linear function of $[\mathbf{X}[n-1], \mathbf{X}[n-2], \dots]$ because in the Gaussian case this MMSE estimator is linear. Since the random sequence is also Markov we have that

$$E[\mathbf{X}[n] | \mathbf{X}[n-1], \dots] = E[\mathbf{X}[n] | \mathbf{X}[n-1]]$$

$$\triangleq \mathbf{A}_n \mathbf{X}[n-1],$$

for some $\mathbf{A}_n$ that may depend on n. In fact, we know the matrix $\mathbf{A}_n$ can be determined by the orthogonality relation

$$(\mathbf{X}[n] - \mathbf{A}_n \mathbf{X}[n-1]) \perp \mathbf{X}[n-1].$$

What remains to be shown is that the prediction-error sequence $\mathbf{X}[n] - \mathbf{A}_n \mathbf{X}[n-1]$ is a white Gaussian random sequence. First, we know it is Gaussian because it is a linear operation on a Gaussian random sequence. Second, we know that the prediction error is (statistically) orthogonal to all previous $\mathbf{X}[k]$ for all $k < n$. Thus

$$(\mathbf{X}[n] - \mathbf{A}_n\mathbf{X}[n-1]) \perp (\mathbf{X}[k] - \mathbf{A}_k\mathbf{X}[k-1])$$

for all $k < n$. Hence it is an orthogonal random sequence, but this is the same as saying it is white and zero-mean. Thus the proof is completed by setting

$$\mathbf{B}_n\sigma^2_{\mathbf{W}}[n]\mathbf{B}^{\dagger}_n \triangleq E[(\mathbf{X}[n] - \mathbf{A}_n\mathbf{X}[n-1])(\mathbf{X}[n] - \mathbf{A}_n\mathbf{X}[n-1])^T].$$

In fact, we can just as well take $\mathbf{B}_n = \mathbf{I}^{\dagger}$ and then

$$\sigma^2_{\mathbf{W}}[n] = E[(\mathbf{X}[n] - \mathbf{A}_n\mathbf{X}[n-1])(\mathbf{X}[n] - \mathbf{A}_n\mathbf{X}[n-1])^T]$$
$$= E[\mathbf{X}[n]\mathbf{X}^T[n]] - \mathbf{A}_nE[\mathbf{X}[n-1]\mathbf{X}^T[n]]$$
$$= \mathbf{K}_{\mathbf{XX}}[n,n] - \mathbf{A}_n\mathbf{K}_{\mathbf{XX}}[n-1,n],$$

but

$$\mathbf{A}_n = \mathbf{K}_{\mathbf{XX}}[n,n-1]\mathbf{K}^{-1}_{\mathbf{XX}}[n-1,n-1] \tag{9.2-5}$$

so

$$\sigma^2_{\mathbf{W}}[n] = \mathbf{K}_{\mathbf{XX}}[n,n] - \mathbf{K}_{\mathbf{XX}}[n,n-1]\mathbf{K}^{-1}_{\mathbf{XX}}[n-1,n-1]\mathbf{K}_{\mathbf{XX}}[n-1,n]. \quad \blacksquare$$

Note that the difference between Theorem 9.2-1 and Theorem 9.2-2 is that in the latter we are assuming that the vector random sequence $\mathbf{X}[n]$ is Gaussian while in the former theorem this condition is not assumed. The sequence $\mathbf{W}[n]$ then is an independent one and is Gaussian in the case where the signal sequence $\mathbf{X}$ is Gaussian as well as Markov.

We also can note that if $\mathbf{X}[n]$ were also stationary in Theorem 9.2-2, then the coefficient matrices $\mathbf{A}_n$ and $\mathbf{B}_n$ would be constants, and the innovations variance matrix would be constant $\sigma^2_{\mathbf{W}}$. Finally note that since we use the expectation operator $\hat{E}$ in Theorem 9.2-2 but only the LLMSE operator $\hat{E}$ in Theorem 9.2-1, the representation in the Gaussian case is really much stronger than in the LMMSE case.

Kalman Predictor and Filter

Here we extend the prediction results developed in the last section by enlarging the class of applications to include prediction from noisy data. This generalization when combined with the Gauss-Markov signal model will result in the celebrated Kalman-Bucy prediction filter. In 1960 R. E. Kalman published the discrete-time theory [9-4]. A year later the continuous-time theory was published by R. E. Kalman and R. S. Bucy [9-5]. Actually, we do not really need the Gauss-Markov assumption. We could just assume the signal is wide-sense Markov and derive the LMMSE filter. The result would be the same as what we will derive here essentially because the MMSE filter is linear for Gaussian data.

---

†Thus ensuring that $\mathbf{B}^{-1}$ exists where it is useful such as for ensuring the innovations property of $\mathbf{W}[n]$.

We will assume that the Gauss-Markov sequence to be predicted is stationary with zero-mean and is defined for $n \geq 0$ with *known* initial condition $\mathbf{X}[-1] = 0$. We will also restrict attention in this section to *real-valued* random vector sequences. By Theorem 9.2-2 we have that the Gauss-Markov signal can be represented as

$$\mathbf{X}[n] = \mathbf{A}\mathbf{X}[n-1] + \mathbf{B}\mathbf{W}[n], \quad n \geq 0, \tag{9.2-6}$$

subject to $\mathbf{X}[-1] = 0$, where $\mathbf{W}[n]$ is white Gaussian noise with zero-mean and variance matrix, $\boldsymbol{\sigma}_{\mathbf{W}}^2$. As earlier, the matrix $\mathbf{A}$ is taken to be nonsingular.

The observations are no longer be assumed noiseless. Instead, we will assume the more practical case where noise has been added to the signal prior to observation,

$$\mathbf{Y}[n] = \mathbf{X}[n] + \mathbf{V}[n],^{\dagger} \quad n \geq 0 \tag{9.2-7}$$

where the random sequence $\mathbf{V}[n]$, called the *observation noise*, is white, Gaussian, and zero-mean. We take the observation noise $\mathbf{V}[n]$ to be stationary with variance matrix $\boldsymbol{\sigma}_{\mathbf{V}}^2 \triangleq E[\mathbf{V}[n]\mathbf{V}[n]^T]$, and we remember that all random sequences are assumed real-valued. Additionally, we assume that $\mathbf{V}$ and $\mathbf{W}$ are orthogonal at all pairs of observation times, that is,

$$\mathbf{V}[n] \perp \mathbf{W}[k] \qquad \text{for all } n, k.$$

Since the two noises are zero-mean, this amounts to saying that $\mathbf{V}$ and $\mathbf{W}$ are uncorrelated at all times n and k. We can write this more compactly as $\mathbf{V} \perp \mathbf{W}$. Furthermore we take $\mathbf{V}$ and $\mathbf{W}$ as jointly Gaussian so that they are in fact *jointly independent* random sequences.

Our method of solution will be to first find the innovation sequence for the noisy observations (Equation (9.2-7)) and then to base our estimate on it. The Kalman predictor and filter are then derived. We will see that they have a convenient predictor–corrector structure. Finally, we will solve for certain error-covariance functions necessary to determine so-called *gain matrices* in the filter.

Now we know that the MMSE prediction of the signal sequence $\mathbf{X}[n]$ based on the observation set $\{\mathbf{Y}[k], \ k < n\}$ is the corresponding conditional mean. Thus we look for $\hat{\mathbf{X}}[n] \triangleq E[\mathbf{X}[n]|\mathbf{Y}[n-1], \mathbf{Y}[n-2], \ldots, \mathbf{Y}[0]]$. We first define an innovations sequence for $\mathbf{Y}[n]$ for the noiseless observations $\mathbf{X}[n]$. Motivated by the requirements of the innovations of Definition 9.2-1, we define the sequence $\tilde{\mathbf{Y}}$ as follows:

$$\tilde{\mathbf{Y}}[0] \triangleq \mathbf{Y}[0] \quad \text{and for} \quad n \geq 1$$

$$\tilde{\mathbf{Y}}[n] \triangleq \mathbf{Y}[n] - E[\mathbf{Y}[n]|\mathbf{Y}[n-1], \mathbf{Y}[n-2], \ldots, \mathbf{Y}[0]]. \tag{9.2-8}$$

We now must show that $\tilde{\mathbf{Y}}[n]$ thus defined is an innovations sequence for $\mathbf{Y}[n]$. To do this we must prove that $\tilde{\mathbf{Y}}[n]$ satisfies the three defining properties (see Definition 9.2-1):

(1) $\tilde{\mathbf{Y}}[n]$ is a causal, linear transformation on $\mathbf{Y}[n]$,

---

†A more general observation model $\mathbf{Y}[n] = \mathbf{C}_n\mathbf{X}[n] + \mathbf{V}[n]$ is treated in Problem 9.10. This generalization allows modeling deconvolution-type problems.

(2) $\mathbf{Y}[n]$ is a causal, linear transformation on $\tilde{\mathbf{Y}}[n]$, and

(3) $\tilde{\mathbf{Y}}[n]$ is an orthogonal (or white) random sequence.

Now (1) is immediate by the definition $\tilde{\mathbf{Y}}[n]$ since $\mathbf{Y}[n]$ is a Gaussian random sequence. To show property (2) we note that we can recursively solve Equation 9.2-8 for $\mathbf{Y}[n]$ as

$$\mathbf{Y}[0] = \tilde{\mathbf{Y}}[0] \quad \text{and}$$

$$\mathbf{Y}[n] = \tilde{\mathbf{Y}}[n] + E[\mathbf{Y}[n]|\mathbf{Y}[n-1], \mathbf{Y}[n-2], \ldots, \mathbf{Y}[0]]$$

$$= \tilde{\mathbf{Y}}[n] + \sum_{k=0}^{n=1} \mathbf{D}_k^{(n)} \tilde{\mathbf{Y}}[k], \quad n \geq 1,$$

where the $\mathbf{D}_k^{(n)}$ are a known sequence of matrices,[†] thus establishing (2). As an additional piece of terminology, when (1) and (2) hold simultaneously we say that $\mathbf{Y}$ and $\tilde{\mathbf{Y}}$ are *causally linearly equivalent*.

To establish (3) we note that, for $k < n$,

$$E[\tilde{\mathbf{Y}}[n]\tilde{\mathbf{Y}}[k]^T] = E[\tilde{\mathbf{Y}}[n](\mathbf{Y}[k] - \tilde{\mathbf{Y}}[k])^T] = \mathbf{0},$$

since $\tilde{\mathbf{Y}}[n] \perp [\mathbf{Y}[n-1], \mathbf{Y}[n-2], \ldots, \mathbf{Y}[0]]$ by the orthogonality principle and hence is orthogonal to any linear combination of the $\mathbf{Y}[k]$ for $k < n$. Similarly, we have $E[\tilde{\mathbf{Y}}[n]\tilde{\mathbf{Y}}[k]^T] = 0$ for $n > k$. Thus combining we have

$$E[\tilde{\mathbf{Y}}[n]\tilde{\mathbf{Y}}[k]^T] = \boldsymbol{\sigma}_{\tilde{\mathbf{Y}}}^2[n] \, \delta[n-k]$$

for some variance matrix $\boldsymbol{\sigma}_{\tilde{\mathbf{Y}}}^2[n]$.[‡] Combining properties (1), (2), and (3) we see that $\tilde{\mathbf{Y}}[n]$ is a desired innovations sequence for the noisy observations $\mathbf{Y}[n]$.

Since $\tilde{\mathbf{Y}}[n]$ and $\mathbf{Y}[n]$ are causally linearly equivalent, we can base our estimate on $\tilde{\mathbf{Y}}[n]$ instead of $\mathbf{Y}[n]$ with expected simplifications due to the orthogonality of the innovations sequence $\tilde{\mathbf{Y}}[n]$. Since the data are Gaussian and the estimate must be linear, that is, $E = \hat{E}$, we can thus write

$$\hat{E}[\mathbf{X}[n]|\mathbf{Y}[n-1], \ldots, \mathbf{Y}[0]] = \hat{E}[\mathbf{X}[n]|\tilde{\mathbf{Y}}[n-1], \ldots, \tilde{\mathbf{Y}}[0]],$$

which by Theorem 9.1-4(b) becomes

$$= \sum_{k=0}^{n-1} \hat{E}[\mathbf{X}[n]|\tilde{\mathbf{Y}}[k]] \tag{9.2-9}$$

$$= \sum_{k=0}^{n-1} \mathbf{C}_k^{(n)} \tilde{\mathbf{Y}}[k]$$

[†]For example, to compute $\mathbf{Y}[1]$ in terms of $\tilde{\mathbf{Y}}[0], \tilde{\mathbf{Y}}[1]$ recall that $\mathbf{Y}[0] = \tilde{\mathbf{Y}}[0]$, and $\mathbf{Y}[1] = \tilde{\mathbf{Y}}[1] + E\{\mathbf{Y}[1]|\mathbf{Y}[0]\}$. But $\mathbf{Y}[0] = \tilde{\mathbf{Y}}[0]$ and for the Gaussian case the MMS estimator is also the LMMS estimator so that $E\{\mathbf{Y}[1]|\mathbf{Y}[0]\} = E\{\mathbf{Y}[1]|\tilde{\mathbf{Y}}[0]\} = \hat{E}\{\mathbf{Y}[1]|\tilde{\mathbf{Y}}[0]\} = \mathbf{D}_0\tilde{\mathbf{Y}}[0]$. Hence $\mathbf{Y}[1] = \tilde{\mathbf{Y}}[1] + \mathbf{D}_0\tilde{\mathbf{Y}}[0]$. As usual, $\mathbf{D}_0$ is found by the orthogonality principle.

[‡]The reader should understand that $\boldsymbol{\sigma}_{\tilde{\mathbf{Y}}}^2[n]$ will not be constant since observations start at $n = 0$, so the innovations may be initially large. In fact $\tilde{\mathbf{Y}}[0] = \mathbf{Y}[0]$.

where

$$E[\mathbf{X}[n]\tilde{\mathbf{Y}}[k]^T] \triangleq \mathbf{C}_k^{(n)}\sigma_{\tilde{\mathbf{Y}}}^2[k], \qquad 0 \le k < n.$$

Assuming that the variance matrix $\sigma_{\tilde{\mathbf{Y}}}^2[k]$ is nonsingular, this equation can be solved for $\mathbf{C}_k^{(n)}$ to yield

$$\mathbf{C}_k^{(n)} = E[\mathbf{X}[n]\tilde{\mathbf{Y}}[k]^T]\sigma_{\tilde{\mathbf{Y}}}^{-2}[k]. \tag{9.2-10}$$

So, we can also write the prediction estimate as

$$\hat{\mathbf{X}}[n] = \sum_{k=0}^{n-1} E[\mathbf{X}[n]\tilde{\mathbf{Y}}[k]^T]\sigma_{\tilde{\mathbf{Y}}}^{-2}[k]\tilde{\mathbf{Y}}[k]. \tag{9.2-11}$$

Since the signal model Equation 9.2-6 is recursive, we suspect there is a way around the growing memory estimate that appears in Equation 9.2-11. Substituting Equation 9.2-6 for $\mathbf{X}[n]$ we can write

$$E[\mathbf{X}[n]\tilde{\mathbf{Y}}[k]^\tau] = \mathbf{A}E[\mathbf{X}[n-1]\tilde{\mathbf{Y}}[k]^T] + \mathbf{B}E[\mathbf{W}[n]\tilde{\mathbf{Y}}[k]^T],$$

but for $k < n$, $\mathbf{W}[n] \perp \mathbf{X}[k]$ and $\mathbf{W}[n] \perp \mathbf{V}[k]$, which implies $\mathbf{W}[n] \perp \mathbf{Y}[k]$ and hence also $\mathbf{W}[n] \perp \hat{\mathbf{Y}}[k]$, so that we have

$$E[\mathbf{X}[n]\tilde{\mathbf{Y}}[k]^\tau] = \mathbf{A}E[\mathbf{X}[n-1]\tilde{\mathbf{Y}}[k]^T] \quad \text{for all } k < n.$$

Hence we can also express the prediction estimate $\hat{\mathbf{X}}[n]$ as

$$\hat{\mathbf{X}}[n] = \mathbf{A}\sum_{k=0}^{n-1} E[\mathbf{X}[n-1]\tilde{\mathbf{Y}}[k]^T]\sigma_{\tilde{\mathbf{Y}}}^{-2}[k]\tilde{\mathbf{Y}}[k]. \tag{9.2-12}$$

But Equation 9.2-11 must hold at $n-1$ as well as n, thus also

$$\hat{\mathbf{X}}[n-1] = \sum_{k=0}^{n-2} E[\mathbf{X}[n-1]\tilde{\mathbf{Y}}[k]^T]\sigma_{\tilde{\mathbf{Y}}}^{-2}[k]\tilde{\mathbf{Y}}[k].$$

Combining this equation with Equation 9.2-12, we finally obtain

$$\hat{\mathbf{X}}[n] = \mathbf{A}\hat{\mathbf{X}}[n-1] + \mathbf{A}E[\mathbf{X}[n-1]\tilde{\mathbf{Y}}[n-1]^T]\sigma_{\tilde{\mathbf{Y}}}^{-2}[n-1]\tilde{\mathbf{Y}}[n-1],$$

which is an efficient way of calculating the prediction estimate of Equation 9.2-12. If we define the *Kalman gain* matrix

$$\mathbf{G}_{n-1} \triangleq E[\mathbf{X}[n-1]\tilde{\mathbf{Y}}[n-1]^T]\sigma_{\tilde{\mathbf{Y}}}^{-2}[n-1], \tag{9.2-13}$$

we can rewrite the preceding result as

$$\hat{\mathbf{X}}[n] = \mathbf{A}(\hat{\mathbf{X}}[n-1] + \mathbf{G}_{n-1}\tilde{\mathbf{Y}}[n-1]), \quad n \ge 0 \tag{9.2-14}$$

with initial condition $\hat{\mathbf{X}}[0] = \mathbf{0}$.

We can eliminate the innovations sequence $\tilde{\mathbf{Y}}[n]$ in Equation 9.2-14 as follows:

$$\tilde{\mathbf{Y}}[n] = \mathbf{Y}[n] - E[\mathbf{Y}[n]|\mathbf{Y}[n-1], \ldots, \mathbf{Y}[0]],$$

so using Equation 9.2-7, we have

$$E[\mathbf{Y}[n]|\mathbf{Y}[n-1], \ldots, \mathbf{Y}[0]] = E[\mathbf{X}[n]|\mathbf{Y}[n-1], \ldots, \mathbf{Y}[0]]$$
$$+ E[\mathbf{V}[n]|\mathbf{Y}[n-1], \ldots, \mathbf{Y}[0]],$$
$$= \hat{\mathbf{X}}[n] + 0.$$

This last step is justified by noting that $\mathbf{V}[n] \perp \mathbf{X}[k]$ and $\mathbf{V}[n] \perp \mathbf{V}[k]$ for all $k < n$ so that we have $\mathbf{V}[n] \perp \mathbf{Y}[k]$ for all $k < n$. Thus we obtain $\tilde{\mathbf{Y}}[n] = \mathbf{Y}[n] - \hat{\mathbf{X}}[n]$. So inserting this into Equation 9.2-14 we finally have

$$\hat{\mathbf{X}}[n] = \mathbf{A}\{\hat{\mathbf{X}}[n-1] + \mathbf{G}_{n-1}(\mathbf{Y}[n-1] - \hat{\mathbf{X}}[n-1])\}, \qquad (9.2\text{-}15)$$

which is the most well known form of the *Kalman predictor*, whose system diagram is shown in Figure 9.2-2.

We can denote the prediction estimate in Equation 9.2-15 more explicitly as

$$\hat{\mathbf{X}}[n|n-1] \triangleq E[\mathbf{X}[n]|\mathbf{Y}[n-1], \mathbf{Y}[n-2], \ldots, \mathbf{Y}[0]] = \hat{\mathbf{X}}[n].$$

On the other hand, we may be interested in calculating the causal estimate

$$\hat{\mathbf{X}}[n|n] \triangleq E[\mathbf{X}[n]|\mathbf{Y}[n], \mathbf{Y}[n-1], \ldots, \mathbf{Y}[0]],$$

which uses all the data up to the present time n. The Kalman predictor can be modified to provide this causal estimate. The resulting recursive formula is called the *Kalman filter*. One method to derive it from Equation 9.2-15 is the following. Consider the prediction

$$\hat{\mathbf{X}}[n|n-1] = E[\mathbf{X}[n]|\mathbf{Y}[n-1], \ldots, \mathbf{Y}[0]],$$

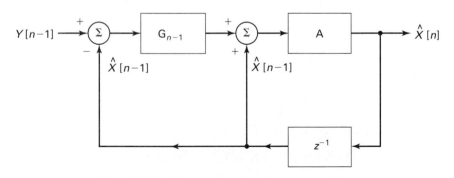

Figure 9.2-2   System diagram of Kalman predictor.

and use the signal model Equation 9.2-6 to obtain

$$\hat{\mathbf{X}}[n|n-1] = \mathbf{A}\hat{\mathbf{X}}[n-1|n-1] + 0,$$

since $\mathbf{W}[n] \perp \mathbf{Y}[k]$ for $k < n$. So

$$\hat{\mathbf{X}}[n-1|n-1] = \mathbf{A}^{-1}\hat{\mathbf{X}}[n|n-1]$$

since $\mathbf{A}$ is nonsingular. Using the result we pre-multiply Equation 9.2-15 by $\mathbf{A}^{-1}$ to get

$$\hat{\mathbf{X}}[n-1|n-1] = \hat{\mathbf{X}}[n-1|n-2] + \mathbf{G}_{n-1}\big(\mathbf{Y}[n-1] - \hat{\mathbf{X}}[n-1|n-2]\big).$$

which can be written equivalently for $n \geq 0$ as

$$\hat{\mathbf{X}}[n|n] = \mathbf{A}\hat{\mathbf{X}}[n-1|n-1] + \mathbf{G}_n\big(\mathbf{Y}[n] - \mathbf{A}\hat{\mathbf{X}}[n-1|n-1]\big), \qquad (9.2\text{-}16)$$

which is known as the Kalman filter equation. Here $\hat{\mathbf{X}}[-1|-1] \triangleq \mathbf{0}$.

By examining either Equation 9.2-15 or Equation 9.2-16 we can recognize a predictor-corrector structure. The first term is the MMSE prediction of $\mathbf{X}[n]$ based on the past observations $\mathbf{Y}[k]$, $k < n$. The second term involving the current data is called the *update*. It is the product of a gain matrix $\mathbf{G}_n$ (which can be precomputed and stored) and the prediction error based on the noisy observations, which we have called the innovations, that is, the new information contained in the current data.

An alternative direct derivation of the Kalman filter (Equation 9.2-16), which avoids the need to invert the system matrix $\mathbf{A}$, proceeds as follows. First we write

$$\hat{\mathbf{X}}[n|n] = \hat{E}[\mathbf{X}[n]|\mathbf{Y}[n], \dots, \mathbf{Y}[0]] \quad n \geq 0$$
$$= \hat{E}[\mathbf{X}[n]|\tilde{\mathbf{Y}}[n], \dots, \tilde{\mathbf{Y}}[0]],$$
$$= \sum_{k=0}^{n} \mathbf{C}_k^{(n)}\tilde{\mathbf{Y}}[k]$$

by use of Theorem 9.1-4 as before. Here $\tilde{\mathbf{Y}}[0] = \mathbf{Y}[0]$, and $\mathbf{C}_k^{(n)} = E[\mathbf{X}[n]\tilde{\mathbf{Y}}^T[k]]\boldsymbol{\sigma}_{\tilde{\mathbf{Y}}}^{-1}[k]$, $0 \leq k \leq n$. Pulling off the $k = n$ term, we write

$$\hat{\mathbf{X}}[n|n] = \mathbf{C}_n^{(n)}\tilde{\mathbf{Y}}[n] + \sum_{k=0}^{n-1} \mathbf{C}_k^{(n)}\tilde{\mathbf{Y}}[k]$$
$$= \mathbf{C}_n^{(n)}\tilde{\mathbf{Y}}[n] + \hat{\mathbf{X}}[n|n-1] \quad \text{by Equation 9.2-9}$$
$$= \hat{\mathbf{X}}[n|n-1] + \mathbf{C}_n^{(n)}\tilde{\mathbf{Y}}[n]$$
$$= \hat{\mathbf{X}}[n|n-1] + \mathbf{G}_n\big(\mathbf{Y}[n] - \hat{\mathbf{X}}[n|n-1]\big), \quad n \geq 1.$$

Figure 9.2-3 shows the system diagram of the Kalman filter.

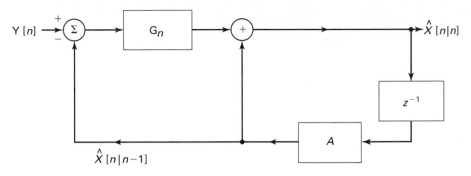

Figure 9.2-3   System diagram of Kalman Filter.

One can go on to derive *Kalman smoothers*, which are fixed-delay estimators of the form

$$\hat{\mathbf{X}}[n|n+k] \triangleq E[\mathbf{X}[n]|\mathbf{Y}[n+k], \mathbf{Y}[n+k-1], \ldots, \mathbf{Y}[0]], \quad k > 0. \tag{9.2-17}$$

These delayed estimators are of importance in the study of various communication and control systems. As yet we have not found an efficient algorithm to calculate the sequence of gain matrices $\mathbf{G}_n$. This is the subject of the next section.

Error-Covariance Equations

We have to find a method for recursively calculating the gain matrix sequence $\mathbf{G}_n$ in Equation 9.2-13. We are also interested in evaluating the mean-square error of the estimate. The *prediction-error variance* matrix is the covariance matrix of $\tilde{\mathbf{X}}[n] \triangleq \hat{\mathbf{X}}[n|n-1] - \mathbf{X}[n]$. We write it as

$$\varepsilon^2[n] \triangleq E[\tilde{\mathbf{X}}[n]\tilde{\mathbf{X}}^T[n]].$$

We start by inserting the observation Equation 9.2-7 into the innovations Equation 9.2-8 to obtain

$$\tilde{\mathbf{Y}}[n] = \mathbf{X}[n] + \mathbf{V}[n] - \hat{\mathbf{X}}[n]^{\dagger}$$
$$= -\tilde{\mathbf{X}}[n] + \mathbf{V}[n]. \tag{9.2-18}$$

But $\mathbf{X}[n] \perp \mathbf{V}[n]$, so upon using $\hat{\mathbf{X}}[n]$ for $\hat{\mathbf{X}}[n|n-1]$,

$$E[\mathbf{X}[n]\tilde{\mathbf{Y}}^T[n]] = -E[\mathbf{X}[n]\tilde{\mathbf{X}}^T[n]]$$
$$= E[(\hat{\mathbf{X}}[n] - \mathbf{X}[n])\tilde{\mathbf{X}}^T[n]] \quad \text{since } \tilde{\mathbf{X}}[n] \perp \hat{\mathbf{X}}[n]$$
$$= E[\tilde{\mathbf{X}}[n]\tilde{\mathbf{X}}^T[n]]$$
$$= \varepsilon^2[n].$$

---

†Remember $\hat{\mathbf{X}}[n]$ is the prediction estimate, $\hat{\mathbf{X}}[n|n-1]$.

Also $\tilde{\mathbf{X}}[n] \perp \mathbf{V}[n]$, so we have from Equation 9.2-18

$$E[\tilde{\mathbf{Y}}[n]\tilde{\mathbf{Y}}[n]^T] = E[\tilde{\mathbf{X}}[n]\tilde{\mathbf{X}}[n]^T] + E[\mathbf{V}[n]\mathbf{V}[n]^T]$$

or

$$\sigma_{\tilde{\mathbf{Y}}}^2[n] = \varepsilon^2[n] + \sigma_{\mathbf{V}}^2[n];$$

thus by Equation 9.2-13 we have

$$\mathbf{G}_n = \varepsilon^2[n](\varepsilon^2[n] + \sigma_{\mathbf{V}}^2[n])^{-1}, \quad n \geq 0. \tag{9.2-19}$$

The problem is now reduced to calculating the prediction error variance matrix $\varepsilon[n]$. From Problem 9.3 we can write

$$\varepsilon^2[n] = E[\mathbf{X}[n]\mathbf{X}^T[n]] - E[\hat{\mathbf{X}}[n]\hat{\mathbf{X}}^T[n]]. \tag{9.2-20}$$

To evaluate the right side of Equation 9.2-20 we use Equation 9.2-6 and $\mathbf{X}[n-1] \perp \mathbf{W}[n]$ to get

$$E[\mathbf{X}[n]\mathbf{X}^T[n]] = \mathbf{A}E[\mathbf{X}[n-1]\mathbf{X}^T[n-1]]\mathbf{A}^T + \mathbf{B}\sigma_{\mathbf{W}}^2\mathbf{B}^T. \tag{9.2-21}$$

Likewise using $\hat{\mathbf{X}}[n] = \mathbf{A}(\hat{\mathbf{X}}[n-1] + \mathbf{G}_{n-1}\tilde{\mathbf{Y}}[n-1])$ and $\hat{\mathbf{X}}[n-1] \perp \tilde{\mathbf{Y}}[n-1]$, we get

$$E[\hat{\mathbf{X}}[n]\hat{\mathbf{X}}^T[n]] = \mathbf{A}E[\hat{\mathbf{X}}[n-1]\hat{\mathbf{X}}^T[n-1]]\mathbf{A}^T + \mathbf{A}\mathbf{G}_{n-1}\sigma_{\tilde{\mathbf{Y}}}^2[n-1]\mathbf{G}_{n-1}^T\mathbf{A}^T$$

$$= \mathbf{A}E[\hat{\mathbf{X}}[n-1]\hat{\mathbf{X}}^T[n-1]]\mathbf{A}^T + \mathbf{A}\varepsilon^2[n-1]\mathbf{G}_{n-1}^T\mathbf{A}^T \tag{9.2-22}$$

where we have used $\mathbf{G}_{n-1}\sigma_{\tilde{\mathbf{Y}}}^2[n-1] = \varepsilon^2[n-1]$.

Substituting Equations 9.2-21 and 9.2-22 into Equation 9.2-20 and simplifying then yields

$$\varepsilon^2[n] = \mathbf{A}\varepsilon^2[n-1](\mathbf{I} - \mathbf{G}_{n-1}^T)\mathbf{A}^T + \mathbf{B}\sigma_{\mathbf{W}}^2\mathbf{B}^T \tag{9.2-23}$$

for $n \geq 0$, where

$$\varepsilon^2[-1] \triangleq E[\tilde{\mathbf{X}}[-1]\tilde{\mathbf{X}}^T[-1]]$$

and

$$E[\tilde{\mathbf{X}}[-1]\tilde{\mathbf{X}}^T[-1]] = \begin{cases} \mathbf{0} & \text{if } \mathbf{X}[-1] = \mathbf{0} \text{ (known)} \\ \sigma_{\mathbf{X}}^2 & \text{if } \mathbf{X} \text{ is WSS.}^\dagger \end{cases}$$

In summary, the Kalman filter for the state equation (Equation 9.2-6) and the observation model (Equation 9.2-7) thus consists of the filtering equation (Equation 9.2-16), the gain equation (Equation 9.2-19), and the prediction-error covariance equation (Equation 9.2-23). The *filtering-error covariance* equation can also be calculated as

$$\mathbf{e}^2[n] \triangleq E[(\mathbf{X}[n] - \hat{\mathbf{X}}[n|n])(\mathbf{X}[n] - \hat{\mathbf{X}}[n|n])^T]$$

$$= \varepsilon^2[n][\mathbf{I} - \mathbf{G}_n^T]$$

$$= \mathbf{A}\varepsilon^2[n-1]\mathbf{A}^T + \mathbf{B}\sigma_{\mathbf{W}}^2\mathbf{B}^T, \quad n \geq 0.$$

The proof of this fact is left as an exercise to the reader.

---

†Strictly speaking, this would require a minor modification of our development, since we have assumed $\mathbf{X}[-1] = \mathbf{0}$. However to model the process in the wide-sense stationary sense, we can just take $\mathbf{X}[-1] = \mathbf{X}$, a Gaussian random vector independent of $\mathbf{W}[n]$ for $n \geq 0$, and with variance matrix $\sigma_{\mathbf{X}}^2$.

Example 9.2-2
(Scalar Kalman filter). Consider the Gauss-Markov signal model

$$X[n] = 0.9X[n-1] + W[n], \qquad n \geq 0,$$

with means equal to zero and $\sigma_W^2 = 0.19$. Also $X[-1] = 0$. Let the scalar observation equation be

$$Y[n] = X[n] + V[n], \qquad n \geq 0,$$

with $\sigma_V^2 = 1$. We have $A = 0.9$ and $B = 1$. The Kalman filter Equation 9.2-16 then becomes

$$\hat{X}[n|n] = 0.9\hat{X}[n-1|n-1] + G_n(Y[n] - 0.9\hat{X}[n-1|n-1]),$$

with initial condition $\hat{X}[-1|-1] = X[-1] = 0$. The Kalman gain Equation 9.2-19 is

$$G_n = \varepsilon^2[n]/(1 + \varepsilon^2[n]),$$

and prediction-error variance Equation 9.2-23 is given as

$$\varepsilon^2[n] = 0.81\varepsilon^2[n-1](1 - G_{n-1}) + 0.19$$

$$= 0.81\varepsilon^2[n-1]\left(1 - \frac{\varepsilon^2[n-1]}{1 + \varepsilon^2[n-1]}\right) + 0.19$$

$$= \frac{0.19 + \varepsilon^2[n-1]}{1 + \varepsilon^2[n-1]}, \qquad n \geq 0,$$

with initial condition $\varepsilon^2[-1] = 0$. We can solve this equation for the steady-state solution $\varepsilon^2[\infty]$

$$\varepsilon^2[\infty] = \frac{0.19 + \varepsilon[\infty]}{1 + \varepsilon[\infty]},$$

and discarding the negative root we obtain

$$\varepsilon^2[\infty] = 0.436.$$

Alternatively, we can use the MATLAB program,

```
eps2(1)
     0.3193
>> for n=2:20
   eps2(n) = ( 0.19 + esp2(n-1) ) / ( 1.0 + eps2(n-1) ) ;
end
>> ,
```

to generate the plot shown in Figure 9.2-4, where we used $\varepsilon^2[0] = 0.19$. We note that convergence to the steady state is monotonic and rapid, essentially occurring by $n = 10$.

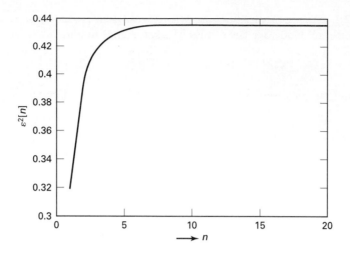

Figure 9.2-4   Prediction-error variance $\varepsilon^2[n]$.

Either way, we see that $\varepsilon^2[n] \to 0.436$ and hence $G_n \to 0.304$ so that the Kalman filter is asymptotically given by

$$\hat{X}[n|n] = 0.9\hat{X}[n-1|n-1] + 0.304(Y[n] - 0.9\hat{X}[n-1|n-1])$$

$$= 0.626\hat{X}[n-1|n-1] + 0.304Y[n]$$

and the steady-state filtering error is $e^2[\infty] = 0.304$.

We have developed the Kalman filter as a computationally attractive solution for estimating a Gauss-Markov signal (Equation 9.2-6) observed in additive white Gaussian noise (Equation 9.2-7) over the *semi-infinite time* interval $0 \le n < \infty$. The filter equations are time-variant because the initial estimates near $n = 0$ effectively have a truncated past. For our constant parameter and constant noise variance assumptions, the estimation error should tend toward an asymptotic or steady-state value as we move away from $n = 0$, at least for stable signal models.

The Kalman filter derivation can easily be generalized to allow time-varying parameters $\mathbf{A}_n$ and $\mathbf{B}_n$ in Equation 9.2-6. We can also permit time-varying noise variances $\boldsymbol{\sigma}^2_{\mathbf{W}}[n]$ and $\boldsymbol{\sigma}^2_{\mathbf{V}}[n]$. In fact, the present derivation will also serve for this time-variant case by just inserting the subscripts on $\mathbf{A}$ and $\mathbf{B}$ as required.

The observation Equation 9.2-7 is also overly restrictive. See Problem 9.10 for a generalization to allow the observation vectors to be of different dimension than the signal vectors,

$$\mathbf{Y}[n] = \mathbf{C}_n\mathbf{X}[n] + \mathbf{V}[n] \tag{9.2-24}$$

In this equation, the rectangular matrix $\mathbf{C}_n$ permits linear combinations of the state vector $\mathbf{X}$ to appear in the observations and hence can model FIR convolution.

Kalman filters have seen extensive use in control systems and automatic target tracking and recognition. Their wide popularity is largely due to the availability of small and modest size computers of great number crunching power. Without such processors. Kalman filters would have remained largely of theoretical interest. There are many books that discuss the Kalman filter, see, for example, [9-3], [9-6], and [9-7].

9.3 WIENER FILTERS FOR RANDOM SEQUENCES

In this section we will investigate optimal linear estimates for signals that are not necessarily Gaussian-Markov of any order. This theory predates the Kalman-Bucy filter of the last section. This optimum linear filter is associated with Norbert Wiener and Andrei N. Kolmogorov, who performed this work in the 1940s. The discrete-time theory that is presented in this section is in fact the work of Kolmogorov [9-8] while Wiener developed the continuous-time theory [9-9] for random processes. Nevertheless, it has become conventional to refer to both types of filters as Wiener filters. These filters are mainly appropriate for WSS random processes (sequences) observed over *infinite time intervals*.

We start with the general problem of finding the LMMSE estimate of the random sequence $X[n]$ from observations of the random sequence $Y[n]$ for all time. We assume both random sequences are WSS and for convenience are zero-mean and *real-valued*. Our approach to find the LMMSE estimate will be based on the innovations sequence. For these infinite time-interval observations, the innovations may be obtained using spectral factorization (cf. Section 6.4).

First we spectrally factor the psd of the observations $\mathsf{S}_{YY}(z)$ into its causal and anti-causal factors,

$$\mathsf{S}_{YY}(z) = \sigma^2 \mathsf{B}(z)\mathsf{B}(z^{-1}), \tag{9.3-1}$$

where $\mathsf{B}(z)$ contains all the poles and zeros of S_{YY} that are inside the unit circle in the Z-plane. Hence B and B^{-1} are stable and causal. We can thus operate on $Y[n]$ with the LSI operator B^{-1} to produce the innovations random sequence $\tilde{Y}[n]$ as shown in Figure 9.3-1. The psd of $\tilde{Y}[n]$ is easily seen to be white,

$$\tilde{\mathsf{S}}_{YY}(z) = \sigma^2.$$

Thus $\tilde{Y}[n]$ satisfies the three defining properties of the innovations sequence as listed in Section 9.2. We can then base our estimate on $\tilde{Y}[n]$.

Unrealizable Case (Smoothing)

Consider an LSI operator G with convolution representation

$$\hat{X}[n] = \sum_{k=-\infty}^{+\infty} g[k]\tilde{Y}[n-k]. \tag{9.3-2}$$

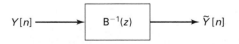

Figure 9.3-1 Whitening filter.

We want to choose the $g[k]$ to minimize the MSE,

$$E[(X(n) - \hat{X}(n))^2].$$

Expanding this expression, we obtain for real-valued system $g[n]$,

$$E[X^2[n]] - 2E\left[\sum_k g[k]X[n]\tilde{Y}[n-k]\right] + E\left[\left(\sum_k g[k]\tilde{Y}[n-k]\right)^2\right]$$

$$= K_{XX}[0] - 2\sum_k g[k]K_{X\tilde{Y}}[k] + \sigma^2 \sum_k g^2[k]$$

$$= K_{XX}[0] + \sum_{k=-\infty}^{+\infty}\left[\sigma g[k] - \frac{K_{X\tilde{Y}}[k]}{\sigma}\right]^2 - \frac{1}{\sigma^2}\sum_{k=-\infty}^{+\infty} K_{X\tilde{Y}}^2[k], \qquad (9.3\text{-}3)$$

where the last line is obtained by completing the square (cf. Appendix A). Examining this equation we see that only the middle term depends on the choice of $g[k]$. The minimum of this term is obviously zero for the choice,

$$g[k] = \frac{1}{\sigma^2}K_{X\tilde{Y}}[k], \qquad -\infty < k < +\infty \qquad (9.3\text{-}4)$$

the LMMSE then being given as

$$\varepsilon_u^2 = K_{XX}[0] - \frac{1}{\sigma^2}\sum_{k=-\infty}^{+\infty} K_{X\tilde{Y}}^2[k]. \qquad (9.3\text{-}5)$$

This result can also be derived directly by using the $\hat{E}$ operator, as shown in Problem 9.11. In the Z-transform domain the operator G is expressed as

$$\mathsf{G}(z) = \frac{1}{\sigma^2}\mathsf{S}_{X\tilde{Y}}(z) = \frac{1}{\sigma^2}\mathsf{S}_{XY}(z)\mathsf{B}^{-1}(z^{-1}).$$

The overall transfer function, including the whitening filter, then becomes

$$\mathsf{H}_u(z) = \mathsf{G}(z)\mathsf{B}^{-1}(z)$$

$$= \mathsf{S}_{XY}(z)/(\sigma^2\mathsf{B}(z)\mathsf{B}(z^{-1}))$$

$$= \mathsf{S}_{XY}(z)/\mathsf{S}_{YY}(z), \qquad (9.3\text{-}6)$$

where the subscript "u" on H_u denotes the *unrealizable* estimator. The MSE is given from Equation 9.3-5 as

$$\frac{1}{2\pi j}\oint_{|z|=1}\left[\mathsf{S}_{XX}(z) - \frac{1}{\sigma^2}\mathsf{S}_{X\tilde{Y}}(z)\mathsf{S}_{X\hat{Y}}(z^{-1})\right]\frac{dz}{z}$$

using Parseval's theorem [9-3], and then simplifies to

$$\frac{1}{2\pi} \int_{-\pi}^{+\pi} (S_{XX}(\omega) - \{|S_{XY}(\omega)|^2/S_{YY}(\omega)\}) d\omega.$$

Example 9.3-1

(Additive noise.) Here we take the special case where the observations consist of a signal plus noise $V[n]$ with $X \perp V$,

$$Y[n] = X[n] + V[n], \qquad -\infty < n + \infty$$

The psd S_{YY} and cross-psd S_{XY} then become

$$S_{YY}(z) = S_{XX}(z) + S_{VV}(z),$$
$$S_{XY}(z) = S_{XX}(z).$$

Thus the optimal unrealizable filter is

$$H_u(z) = \frac{S_{XX}(z)}{S_{XX}(z) + S_{VV}(z)}.$$

The MSE expression becomes

$$\frac{1}{2\pi} \int_{-\pi}^{+\pi} \frac{S_{XX}(\omega) S_{VV}(\omega)}{S_{XX}(\omega) + S_{VV}(\omega)} d\omega.$$

Examining the frequency response

$$H_u(\omega) = \frac{S_{XX}(\omega)}{S_{XX}(\omega) + S_{VV}(\omega)},$$

we see the interpretation of $H_u(\omega)$ as a frequency-domain weighting function. When the signal-to-noise (SNR) ratio is high at a given frequency, $H_u(\omega)$ is close to 1, that is,

$$H_u(\omega) \simeq 1 \quad \text{when} \quad S_{XX}(\omega)/S_{VV}(\omega) \gg 1.$$

Similarly, when the SNR is low at ω, then reasonably enough $H_u(\omega)$ is near zero.

Causal Wiener Filter

If we add the constraint that the LMMSE estimate must be causal, we get a filter that can be approximated more easily in the time domain. We can proceed as before up to Equation 9.3-3 at which point we must apply the constraint that $g[k] = 0$ for $k < 0$. Thus the optimal solution is to set

$$g[n] = \frac{1}{\sigma^2} K_{X\tilde{Y}}[n] u[n]. \tag{9.3-7}$$

The overall error then becomes

$$\varepsilon_c^2 = K_{XX}[0] - \frac{1}{\sigma^2} \sum_{k=0}^{\infty} K_{X\tilde{Y}}^2[k], \tag{9.3-8}$$

which is seen larger than the unrealizable error in Equation 9.3-5.

To express the optimal filter in the transform domain, we introduce the notation

$$[\mathsf{F}(z)]_+ \overset{\Delta}{=} \sum_{n=0}^{\infty} f[n]z^{-n} \tag{9.3-9}$$

and then write the Z-transform of Equation 9.3-7 as

$$\mathsf{G}(z) = \frac{1}{\sigma^2}[\mathsf{S}_{X\tilde{Y}}(z)]_+.$$

The overall LMMSE causal filter then becomes

$$\mathsf{H}_c(z) = \frac{1}{\sigma^2\mathsf{B}(z)}\left[\frac{\mathsf{S}_{XY}(z)}{\mathsf{B}(z^{-1})}\right]_+, \tag{9.3-10}$$

where the subscript c denotes a *causal* estimator. The causal filtering MSE is expressed as

$$\varepsilon_c^2 = \frac{1}{2\pi}\int_{-\pi}^{+\pi}(S_{XX}(\omega) - H_c(\omega)S_{XY}^*(\omega))d\omega.$$

Example 9.3-2 _____

(First-order signal in white noise.) Let the signal psd be given as

$$\mathsf{S}_{XX}(z) = \frac{0.19}{(1 - 0.9z^{-1})(1 - 0.9z)}.$$

Let the observations be given as

$$Y[n] = X[n] + V[n],$$

where $V[n]$ is white noise with variance $\sigma_V^2 = 1$. Then the psd of the observations is

$$\mathsf{S}_{YY}(z) = \mathsf{S}_{XX}(z) + \mathsf{S}_{VV}(z)$$

$$= 1.436\left[\frac{1 - 0.627z^{-1}}{1 - 0.9z^{-1}}\right]\left[\frac{1 - 0.627z}{1 - 0.9z}\right]$$

$$= \sigma^2\,\mathsf{B}(z)\cdot\mathsf{B}(z^{-1}),$$

so that

$$\left[\frac{\mathsf{S}_{XX}(z)}{\mathsf{B}(z^{-1})}\right]_+ = \left[\frac{0.19}{(1 - 0.9z^{-1})(1 - 0.627z)}\right]_+$$

$$= \left[\frac{0.436}{1 - 0.9z^{-1}} + \frac{0.273}{z^{-1} - 0.627}\right]_+$$

$$= \frac{0.436}{1 - 0.9z^{-1}},$$

where we have used the partial fraction expansion to recover the causal part as required by Equation 9.3-9. Thus by Equation 9.3-10, the optimal causal filter is

$$H_c(z) = \frac{1}{\sigma^2 B(z)} \frac{0.436}{1 - 0.9z^{-1}}$$

$$= \frac{0.304}{1 - 0.627z^{-1}}.$$

While Wiener filters do not require a signal model such as Gauss–Markov, as required in the Kalman filter, they suffer in comparison with the latter in that their memory storage requirements are greater and they are less conveniently adaptive to changes in system or noise parameters.

9.4 EXPECTATION-MAXIMIZATION ALGORITHM

The *expectation-maximization* (E-M) algorithm is an iterative method of obtaining a *maximum-likelihood estimator* (MLE) of a parameter θ or parameter vector $\boldsymbol{\theta}$ of a probability function such as the PMF (probability mass function), pdf (probability density function), and so forth. The MLE is obtained by forming the likelihood function $l_{\mathbf{X}}(\theta) = \prod_{i=1}^{N} f_{X_i}(X_i; \theta)$, or log-likelihood function $L_{\mathbf{X}}(\theta) = \log \prod_{i=1}^{N} f_{X_i}(X_i; \theta)$, and, most often, (but not always) differentiating with respect to θ to find an estimator $\hat{\theta}$ (a random variable) that maximizes the likelihood function. The MLE $\hat{\theta}$ depends only on the data, that is, $\hat{\theta} = d(X_1, \ldots, X_N)$. The principle is the same if θ is a vector parameter, that is, $\boldsymbol{\theta} = (\theta_1, \ldots, \theta_k)$; in that case the MLE of $\boldsymbol{\theta}$ would involve k functions $\{d_i\}$ of the data, that is, $\hat{\theta}_1 = d_1(X_1, \ldots, X_N)$, $\hat{\theta}_2 = d_2(X_1, \ldots, X_N), \ldots, \hat{\theta}_K = d_K(X_1, \ldots, X_N)$. However, (and this is the crux of the problem) what happens when the data $X_1, \ldots, X_N$ are not directly observable? For example, suppose we observe *not* the vector of random variables $\mathbf{X} = (X_1, \ldots, X_N)$ but, instead, $\mathbf{Y} = (Y_1, \ldots, Y_M)$ where $Y_1 = T_1(\mathbf{X}), \ldots, Y_M = T_M(\mathbf{X})$ and $M < N$. Here the functions $\{T_i\}$ are often many-to-one and describe the physical process by which the *unobserved*, but so-called *complete* data $\mathbf{X}$ gets transformed into the *observed* but *incomplete* data $\mathbf{Y}$. For example, in computer-aided tomography (CAT), we measure $Y_i = \sum_{j=1}^{N} a_{ij} X_j, i = 1, \ldots, D$, where the $\{Y_i\}$ are the D detector readings (the observable but incomplete data), the $\{a_{ij}\}$ relate to the geometry of the configuration, and the $\{X_j\}$ are the pixel opacities (the desired and complete data but not directly observable). If we wish to determine the parameters governing the distribution of the $\{X_j\}$, it seems that we must do so through the incomplete data $\{Y_j\}$. Unfortunately this is hard to do in many cases because of the nature of the transformation from $\mathbf{X}$ to $\mathbf{Y}$. The E-M algorithm enables us to find the MLE of the unknown parameter(s) by a series of iterations in which only the incomplete data $\mathbf{Y}$ are involved. Each iteration consists of two steps: the *expectation* or so-called E-step, and the *maximization* or so-called M-step. The expectation is with respect to the underlying variables $\mathbf{X}$, using the current estimate of the unknown parameters and conditioned upon the observations $\mathbf{Y}$. The maximization step provides a new estimate of the unknown parameters.

There are many examples of complete and incomplete data. For example, suppose we want to estimate the mean of a Normal random variable X by making N i.i.d. observations $X_1, X_2, \ldots, X_N$ on X. The data, sometimes called *the complete data*, allows us to obtain the MLE as $\hat{\theta} = \frac{1}{n} \sum_{i=1}^{n} X_i$. However, suppose our data is corrupted by independent Normally distributed noise. Then instead of the clean $X_i's$, we now get the corrupted data $Y_i = X_i + N_i$, $i = 1, \ldots, n$. Here the $\{Y_i\}$, being a many-to-one transformation on the $\{X_i\}$, represent the incomplete data.

The E-M algorithm cannot work miracles. The MLE obtained from the $\{Y_i\}$, whether by direct methods or by iterations as in the E-M algorithm, will generally have a higher variance than the MLE obtained from the complete data. We illustrate with an example.

Example 9.4-1

We wish to obtain the MLE of the mean, μ, of a real, Normal random variable X with variance σ^2 from two independent observations on X, namely X_1 and X_2. The joint pdf of the data is $f_{X_1 X_2}(x_1, x_2) = (2\pi\sigma^2)^{-1} \exp\{-\frac{1}{2}[(x_1 - \mu)^2 + (x_2 - \mu)^2]/\sigma^2\}$. By differentiating with respect to μ we obtain the MLE as $\hat{\theta}_X = \hat{\mu}_1 = \frac{1}{2} \sum_{i=1}^{2} X_i$. Suppose, however, that our measurement is the datum $Y = T(\mathbf{X}) = 2X_1 + 3X_2$ which is a many-to-one linear transformation. Clearly we cannot form the MLE $\hat{\theta}_X = \hat{\mu}_1 = \frac{1}{2} \sum_{i=1}^{2} X_i$ from Y alone. However, if is easy to compute the pdf of Y as

$$f_Y(y) = \frac{1}{\sigma\sqrt{26\pi}} \exp\left\{ -\frac{1}{2} \left[\frac{y - 5\mu}{\sigma\sqrt{13}} \right]^2 \right\}.$$

The log-likelihood function of the random variable Y is

$$\log f_Y(Y) = -1/2 \log(26\pi\sigma^2) + 1/2 \left\{ \frac{Y - 5\mu}{\sigma\sqrt{13}} \right\}^2$$

and maximizing with respect to μ yields the MLE, based on Y, as

$$\hat{\theta}_Y = \hat{\mu}_2 = \frac{1}{5} Y.$$

While both $\hat{\theta}_X$ *and* $\hat{\theta}_Y$ yield unbiased estimates of μ, $\text{Var}(\hat{\theta}_Y)$ is greater than $\text{Var}(\tilde{\theta}_X)$

In Example 9.4-1 the random variable Y is an example of *incomplete* data. But this implies that there might exist other so-called *left-out data* that, when combined with the incomplete data, can yield an MLE with the variance associated with the complete data.

We illustrate with an example.

Example 9.4-2

Suppose we could get a second measurement in Example 9.4-1 that was functionally independent from the first, for example, say, $W = X_1 - 4X_2$. Then we can easily obtain the joint pdf of $f_{WY}(w, y)$ in terms of $f_{X_1 X_2}(x_1, x_2)$ as

$$f_{WY}(w, y; \theta) = 11 \cdot f_{X_1 X_2}(x_1, x_2; \theta),$$

where $x_1 = (4y + 3w)/11$, $x_2 = (y - 2w)/11$, and the factor of 11 is the required Jacobian scaling. Here, we added the argument θ to emphasize the dependence of the pdf's on the unknown parameter θ. The MLE of μ is $\hat{\theta}_{WY} = (5Y + W)/11$ which has the same variance as that of $\hat{\mu}_1$. We can think of W as one form of the left-out data needed to make the observations complete. For a given transformation T there are many choices for W. Both Y and W are functions on the Euclidean space R^1 while the vector $\mathbf{X} = (X_1, X_2)$ is a function onto R^2. Thus the measurement and left-out data spaces are subspaces of the nonobservable, but complete, data space.

The point of this example is to introduce the notion of left-out data and to show that we can write the pdf of $\mathbf{X}$ in terms of the pdfs of Y and W. Likewise, we can write the likelihood function of $\mathbf{X}$ in terms of those of Y and W. Indeed, in the more general case, with $\mathbf{X}, \mathbf{Y}, \mathbf{W}$ representing the complete, incomplete, and left-out data, respectively we can write that

$$K' f_{\mathbf{X}}(\mathbf{x}; \theta) = f_{\mathbf{WY}}(\mathbf{w}, \mathbf{y}; \theta) = f_{\mathbf{W}|\mathbf{Y}}(\mathbf{w}|\mathbf{y}; \theta) f_{\mathbf{Y}}(\mathbf{y}; \theta),$$

where K' is related to the Jacobian of the transformation and is 11 in this example.

The next example illustrates a situation where obtaining the MLE by the direct method does not seem feasible and the E-M algorithm is required.

Example 9.4-3 _____

Consider an image $\{F(i, j)\}_{N \times N}$ represented by a (possibly long) vector $\mathbf{F}$ with N^2 components. It is shown in the research literature [9-10] that $\mathbf{F}$ can be decomposed into two parts: $\mathbf{F} = \mathbf{AF} + \mathbf{V}$, where $\mathbf{A}$ is an autoregressive image model matrix and $\mathbf{V}$ is a zero-mean, stationary Gaussian process with diagonal covariance $\mathbf{K_{VV}} = \sigma_v^2 \mathbf{I}$. The *observed image* $\mathbf{G}$ is often a blurred and noisy version of $\mathbf{F}$. It is modeled by a linear, shift-invariant operation on $\mathbf{F}$ by a matrix $\mathbf{D}$ followed by the addition of independent, additive Gaussian noise, that is $\mathbf{G} = \mathbf{DF} + \mathbf{W}$. The covariance matrix of $\mathbf{W}$ is assumed diagonal, that is $\mathbf{K_{WW}} = \sigma_W^2 \mathbf{I}$. The unknown parameter vector $\boldsymbol{\theta} \triangleq \{d(m, n), a(l, k), \sigma_W^2, \sigma_V^2\}$, where the $\{d(m, n)\}$ and the $\{a(k, l)\}$ are the coefficients of the $\mathbf{D}$ and $\mathbf{A}$ matrices, respectively, is to be determined from the observed image, the known structure of the image and blurring models, and the *a priori* known forms of the probability functions for $\mathbf{V}$ and $\mathbf{W}$. With $\mathbf{K_{GG}}$ defined as the covariance matrix of $\mathbf{G}$ and given by

$$\mathbf{K_{GG}} = \mathbf{D}(\mathbf{I} - \mathbf{A})^{-1} \mathbf{K_{vv}} (\mathbf{I} - \mathbf{A})^{-1} \mathbf{D} + \mathbf{K_{WW}},$$

the MLE of $\boldsymbol{\theta}$ is obtained from

$$\boldsymbol{\theta}_{ML} = \arg\max_{\boldsymbol{\theta}} \{-\log(\det |\mathbf{K_{GG}}|) - \mathbf{G}^T \mathbf{K_{GG}} \mathbf{G}\}, \qquad (9.4\text{-}1)$$

where the superscript ^{T} denotes transpose. Because finding the solution to Equation 9.4-1 involves a complicated non-linear optimization in many variables, a direct solution does not seem possible. However, the problem has been solved by the E-M method and the solution is described in the published literature.

Log-likelihood for the Linear Transformation

We now extend the results of Example 9.4-2 to a general many-to-one linear transformation and write

$$W_1 = L_1(X_1, X_2, \ldots, X_N)$$
$$W_2 = L_2(X_1, X_2, \ldots, X_N)$$
$$\vdots$$
$$W_K = L_K(X_1, X_2, \ldots, X_N)$$
$$Y_1 = L_{K+1}(X_1, X_2, \ldots, X_N)$$
$$\vdots$$
$$Y_M = L_{K+M}(X_1, X_2, \ldots, X_N),$$

where the L_i, $i = 1, \ldots, K + M = N$ are linear operators. The appropriate transformation for the pdf's is

$$K' f_{\mathbf{X}}(\mathbf{x}; \theta) = f_{\mathbf{WY}}(\mathbf{w}, \mathbf{y}; \theta) = f_{\mathbf{W}|\mathbf{Y}}(\mathbf{w}|\mathbf{y}; \theta) f_{\mathbf{Y}}(\mathbf{y}; \theta), \qquad (9.4\text{-}2)$$

where K' is the Jacobian scaling of the transformation and is of little interest in what follows. Written as a *log-likelihood-function*, Equation 9.4-2 can be converted to

$$\log f_{\mathbf{Y}}(\mathbf{Y}; \theta) = \log f_{\mathbf{X}}(\mathbf{X}; \theta) - \log f_{\mathbf{W}|\mathbf{Y}}(\mathbf{W}|\mathbf{Y}; \theta) + K, \qquad (9.4\text{-}3)$$

where $K \triangleq \log K'$. With the exception of the constant term, each of the terms in Equation 9.4-3 can be interpreted as a likelihood function. If we take the conditional expectation of Equation 9.4-3, term by-term, for some value of the parameter θ, say θ', conditioned on both $\mathbf{Y}$ and the current estimator of θ, say $\theta^{(k)}$, we obtain

$$\log f_{\mathbf{Y}}(\mathbf{Y}; \theta') = E\{\log f_{\mathbf{X}}(\mathbf{X}; \theta')|\mathbf{Y}; \theta^{(k)}\} - E\{\log f_{\mathbf{W}|\mathbf{Y}}(\mathbf{W}|\mathbf{Y}; \theta')|\mathbf{Y}; \theta^{(k)}\} + K, \quad (9.4\text{-}4)$$

where the left-hand side (LHS) of Equation 9.4-4 is a constant with respect to the expectation, since the conditioning is on $\mathbf{Y}$.

To shorten the notation, define

$$U(\theta', \theta^{(k)}) \triangleq E\{\log f_{\mathbf{X}}(\mathbf{X}; \theta')|\mathbf{Y}; \theta^{(k)}\} \qquad (9.4\text{-}5)$$

and

$$V(\theta', \theta^{(k)}) \triangleq E\{\log f_{\mathbf{W}|\mathbf{Y}}(\mathbf{W}|\mathbf{Y}; \theta')|\mathbf{Y}; \theta^{(k)}\} \qquad (9.4\text{-}6)$$

so that Equation 9.4-4 can be rewritten as

$$\log f_{\mathbf{Y}}(\mathbf{Y}; \theta') = U(\theta', \theta^{(k)}) - V(\theta', \theta^{(k)}) + K, \qquad (9.4\text{-}7)$$

Now if it can be shown that the function $V(\cdot,\cdot)$ has the property that

$$V(\theta', \theta^{(k)}) \leq V(\theta^{(k)}, \theta^{(k)}), \; Condition \; 1, \tag{9.4-8}$$

and if θ' is chosen such that

$$U(\theta', \theta^{(k)}) \geq U(\theta^{(k)}, \theta^{(k)}), \; Condition \; 2, \tag{9.4-9}$$

then, it follows from Equation 9.4-7 that

$$\log f_{\mathbf{Y}}(\mathbf{Y}; \theta') \geq \log f_{\mathbf{Y}}(\mathbf{Y}; \theta^{(k)}) \tag{9.4-10}$$

or, equivalently,

$$f_{\mathbf{Y}}(\mathbf{Y}; \theta') \geq f_{\mathbf{Y}}(\mathbf{Y}; \theta^{(k)}). \tag{9.4-11}$$

Equations 9.4-8, 9.4-9, and 9.4-11 are the basis for the E-M algorithm. For Equation 9.4-10 to be true, we must first show that Equation 9.4-8 is true. To this end define

$$Z \triangleq f_{\mathbf{W}|\mathbf{Y}}(\mathbf{W}|\mathbf{Y}; \theta') / f_{\mathbf{W}|\mathbf{Y}}(\mathbf{W}|\mathbf{Y}; \theta^{(k)}) \tag{9.4-12}$$

and use the fact that $\log Z \leq Z - 1$. Since this is true for every realization of Z, it must also be true for the expectations, conditional or otherwise. Thus

$$E\{\log\{f_{\mathbf{W}|\mathbf{Y}}(\mathbf{W}|\mathbf{Y}; \theta') / f_{\mathbf{W}|\mathbf{Y}}(\mathbf{W}|\mathbf{Y}; \theta^{(k)})\}|\mathbf{Y}; \theta^{(k)}\}$$

$$\leq E\{[f_{\mathbf{W}|\mathbf{Y}}(\mathbf{W}|\mathbf{Y}; \theta') / f_{\mathbf{W}|\mathbf{Y}}(\mathbf{W}|\mathbf{Y}; \theta^{(k)})]|\mathbf{Y}; \theta^{(k)}\} - 1. \tag{9.4-13}$$

But, by definition of the expectation, the right-hand side (RHS) is merely

$$\int_{-\infty}^{\infty} \frac{f_{\mathbf{W}|\mathbf{Y}}(w|y; \theta')}{f_{\mathbf{W}|\mathbf{Y}}(w|y; \theta^{(k)})} f_{\mathbf{W}|\mathbf{Y}}(w|y; \theta^{(k)}) dw - 1 = 0,$$

from which it follows that

$$E\{\log\{f_{\mathbf{W}|\mathbf{Y}}(\mathbf{W}|\mathbf{Y}; \theta') / f_{\mathbf{W}|\mathbf{Y}}(\mathbf{W}|\mathbf{Y}; \theta^{(k)})\}|\mathbf{Y}; \theta^{(k)}\} \leq 0 \tag{9.4-14}$$

or, equivalently,

$$V(\theta', \theta^{(k)}) \triangleq E\{\log f_{\mathbf{W}|\mathbf{Y}}(\mathbf{W}|\mathbf{Y}; \theta')|\mathbf{Y}; \theta^{(k)}\}$$

$$\leq E\{\log f_{\mathbf{W}|\mathbf{Y}}(\mathbf{W}|\mathbf{Y}; \theta^{(k)})|\mathbf{Y}; \theta^{(k)}\} \triangleq V(\theta^{(k)}, \theta^{(k)}). \tag{9.4-15}$$

Thus Condition 1 is met. To meet Condition 2, it is merely required to compute

$$U(\theta', \theta^{(k)}) \triangleq E\{\log f_{\mathbf{X}}(\mathbf{X}; \theta')|\mathbf{Y}; \theta^{(k)}\}$$

and update the estimate of $\theta^{(k)}$ by finding $\theta^{(k+1)}$ as

$$\theta^{(k+1)} = \arg\max_{\theta'} U(\theta', \theta^{(k)}). \tag{9.4-16}$$

The operation $\theta^{(k+1)} = \arg\max_{\theta'} U(\theta', \theta^{(k)})$ is short for "the value of θ' that maximizes $U(\theta', \theta^{(k)})$ and is called $\theta^{(k+1)}$".

The net effect of these operations is to achieve the goal of Equation 9.4-11, namely,

$$f_{\mathbf{Y}}(\mathbf{Y}; \theta^{(k+1)}) \geq f_{\mathbf{Y}}(\mathbf{Y}; \theta^{(k)}).$$

The E-M algorithm is not *guaranteed* to converge to the MLE of θ. The algorithm, which we summarize below, can stagnate at a local minimum. Nevertheless the E-M algorithm has been used with good success in tomography, speech recognition, active noise cancellation, spread-spectrum communications, and still others [9-11].

Summary of the E-M algorithm

Start with an arbitrary initial estimate $\theta^{(0)}$ of the unknown parameter θ. For $k = 0, 1, 2, \ldots$ compute

1. **The E-step:** compute $U(\theta', \theta^{(k)}) \triangleq E\{\log f_{\mathbf{X}}(\mathbf{X}; \theta') | \mathbf{Y}; \theta^{(k)}\}$ as a function of θ'. (This is often the hard part.)
2. **The M-step:** compute $\theta^{(k+1)} \triangleq \arg\max_{\theta'} U(\theta', \theta^{(k)})$.

Stop when $\theta^{(k)}$ stops changing, or is changing so slowly that an imposed convergence criterion is met.

E-M Algorithm for Exponential Probability Functions

A simplification of the E-step occurs when the pdf or PMF of the data belongs to the exponential family of probability functions. Such functions can be written in the somewhat general form as

$$f_{\mathbf{X}}(\mathbf{x}; \theta) = b(\mathbf{x})c(\theta)\exp\{\mathbf{t}(\mathbf{x})\mathbf{\Gamma}^T(\theta)\} \tag{9.4-17}$$

where $b(\mathbf{x})$ is a function that depends only on $\mathbf{x}$, $c(\theta)$ is a constant, $\mathbf{\Gamma}(\theta)$ is a (possibly vector) function of the unknown parameter θ, and $\mathbf{t}(\mathbf{x})$ is a (possibly row vector) function that depends only on the realizations and is independent of the parameter θ. When used in a likelihood function, $\mathbf{t}(\mathbf{X})$ is called a *sufficient statistic* for θ because, for the exponential family of pdf's (or PMF's), it aggregates the data in a fashion that is sufficient to form an estimator for θ. Being essentially an estimator, a sufficient statistic cannot depend on θ. A key requirement on a sufficient statistic is that the pdf (or PMF) of the data $\mathbf{X}$, conditioned on the sufficient statistic, must not depend on θ.

Example 9.4-4

Consider the joint PMF of N independent Poisson random variables $X_1, \ldots, X_N$ with Poisson parameters $\theta_1, \ldots, \theta_N$, respectively. The joint PMF can be written as

$$P_{\mathbf{X}}[\mathbf{x}; \boldsymbol{\theta}] = \prod_{i=1}^{N} e^{-\theta_i} \theta_i^{x_i} / x_i!, \tag{9.4-18}$$

where $\mathbf{X} \triangleq (X_1, \ldots, X_N)$, $\mathbf{x} \triangleq (x_1, \ldots, x_N)$, and $\boldsymbol{\theta} \triangleq (\theta_1, \ldots, \theta_N)$. Using the fact that $\theta^x \triangleq e^{x \log \theta}$, we can rewrite Equation 9.4-18 as

$$P_{\mathbf{X}}[\mathbf{x}; \boldsymbol{\theta}] = \left(\prod_{i=1}^{N} \frac{1}{x_i!}\right) \times \exp\left\{-\sum_{i=1}^{N} \theta_i\right\} \times \exp\{\mathbf{x}\mathbf{\Gamma}^T(\theta)\}, \tag{9.4-19}$$

Thus we associate $b(\mathbf{x})$ with $(\prod_{i=1}^{N} \frac{1}{x_i!})$, $c(\boldsymbol{\theta})$ with $\exp\{-\sum_{t=1}^{N} \theta_i\}$, $\mathbf{t}(\mathbf{x})$ with $\mathbf{x}$, and $\boldsymbol{\Gamma}(\boldsymbol{\theta})$ with the (row) vector $(\log\theta_1, \ldots, \log\theta_N)$.

Still restricting ourselves to the exponential family of probability functions, we consider now the log-likelihood function associated with estimating the vector parameter $\boldsymbol{\theta}$. This yields

$$L_{\mathbf{X}}(\boldsymbol{\theta}) = \log b(\mathbf{X}) + \log c(\boldsymbol{\theta}) + \boldsymbol{\Gamma}(\boldsymbol{\theta})\mathbf{t}^T(\mathbf{X}). \tag{9.4-20}$$

The expression for the E-step is then

$$U(\boldsymbol{\theta}', \boldsymbol{\theta}^{(k)}) = E\{\log[b(\mathbf{X})|\mathbf{Y}; \boldsymbol{\theta}^{(k)}]\} + \log c(\boldsymbol{\theta}') + \boldsymbol{\Gamma}(\boldsymbol{\theta}')E\{\mathbf{t}^T(\mathbf{X})|\mathbf{Y}; \boldsymbol{\theta}^{(k)}\}$$

However, recall that in the M-step, any expression *not* containing θ' becomes irrelevant in the maximization routine. Hence we may drop the term $E\{\log[b(\mathbf{X}|\mathbf{Y}; \theta^{(k)})]\}$. Therefore for the family of exponential pdf's or PMF's, the E-M algorithm takes the somewhat simpler form.

Start with an arbitrary initial estimate $\boldsymbol{\theta}^{(0)}$ of the unknown parameter $\boldsymbol{\theta}$. For $k = 0, 1, 2, \ldots$ compute.

1. **The E-step:**
$$\mathbf{t}^{(k+1)} = E\{\mathbf{t}(\mathbf{X})|\mathbf{Y}; \boldsymbol{\theta}^{(k)}\};$$

2. **The M-step:**
$$\boldsymbol{\theta}^{(k+1)} = \arg\max_{\boldsymbol{\theta}'}\{\log c(\boldsymbol{\theta}') + \mathbf{t}^{(k+1)}\boldsymbol{\Gamma}^T(\boldsymbol{\theta}')\}.$$

Repeat until the convergence criterion is met.

For the important Poisson case the E-M algorithm takes a special form.

1. **The E-step:**
$$\mathbf{X}^{(k+1)} = E[\mathbf{X}^{(k)}|\mathbf{Y}; \boldsymbol{\theta}^{(k)}].$$

2. **The M-step:**

$$\boldsymbol{\theta}^{(k+1)} = \arg\max_{\boldsymbol{\theta}'}\left\{-\left(\sum_{i=1}^{N}\theta_i'\right) + \mathbf{X}^{(k+1)} \cdot (\log\theta_1', \log\theta_2', \ldots, \log\theta_N')^T\right\}.$$

Repeat until the convergence criterion is met.

Application to Emission Tomography

Emission tomography (ET) is a medical imaging technique in which body tissue is stimulated to emit photons. Typically, a radioactive positron-emitting substance is attached to glucose and injected into the body of the patient. Areas of the body where the glucose is rapidly metabolized show up as "hot spots," that is, regions of strong photon emissions. For example, in metastatic cancer, the tumors, because of rapid cell division and growth, exhibit above-average metabolic activity and therefore emit strong streams of photons. Another application of ET is in imaging the areas in the brain exhibiting the metabolism of glucose while the patient is engaged in various activities such as reading, playing chess, watching a movie, and so on. In this way, researchers can determine which part of the brain is involved while doing a particular activity. (See for example, the article by T.K. Moon [9-11].)

Detectors located around the body collect the photons and the number of photons collected at each detector constitutes the data $\mathbf{Y}$ (Figure 9.4-1). The fundamental problem in ET is, then, to reconstruct an image of the spatial distribution of the photon emission vector Λ from the data $\mathbf{Y}$.

It should be noted that the detector data is often incomplete to begin with. We illustrate with an example.

Example 9.4-5

A greatly simplified tomographic configuration is shown in Figure 9.4-2. Each cell emits photons in a Poisson mode. The photons emitted during a certain interval of time are represented by the components of the vector $\mathbf{X} = (X_1, X_2, X_3, X_4)$, which denote the number of photons emitted from cells 1, 2, 3, 4, respectively.

The vector $\mathbf{X}$ can be said to constitute the complete data. The MLE of the photon emission vector $\Lambda = (\lambda_1, \lambda_2, \lambda_3, \lambda_4)$ is easily shown to be $\hat{\lambda}_i = X_i$, $i = 1, 4$. However, the

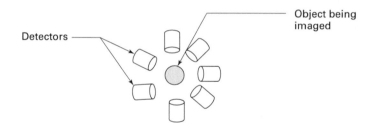

Figure 9.4-1   Emission tomography configuration.

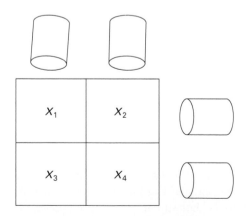

Figure 9.4-2   In ET, the data consist of the sum of the aggregate outputs of the individual cells. In this illustrative example, the summations are along the rows and the columns.

data collected at the four detectors are represented by

$$\mathbf{Y} = (X_1 + X_3, X_2 + X_4, X_3 + X_4, X_1 + X_2)$$

which is *incomplete*. For example, exactly the same data $\mathbf{Y}$ would be collected if the complete data were given instead by $\mathbf{X}' = (X_1 + \delta, X_2 - \delta, X_3 - \delta, X_4 + \delta)$. Hence, there is a *many-to-one transformation* in going from $\mathbf{X}$ to $\mathbf{Y}$.

Following the example in the previously cited article by Moon, we assume that it is desired to estimate Λ, the Poisson spatial emission function. The object to be imaged consists of B cells and the Poisson law governs the emission of photons from each cell. The components of Λ are the Poisson emission parameters of the cells, that is, $\Lambda = (\lambda_1, \lambda_2, \ldots, \lambda_3)$. Thus during T seconds, the probability of n photon emissions from cell b is $P[X_b = n | \lambda_b] = \exp(-\lambda_b T) \cdot (\lambda_b T)^n / n!$, where X_b is the number of emitted photons from cell b, and λ_b is the Poisson rate parameter from cell b. Without loss of generality, we take $T = 1$. Let p_{bd}, $b = 1, \ldots, B$; $d = 1, \ldots, D$ denote the probability that a photon from cell b is detected by detector tube d. The $\{p_{bd}\}$ can be determined from the geometry of the sensors with respect to the body. It is assumed that no photons get lost: $\sum_{d=1}^{D} p_{bd} = 1$. The joint probability of the number of photons captured at each of the D detectors obeys the *multinomial law* if we condition this probability on the total number of emitted photons. Without such conditioning, the number of photons, Y_d, collected at collector d is Poisson (we leave the demonstration of this result to the reader) with PMF

$$P[Y_d = y] = e^{-\lambda_d} (\lambda_d)^y / y! \qquad (9.4\text{-}21)$$

Let X_{bd} denote the number of emissions from cell b detected by detector d. The set $\mathbf{X} = \{X_{bd}, b = 1, \ldots, B; d = 1, \ldots, D\}$ represents the complete but unobserved data. The incomplete data are the detector readings $\{Y_d : d = 1, \ldots, D\}$. The many-to-one map is implied by the system of equations:

$$Y_d = \sum_{b=1}^{B} X_{bd}, d = 1, \ldots, D, \qquad (9.4\text{-}22)$$

where we assume that $B > D$.

The expected value of Y_d is given by

$$E[Y_d] = \lambda_d \qquad (9.4\text{-}23)$$

$$= \sum_{b=1}^{B} E_{X_b}\{E[X_{bd} | X_b]\}$$

$$= \sum_{b=1}^{B} E_{X_b}\{p_{bd} X_b\}$$

$$= \sum_{b=1}^{B} \lambda_b p_{bd}. \qquad (9.4\text{-}24)$$

Hence, $\lambda_d = \sum_{b=1}^{B} \lambda_b p_{bd}$ and each random variable X_{bd} is Poisson with $\lambda_{bd} = \lambda_b p_{bd}$. We now demonstrate the use of the E-M algorithm to estimate $\Lambda = \{\lambda_1, \ldots, \lambda_B\}$.

Log-likelihood Function of Complete Data

A basic assumption is that each cell emits independently of any other cell and each detector operates independently of any other detector. Under this assumption the likelihood function $l_X(\Lambda)$ is given by

$$l_{\mathbf{X}}(\Lambda) = \prod_{b,d} P_{\mathbf{X}}[X_{bd}]$$

$$= \prod_{b,d} e^{-\lambda_b p_{bd}} (\lambda_b p_{bd})^{X_{bd}} \Big/ X_{bd}! \qquad (9.4\text{-}25)$$

where, for convenience, we omit the explicit dependence of the $\{X_{bd}\}$ on $l_X(\Lambda)$.

As usual, we prefer to deal with the log-likelihood function, $L_{\mathbf{X}}(\Lambda)$, by taking the natural logarithm of $l_{\mathbf{X}}(\Lambda)$. This yields

$$L_{\mathbf{X}}(\Lambda) = \sum_{b,d} -\lambda_b p_{bd} + X_{bd} \log \lambda_b + X_{bd} \log p_{bd} - \log X_{bd}! \qquad (9.4\text{-}26)$$

We wish to find the vector Λ that maximizes this expression. The E-M algorithm will be used to estimate the unobserved data using the current best estimate, $\Lambda^{(k)}$, of Λ. Then the estimated unobserved data $\mathbf{X}^{(k+1)}$ will be used to improve the estimate of $\Lambda^{(k)}$ to $\Lambda^{(k+1)}$. The procedure will be repeated until the convergence criteria established by the user are met.

E-step

Assume that we are at the kth step and our current estimate of Λ is $\Lambda^{(k)}$. We compute the E-step as

$$X_{bd}^{(k+1)} = E[X_{bd}^{(k)}|\mathbf{Y}, \Lambda^{(k)}] = E[X_{bd}^{(k)}|Y_d, \Lambda^{(k)}]$$

(independent photon aggregate at the detectors)

$$= \sum_{x_{bd}} x_{bd} P\left[X_{bd}^{(k)} = x_{bd} \,\middle|\, \sum_{i=1}^{B} X_{id}^{(k)} = y_d, \Lambda^{(k)}\right] \left(\text{recall that } Y_d = \sum_{i=1}^{B} X_{id}\right)$$

$$= \sum_{x_{bd}} x_{bd} \frac{P\left[X_{bd}^{(k)} = x_{bd}, \sum_{i \neq b}^{B} X_{id}^{(k)} = y_d - x_{bd}\right]}{P[Y_d = y_d]} \quad \text{(using Bayes' rule, rewriting the}$$

second of the joint events, and submerging the condition on $\Lambda^{(k)}$ to save space)

$$= \sum_{x_{bd}} x_{bd} \frac{P[X_{bd}^{(k)} = x_{bd}] \cdot P\left[\sum_{i=b}^{B} X_{id}^{(k)} = y_d - x_{bd}\right]}{P[Y_d = y_d]}$$

(using the independence of the $X_{bd}^{(k)}$).   (9.4-27)

Now use

$$P[X_{bd}^{(k)} = x_{bd}] = \exp(-\lambda_{bd}^{(k)}) \cdot (\lambda_{bd}^{(k)})^{x_{bd}} / x_{bd}! \qquad \text{(Line 1)}$$

$$P\left[\sum_{i \neq b} X_{id}^{(k)} = y_d - x_{bd}\right] = \exp\left(-\sum_{i \neq b}^{B} \lambda_{id}^{(k)}\right) \cdot \left(\sum_{i \neq b}^{B} \lambda_{id}^{(k)}\right)^{y_d - x_{bd}} / (y_d - x_{bd})! \qquad \text{(Line 2)}$$

$$P\left[Y_d \text{ and } \lambda_{bd}^{(k)}\right] = \exp\left(-\sum_{i=1}^{B} \lambda_{id}^{(k)}\right) \cdot \left(\sum_{i=1}^{B} \lambda_{id}^{(k)}\right)^{y_d} / y_d!$$

$$\text{and } \lambda_{bd}^{(k)} = \lambda_b^{(k)} \times p_{bd}. \qquad \text{(Line 3)}$$

In line 2 we used the result that the sum of Poisson random variables is Poisson with the parameter being the sum of the Poisson parameters of the random variables.

After some elementary algebraic manipulations, we obtain

$$X_{bd}^{(k+1)} = \frac{Y_d \lambda_b^{(k)} p_{bd}}{\sum_{i=1}^{B} \lambda_i^{(k)} p_{id}}, \, b = 1, \ldots, B; d = 1, \ldots, D \qquad (9.4\text{-}28)$$

M-step

The M-step is usually easier to realize computationally than the E-step. We merely maximize the log-likelihood function with respect to λ_b using the updated estimate of the complete data $X_{bd}^{(k+1)}$.

Thus set

$$0 = \frac{\partial}{\partial \lambda_b} L_{X^{(k+1)}}(\Lambda)$$

which yields the desired result as

$$\lambda_b^{(k+1)} = \sum_{d=1}^{D} X_{bd}^{(k+1)} \qquad (9.4\text{-}29)$$

In obtaining Equation 9.4-29 we used the fact that $\sum_{d=1}^{D} p_{bd} = 1$.

Finally, Equation 9.4-28 can be combined with Equation 9.4-29 to yield the single update equation:

$$\lambda_b^{(k+1)} = \lambda_b^{(k)} \sum_{d=1}^{D} \left(\frac{Y_d p_{bd}}{\sum\limits_{i=1}^{B} \lambda_i^{(k)} p_{id}} \right). \tag{9.4-30}$$

We then repeat the steps of the EM algorithm until the convergence criterion has been met.

Clearly the EM algorithm is non-trivial to execute. Without computers its application to all but the simplest problems would not feasible. It has become a powerful technique for MLE-type problems in engineering and medicine. For further reading on the EM algorithm, see [9-12, 9-13].

9.5   HIDDEN MARKOV MODELS (HMM)

In ordinary Markov models the underlying stochastic mechanism is the transition between states. The sequences of states, as time progresses, are the *observations*. Each state corresponds to a physical event. The evolution of the states as a function of time (or displacement, or volume) forms a Markov random sequence. The observer knows the Markov model itself.

For a class of engineering problems of growing importance such as speech and image processing by computer, the known model assumption is too restrictive. For this reason certain classes of Markov models that exhibit a second degree of randomness have been introduced. These models are called *hidden Markov models* (HMM's) because the data available to the observer are not the evolution of the states but a second stochastic process that is a probabilistic function of the states. Thus it is not known which sequence of states produced the observed data. Indeed there are many state sequences that could have produced the same observations. An important question in this regard is, *given a vector of observations, which state sequence was most likely to have produced it?*

Excellent tutorial articles have been written on HMM's. In particular the tutorial articles by Lawrence Rabiner and his colleagues [9-14, 9-15] have facilitated the understanding of HMM's by the nonspecialist in statistics. In what follows, we shall closely follow the discussion and style of these tutorial articles.

In certain types of problems, not only is it unknown which sequence of states produced a given observation recognition but it is also not even known which of several competing Markov models was the most likely generator of the given observations. We illustrate with an example adapted from [9-15].

Example 9.5-1 _____

(Coin tossing experiment.) We are given the results of a coin tossing experiment in which each observation is either a head $\{H\}$ or a tail $\{T\}$. We do not know which of two models is responsible for the observation sequence $\mathbf{O} = \{HTT\}$. The two models, M_1, M_2 are (1) two biased coins with a stochastic method of choosing between them for the next coin flip; or (2) three biased coins with a stochastic mechanism for choosing among them for the next flip. We use the notation $P^i[H|C_j]$ to mean the probability of getting a head in model i

using coin C_j in that model (note that $j = 1, 2$ in the first model and $j = 1, 2, 3$ in the second model). We let the choice of coins be the state. Suppose we know or estimate the following probabilities:

$$P^1[H|C_1] = 0.3, \quad P^1[T|C_1] = 0.7, \quad P^1[H|C_2] = 0.6, \quad P^1[T|C_2] = 0.4 \quad \text{and}$$

$$P^2[H|C_1] = 0.5, \quad P^2[T|C_1] = 0.5, \quad P^2[H|C_2] = 0.8, \quad P^2[T|C_2] = 0.2, \quad \text{and}$$

$$P^2[H|C_3] = 0.4, \quad P^2[T|C_3] = 0.6.$$

These probabilities are known as the *state conditional output symbol probabilities*. The state $S[n]$ at time n is the particular coin in use at time n. The event $\{S[n] = C_1, M_2\}$ means that the state in model *two* at time n is C_1. We next need to specify the stochastic mechanism by which coins are selected that represents state transitions. Specifying a set of *state transition probabilities* does this. We use the notation $a_{ij}^1[k] = P[S[k] = C_j|S[k-1] = C_i, M_1]$ to denote these state transition probabilities, conditioned on model *one,* and likewise for model *two*: $a_{ij}^2[k] = P[S[k] = C_j|S[k-1] = C_i, M_2]$. In the models we consider, the state transition probabilities are time-invariant. That is, the probability of reaching one state from another remains the same whatever the time might be. For the sake of illustration we assign values to the state transition probabilities as in Figure 9.5-1.

To finish the description of these models we need to specify the initial state probabilities, that is, the probabilities of being in the various states at time $n = 1$. We denote these probabilities by $p_j^1[1]$ or $p_j^2[1]$, $j = 1, 2$ for model *one* and $j = 1, 2, 3$ for model *two*. It is convenient to define a set of initial state probability vectors $\mathbf{p}^1[1] = (p_1^1[1], p_2^1[1])$ for model *one* and $\mathbf{p}^2[1] = (p_1^2[1], p_2^2[1], p_3^2[1])$ for model *two*. For the sake of specificity we assign $\mathbf{p}^1[1] = (0.7 \ 0.3)$ and $\mathbf{p}^2[1] = (0.8 \ 0.1 \ 0.1)$. If the initial state probability vector has zeros everywhere except a one in position j, then the *initial state must be* coin C_j. The two models are shown in Figure 9.5-1

Hidden Markov models have considerable versatility in describing and recognizing complex random phenomena such as natural speech. Indeed HMM's are used in natural speech recognition by computer. An HMM is typically a doubly stochastic construct and its complexity and versatility depend on the number of states and whether the states are massively or sparsely interconnected.

Specification of an HMM

We limit ourselves to HMM's where output symbols come from a discrete alphabet with a finite number of symbols. To fully specify an HMM we need to specify the following parameters:

(1) *The number of states N.* In speech processing, each state might represent a different position of the vocal organs. Then the underlying stochastic mechanism of moving between states is associated with the dynamics of the vocal organs as a spoken word is being produced. The inertia of the mass of the vocal organs as well as the limited speed of nerve impulses constrains which states can be reached from

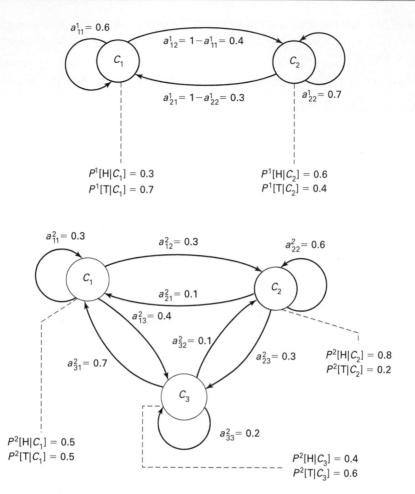

Figure 9.5-1   Two possible models that might have produced the observation sequence $\mathbf{O} = \{HTT\}$. In each of the two models there are obviously many state sequences that could have produced this observation vector. A basic question is, *which of the two models was most likely to have produced this sequence?* Another basic question is, *how do we efficiently compute the probability that a model produced the observation vector?*

other states. Also if the number of different positions of the vocal organs is very large, then the number of states will be very large and, indeed, might be too large for computational purposes. If the number of states has to be kept small, again for computational reasons, then the one-to-one relation between states and the position of the vocal organs becomes blurred. This does not mean, however, that the HMM cannot be used effectively. All states must have realizations in the vector $\mathbf{Q} = (q_1, \ldots, q_N)$. Thus the statement that at time n the HMM is in state q_j is written as $\{S[n] = q_j\}$.

(2) *The $N \times N$ state-transition probability matrix* $\mathbf{A} = \{a_{ij}\}$. In more complex models the elements $a_{ij}[n] \overset{\Delta}{=} P[S[n] = q_j | S[n-1] = q_i]$ might depend on the time index n. Here we assume that the state transition probabilities do not depend on time. Therefore we write a_{ij} and not $a_{ij}[n]$. Note that if $a_{ij} = 0$, it is not possible to go to state q_j from state q_i in one step. If $a_{ij} > 0$ for all i, j, then every state can be reached from every other state in one step. Note that the row probabilities in $\mathbf{A}$ must add to unity, that is, $\sum_{j=1}^{N} a_{ij} = 1$, $j = 1, 2, \ldots, N$ since the HMM must be in one of the N states.

(3) *The set of discrete observation symbols* $\mathbf{V} = \{v_1, v_2, \ldots, v_M\}$. The number M is called the *discrete alphabet size*. It is elements of the set $\mathbf{V}$ that are actually observed, not the states which produced them. In the case of speech or music; the actual physical output is typically considered a continuous-time, continuous-amplitude process. This means that to model such a process with a discrete-time, finite set of symbols, sampling and quantizing are required.

(4) *The state conditional output probabilities.* These are the probabilities b_{jk} of observing the output symbol v_k while in state q_j. By definition $b_{jk} = P[X[n] = v_k | S[n] = q_j]$ for $k = 1, \ldots, M$ and $j = 1, \ldots, N$. The random variable $X[n]$ is the observation at time n. In a more sophisticated model these probabilities could depend on the time index n, but here we assume time independence, hence we write b_{jk} and not $b_{jk}[n]$.

The state conditional output probability matrix $\mathbf{B} \overset{\Delta}{=} [b_{jk}]$ is $N \times M$.

(5) *The initial state probability vector.* This is the vector $\mathbf{p}[1] = (p_1[1], p_2[1], \ldots, p_N[1])$ whose components $p_j[1]$ are the probabilities of starting the observation sequence in state q_j, $j = 1, 2, \ldots, N$. As in Example 9.5-1 if for some j, $p_j[1] = 1$, the sequence must begin in state q_j.

For convenience the HMM can be defined by the six-tuple

$$\mathbf{M} \overset{\Delta}{=} (N, M, \mathbf{V}, \mathbf{A}, \mathbf{B}, \mathbf{p}[1])$$

although it is customary in the applications literature [9-11] to use the more compact notation $\mathbf{M} = (\mathbf{A}, \mathbf{B}, \mathbf{p}[1])$.

Example 9.5-2 ⎯⎯⎯⎯⎯⎯⎯⎯⎯⎯⎯⎯⎯⎯⎯⎯⎯⎯⎯⎯⎯⎯⎯⎯⎯
Describe the two models in Example 9.5-1 using the parameters of the six-tuple $\mathbf{M}$,

Solution For model *one* we obtain $N = 2$ (two states), $M = 2$ (two characters H and T), and,

$$\mathbf{V} = (H, T), \mathbf{A} = \begin{bmatrix} 0.6 & 0.4 \\ 0.3 & 0.7 \end{bmatrix}, \mathbf{B} = \begin{bmatrix} 0.3 & 0.7 \\ 0.6 & 0.4 \end{bmatrix}, \mathbf{p}[1] = [0.7 \; 0.3].$$

For model *two* we obtain $N = 3$, $M = 2$, and

$$\mathbf{V} = (H, T), \mathbf{A} = \begin{bmatrix} 0.3 & 0.3 & 0.4 \\ 0.1 & 0.6 & 0.3 \\ 0.7 & 0.1 & 0.2 \end{bmatrix}, \mathbf{B} = \begin{bmatrix} 0.5 & 0.5 \\ 0.8 & 0.2 \\ 0.4 & 0.6 \end{bmatrix}, \mathbf{p}[1] = [0.8, \quad 0.1, \quad 0.1].$$

In digital signal processing an HMM can be designed for simulation by using the prescribed parameters to generate realizations of the random observation sequence $X[1], X[2], \ldots, X[n]$, $\ldots, X[L]$. Here L is an integer that denotes the maximum number of observations. A procedure to realize this design might proceed as follows:

(1) Choose a realization of the initial state $S[1]$ according to the initial-state distribution vector. For example, if $N = 3$ and $p_1[1] = 0.6$, $p_2[1] = 0.3$, and $p_3[1] = 0.1$, use a uniform random number (RN) generator to determine the initial state. If $0 \leq \mathrm{RN} < 0.6$, $S[1] = q_1$; if $0.6 \leq \mathrm{RN} < 0.9$, $S[1] = q_2$; and if $0.9 \leq \mathrm{RN} < 1.0$, $S[1] = q_3$.

(2) Set $n = 1$

(3) Obtain a realization of $X[n] = v_{p(n)} \in \mathbf{V}$ according to the state conditional output probabilities in $\mathbf{B}$. Again a random number generator can be used for this purpose.

(4) Use a random number generator (or other pseudo-stochastic or stochastic method) to transfer to the next state in accordance with state transition probabilities in $\mathbf{A}$.

(5) For $n < L$ set $n = n + 1$ and go to step (3). Otherwise terminate the procedure.

Thus an HMM can be used as a generator of observations. It can also be used in reverse fashion, that is, given a sequence of observations, determine which of several competing models was most likely responsible for the observations.

Application to Speech Processing

As a review of the literature will show, HMM's are extensively used in speech recognition. Here, we only briefly review the most basic aspects of isolated word recognition. Indeed, a description of the procedure to merely obtain suitable observation vectors could easily take more than a chapter of a book. We mention only that the observation vectors are obtained from the spectral content of the speech sample through a process called *linear predictive coding* (LPC) which is extensively discussed in the literature on speech processing [9-16]. In speech recognition each of, say, K words is modeled by an HMM and the totality of all K words are then modeled by K HMM's that we denote as $\mathbf{M}_1, \mathbf{M}_2, \ldots, \mathbf{M}_k$, where $\mathbf{M}_i = (\mathbf{A}_i, \mathbf{B}_i, \mathbf{p}_i[1])$. The $\mathbf{M}_i$, $i = 1, \ldots, K$ are the so-called *word models*. The problem of designing an appropriate word model for a given word is often considered the most difficult of the various tasks associated with speech recognition. It requires extensive training involving many human talkers if it is to be speaker independent. Fortunately, it need be done only once for each word and is done off-line, that is, before the model is used as an automatic word recognizer.

Example 9.5-3

(The left-to-right model.) As an example of an HMM that is useful in isolated word recognition is the left-to-right HMM shown in Figure 9.5-2.

The left-to-right HMM has the property that the $\mathbf{A}$ matrix is upper-triangular and that there are single initial and final states. Once the process enters a new state it can never return to an earlier state. The initial state distribution vector is $\mathbf{p}[1] = (1, 0, 0, \ldots, 0)$. When in the current state, the model can repeat the current state or advance one or two states.

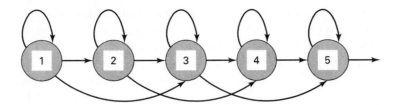

Figure 9.5-2   A five state left-to-right HMM used in word recognition.

The basic problem in spoken word recognition is to determine which from among the K word models $\mathbf{M}_1, \mathbf{M}_2, \ldots, \mathbf{M}_k$ is the source of the event $\mathbf{E} = \{X[1] = \nu_{p(1)}, X[2] = \nu_{p(2)}, \ldots, X[L] = \nu_{p(L)}\}$. In keeping with the literature, we call $\mathbf{E}$. the complete observation sequence. We remind the reader that $\nu_{p(i)} \in \mathbf{V}$ for all $i = 1, \ldots, L$ and $p(i) \in \{1, 2, \ldots, M\}$. A *maximum a posteriori* (MAP) estimator, $\mathbf{M}^*$, would require solving $\mathbf{M}^* = \arg\max_i P[\mathbf{M}_i|\mathbf{E}]$. This, in turn, would require *a priori* knowledge of the distribution of the words which is not generally known. If, on the other hand, we assume that all the $\mathbf{M}_i$ are equally likely, then the MAP estimator is equivalent to the *maximum-likelihood estimator* (MLE) $\mathbf{M}^* = \arg\max_i P[\mathbf{E}|\mathbf{M}_i]$. The efficient computation of $P[\mathbf{E}|\mathbf{M}_i]$ for each word is an important consideration. A brute force approach, that is, one following directly from the definition, is not feasible.

Example 9.5-4
Assume a massively interconnected HMM with N states and L observations. Show that the number of operations required to compute $P[\mathbf{E}|\mathbf{M}]$ from the definition is of the order of N^L.

Solution: Computing $P[\mathbf{E}|\mathbf{M}]$ directly from the definition requires summing the probabilities over all possible state sequences. A particular state sequence is sometimes called a *path*. If we use the symbol Q_i to denote the event that the observation vector is associated with the ith path, then $P[\mathbf{E}|\mathbf{M}] = \sum_i P[\mathbf{E}, Q_i|\mathbf{M}] = \sum_i P[\mathbf{E}|Q_i, \mathbf{M}]P[Q_i|\mathbf{M}]$.

For each of the observations, starting at $n = 1$ and ending at $n = L$, there is a choice of N states and, therefore, for all L observations there are $\sim N^L$ possible state sequences. For $N = 5$ and $L = 100$, there are of the order of 5^{100} sums of products. A more precise calculation [9-14] shows that there are $2LN^L$ operations required. For the numbers given above these amount to $\sim 10^{72}$ operations. We consider next a more efficient approach to this problem by capitalizing on the structure of the computations.

Efficient Computation of $P[\mathrm{E}|\mathrm{M}]$ with a Recursive Algorithm

As Example 9.5-4 demonstrated, computing $P[\mathbf{E}|\mathbf{M}]$ without considering the underlying lattice structure of the state transition sequence is a hopeless task. A recursive algorithm that does take advantage of this structure is the so-called *forward-backward* procedure. The forward procedure refers to an iteration that begins at $n = 1$ and proceeds to the present, while the backward procedure refers to a very similar algorithm that proceeds to the present from $n = L$. We shall focus primarily on the forward procedure. Either approach yields the same results. The forward procedure does not require the availability of all the observations before it can proceed. We shall first describe the algorithms and then explain why they are so much more efficient than the direct approach.

Define the event $\mathbf{E}_n \triangleq \{X[1] = \nu_{p(1)}, \ldots, X[n] = \nu_{p(n)}\}$ and the forward variable $\alpha_n[i] = P[\mathbf{E}_n, S[n] = q_i | \mathbf{M}]$. Here i refers to the state q_i and n, whether as an argument or subscript, refers to time. In words $\alpha_n[i]$ is the probability of the joint event $\mathbf{E}_n$ *and* that the present state is q_i. We require the initialization $\alpha_1[i] \triangleq p_i[1] \times P[X[1]] = \nu_{p(1)} | S[1] = q_i, \mathbf{M}]$. Note that $\alpha_{n+1}[j] = P[\mathbf{E}_{n+1}, S[n+1] = q_j | \mathbf{M}]$. Had we known that the previous state had been q_i, then we could have written that $\alpha_{n+1}[j] = \alpha_n[i] a_{ij} b_{j,p(n)}$. (Recall that α_{ij} is the transition probability of going to state q_j from state q_i and $b_{j,p(n)}$ is the probability of observing the symbol $\nu_{p(n)}$ while in state q_j.) However, since the transition to state q_j could have come from any state for which $a_{ij} \neq 0$, the correct recursion is

$$\alpha_{n+1}[j] = \sum_{i=1}^{N} \alpha_n[i] a_{ij} b_{j,p(n)} \tag{9.5-1}$$

for $j = 1, \ldots, N$ and $n = 1, \ldots, L-1$ When $n = L-1$, we obtain $\alpha_L[j] = P[\mathbf{E}, S[L] = q_j | \mathbf{M}]$, where we implicitly let $\mathbf{E}_L = \mathbf{E}$. Now recalling from basic probability theory that if a sequence of events $\{A_j\}$, $j = 1, \ldots, N$ has the property that $A_i A_j = \phi$ (the empty set) for $i \neq j$ and $\cup_i A_i = \Omega$ (the certain event), then for any event B, $P[B] = \sum_{i=1}^{N} P[BA_i]$. Associating the $\{A_j\}$ with $\{S[L] = q_j\}$ we obtain the important result that

$$P[\mathbf{E}|\mathbf{M}] = \sum_{j=1}^{N} \alpha_L[j]. \tag{9.5-2}$$

An estimate of the number of computations for the forward algorithm is easily obtained as follows: From Equation 9.5-1 we see that for a fixed j and n, the computation of $\alpha_{n+1}[j]$ requires $\sim N$ operations. Repeating the process for $j = 1, 2, \ldots, N$ requires a total of N^2 operations. Finally repeating the calculation for $n = 1, 2, \ldots, L-1$ yields a total of $\sim N^2 L$ multiplication/addition type of operations, to be contrasted with $2LN^L$ using the definitions (or ~ 2500 versus $\sim 200 \times 5^{100}$). To what is owed this saving of roughly 69 orders of magnitude?

As an analogy consider a surveyor who is assigned to measure the road distances from a distant city, say, A to three cities B, C, and D that are quite near each other. The road map is shown in Figure 9.5-3.

An efficient way to measure the distances d_{AB}, d_{AC}, d_{AD} is as follows: Measure d_{AB}. Then measure d_{BC} and d_{CD} and compute $d_{AC} = d_{AB} + d_{BC}$, and $d_{AD} = d_{AC} + d_{CD}$. Alternatively, the surveyor could measure d_{AB}, return to A, and measure the total distance d_{AC} without considering that to get to C he must pass through B, return to A once again,

Figure 9.5-3 A road map from city A to cities B, C, and D.

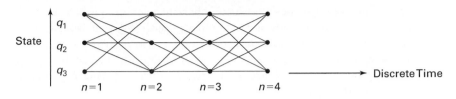

Figure 9.5-4 Lattice structure of the implementation of the computation of $\alpha_n[j]$.

and measure the total distance d_{AD}. In the former method the surveyor builds on earlier measurements and knowledge of the road map to get subsequent distances. In the second case, the surveyor makes each measurement without building up from earlier results. It is essentially the same with computing $P[\mathbf{E}|\mathbf{M}]$ from the definition. To do so ignores the lattice structure of the computations and ignores the fact that different state sequences share the same subsequences. In the recursive approach, the computations build on each other. As n increases, the updated probability computation builds on the previous ones. The lattice structure of the algorithm is shown in Figure 9.5-4 for three states and for four time increments. It should be clear from the diagram that as time increases, different state sequences remerge into the same three states, sharing many of the previous subsequences.

One can also define an event $\mathbf{E}'_n \overset{\Delta}{=} \{X[n+1] = v_{p(n+1)}, \dots, X[L] = v_{p(L)}\}$ so that $\mathbf{E}_n \cup \mathbf{E}'_n = \mathbf{E}$. In terms of this event one can define the so-called backward variable $\beta_n[i] \overset{\Delta}{=} P[\mathbf{E}'_n, S[n] = q_i|\mathbf{M}]$ with the arbitrary initialization $\beta_L[i] = 1$, $i = 1, \dots, N$ to obtain the recursion

$$\beta_n[i] = \sum_{j=1}^{N} \beta_{n+1}[j] a_{ij} b_{i,p(n)}. \tag{9.5-3}$$

The recursive algorithm to compute $P[\mathbf{E}|\mathbf{M}]$ using Equation 9.5-3 has the same computational complexity as the one using the *forward variable* $\alpha_n[j]$. We leave the details as a exercise for the reader.

Viterbi Algorithm and the Most Likely State Sequence for the Observations

In the algorithm based on the forward variable (a similar statement applies to the backward variable recursion) we computed $P[\mathbf{E}|\mathbf{M}]$ by averaging over all state transitions. An alternative procedure, which is often more efficient, is to find, for each model, the path that was most likely to have produced the observations. Sometimes this path is called the *minimum cost* path in the event that a criterion other than *most probable* is used. The algorithm for finding such a path is based on the principle of *dynamic programming* and the algorithm itself is called the *Viterbi algorithm* [9-17]. The principle of dynamic programming can be illustrated as follows. Suppose we wish to find the shortest path from a point A to B in a connected graph with many links and nodes and many possible paths from point A to B. Consider a node η_i; then among all paths going from A to B via η_i, only the shortest path I_i^* from η_i to B needs to be stored. All other paths from η_i to B can be discarded. This follows from there being only two possibilities: Either the shortest path from A to B takes us through η_i or not. If it takes us through η_i, then any path different from I_i^* will increase the overall path length. If the shortest path does not go through η_i, it must go through some other node, say, η_j. Then we repeat the process of finding the shortest path from η_j to B. In this way we can backtrack back to A from B always storing the subsequence of nodes that yield the shortest paths.

Consider now the application of this reasoning to finding the sequence of states that was most likely to have produced the given partial observation vector, that is, the event $\mathbf{E}_n$. Equation 9.5-1 is the recursion that averages over all paths. If we replace this recursion with the most the most likely path algorithm, we need to replace the summation by a maximization over all path sequences leading to the present state. To find the single most likely path $\mathbf{Q}^* \triangleq (q_{p(1)}^*, q_{p(2)}^*, \ldots, q_{p(L)}^*)$ we proceed as follows. Define the event $\mathbf{S}_n[i] \triangleq \{S[1] = q_{p(1)}, S[2] = q_{p(2)}, \ldots, S[n] = q_i\}$. From elementary probability theory we can write

$$P[\mathbf{S}_n[i]|\mathbf{E}_n, \mathbf{M}] = \frac{P[\mathbf{S}_n[i], \mathbf{E}_n|\mathbf{M}]}{P[\mathbf{E}_n|\mathbf{M}]}.$$

Since the denominator does not depend on $\mathbf{S}_n[i]$, maximizing the left-hand side is equivalent to maximizing the right-hand side over the state sequences leading up to state q_i. For the Viterbi algorithm we define the variable

$$\varphi_n[i] = \max_{q_{p(1)}, q_{p(2)}, \ldots, q_{p(n-1)}} P[\mathbf{S}_n[i], \mathbf{E}_n|\mathbf{M}] \tag{9.5-4}$$

and observe that

$$\varphi_{n+1}[j] = \max_i \varphi_n[i] a_{ij} b_{j,p(n+1)}. \tag{9.5-5}$$

The interpretation of Equations 9.5-4 and 9.5-5 is as follows. Equation 9.5-4 finds the most probable path to state q_i, at time n, that accounts for the observation vector $\mathbf{E}_n$. Now to find the most probable path at time $n + 1$ to state q_j, we need only consider the most probable paths to states $q_1, q_2, \ldots, q_N$ since the overall most probable path must transit through one of these states at time n. This is a forward application of the dynamio programming principle if we replace "shortest path" in the earlier discussion on dynamic programming by "most likely path." Figure 9.5-5 shows how the recursion works.

Below, we furnish the entire Viterbi procedure for finding the most likely state sequence consistent with the observations:

1. *Initialization*

$$\varphi_1[i] = p_i[1] b_{i,p(1)}, i = 1, \ldots, N;$$
$$\psi_1[i] = 0, i = 1, \ldots, N.$$

The $\psi_n[i]$ are the path tracking functions.

2. *Recursion*

$$\varphi_{n+1}[j] = \max_i \varphi_n[i] a_{ij} b_{j,p(n+1)}], n = 1, \ldots, L - 1; j = 1, \ldots, N;$$
$$\psi_{n+1}[j] = \arg \max_i [\varphi_n[i] a_{ij}], n = 1, \ldots, L - 1; j = 1, \ldots, N.$$

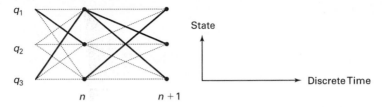

Figure 9.5-5   The heavy black lines leading to the state nodes at time n represent the overall most probable path to these states. For simplicity's sake, think of these computations as having been done according to Equation 9.5-4. The dotted lines indicate tested paths that have been rejected. The most likely paths to the various states at $n+1$ are then computed as in Equation 9.5-7; these are shown as heavy black lines leading to the state nodes at $n+1$. At each instant of time, we must store the N values of $\varphi_n[j]$, $j = 1, \ldots, N$. For L observations we must store, therefore, $N\,L$ such values in addition to the observations themselves.

3. *Termination*

$$P^* \triangleq P[\mathbf{E}, \mathbf{S}^*[L]|\mathbf{M}] = \max_i \varphi_L[i];$$

$$q_L^* = \arg\,\max_i \varphi_L[i].$$

4. *State sequence backtracking*

$$q_n^* = \psi_{n+1}(q_{n+1}^*), n = L - 1, L - 2, \ldots, 1.$$

For any but small values of N, T these recursions require the aid of a computer. We illustrate with a hand-computable example.

Example 9.5-5 _____
In Example 9.5-1, the sequence HTT is observed. For model *one,* compute the most likely path and its probability using the Viterbi algorithm.

Solution   For model *one,* the parameters are, from Example 9.5-2,

$$\mathbf{A} = \begin{bmatrix} 0.6 & 0.4 \\ 0.3 & 0.7 \end{bmatrix} \quad \mathbf{p}[1] = [0.7 \quad 0.3] \quad \mathbf{V} = (H, T) \text{ and } \mathbf{B} = \begin{bmatrix} 0.3 & 0.7 \\ 0.6 & 0.4 \end{bmatrix}.$$

Step 1: *Initialization* $(n = 1)$

$$\varphi_1[1] = p_1[1] \times P[H|q_1] = 0.7 \times 0.3 = 0.21;$$

$$\varphi_1[2] = p_1[2] \times P[H|q_2] = 0.3 \times 0.6 = 0.18;$$

Step 2: *Recursion*

$$n = 2$$

$$\varphi_2[1] = \max\{\varphi_1[1]a_{11} \times P[T|q_1], \varphi_1[2]a_{21} \times P[T|q_1]\}; \text{ state } q_1;$$

$$= \max\{0.0882, 0.0378\} = 0.0882;$$

$$\psi_2[1] = 1;$$

$$\varphi_2[2] = \max\{\varphi_1[1]a_{12} \times P[T|q_2], \varphi_1[2]a_{22} \times P[T|q_2]\}; \text{ state } q_2;$$

$$= \max\{0.0336, 0.0504\} = 0.0504;$$

$$\psi_2[2] = 2;$$

$$n = 3;$$

$$\varphi_3[1] = \max\{\varphi_2[1]a_{11} \times P[T|q_1], \varphi_2[2]a_{21} \times P[T|q_1]\}; \text{ state } q_1;$$

$$= \max\{0.0370, 0.0106\} = 0.0370;$$

$$\psi_3[1] = 1;$$

$$\varphi_3[2] = \max\{\varphi_2[1]a_{12} \times P[T|q_2], \varphi_2[2]a_{22} \times P[T|q_2]\};$$

$$= \max\{0.0141, 0.0141\} = 0.0141;$$

$$\psi_3[2] = 1 \text{ or } 2.$$

Step 3: *Termination*

$$P^* = \max\{\varphi_3[1], \varphi_3[2]\} = \max\{0.0370, 0.0141\} = 0.0370;$$

$$q_3^* = \arg\max\{\varphi_3[1], \varphi_3[2]\} = 1;$$

Step 4: *State sequence backtracking*

$$q_n^* = \psi_{n+1}(q_{n+1}^*); n = L - 1, L - 2, \ldots, 1;$$

Hence

$$q_2^* = \psi_3(1) = 1;$$

$$q_1^* = \psi_2(1) = 1.$$

So the most likely state sequence for observing HTT is 1, 1, 1.

Were we to repeat the calculation for the observation HHH, we would have found the most likely sequence to be 2, 2, 2. We leave the verification of this result as an exercise for the reader.

9.6 SPECTRAL ESTIMATION

In Chapter 5, we estimated the means and correlations (or covariances) of both random vectors and random sequences (cf. Sections 5.8-9 and also Section 6.8). In the special case of a wide-sense stationary (WSS) random sequence, we employed the ergodic hypothesis (cf. Section 8.4) to provide an asymptotically exact estimate of the mean and correlation function as the number of samples tends to infinity. Here, in this section, we show how to make correspondingly good estimates of the power spectral densities (psd) of random sequences.

Basically there are two classical approaches to power spectral density estimation. One can either first estimate the correlation function and then calculate the psd estimate, most notably with the discrete Fourier transform, as the Fourier transform of the correlation function estimate. Or, one can estimate the psd directly from the Fourier transform of the data. The conceptually simplest approach here is to take the magnitude squared of the Fourier transform of the data record, also known as the *periodogram*. While this simple periodogram estimate is often used in practice, we will see below that its large variance makes it quite a noisy psd estimator.

The Periodogram

We start with a WSS and *zero-mean* random sequence $X[n]$ observed over $-\infty \leq n \leq +\infty$. We must additionally assume the sequence is ergodic in correlation. We use the window function $w_N[n] \triangleq 1$ for $0 \leq n \leq N-1$ and 0 elsewhere, to cut out a finite-length section:

$$X_N[n] \triangleq w_N[n]X[n].$$

We then estimate the correlation function $R_{XX}[m]$ by the time average

$$\hat{R}_N[m] \triangleq \frac{1}{N} X_N[m] * X_N[-m]. \tag{9.6-1}$$

Since the random sequence is ergodic in correlation (cf. Definition 8.4-3 for the random process version), say in the mean-square (m.s.) sense, then this estimate should converge to the ensemble average correlation function $R_{XX}[m]$ in the m.s. sense, that is,

$$\lim_{N \to \infty} \hat{R}_N[m] = R_{XX}[m], \qquad \text{(m.s.)} \tag{9.6-2}$$

based on discrete-time versions of the ergodic arguments used in Section 8.4 in particular see Theorem 8.4-2. Problem 9.12 which works out the details in the case of a Gaussian random sequence.

Since the psd $S_{XX}(\omega)$ is the Fourier transform of the correlation function $R_{XX}[m]$, one might hope that the Fourier transform of $\hat{R}_N[m]$ would be a good estimate of $S_{XX}(\omega)$. We investigate this possibility by evaluating the mean and variance of this psd estimate. We hope that the mean of this psd estimate, called an *unbiased estimate*, will be correct and that the variance of the psd estimator will decrease toward zero as $N \to \infty$, a so-called *consistent* estimate.

Denote the Fourier transform of $\hat{R}_N[m]$ as $I_N(\omega)$; then we have

$$I_N(\omega) = FT\{\hat{R}_N[m]\}$$

$$= \sum_{m=-N}^{+N} \hat{R}_N[m] \exp(-j\omega m)$$

$$= \sum_m \frac{1}{N} \left(\sum_n X[n+m]X^*[n] \right) \exp(-j\omega m)$$

$$= \frac{1}{N} |X_N(\omega)|^2, \tag{9.6-3}$$

where

$$X_N(\omega) \triangleq FT\{X_N[n]\}.$$

In words $X_N(\omega)$ is the Fourier transform of N samples of the WSS random sequence $X[n]$. This quantity can thus be well approximated by any of the fast DFT routines [9-18] normally available on scientific workstations.

Calculating the mean value of the periodogram $I_N(\omega)$, we obtain

$$E\{I_N(\omega)\} = E \left\{ \sum_{m=-(N-1)}^{+(N-1)} \hat{R}_N[m] \exp(-j\omega m) \right\}$$

$$= \sum_{m=-(N-1)}^{+(N-1)} E\{\hat{R}_N[m]\} \exp(-j\omega m)$$

$$= \sum_{m=-(N-1)}^{+(N-1)} \left(\frac{N - |m|}{N} \right) R_{XX}[m] \exp(-j\omega m), \tag{9.6-4}$$

so, as $N \to \infty$, we would expect the mean to be asymptotically correct, that is $S_{XX}(\omega)$, if $|R_{XX}[m]|$ tends to zero fast enough as $|m| \to \infty$. It is easy to see that the precise criteria needed is $\sum |m| |R_{XX}[m]| < \infty$. Then the mean of the periodogram $I_N(\omega)$ will be asymptotically correct, that is the periodogram estimate will be *asymptotically unbiased*:

$$\lim_{N \to \infty} E\{I_N(\omega)\} = S_{XX}(\omega).$$

We note in passing, that the random sequence $X[n]$ must have zero-mean to satisfy the convergence criteria $\sum |m| |R_{XX}[m]| < \infty$. If the mean is not zero, then there is an impulse at $\omega = 0$, which must be handled separately.

We can also look at Equation 9.6-4 as introducing a *correlation window* as a multiplier on R_{XX}; then by the well-known correspondence of convolution and multiplication, we obtain

$$FT\{t_N[m] \cdot R_{XX}\} = \frac{1}{2\pi} \int_{-\pi}^{+\pi} S_{XX}(v) \frac{1}{N} \left(\frac{\sin[(\omega-v)N/2]}{\sin[(\omega-v)/2]} \right)^2 dv$$

where the triangular window function,

$$t_N[m] = w_N[m] @ w_N[m]^\dagger$$

has as its Fourier transform, the square of the periodic sinc function, as shown.

We now turn to the evaluation of the variance of the periodogram $I_N(\omega)$. Unfortunately, we find that the variance does not tend to zero as N approaches infinity. In fact the variance does not even get small. Under certain assumptions to be detailed below, the asymptotic variance of the periodogram estimator is lower bounded by the square of the power spectral density! Thus $I_N(\omega)$ turns out to be quite a bad estimate of the psd.

To calculate the variance we proceed as follows: Employing $\text{Var}\{I_N(\omega)\} = E\{I_N^2(\omega)\} - E^2\{I_N(\omega)\}$, we first calculate

$$E\{I_N^2(\omega)\} = \frac{1}{N^2} \sum_n \sum_m \sum_p \sum_q w_N[n]w_N[m]w_N[p]w_N[q]$$

$$\times E\{X[n]X^*[m]X[p]X^*[q]\}e^{-j\omega(n-m)}e^{-j\omega(p-q)}.$$

Next we invoke the Gaussian assumption and use the fourth-moment property of the Gaussian distribution (ref. Problem 5.20) to obtain

$$E\{I_N^2(\omega)\} = E^2\{I_N(\omega)\} + \left| \frac{1}{N} \sum_n \sum_p w_N[n]w_N[p]R_{XX}[n-p] = e^{-j\omega(n+p)} \right|^2$$

$$+ \left| \frac{1}{N} \sum_n \sum_q w_N[n]w_N[q]R_{XX}[n-q]e^{-j\omega(n-q)} \right|^2.$$

Thus the variance is just given by the last two terms. Converting them to the frequency domain, we can then write

$$\text{Var}\{I_N(\omega)\} = \left| \frac{1}{2\pi} \int_{-\pi}^{+\pi} S_{XX}(\lambda)\frac{1}{N}W_N(\lambda+\omega)W_N^*(\lambda-\omega)d\lambda \right|^2$$

$$+ \left(\frac{1}{2\pi} \int_{-\pi}^{+\pi} S_{XX}(\lambda)\frac{1}{N}|W_N(\lambda-\omega)|^2 d\lambda \right)^2, \qquad (9.6\text{-}5)$$

where $W_N(\omega) \triangleq FT\{w_N[n]\}$. For sufficiently large N, the first term will be near zero when ω is not near zero, while the second term will tend to $E^2\{I_N(\omega)\} \approx S_{XX}^2(\omega)$. We thus conclude that the periodogram is not a very accurate estimator of the psd, with its asymptotic variance satisfying

$$\text{Var}\{I_N(\omega)\} \approx S_{XX}^2(\omega). \qquad (9.6\text{-}6)$$

While the mean of the periodogram is asymptotically correct, we see that the standard deviation of this spectral estimate is as big as the psd!

[†]The deterministic autocorrelation operator is denoted by the '@' symbol $x[n] @ y[n] \triangleq \sum_{k=-\infty}^{+\infty} x[k]\,y[n+k]$.

Bartlett's Procedure–Averaging Periodograms[†]

Since we assume that the random sequence $X[n]$ is available time $[0, N-1]$, we can form several or many periodogram estimates from different segments of the random sequence, that is, let M divide N and set $K \triangleq N/M$. Then form K segments of length M as follows:

$$X^i[n] \triangleq X[n + iM - M] \quad \text{where} \quad 0 \le n \le M-1,$$

$$\text{and} \quad 1 \le i \le K.$$

We can now form K periodograms whose statistical errors will be largely uncorrelated; then by performing simple averaging on these periodograms we should obtain a usable estimate of the power spectral density $S_{XX}(\omega)$. We define the K periodograms as follows:

$$I^i_M[\omega] \triangleq \frac{1}{M} \left| \sum_{n=0}^{M-1} X^i[n] \exp(-j\omega n) \right|^2, \quad 1 \le i \le K.$$

Then define the averaged periodogram estimate of the psd as

$$B_X(\omega) \triangleq \frac{1}{K} \sum_{i=1}^{k} I^{(i)}_M(\omega).$$

Clearly this averaging will not disturb the asymptotic unbiasedness of the individual periodograms, although much more data may now be required to ameliorate the windowing effects due to the smaller window length M used here. However a real advantage will occur in the variance reduction of the Bartlett estimate, due to its averaging of individual periodogram estimates of nonoverlapping data segments.

Calculating the mean of $B_X(\omega)$, we obtain

$$E\{B_X(\omega)\} = E\{I^{(1)}_M(\omega)\}$$

$$= \frac{1}{2\pi} \int_{-\pi}^{+\pi} S_{XX}(\omega) \frac{1}{M} \left(\frac{\sin[(\omega - \upsilon)M/2]}{\sin[(\omega - \upsilon)/2]} \right)^2 d\upsilon$$

$$\to S_{XX}(\omega) \quad \text{as} \quad M \to \infty.$$

Using Equation 9.6-6, the approximate variance of the Bartlett estimator of the psd can be found as

$$\text{Var}B_X(\omega)] \approx \frac{1}{K} \text{Var}I^{(1)}_M(\omega)]$$

$$\approx \frac{1}{K} S^2_{XX}(\omega), \tag{9.6-7}$$

as long as the segment length $M = N/K$ is still large enough for Equation 9.6-6 to hold with M in place of N.

[†]See reference [9-19].

Example 9.6-1 _____

This example uses MATLAB to illustrate the periodogram and Bartlett estimator for a 1024-element sample of simulated white Gaussian noise $W[n]$. The plots in this example are generated by the *m-file* titled `Bartlett.m` that is located at this book's Web site.

The function rand is first called to generate a 1000-element vector with the Gaussian or *Normal* pdf. The mean is zero and the variance is one, that is, $\sigma_W^2 = 1$. The periodogram is calculated via Equation 9.6-3 using the MATLAB `fft` function with $N = 1024$. The resulting periodogram is shown in Figure 9.6-1. The horizontal axis is numbered corresponding to uniform samples of ω going from 0 to 2π, thus the highest discrete-time frequency $\omega = \pi$ is in the middle of the plot. We note that the variance of this periodogram seems quite large, with spikes going up almost to seven times the true value of the psd $S_{WW}(\omega) = \sigma_W^2 = 1$.

Next the 1024-point data vector is broken down successively, first into four 256-point data vectors and then sixteen 64-point data vectors. The Bartlett procedure is then carried out for $K = 4$ and $K = 16$, respectively, by `Bartlett.m`. The smoother $K = 16$ result is shown in Figure 9.6-2. In these cases the value of M decreases to $M = 256$ and $M = 64$, respectively. This is also the number of samples of ω provided by the numerical routine `fft` in these cases, again with the Nyquist frequency sample in the middle of the range. As is common with MATLAB plots, the vertical range has automatically re-sized to match the range of the data. However, with reference to the vertical axis we can see that the Bartlett estimate has the expected reduction in variance (cf. Equation 9.6-7). The reader should note that we are trading *frequency resolution* in some sense for this improved statistical behavior. If the true psd were not white, that is, not flat, then this disadvantage of lost resolution would show up.

Example 9.6-2 _____

Here the same Bartlett procedure is used to estimate a nonwhite power spectral density. (cf. Fig. 9.6-5) A 1024-point sequence generated by MATLAB is used to stimulate an

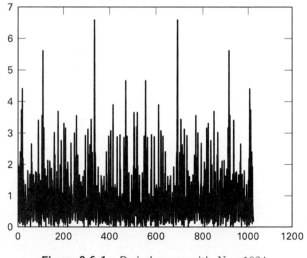

Figure 9.6-1   Periodogram with $N = 1024$.

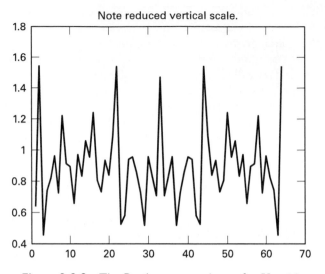

Figure 9.6-2   The Bartlett type estimate for $K = 16$.

autoregressive moving average (ARMA) random sequence. This sequence is the output of a filter whose input was Normal white noise, generated by the routine **randn**. Then filter was applied with numerator coefficient vector **b** $= [1.0 - 0.8 - 0.1]$ and denominator coefficient vector **a** $= [1.0 - 1.2 + 0.4]$. The corresponding transfer function, $\mathsf{H}(z) = (1 - 0.8z^{-1} - 0.1z^{-2})/(1 - 1.2z^{-1} + 0.4z^{-2})$, then gives the ARMA power spectral density:

$$S(\omega) = \sigma^2 |\mathsf{H}(e^{j\omega})|^2$$

$$= \frac{1.65 - 1.44 \cos \omega - 0.2 \cos 2\omega}{2.60 - 3.36 \cos \omega + 0.8 \cos 2\omega}.$$

The periodogram and Bartlett ($K = 16$) spectral estimates are plotted in Figures 9.6-3 and 9.6-4. Note the reduced noise in the Bartlett $K = 16$ estimate, as well as the reduced resolution evident in the broadening of the spectral density peak in comparison with the true psd shown in Figure 9.6-5.

     This example is programmed in the MATLAB file **specest.m** located at this book's Web site. You can edit this file and run other cases with different ARMA coefficient values. Also by repeatedly calling this *.m* file, with the "state $= 0$" command removed, you can see the effect on these spectral estimates of different sample sequences of the driving white noise at the input to the filter.

Parametric Spectral Estimate

The Markov random sequence was studied in Section 6.5. There in Example 6.5-2, we saw that a first-order linear difference equation, driven by an i.i.d. random sequence, generated a Markov random sequence. A Markov-p, or pth order Markov sequence, likewise

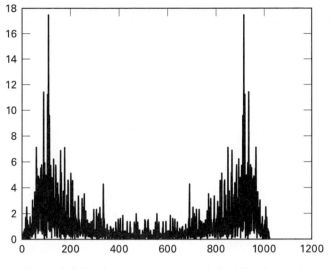

Figure 9.6-3   Periodogram estimate for ARMA model.

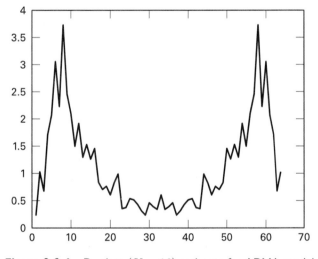

Figure 9.6-4   Bartlett ($K = 16$) estimate for ARMA model.

can be generated by passing an i.i.d. sequence through a pth order difference equation (ref. Definition 6.5-2). If we are willing to model our observed random sequence by such a model, then we can estimate its psd by estimating the parameters of the model. For this model, the best linear predictor will yield the model parameters. In Section 9.1 it is shown how to determine this predictor.

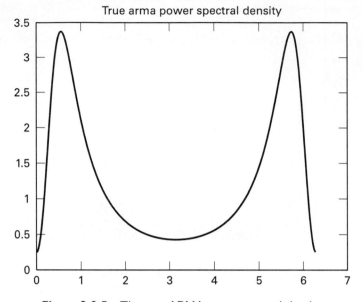

Figure 9.6-5 The true ARMA power spectral density.

Consider the Markov-p model

$$X[n] = \sum_{k=1}^{p} a_k X[n-k] + W[n].$$

Take X and W as zero-mean and write the variance $\text{Var}\{W[n]\}$ as σ_W^2. Let the vector $\mathbf{X}$ in Theorem 9.1-2 be given as the p-dimensional vector $\mathbf{X} \triangleq (X[n-1], X[n-2], \ldots, X[n-p])^T$. Then Equation 9.1-9 provides the linear prediction estimate of the scalar random variable $Y = X[n]$ in terms of the a_k. Expressed in vector form as $\mathbf{a} \triangleq (a_1, a_2, \ldots, a_p)^T$, these coefficients are then determined as the solution to the orthogonality equations, given by Equation 9.1-10, repeated here for convenience,

$$\mathbf{a}_o^T = \mathbf{k}_{Y\mathbf{X}} \mathbf{K}_{\mathbf{X}\mathbf{X}}^{-1}. \tag{9.6-8}$$

The cross-covariance vector and covariance matrix then become

$$\mathbf{k}_{Y\mathbf{X}}^T = (K_{XX}[1], K_{XX}[2], \ldots, K_{XX}[p])^T,$$

and

$$\mathbf{K_{XX}} = \begin{bmatrix} K_{XX}[0] & K_{XX}[1] & \cdots & K_{XX}[p-1] \\ K_{XX}[-1] & K_{XX}[0] & \ddots & \vdots \\ \vdots & \ddots & \ddots & K_{XX}[1] \\ K_{XX}[-(p-1)] & \cdots & K_{XX}[-1] & K_{XX}[0] \end{bmatrix}.$$

To obtain a simple parametric psd estimate, we can just replace the covariance function $K_{XX}[m]$ in the above equations with its estimate provided by Equation 9.6-1.[†] The solution will then yield parameter estimates $\widehat{a_1}, \widehat{a_2}, \ldots, \widehat{a_p}$, so that the psd estimate can be written as:

$$\hat{S}_{XX}(\omega) = \frac{\widehat{\sigma_W^2}}{|1 - \sum_{k=1}^{p} \widehat{a_k} \exp(-j\omega k)|^2}. \tag{9.6-9}$$

It is interesting to note that the AR parametric spectral estimate of the psd has the *covariance (correlation) matching property*, that is,

$$IFT\{\hat{S}_{XX}(\omega)\} = \hat{R}_{XX}[m], \qquad |m| \leq p.$$

We note that if the $\hat{R}_{XX}[m]$ are sufficiently accurate, then $\hat{S}_{XX}(\omega)$ can be quite close to the true psd $S_{XX}(\omega)$, if the random sequence is Markov-p. However, if the true random sequence $X[n]$ is not Markov-p, than the opposite can occur.

Such parametric spectral estimates can provide greater resolution of closely spaced spectral components than can the classical methods. This may be an advantage when the amount of data is small. On the other hand, one pays the price of greater sensitivity to model assumptions.

Example 9.6-3

This example uses the same ARMA random sequence as was used in Example 9.6-2. Here, however, we use an AR spectral density estimate and investigate the effect of increasing the predictor order for $p = 2$, 3, and 4. Since the data is not Markov (i.e., the psd is not AR), we do not expect to get the precise psd as the data length $N \to \infty$. However, for large N, and here we use $N = 512$, we do expect that for a sufficiently large AR predictor order p, we will get an accurate estimate.

The correlation function estimate (or covariance function since the involved data is generated with zero-mean) is given as

$$\hat{R}_N[m] = \frac{1}{N} \sum_{n=0}^{N-1} y[n]y[n+m].$$

Then the estimated value of parameter vector **a** is determined via Equation 9.6-8. These vectors for $p = 2$, and 4, are then plugged into Equation 9.6-9 to yield the resulting psd plots of Figures 9.6-6 and 9.6-7, respectively. Each plot also shows the true ARMA psd plotted as dashed line for ease of comparison. Note that as the predictor order p increases, the approximation to the true ARMA psd gets better. Note also that this estimate appears to have better resolution than the Bartlett estimator in that it can better match the narrow spectral peaks of the true psd.

[†]Since the mean is zero here, the correlation and covariance functions are equal.

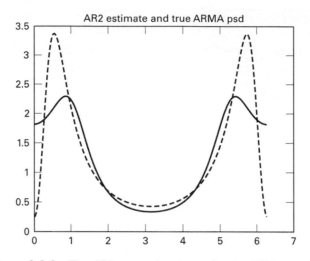

Figure 9.6-6 The AR2 spectral estimate for the ARMA model.

Maximum Entropy Spectral Density

The maximum entropy spectral density is not really a spectral estimate at all! It is simply the psd that has maximum entropy while agreeing with a certain number of given correlation values. Entropy is a concept borrowed from information theory[†] that measures the *uncertainty* in a random quantity. This uncertainty manifests itself in relatively flat spectral densities with a minimum of narrow peaks. Thus, we can think of the maximum entropy psd as the flattest one that agrees with the measured correlations. Certainly, this is a conservative choice, if we are looking for high resolution of closely spaced spectral components. The surprising result of such an approach, however, is that the maximum entropy psd has much higher resolution for closely spaced spectral peaks than do the classical methods. The resolution of this seeming paradox is that the classical techniques are not *correlation matching*. The maximum entropy psd is the flattest spectra that satisfies the constraint of matching the known (exact) correlation data.

It turns out that the Gaussian random sequence has the largest entropy of all random sequences with a given correlation function, so we can restrict attention to Gaussian random sequences. For such a sequence $X[n]$ with psd $S_{XX}(\omega)$, the continuous entropy can be expressed as

$$H(X) = \frac{1}{2\pi} \int_{-\pi}^{+\pi} \log S_{XX}(\omega) d\omega.$$

We wish to maximize this quantity subject to the constraint that the inverse Fourier transform of $S_{XX}(\omega)$ agree with the correlation values $R_{XX}[m] = r_m$, for $m = -p, \ldots, +p$. We can write the resulting constrained maximization problem in Lagrange multiplier form

---

[†]Actually, in information theory, this quantity is called *continuous entropy*. Simple entropy would be infinite in this case of continuous random variables.

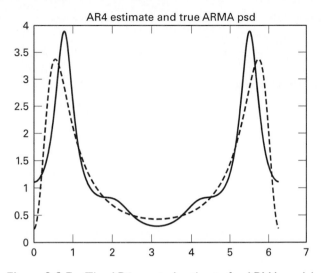

Figure 9.6-7   The AR4 spectral estimate for ARMA model.

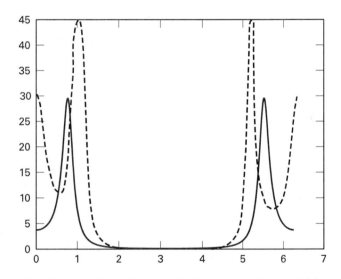

Figure 9.6-8   True (dashed) and MEM2 (solid) psd's for AR(3) model.

by introducing the variables $\lambda_{-p}, \ldots, \lambda_{+p}$ as follows:

$$\Lambda = \frac{1}{2\pi} \int_{-\pi}^{+\pi} \log S_{XX}(\omega) d\omega - \sum_{m=-p}^{+p} \lambda_m r_m.$$

Consider the partial derivatives of Λ with respect to the r_m,

$$\frac{\partial \Lambda}{\partial r_m} = \frac{1}{2\pi} \int_{-\pi}^{+\pi} \frac{1}{S_{XX}(\omega)} \exp(-j\omega m) d\omega - \lambda_m.$$

Upon setting these partial derivatives to zero for all $m \geq p$, we obtain the result that the inverse Fourier transform of $S_{XX}^{-1}(\omega)$ equals the FIR sequence $\lambda_{-p}, \ldots, \lambda_0, \ldots, \lambda_p$, that is,

$$IFT\{S_{XX}^{-1}(\omega)\} = \begin{cases} \lambda_m, & |m| \leq p \\ 0, & \text{else} \end{cases}.$$

Then, by taking the Fourier transform of the $2p + 1$ point sequence λ_m, and solving for $S_{XX}(\omega)$, we can obtain

$$S_{XX}(\omega) = \frac{1}{\sum_{m=-p}^{+p} \lambda_m \exp(-j\omega m)} \quad |\omega| \leq \pi.$$

Incidentally, we assume that $\lambda_m = \lambda_{-m}^*$ as required to make $S_{XX}(\omega)$ real-valued. Finally we must solve for the λ_m to match the known correlation values r_m. To see when this is possible, remember that the Markov-p random sequence has an all-pole power spectral density of this same form. Further the coefficients were related to the correlation values by the so-called *Normal* or Yule–Walker Equation 9.1-10; see also Equation 9.2-5. Thus by writing

$$\sum_{m=-p}^{+p} \lambda_m \exp(-j\omega m) = \frac{1}{\sigma^2} \left| 1 - \sum_{m=1}^{+p} a_m \exp(-j\omega m) \right|^2,$$

we can find the required coefficients a_m and σ_e^2 as the solution to the equations

$$\begin{bmatrix} r_0 & r_1 & r_2 & \cdot & \cdot & r_{p-1} \\ r_{-1} & r_0 & r_1 & r_2 & \cdot & \cdot \\ r_{-2} & r_{-1} & r_0 & r_1 & \cdot & \cdot \\ \cdot & r_{-2} & r_{-1} & \cdot & \cdot & r_2 \\ \cdot & \cdot & \cdot & \cdot & r_0 & r_1 \\ r_{-(p-1)} & \cdot & \cdot & r_{-2} & r_{-1} & r_0 \end{bmatrix} \begin{bmatrix} a_1 \\ a_2 \\ \cdot \\ \cdot \\ \cdot \\ a_p \end{bmatrix} = \begin{bmatrix} r_1 \\ r_2 \\ \cdot \\ \cdot \\ \cdot \\ r_p \end{bmatrix},$$

$$\sigma_e^2 = r_0 - \sum_{m=1}^{p} a_m r_m > 0.$$

The solution,

$$\mathbf{a} = \mathbf{r}\mathbf{R}_p^{-1},$$

is guaranteed to exist whenever the correlation matrix $\mathbf{R}_p$ is positive definite.

We see that, practically speaking, the maximum entropy psd is just the AR spectral estimate that would be generated if the correlation (or covariance) estimated values were the exact values. The following example shows how the maximum entropy spectral estimate varies with assumed order p. It is very similar to the previous example on AR spectral estimates, but here the true values are used for the correlation (or covariance) values needed.

Example 9.6-4 _____

The purpose of this example is to investigate the effect of increasing assumed model order p on the maximum entropy spectral density estimate. For each p, we denote the resulting spectral density as MEM(p). This example was computed using the four MATLAB program files mem_p.m, for $p = 1$ through 4, located at this book's Web site. We use the ar(3) model with parameters indicated by the MATLAB vector a = $[1. \; -1.7 \; 1.53 \; -.648]$. You can run these four .m files yourself to experiment with other AR and ARMA correlation models. We determined the following values for

$$\mathbf{a}_0 = 0.7026, \begin{bmatrix} 1.2215 \\ -0.7385 \end{bmatrix}, \quad \begin{bmatrix} 1.700 \\ -1.53 \\ 0.648 \end{bmatrix}, \quad \begin{bmatrix} 1.700 \\ -1.53 \\ 0.648 \\ -0.00 \end{bmatrix},$$

$$p = \quad 1, \qquad 2, \qquad\qquad 3, \qquad\qquad 4.$$

Note that starting with MEM(3), the a_0 values are numerically correct. Figure 9.6-8 shows the corresponding psd plots for $p = 2$ and 3. We see the MEM(2) psd only detects the larger peak at $\omega = \pi/3$, while the MEM(3) of Figure 9.6-8 psd just overplots the true ar(3) spectral density at this plotting resolution. Thus, in this example at least, there is no numerical instability associated with choosing too high a model order. In real cases, and where the correlation data is not so exact, there are practical problems encountered when the model order is chosen too high for the data.

9.7 SIMULATED ANNEALING

Simulated annealing is a powerful stochastic technique for optimization. It can be applied to both deterministic and random optimization problems, seeking either maxima or minima. Here we look at its use for the problem of maximum _a posteriori_[†] probability (MAP) estimation for Markov random sequences, where the goal is to maximize the conditional probability of the signal given noisy observations. In general, such optimization problems have many local maxima so that a simple hill climbing algorithm like steepest ascent will not be effective. Simply speaking, simulated annealing (SA) avoids ending at a local maxima, by following an iterative stochastic procedure that samples the _a posteriori_ pdf (or pmf) at a huge number of points, near both local maxima and the global maximum. As the iteration proceeds, a parameter termed _temperature_ which governs the randomness in the _a posteriori_ pdf is slowly reduced. This causes the estimator to spend more and more of its time near the global maximum. If the temperature is reduced towards zero slowly enough, the MAP estimate is obtained.

SA is a stochastic procedure because the steps are random and are obtained as samples of the conditional pdf at each time (or spatial location). SA is not needed for the Gaussian

_____

[†]The Latin phrase _a posteriori_ simply means "afterwards," in this case after the noisy observations. The complimentary term _a priori_ means "before." The _a priori_ pdf would then be the original pdf of the noise-free signal.

Markov random sequences studied in earlier sections, where Wiener and Kalman filters yield the MMSE estimates very efficiently. In the Gaussian case, the MMSE and MAP estimate are the same since the MMSE estimate is the conditional mean of the a posteriori pdf, and being Gaussian this conditional mean is the peak. For more complicated, especially compound (sometimes called doubly stochastic) Markov models, the solution is non-linear and usually cannot be obtained in closed form, so that some form of iterative procedure is then required to find either the MMSE or the MAP estimate.

Physical *annealing* describes the process of gradually cooling a liquid in such as way that large scale crystal-like regions form as the *temperature* is reduced, i.e., the material *freezes*. The freezing occurs at a critical temperature called the *Curie point*. A similar effect is observed for so-called ferromagnetic material when placed in a strong magnetic field. In this material, the internal magnetic dipoles align with an imposed magnetic field, more and more, as the temperature is gradually reduced, and the material becomes magnetized. The *Ising* model [9-20] is a simple Markov chain model that has been able to capture the essential freezing characteristics of magnetization in ferromagnetic solids. SA is an optimization method that has, because of some similarities, adopted the terminology of physical annealing. Another name for SA is stochastic relaxation. It is a modern and efficient way of doing Monte Carlo simulation.

We will apply SA to the problem of finding the MAP estimate for Markov random sequences. However, its main application area has been in image processing, to two-dimensional random sequences called *random fields*. In particular Markov random fields often exhibit two properties that require SA. First they are often noncausal, so that the Kalman recursive-in-time scheme cannot be used to get the MMSE estimate even if the data is Gaussian. Secondly, the Markov random fields of interest in image processing, are often compound with resultant nonGaussianity, so that the MMSE estimator is not linear, ruling out a simple two-dimensional extension of the Wiener filter. On the other hand, extension of the one-dimensional SA method for noncausal random sequences, to such spatial and even spatiotemporal processes is straightforward.

Gibbs Sampler

The Gibbs sampler [9-21] is a stochastic iterative procedure which *samples* the conditional pdf (or pmf) of the signal given both the noisy observations and neighboring, but prior signal estimates. Assume there are noisy observations of the N-dimensional vector $\mathbf{X} \triangleq (X[0], X[2], \ldots, X[N-1])^T$ available in the observation vector $\mathbf{Y} \triangleq (Y[0], Y[2], \ldots, Y[N-1])^T$. We define the MAP estimate of $\mathbf{X}$ then as

$$\widehat{\mathbf{X}}_{MAP}(\mathbf{Y}) \triangleq \arg \max_{\mathbf{x}} f(\mathbf{x}|\mathbf{Y}).$$

The Gibbs sampler does not sample the a posteriori pdf $f(\mathbf{x}|\mathbf{Y})$ directly, as this would be difficult for large N. Rather it samples the related conditional pdf $f(x[n]|\mathbf{x}_n, \mathbf{Y})$, where $\mathbf{x}_n \triangleq (x[0], \ldots, x[n-1], x[n+1], \ldots, x[N-1])^T$ is the so-called *deleted vector* obtained be removing $x[n]$. The sampling is usually implemented in sequence, sweeping through the N lattice points (time or space) to complete one step of the iteration. Repeated sweeps with the Gibbs sampler thus generate a sequence of estimates $\widehat{\mathbf{X}}[k]$ for increasing k, where k is the iteration number that under certain conditions can be shown to converge to the MAP estimate $\widehat{\mathbf{X}}_{MAP}$. In fact, the sequence of vector estimates $\widehat{\mathbf{X}}[k]$ can be shown to be a Markov random sequence itself [9-21]. As part of the Gibbs sampling procedure, a relaxation or annealing parameter T is introduced, called the *temperature*, which gradually reduces the variance of this conditional density as $T \to 0$, finally resulting in a *freezing* at hopefully the global optimum value of the objective function. We denote the modified conditional pdf as f_T. Since the variance of the distribution is decreasing with temperature, the local maximas of $f_T(\mathbf{x}|\mathbf{Y})$ are decreasing with respect to the global maximum $f_T(\widehat{\mathbf{X}}_{MAP}|\mathbf{Y})$. This means that near $T = 0$, the conditional pdf should be nearly an impulse centered at $\mathbf{x} = \widehat{\mathbf{X}}_{MAP}(\mathbf{Y})$.

Here is the basic algorithm for the Gibbs sampler.

1. Set *iteration number* $k = 1$ and temperature $T[k] = 1$, max number of iterations k_f, and initialize the signal estimate $\widehat{\mathbf{X}}[0]$.
2. Using temperature $T[k]$, at each site n, sample the conditional pdf $f_T(x[n]|\mathbf{x}_n, \mathbf{Y})$, where $\mathbf{x}_n$ is the vector of samples of estimates computed up to this time. After we complete this sweep of all N samples, call the new signal estimate vector $\widehat{\mathbf{X}}[k]$.
3. Set $k = k + 1$ and go to step 1 until $k > k_f$ or convergence criteria are met.

Below, we will see how to obtain the conditional pdf $f_T(x[n]|\mathbf{x}_n, \mathbf{Y})$ from which we obtain the new signal estimate $\widehat{\mathbf{X}}[k]$ from the prior estimate $\widehat{\mathbf{X}}[k-1]$ in the important case of compound Gauss–Markov random sequence. The rate at which the temperature $T[k]$ is reduced over time is crucial and is called the *annealing schedule*. Proofs of convergence of SA algorithms [9-21], [9-22] depend on how slowly the temperature reduces. Unfortunately, practical applications usually require a faster reduction than the logarithmic annealing schedule appearing in key proofs. Still simulated annealing with the Gibbs sampler often results in much improved estimates in practice for compound Gauss–Markov models. A concrete example of how this is done appears below after introduction of noncausal Gauss-Markov random sequences.

Noncausal Gauss–Markov Models

In earlier work in this book, we have looked at the Gauss–Markov model and seen that it is modeled by white Gaussian noise input to an all-pole filter, so that the resulting difference equation, in the WSS case, is

$$X[n] = \sum_{k=1}^{p} a_k X[n-k] + W[n], \qquad -\infty < n < +\infty, \qquad (9.7\text{-}1)$$

with power spectral density $S_{XX}(\omega) = \sigma_W^2 / |1 - \sum_{k=1}^{p} a_k e^{-j\omega k}|^2$. Here $W[n]$ has the interpretation of the MMSE prediction error. This is a *causal* model, meaning that the sequence $X[n]$ can be recursively generated with increasing n, from some starting point (here assumed at $-\infty$). We can also generate the same random sequence from the *noncausal* model[†]

$$X[n] = \sum_{k=-p,\,k\neq0}^{p} c_k X[n-k] + U[n], \qquad -\infty < n < +\infty, \tag{9.7-2}$$

where the c_k's are the linear interpolator coefficients and the random sequence $U[n]$ is the linear MMSE interpolation error. $U[n]$ has a neighborhood limited (i.e. finite support) correlation function

$$R_{UU}[m] = \begin{cases} \sigma_U^2, & m = 0, \\ -c_m \sigma_U^2, & 0 < |m| \leq p, \\ 0, & \text{else.} \end{cases} \tag{9.7-3}$$

We can derive Equation 9.7-2, by multiplying out the denominator of $S_{XX}(\omega)$ above to get the noncausal model and the c_k values. We find explicitly that

$$S_{UU}(\omega) = \left(1 - \sum_{k=-p,\,k\neq0}^{p} c_k e^{-j\omega k}\right)\sigma_U^2 = \left|1 - \sum_{k=1}^{p} a_k e^{-j\omega k}\right|^2 \Big/ \sigma_W^2. \tag{9.7-4}$$

Correlation models of the form in Equation 9.7-3 have finite support, so that beyond a distance p, there is exactly zero correlation. Thus, unlike the input $W[n]$ in a causal Markov model, the input $U[n]$ in a noncausal Markov random sequence is not a white noise.

The reader may wonder why we introduce equivalent, but noncausal models. The reason is so that we can demonstrate SA, in particular the Gibbs sampler, without going to two-dimensions. In two and higher dimensions, the noncausal model equation does not factor simply as in Equation 9.7-4. It also turns out that the noncausal model is the more general of the two, and the basis for defining the Markov random field in two and higher dimensions.

The general definition of Markov random sequences (and fields) is given in terms of their interpolative conditional pdf $f_X(x_n|x_{n+p}, \ldots, x_{n+1}, x_{n-1}, \ldots, x_{n-p})$.

Definition 9.7-1   A random sequence $X[n]$ defined on $[0, N-1]$ is said to be *Markov-p*, if for all n,

$$f_X(x_n|x_0, \ldots, x_{n+1}, x_{n-1}, \ldots, x_N) = f_X(x_n|x_{n+p}, \ldots, x_{n+1}, x_{n-1}, \ldots, x_{n-p}).$$

If $n \pm p$ are outside $[0, N-1]$ then use boundary conditions or reduce the model order near these boundaries. ∎

We will use this general definition of Markov random sequence in the sequel. In the two-dimensional, finite lattice case, where factorization of the noncausal representation is not possible, there was the problem of how to relate this conditional representation to the joint pdf $f_X(\mathbf{x})$, where $\mathbf{X}$ is the random vector of all the sample random variables $X[n]$. In [9-23], Besag showed how to relate the two representations, i.e. conditional and joint probabilities,

---

[†]This is the same representation for Markov random sequences that was found in Section 9.6 for the Maximum Entropy spectrum.

in a theorem of Hammersley and Clifford. For the discrete-valued case on a finite lattice (or bounded region), the Hammersley–Clifford theorem expresses the joint Markov pmf $P_X(\mathbf{x})$, in terms of an *energy function* $U(\mathbf{x})$ (not related to $U[n]$ above!) as

$$P_X(\mathbf{x}) = k \exp\{-U(\mathbf{x})\},$$

with the assumption that $P_X(\mathbf{x}) > 0$ for all $\mathbf{x}$, and with k being a normalizing constant. The *non-negative* energy function $U(\mathbf{x})$ is in turn defined in terms of *potential functions* that are summed over *cliques*, which are single sites and neighbor pairs in the local neighborhood regions of the Markov random sequence,

$$U(\mathbf{x}) \triangleq \sum_{c \epsilon C} V_c(\mathbf{x}).$$

Here the sum is over each site or location n, and potentials associated with that site. This general Markov random sequence representation also applies for continuous valued random sequences to the pdf's on finite lattices, again with the constraint that $f_X(\mathbf{x}) > 0$ for all $\mathbf{x}$. Note that the energy function is a sum over local characteristics expressed in terms of the potential functions. We see that the maximum probability occurs at the minimum energy.

Example 9.7-1

For the noncausal Gauss–Markov random sequence of order p given in Equation 9.7-2, the cliques c are: the single sites $[n]$ and all the site pairs $[n], [k]$ with $|n - k| \le p$, with corresponding potential functions

$$V_c(\mathbf{x}) = \frac{x^2[n]}{2\sigma_U^2} \quad \text{and} \quad V_c(\mathbf{x}) = -\frac{c_{n-k} x[n] x[k]}{\sigma_U^2} \quad \text{for } k \ne n \text{ and } |n - k| \le p.$$

Here σ_U^2 is the minimum *mean-square interpolation error*, that is obtained with *interpolation coefficients* c_k. The overall energy $U(\mathbf{x})$ is then obtained by summing the above potential functions over all n, and for each n, then for all $k \ne n$ in a local neighborhood such that $|n - k| \le p$. In summing over clique pairs $[n], [k]$, we multiply by $1/2$ so as not to count these site pairs twice. We thus get

$$U(\mathbf{x}) = \sum_{n=0}^{N-1} \left(\frac{x^2[n]}{2\sigma_U^2} - \sum_{k \ne n, k=n-p}^{n+p} \frac{c_{n-k} x[n] x[k]}{2\sigma_U^2} \right),$$

which we recognize as the exponential argument of a multidimensional Gaussian random vector of Chapter 5. In this case the $N \times N$ correlation matrix $\mathbf{R}_{XX}$ is given in inverse form, for first order $p = 1$ and $N = 6$ case for example, as

$$\mathbf{R}_{XX}^{-1} = \frac{1}{\sigma_U^2} \begin{bmatrix} 1 & -c_1 & 0 & 0 & 0 & 0 \\ -c_{-1} & 1 & -c_1 & 0 & 0 & 0 \\ 0 & -c_{-1} & 1 & -c_1 & 0 & 0 \\ 0 & 0 & -c_{-1} & 1 & -c_1 & 0 \\ 0 & 0 & 0 & -c_{-1} & 1 & -c_1 \\ 0 & 0 & 0 & 0 & -c_{-1} & 1 \end{bmatrix}.$$

We note that this implies something about the boundary conditions, in particular the boundary condition $x[-1] = x[N] = 0$ is needed here. The fact that different boundary conditions are needed for the causal and noncausal Markov random sequences has been pointed out by Derin in [9-20].

The next example uses the Gibbs sampler to perform an iterative MAP estimate for a Gauss–Markov random sequence. While this problem could be more efficiently solved by a linear Kalman or Wiener filter, the example serves to illustrate the SA approach. This same approach is then used in a later example for a compound random process, for which linear estimates are not optimal.

Example 9.7-2

(Interpolative (conditional) representation.) Consider the following estimation problem. For some $N > 1$, we have observations

$$Y[n] = X[n] + V[n] \text{ for } n = 0, \ldots, N - 1,$$

where the signal vector $\mathbf{X} \triangleq (X[0], X[1], \ldots, X[N-1])^T$ is non-causal Gauss–Markov, and the additive noise $V[n]$ is independent white Gaussian noise of variance σ_V^2. We seek the MAP estimate $\widehat{\mathbf{X}}_{MAP} = \arg\max_{\mathbf{x}} f(\mathbf{x}|\mathbf{Y})$ using the Gibbs sampler, where $\mathbf{Y} = (Y[0], Y[1], \ldots, Y[N-1])^T$.

To run the Gibbs sampler, and implement the simulated annealing schedule, we modify the variances $\sigma_U^2 \to T\sigma_U^2$ and $\sigma_V^2 \to T\sigma_V^2$. With $T = 1$, we have the true variances, but as $T \to 0$, we simulate annealing to converge to the global MAP estimate $\widehat{\mathbf{X}}_{MAP}(\mathbf{Y})$. We need the conditional pdf $\pi_X[n] \triangleq f_T(x[n]|\mathbf{x}_n, \mathbf{Y})$, where $\mathbf{x}_n$ is the deleted version of $\mathbf{x}$, and all the variances are multiplied by T. Starting out, we write

$$
\begin{aligned}
\pi_X[n] &= f_T(x[n]|\mathbf{x}_n, \mathbf{Y}) \\
&= \frac{f_T(x[n], \mathbf{x}_n, \mathbf{Y})}{\int f_T(z, \mathbf{x}_n, \mathbf{Y})dz}, \quad \text{where the integration is over all values of } z = x[n], \\
&= \frac{f_T(\mathbf{x})f_T(\mathbf{Y}|\mathbf{x})}{\int f_T(z, \mathbf{x}_n)f_T(\mathbf{Y}|\mathbf{x})dz}, \\
&= \frac{f_T(x[n]|\mathbf{x}_n)f_T(\mathbf{x}_n)f_T(\mathbf{Y}|\mathbf{x})}{\int f_T(z|\mathbf{x}_n)f_T(\mathbf{x}_n)f_T(\mathbf{Y}|\mathbf{x})dz}, \\
&= kf_T(x[n]|\mathbf{x}_n)f_T(\mathbf{Y}|\mathbf{x}), \\
&= k\exp\left(-\frac{(x[n] - (c*x)[n])^2}{2T\sigma_U^2} - \frac{(Y[n] - x[n])^2}{2T\sigma_V^2}\right),
\end{aligned}
\tag{9.7-5}
$$

where many terms not involving $x[n]$ have been absorbed into the constant normalizing k, and where the term[†]

$$(c * x)[n] \triangleq \sum_{k=-p,\ k\neq 0}^{p} c_k x[n-k].$$

Now recognizing that this is a Gaussian pdf, and completing the square, we finally arrive at a Gaussian sampling pdf $\pi_X[n] \sim N(\mu, \sigma^2)$ with:

$$\mu \triangleq (c * x)[n]\frac{\sigma_V^2}{\sigma_U^2 + \sigma_V^2} + Y[n]\frac{\sigma_U^2}{\sigma_U^2 + \sigma_V^2}$$

$$\sigma^2 \triangleq \frac{\sigma_U \sigma_V}{\sqrt{\sigma_U^2 + \sigma_V^2}} T.$$

We can recognize the conditional mean μ as a weighted average of the observed value of $Y[n]$, and the interpolated value, based on the current estimate of the neighborhood of site n. Similarly σ^2 is the variance of the error when μ is used as a summary (estimator) of this distribution.

While this is an interesting exercise, and as we have said before, SA is not needed for one-dimensional Gauss–Markov random sequences, because the MMSE and MAP estimates can be calculated recursively by the Kalman filter as we saw in Section 9.2. Since we cannot factor general noncausal Markov models in two and higher dimensions, an iterative solution is often appropriate, but since there is only one maximum in the *a posteriori* pdf in the Gauss–Markov case, a deterministic version of simulated annealing can be used [9-20], or an iterative Wiener filter (cf. Section 9.3) can be used. However, the real importance of SA lies in finding the MAP estimate for compound stochastic models, with their characteristic of many local maxima of the *a posteriori* pdf. Below, we consider the case of compound Markov random sequences that have been used with success in image processing. To simplify the discussion, we will continue to treat only the one-dimensional case. Generalization of the SA method to two and higher dimensions is straight forward.

Compound Markov Models

Returning to Equation 9.7-2, we now write the coefficients as randomly selected by some underlying Markov chain $L[n]$, called a *line sequence* (also commonly called *line process*):

$$X[n] = \sum_{k=-p,\ k\neq 0}^{p} c_k^{\mathbf{L}[n]} X[n-k] + U_{\mathbf{L}[n]}[n], \qquad 1 \le n \le N, \qquad (9.7\text{-}6)$$

Here the line process $L[n]$ is interpreted as "bonds" between the data values $X[n]$, and is located between the data points, in the time (or spatial) domain. If $L = 0$, the bond

---

[†]We note that for a given set of realizations $\mathbf{X}_n = \mathbf{x}_n$ (recall that $\mathbf{X}_n = (X[n-p], \ldots, X[n-1], X[n+1], \ldots, X[n+p])^T$ and $\mathbf{x}_n = (x[n-p], \ldots, x[n-1], x[n+1], \ldots, x[n+p])^T$) we have from Equation 9.7-2 that $E\{X[n]|\mathbf{x}_n\} = \sum_{k=-p,k\neq 0}^{p} c_k x[n-k]$ since $E\{U[n]\} = 0$ i.e., $R_{UU}[m] \xrightarrow{m\to\infty} \mu_U^2 = 0$.

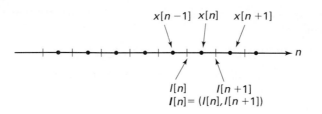

Figure 9.7-1   Placement of line process with regard to data samples.

is present and there is normal (high) correlation between the neighboring values on either side of the bond, but if $L = 1$, then the bond is broken and little correlation exists. In the interpolative model Equation 9.7-6 then, the interpolation coefficients $c_k^{\mathbf{L}[n]}$ are either big or small, based on the value of $\mathbf{L}[n] \triangleq (L[n], L[n+1])$, with temporal (spatial) arrangement as shown in Figure 9.7-1. In this Figure, the black dots, indicate the locations (sites) for the observed random variables $X[n]$, while the vertical bars indicate the interstitial locations or sites of the unobserved bonds or line variables $L[n]$. We note that the binary random vector $\mathbf{L}[n]$ contains the two bonds on either side of $X[n]$, so that the value of the $c_k^{\mathbf{L}[n]}$ will directly depend on whether these bonds are intact. Note also, that the interpolation error $U_{\mathbf{L}[n]}[n]$ would be expected to be large or small, based on these local bonds. For example if all the $L[n] = 1$ then the interpolation error will be larger since there is no "bonding" between points. We model this error here as a conditionally Gaussian pdf, with zero mean, and a variance dependent on the line process nearest neighbors $\mathbf{L}[n]$.

It is convenient to consider the *joint* MAP estimate here

$$(\widehat{\mathbf{X}}, \widehat{\mathbf{L}}) \triangleq \arg\max_{\mathbf{x},\mathbf{l}} f(\mathbf{x}, \mathbf{l}|\mathbf{Y}).$$

As in the previous example, we consider only the first-order case here where the only coefficients are $c_1^{\mathbf{l}} = c_{-1}^{\mathbf{l}}$. We can write the joint mixed[†] pdf $f(\mathbf{x}, \mathbf{l}, \mathbf{Y}) = P(\mathbf{l}) f(\mathbf{x}|\mathbf{l}) f(\mathbf{Y}|\mathbf{x}, \mathbf{l})$ and then we have

$$f(\mathbf{x}, \mathbf{l}|\mathbf{Y}) = k P(\mathbf{l}) f(\mathbf{x}|\mathbf{l}) f(\mathbf{Y}|\mathbf{x}, \mathbf{l}),$$

where the normalization constant k is not a function[‡] of $\mathbf{x}$ or $\mathbf{l}$. To proceed further we need the pmf of the discrete valued line sequence $\mathbf{L}$, for which we take a one-dimensional version of the Gemans' line sequence [9-21], as shown below.

Gibbs Line Sequence

We write the joint probability of the line sequence $\mathbf{L} = \{L[n]\}$ as

$$P_L(\mathbf{l}) = k \exp\{-U(\mathbf{l})\}$$

[†]Note that this probability is part 'density' and part 'mass function,' hence the word 'mixed.' Thus it must be integrated over $\mathbf{x}$ and $\mathbf{r}$, but summed over $\mathbf{l}$, to give actual probabilities.

[‡]Since $f(\boldsymbol{x}, \boldsymbol{l}, |\mathbf{Y}) = f(\boldsymbol{x}, \boldsymbol{l}, \mathbf{Y})/f(\mathbf{Y})$ we find that $k \triangleq 1/f(\mathbf{Y})$.

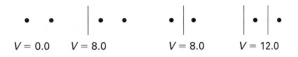

$V = 0.0 \qquad V = 8.0 \qquad\qquad V = 8.0 \qquad V = 12.0$

Figure 9.7-2 Assignment of Line sequence potential function to line cliques.

with energy function

$$U(\mathbf{l}) \triangleq \sum_{c_l \in C_l} V_{c_l}(\mathbf{l}),$$

where $V_{c_l}(\mathbf{l})$ is the potential function at clique c_l. We consider four such values illustrated by Figure 9.7-2, where the black dots are data value sites, and the line sequence is shown between the samples. Four separate cases are shown here, with increasing energy (lower probability) from right to left. Here the black line indicates a broken bond, i.e. $l = 1$, while its absence indicates an intact bond, i.e. $l = 0$. Note that the potential function values V_c increase with the number of broken bonds, with two neighboring broken bonds being given the highest potential, leading to highest energy, and hence lowest probability. The numerical values given were experimentally determined from the data, and have no general significance. For a new problem, another set of increasing potential function values would have to be carefully selected.

To run the Gibbs sampler for the compound Gauss–Markov sequence, we need the conditional pmf of the line sequence, as in Equation 9.7-5

$$\pi_l[n] \triangleq P\{l[n]|\mathbf{l}_n, \mathbf{x}, \mathbf{Y}\}$$
$$= \frac{p\{\mathbf{x}, \mathbf{l}, \mathbf{Y}\}}{\sum_l p\{\mathbf{x}, \mathbf{l}_n, l, \mathbf{Y}\}}$$

where the deleted vector $\mathbf{l}_n \triangleq (l[0], \dots, l[n-1], l[n+1], \dots, l[N-1])^T$ excludes $l[n]$, and the sum in the denominator is over the (two) values of $l[n]$. After some manipulations, we come up with the following expression for $\pi_l[n]$ in terms of $V_{c_l}(\mathbf{l}[n])$ and the neighboring data values $x[n]$ and $x[n+1]$, which is valid for first order models, i.e. $p = 1$,

$$\pi_l[n] = k \exp\left(-\frac{\left(\frac{x^2[n]}{2\sigma^2_{1[n]}} - \frac{c_1^{1[n]}x[n]x[n+1]}{\sigma^2_{1[n]}} + \frac{x^2[n+1]}{2\sigma^2_{1[n+1]}}\right) + V_{c_l}(\mathbf{l}[n])}{T}\right),$$

where we have inserted the annealing temperature variable T, which again is $T = 1$ for the actual stochastic model.

Example 9.7-3

(Simulated annealing for noisy pulse sequence.) Here we apply the Gibbs sampler to the estimation of 100 points of a somewhat random pulse sequence $X[n]$ shown in Figure 9.7-3 corrupted by an independent, additive, white, Gaussian noise $W[n]$. Figure 9.7-4 shows the

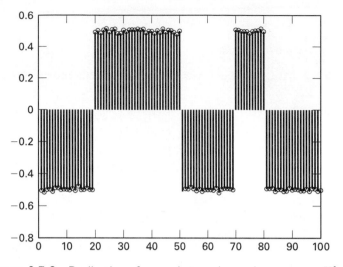

Figure 9.7-3   Realization of somewhat random pulse sequence $X[n]$.

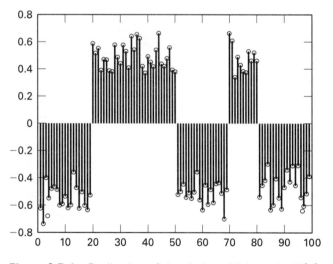

Figure 9.7-4   Realization of signal plus additive noise $Y[n]$.

received noisy pulse sequence. The approximate signal variance is 0.245 and the input noise variance is 0.009. We see that the additive noise has significantly increased the randomness of the height of the pulses. After SA processing for 500 iterations the mean square error is reduced to approximately 0.004, while after linear iterative processing the MSE is approximately 0.0065.

Figure 9.7-5 shows the SA estimate of $L[n]+1$, and we note that it has correctly detected the jumps in the input random pulse train. Because the bonds are now broken across the

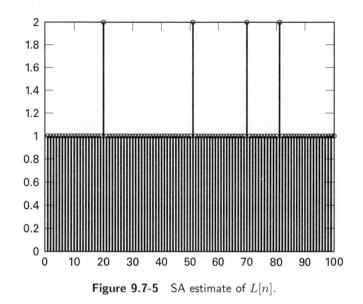

Figure 9.7-5 SA estimate of $L[n]$.

pulse edges, we expect the SA estimate of the signal $X[n]$ will be smoothed *within* the pulses, but not across the pulse edges. This should result in improved performance.

Figure 9.7-6 shows the SA estimate obtained with 500 iterations. Note that the observation noise $W[n]$ has been attenuated without smoothing the edges of the pulses. Figure 9.7-7 shows a linear Wiener estimate obtained without aid from the line sequence. Note the oversmoothing or blurring at the pulse edges seen here.

In a 2-D version of the signal-in-noise estimation problem of the last example, where Markov random fields and SA are used for image processing, the use of the compound Markov model is essential to avoid, the otherwise visually annoying, blurring of image edges. Like the non-compound or Gauss–Markov model in the above 1-D example, a simple 2-D Gauss–Markov model would give substantial blurring of image edges and an overall out-of-focus effect [9-24]. Other applications of SA in image processing include object detection and motion estimation between two frames of a video or movie.

9.8 SUMMARY

This chapter has presented several applications of random sequences and processes in the area generally known as statistical signal processing. The reader should note that there are many more applications that we did not have room to mention. Notable among these is the application of probabilistic and statistical theory to communications, data compression, and networking. Having completed this book, it is our belief and hope that the student will find other important applications of random sequences and processes in his or her future work.

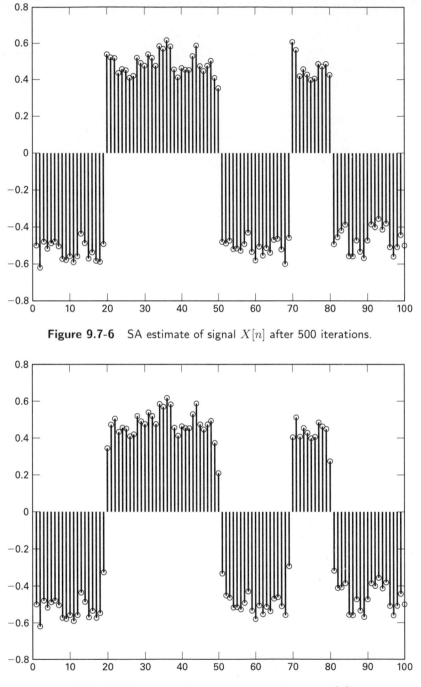

Figure 9.7-6   SA estimate of signal $X[n]$ after 500 iterations.

Figure 9.7-7   Linear iterative estimate of signal $X[n]$.

PROBLEMS

9.1. Let $\mathbf{X}$ and $\mathbf{Y}$ be real-valued random vectors with $E[\mathbf{X}] = E[\mathbf{Y}] = \mathbf{0}$ and $\mathbf{K}_1 \triangleq E[\mathbf{X}\mathbf{X}^T], \mathbf{K}_2 \triangleq E[\mathbf{Y}\mathbf{Y}^T]$, and $\mathbf{K}_{12} \triangleq E[\mathbf{X}\mathbf{Y}^T] = \mathbf{K}_{21}^T \triangleq (E[\mathbf{Y}\mathbf{X}^T])^T$. It is desired to estimate the value of $\mathbf{Y}$ from observing the value of $\mathbf{X}$ according to the rule

$$\widehat{\mathbf{Y}} = \mathbf{A}\mathbf{X}.$$

Show that with

$$\mathbf{A} = \mathbf{A}_o \triangleq \mathbf{K}_{21}\mathbf{K}_1^{-1}$$

the estimator $\widehat{\mathbf{Y}}_o$ is a minimum-variance estimator of $\mathbf{Y}$. By minimum variance is meant that the diagonal terms of $E[(\widehat{\mathbf{Y}} - \mathbf{Y})(\widehat{\mathbf{Y}} - \mathbf{Y})^T]$ are at a minimum.

9.2. Let $\mathbf{X}$ and $\mathbf{Y}$ be real-valued random vectors with $E[\mathbf{X}] = \boldsymbol{\mu}_1, E[\mathbf{Y}] = \boldsymbol{\mu}_2, E[(\mathbf{X} - \boldsymbol{\mu}_1)(\mathbf{X} - \boldsymbol{\mu}_1)^T] \triangleq \mathbf{K}_1, E[(\mathbf{Y} - \boldsymbol{\mu}_2)(\mathbf{Y} - \boldsymbol{\mu}_2)^T] \triangleq \mathbf{K}_2$ and $E[(\mathbf{X} - \boldsymbol{\mu}_1)(\mathbf{Y} - \boldsymbol{\mu}_2)^T] \triangleq \mathbf{K}_{12}$. Show that

$$\widehat{\mathbf{Y}} = \boldsymbol{\mu}_2 + \mathbf{K}_{21}\mathbf{K}_1^{-1}(\mathbf{X} - \boldsymbol{\mu}_1)$$

is a linear minimum variance for $\mathbf{Y}$, based on $\mathbf{X}$.

9.3. Use the orthogonality principle to show that the MMSE

$$\varepsilon^2 \triangleq E[(X - E[X|Y])^2],$$

for real-valued random variables can be expressed as

$$\varepsilon^2 = E[X(X - E[X|Y])]$$

or as

$$\varepsilon^2 = E[X^2] - E[E[X|Y]^2].$$

Generalize to the case where $\mathbf{X}$ and $\mathbf{Y}$ are real-valued random vectors, that is, show that the MMSE matrix is

$$\varepsilon^2 = E[(\mathbf{X} - E[\mathbf{X}|\mathbf{Y}])(\mathbf{X} - E[\mathbf{X}|\mathbf{Y}])^T]$$
$$= E[\mathbf{X}(\mathbf{X} - E[\mathbf{X}|\mathbf{Y}])^T]$$
$$= E[\mathbf{X}\mathbf{X}^T] - E[E[\mathbf{X}|\mathbf{Y}]E^T[\mathbf{X}|\mathbf{Y}]].$$

9.4. Conclude that the limit in Equation 9.1-17 exists with probability-1 by invoking the Martingale Convergence Theorem 6.8-4 applied to the random sequence $G[N]$ with parameter N defined in Equation 9.1-18. Specifically show that $\sigma_G^2[N]$ remains uniformly bounded as $N \to \infty$.

9.5. Modify Theorem 9.1-3 to specify the LMMSE estimate of the zero-mean random sequence $X[n]$ based upon the most recent p observations of the zero-mean random sequence $Y[n]$, i.e.

$$\widehat{E}[\mathbf{X}[n]|\mathbf{Y}[n], \ldots, \mathbf{Y}[n-p]] = \sum_{i=0}^{p} a_i^{(p)}\mathbf{Y}[n-i].$$

      (a) Write equations analogous to Equations 9.1-25 and 9.1-26.

      (b) Derive the corresponding equation for $\varepsilon^2_{\min}$ as in Equation 9.1-27.

9.6.   (Larson & Shubert [9-25]). A Gaussian random sequence $X[n], n = 0, 1, 2, \ldots$ is defined by the equation

$$X[n] = -\sum_{k=1}^{n} \binom{k+2}{2} X[n-k] + W[n] \quad n = 1, 2, \ldots,$$

where $X[0] = W[0]$, and $W[n], n = 0, 1, 2, \ldots$ is a Gaussian white noise sequence with zero mean and variance of unity.

      (a) Show that $W[n]$ is the innovations sequence for $X[n]$.

      (b) Show that $X[n] = W[n] - 3W[n-1] + 3W[n-2] - W[n-3]$, for $n = 0, 1, 2, \ldots$ where $W[-3] = W[-2] = W[-1] = 0$.

      (c) Use the preceding result to obtain the best two-step predictor of $X[12]$ as a linear combination of $X[0], \ldots, X[10]$. Also calculate the resulting mean-square prediction error.

9.7.   Let $W[n]$ be a sequence of independent, identically distributed Gaussian random variables with zero mean and unit variance. Define

$$X[n] \triangleq W[1] + \ldots + W[n] \quad n = 1, 2, \ldots$$

      (a) Find the innovations sequence for $X[n]$.

      (b) Let there be noisy observations available:

$$Y[n] = X[n] + V[n], \quad n = 1, 2, \ldots,$$

where $V[n]$ is also a white Gaussian random sequence with variance σ_V^2, and $V[n]$ is orthogonal to $W[n]$.

Find the recursive filtering structure for computing the MMSE estimate

$$\widehat{X}[n|n] \triangleq E\big[X[n]|Y[1], \ldots, Y[n]\big].$$

      (c) Find the recursive equations specifying any unknown constants in the filter of (b). Specify the initial conditions.

9.8.   A random sequence $Y[n], n = 0, 1, 2, \ldots$, satisfies a second-order linear-difference equation

$$2Y[n+2] + Y[n+1] + Y[n] = 2W[n], \quad Y[0] = 0, Y[1] = 1,$$

with $W[n], n = 0, 1, \ldots$, a standard white Gaussian random sequence (that is, $N(0, 1)$). Transform this equation into the state-space representation and evaluate the mean function $\boldsymbol{\mu}_{\mathbf{X}}[n]$ and the correlation function $\mathbf{R}_{\mathbf{XX}}[n_1, n_2]$ at least for the first few values of n.

Hint: Define the state vector $\mathbf{X}[n] = (Y[n+2], Y[n+1])^T$.

9.9. In our derivation of the Kalman filter in Section 9.2, we assumed that the Gauss-Markov signal model (Equation 9.2-6) was zero-mean. Here we modify the Kalman filter to permit the general case of nonzero mean for $\mathbf{X}[n]$. Let the Gauss-Markov signal model be

$$\mathbf{X}[n] = \mathbf{A}\mathbf{X}[n-1] + \mathbf{B}\mathbf{W}[n], \quad n \geq 0$$

where $\mathbf{X}[-1] = \mathbf{0}$ and the centered noise $\mathbf{W}_c[n] \triangleq \mathbf{W}[n] - \boldsymbol{\mu}_{\mathbf{W}}[n]$ is white Gaussian with variance $\sigma_{\mathbf{W}}^2$, and $\boldsymbol{\mu}_{\mathbf{W}}[n] \neq \mathbf{0}$. The observation equation is still Equation 9.2-7 and $\mathbf{V} \perp \mathbf{W}_c$.

(a) Find an expression for $\boldsymbol{\mu}_{\mathbf{X}}[n]$ and $\boldsymbol{\mu}_{\mathbf{Y}}[n]$.

(b) Show that the MMSE estimate of $\mathbf{X}[n]$ equals the sum of $\boldsymbol{\mu}_{\mathbf{X}}[n]$ and the MMSE estimate of $\mathbf{X}_c[n] \triangleq \mathbf{X}[n] - \boldsymbol{\mu}_{\mathbf{X}}[n]$ based on the centered observations $\mathbf{Y}_c[n] \triangleq \mathbf{Y}[n] - \boldsymbol{\mu}_{\mathbf{Y}}[n]$.

(c) Extend the Kalman filtering Equation 9.2-16 to the nonzero mean case by using the result of (b).

(d) How do the gain and error-covariance equations change?

9.10. (Larson & Shubert [9-25]). Suppose that the observation equation of the Kalman predictor is generalized to

$$\mathbf{Y}[n] = \mathbf{C}_n \mathbf{X}[n] + \mathbf{V}[n],$$

where the $\mathbf{C}_n, n = 0, 1, 2, \ldots$ are $(M \times N)$ matrices, $\mathbf{X}[n]$ is a $(N \times 1)$ random vector sequence, and $\mathbf{Y}[n]$ is a $(M \times 1)$ random vector sequence. Let the time varying signal model be given as

$$\mathbf{X}[n] = \mathbf{A}_n \mathbf{X}[n-1] + \mathbf{B}_n \mathbf{W}[n].$$

Repeat the derivation of the Kalman predictor to show that the prediction estimate now becomes

$$\widehat{\mathbf{X}}[n] = \mathbf{A}_n[(\mathbf{I} - \mathbf{G}_{n-1}\mathbf{C}_{n-1})\widehat{\mathbf{X}}[n-1] + \mathbf{G}_{n-1}\mathbf{Y}[n-1]],$$

with Kalman gain

$$\mathbf{G}_n = \varepsilon^2[n]\mathbf{C}_n^T[\mathbf{C}_n\varepsilon^2[n]\mathbf{C}_n^T + \sigma_{\mathbf{V}}^2[n]]^{-1}.$$

What happens to the equation for the prediction MSE matrix $\varepsilon^2[n]$?

9.11. Here we show how to derive Equation 9.3-5 using property (b) of the $\widehat{E}$ operator of Theorem 9.1-4.

(a) Use property (b) in Theorem 9.1-4 iteratively to conclude,

$$\widehat{E}[X[n]|\tilde{Y}[-N], \ldots, \tilde{Y}[+N]] = \sum_{K=-N}^{+N} g[k]\tilde{Y}[k],$$

with $g[k]$ given by Equation 9.3-4.

(b) Use the result of Problem 9.4 to show that $\lim_{N\to\infty} \sum_{k=-N}^{+N} g[k]\tilde{y}[k]$ exists with probability-1 when X and Y are jointly Gaussian.

9.12. Let $X[n], -\infty < n < +\infty$, be a WSS random sequence with zero mean. Further assume that X is Gaussian distributed. Let $\widehat{R}_N[m]$ be defined as in Equation 9.6-1.

(a) Show that $\widehat{R}_N[m]$ is an unbiased estimator of $R_N[m]$, i.e. $E\{\widehat{R}_N[m]\} = R_N[m], -\infty < m < +\infty$.

(b) Show that $\widehat{R}_N[m] \to R_{XX}[m]$ in mean-square. Hint: Consider $E\{\widehat{R}_N[m] \widehat{R}_N^*[m]\}$, and use the fourth order moment property of the Gaussian distribution.

9.13. Let the correlation function

$$R_{XX}[m] = 10e^{-\lambda_1|m|} + 5e^{-\lambda_2|m|}$$

for $\lambda_1 > 0$ and $\lambda_2 > 0$.

(a) Find $S_{XX}(\omega)$ for $|\omega| \le \pi$.

(b) Evaluate Equation 9.6-4 and show that

$$\lim_{N\to\infty} E\{I_N(\omega)\} = S_{XX}(\omega)$$

in this example.

9.14. Let covariance values for the zero-mean, WSS random sequence $X[n]$ be known for $m = 0, \pm 1$ and given as $K_{XX}[0] = \sigma_X^2$ and $K_{XX}[\pm 1] = \sigma_X^2 \rho$ with $|\rho| < 1$. Find the maximum entropy power density spectra (psd) for this covariance data and corresponding to $p = 1$.

9.15. Using the MATLAB file ar_ar.m, investigate the effect of increasing data length N on the parametric ar(3) spectral density estimate. Plot the resulting psd estimates and the true psd to compare for $N = 25, 100$, and 512. Also plot the correlation function estimate for $N = 100$.

9.16. (Hidden Markov Model) In Example 9.5-5 assume that the observation at $n = 1, 2, 3$ are, respectively H, H, H. Use the Viterbi algorithm to show that the optimum state-sequence is

$$q_1^* = q_2^* = q_3^* = 2.$$

9.17. Write a MATLAB.m file for computing the optimal state-sequence for model *one* in Example 9.5-1 and the parameters given in Example 9.5-5. Let the model allow for 5 observations, where a 1 represents a *Head* and a zero represents a *Tail*. Thus given the observation $\{H, T, T, H, H\}$; which is represented by $\{1, 0, 0, 1, 1\}$, the program should compute the state sequence most likely to have produced it.

9.18. (Expectation Maximization Algorithm) Assume a tomographic configuration consisting of two Poisson-emitting cells. Let the emission from cells 1 and 2 in one second *be* denoted by X_1, X_2 respectively. The detector readings are denoted by $Y_1 = \frac{2}{3}X_1 + \frac{1}{3}X_2$ and $Y_2 = \frac{1}{3}X_1 + \frac{2}{3}X_2$. Use the E-M algorithm to find the ML estimates of θ_1 and θ_2, the Poisson parameters for cells 1 and 2 respectively. (Note that this problem can be solved without using the E-M algorithm.)

9.19. Solve Equation 9.7-4 for the mean square interpolation error of in terms of the Gaussian model intepolator coefficients c_k and the mean square prediction error σ_W^2. Assume the equation holds for all time n and use psd's in the solution.

9.20. Returning to Equation 9.7-2, consider the deterministic iterative estimate

$$X^{(k+1)}[n] \triangleq (c * X^{(k)})[n] \frac{\sigma_V^2}{\sigma_U^2 + \sigma_V^2} + Y[n] \frac{\sigma_U^2}{\sigma_U^2 + \sigma_V^2},$$

starting at $X^{(0)}[n] = 0$ for all n. Show that under the condition $\sum |c_k| < 1$, this iteration should in the limit achieve the stationary point,

$$X[n] \triangleq (c * X)[n] \frac{\sigma_V^2}{\sigma_U^2 + \sigma_V^2} + Y[n] \frac{\sigma_U^2}{\sigma_U^2 + \sigma_V^2},$$

which yields the non-causal. Wiener filter solutions for this problems.

REFERENCES

9-1. K. S. Miller, *Complex Stochastic Processes*, Reading, Mass.; Addison-Wesley, 1974, pp. 76–80.

9-2. J. G. Proakis et al. *Advanced Digital Signal Processing*, New York: Macmillan, 1992.

9-3. M. S. Grewal and A. P. Andrews, *Kalman Filtering*, Upper Saddle River, N.J.: Prentice-Hall, 1993.

9-4. R. E. Kalman, "A New Approach to Linear Filtering and Prediction Problems," *Journal of Basic Eng.*, Vol. 82, March 1960, pp. 35–45.

9-5. R. E. Kalman and R. S. Bucy, "New Results in Linear Filtering and Prediction Theory," *Journal of Basic Eng.*, Vol. 83, Dec. 1961, pp. 95–107.

9-6. S. M. Kay, *Fundamentals of Statistical Signal Processing: Estimation*, Upper Saddle River, N.J.: Prentice-Hall, 1993.

9-7. T. K. Moon and W. C. Stirling, *Mathematical Methods and Algorithms for Signal Processing*, Upper Saddle River, N.J.: Prentice-Hall, 2000, Chapter 13.

9-8. A. A. Kolmogorov, "Über die analytichen Methoden in der Wahrscheinlichkeitsrechnung," *Mathematische Annelen*, Vol. 104, pp. 415–458, 1931.

9-9. N. Wiener, *The Extrapolation, Interpolation, and Smoothing of Stationary Time Series with Engineering Application*, New York: Wiley, 1949.

9-10. R. L. Lagendijk et al. "Identification and Restoration of Noisy Blurred Images Using the Expectation–Maximization Algorithm," *IEEE Trans. Acoust., Speech, Signal Processing*, Vol. 38, 1990, pp. 1180–91.

9-11. T. K. Moon, "The EM Algorithm in Signal Processing," *IEEE Signal Processing Magazine*, Vol. 13, November 1996, pp. 47–60.

9-12. A. K. Katsaggelos ed., *Digital Image Restoration*, New York: Springer-Verlag, 1989, Chapter 6.

9-13. See Chapter 17 in [9-7].

9-14. L. R. Rabiner, "A Tutorial on Hidden Markor Models and Selected Applications in Speech Recognition," Proc. IEEE, Vol. 77, February 1989, pp. 257–86.

9-15. L. R. Rabiner and B-H Juang, *Fundamentals of Speech Recognition*, Upper Saddle River, N.J.: Prentice-Hall, 1993.

9-16. G. A. Frantz and R. H. Wiggins, "Design Case History: *Speak and Spell* learns to talk," *IEEE Spectrum*, February 1982, pp. 45–49.

9-17. G. D. Forney Jr., "The Viterbi Algorithm," *Proc. IEEE*, Vol. 61, March 1978, pp. 268–78.

9-18. A. V. Oppenheim and R. W. Schafer, *Discrete-Time Signal Processing*, Englewood Cliffs, N.J.: Prentice-Hall, 1989, Chapter 3.

9-19. G. E. P. Box and G. M. Jeukins, *Time Series Analysis: Forecasting and Control*, San Francisco: Holden Day, 1978.

9-20. H. Derin and P. A. Kelly, "Discrete-Index Markov-Type Random Processes," Proceedings of the IEEE, Vol. 77, October 1989, pp. 1485–1510.

9-21. S. Geman and D. Geman, "Stochastic Relaxation, Gibbs Distributions, and the Bayesian Restoration of Images," *IEEE Trans. Pattern Analysis and Machine Intelligence*, Vol. PAMI-6, Nov. 1984, pp. 721–741.

9-22. F.-C. Jeng and J. W. Woods, "Simulated Annealing in Compound Gaussian Random Fields," *IEEE Trans. Information Theory*, Vol. IT-36, Jan. 1990, pp. 94–107.

9-23. J. Besag, "Spatial Interaction and the Statistical Analysis of Lattice Systems," *J. Royal Statistical Society*", series B, Vol. 34, 1974, pp. 192–236.

9-24. F.-C. Jeng and J. W. Woods, "Compound Gauss-Markov Random Fields for Image Estimation," *IEEE Trans. Signal Processing*, Vol. 39, March 1991.

9-25. H. J. Larson and B. O. Shubert, *Probabilistic Models in Engineering Sciences: Vol. II.*, New York: John Wiley, 1979.

Appendix A

Review of Relevant Mathematics

This section will review the mathematics needed for the study of probability and random processes. We start with a review of basic discrete and continuous mathematical concepts.

A.1 BASIC MATHEMATICS

We review the concept of sequence and present several examples. We then look at summation of sequences. Next the Z-transform is reviewed.

Sequences

A sequence is simply a mapping of a set of integers into the set of real or complex numbers. Most often the set of integers is the nonnegative integers $\{n \geq 0\}$ or the set of all integers $\{-\infty < n < +\infty\}$.

An example of a sequence often encountered is the exponential sequence a^n for $\{n \geq 0\}$, which is plotted in Figure A.1-1 for several values of the real number a. Note that for $|a| > 1$, the sequence diverges, while for $|a| < 1$, the sequence converges to 0. For $a = 1$, the sequence is the constant 1, and for $a = -1$, the sequence alternates between $+1$ and -1.

A related and important sequence is the complex exponential $\exp(j\omega n)$. These sequences are eigenfunctions of linear time-invariant systems, which just means that for such a

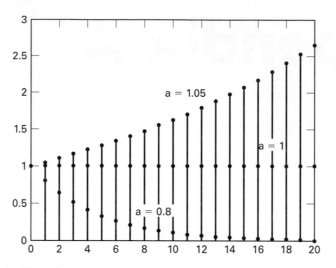

Figure A.1-1   Plot of exponential sequence for three values of $a := 1.05, 1.0,$ and 0.8.

system with frequency response function $H(\omega)$, the response to the input $\exp(j\omega n)$ is just $H(\omega)\exp(j\omega n)$, a scaled version of the input.

Convergence

A sequence, denoted $x[n]$ or x_n, which is defined on the positive integers $n \geq 1$, converges to a limiting value x if the values $x[n]$ become nearer and nearer to x as n becomes large. More precisely, we can say that for any given $\varepsilon > 0$, there must exist a value $N_0(\varepsilon)$ such that for all $n \geq N_0$, we have $|x[n] - x| < \varepsilon$. Note that N_0 is allowed to depend on ε.

Example A.1-1 _____

Let the sequence a_n be given as

$$a_n = 2^n/(2^n + 3^n),$$

and find the limit as $n \to \infty$. From observation, we see that the limit is $a = 0$. To complete the argument, we can then express $N_0(\varepsilon)$ from the equation

$$\left| \frac{2^n}{2^n + 3^n} \right| < \varepsilon$$

as

$$N_0(\varepsilon) = \frac{\ln\left(\frac{1-\varepsilon}{\varepsilon}\right)}{\ln\frac{3}{2}},$$

where we assume that $0 < \varepsilon < 1$. We note that for any fixed $0 < \varepsilon < 1$, the value N_0 is finite as required.

Summations

Summations of sequences arise quite often in our work. A common sequence used to illustrate summation concepts is the geometric sequence a^n. The following summation formula can be readily derived: Take $n_2 \geq n_1$.

$$\sum_{n=n_1}^{n_2} a_n = \frac{a^{n_1} - a^{n_2+1}}{1-a} \quad \text{for} \quad a \neq 1. \tag{A.1-1}$$

Of course, when $a = 1$, the summation is just $n_2 - n_1 + 1$. A simple way to see the validity of Equation A.1-1 is to first define $S = \sum_{n_1}^{n_2} a^n$ and then note that, by the special property of the geometric sequence,

$$aS = S + a^{n_2+1} - a^{n_1}.$$

Then, by solving for S, we derive Equation A.1-1 when $a \neq 1$.

When $|a| \leq 1$, the upper limit of summation can be extended to ∞ to yield

$$\sum_{n=n_1}^{\infty} a^n = \frac{a^{n_1}}{1-a} \quad \text{for} \quad |a| < 1. \tag{A.1-2}$$

Another useful related summation is:

$$\sum_{n=n_1}^{\infty} na^n = \frac{n_1 a^{n_1}(1-a) + a^{n_1+1}}{(1-a)^2} \quad \text{for} \quad |a| < 1. \tag{A.1-3}$$

Equations A.1-2 and A.1-3 most often occur with $n_1 = 0$.

Z-Transform

This transform is very helpful in solving for various quantities in a linear time-invariant system and also for the solution of linear constant-coefficient difference equations. The Z-transform is defined for a deterministic sequence $x[n]$ as follows:

$$\mathsf{X}(z) = \sum_{n=-\infty}^{+\infty} x[n] z^{-n}, \text{ for } z \in \mathscr{R}.$$

In this equation, the region $\mathscr{R}$ is called the *region of convergence* and denotes the set of complex numbers z for which the transform is defined. This set $\mathscr{R}$ is further specified as those z for which the relevant sum converges absolutely, i.e.,

$$\sum_{n=-\infty}^{+\infty} |x[n]||z|^{-n} < \infty.$$

This region $\mathscr{R}$ can be written in general as $\mathscr{R} = \{z : R_- < |z| < R_+\}$, an annular shaped region. The set $\{z | R_- < |z| < R_+\}$ is to read as "the set of all points z whose magnitude (length) is greater than R_- and less than R_+."

Example A.1-2 _____

Let the discrete-time sequence $x[n]$ be given as the damped exponential

$$x[n] = a^n \exp(j\omega_0 n)u[n],$$

where $u[n]$ denotes the unit step sequence, $u[n] = 1$ for $n \geq 0$ and $u[n] = 0$ for $n < 0$.

Calculating the Z-transform, we get

$$\mathsf{X}(z) = \sum_{n=0}^{\infty} a^n \exp(j\omega_0 n)z^{-n}$$

$$= \sum_{n=0}^{\infty} (ae^{j\omega_0}z^{-1})^n$$

$$= \frac{1}{1 - ae^{j\omega_0}z^{-1}} \quad \text{for} \quad |z| > |a|. \quad \mathscr{R} = \{z : |a| < |z|\}.$$

The Z-transform is quite useful in discrete-time signal processing because of the following fundamental theorem relating convolution and multiplication of the corresponding Z-transforms.

Theorem A.1-1   Consider the convolution of two absolutely summable sequences $x[n]$ and $h[n]$, which generates a new sequence $y[n]$ as follows:

$$y[n] = \sum_{m=-\infty}^{+\infty} x[m]h[n-m]$$

which we denote operationally as $y = h * x$. Then the Z-transform of y is given in terms of the corresponding Z-transforms of x and h as

$$\mathsf{Y}(z) = \mathsf{H}(z)\mathsf{X}(z) \quad \text{for} \quad z \in \mathscr{R}_h \cap \mathscr{R}_x.$$

Because the two sequences h and x are absolutely summable, their regions of convergence $\mathscr{R}_h$ and $\mathscr{R}_x$ will both include the unit circle of the z-plane, that is, $\{|z| = 1\}$. Then the Z-transform $\mathsf{Y}(z)$ will exist for $z \in \mathscr{R}_h \cap \mathscr{R}_x$, which is then guaranteed to be nonempty.   ■

After obtaining the Z-transform of a convolution using this result, one can often take the inverse Z-transform to get back the output sequence $y[n]$. There are several ways to do this, including expansion of the Z-transform $\mathsf{Y}(z)$ in a power series, doing long division in the typical case when $\mathsf{Y}(z)$ is a ratio of polynomials in z, and the most powerful method, the method of residues. This last method, along with the residue method for inverse Laplace transforms, is the topic of Section A.3 of this appendix.

A.2 CONTINUOUS MATHEMATICS

The intent here is to review some ideas from the integral calculus of one- and two-dimensional functions of real variables.

Definite and Indefinite Integrals

In a basic calculus course, we study two types of integrals, definite and indefinite:

$$\int x^2 dx = \frac{1}{3}x^3 + C \qquad \text{indefinite}$$

$$\int_a^b x^2 dx = \frac{1}{3}b^3 - \frac{1}{3}a^3 \qquad \text{definite.}$$

In this course we will most always write the definite integral, almost never the indefinite integral. This is because we will use integrals to measure specific quantities, not merely to determine the class of functions that have a given derivative. Please note the difference between these two integrals. Unlike the indefinite integral, the definite integral is a function of its upper and lower limits, but not of x itself! Sometimes we refer to x in our definite integrals as a "dummy variable" for this reason, that is, x could just as well be replaced by another variable, say y, with no change resulting to our definite integral, i.e.,

$$\int_a^b x^3 dx = \int_a^b y^3 dy.$$

To compute the definite integral we first compute the indefinite integral and then subtract its evaluation at the lower limit from its evaluation at the upper limit.

Differentiation of Integrals

From time to time, it becomes necessary to differentiate an integral with respect to a parameter which appears either in the upper limit, the lower limit, or the integrand itself:

$$\frac{d}{dy}\int_{a(y)}^{b(y)} f(x,y)dx = f(b(y),y)\frac{db(y)}{dy} - f(a(y),y)\frac{da(y)}{dy} + \int_{a(y)}^{b(y)} \frac{\partial f(x,y)}{\partial y}dx. \qquad \text{(A.2-1)}$$

This important formula is easily derived by recalling that for a function $I(b,a,y)$ where in turn, $b = b(y)$ and $a = a(y)$ are two functions of y, we have

$$\frac{dI}{dy} = \frac{\partial I}{\partial b}\frac{db}{dy} + \frac{\partial I}{\partial a}\frac{da}{dy} + \frac{\partial I}{\partial y}.$$

If we denote

$$I \triangleq \int_{a(y)}^{b(y)} f(x,y)dx$$

and define a function $F(x,y)$ such that

$$f(x,y) \triangleq \frac{\partial F(x,y)}{\partial x},$$

then clearly

$$\frac{\partial I}{\partial b} = f(b(y), y)$$

$$\frac{\partial I}{\partial a} = f(a(y), y)$$

$$\frac{\partial I}{\partial y} = \frac{\partial}{\partial y} \int_{a(y)}^{b(y)} f(x, y) dx = \int_{a(y)}^{b(y)} \frac{\partial}{\partial y} f(x, y) dx.$$

The last step on the right follows from treating $b(y)$ and $a(y)$ as *constants*, since the variation of I arising from its upper and lower limits is already counted by the first two terms.

An example of use of this formula, which arises in the study of how systems transform probability functions, is shown next.

Example A.2-1 _____

Consider the example where the function $f(x, y) = x + 2y$,

$$\frac{\partial}{\partial y} \int_0^y (x + 2y)^2 dx = (y + 2y)^2 1 - (0 + 2y)^2 0 + \int_0^y 4(x + 2y) dx$$

$$= (3y)^2 + 4 \left(\frac{1}{2} x^2 + 2yx \right) \Big|_0^y$$

$$= (3y)^2 + 2y^2 + 8y^2 = 19y^2$$

Integration by Parts

Integration by parts is a useful technique for explicit calculation of integrals. We write the formula as follows:

$$\int_a^b u(x) dv(x) = u(x) v(x) \Big|_a^b - \int_a^b v(x) du(x), \tag{A.2-2}$$

where u and v denote functions of the variable x with the integral extending over the range $a \geq x \geq b$. This formula is derived using the chain rule for derivatives, applied to the derivative of the product function $u(x) v(x)$. An example is shown below. Integration by parts is useful to extend the class of integrals that are doable analytically.

Example A.2-2 _____

Consider the following integration problem:

$$\int_0^\infty x e^{-2x} dx$$

Let $u(x) = x$ and $dv(x) = e^{-2x} dx$; then using the above integration by parts formula we obtain

$$\int_0^\infty x e^{-2x} dx = x \left(-\frac{1}{2} e^{-2x} \right) \Big|_0^\infty - \int_0^\infty \left(-\frac{1}{2} e^{-2x} \right) dx$$

$$= \frac{1}{2} \int_0^\infty e^{-2x} dx = \frac{1}{2} \left(-\frac{1}{2} e^{-2x} \right) \Big|_0^\infty$$

$$= \frac{1}{4}.$$

Completing the Square

The method of completing the square is applied to the calculation of integrals by transforming an unknown integral into a known one by turning the argument of its integrand into a perfect square. For example, consider making a perfect square out of $x^2 + 4x$. We can transform it into the perfect square $(x+2)^2$ by adding and subtracting 4, that is,

$$x^2 + 4x = (x+2)^2 - 4.$$

To see how this polynomial concept can be used to calculate integrals, consider the well-known Gaussian integral that we often encounter in this course:

$$\int_{-\infty}^{+\infty} e^{-\frac{1}{2}x^2} dx = \sqrt{2\pi}.$$

If, instead we need to calculate

$$\int_{-\infty}^{+\infty} e^{-\frac{1}{2}(x^2 + 4x)} dx = ?,$$

we can do so by completing the square as follows:

$$e^2 \int_{-\infty}^{+\infty} e^{-\frac{1}{2}(x^2 + 4x + 4)} dx$$

where we have multiplied by e^{-2} inside the integral and by e^2 outside. Then we continue,

$$= e^2 \int_{-\infty}^{+\infty} e^{-\frac{1}{2}(x+2)^2} dx.$$

With the change of variables $y = x + 2$, this then becomes

$$= e^2 \int_{-\infty}^{+\infty} e^{-\frac{1}{2}y^2} dy$$

$$= e^2 \sqrt{2\pi}.$$

Double Integration

Integrals on the (x, y) plane are properly called double integrals. The infinitesimal element is an area, written as $dxdy$. We often evaluate these integrals in some order, say x first and then y, or vice versa. Then the integral is called an iterated integral. We can write the three possible situations as follows:

$$\int_{y_1}^{y_2} \left(\int_{x_1}^{x_2} f(x,y)dx \right) dy = \int_{x_1}^{x_2} \int_{y_1}^{y_2} f(x,y)dx\,dy = \int_{x_1}^{x_2} \left(\int_{y_1}^{y_2} f(x,y)dy \right) dx,$$

where the integral in the middle is the true double or area integral. Since limiting operations are the basis for any integral, there is actually a question of whether the three two-dimensional integrals are always equal. Fortunately, an advanced result in measure theory [7-1] shows that when the integrals are defined in the modern Lebesgue sense, then all three either exist and are equal, or do not exist. We will stick with the ordinarily occurring case where the above three integrals exist and are equal.

Note that on the left, when we integrate on x first, that the limits are interchanged versus the situation on the right where we integrate in the y direction first. The double or area integral in the middle, adopts the notation that one reads the limits in x, y order, just as in the function arguments and the area differential $dxdy$. Thus there should be no confusion in interpreting such expressions as

$$\int_1^3 \int_0^5 xe^{-y}dx\,dy,$$

since we would read this correctly as an integral over the rectangle with opposite corners $(x,y) = (1,0)$ and $(x,y) = (3,5)$.

Functions

A function is a unique mapping from a domain space $\mathcal{X}$ to a range space $\mathcal{Y}$. The only condition is uniqueness which means that only one y goes with each x, that is, $f(x)$ has one and only one value. An example is $f(x) = x^2$. A counterexample is $f(x) = \pm\sqrt{x}$.

Monotone Functions.   A monotone function of the real variable x is one that *always* increases as x increases or *always* decreases as x increases. The former, with the positive slope, is called *monotone increasing*, while the latter, with the negative slope, is called *monotone decreasing*, as illustrated in Figures A.2-1 and A.2-2. If a function is monotone

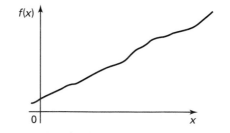

Figure A.2-1   Example of a *monotone increasing* function.

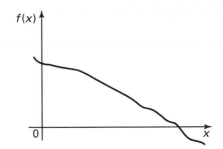

Figure A.2-2   Example of a *monotone decreasing* function.

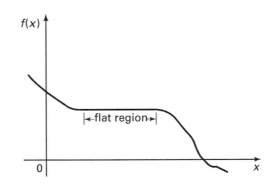

Figure A.2-3   Example of a monotone *nonincreasing* function.

except for some flat regions of zero slope, then we use the terms monotone nondecreasing or monotone nonincreasing to describe them, as illustrated in Figure A.2-3.

Inverse Functions.   A function may or may not have an inverse. The inverse function exists when the original function has the additional uniqueness property that to each y in $\mathcal{Y}$, there corresponds only one x (in $\mathcal{X}$). This allows us to define an inverse function $f^{-1}(y)$ to map back from $\mathcal{Y}$ to $\mathcal{X}$. We note that a sufficient condition for the inverse function to exist is that the original function $f(x)$ is monotone increasing *or* monotone decreasing. The function sketched in Figure A.2-3 does not have an inverse due to the flat section of zero slope.

A.3 RESIDUE METHOD FOR INVERSE FOURIER TRANSFORMATION

In Chapters 6 and 7, we defined the power spectral density (psd) $S(\omega)$ for both discrete and continuous time and showed that the psd is central to analyze LSI systems with random sequence and process inputs. We often want to take an inverse transform to find the correlation function corresponding to a given psd to obtain a time-domain characterization. This section summarizes the powerful residue method for accomplishing the necessary inverse Fourier transformation.

We start by recalling the relation between the psd and correlation function for a WSS random process,

$$S(\omega) = \int_{-\infty}^{+\infty} R(\tau)e^{-j\omega\tau}d\tau,$$

$$R(\tau) = \frac{1}{2\pi}\int_{-\infty}^{+\infty} S(\omega)e^{+j\omega\tau}\,d\tau.$$

To apply the residue method of complex variable theory [A-3] to the evaluation of the above IFT, we must first express this integral as an integral along a contour in the *complex s-plane*. We define a new function S of the *complex variable* $s = \sigma + j\omega$ as follows.

First we define $S(s)$ on the imaginary axis in terms of the function of a *real variable* $S(\omega)$ as

$$\mathsf{S}(s)|_{s=j\omega} \overset{\Delta}{=} S(\omega).$$

Then we replace $j\omega$ by s to extend the function $\mathsf{S}(j\omega)$ to the entire complex plane. Thus

$$\mathsf{S}(s)|_{s=j\omega} = S(\omega) = \int_{-\infty}^{+\infty} R(\tau)e^{-j\omega\tau}\,d\tau$$

so

$$\mathsf{S}(s) = \int_{-\infty}^{+\infty} R(\tau)e^{-s\tau}\,d\tau, \tag{A.3-1}$$

which is the two-sided Laplace transform of the correlation function R. Also by inverse Fourier transform,

$$R(\tau) = \frac{1}{2\pi j}\int_{-\infty}^{+\infty} \mathsf{S}(s)|_{s=j\omega}e^{j\omega\tau}d(j\omega),$$

$$= \frac{1}{2\pi j}\int_{-j\infty}^{+j\infty} \mathsf{S}(s)e^{s\tau}\,ds, \tag{A.3-2}$$

which is an integral along the imaginary axis of the s-plane.

The integral in Equation A.3-2 is called a *contour integral* in the theory of functions of a complex variable [A-3], where it is shown that one can evaluate such an integral over a closed contour by the *method of residues*. This method is particularly easy to apply when the functions are rational; that is, the function is the ratio of two polynomials in s. Since this situation often occurs in linear systems whose behavior is modeled by differential equations, this method of evaluation can be very useful. We state the main result as a fact from the theory of complex variables.

Fact

Let $\mathsf{F}(s)$ be a function of the complex variable s, which is analytic inside and on a closed *counterclockwise* contour C except at P poles located inside C. The contour C encircles the origin. The P poles are located at $s = p_i, i = 1, \ldots, P$. Then

$$\frac{1}{2\pi j} \oint_C F(s)ds = \sum_{\substack{p_i \text{ inside} \\ C}} \text{Res}[F(s); s = p_i], \qquad (A.3\text{-}3)$$

where

1. at a first-order pole, $\text{Res}[F(s); s = p] = [F(s)(s - p)]|_{s=p}$;
2. at a second-order pole, $\text{Res}[F(s); s = p] = \frac{d}{ds}[F(s)(s - p)^2]|_{s=p}$; and at an nth order pole
3. $\text{Res}[F(s); s = p] = \frac{1}{(n-1)!} \left(\frac{d^{(n-1)}}{ds^{(n-1)}}[F(s)(s - p)^n] \right)\Big|_{s=p}$.

In applying these results to our problem we first have to close the contour in some fashion. If we close the contour with a half-circle of infinite radius C_L as shown in Figure A.3-1, then provided that the function being integrated, $S(s)e^{s\tau}$, tends to zero fast enough as $|s| \to +\infty$, the value of the integral will not be changed by this closing of the contour. In other words, the integral over the semicircular part of the contour will be zero. The conditions for this are $|S(s)|$ stage bounded as $|s| \to +\infty$, and

$$|e^{s\tau}| \to 0 \quad \text{as} \quad Re(s) \to -\infty,$$

the latter of which is satisfied for all $\tau > 0$. Thus for positive τ we have

$$R(\tau) = \frac{1}{2\pi j} \oint_{C_L} S(s)e^{s\tau}ds = \sum_{\substack{p_i \\ \text{inside } C_L}} \text{Res}[S(s)e^{s\tau}; s = p_i],$$

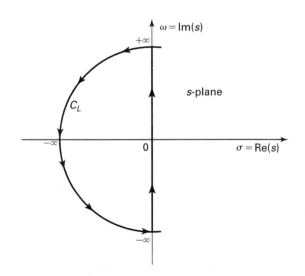

Figure A.3-1   Closed contour in left-half of s-plane for $\tau > 0$.

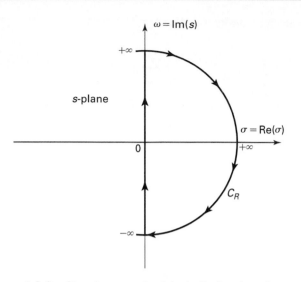

Figure A.3-2   Closed contour in right-half of s-plane for $\tau < 0$.

Similarly, for $\tau < 0$ one can show that it is permissible to close the contour to the right as shown in Figure A.3-2, in which case we have

$$|e^{s\tau}| \to 0 \quad \text{as} \quad Re(s) \to +\infty,$$

so that we get

$$R(\tau) = \frac{1}{2\pi j} \oint_{C_R} S(s)e^{s\tau} ds = - \sum_{\substack{p_i \\ \text{inside } C_R}} \text{Res}[S(s)e^{s\tau}; s = p_i]$$

for $\tau < 0$, the minus sign arising from the clockwise traversal of the contour.

Example A.3-1 _____

(First-order psd.) Let

$$S(\omega) = 2\alpha/(\alpha^2 + \omega^2), \qquad 0 < \alpha < 1.$$

Then

$$S(s)|_{s=j\omega} = S(\omega) = 2\alpha/(\alpha^2 + \omega^2) = 2\alpha/(j\omega + \alpha)(-j\omega + \alpha),$$

so

$$S(s) = \frac{2\alpha}{(s + \alpha)(-s + \alpha)},$$

where the configuration of the poles in the s-plane is shown in Figure A.3-3.

Evaluating the residues for $\tau > 0$, we get

$$R(\tau) = \text{Res}[S(s)e^{s\tau}; s = -\alpha] = \frac{2\alpha e^{s\tau}}{(-s + \alpha)}\Bigg|_{s=-\alpha}$$

$$= \frac{2\alpha}{2\alpha}e^{-\alpha\tau},$$

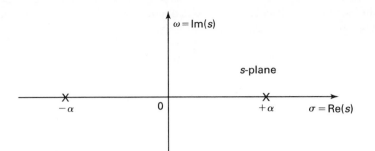

Figure A.3-3   Pole-zero diagram.

while for $\tau < 0$ we get

$$R(\tau) = -\text{Res}[S(s)e^{s\tau}; s = +\alpha]$$

$$= -\left.\frac{2\alpha e^{s\tau}(s - \alpha)}{(s + \alpha)(-s + \alpha)}\right|_{s=+\alpha}$$

$$= \left.\frac{-2\alpha e^{s\tau}}{(s + \alpha)(-1)}\right|_{s=\alpha} = \frac{2\alpha}{2\alpha}e^{\alpha\tau}.$$

Combining the results into a single formula, we get

$$R(\tau) = \exp(-\alpha|\tau|), \qquad -\infty < \tau < +\infty.$$

Inverse Fourier Transform for psd of Random Sequence

In the case of a random sequence one can do a similar contour integral evaluation in the complex z-plane. We recall the transform and inverse transform for a sequence:

$$S(\omega) = \sum_{m=-\infty}^{+\infty} R[m]e^{-j\omega m},$$

$$R[m] = \frac{1}{2\pi}\int_{-\pi}^{+\pi} S(\omega)e^{+j\omega m}d\omega.$$

We rewrite the latter integral as a contour integral around the unit circle in a complex plane by defining the function of a complex variable, $S(z)|_{z=e^{j\omega}} \triangleq S(\omega)$, and then substituting z for $e^{j\omega}$ into this new function to obtain the psd as a z-transform,

$$S(z) = \sum_{m=-\infty}^{+\infty} R[m]z^{-m} \quad \text{and}$$

$$R[m] = \frac{1}{2\pi j}\oint_C S(z)z^{m-1}dz \quad \text{where } C = \{|z| = 1\}. \tag{A.3-4}$$

In this case the contour is already closed and it encircles the origin in a counterclockwise direction, so we can apply Equation A.3-3 directly to obtain

$$R[m] = \sum_{\substack{p_i \\ \text{inside } C}} \text{Res}[S(z)z^{m-1}; z = p_i],$$

where the sum is over the residues at the poles inside the unit circle. This formula is valid for all values of the integer m; however, it is awkward to evaluate for negative m due to the variable-order pole contributed by z^{m-1} at $z = 0$. Fortunately, a transformation mapping z to $1/z$ conveniently solves this problem, and we have [A-1],

$$R[m] = \frac{1}{2\pi j} \oint_C S(z^{-1})z^{-m+1}(-z^{-2}dz),$$

$$= \frac{1}{2\pi j} \oint_C S(z^{-1})z^{-m-1}dz,$$

avoiding the variable-order pole for $m < 0$. We thus arrive at the prescription:
For $m \geq 0$

$$R[m] = \sum_{\substack{i:\text{poles} \\ \text{inside unit circle}}} \text{Res}[S(z)z^{m-1}; z = p_i],$$

and for $m < 0$

$$R[m] = \sum_{\substack{i:\text{poles} \\ \text{outside unit circle}}} \text{Res}[S(z^{-1})z^{-m-1}; z = p_i^{-1}].$$

Example A.3-2

(First-order psd of random sequence.) We consider a psd given as

$$S(\omega) = \frac{2(1 - \rho^2)}{(1 + \rho^2) - 2\rho\cos\omega}, \qquad |\omega| \leq \pi, \tag{A.3-5}$$

which is plotted in Figure A.3-4.

Using the identify $\cos\omega = 0.5(\exp j\omega + \exp -j\omega)$, we can make this substitution in Equation A.3-5 to obtain the function of a complex variable,

$$S(z)|_{z=e^{j\omega}} = S(\omega) = \frac{2(1 - \rho^2)}{(1 + \rho^2) - 2\rho\cos\omega}$$

$$= \frac{2(1 - \rho^2)}{(1 + \rho^2) - \rho(e^{+j\omega} + e^{-j\omega})}.$$

Then we replace $e^{j\omega}$ by z to obtain the function of z,

$$S(z) = \frac{2(1 - \rho^2)}{(1 + \rho^2) - \rho(z + z^{-1})}$$

$$= -2(\rho^{-1} - \rho)\frac{z}{(z - \rho)(z - \rho^{-1})}.$$

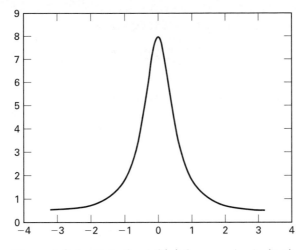

Figure A.3-4   Plot of psd $S(\omega)$ for a ρ value in $(0,1)$.

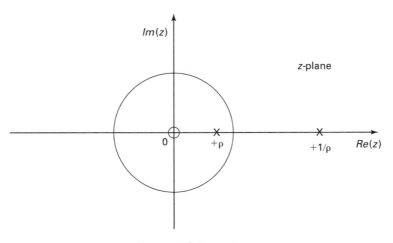

Figure A.3-5 z-plane.

The z-plane pole-zero configuration of this function is shown in Figure A.3-5. The overall transformation from $S(\omega)$ to $\mathsf{S}(z)$ is thus given by the replacement

$$\cos\omega \leftarrow \tfrac{1}{2}(z + z^{-1}). \tag{A.3-6}$$

For $m \geq 0$ we get

$$R[m] = \mathrm{Res}[\mathsf{S}(z)z^{m-1}; z = \rho]$$

$$= -2(\rho^{-1} - \rho)\frac{\rho\rho^{m-1}}{(\rho - \rho^{-1})}$$

$$= 2\rho^{m}.$$

For $m < 0$ we have

$$R[m] = \text{Res}[\mathsf{S}(z^{-1})z^{-m-1}; z = \rho],$$

since $z = p^{-1}$ is the one pole outside the unit circle.
Now

$$\mathsf{S}(1/z) = -2(\rho^{-1} - \rho)\frac{z^{-1}}{(z^{-1} - \rho)(z^{-1} - \rho^{-1})}$$

$$= -2(\rho^{-1} - \rho)\frac{z}{(z - \rho^{-1})(z - \rho)},$$

which could easily have been foretold from the symmetry evident in Equation A.3-6. Then

$$\text{Res}[\mathsf{S}(z^{-1})z^{-m-1}; z = \rho] = -2(\rho^{-1} - \rho) \left. \frac{z^{-m}(z - \rho)}{(z - \rho^{-1})(z - \rho)}\right|_{z=p}$$

$$= -2\frac{(\rho^{-1} - \rho)\rho^{-m}}{(\rho - \rho^{-1})}$$

$$= 2\rho^{-m}.$$

Combining, we get the overall answer

$$R[m] = 2\rho^{|m|}, \qquad -\infty < m < +\infty.$$

A.4 MATHEMATICAL INDUCTION[†] [A-4]

Many proofs in probability are obtained by *mathematical induction*. Mathematical induction is a method for obtaining results, especially proving theorems, which are difficult if not impossible to get by any other method. For example: It is claimed that the set S contains all the positive integers. How do we verify this? We could show that $1 \in S$, $2 \in S$, $3 \in S$ etc. But using this procedure would not allow us to finish in finite time. Instead we use the following Axiom.

Axiom of Induction

Let S be a set of integers. If

(i) $1 \in S$; and
(ii) $k \in S$ implies that $k + 1 \in S$

then all positive integers are in S.   ■

Suppose it is known that S satisfies conditions (i) and (ii). Then $1 \in S$ implies that $2 \in S$, and $2 \in S$ implies that $3 \in S$, etc. Thus any positive integer $k+1$ can be obtained by adding 1 successively (k times) to 1. Thus $k + 1 \in S$ for any positive integer k.

[†]See Calculus, 3rd edition by S. L. Salas and Einar Hille, John Wiley & Sons, New York, 1978.

Example A.4-1 _____

Show that if $0 < a < b$ then $a^n < b^n$ for all positive integers n.

Solution   Suppose we have two numbers a, b such that $0 < a < b$ and let S be the set of positive integers for which $a^n < b^n$. So $1 \in S$. Now assume that $k \in S$, meaning $a^k < b^k$. Then $a^{k+1} = a \times a^k < a \times b^k < b \times b^k = b^{k+1}$. Thus we have shown that $1 \in S$ and that $k \in S$ implies that $k + 1 \in S$. By the Axiom of Induction we conclude that all positive integers are in S

_____

REFERENCES

A-1.   A. V. Oppenheim and R. W. Schafer, _Discrete-Time Signal Processing_. Upper Saddle River, NJ: Prentice Hall, 1989, Chapter 3.

A-2.   T. Kailath, _Linear Systems_. Upper Saddle River, NJ: Prentice Hall, 1980, pp. 161–166.

A-3.   E. Hille, _Analytic Function Theory_, Vol. I. New York: Blaisdell, 1965, Chapters 7 and 9.

A-4.   S. L. Salas & Einar Hille, _Calculus_, 3rd ed., John Wiley & Sons: New York, 1978.

Appendix B

Gamma and Delta Functions

B.1 GAMMA FUNCTION

The Gamma function $\Gamma(\alpha)$ is defined by the integral

$$\Gamma(\alpha) \triangleq \int_0^\infty x^{\alpha-1} e^{-x} dx,$$

where $\alpha > 0$. The Gamma function is sometimes extended for $\alpha \leq 0$ as follows: For any integer $n = 1, 2, 3, \cdots, \Gamma(\alpha) = \Gamma(\alpha + 1)/\alpha$ whenever $-(n+1) < \alpha < -n$. Also $\Gamma(0) = \infty$, and

$$\Gamma(-n) = \frac{(-1)}{n!} \Gamma(0) = \pm\infty.$$

We are primarily interested in the Gamma function for $\alpha > 0$. If we integrate by parts, we obtain

$$\Gamma(\alpha) = \int_0^\infty x^{\alpha-1} d(-e^{-x}) = -x^{\alpha-1} e^{-x}\big|_0^\infty + (\alpha - 1) \int_0^\infty x^{\alpha-2} e^{-x} dx = 0 + (\alpha - 1)\Gamma(\alpha - 1),$$

where the term on the far right is valid only for $\alpha > 1$, since the integral won't converge otherwise. Likewise, $\Gamma(\alpha-1) = (\alpha-2)\Gamma(\alpha-2)$ for $\alpha > 2$. Thus $\Gamma(\alpha) = (\alpha-1)(\alpha-2)\Gamma(\alpha-2)$, etc. If α is an integer, then repeating the integration by parts recursively, we obtain

$$\Gamma(\alpha) = (\alpha - 1)!\Gamma(1).$$

But

$$\Gamma(1) = \int_0^\infty e^{-x}dx = 1.$$

Hence, for $\alpha = 1, 2, 3, \cdots, \Gamma(\alpha) = (\alpha - 1)!$

Therefore, note that $0! = 1$. We leave it to the reader to show that $\Gamma(0.5) = \sqrt{\pi}$ and $\Gamma(1.5) = \sqrt{\pi}/2$. The Gamma function is sometimes called the *generalized factorial function*.

B.2 DIRAC DELTA FUNCTION

The Dirac delta function $\delta(x)$ is often defined as a "function" that is zero everywhere except at $x = 0$, where it is infinite such that

$$\int_{-\infty}^\infty \delta(x)dx = 1.$$

Another definition is to regard $\delta(x)$ as the limit of one of several pulses. For example, with rectangular window,

$$w\left(\frac{x}{b}\right) \triangleq \begin{cases} 1, & -b/2 \le x \le b/2 \\ 0, & \text{else,} \end{cases}$$

we can define $\delta(x)$ as

$$\delta(x) \triangleq \lim_{a \to \infty} \{aw(ax)\}.$$

Another possibility is to define $\delta(x)$ as

$$\delta(x) \triangleq \lim_{a \to \infty} \{a\exp(-\pi a^2 x^2)\}.$$

The rectangular and Gaussian shaped pulses are shown in Figure B.2-1. The function $aw(ax)$ has discontinuous derivatives, whereas $a\exp(-\pi a^2 x^2)$ has continuous derivatives. The exact shape of these functions is immaterial. Their important features are (1) unit area and (2) rapid decrease to zero for $x \ne 0$.

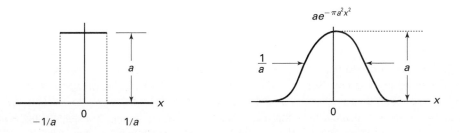

Figure B.2-1   Rectangular and Gaussian shaped pulses of unit area.

Still another defintion is to call any object a delta function if for any function $f(\cdot)$ continuous at x it satisfies the integral equation[†]

$$\int_{-\infty}^{\infty} f(y)\delta(y-x)\,dy = f(x). \tag{B.2-1}$$

This definition can, of course, be related to the previous one, since either of the pulses when substituted for $\delta(x)$ in Equation (B.2-1) will essentially furnish the same result when a is large. This follows because the integrand is significantly nonzero only for $x \simeq y$. The integral can, therefore, be approximately evaluated by replacing $f(y)$ by $f(x)$ and moving it outside the integral. Then, since both pulses have unit-area, the result follows. Note that $\delta(x) = \delta(-x)$.

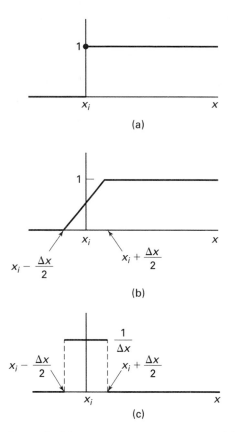

Figure B.2-2   (a) unit step $u(x - x_i)$; (b) approximation to unit step; (c) derivative of function in (b).

[†] A word of caution is in order here. Since $\delta(x)$ is zero everywhere except at a single point, its integral (in the Riemann sense) is not defined. Hence, Equation B.2-1 is essentially symbolic, that is, it implies a limiting operation as was done with the rectangular and Gaussian pulses.

Consider now the unit step $u(x - x_i)$, which is discontinuous at $x = x_i$ with $u(0) \triangleq 1$ (Figure B.2-2a). The discontinuity can be viewed as the limit of the function shown in Figure B.2-2b. The derivative is shown in Figure B.2-2c.

The derivative of the function shown in Figure B.2-2b is given by

$$\left. \frac{dF}{dx} \right|_{x_i} \triangleq \frac{dF(x_i)}{dx_i} = \lim_{\Delta x \to 0} \frac{1}{\Delta x_i} w \left[\frac{x - x_i}{\Delta x} \right]$$

$$= \delta(x - x_i). \tag{B.2-2}$$

Thus, formally, the derivative at a step discontinuity is a delta function with weight[†] proportional to the height of the jump. It is not uncommon to call $\delta(x - x_i)$ the delta function at "x_i."

Returning now to Equation 2.5-7 in Chapter 2, which can be written as

$$F(x) = \sum_i P_x(x_i) u(x - x_i)$$

and using the result of Equation B.2-2 enables us to write for a discrete r.v.:

$$f(x) = \frac{dF(x)}{dx} = \sum_i P_X(x_i) \delta(x - x_i), \tag{B.2-3}$$

where we recall that $P_X(x) \triangleq F(x_i) - F(x_i^-)$ and the unit step assures that the summation is over all i such that $x_i \leq x$.

[†]It is also called the area of the delta function.

Appendix C

Functional Transformations and Jacobians

C.1 INTRODUCTION

Functional transformations play an important role in probability theory as well as many other fields. In this appendix, we shall review the theory of Jacobians, beginning with a two-function-to-two function transformation and extending the result to the n-function-to-n-function case. First, we should recall two basic results from advanced calculus:

Theorem C.1-1 Consider a bounded linear transformation L from E^n to E^n. If D is a bounded set in E^n with n-dimensional volume $V(D)$, then the volume of $L(D)$ is merely $k \times V(D)$, where k is a constant *independent* of D. ■

Theorem C.1-2 If T is a transformation of class C^1 from E^n to E^n in an open set D then, at every point $p \in D$, dT is a *linear transformation* from E^n to E^n. ■

The first theorem states that the effect of L is merely to multiply the volume by a constant that doesn't depend on the shape of D. The second theorems states that, *at the differential level*, even nonlinear transformations become linear, provided that the transformations consist of differential functions. Both theorems will find application in this development.

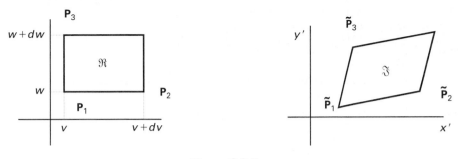

Figure C.2-1

C.2 JACOBIANS FOR $n = 2$

Consider the pair of one-to-one[†] differentiable functions $v = g(x, y)$, $w = h(x, y)$ with the unique inverse $x = \phi(v, w)$, $y = \varphi(v, w)$. As the vector $\mathbf{z} = (v, w)$ traces out the infinitesimal rectangle $\Re$ in the v'–w' plane, the vector $\mathbf{u} = (x, y)$ traces out the infinitesimal parallelogram $\Im$ in the x'–y' plane. By Theorem C.1-1, this differential transformation is linear, and by Theorem C.1-2, the ratio of the areas $A(\Im)/A(\Re)$ is a constant. We shall denote this constant by $|\tilde{J}|$ and compute its value.

We can easily compute the constant $\tilde{J}$ with the aid of Figure C.2-1. Recalling that $x = \phi(v, w)$, $y = \varphi(v, w)$, we compute the image points $\tilde{\mathbf{P}}_1$, $\tilde{\mathbf{P}}_2$, $\tilde{\mathbf{P}}_3$ of the vertices at $\mathbf{P}_1$, $\mathbf{P}_2$, $\mathbf{P}_3$ as:

$$\tilde{\mathbf{P}}_1 = (x, y), \ \tilde{\mathbf{P}}_2 = \left(x + \frac{\partial \phi}{\partial v} dv, y + \frac{\partial \varphi}{\partial v} dv \right), \ \tilde{\mathbf{P}}_3 = \left(x + \frac{\partial \phi}{\partial w} dw, y + \frac{\partial \varphi}{\partial w} dw \right).$$

These results are directly obtained by a Taylor series expansion about (x, y). Thus, for example, the coordinates (x_2, y_2) of $\tilde{\mathbf{P}}_2$ are obtained from

$$x_2 = \phi(v + dv, w) \approx \phi(v, w) + \frac{\partial \phi}{\partial v} dv \text{ and } y_2 = \varphi(v + dv, w) \approx \varphi(v, w) + \frac{\partial \varphi}{\partial v} dv.$$

There are no finite derivatives with respect to w because w is held constant in going from $\mathbf{P}_1$ to $\mathbf{P}_2$. From vector analysis, it is well known that the area of a parallelogram spanned by the vectors $\mathbf{v}_1$ and $\mathbf{v}_2$ is given by the magnitude of the cross-product, that is,

$$A(\Im) = |\mathbf{v}_1 \times \mathbf{v}_2| = \left| \left(\frac{\partial \phi}{\partial v} dv \mathbf{i} + \frac{\partial \varphi}{\partial v} \mathbf{j} dv \right) \times \left(\frac{\partial \phi}{\partial w} \mathbf{i} dw + \frac{\partial \varphi}{\partial w} \mathbf{j} dw \right) \right|$$

where we used the fact that $\mathbf{v}_1 = \tilde{\mathbf{P}}_2 - \tilde{\mathbf{P}}_1$ and $\mathbf{v}_2 = \tilde{\mathbf{P}}_3 - \tilde{\mathbf{P}}_1$. The unit vectors $\mathbf{i}, \mathbf{j}$ satisfy $\mathbf{i} \times \mathbf{j} = \mathbf{k}, \mathbf{j} \times \mathbf{i} = -\mathbf{k}, \mathbf{i} \times \mathbf{i} = \mathbf{j} \times \mathbf{j} = 0$, where $\mathbf{k} \perp \mathbf{i}, \mathbf{j}$. Thus

$$A(\Im) = \left| \frac{\partial \phi}{\partial v} \frac{\partial \varphi}{\partial w} - \frac{\partial \phi}{\partial w} \frac{\partial \varphi}{\partial v} \right| dv\, dw.$$

[†]This means that every point (x, y) maps into a unique (u, v) and vice versa.

Since $A(\Re) = dv\,dw$, we find that the ratio of the areas is

$$A(\Im)/A(\Re) = \left| \frac{\partial \phi}{\partial v} \frac{\partial \varphi}{\partial w} - \frac{\partial \phi}{\partial w} \frac{\partial \varphi}{\partial v} \right| \triangleq |\tilde{J}|.$$

In higher dimensions it is easier to write $\tilde{J}$ as a determinant. Indeed, even in this two-dimensional case we can write:

$$\tilde{J} = \begin{vmatrix} \dfrac{\partial \phi}{\partial v} & \dfrac{\partial \phi}{\partial w} \\[2mm] \dfrac{\partial \varphi}{\partial v} & \dfrac{\partial \varphi}{\partial w} \end{vmatrix} = \frac{\partial \phi}{\partial v} \frac{\partial \varphi}{\partial w} - \frac{\partial \phi}{\partial w} \frac{\partial \varphi}{\partial v}.$$

The quantity $\tilde{J}$ is called the Jacobian of the transformation $x = \phi(v, w)$, $y = \varphi(v, w)$.

Among other things, the Jacobian is necessary to preserve probability measure (sometimes called the probability mass or probability volume). For example, consider a pdf $f_{XY}(x, y)$ and the transformation $x = \phi(v, w)$, $y = \varphi(v, w)$. Consider the event $B \triangleq \{\zeta : (X, Y) \in \wp \subset E^2\}$. Then

$$P(B) = \int\!\!\int_{\wp} f_{XY}(x, y)dx\,dy \neq \int\!\!\int_{\wp} f_{XY}(\phi(v, w), \varphi(v, w))dv\,dw$$

because the volume $dx\,dy \neq dv\,dw$. What is needed is the Jacobian to create the equality among the integrals as

$$\int\!\!\int_{\wp} f_{XY}(x, y)dx\,dy = \int\!\!\int_{\wp} f_{XY}(\phi(v, w), \varphi(v, w))|\tilde{J}|dv\,dw$$

Sometimes it may be easier to deal with the original functions $v = g(x, y)$, $w = h(x, y)$ than the inverse functions $x = \phi(v, w)$, $y = \varphi(v, w)$. To get the desired result, we recompute the ratio of areas by considering the image, $\Re'$, in the v–w system, of an infinitesimal rectangle, $\Im'$, in the x–y system (Figure C.2-2). Following the same procedure as before, we obtain $A(\Im')/A(\Re') \triangleq 1/|J|$, where the primes help indicate the regions in the two systems and J is given by

$$J = \begin{vmatrix} \dfrac{\partial g}{\partial x} & \dfrac{\partial g}{\partial y} \\[2mm] \dfrac{\partial h}{\partial x} & \dfrac{\partial h}{\partial y} \end{vmatrix}.$$

But, by Theorem C.1-1, $A(\Im')/A(\Re') = A(\Im)/A(\Re)$ and, hence, $|\tilde{J}| = 1/|J|$ or $|\tilde{J}J| = 1$. We leave the details of the computation as an exercise for the reader.

C.3 JACOBIAN FOR GENERAL n

The general case is easy to deal with if we allow ourselves to use matrix and vector notation and some results from linear algebra. First, it is not convenient to use the unit vectors $\mathbf{i}, \mathbf{j}, \mathbf{k}$

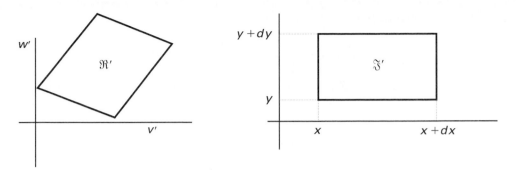

Figure C.2-2

in higher dimensions. Instead, we use unit vectors that are represented by column vectors. Thus, in E^2 we use $\mathbf{e}_1 = [1, 0]^T$ and $\mathbf{e}_2 = [0, 1]^T$. Then

$$\mathbf{v}_1 = \frac{\partial \phi}{\partial v} dv \, \mathbf{e}_1 + \frac{\partial \vartheta}{\partial v} dv \, \mathbf{e}_2 = \left[\frac{\partial \phi}{\partial v} dv, \frac{\partial \vartheta}{\partial v} dv \right]^T$$

and

$$\mathbf{v}_2 = \frac{\partial \phi}{\partial w} dw \, \mathbf{e}_1 + \frac{\partial \vartheta}{\partial w} dw \, \mathbf{e}_2 = \left[\frac{\partial \phi}{\partial w} dw, \frac{\partial \vartheta}{\partial w} dw \right]^T$$

Next, we form the 2×2 matrix $\mathbf{V}_2 = [\mathbf{v}_1 \ \mathbf{v}_2]$, where the subscript 2 on $\mathbf{V}_2$ refers to two-dimensional Euclidean space.

Then, for the special case of $n = 2, A(\Im)$ is given by $|\det \mathbf{V}_2|$. As we go to higher dimensions we drop the term "area of the parallelepiped" in favor of "volume of the parallelepiped," although purists would argue that for spaces of dimensions higher than three we should use "hypervolume." Also in higher dimensions, it is easier to use different subscripts rather than different symbols for functions and arguments. In n-dimensional space, the volume of a parallelepiped is always given by the height times the base area, where the base area is the volume of the parallelepiped in $n - 1$ dimensional space and the height is the length of the component of $\mathbf{v}_n$, which is orthogonal to the vectors that span E^{n-1}. Thus, in E^2 the base area is the length of the chosen base vector and the height is the length of the orthogonal component of the second vector. In E^3, the base area is the area of the parallelogram spanned by any two of the three vectors and the height is the length of the component of the third vector orthogonal to the plane containing the first two vectors.

We wish to compute the volume of an infinitesimal parallelepiped in n-dimensional space. Motivated by the fact that the volume, V_2, in two-dimensional space is given by $V_2 = |\det \mathbf{V}_2|$, we are tempted to write that $V_n = |\det \mathbf{V}_n|$. Is this true? The answer is yes and the proof is furnished by *induction*. Thus, we assume that $V_n = |\det \mathbf{V}_n|$ is true and we must prove that $V_{n+1} = |\det \mathbf{V}_{n+1}|$. Now in terms of the vectors $\mathbf{v}_1, \mathbf{v}_2, \cdots, \mathbf{v}_n, \mathbf{v}_{n+1}$, the matrix $\mathbf{V}_{n+1}$ can be written as

$$
\mathbf{V}_{n+1} =
\begin{bmatrix}
\mathbf{v}_1 \cdots \mathbf{v}_n & v_{n+1,1} \\
\vdots \; \vdots \; \vdots & v_{n+1,2} \\
\vdots \; \vdots \; \vdots & \vdots \\
0 \quad 0 \quad 0 & v_{n+1,n+1}
\end{bmatrix}
$$

$$
=
\begin{bmatrix}
\mathbf{V}_n & v_{n+1,1} \\
 & \vdots \\
0 \;\; 0 & v_{n+1,n+1}
\end{bmatrix}
$$

To compute $|\det \mathbf{V}_{n+1}|$ we expand by the bottom row to obtain $|\det \mathbf{V}_{n+1}| = |v_{n+1,v+1}||\det \mathbf{V}_n|$, since all other terms in the expansion are zero. Now consider the vector $\mathbf{v}_{n+1}$ in more detail. In terms of the unit vectors $\mathbf{e}_1, \mathbf{e}_2, \ldots, \mathbf{e}_{n+1}$, it can be written as

$$
\mathbf{v}_{n+1} = v_{n+1,n+1}\mathbf{e}_{n+1} + \sum_{i=1}^{n} v_{n+1,i}\,\mathbf{e}_i,
$$

where $\mathbf{e}_i$ has a 1 in the ith position (row) and 0's in the remaining n position. But $\mathbf{e}_{n+1}$ is the unit vector orthogonal to the $\mathbf{e}_1, \mathbf{e}_2, \ldots, \mathbf{e}_n$, and hence is orthogonal to the space spanned by them, and $|v_{n+1,n+1}|$ is its height. Also recall that $|\det \mathbf{V}_n|$ is the volume of the parallelepiped in n-dimensions and therefore represents the base area in $n+1$ dimensions. Hence $|\det \mathbf{V}_{n+1}| = |v_{n+1,v+1}||\det \mathbf{V}_n|$ is indeed height times base area and the proof is complete.

Readers familiar with Hadamard's inequality and the Gram–Schmidt orthogonalization procedure can furnish a faster, more direct, proof that avoids induction, but is less intuitive.

In Chapter 5 we considered the transformation

$$
y_1 = g_1(x_1, x_2, \ldots, x_n)
$$
$$
y_2 = g_2(x_1, x_2, \ldots, x_n)
$$
$$
y_n = g_n(x_1, x_2, \ldots, x_n),
$$

with unique inverse

$$
x_1 = \phi_1(y_1, y_2, \ldots, y_n)
$$
$$
x_2 = \phi_2(y_1, y_2, \ldots, y_n)
$$
$$
\vdots
$$
$$
x_n = \phi_n(y_1, y_2, \ldots, y_n)
$$

Then, a rectangular parallelepiped in the $(y_1, y_2, \ldots, y_n)$ system with volume $\prod_{i=1}^{n} |dy_i|$ maps into a parallelepiped in the $(x_1, x_2, \ldots, x_n)$ system with volume $|\det \mathbf{V}_n| = |\det[\mathbf{v}_1, \mathbf{v}_2, \ldots, \mathbf{v}_n]|$. Here, by computing the differentials of the transformation, we obtain for the $\mathbf{v}_i, i = 1, \ldots, n$:

$$\mathbf{v}_i = \left(\frac{\partial \phi_1}{\partial y_i} dy_i, \ldots, \frac{\partial \phi_n}{\partial y_i} dy_i \right)^T, \qquad i = 1, \ldots, n.$$

Appendix D

Measure and Probability

D.1 INTRODUCTION AND BASIC IDEAS

Some mathematicians describe probability theory as a special case of *measure theory*. Indeed, random variables are said to be *measurable* functions; the distribution function is said to be a *measure*; events are *measurable* sets; the sample description space together with the field of events is a *measurable space*; and a probability space is a *measure* space. In this appendix, we furnish some results for readers not familiar with the basic ideas of measure theory. We assume that the reader has read Chapter 1 and is familiar with set operations, fields, and sigma fields. The bulk of the material in this appendix is adapted from the classic work by Billingsley.[†]

Let Ω be a space (a universal set) and let $A, B, C, \ldots$ be elements (subsets) of Ω. Also, as in the text, let ϕ denote the empty set. Let $\Im$ be a field of sets on Ω. Then the pair $(\Omega, \Im)$ is a *measurable space* if $\Im$ is an σ–*field* on Ω. Let μ be a *set function*[‡] on $\Im$. Then μ is *a measure* if it satisfies these conditions:

(i) Let $A \in \Im$, then $\mu[A] \in [0, \infty)$;
(ii) $\mu[\phi] = 0$;

[†]Patrick Billingsley: *Probability and Measure.* John Wiley & Sons, New York, 1978.
[‡]A set function is a real valued function defined on the field $\Im$ of subsets of the space Ω.

(iii) if $A_1, A_2, \ldots$ is a disjoint sequence of sets in $\Im$ and if $\cup_{k=1}^{\infty} A_k \in \Im$, then

$$\mu\left[\bigcup_{k=1}^{\infty} A_k\right] = \sum_{k=1}^{\infty} \mu[A_k].$$

This property is called *countable additivity*. A measure μ is called *finite* if $\mu[\Omega] < \infty$; it is *infinite* if $\mu[\Omega] = \infty$. It qualifies as a probability measure if $\mu[\Omega] = 1$, as denoted in Chapter 1. If $\Im$ is an σ-field in Ω, the triplet $(\Omega, \Im, \mu)$ is a *measure space*.

Countable additivity implies *finite* additivity, that is,

$$\mu\left[\bigcup_{k=1}^{n} A_k\right] = \sum_{k=1}^{n} \mu[A_k]$$

if the sets are disjoint. A measure μ is *monotone*, that is $\mu[A] \le \mu[B]$ whenever $A \subset B$. The proof of this statement is easy. Write, as is customary in the literature on measure theory, $BA^c \triangleq B - A$ *and* $B = (B - A) \bigcup AB = (B - A) \bigcup A$. Then, since A and B–A are disjoint, it follows that $\mu[B] = \mu[B - A] + \mu[A] \ge \mu[A]$. Also, since $A \bigcup B = (A - B) \bigcup (B - A) \bigcup AB$, it follows that $\mu[A \bigcup B] = \mu[A - B] + \mu[B - A] + \mu[AB]$. This result can be extended to many sets in a σ-field, (sets in a σ-field are called σ-sets), that is,

$$\mu\left[\bigcup_{k=1}^{n} A_k\right] = \sum_{k=1}^{n} \mu[A_k] - \sum_{i<j} \mu[A_i A_j] + \ldots + (-1)^{n+1} \mu[A_1 A \ldots A_n]$$

Of course, this equation makes sense only if the sets have finite measure. It is also straightforward to show that $\mu[\cdot]$ has the property of *subadditivity*:

$$\mu\left[\bigcup_{k=1}^{n} A_k\right] \le \sum_{k=1}^{n} \mu[A_k].$$

Example D.1-1 _____

Lebesgue measure. Consider the σ-field, $\Im$, of intervals on $\Omega = (0, 1)$. The elements of $\Im$ are called linear *Borel* sets and the σ-field of intervals is called the Borel field $\mathscr{B}$. We shall use this notation for any σ-field on the real line. A measure $\mu[\cdot]$ on $\Im$ is $\mu = \lambda(a, b) \triangleq b - a$, where $b \ge a$. This measure is called the *Lebesgue measure* on $(a, b]$. It can be directly generalized to the real line R^1. An extension of the Lebesgue measure to k-dimensional Euclidean space is:

$$\mu = \lambda_k[x : a_i < x_i \le b_i, i = 1, \ldots, k] \triangleq \prod_{i=1}^{k} (b_i - a_i)$$

Thus the Lebesgue measures are *length* $(k = 1)$, *area* $(k = 2)$, *volume* $(k = 3)$, and *hypervolume* $(k > 3)$. We denote the associated σ-field generated by these *generalized rectangles* by the symbol $\mathscr{B}^k$.

There are many important theorems regarding measures. We cite several below.

Theorem D.1-1   (Translation invariance.) Let $A \in \mathscr{B}^k$ and define $A + x \triangleq \{a + x : a \in A\}$. Then $\lambda_k(A) = \lambda_k(A + x)$ for all translation vectors x.   ■

Theorem D.1-2   (Lebesgue measure of transformation.) Let $T : R^k \to R^k$ denote a linear and nonsingular transformation from the Euclidean space R^k to R^k. Then $A \in \mathscr{B}^k$ implies that $TA \in \mathscr{B}^k$ and $\lambda_k(TA) = |\det T| \cdot \lambda_k(A)$. For example, if T is a rotation, or reflection, that is, an orthogonal or unitary transformation, then $|\det T| = 1$ and $\lambda_k(TA) = \lambda_k(A)$.   ■

Theorem D.1-3   (Lebesgue Measure of Subspaces of R^k). Every $(k-1)$ dimensional hyperplane has k-dimensional Lebesgue measure zero.   ■

Theorem D.1-4   (Continuity of measure.) (i) Let μ be a measure on a field $\mathfrak{F}$. Then if A_n and A lie in $\mathfrak{F}$ and $A_n \uparrow A$, then $\mu[A_n] \uparrow \mu[A]$. This is called *continuity of measure from below*. $A_n \uparrow A$ means that $A_{n-1} \subset A_n \subset A_{n+1} \subset \cdots$ *and*

$$A = \bigcup_{n=1}^{\infty} A_n$$

Likewise, $\mu[A_n] \uparrow \mu[A]$ means that $\mu[A_n] \le \mu[A_{n+1}] \le \mu[A]$ and $\lim\limits_{n \to \infty} \mu[A_n] = \mu[A]$.

(ii) Let μ be a measure on a field $\mathfrak{F}$. Then if A_n and A lie in $\mathfrak{F}$ and $A_n \downarrow A$, then $\mu[A_n] \downarrow \mu[A]$. This is called *continuity of measure from above*. $A_n \downarrow A$ means that $A_{n-1} \supset A_n \supset A_{n+1} \supset \cdots$ *and*

$$A = \bigcap_{n=1}^{\infty} A_n$$

Likewise, $\mu[A_n] \downarrow \mu[A]$ means that $\mu[A_n] \ge \mu[A_{n+1}] \ge \mu[A]$ and $\lim\limits_{n \to \infty} \mu[A_n] = \mu[A]$.   ■

Measurable Mappings and Functions

Let $(\Omega, \mathfrak{F})$ *and* $(\Omega', \mathfrak{F}')$ be two measurable spaces with two sets $A \in \mathfrak{F}$ and $A' \in \mathfrak{F}'$. For a mapping $T : \Omega \to \Omega'$, consider the inverse image $T^{-1}A' = \{\omega \in \Omega : T\omega \in A'\}$ *for* $A' \subset \Omega'$. The *mapping is measurable* if $T^{-1}A' \in \mathfrak{F}$ *for every* $A' \in \mathfrak{F}'$. For example, consider the unit interval $\Omega = (0, 1]$ with $\mathfrak{F} = \mathscr{B}$ and the mapping $Tx = x^2$. Here, $\Omega' = \Omega$ *and* $\mathfrak{F}' = \mathscr{B}$. Clearly, the inverse image of every Borel interval in Ω' is a Borel interval in Ω. Hence, T is a measurable mapping.

A real function X on Ω, with image space R^1, is said to be *measurable* if its inverse image $X^{-1}B = \{\omega : X(\omega) \in B\} \in \mathfrak{F}$ for every $B \in \mathfrak{F}$.

D.2 APPLICATION OF MEASURE THEORY TO PROBABILITY

A set function P on a σ-field $\mathfrak{F}$ is a *probability measure* if:

(i) $0 \le P[A] \le 1$ *for every* $A \in \mathfrak{F}$;

(ii) $P[\phi] = 0$, $P(\Omega) = 1$;

(iii) if $A_1, A_2, \ldots, A_k, \ldots$ is a disjoint sequence of $\Im$-sets such that

$$\bigcup_{k=1}^{\infty} A_k \in \Im$$

then

$$P\left[\bigcup_{k=1}^{\infty} A_k\right] = \sum_{k=1}^{\infty} P[A_k].$$

(This is the countable additivity property of the probability measure.)

Distribution Measure

In keeping with the notation in the main text, we replace ω with ζ to denote the elements of Ω. Recall that this was done to save ω for the Fourier transform variable needed throughout the text. Let $B \in \mathscr{B}$, the Borel σ-field of intervals on the real line. Consider a (probability) measure μ on $(R^1, \mathscr{B})$ defined by $\mu[B] \triangleq P[\zeta : X(\zeta) \in B] = P_X[B]$. This measure is called the *distribution* or *law* of a random variable. The *distribution function* of X is defined by

$$F_X(x) \triangleq \mu(-\infty, x] = P[X \leq x],$$

where $P[X \leq x]$ is short for $P[\zeta : X(\zeta) \leq x]$. By the *continuity from above* part of the *continuity of measure theorem*, $F_X(x)$ is continuous from the right.

Since the field of events is a *σ-field*, and the distribution function is generated by a measure, all of the properties of measures apply in probability. It is for this reason that probability and measure theories are so closely related. However, to look at probability theory just from the point of view of measure theory is to ignore its rich calculus which enables the solution of engineering, scientific, and statistical problems.

Appendix E

Sampled Analog Waveforms and Discrete-time Signals

Discrete-time signals are often realized by sampling continuous-time analog wave forms. Here, we briefly review the relationship between the two types of signals. The reconstruction of a continuous-time signal from its equally-spaced samples is governed by the famous Whittaker–Nyquist–Shannon sampling theorem, which states the following.

Theorem E.1-1   *A continuous signal $x(t)$ with real frequencies no higher than $v_{\max}$ can be reconstructed exactly from its samples $x(nT)$ if the sampling interval T satisfies $T < \frac{1}{2v_{\max}}$.* ∎

The proof of this important theorem is given in many places, for example, *Principles of Communication Engineering* by John M. Wozencraft and Irwin M. Jacobs, John Wiley and Sons, N.Y, 1965. Let $x(t)$, $y(t)$, and $h(t)$ denote the input signal, output signal, and impulse response of a linear, shift-invariant (LSI) system respectively. Let B, in Hertz, denote a bandwidth that is greater than any signal or system bandwidth encountered in the system and let $\Delta \triangleq 1/(2B)$. For ease of notation define

$$\operatorname{sinc}\,(x) \triangleq \frac{\sin \pi x}{\pi x}.$$

The relationship between input and output for an LSI system is

$$y(t) = \int_{-\infty}^{\infty} h(s)x(t-s)ds$$

and from the sampling theorem:

$$y(t) = \sum_l y(l\Delta) \text{ sinc } (2B[t - l\Delta])$$

$$x(t) = \sum_l x(l\Delta) \text{ sinc } (2B[t - l\Delta])$$

$$h(t) = \sum_l h(l\Delta) \text{ sinc } (2B[t - l\Delta])$$

If we insert the top three lines into the input–output integral and evaluate at $y(t)$ at $t = l\Delta$, we obtain

$$y(l\Delta) = \sum_n \sum_m h(n\Delta)x(m\Delta)I(l, m, n),$$

where

$$I(l, m, n) \triangleq \int_{-\infty}^{\infty} \text{ sinc } (2B[s - n\Delta]) \text{ sinc } (2B[s - (l - m)\Delta])ds = 0,$$

for all real integers l, m, n except when $l - m = n$, whereupon it assumes the value Δ. Hence, we obtain the important result that

$$y(l\Delta) = \sum_n h(n\Delta)x([l - n]\Delta)\Delta,$$

Often the factor Δ is submerged into $h(n\Delta)$. In a computer the sampled values of the functions become mere sequences of numbers as $y(l\Delta) \triangleq y[l]$, $x(l\Delta) \triangleq x[l]$, and $h(n\Delta) \triangleq h[n]$. Then, we obtain

$$y[n] = \sum_n h[n]x[l - n]$$

that we recognize as a discrete convolution. The important fact to remember is that the processing of analog signals can be done by operating on their samples and then reconstructing an analog waveform by filtering.

Another point to consider is that the sequence of numbers $\{x[n]\}$ does not contain information about the sampling period, For example, consider the sinusoid $x(t) = A\cos(\omega_r t + \theta)$. If we sample at $t = n\Delta$, $n = \ldots, -2, -1, 0, 1, 2, \ldots$, we obtain the samples $x(n\Delta) = A\cos(n\Delta\omega_r + \theta) = A\cos(n\omega + \theta) \triangleq x[n]$, where $\omega \triangleq \Delta\omega_r$. The radian "frequency" ω is dimensionless, which is consistent with the dimensionless "time" n. It is well to remember that to convert to analog frequencies ω_r (radians/sec) or v_r (Hertz) we must use $\omega_r = \omega\Delta$ or $v_r = v\Delta$. For example, the Fourier transform of a sequence of numbers $\{x[n]\}$ will yield a spectrum of sinusoids at normalized frequencies ω that lie in the interval $[-\pi, \pi]$. If we convert to analog radian frequencies, then the spectrum will lie in the interval $[-2\pi B, 2\pi B]$.

Index

Q1 a) Compute the pdf of $Z = \max(X,Y)$ if X and Y are independent r.v.s

b) Specialize your result to $f_X(x) = f_Y(y) \sim U(0,1)$.

S1 a) $\qquad F_Z(z) = P(\max(X,Y) \le z)$

But this event is equivalent to
$$\{X \le z, Y \le z\}$$

So, $P(Z \le z) = P(X \le z, Y \le z) = F_X(z) \, F_Y(z)$

due to independence. It follows that
$$f_Z(z) = f_Y(z) \, F_X(z) + f_X(z) \, F_Y(z)$$

b) $\qquad f_Z(z) = 2z \left(u(z) - U(z-1) \right)$

Q2 Let X, Y be independent, identically distributed r.v.s with $f_X(x) = e^{-x} u(x)$. Let $Z = \max(X,Y)$. Compute $f_Z(z)$ and $P\{Z \le 1\}$.

S2 From previous problem, $F_Z(z) = F_X(z) \, F_Y(z) = \left(1 - z^{-2}\right)^2 u(z)$

and $f_Z(z) = \dfrac{dF_Z(z)}{dz} = 2e^{-z}(1 - e^{-z}) u(z)$.

$F_Z(1) = (1 - e^{-1})^2 u(1) \cong 0.4$.

Q3 Given two functions $\quad v = g(x,y) = 3x + 5y$
$$w = h(x,y) = x + 2y$$
and $f_{XY}(x,y) = \dfrac{1}{2a} \exp\{-\tfrac{1}{2}(x^2 + y^2)\}$. What is $f_{VW}(v,w)$?

S3 Inverse mappings are: $X = 2v - 5w, \quad y = -v + 3w$

$|\det J| = \text{mag} \begin{vmatrix} 2 & -5 \\ -1 & 3 \end{vmatrix} = 1$ & $f_{VW}(v,w) = \dfrac{1}{2\pi} \exp\{-\tfrac{1}{2}(5v^2 - 26vw + 34w^2)\}$

Q4 Calculate $f_{VW}(v,w)$ for $\quad V = X\cos\theta + Y\sin\theta$

$$W = X\sin\theta - Y\cos\theta$$

with $\quad f_{XY}(x,y) = \dfrac{1}{2\pi\sigma^2} e^{-(x^2+y^2)/2\sigma^2}$

S4 Inverse mapping is: $\quad x = v\cos\theta + w\sin\theta$

$$y = v\sin\theta - w\cos\theta$$

$$|\det J| = \text{mag} \begin{vmatrix} \cos\theta & \sin\theta \\ \sin\theta & -\cos\theta \end{vmatrix} = 1.$$

Hence $\quad f_{VW}(v,w) = \dfrac{1}{2\pi\sigma^2} \exp\left\{ -(v^2 + w^2)/2\sigma^2 \right\}$

Q5 If X and Y are independent r.v.s identically distributed as $N(0,1)$, find the mean and variance of

$$Z = \sqrt{X^2 + Y^2}$$

S5 It has been found out that $\quad f_Z(z) = z\, e^{-\frac{1}{2}z^2} u(z)$

So $\quad E(Z) = \displaystyle\int_0^\infty z\, f_Z(z)\,dz = \int_0^\infty z^2 e^{-\frac{1}{2}z^2}\,dz = \sqrt{\dfrac{\pi}{2}}$

$$E(Z^2) = \int_0^\infty z^3 e^{-\frac{1}{2}z^2}\,dz = 2$$

$$\sigma_Z^2 = E(Z^2) - \left(\bar{Z}\right)^2 = 2 - \dfrac{\pi}{2} \cong 0.43$$

Q6 Consider the pmf:

Define

$$X(k) = k, \quad Y(k) = k^2$$

a) Show that X and Y are independent
b) Show that X and Y are uncorrelated

S6 a) $P(X = -1) = P(k = -1) = \dfrac{1}{5}, \quad P(Y=1) = P(k=-1) + P(k=1) = \dfrac{2}{5}$

$$P(X = -1, Y=1) = P(k=-1) = \dfrac{1}{5}$$

$$P(X=-1)\, P(Y=1) = \dfrac{1}{5} \cdot \dfrac{2}{5} = \dfrac{2}{25} \neq P(X=-1, Y=1)$$

So dependent

b) $E(X) = \displaystyle\sum_{k=1}^{5} k_i\, P(k_i) = \dfrac{1}{5}\left(-1 - \dfrac{1}{2} + 0 + \dfrac{1}{2} + 1\right) = 0$

$$E(Y) = \dfrac{1}{5}\left(1 + \dfrac{1}{4} + 0 + \dfrac{1}{4} + 1\right), \quad E(XY) = \sum_5 xy\, P(x,y)$$

$$E(X)\,E(Y) = 0 \cdot \dfrac{1}{2} = 0 = E(XY) \qquad = \sum_{i=1}^{5} k^3 P(k) = 0$$

Uncorrelated